Oscar Zariski: Collected Papers
Volume II
Holomorphic Functions and Linear Systems

Mathematicians of Our Time

Gian-Carlo Rota, series editor

1. Charles Loewner: Theory of Continuous Groups
   notes by Harley Flanders and Murray H. Protter

2. Oscar Zariski: Collected Papers
   Volume I
   Foundations of Algebraic Geometry and Resolution of Singularities
   edited by H. Hironaka and D. Mumford

3. Collected Papers of Hans Rademacher
   Volume I
   edited by Emil Grosswald

4. Collected Papers of Hans Rademacher
   Volume II
   edited by Emil Grosswald

5. Paul Erdös: The Art of Counting
   edited by Joel Spencer

6. Oscar Zariski: Collected Papers
   Volume II
   Holomorphic Functions and Linear Systems
   edited by M. Artin and D. Mumford

Oscar Zariski

Collected Papers

Volume II
Holomorphic Functions and Linear Systems

Edited by
M. Artin and D. Mumford

The MIT Press
Cambridge, Massachusetts, and London, England

This book was printed and bound
by Semline Inc.
in the United States of America.

Library of Congress Cataloging in Publication Data

Zariski, Oscar, 1899–
Collected papers.

(Mathematicians of our time, v. 2, 6)
CONTENTS: v. 1. Foundations of algebraic geometry and resolution of singularities.—v. 2. Holomorphic functions and linear systems.
Bibliography: p.
1. Mathematics—Collected works. I. Series.
QA3.Z37 510′.8 73-171558
ISBN 0-262-01038-0

Oscar Zariski

Contents

(Bracketed numbers are from the Bibliography)

Part II
Linear Systems, the Riemann-Roch Theorem and Applications

Introduction by D. Mumford
175

Reprints of Papers

Preface

The series "Mathematicians of Our Time" embraces, at least in principle, the works of living mathematicians; therefore, the term "collected works," as applied to this series stands of necessity for an open-ended entity, because the author—contrary to the old cliché that "mathematics is a young man's game" —may still be actively engaged in research and therefore continue to produce papers while the "collected" works is being printed. Thus, in my case, the bibliography of papers that appeared in the first volume did not include two papers, one of which was in course of publication and another of which was in preparation. (These two papers form, together with the last paper [89] of the list printed in the first volume, a sequence of three papers under the common title "General theory of saturation and of saturated local rings.") The paper that was then "in course of publication" appears now, in this second volume, as the last published paper [90] in my bibliography, while the third paper of that series is now in course of publication. At any rate, the present plans envisage the publication of four volumes. These will include all my published works, with the following exceptions (the numbers in brackets refer to the bibliography as printed in this volume):

1. Books [6,25,72,75].
2. Lecture notes [87].
3. Expository articles in fields to which I have made no original contribution myself [1,3,5,11]. All these articles deal with the foundations of set theory, and I wrote them in my early postgraduate years in Rome at the urging and with the encouragement of my teacher F. Enriques, whose primary interest at that time was in the philosophy and history of science and who was editor of a series of books entitled "Per la Storia e la Filosofia delle Matematiche." As the reader can see, the book listed under [6] was published in that series.

The editorial preparation and the writing of introductions to each volume is entrusted to the capable hands of younger men, who are experts in the field of algebraic geometry and who at one time or another have been either my students at Harvard or have been closely associated with me in some capacity at Harvard or elsewhere. Thus, the editors of the first volume, H. Hironaka and D. Mumford, as well as M. Artin, who joined D. Mumford as editor of this second volume, are truly leaders in the field of algebraic geometry and have studied at Harvard.

While all the papers printed in these collected works belong, without exception, to algebraic geometry, the reader will undoubtedly notice that beginning with the year 1937 the nature of my work underwent a radical change. It became strongly *algebraic* in character, both as to methods used and as to the very formulation of the problem studied (these problems, nevertheless, always have had, and never ceased to have in my mind, their origin and motivation in algebraic *geometry*). A few words on how this change came about may be of some interest to the reader. When I was nearing the age of 40, the circumstances that led me to this radical change of direction in my research (a change that marked the beginning of what was destined to become my chief contribution to algebraic geometry) were in part personal in character, but chiefly they had to do with the objective situation that prevailed in algebraic geometry in the 1930s.

In my early studies as a student at the University of Kiev in the Ukraine, I was interested in algebra and also in number theory (by tradition, the latter subject is strongly cultivated in Russia). When I became a student of the University of Rome in 1921, algebraic geometry reigned supreme in that university. I had the great fortune of finding there on the faculty three great mathematicians, whose very names now symbolize and are identified with classical algebraic geometry: G. Castelnuovo, F. Enriques, and F. Severi. Since even within the classical framework of algebraic geometry the algebraic background was clearly in evidence, it was inevitable that I shoud be attracted to that field. For a long time, and in fact for almost ten years *after* I left Rome in 1927 for a position at the Johns Hopkins University in Baltimore, I felt quite happy with the kind of "synthetic" (an adjective dear to my Italian teachers) geometric proofs that constituted the very life stream of classical algebraic geometry (Italian style). However, even during my Roman period, my algebraic tendencies were showing and were clearly perceived by Castelnuovo, who once told me: "You are here with us but are not one of us." This was said not in reproach but good-naturedly, for Castelnuovo himself told me time and time again that the methods of the Italian geometric school had done all they could do, had reached a dead end, and were inadequate for further progress in the field of algebraic geometry. It was with this perception of my algebraic inclination that Castelnuovo suggested to me a problem for my doctoral dissertation, which was closely related to Galois theory (see [2 and 12]).

Both Castelnuovo and Severi always spoke to me in the highest possible terms of S. Lefschetz's work in algebraic geometry, based on topology; they both were of the opinion that topological methods would play an increasingly important role in the development of algebraic geometry. Their views, very amply justified by future developments, have strongly influenced my own work for some time. This explains the topological trend in my work during the period 1929 to 1937 (see [15,16,17,20,22,27,28,29]). During that period I made frequent trips from Baltimore to Princeton to talk to and consult with Lefschetz, and I owe a great deal to him for his inspiring guidance and encouragement.

The breakdown (or the breakthrough, depending on how one looks at it) came when I wrote my Ergebnisse monograph *Algebraic Surfaces* [25]. At that time (1935) modern algebra had already come to life (through the work of Emmy Noether and the important treatise of B. L. van der Waerden), but while it was being applied to some aspects of the foundations of algebraic geometry by van der Waerden, in his series of papers "*Zur algebraischen Geometrie*," the deeper aspects of *birational* algebraic geometry (such as the problem of reduction of singularities, the properties of fundamental loci and exceptional varieties of birational transformations, questions pertaining to complete linear systems and complete "continuous" systems of curves on surfaces, and so forth) were largely, or even entirely, virgin territory as far as algebraic exploration was concerned. In my Ergebnisse monograph I tried my best to present the underlying ideas of the ingenious geometric methods and proofs with which the Italian geometers were handling these deeper aspects of the whole theory of surfaces, and in all probability I succeeded, but at a price. The price was my own personal loss of the geometric paradise in which I so happily had been living. I began to feel distinctly unhappy about the rigor of the original proofs I was trying to sketch (without losing in the least my admiration for the imaginative geometric spirit that permeated these proofs); I became convinced that the whole structure must be done over again by purely algebraic methods. After spending a couple of years just studying modern algebra, I had to begin somewhere, and it was not by accident that I began with the problem of local uniformization and reduction of singularities. At that time there appeared the Ergebnisse monograph *Idealtheorie* of W. Krull, emphasizing valuation theory and the concept of integral dependence and integral closure. Krull said somewhere in his monograph that the *general* concept of valuation (including, therefore, nondiscrete valuations and valuations of rank > 1) was not likely to have applications in algebraic geometry. On the contrary, after some trial tests (such as the valuation-theoretic analysis of the notion of infinitely near base points; see title [35]), I felt that this concept could be extremely useful for the analysis of singularities and for the problem of reduction of singularities. At the same time I noticed some promising connections between integral closure and complete linear systems; a systematic study of these latter connections later led me to the notions of normal varieties and normalization. However, I also concluded that this program could be successful only provided that much of the preparatory work be done for ground fields that are not algebraically closed. I restricted myself to characteristic zero: for a short time, the quantum jump to $p \neq 0$ was beyond the range of either my intellectual curiosity or my newly acquired skills in algebra; but it did not take me too long to make that jump; see for instance [48,49,50] published in 1943 to 1947.

I carried out this initial program of work primarily in the four papers [37,39,40,41] published in 1939 and 1940. From then on, for more than 30 years, my work ranged over a wide variety of topics in algebraic geometry. It is not my intention here, nor is it the purpose of this preface, to brief the reader on the nature of these topics and the results obtained or the manner in which my papers can be grouped together in various categories, according to the principal topics treated. This is the task of the editors of the various volumes. I will say only a few words about the first and the present (second) volumes.

The papers collected in the first volume are divided in two groups: (1) *foundations*, meaning primarily properties of normal varieties, linear systems, birational transformations, and so on, and (2) *local uniformization and resolution of singularities*. These two subdivisions correspond precisely to the twofold aim I set to myself in my first concerted attack on algebraic geometry by purely algebraic methods—an undertaking and a state of mind about which I have already said a few words earlier. As a matter of fact, of the four main papers that I mentioned earlier as being the chief fruit of my first huddle with modern algebra and its applications to algebraic geometry, exactly two [37 and 40] belong to "foundations," while the other two [39,41] belong to the category "resolution of singularities and local uniformization."

The papers collected in this second volume are also divided in two groups: (1) *theory of formal holomorphic functions on algebraic varieties* (in any characteristic), meaning primarily analytic properties of an algebraic variety V, either in the neighborhood of a point (strictly *local* theory) or—and this is the deeper aspect of the theory—in the neighborhood of an algebraic subvariety of V (semi-global theory); (2) *linear systems, the Riemann-Roch theorem and applications* (again in any characteristic), the applications being primarily to algebraic surfaces (minimal models, characterization of rational or ruled surfaces, etc.).

My work on formal holomorphic functions was a natural outgrowth of my previous work on the local theory of singularities and their resolution. In the course of this previous work I developed an absorbing interest in the formal aspects of Krull's theory of local rings and their completions. In particular, I gave much thought to the possibility of extending to varieties V over arbitrary ground fields the classical notion of analytic continuation of a holomorphic function defined in the neighborhood of a point P of V. I sensed the probable existence of such an extension provided the analytic continuation were carried out along an algebraic subvariety W of V passing through P. It was wartime, and my heavy teaching load at Johns Hopkins University (18 hours a week) left me with little time for developing these ideas. Fortunately I was invited in January 1945 to spend at least one year at the University of São Paolo, Brazil, as exchange professor under the auspices of our Department of State. My light teaching schedule at São Paolo gave me the necessary leisure time to concentrate on an abstract theory of holomorphic functions. The year spent at São Paolo also presented me with a superlative audience consisting of one person—André Weil (who spent two or three years in São Paolo)—to whom I could speak about these ideas of mine during our frequent walks. The full theory of holomorphic functions—in the difficult case of complete (projective) varieties—was developed by me in my 1951 Transactions memoir [58]. However, the germ of this theory, in the easier case of affine varieties, appears already in my 1946 paper [49] written and published in Brazil. The key ingredient of the theory developed in this earlier paper is the concept of certain special rings, which later were named "Zariski rings," and properties of the completion of these rings. It is also this earlier Brazilian paper that led me to the discovery of a connection between the general theory of holomorphic functions and the connectedness theorem on algebraic varieties (and, in particular, the so-called principle of degeneration of Enriques). This connection was fully developed in my memoir [58] mentioned above. To a more strictly local frame of reference belong such papers as [52], [53], and [59] which deal with analytic properties of normal points of a variety.

In my student days in Rome algebraic geometry was almost synonymous with the theory of algebraic surfaces. This was the topic on which my Italian teachers lectured most frequently and in which arguments and controversy were also most frequent. Old proofs were questioned, corrections were offered, and these corrections were—rightly so—questioned in their turn. At any rate, the general theory of algebraic surfaces was very much on my mind in subsequent years, as witnessed—on a expository level—by my monograph [25] on algebraic surfaces, and—on a more significant research level—by the connection which I have found exists for varieties V of any dimension between normal (respect., arithmetically normal) varieties and the property that the hypersurfaces of a sufficiently high order (respect., of all orders) cut out on V complete linear systems. With this result as a starting point and with the conviction, indelibly impressed in my mind by my Italian teachers, that the theory of algebraic surfaces is the apex of algebraic geometry, it is no wonder that as soon as I realized that further progress in the problem of resolution of singularities would probably take years and years of further effort on my part, I decided that it was time for me to come to grips with the theory of algebraic surfaces. I felt that this would be the real testing ground for the algebraic methods which I had developed earlier. In his introduction to the second part of this volume Mumford says that he believes that my research on linear systems was to me "something like a dessert" (after the arduous efforts of the previous phase of my work). Objectively, Mumford may be right, but to me, subjectively, the proposed new work on linear systems felt more like the "main course." This work was also, in part, an answer to the following challenge sounded by Castelnuovo in his 1949 introduction to the treatise "Le superficie algebriche" of Enriques: "Verrà presto il continuatore dell'opera delle scuole italiana e francese il quale riesca a dare alla teoria delle superficie algebriche la perfezione che ha raggiunto la teoria delle curve algebriche?" (Note Castelnuovo's answer: "Lo spero ma ne dubito".) With this challenging question of Castelnuovo in mind, the reader will read with particular interest Mumford's analysis of my papers on linear systems and of later work done by others in the theory of algebraic surfaces. The reader will then realize that the theory of surfaces is still a very lively topic of research and that everything points to the likelihood that this theory will reach the degree of perfection dreamed of by Castelnuovo, except that this will not be the work of one "continuatore," but of many.

In 1950 I gave a lecture at the International Congress of Mathematicians at Harvard; the title of that lecture was "The fundamental ideas of abstract algebraic geometry" [60]. This is a good illustration of how relative in nature is what we call "abstract" at a given time. Certainly that lecture was very "abstract" for *that* time when compared with the reality of the Italian geometric school. Because it dealt only with projective varieties, that lecture, viewed at the present time, however, after the great generalization of the subject due to Grothendieck, appears to be a very, very concrete brand of mathematics. There is no doubt that the concept of "schemes" due to Grothendieck was a sound and inevitable generalization of the older concept of "variety" and that this generalization has introduced a new dimension into the conceptual content of algebraic geometry. What is more important is that this generalization has met what seems to me to be the true test of any generalization, that is, its effectiveness in solving, or throwing new light on, old problems by generalizing the terms of the problem (for example: the Riemann-Roch theorem for varieties of any dimension; the problem of the completeness of the characteristic linear series of a complete algebraic system of curves on a surface, both in characteristic zero and especially in characteristic $p \neq 0$; the computation of the fundamental group of an algebraic curve in characteristic $p \neq 0$).

But a mathematical theory cannot thrive indefinitely on greater and greater generality. A proper balance must ultimately be maintained between the generality and the concreteness of the structure studied, and usually this balance is restored after a period in which it was temporarily (and understandably) lost. There are signs at the present moment of the pendulum swinging back from "schemes," "motives," and so on toward concrete but difficult unsolved questions concerning the old pedestrian concept of a projective variety (and even of algebraic surfaces). There is no lack of such problems. It suffices to mention such questions as (1) criteria of rationality of higher varieties; (2) the study of cycles of codimension > 1 on any given variety; (3) even for divisors D on a variety there is the question of the behavior of the numerical function of n: dim $|nD|$ and finally (4) problems, such as reduction of singularities or the behavior of the zeta function, which are still unsolved when the ground field is of characteristic $p \neq 0$ (and is respectively algebraically closed or a finite field). These are new tasks that face the younger generation; I wholeheartedly wish that generation good speed and success.

Oscar Zariski

Cambridge, Massachusetts
March 1973

Bibliography of Oscar Zariski

(Entries preceded by one asterisk are reprinted in volume I; entries preceded by two asterisks are reprinted in this volume.)

[1] *I fondamenti della teoria degli insiemi di Cantor*, Period. Mat., serie 4, vol. 4 (1924) pp. 408–437.

[2] *Sulle equazioni algebriche contenenti linearmente un parametro e risolubili per radicali*, Atti Accad. Naz. Lincei Rend., Cl. Sci. Fis. Mat. Natur., serie V, vol. 33 (1924) pp. 80–82.

[3] *Gli sviluppi più recenti della teoria degli insiemi e il principio di Zermelo*, Period. Mat., serie 4, vol. 5 (1925) pp. 57–80.

[4] *Sur le développement d'une fonction algébroide dans un domaine contenant plusieurs points critiques*, C. R. Acad. Sci., Paris, vol. 180 (1925) pp. 1153–1156.

[5] *Il principio de Zermelo e la funzione transfinita di Hilbert*, Rend. Sem. Mat. Roma, serie 2, vol. 2 (1925) pp. 24–26.

[6] *R. Dedekind, Essenza e Significato dei Numeri. Continuità e Numeri Irrazionali, Traduzione dal tedesco e note storico-critiche di Oscar Zariski* ("Per la Storia e la Filosofia delle Matematiche" series), Stock, Rome, 1926, 306 pp. The notes fill pp. 157–300.

[7] *Sugli sviluppi in serie delle funzioni algebroidi in campi contenenti più punti critici*, Atti Accad. Naz. Lincei Mem., Cl. Sci. Fis. Mat. Natur., serie VI, vol. 1 (1926) pp. 481–495.

[8] *Sull'impossibilità di risolvere parametricamente per radicali un'equazione algebrica* $f(x,y) = 0$ *di genere* $p > 6$ *a moduli generali*, Atti Accad. Naz. Lincei Rend., Cl. Sci. Fis. Mat. Natur., serie VI, vol. 3 (1926) pp. 660–666.

[9] *Sulla rappresentazione conforme dell'area limitata da una lemniscata sopra un cerchio*, Atti Accad. Naz. Lincei Rend., Cl. Sci. Fis. Mat. Natur., serie VI, vol. 4 (1926) pp. 22–25.

[10] *Sullo sviluppo di una funzione algebrica in un cerchio contenente più punti critici*, Atti Accad. Naz. Lincei Rend., Cl. Sci. Fis. Mat. Natur., serie VI, vol. 4 (1926) pp. 109–112.

[11] *El principio de la continuidad en su desarrolo histórico*, Rev. Mat. Hisp.-Amer., serie 2, vol. 1 (1926) pp. 161–166, 193–200, 233–240, 257–260.

[12] *Sopra una classe di equazioni algebriche contenenti linearmente un parametro e risolubili per radicali*, Rend. Circolo Mat. Palermo, vol. 50 (1926) pp. 196–218.

[13] *On a theorem of Severi*, Amer. J. Math., vol. 50 (1928) pp. 87–92.

[14] *On hyperelliptic θ-functions with rational characteristics*, Amer. J. Math., vol. 50 (1928) pp. 315–344.

[15] *Sopra il teorema d'esistenza per le funzioni algebriche di due variabili*, Atti Congr. Internaz. Mat. 2, Bologna, vol. 4 (1928) pp. 133–138.

[16] *On the problem of existence of algebraic functions of two variables possessing a given branch curve*, Amer. J. Math., vol. 51 (1929) pp. 305–328.

[17] *On the linear connection index of the algebraic surfaces* $z^n = f(x,y)$, Proc. Nat. Acad. Sci. U.S.A., vol. 15 (1929) pp. 494–501.

[18] *On the moduli of algebraic functions possessing a given monodromie group*, Amer. J. Math., vol. 52 (1930) pp. 150–170.

[19] *On the non-existence of curves of order 8 with 16 cusps*, Amer. J. Math., vol. 53 (1931) pp. 309–318.

[20] *On the irregularity of cyclic multiple planes*, Ann. of Math., vol. 32 (1931) pp. 485–511.

[21] *On quadrangular 3-webs of straight lines in space*, Abh. Math. Sem. Univ. Hamburg, vol. 9 (1932) pp. 79–83.

[22] *On the topology of algebroid singularities*, Amer. J. Math., vol. 54 (1932) pp. 453–465.

[23] *On a theorem of Eddington*, Amer. J. Math., vol. 54 (1932) pp. 466–470.

[24] *Parametric representation of an algebraic variety*, Symposium on Algebraic Geometry, Princeton University, 1934–1935, mimeographed lectures, Princeton, 1935, pp. 1–10.

[25] *Algebraic Surfaces*, Ergebnisse der Mathematik, vol. 3, no. 5, Springer-Verlag, Berlin, 1935, 198 pp; second supplemented edition, with appendices by S. S. Abyankar, J. Lipman, and D. Mumford, Ergebnisse der Mathematik, vol. 61, Springer-Verlag, Berlin-Heidelberg-New York, 1971, 270 pp.

**[26] (with S. F. Barber) *Reducible exceptional curves of the first kind*, Amer. J. Math., vol. 57 (1935) pp. 119–141.

[27] *A topological proof of the Riemann-Roch theorem on an algebraic curve*, Amer. J. Math., vol. 58 (1936) pp. 1–14.

[28] *On the Poincaré group of rational plane curves*, Amer. J. Math., vol. 58 (1936) pp. 607–619.

[29] *A theorem on the Poincaré group of an algebraic hypersurface*, Ann. of Math., vol. 38 (1937) pp. 131–141.

[30] *Generalized weight properties of the resultant of* $n + 1$ *polynomials in* n *indeterminants*, Trans. Amer. Math. Soc., vol. 41 (1937) pp. 249–265.

[31] *The topological discriminant group of a Riemann surface of genus* p, Amer. J. Math., vol. 59 (1937) pp. 335–358.

[32] *A remark concerning the parametric representation of an algebraic variety*, Amer. J. Math., vol. 59 (1937) pp. 363–364.

[33] (In Russian) *Linear and continuous systems of curves on an algebraic surface*, Progress of Mathematical Sciences, Moscow, vol. 3 (1937).

*[34] *Some results in the arithmetic theory of algebraic functions of several variables*, Proc. Nat. Acad. Sci. U.S.A., vol. 23 (1937) pp. 410–414.

*[35] *Polynominal ideals defined by infinitely near base points*, Amer. J. Math., vol. 60 (1938) pp. 151–204.

*[36] (with O. F. G. Schilling) *On the linearity of pencils of curves on algebraic surfaces*, Amer. J. Math., vol. 60 (1938) pp. 320–324.

*[37] *Some results in the arithmetic theory of algebraic varieties*, Amer. J. Math., vol. 61 (1939) pp. 249–294.

*[38] (with H. T. Muhly) *The resolution of singularities of an algebraic curve*, Amer. J. Math., vol. 61 (1939) pp. 107–114.

*[39] *The reduction of the singularities of an algebraic surface*, Ann. of Math., vol. 40 (1939) pp. 639–689.

*[40] *Algebraic varieties over ground fields of characteristic zero*, Amer. J. Math., vol. 62 (1940) pp. 187–221.

*[41] *Local uniformization on algebraic varieties*, Ann. of Math., vol. 41 (1940) pp. 852–896.

*[42] *Pencils on an algebraic variety and a new proof of a theorem of Bertini*, Trans. Amer. Math. Soc., vol. 50 (1941) pp. 48–70.

*[43] *Normal varieties and birational correspondences*, Bull. Amer. Math. Soc., vol. 48 (1942) pp. 402–413.

*[44] *A simplified proof for the resolution of singularities of an algebraic surface*, Ann. of Math., vol. 43 (1942) pp. 583–593.

*[45] *Foundations of a general theory of birational correspondences*, Trans. Amer. Math. Soc., vol. 53 (1943) pp. 490–542.

*[46] *The compactness of the Riemann manifold of an abstract field of algebraic functions*, Bull. Amer. Math. Soc., vol. 45 (1944) pp. 683–691.

*[47] *Reduction of the singularities of algebraic three dimensional varieties*, Ann. of Math., vol. 45 (1944) pp. 472–542.

*[48] *The theorem of Bertini on the variable singular points of a linear system of varieties*, Trans. Amer. Math. Soc., vol. 56 (1944) pp. 130–140.

**[49] *Generalized semi-local rings*, Summa Brasiliensis Mathematicae, vol. 1, fasc. 8 (1946) pp. 169–195.

*[50] *The concept of a simple point of an abstract algebraic variety*, Trans. Amer. Math. Soc., vol. 62 (1947) pp. 1–52.

[51] *A new proof of Hilbert's Nullstellensatz*, Bull. Amer. Math. Soc., vol. 53 (1947) pp. 362–368.

**[52] *Analytical irreducibility of normal varieties*, Ann. of Math., vol. 49 (1948) pp. 352–361.

**[53] *A simple analytical proof of a fundamental property of birational transformations*, Proc. Nat. Acad. Sci. U.S.A., vol. 35 (1949) pp. 62–66.

**[54] *A fundamental lemma from the theory of holomorphic functions on an algebraic variety*, Ann. Mat. Pura Appl., series 4, vol. 29 (1949) pp. 187–198.

**[55] *Quelques questions concernant la théorie des fonctions holomorphes sur une variété algébrique*, Colloque d'Algèbre et Théorie des Nombres, Paris, 1949, pp. 129–134.

**[56] *Postulation et genre arithmétique*, Colloque d'Algèbre et Théorie des Nombres, Paris, 1949, pp. 115–116.

**[57] (with H. T. Muhly) *Hilbert's characteristic function and the arithmetic genus of an algebraic variety*, Trans. Amer. Math. Soc., vol. 69 (1950) pp. 78–88.

**[58] *Theory and applications of holomorphic functions on algebraic varieties over arbitrary ground fields*, Mem. Amer. Math. Soc., no. 5 (1951) pp. 1–90.

**[59] *Sur la normalité analytique des variétés normales*, Ann. Inst. Fourier (Grenoble), vol. 2 (1950) pp. 161–164.

[60] *The fundamental ideas of abstract algebraic geometry*, Proc. Internat. Cong. Math., Cambridge, Massachusetts, 1950, pp. 77–89.

**[61] *Complete linear systems on normal varieties and a generalization of a lemma of Enriques-Severi*, Ann. of Math., vol. 55 (1952) pp. 552–592.

*[62] *Le problème de la réduction des singularités d'une variété algébrique*, Bull. Sci. Mathématiques, vol. 78 (January-February 1954) pp. 1–10.

**[63] *Interprétations algébrico-géométriques du quatorzième problème de Hilbert*, Bull. Sci. Math., vol. 78 (July-August 1954) pp. 1–14.

[64] *Applicazioni geometriche della teoria delle valutazioni*, Rend. Mat. e Appl., vol. 13, fasc. 1–2, Roma (1954) pp. 1–38.

[65] (with S. Abhyankar) *Splitting of valuations in extensions of local domains*, Proc. Nat. Acad. Sci. U.S.A., vol. 41 (1955) pp. 84–90.

**[66] *The connectedness theorem for birational transformations*, Algebraic Geometry and Topology (Symposium in honor of S. Lefschetz), edited by R. H. Fox, D. C. Spencer, and A. W. Tucker, Princeton University Press, 1955, pp. 182–188.

[67] *Algebraic sheaf theory* (Scientific report on the second Summer Institute), Bull. Amer. Math. Soc., vol. 62 (1956) pp. 117–141.

[68] (with I. S. Cohen) *A fundamental inequality in the theory of extensions of valuations*, Illinois J. Math., vol. 1 (1957) pp. 1–8.

**[69] *Introduction to the problem of minimal models in the theory of algebraic surfaces*, Publ. Math. Soc. Japan, no. 4 (1958) pp. 1–89.

**[70] *The problem of minimal models in the theory of algebraic surfaces*, Amer. J. Math., vol. 80 (1958) pp. 146–184.

**[71] *On Castelnuovo's criterion of rationality* $p_a = P_2 = 0$ *of an algebraic surface*, Illinois J. Math., vol. 2 (1958) pp. 303–315.

[72] (with Pierre Samuel and cooperation of I. S. Cohen) *Commutative Algebra*, vol. I, D. Van Nostrand Company, Princeton, N.J., 1958.

[73] *On the purity of the branch locus of algebraic functions*, Proc. Nat. Acad. Sci. U.S.A., vol. 44 (1958) pp. 791–796.

**[74] *Proof that any birational class of non-singular surfaces satisfies the descending chain condition*, Mem. Coll. Sci., Kyoto Univ., series A, vol. 32, Mathematics no. 1 (1959) pp. 21–31.

[75] (with Pierre Samuel) *Commutative Algebra*, vol. II, D. Van Nostrand Company, Princeton, N.J., 1960.

[76] (with Peter Falb) *On differentials in function fields*, Amer. J. Math., vol. 83 (1961) pp. 542–556.

**[77] *On the superabundance of the complete linear systems* $|nD|$ *(n-large) for an arbitrary divisor D on an algebraic surface*, Atti del Convegno Internazionale di Geometria Algebrica tenuto a Torino, Maggio 1961, pp. 105–120.

*[78] *La risoluzione delle singolarità delle superficie algebriche immerse*, Nota I e II, Atti Accad. Naz. Lincei Rend., Cl. Sci. Fis. Mat. Natur., serie VIII, vol. 31, fasc. 3–4 (Settembre-Ottobre 1961) pp. 97–102; e fasc. 5 (Novembre 1961) pp. 177–180.

**[79] *The theorem of Riemann-Roch for high multiples of an effective divisor on an algebraic surface*, Ann. Math., vol. 76 (1962) pp. 560–615.

[80] *Equisingular points on algebraic varieties*, Seminari dell'Istituto Nazionale di Alta Matematica, 1962–1963, Edizioni Cremonese, Roma, 1964, pp. 164–177.

[81] *Studies in equisingularity I. Equivalent singularities of plane algebroid curves*, Amer. J. Math., vol. 87 (1965) pp. 507–536.

[82] *Studies in equisingularity II. Equisingularity in co-dimension 1 (and characteristic zero)*, Amer. J. Math., vol. 87 (1965) pp. 972–1006.

[83] *Characterization of plane algebroid curves whose module of differentials has maximum torsion*, Proc. Nat. Acad. Sci. U.S.A., vol. 56 (1966) pp. 781–786.

*[84] *Exceptional singularities of an algebroid surface and their reduction*, Atti Accad. Naz. Lincei Rend., Cl. Sci. Fis. Mat. Natur., serie VIII, vol. 43, fasc. 3–4 (Settembre-Ottobre 1967) pp. 135–146.

[85] *Studies in equisingularity III. Saturation of local rings and equisingularity*, Amer. J. Math., vol. 90 (1968) pp. 961–1023.

[86] *Contributions to the problem of equisingularity*, Centro Internazionale Matematico Estivo (C.I.M.E.), Questions on Algebraic varieties. III ciclo, Varenna, 7–17 Settembre 1969, Edizioni Cremonese, Roma, 1970, pp. 261–343.

[87] *An Introduction to the Theory of Algebraic Surfaces*, Lecture Notes in Mathematics, No. 83, Springer-Verlag, Berlin, 1969.

[88] *Some open questions in the theory of singularities*, Bull. Amer. Math. Soc., vol. 77 (1971) pp. 481–491.

[89] *General theory of saturation and of saturated local rings, I. Saturation of complete local domains of dimension one having arbitrary coefficient fields (of characteristic zero)*, Amer. J. Math., vol. 93 (1971) pp. 573–648.

[90] *General theory of saturation and of saturated local rings, II. Saturated local rings of dimension 1*, Amer. J. Math., vol. 93 (1971) pp. 872–964.

Part I

Holomorphic Functions

Introduction by M. Artin

Zariski's work on holomorphic functions occupied him almost exclusively from the mid-forties through 1950, culminating in his famous Memoir [58].† It was published in 1951, but Zariski had the main results several years earlier, and in fact he alluded to them already in 1946, in the Summa Brasiliensis paper [49]. The use of holomorphic functions in algebraic geometry was slowly emerging at this time, largely through the work of Weil[27] and Chevalley[5,6] on local rings, and of Zariski himself. Because of his interest in geometry in the large, Zariski quickly saw the importance of holomorphic functions along subvarieties, which he thought of as analogues of classical analytic functions. (This point of view was suppressed later in favor of viewing them as inverse limits, but we hope it may be re-emerging.)

Let W be a subvariety of a variety V, defined by a sheaf of ideals I on V. In the terminology of the Memoir, a *strongly holomorphic function* f along W is an equivalence class of sequences $\{f_i\}$ of functions regular along W, which are Cauchy sequences with respect to the I-adic metric. Zariski identifies such a function by means of its "branches" f_w (its images) in the completions of the local rings of the points $w \in W$. A *holomorphic function* f along W is given by a collection of branches $f_w \in \mathcal{O}^*_{V,w}$, one for each $w \in W$, having the following property: for some affine open covering $\{V_i\}$ of V, there are strongly holomorphic functions on each V_i whose branches are the given ones f_w if $w \in W \cap V_i$.

Thus a strongly holomorphic function may be viewed as an element of the I-adic completion of the ring of functions on V, regular along W. This led Zariski to introduce from the start the notion of the completion R^* of an essentially *arbitrary* noetherian ring R with respect to an ideal $\mathfrak{m}$.[18] The exposition in [49] is very readable and makes an excellent introduction to completions in general, as well as to his Memoir. A nice feature of both of these papers is that the analytical nature of his theory is brought out clearly. In more recent treatments of completions, such as one by Bourbaki,[4] the major novelty is their emphasis on the flatness of R^* over R, a concept which was introduced in 1956 by Serre.[26]

It may be worthwhile to review the terminology of [49], for it differs from present usage. Mainly, the term *semi-local ring*, as used by Zariski, does not mean a ring with finitely many maximal ideals; this is Chevalley's terminology and is the common one at present. Instead, Zariski uses the adjective "semi-

†Bracketed numbers refer to the Bibliography found at the front of this volume. This list includes all of Zariski's published work.

local" more generally, to refer to an $\mathfrak{m}$-adic ring R having the following property:

If $s \equiv 1$ (modulo $\mathfrak{m}$), then s is a unit in R.

He shows that this is equivalent with the condition

$\mathfrak{m} \subset \text{rad } R$,

where rad R is the Jacobson radical, called the *kernel* by Zariski. Such "semilocal" rings are now usually called *Zariski rings*. The localization of any noetherian affine scheme along a closed subscheme is a Zariski ring, and so the study of their completions leads to an essentially general theory.

In the Memoir, Zariski lays the foundations of a global theory of holomorphic functions and proves the following theorem on the invariance of the ring of holomorphic functions under rational transformations:

Let $T : V' \to V$ be a projective morphism of varieties. Assume that V is normal, and that the function field $K(V)$ is algebraically closed in $K(V')$. Let W be a closed subvariety of V, and let $W' = T^{-1}(W)$. Then the rings of holomorphic functions $\mathcal{O}^*_{V,W}$ and $\mathcal{O}^*_{V',W'}$ are isomorphic.

Zariski applied this theorem to a "principle of degeneration" which extended his Main Theorem [45] (see Mumford's introduction to Volume I) to a very general connectedness theorem. His formulation of this principle of degeneration is explained in the survey articles [55] and [60], as well as in the introduction of the Memoir. It has been translated into the language of schemes and generalized slightly by Grothendieck,[13] to give the following assertion:

Let f: $V' \to V$ be a proper morphism of noetherian schemes such that $f_*\mathcal{O}_{V'} = \mathcal{O}_V$. Then the geometric fibres of f are connected.

The method of holomorphic functions was immediately recognized as having great potential. (It was called "bahnbrechend" by Krull in his review[19] of the Memoir.) But the principle of degeneration continued to be its main application for several years, and it was not until Grothendieck's work that its full power was realized. Important as the connectedness results are, they are now only one aspect of a method which has become a main tool of modern algebraic geometry.

Basically, Grothendieck has enriched the theory in two ways:

(1) He generalized the principal theorem, replacing the structure sheaf $\mathcal{O}$ by a coherent sheaf of modules, and f_* by the higher direct images $R^q f_*$. This extension was quite natural. It was at least partly foreseen by Zariski [67, p. 120] when he learned of Serre's work[25] on coherent sheaves, and was proved by Grothendieck a year or two later.

(2) More importantly, using techniques introduced by Serre,[26] Grothendieck has proved an *existence* theorem for coherent sheaves given formally,[13] and has applied it to many problems, such as computing the tame fundamental group of a curve in characteristic p.[14]

As an aid to a casual reader wishing to browse in the Memoir, here are a few salient features. There is an interesting discussion of some open questions on page 24, on which considerable work has been done in the meantime. Problems B and C have been answered negatively in their strict interpretations, but many positive results have been obtained by Grauert,[11] Hartshorne,[16] Hironaka,[18] Matsumura,[17] and Artin.[2] Problem A, which is still open, has also been clarified by all of this work.

The first main result is Theorem 10 on page 40, which asserts that, on an affine variety, a function which is holomorphic is strongly holomorphic, i.e., is in the completion of the affine coordinate ring. This appears as the coherence property of holomorphic functions in Grothendieck's elements.[13]

Then comes Part II, with the proof of the principal theorem. In contrast with Grothendieck's more general proof for the higher direct images $R^q T_*$, which uses descending induction on q and gives little intuition as to why the theorem is true, Zariski's approach is quite constructive. It is based on the classical method of reduction to relative dimension 1, which he uses also in his first proof of the Main Theorem [45].

Let T: $V' \to V$ be the given map, which we assume to be birational. There exists a normal variety V'' dominating V' such that the composed transformation T'': $V'' \to V$ is a composition of regular maps ϕ_i: $V_{i-1} \to V_i$, and each V_{i-1} is a closed subset of $\mathbf{P}^1 \times V_i$. It is clear that V' may be replaced by V''. Then since he assumes the range normal, Zariski has to consider the normalization of each V_i, and so he is left with two cases:

(1) V' is a closed subset of $\mathbf{P}^1 \times V$;

(2) V' is the normalization of a closed subset of $\mathbf{P}^1 \times V$.

Case 2 is reduced to case 1 by a conductor argument, and the crucial step 1, the "blowing-up," is done on pages 61–64.

Very briefly, the argument goes this way: Say that V = Spec R, where R is an integrally closed domain. Then V' will be the union of the spectra of two rings $R[t]$ and $R[t^{-1}]$, for some t in the field of fractions K. A holomorphic function ϕ is given by convergent sequences

$$f_i(t) \in R[t] \quad \text{and} \quad g_i(t) \in R[t^{-1}],\ i = 1, 2, \ldots,$$

the f_i and g_i being polynomials with coefficients in R. These sequences must have the same limit on Spec $R[t, t^{-1}]$, i.e., $f_i(t) - g_i(t^{-1}) = 0$ (modulo $\mathfrak{m}^{\rho_i} R[t, t^{-1}]$), where $\rho_i \to \infty$. We may therefore write this difference in the form $h_i(t) + h_i'(t^{-1})$, where the polynomials h_i and h_i' have coefficients in $\mathfrak{m}^{\rho_i}$. Let ξ_i be the rational function

$$f_i(t) - h_i(t) = g_i(t^{-1}) + h_i'(t^{-1}).$$

Zariski notes that this function has no poles on V'! For, if D is a prime divisor and t has no pole on D, the left side of the equation shows that ξ_i has no pole on D. If t has a pole on D, the right side shows that ξ_i has no pole there. Thus, $\xi_i \in R$ since R is integrally closed, and it is immediately seen that the sequence ξ_i converges to the holomorphic function ϕ.

Zariski returned to the connectedness theorem two more times, to give proofs using valuation rings [53], [66]. Both papers involve the notion of *prime divisor*[1] in a d-dimensional local domain $(\mathfrak{o},\mathfrak{m})$. This is a discrete rank 1 valuation ring (R,P) centered at $\mathfrak{o}$, such that the transcendence degree of R/P over $\mathfrak{o}/\mathfrak{m}$ is $d-1$. In [53] he proves the following analytic result:

Theorem 2: If the local domain $\mathfrak{o}$ is analytically irreducible, then it is a *subspace* of R, i.e., $\mathfrak{p}^i \cap \mathfrak{o} \subset \mathfrak{m}^{\nu(i)}$, where $\nu(i) \to \infty$.

He concludes that if $F : V' \to V$ is a birational morphism and V is normal, then the local ring $\mathcal{O}_{V,f(p)}$ is a subspace of $\mathcal{O}_{V',p'}$. Zariski's Main Theorem follows easily from this fact.

Assuming that the singularities of the local domain can be resolved by monoidal transformations, one can deduce Theorem 2 from another important fact:

Theorem: A prime divisor (R,P) becomes of the first kind after a finite succession of monoidal transformations.

In other words, the procedure of replacing the local ring $(\mathfrak{o},\mathfrak{m})$ by the center of the valuation (R,P) on the monoidal transform of $(\mathfrak{o},\mathfrak{m})$ results, after a finite number of steps, in $(R,P) = (\mathfrak{o},\mathfrak{m})$. This theorem was extended to arbitrary regular local rings by Abhyankar.[1] It was used by Zariski several times: first in his proof [39], [69] that a birational transformation of smooth surfaces can be factored into quadratic transformations, and again in his paper [47] on resolution of singularities for three-folds. In [66], Zariski shows that the connectedness theorem for smooth varieties can also be derived from it.

A quite different and interesting approach to the principle of degeneration has been used by Chow.[7,8] He proves a completely general version constructively, using Chow coordinates and a generalized form of Hensel's lemma. Other proofs of the Main Theorem are discussed in Mumford's introduction to Volume I.

In terms of stimulus to further research, two of Zariski's most influential papers are the ones on analytic irreducibility [52] and analytic normality [59] of normal varieties. Together with the work of Chevalley[5,6] and Cohen,[9] they form the beginning of the finer structure theory of local rings. Chevalley[6] proved that the local ring $\mathfrak{o}$ of an algebraic variety is analytically unramified, i.e., that its completion $\mathfrak{o}^*$ has no nilpotent elements. In [52], Zariski reproves Chevalley's theorem and uses it to show that if $\mathfrak{o}$ is normal, $\mathfrak{o}^*$ is an integral domain. In [59] he proves the stronger assertion that $\mathfrak{o}^*$ is in fact normal, again basing the argument on Chevalley's theorem.

At the end of paper [52], Zariski considers the problem of extending his results to more general local rings and poses several questions, among them the following: Suppose $\mathfrak{o}$ is a local integral domain whose normalization $\bar{\mathfrak{o}}$ is a finite $\mathfrak{o}$-module. Is $\mathfrak{o}$ analytically unramified? This property of finiteness of integral closure was taken up by Nagata[23] and has become the central point of his theory. Nagata gave an example answering Zariski's question negatively, but he proved that if the normalization of $\mathfrak{o}'$ is a finite module for every $\mathfrak{o}'$ finite over $\mathfrak{o}$, then $\mathfrak{o}$ is analytically unramified. This allowed him to extend Zariski's result to a wide class of rings. He also delineated the theory with several other important examples. Further examples have been given recently by Ferrand and Raynaud.[10]

The most important development in this area since Nagata's book appeared is Grothendieck's use of the concept of *formal smoothness* of $\mathfrak{o}^*$ over $\mathfrak{o}$, and his definition of *excellent* rings.[15] A systematic exposition combining Nagata's and Grothendieck's ideas has been given recently by Matsumura.[21] In a related direction the notions of *henselian ring* and of *henselization* of a ring along an ideal have been introduced by Azumaya[3] and Nagata,[22] and studied by Lafon,[20] Greco,[12] Raynaud,[24] and others. The henselization is an intermediate ring between R and R^* and has many of the properties of the completion.

For non-local rings, the first result on analytic irreducibility was obtained by Zariski in 1946 ([49], Theorem 9). The theory of completions of general rings is still far from definitive, and this is an active research area at the present time.

References

1. S. Abhyankar, *On the valuations centered in a local domain*, Amer. J. Math., vol. 78 (1956) pp. 332–336.

2. M. Artin, *Algebraization of formal moduli II, existence of modifications*, Ann. of Math., vol. 91 (1970) pp.

3. G. Azumaya, *On maximally central algebras*, Nagoya Math. J., vol. 2 (1950) pp. 119–150.

4. N. Bourbaki, *Algèbre Commutative*, Ch. 3, Hermann, Paris, 1961.

5. C. Chevalley, *On the theory of local rings*, Ann. of Math., vol. 44 (1943) pp. 690–708.

6. C. Chevalley, *Intersections of algebraic and algebroid varieties*, Trans. Amer. Math. Soc., vol. 57 (1945) pp. 1–85.

7. W. -L. Chow, *On the principle of degeneration in algebraic geometry*, Ann. of Math., vol. 66 (1957) pp. 70–79.

8. W. -L. Chow, *On the connectedness theorem in algebraic geometry*, Amer. J. Math., vol. 81 (1959) pp. 1033–1074.

9. I. S. Cohen, *On the structure and ideal theory of complete local rings*, Trans. Amer. Math. Soc., vol. 59 (1946) pp. 54–106.

10. D. Ferrand and M. Raynaud, *Fibres formelles d'un anneau local noethérien*, Ann. Sci. École Norm. Sup. 4e série, vol. 3 (1970) pp. 295–311.

11. H. Grauert, *Über Modifikationen und exzeptionelle analytische Mengen*, Math. Ann., vol. 146 (1962) pp. 331–368.

12. S. Greco, *Algebras over non-local Hensel rings I, II*, J. Algebra, vol. 8 (1968) pp. 45–59; and vol. 13 (1969) pp. 48–56.

13. A. Grothendieck, *Éléments de géométric algébrique* III_1, Inst. Hautes Études Sci., Publ. Math., vol. 11 (1961) pp. 130, 122, 149.

14. A. Grothendieck, *Géométrie formelle et géométrie algébrique*, Séminaire Bourbaki, vol. 11 (1958–1959) exp. 182, Benjamin, New York, 1966.

15. A. Grothendieck, *Éléments de géométrie algébrique* IV_2, Inst. Hautes Études Sci., Publ. Math., vol. 24 (1965) p. 182ff.

16. R. Hartshorne, *Cohomological dimension of algebraic varieties*, Ann. of Math., vol. 88 (1968) pp. 403–450.

17. H. Hironaka and H. Matsumura, *Formal functions and formal embeddings*, J. Math. Soc. Japan, vol. 20 (1968) pp. 52–82.

18. H. Hironaka, *Formal line bundles along exceptional loci*, Algebraic Geometry, Papers Presented at the Bombay Colloquium, Oxford, Bombay, 1969, pp. 201–218.

19. W. Krull, Review of Zariski's Memoir [58], Zentralblatt Math., vol. 45 (1953) pp. 240–241.

20. J. -P. Lafon, *Anneau henséliens*, Bull. Soc. Math. France, vol. 91 (1963) pp. 77–107.

21. H. Matsumura, *Commutative algebra*, Benjamin, New York, 1970.

22. M. Nagata, *On the theory of Henselian rings, I, II*, Nagoya Math. J., vol. 5 (1953) pp. 45–57, and vol. 7 (1954) pp. 1–19.

23. M. Nagata, *Local Rings*, Interscience, New York, 1962.

24. M. Raynaud, *Anneaux locaux henséliens*, Lecture Notes in Math No. 169, Springer, Berlin, 1970.

25. J. -P. Serre, *Faisceaux algébriques cohérents*, Ann. of Math., vol. 61 (1955) pp. 197–278.

26. J. -P. Serre, *Géométrie algébrique et géométrie analytique*, Ann. Inst. Fourier (Grenoble), vol. 6 (1956) pp. 1–42.

27. A. Weil, *Foundations of algebraic geometry*, American Mathematical Society, Providence, R.I., 1962.

Reprints of Papers

GENERALIZED SEMI-LOCAL RINGS (*)

By Oscar Zariski

Contents:

§ 1. Introduction

The theory of local rings developed by W. Krull (1) and C. Chevalley (2) provides some of the algebraic tools which are used in the local theory of algebraic varieties. To this local theory, where one studies the properties of an algebraic variety in the neighborhood of a point, belong such analytic and algebro-geometric concepts as "holomorphic function", "analytical branch", "simple point", "local uniformizing parameters", "intersection multiplicity" etc.

(1) Dimensionstheorie in Stellenringen, J. Reine Angew. Math., vol. 179 (1938).

(2) On the theory of local rings, Ann. of Math., vol. 44 (1943).

(*) Trabalho apresentado à Sociedade de Matematica de S. Paulo.
Manuscrito recebido pela F. G. V. em Outubro de 1945.

By a suitable transcendental extension of the ground field it is possible to incorporate in the local theory certain questions pertaining to the behaviour of an algebraic variety V along a subvariety Γ of dimension >0. However, this can be done only at the cost of treating Γ as a "point" (or as a finite set of points, if Γ is reducible), over the new ground field. In this "reduction to the zero-dimensional case" some of the finer aspects of the geometric and analytical configuration represented by the pair (V, Γ) are entirely lost. For instance, the study of the class of functions which are holomorphic at *each* point of Γ requires a good deal more than the strictly local theory. Another example is the following. The "analytical irreducibility" of a variety V in the neighborhood of a point corresponds, arithmetically speaking, to the fact that the completion of the quotient ring of the point is an integral domain. If we have a finite set E of n points on V, then in the neighborhood of E, V is analytically reducible and consists of at least n analytical branches. In the case of a subvariety Γ, of dimension >0, the above reduction to the zero-dimensional case leads therefore to the untenable conclusion that V is always analytically reducible along a composite Γ, whereas in point of fact V will be analytically irreducible along Γ if it is analytically irreducible at each point of Γ *and if* Γ *is connected* (see Theorem 9, § 8).

The object of this paper is to develop a semi-local theory which should make it possible to treat such questions as above, which are halfway between the local theory and the theory in the large. We generalize some of the results of the local theory and we add some new results. This is a preliminary treatment of the subject. In a subsequent paper we shall develop further the semi-local theory and we shall apply it to the derivation of a "principle of degeneration" for abstract algebraic varieties.

§ 2. Metrics derived from null sequences of ideals.

Let R be a commutative ring and let

$$(1) \qquad \mathfrak{A}_0 \supseteqq \mathfrak{A}_1 \supseteqq \mathfrak{A}_2 \supseteqq \ldots, \qquad \mathfrak{A}_0 = R,$$

be a descending sequence of ideals $\mathfrak{A}_i$ in R. We say that (1) is a *null sequence* if

(2) $$\bigcap_{i=0}^{\infty} \mathfrak{A}_i = (0).$$

We assume that (1) is a null sequence and we introduce a corresponding metric d in R as follows: if $x, y \in R$, $x-y \neq 0$, we set

(3) $$d(x, y) = e^{-n}, \quad e \text{ real}, \quad e > 1,$$

where the integer n is defined by the condition

(4) $$x-y \equiv 0 \pmod{\mathfrak{A}_n}, \quad x-y \not\equiv 0 \pmod{\mathfrak{A}_{n+1}}.$$

If $x=y$ we set $d(x, y)=0$. The triangle axiom follows from $\mathfrak{A}_i \supseteq \mathfrak{A}_{i+1}$, and as a matter of fact the following stronger relation holds:

$$d(x, y) \leqq \max \{d(x, z), d(y, z)\}.$$

We denote by $W_{x,n}$ the closed sphere of center x and radius e^{-n}. (n, an integer). Every open sphere of center x and radius $r>0$ coincides with the closed sphere $W_{x,n}$, $n=[-\log_e r]+1$ (if $r>1$ then the sphere is the entire space R). It follows that the space is totally disconnected.

The sphere $W_{x,n}$ is the residue class mod $\mathfrak{A}_n$ which contains the element x: $W_{x,n}=x+\mathfrak{A}_n$. Hence $W_{x,n}=W_{y,n}$ if $x\equiv y \pmod{\mathfrak{A}_n}$ and $W_{x,n} \cap W_{y,n}=\emptyset$ if $x \not\equiv y \pmod{\mathfrak{A}_n}$, where $\emptyset$ denotes the empty set. Similarly, if $n<m$ we find: $W_{x,n} \supseteq W_{y,m}$ if $x\equiv y \pmod{\mathfrak{A}_n}$ and $W_{x,n} \cap W_{y,m}=\emptyset$ if $x \not\equiv y \pmod{\mathfrak{A}_n}$.

If X, Y are subsets of R, we denote by $-X$, $X-Y$ and XY the sets consisting respectively of all the elements $-x$, $x-y$, and xy, where $x \in X$ and $y \in Y$.

We have: $-W_{x,n}=W_{-x,n}$, $W_{x,n}-W_{y,n}=W_{x-y,n}$ and $W_{x,n} \cdot W_{y,n} \subseteq W_{xy,n}$, and therefore the ring operations in R are continuous. *Hence R is a metric ring.*

Given another null sequence of ideals:

$$\mathfrak{A}_0' \supseteq \mathfrak{A}_1' \supseteq \mathfrak{A}_2' \supseteq \ldots, \qquad \mathfrak{A}_0' = R,$$

$$\bigcap_{i=0}^{\infty} \mathfrak{A}_i' = (0),$$

we shall say that the two ideal sequences $\{\mathfrak{A}_i\}$ and $\{\mathfrak{A}_i'\}$ are *equivalent* if *each* ideal $\mathfrak{A}_i$ is a divisor of *some* ideal $\mathfrak{A}_j'$ *and* if *each* ideal $\mathfrak{A}_i'$ is a divisor of *some* ideal $\mathfrak{A}_j$. If we call d' the metric defined in R by the null sequence $\{\mathfrak{A}_i'\}$, then it is clear that the two metrics d and d' define in R the same topology if and only if the two given ideal sequences are equivalent. We are interested primarily in the *topological* ring R: the metric is used by us only as a convenient tool.

We shall say that the ring R is *complete* with respect to the given sequence $\{\mathfrak{A}_i\}$ if R is a complete space with respect to the metric d defined by this sequence. If we have our two equivalent sequences $\{\mathfrak{A}_i\}$ and $\{\mathfrak{A}_i'\}$, then it is seen immediately that given any positive real number ε there exists a positive number δ such that $d'(x, y) < \varepsilon$ whenever $d(x, y) < \delta$ (and viceversa, with d and d' interchanged). From this it follows that if R is complete with respect to a given null sequence $\{\mathfrak{A}_i\}$ it is also complete with respect to any null sequence $\{\mathfrak{A}_i'\}$ which is equivalent to $\{\mathfrak{A}_i\}$.

§ 3. The completion of a ring.

Following the procedure of completion of metric spaces, we can embedd the metric space R of the preceding section in a complete metric space R^* such that R is everywhere dense in R^*. Every Cauchy sequence in R^* has a (unique) limit in R^*, and every element of R^* is the limit of a Cauchy sequence in R.

A sequence $\{x_i\}$ of elements of R is a Cauchy sequence if and only if $x_i - x_{i-1}$ belongs to an ideal $\mathfrak{A}_{\nu(i)}$ whose subscript $\nu(i)$ increases indefinitely with i. The space R^* is uniquely determined, to within isometries.

The ring operations in R can be extended to R^* by the usual definitions: if $x^* = \lim x_i$ and $y^* = \lim y_i$, then $x^* + y^* = \lim (x_i + y_i)$ and $x^* y^* = \lim x_i y_i$, With these definitions, the metric space R^* becomes a metric ring which contains R as a subring and as a metric subspace.

Let $\{\mathfrak{A}_i'\}$ be a null sequence of ideals in R which is equivalent to the given sequence $\{\mathfrak{A}_i\}$ (see § 2), and let d' be the distance function in R defined by the sequence $\{\mathfrak{A}_i'\}$. It follows readily from the considerations at the end of § 2 that the completion R'^* of R with respect to the metric d' is algebraically R–isomorphic and topologically R–equivalent with R^*. Therefore the ring R^* is uniquely determined as an *abstract topological ring* by the topology defined in R by the class of null ideal sequences which are equivalent to the given sequence $\{\mathfrak{A}_i\}$. We shall call R^* the completion of R *with respect to the sequence* $\{\mathfrak{A}_i\}$.

We shall now show that the ring R^* is complete with respect to a suitable null sequence of ideals in R^* (in the sense of § 2).

Let $\{x_i\}$ be a Cauchy sequence in R and let $x^* = \lim x_i$. Let us suppose that if i is sufficiently high, the elements x_i belong to a given ideal $\mathfrak{A}_n$ of our null sequence $\{\mathfrak{A}_j\}$:

(5) $x^* = \lim x_i$; $x_i \equiv 0 \pmod{\mathfrak{A}_n}$ if i is sufficiently high.

Under this assumption we will have by (3): $d(x_i, 0) \leqq e^{-n}$ if $i \geqq i_0$, and hence $d(x^*, 0) = \lim d(x_i, 0) \leqq e^{-n}$. Conversely, if $d(x^*, 0) \leqq e^{-n}$, then $d(x_i, 0) \leqq e^{-n}$ for i sufficiently high (because each distance $d(x_i, 0)$ is a real number of the form $e^{-\nu_i}, \nu_i$-an integer), and therefore $x_i \equiv 0$ (mod $\mathfrak{A}_n$) for i sufficiently high. Consequently, if for a given n, (5) hold for one Cauchy sequence $\{x_i\}$ which converges to x^* then it holds for any Cauchy sequence which converges to x^*.

We denote by $\mathfrak{A}^*_n$ the set of elements x^* in R^* which satisfy condition (5):

$$\mathfrak{A}^*_n = \{x^* \in R^* \mid x^* \text{ satisfies } (5)\} \tag{6}$$

From the definitions of the ring operations in R^* it follows directly that $\mathfrak{A}^*_n$ is an ideal. Moreover we have obviously:

$$\mathfrak{A}^*_0 \supseteqq \mathfrak{A}^*_1 \supseteqq \mathfrak{A}^*_2 \supseteqq \ldots, \quad \mathfrak{A}^*_0 = R^*. \tag{7}$$

If $y^* = \lim y_i$ and if y^* belongs to all the ideals $\mathfrak{A}^*_n$, then $d(y^*, 0) \leqq e^{-n}$ for all integers n, and therefore $y^* = 0$. *Hence* $\{\mathfrak{A}^*_n\}$ *is a null sequence of ideals in* R^*.

Suppose that the element x^* in (5) does not belong to $\mathfrak{A}^*_{n+1}$. Since $\{x_i\}$ is a Cauchy sequence, there exists an integer N such that

$x_i - x_j \equiv 0 \pmod{\mathfrak{A}_{n+1}}$ if $i, j \geqq N$. Since $x^* \notin \mathfrak{A}^*_{n+1}$ we can find an integer $N_0 \geqq N$ such that $x_N \not\equiv 0 \pmod{\mathfrak{A}_{n+1}}$. Hence $x_i \not\equiv 0 \pmod{\mathfrak{A}_{n+1}}$ if $i \geqq N_0$, i.e.: if the element x^* belongs to $\mathfrak{A}^*_n$ but does not belong to $\mathfrak{A}^*_{n+1}$, then $x_i \equiv 0 \pmod{\mathfrak{A}_n}$, $x_i \not\equiv 0 \pmod{\mathfrak{A}_{n+1}}$, for i sufficiently high. In that case we will have $d(x_i, 0) = e^{-n}$ for all sufficiently high values of i, and consequently $d(x^*, 0) = e^{-n}$. *This shows that the metric of R^* coincides with the metric defined in the ring R^* (in the sense of § 2) by the null sequence $\{\mathfrak{A}^*_n\}$.* R^* is of course complete with respect to this sequence.

We observe that from the definition (6) of the ideals $\mathfrak{A}^*_n$ follows directly that

$$\mathfrak{A}^*_n \cap R = \mathfrak{A}_n. \tag{8}$$

§ 4. The closure of an ideal in the topological ring R.

If x is an element of R then the spheres $W_{x,n} = x + \mathfrak{A}_n$ form a local base at x for the topological ring R. Hence if E is an arbitrary subset of R and if $\bar{E}$ is the closure of E, then $x \in \bar{E}$ if and only if $(x+\mathfrak{A}_n) \cap E \neq \emptyset$, $n = 0, 1, 2, \ldots$, i.e. if and only if $x \in \bigcap_{n=0}^{\infty} (E+\mathfrak{A}_n)$. Hence

$$\bar{E} = \bigcap_{n=0}^{\infty} (E+\mathfrak{A}_n). \tag{9}$$

In particular, if E is an ideal in R then it follows from (9) that also $\bar{E}$ is an ideal.

LEMMA 1. *Let $\mathfrak{B}$ be an ideal in R and let $\bar{\mathfrak{B}}$ and $\bar{\mathfrak{B}}^*$ denote the closures of $\mathfrak{B}$ in R and R^* respectively. If $R^*\mathfrak{B}$ denotes the extended ideal in the complete ring R^*, then*

$$R^*\mathfrak{B} \subseteq \bar{\mathfrak{B}}^*, \tag{10}$$

$$R^*\mathfrak{B} \cap R \subseteq \bar{\mathfrak{B}}. \tag{10'}$$

Proof. If x^* is any element of $R^*\mathfrak{B}$, then x^* can be written in the form: $x^* = b_1x^*_1 + b_2x^*_2 + \ldots + b_sx^*_s$, where $b_i \in \mathfrak{B}$, $x^*_i \in R^*$, $x^*_i = \lim_{j \to \infty} x_{ij}$, $x_{ij} \in R$. For any given integer n we can find an integer $N = N(n)$ such that

(11) $$x^*_i - x_{ij} \equiv 0 \pmod{\mathfrak{A}^*_n}, \quad 1 \leqq i \leqq s, \quad j \geqq N.$$

If we put $x_N = b_1 x_{1N} + b_2 x_{2N} + \ldots + b_s x_{sN}$, then $x_N \in \mathfrak{B}$ and on the other hand we have, by (11), $x^* - x_N \in \mathfrak{A}^*_n$. Hence $x^* = \lim x_N$, and since $x_N \in \mathfrak{B}$, it follows that $x^* \in \bar{\mathfrak{B}}^*$, which establishes (10) and consequently also (10′)

§ 5. $\mathfrak{m}$-adic completions.

From now on we shall suppose that R is a Noetherian ring, i.e. (see Chevalley, loc. cit. Note 2, p. 690) a commutative ring with an element 1 and satisfying the maximal condition for ideals. We fix in R an ideal $\mathfrak{m}$ satisfying the condition $\bigcap_{i=1}^{\infty} \mathfrak{m}^i = (0)$. It is known that an equivalent condition on $\mathfrak{m}$ is that no element in R which is $\equiv 1 \pmod{\mathfrak{m}}$ be a zero divisor (Krull, loc. cit, Note 1; Chevalley, loc. cit., Note 2). The considerations of the preceding sections are then applicable to the special case

$$\mathfrak{A}_i = \mathfrak{m}^i, \quad (\mathfrak{A}_0 = \mathfrak{m}^0 = R).$$

The ring R, with the topology defined in it by the null sequence $\{\mathfrak{m}^i\}$, shall be called an *$\mathfrak{m}$-adic ring*. The corresponding complete ring R^* shall be referred to as *the $\mathfrak{m}$-adic completion of R*. If $R^* = R$ then we say that R is a *complete $\mathfrak{m}$-adic ring*.

Let $\mathfrak{m}'$ be the radical (3) of $\mathfrak{m}$. We have $\mathfrak{m}'^i \supseteqq \mathfrak{m}^i$, $i = 1, 2, \ldots$ On the other hand, the maximal condition for ideals in R implies that some power of $\mathfrak{m}'$ is contained in $\mathfrak{m}$. If $\mathfrak{m} \supseteqq \mathfrak{m}'^\rho$, then $\mathfrak{m}^i \supseteqq \mathfrak{m}'^{i\rho}$, and therefore the two sequences $\{\mathfrak{m}^i\}$, $\{\mathfrak{m}'^i\}$ are equivalent (§ 2). It follows that *there is no loss of generality if we assume that $\mathfrak{m}$ is an intersection of prime ideals*. Accordingly we shall assume from now on that $\mathfrak{m} =$ Radical of $\mathfrak{m}$.

The local theory of Krull corresponds to the special case where $\mathfrak{m}$ is a maximal prime ideal in R.

(3) The radical of $\mathfrak{m}$ is defined as the set of elements ω in R such that some power of ω belongs to $\mathfrak{m}$. An equivalent definition is that the radical $\mathfrak{m}'$ is the intersection of the isolated prime ideals of $\mathfrak{m}$.

In the semi-local theory of Chevalley $\mathfrak{m}$ is the intersection of a finite number of maximal prime ideals.

Let $\omega_1,\omega_2,\ldots,\omega_s$ be a finite base of $\mathfrak{m}$. We denote by $R<z_1,z_2,\ldots z_s>$, or also by $R<z>$, the ring of formal power series in s symbols $z_1,z_2,\ldots z_s$, with coefficients in R. Let $F^*(z)=\sum_{j=0}^{\infty} F_j(z)$ be a power series in $R<z>$; here $F_j(z)=F_j(z_1,z_2,\ldots,z_s)$ is a form of degree j with coefficients in R. By the formal substitution $z_i\rightarrow\omega_i$ we associate with each partial sum $\sum_{j=o}^{n} F_j(z)$ an element x_n in R:

$$x_n=\sum_{j=o}^{n} F_j(\omega), \quad n=0,\ 1,\ 2,\ \ldots.$$

Since $F_j(\omega)\equiv 0(\mathfrak{m}^j)$, the sequence $\{x_n\}$ converges to an element x^* in the $\mathfrak{m}$-adic completion R^* of R. We write formally

$$x^*=\sum_{j=o}^{\infty} F_j(\omega)=F^*(\omega).$$

We have thus a transformation τ of $R<z>$ into R^*:

$$\tau:\quad F^*(z)\rightarrow F^*(\omega), \tag{12}$$

and it is clear that τ is a homomorphism.

It is not difficult to see that τ is a homomorphism of $R<z>$ *onto* R. For let x^* be any element of R^* and let $x^*=\lim x_i$, $x_i\in R$. We have $x_i-x_{i-1}\equiv 0 \pmod{\mathfrak{m}^{\nu(i)}}$, $\nu(i)\rightarrow+\infty$. Replacing $\{x_i\}$ by a suitable subsequence we may arrange matters so as to have $x_i-x_{i-1}\equiv 0 \pmod{\mathfrak{m}^i}$. Then the difference x_i-x_{i-1} can be written as a form of degree i in $\omega_1,\omega_2,\ldots,\omega_s$ with coefficients in R: $x_i-x_{i-1}=F_i(\omega_1,\omega_2,\ldots,\omega_s)$. We put $x_o=F_o(\omega_1,\omega_2,\ldots,\omega_s)$, whence $x_i=\sum_{j=o}^{i} F_j(\omega)$ and therefore $x^*=\tau(F^*(z))$, where $F^*(z)=\sum_{j=o}^{\infty} F_j(z)$.

The ring $R<z>$ is a Noetherian ring (Chevalley, loc. cit. note 2, p. 696). Therefore also R^*, a homomorphic image of $R<z>$, is a Noetherian ring.

In the ring R^* we have the null sequence of ideals $\mathfrak{A}^*_n$ (see § 3, (5) and (6)), and we know that R^* is complete with respect to this sequence. We also have by (8)

(13) $$\mathfrak{A}^*_n \cap R = \mathfrak{m}^n.$$

Let x^* be any element of $\mathfrak{A}^*_n$: $x^* = \lim x_i$, $x_i \equiv 0 \pmod{\mathfrak{m}^n}$ for all $i \geqq n$, and $x_i - x_{i-1} \equiv 0 \pmod{\mathfrak{m}^i}$. In regard to the above construction of a power series $F^*(z)$ satisfying the condition $x^* = \tau(F^*(z))$ we observe that x_n can be expressed as a form $G_n(\omega_1, \omega_2, \ldots, \omega_s)$ of degree n, with coefficients in R. We set: $G_i(\omega) = F_i(\omega)$ for $i > n$ and

$$G^*(z) = G_n(z) + G_{n+1}(z) + \ldots$$

Then it is clear that $\tau(G^*(z)) = x^*$, but this time the power series $G^*(z)$ begins with term of degree n. Hence $G^*(z)$ belongs to the nth power of the ideal which is generated in $R<z>$ by the elements $z_1, z_2, \ldots z_s$. Since $\tau z_i = \omega_i$, it follows that x^* belongs to the nth power of the ideal $R^*\mathfrak{m}$. Hence $\mathfrak{A}^*_n \subseteqq R^*. \mathfrak{m}^n$, and consequently by (13) we conclude that

(14) $$\mathfrak{A}^*_n = R^*. \ \mathfrak{m}^n.$$

We put $\mathfrak{m}^* = R^* \mathfrak{m}$. Then we have, by (13) and (14):

(15) $$\mathfrak{A}^*_n = \mathfrak{m}^{*n}; \ \mathfrak{m}^{*n} \cap R = \mathfrak{m}^n.$$

We conclude that R^* *is a complete* $\mathfrak{m}^*$*-adic ring*

LEMMA 2. *Let R be an $\mathfrak{m}$-adic ring and let R_1 be an $\mathfrak{m}_1$-adic ring containing R as a subring. The $\mathfrak{m}$-adic ring R is a subspace of the $\mathfrak{m}_1$-adic ring R_1 if and only if* (a) $\mathfrak{m} \equiv 0 \pmod{\mathfrak{m}_1}$ *and* (b) $\mathfrak{m}_1^n \cap R \equiv 0 \pmod{\mathfrak{m}^{\nu(n)}}$, $\nu(n) \rightarrow +\infty$.

Proof. The ideals $\mathfrak{m}^n$ form a local base of the zero element in R. Similarly the ideals $\mathfrak{m}_1^n$ form a local base of the zero in R_1. Hence R is a subspace of R_1 if and only if the ideal sequences $\{\mathfrak{m}^n\}$ and

$\{\mathfrak{m}_1^n \cap R\}$ are equivalent. The condition (b) expresses the fact that every ideal in the first sequence contains some ideal in the second sequence. Condition (a) is sufficient in order that every ideal of the second sequence contain some ideal in the first sequence, for if $\mathfrak{m} \subseteq \mathfrak{m}_1$ then $\mathfrak{m}^n \subseteq \mathfrak{m}_1^n \cap R$. But (a) is also a necessary condition, for if $\mathfrak{m}_1$ contains some power of $\mathfrak{m}$ then $\mathfrak{m}_1$ necessarily contains $\mathfrak{m}$ because $\mathfrak{m}_1$ = Radical of $\mathfrak{m}_1$.

THEOREM 1. *If an $\mathfrak{m}$-adic ring R is a subring and a subspace of an $\mathfrak{m}_1$-adic ring R_1, then the $\mathfrak{m}$-adic completion R^* of R is a subring and a subspace of the $\mathfrak{m}_1$-adic completion R_1^* of R_1.* (4)

Proof. Since $\mathfrak{m} \equiv 0 \pmod{\mathfrak{m}_1}$, every Cauchy sequence $\{x_i\}$ in the $\mathfrak{m}$-adic ring R is also a Cauchy sequence in the $\mathfrak{m}_1$-adic ring R_1, and if the sequence converges to 0 in R it also converges to zero in R_1. Moreover, if $\{x_i\}$ is a Cauchy sequence in R and if its limit in R_1 is zero, then the limit of the sequence is zero also in R, in view of condition (b) of the preceding lemma. This shows that R^* is a subring of R_1^*. It is obvious that the ring R^*, regarded as the $\mathfrak{m}$-adic completion of R, has the same topology as the one which it possesses if it is regarded as the closure of the *subspace* R of the topological ring R_1^*. Hence R^* is a subspace of R_1^*.

§ 6. Closed ideals in $\mathfrak{m}$-adic rings.

If $\mathfrak{B}$ is an ideal in an $\mathfrak{m}$-adic ring R, its closure $\overline{\mathfrak{B}}$ is given according to (9) (§ 4) by the formula:

$$\overline{\mathfrak{B}} = \bigcap_{n=1}^{\infty} (\mathfrak{B} + \mathfrak{m}^n). \tag{16}$$

LEMMA 3. *An ideal $\mathfrak{B}$ in an $\mathfrak{m}$-adic ring R is a closed subset of R, or equivalently:*

$$\mathfrak{B} = \bigcap_{n=1}^{\infty} (\mathfrak{B} + \mathfrak{m}^n), \tag{17}$$

---

(4) Compare with theorem 1 in Chevalley, loc. cit., note 3. p. 698.

if an only if either of the following conditions is satisfied:

a) *Each element* ω *in* R *such that* $\omega \equiv 1$ (mod $\mathfrak{m}$) *is prime to* $\mathfrak{B}$,: *i.e.* $\mathfrak{B}: (\omega) = \mathfrak{B}$

b) $\mathfrak{p} + \mathfrak{m} \neq (1)$ *for each prime ideal* $\mathfrak{p}$ *of* $\mathfrak{B}$.

Proof. Let $\bar{R} = R/\mathfrak{B}$ and let σ denote the natural homomorphism $R \sim \bar{R}$. If $\bar{\mathfrak{m}} = \sigma\mathfrak{m} = (\mathfrak{m} + \mathfrak{B})/\mathfrak{B}$, then $\bar{\mathfrak{m}}^n = (\mathfrak{m}^n + \mathfrak{B})/\mathfrak{B}$. Hence (17) is equivalent to

$$(17') \qquad \bigcap_{n=1}^{\infty} \bar{\mathfrak{m}}^n = (\bar{0}),$$

where $\bar{0}$ denotes the element zero of $\bar{R}$. We know that (17') is equivalent to the following condition:

a') no element $\bar{\omega}$ of $\bar{R}$ which is $\equiv \bar{1}$ (mod $\bar{\mathfrak{m}}$) is a zero divisor, i.e. $\bar{\omega} \equiv \bar{1}$ (mod $\bar{\mathfrak{m}}$) implies $(\bar{0}) : (\bar{\omega}) = (\bar{0})$.

On the other hand, if ω is any element of R and if $\bar{\omega} = \sigma\,\omega$, then the conditions $\mathfrak{B}: (\omega) = \mathfrak{B}$ and $(\bar{0}): (\bar{\omega}) = (\bar{0})$ are equivalent. Hence condition a) is equivalent to (17).

If b) holds and if $\omega \equiv 1$ (mod $\mathfrak{m}$), then ω does not belong to any of the prime ideal $\mathfrak{p}$ of $\mathfrak{B}$, and hence ω is prime to $\mathfrak{B}$. On the other hand, if b) does not hold and if $\mathfrak{p}$ is a prime ideal of $\mathfrak{B}$ such that $\mathfrak{p} + \mathfrak{m} = (1)$, then $\mathfrak{p}$ contains an element ω satisfying the congruence $\omega \equiv 1$ (mod $\mathfrak{m}$), and this element is certainly not prime to $\mathfrak{B}$, for $\omega \equiv 0$ (mod $\mathfrak{p}$). We have shown that conditions a) and b) are equivalent, and this completes the proof of the lemma.

COROLLARY. *If* R *is a complete* $\mathfrak{m}$*-adic ring then every ideal in* R *is a closed subset of* R.

For if $\omega = 1 - \pi$, $\pi \in \mathfrak{m}$, is an element of R which is $\equiv 1$ (mod $\mathfrak{m}$), then $\pi^n \in \mathfrak{m}^n$, and hence the infinite series $1 + \pi + \pi^2 + \ldots$ converges to an element x of R. Since $x\,\omega = 1$, ω is a unit in R, and therefore ω is prime to any ideal in R.

---

(5) We observe that the proof makes no use of the hypothesis that $\bigcap_{n=1}^{\infty} \mathfrak{m}^n = (0)$. Hence it is true that in an arbitrary Noetherian ring R conditions (17), a and b) are all equivalent.

THEOREM 2. *If $\bar{\mathfrak{B}}$ is the closure of an ideal $\mathfrak{B}$ in an $\mathfrak{m}$-adic ring R and if $\bar{\mathfrak{B}}^*$ denotes the closure of $\mathfrak{B}$ in the $\mathfrak{m}$-adic completion R^* of R, then*

$$\bar{\mathfrak{B}}^* = R^*\mathfrak{B}, \tag{18}$$

$$\bar{\mathfrak{B}} = R^*\mathfrak{B} \cap R. \tag{18'}$$

Proof. We know [see (15), § 5] that R^* is a complete $\mathfrak{m}^*$-adic ring, where $\mathfrak{m}^* = R^*\mathfrak{m}$. By the above corollary the ideal $R^*\mathfrak{B}$ is a closed subset of R^*. Since $\mathfrak{B} \subseteqq R^*\mathfrak{B}$, we have therefore $\bar{\mathfrak{B}}^* \subseteqq R^*\mathfrak{B}$, and from this inclusion the equality (18) follows in view of (10), Lemma 1. As to (18′) it is a direct consequence of (18).

The above theorem confers a purely arithmetic meaning to the topological operation of closure as applied to ideals in an $\mathfrak{m}$-adic ring. We shall give now another arithmetic determination of the closure of an ideal $\mathfrak{B}$ which has the advantage of depending only on operations performed *within* the ring R (whereas in (18) we have to pass from R to its $\mathfrak{m}$-adic completion R^*). It will be merely a generalization of the criterion b) of lemma 3 for closed ideals.

THEOREM 3. *Let $\bar{\mathfrak{B}}$ be the closure of an ideal $\mathfrak{B}$ in an $\mathfrak{m}$-adic ring R and let $\mathfrak{B} = [\mathfrak{q}_1, \mathfrak{q}_2, \ldots, \mathfrak{q}_g]$ be a normal decomposition of $\mathfrak{B}$ into primary components. If $\mathfrak{p}_1, \mathfrak{p}_2, \ldots, \mathfrak{p}_g$ denote the prime ideals of $\mathfrak{B}$ and if*

$$(\mathfrak{p}_i, \mathfrak{m}) \neq (1) \text{ for } i = 1, 2, \ldots, h; \tag{19}$$

$$(\mathfrak{p}_i, \mathfrak{m}) = (1) \text{ for } i = h+1, h+2, \ldots, g, \tag{19'}$$

then

$$\bar{\mathfrak{B}} = [\mathfrak{q}_1, \mathfrak{q}_2, \ldots, \mathfrak{q}_h], \tag{20}$$

or equivalently [see (16)]:

$$\bigcap_{n=1}^{\infty} (\mathfrak{B} + \mathfrak{m}^n) = [\mathfrak{q}_1, \mathfrak{q}_2, \ldots \mathfrak{q}_h]. \tag{20'}$$

Proof. By (19′) there exists an element ω_i in $\mathfrak{p}_i$ such that $\omega_i \equiv 1$ (mod $\mathfrak{m}$), $i=h+1$, $h+2$, ..., g. Since $\mathfrak{p}_i$ is the prime ideal of the primary ideal $\mathfrak{q}_i$, a certain power of ω_i belongs to $\mathfrak{q}_i$, say $\omega_i^{\rho(i)} \equiv 0$ (mod $\mathfrak{q}_i$). We have $\omega_i^{\rho(i)} \equiv 1$ (mod $\mathfrak{m}$) and hence if we put

$$\omega = \prod_{i=h+1}^{g} \omega_i^{\rho(i)},$$

then

(21) $\omega \equiv 0$ (mod $[\mathfrak{q}_{h+1}, \mathfrak{q}_{h+2}, \ldots, \mathfrak{q}_g]$),

(21′) $\omega \equiv 1$ (mod $\mathfrak{m}$).

We denote by $\mathfrak{C}$ the ideal $[\mathfrak{q}_1, \mathfrak{q}_2, \ldots, \mathfrak{q}_h]$, and we consider an arbitrary element x of $\mathfrak{C}$. By (21′) we can write $\omega = 1 - \pi$, $\pi \equiv 0$ (mod $\mathfrak{m}$), and hence we have $\omega x = x - x\pi$. By (21) the product ωx belongs to $\mathfrak{B}$. Hence $x \equiv x\pi$ (mod $\mathfrak{B}$), $x\pi \equiv x\pi^2$ (mod $\mathfrak{B}$), ..., $x\pi^{n-1} \equiv x \pi^n$ (mod $\mathfrak{B}$), and consequently we have $x \equiv x\pi^n$ (mod $\mathfrak{B}$), for $n=1$, 2, ... Since $\pi \in \mathfrak{m}$, it follows that x belongs to all the ideals $\mathfrak{B}+\mathfrak{m}^n$, $n=1, 2, \ldots$, Since x was any element of $\mathfrak{C}$, we conclude that $\mathfrak{C} \subseteq \bigcap_{n=1}^{\infty} (\mathfrak{B}+\mathfrak{m}^n)$.

On the other hand, by the definition of $\mathfrak{C}$, and in view of (19) it follows from the criterion b) of lemma 3 that $\bigcap_{n=1}^{\infty} (\mathfrak{C}+\mathfrak{m}^n) = \mathfrak{C}$. Since $\mathfrak{B} \subset \mathfrak{C}$, we conclude that $\bigcap_{n=1}^{\infty} (\mathfrak{B}+\mathfrak{m}^n) \subseteq \mathfrak{C}$, and this completes the proof of the theorem [6].

Lemma 4. *If R^* is the $\mathfrak{m}$-adic completion of an $\mathfrak{m}$-adic ring R and if $\mathfrak{p}$ is a prime ideal in R which divides $\mathfrak{m}$, then $\mathfrak{p}^* = R^*\mathfrak{p}$ is a prime ideal is R^*. More generally, if an ideal $\mathfrak{B}$ in R is an intersection of prime ideals, $\mathfrak{B} = [\mathfrak{p}_1, \mathfrak{p}_2, \ldots, \mathfrak{p}_g]$, and if each of the prime ideals $\mathfrak{p}_1, \mathfrak{p}_2, \ldots, \mathfrak{p}_g$ divides $\mathfrak{m}$, then $R^*\mathfrak{B} = [\mathfrak{p}^*_1, \mathfrak{p}^*_2, \ldots, \mathfrak{p}^*_g]$, where $\mathfrak{p}^*_i = R^*\mathfrak{p}_i$.*

Proof. Let $\mathfrak{B}$ be an ideal in R which divides $\mathfrak{m}$. The relation $R^*\mathfrak{B} \cap R = \mathfrak{B}$ follows directly from (16) and (18′). If $x^* = \lim x_i$ is an

(6) Also this proof makes no use of the condition $\bigcap_{n=1}^{\infty} \mathfrak{m}^n = (0)$ which is implicit in our assumption that R is an $\mathfrak{m}$-adic ring; compare with note 5.

element of R^*, where $x_0, x_1, \ldots$ are elements of R and where we may assume that $x_n \equiv x_{n-1}$ (mod $\mathfrak{m}^n$), then x^* belongs to $R^*\mathfrak{B}$ *if and only if* x_0 *belongs to* $\mathfrak{B}$. For if x_0 belongs to $\mathfrak{B}$, then all the elements of the sequence $\{x_n\}$ belong to $\mathfrak{B}$ (since $\mathfrak{m}^n \subseteq \mathfrak{B}$, $n=1, 2, \ldots$), and consequently x^* belongs to $R^*\mathfrak{B}$, because $R^*\mathfrak{B}$ is a closed subset of R (see Corollary to Lemma 3). Conversely, if $x^* \in R^*\mathfrak{B}$ then it follows from Lemma 1 (§ 4) that x^* is the limit of a sequence of elements y_i *belonging to* $\mathfrak{B}$ [see (10)]. Since the two sequences $\{x_n\}$ and $\{y_n\}$ have the same limit we must have $x_n - y_n \equiv 0$ $(\mathfrak{m}) \equiv 0$ (mod $\mathfrak{B}$) for $n \geqq n_0$, where n_0 is a suitable integer. Consequently $x_n \equiv 0$ (mod $\mathfrak{B}$) for $n \geqq n_0$, and since $x_0 - x_n \equiv 0$ (mod $\mathfrak{m}) \equiv 0$ (mod $\mathfrak{B}$), it follows that $x_0 \in \mathfrak{B}$.

The above result shows that if $\mathfrak{B}$ is a prime ideal which divides $\mathfrak{m}$, then also $R^*\mathfrak{B}$ is a prime ideal. If $\mathfrak{B} = [\mathfrak{p}_1, \mathfrak{p}_2, \ldots, \mathfrak{p}_g], \supseteq \mathfrak{m}$, then x^* belongs to $R^*\mathfrak{B}$ if and only if $x_0 \in \mathfrak{p}_i$, $i=1, 2, \ldots, g$, i.e. if and only if $x^* \in R^*\mathfrak{p}_i$. This completes the proof of the lemma.

COROLLARY. *If* $\mathfrak{m} = [\mathfrak{p}_1, \mathfrak{p}_2, \ldots, \mathfrak{p}_s]$, *where the* $\mathfrak{p}_i$ *are prime ideal in* R, *then the ideals* $\mathfrak{p}^*_i = R^*\mathfrak{p}_i$, $i=1, 2, \ldots, s$, *are prime ideals in the* $\mathfrak{m}$*-adic completion* R^* *of* R, *and we have:*

$$\mathfrak{m}^* = R^*\mathfrak{m} = [\mathfrak{p}^*_1, \mathfrak{p}^*_2, \ldots, \mathfrak{p}^*_s],\ \mathfrak{p}^*_i \cap R = \mathfrak{p}_i.$$

§ 7. **Semi-local rings.**

Let R be an $\mathfrak{m}$-adic ring and let S denote the set of elements s of R such that $s \equiv 1$ (mod $\mathfrak{m}$). The set S is a multiplicatively closed system, and no element of S is a zero divisor since $\bigcap_{n=1}^{\infty} \mathfrak{m}^n = (0)$. Hence we can form the ring of quotients of S with respect to R. The elements of this ring are the quotients of the form a/s; $a, s \in R$, $s \in S$. Let this ring be denoted by R_S.

Let $\mathfrak{B} = [\mathfrak{q}_1, \mathfrak{q}_2, \ldots, \mathfrak{q}_g]$ be a normal decomposition of an ideal $\mathfrak{B}$ in R into primary components, and let $\mathfrak{p}_i$ be the associated prime ideal of $\mathfrak{q}_i$, $1 \leqq i \leqq g$. We assume that $\mathfrak{p}_i \cap S = \varnothing$ for $1 \leqq i \leqq h \leqq g$ and that $\mathfrak{p}_i \cap S \neq \varnothing$ for $h+1 \leqq i \leqq g$. From the theory of quotient rings it is then well known that

$$R_S . \mathfrak{B} \cap R = [\mathfrak{q}_1, \mathfrak{q}_2, \ldots, \mathfrak{q}_h].$$

On the other hand it is clear that the condition $\mathfrak{p} \cap S \neq \emptyset$ is equivalent to the condition $\mathfrak{p}+\mathfrak{m}=(1)$. Taking into account Theorem 3 (§ 6) we find therefore still another expression for the closure $\overline{\mathfrak{B}}$ of an ideal $\mathfrak{B}$ in an $\mathfrak{m}$-adic ring:

$$(22) \qquad \overline{\mathfrak{B}}=R_S \mathfrak{B} \cap R, \qquad S=\{s \mid s\equiv 1 \pmod{\mathfrak{m}}\}$$

Let $\mathfrak{M}=R_S.\,\mathfrak{m}$, and let $\frac{a}{s}$ be an element of $\mathfrak{M}^n$, $(a, s \in R, s \in S)$. Then $\frac{a}{s} = \frac{\pi'}{s'}$, where $\pi' \in \mathfrak{m}^n$ and $s' \in S$. We have $as' = s\pi'$ and $s'=1-\pi$, $\pi \in \mathfrak{m}$. Hence $a \equiv a\pi \pmod{\mathfrak{m}^n}$, $a\pi \equiv a\pi^2 \pmod{\mathfrak{m}^n}, \ldots,$ $a\pi^i = a\pi^{i+1} \pmod{\mathfrak{m}^n}$, and consequently $a \equiv a\pi^i \pmod{\mathfrak{m}^n}$, $i=1, 2, \ldots$ Since $\pi \in \mathfrak{m}$ it follows, for $i=n$, that $a\equiv 0\ (\mathfrak{m}^n)$. Hence $\bigcap_{n\ 1}^{\infty} \mathfrak{M}^n=(0)$ and $\mathfrak{M}^n \cap R=\mathfrak{m}^n$. This shows that R_S *is an* $\mathfrak{M}$*-adic ring and that the* $\mathfrak{m}$*-adic ring* R *is a subspace of* R_S, and therefore the $\mathfrak{m}$-adic completion R^* of R is a subring and subspace of the $\mathfrak{M}$-adic completion $R_S{}^*$ of R_S (Theorem 1, § 5). If $\frac{a}{s}$ is any element of R_S $(a \in R, s \in S)$, then $\frac{a}{s} = a+a\pi+a\pi^2+\ldots$, where $\pi=1-s\equiv 0 \pmod{\mathfrak{m}}$. This shows that R is everywhere dense in R_S. *Consequently* $R^*=R_S{}^*$.

It is thus seen that for the purpose of completion it is permissible to replace the ring R by the quotient ring R_S.

The $\mathfrak{M}$-adic ring R_S has the following property: *if an element* ω *in* R_S *is* $\equiv 1 \pmod{\mathfrak{M}}$ *then* ω *is a unit in* R_S. We shall use this property in order to define *semi-local* rings. However, we first wish to exclude the trivial case $\mathfrak{m}=(0)$. The condition "$\omega\equiv 1 \pmod{\mathfrak{m}}$ implies that ω is a unit in R" is trivially satisfied by every Noetherian ring R if we take for $\mathfrak{m}$ the zero ideal. But when $\mathfrak{m}=0$ then the $\mathfrak{m}$-adic ring R is a discrete space, and the operation of completion is vacuous (R always coincides with R^*). For this reason we shall define semi-local rings as follows:

Definition 1. *An* $\mathfrak{m}$*-adic ring* R *is semi-local if* $\mathfrak{m} \neq (0)$ *and if every element of* R *which is* $\equiv 1 \pmod{\mathfrak{m}}$ *is a unit in* R.

The ring of quotients R_S considered above is a semi-local $\mathfrak{M}$-adic ring if $\mathfrak{m} \neq (0)$. The completion of any $\mathfrak{m}$-adic ring R ($\mathfrak{m} \neq (0)$) is also the completion of a semi-local ring (of the ring R_S).

From now on when we speak of an $\mathfrak{m}$-adic ring *we shall always assume that* $\mathfrak{m} \neq (0)$.

THEOREM 4. *Each of the following two conditions is necessary and sufficient in order that an* $\mathfrak{m}$*-adic ring* R *be semi-local:*

a) $R = R_S$, where $S = \{s \mid s \equiv 1 \pmod{\mathfrak{m}}\}$.

b) *All ideals in* R *are closed sets.*

Proof. We have just proved that a) is a sufficient condition. On the other hand, if R is semi-local, then all elements of S are units in R, and hence $R = R_S$. From a) follows b) in view of (22). It remains to show that b) is a sufficient condition. If $s \in S$, then s is a unit in R_S, and hence by (22) the closure (in R) of the principal ideal $R.\,s$ is the unit ideal. Therefore if b) holds we must have $R.\,s = (1)$, i.e. s is a unit in R, and this shows that the $\mathfrak{m}$-adic ring R is semi-local.

COROLLARY. *Every complete* $\mathfrak{m}$*-adic ring is semi-local* (see § 6, Lemma 3, Corollary).

Let $\mathfrak{K}$ denote the intersection of all the *maximal prime ideals* of R. We shall call $\mathfrak{K}$ the *kernel* of R.

THEOREM 5. *If* $\mathfrak{m}$ *is an ideal in a Noetherian ring* R *and if* $\mathfrak{m} \neq (0)$ *then* R *is an* $\mathfrak{m}$*-adic semi-local ring if and only if* $\mathfrak{m} \equiv 0 \pmod{\mathfrak{K}}$.

Proof. Suppose that R is a semi-local $\mathfrak{m}$-adic ring, and let $\mathfrak{p}$ be a maximal prime ideal in R. The ideal $\mathfrak{p}+\mathfrak{m}$ is either $\mathfrak{p}$ or the unit ideal. If $\mathfrak{p}+\mathfrak{m} = (1)$ then $\mathfrak{p}$ contains an element s such that $s \equiv 1$ (mod $\mathfrak{m}$), and this is impossible since any such element s is a unit in R, in view of the assumption that R is semi-local. Hence we must have $\mathfrak{p}+\mathfrak{m} = \mathfrak{p}$, i.e. $\mathfrak{m} \equiv 0$ ($\mathfrak{p}$). Since this congruence must hold for any maximal prime ideal $\mathfrak{p}$, it follows that $\mathfrak{m}$ is contained in the kernel $\mathfrak{K}$.

Conversely, suppose that $(0) \neq \mathfrak{m} \subseteqq \mathfrak{K}$. If s is an element of R such that $s \equiv 1$ (mod $\mathfrak{m}$), then s is not contained in any maximal prime ideal $\mathfrak{p}$, for we have always $\mathfrak{m} \equiv 0$ ($\mathfrak{p}$). Therefore s

is necessarily a unit in R. This shows in the first place that $\bigcap_{n=1}^{\infty} \mathfrak{m}^n = (0)$, i. e that R is an $\mathfrak{m}$-adic ring, and also that R is semi-local.

The preceding theorem shows that *semi-local rings are characterized by the condition of having a non zero kernel* $\mathfrak{K}$ and that if a ring R satisfies this condition then R is a semi-local $\mathfrak{K}$-adic ring.

The local rings of Krull are those Noetherian rings in which the kernel is itself a maximal prime ideal. In the special semi-local rings studied by Chevalley the kernel is a finite intersection of maximal prime ideals.

We shall devote the rest of this section to the derivation of some theorems concerning the completions of semi-local rings and homomorphic images of these rings.

THEOREM 6. *Let R be a semi-local $\mathfrak{m}$-adic ring and let R^* be the $\mathfrak{m}$-adic completion of R. If we associate with each ideal $\mathfrak{B}$ in R its extended ideal $\mathfrak{B}^* = R^* \mathfrak{B}$ in R^*, then this association induces a* (1.1) *correspondence between the ideals $\mathfrak{B}$ in R which divide $\mathfrak{m}$ and the ideals $\mathfrak{B}^*$ in R^* which divide $\mathfrak{m}^* = R^* \mathfrak{m}$. If $\mathfrak{B}$ and $\mathfrak{B}^*$ are two corresponding ideals* ($\mathfrak{B} \supseteq \mathfrak{m}$) *then the residue class rings $R/\mathfrak{B}$, $R^*/\mathfrak{B}^*$ coincide.*

Proof. Since every ideal in a semi-local ring is a closed set (Theorem 4), it follows that $R^* \mathfrak{B} \cap R = \mathfrak{B}$. Hence the association $\mathfrak{B} \leftrightarrow \mathfrak{B}^* = R^*\mathfrak{B}$ defines a (1, 1) correspondence between the ideals in R and certain ideals in R^*. If $\mathfrak{B}$ is a divisor of $\mathfrak{m}$ then $R^* \mathfrak{B}$ is divisor of $\mathfrak{m}^*$. Conversely, let $\mathfrak{B}^*$ be an ideal in R^* which divides $\mathfrak{m}^*$ and let $\mathfrak{B}^* \cap R = \mathfrak{B}$. Then $\mathfrak{B}$ is a divisor of $\mathfrak{m}$. Every element x^* of R^* is congruent modulo $\mathfrak{m}$ to an element x_0 of R. Since $\mathfrak{m}^* \subseteq \mathfrak{B}^*$, we have therefore.

$$x \equiv x_0 \pmod{\mathfrak{B}^*}, \qquad x_0 \in R. \tag{23}$$

If $x^* \in \mathfrak{B}^*$, then (23) shows that $x_0 \in \mathfrak{B}^* \cap R$, i. e. $x_0 \in \mathfrak{B}$. Since $x^* \equiv x_0 \pmod{\mathfrak{m}^*}$, we conclude that $\mathfrak{B}^* \subseteq \mathfrak{B} + R^*\mathfrak{m} \subseteq \mathfrak{B} + R^*\mathfrak{B} = R^*\mathfrak{B}$, and consequently $\mathfrak{B}^* = R^*\mathfrak{B}$. We have thus shown that every divisor of $\mathfrak{m}^*$ is the extension of a divisor of $\mathfrak{m}$. The equality of the rings $R/\mathfrak{B}$ and $R^*/\mathfrak{B}^*$ follows from the relation $\mathfrak{B}^* \cap R = \mathfrak{B}$ and from (23).

COROLLARY *1. The correspondence* $\mathfrak{B} \leftrightarrow R^* \mathfrak{B}$ *induces a* (1, 1) *correspondence between the maximal prime ideals in R and the maximal prime ideals in* R^*. This corollary follows from the preceding theorem, from Theorem 5 and from the fact that $\mathfrak{p}$ is maximal in R if and only if $R/\mathfrak{p}$ is a field.

COROLLARY 2. *If* $\mathfrak{K}$ *is the kernel of a semi-local* $\mathfrak{m}$*-adic ring then* $R^* \mathfrak{K}$ *is the kernel* $\mathfrak{K}^*$ *of the* $\mathfrak{m}$*-adic completion* R^* *of* R. For by Corollary 1 we have $\mathfrak{K}^* \cap R = \mathfrak{K}$, and since $\mathfrak{K}^*$ is a divisor of $\mathfrak{m}^*$ it follows that $\mathfrak{K}^* = R^* \mathfrak{K}$.

THEOREM 7. *Let* $\mathfrak{m}$ *and* $\mathfrak{m}'$ *be two ideals in a Noetherian ring R such that R is semi-local both as an* $\mathfrak{m}$*-adic and as an* $\mathfrak{m}'$*-adic ring, and let* R^* *and* R'^* *be respectively the* $\mathfrak{m}$*-adic and the* $\mathfrak{m}'$*-adic completion of R. If* $\mathfrak{m}' \equiv 0 \pmod{\mathfrak{m}}$ *then* R'^* *is a subring of* R^*. *Moreover, if* $\mathfrak{m}_1$ *denotes the ideal* $R'^* . \mathfrak{m}$, *then* R^* *is the* $\mathfrak{m}_1$*-adic completion of* R'^*.

Proof. From the condition $\mathfrak{m}' \equiv 0$ ($\mathfrak{m}$) follows that every Cauchy sequence, and in particular every null sequence, $\{x_i\}$ in the $\mathfrak{m}'$-adic ring R is respectively a Cauchy sequence or a null sequence in the $\mathfrak{m}$-adic ring R. Hence R^* contain a homomorphic image of R'^*. To show that this homomorphism is an isomorphism, we have to show that if a Cauchy sequence $\{x_n\}$ in the $\mathfrak{m}'$-adic ring R is a null sequence in the $\mathfrak{m}$-adic ring R, then $\{x_n\}$ is also a null sequence in the $\mathfrak{m}'$-adic ring R. We have, by hypothesis, $x_n \equiv x_i$ (mod $\mathfrak{m}'^{\rho(n)}$), $i \geqq n$, $\lim \rho(n) = +\infty$, and also $x_i \equiv 0$ (mod $\mathfrak{m}^{\nu(i)}$), $\lim \nu(i) = +\infty$. Hence $x_n \in \mathfrak{m}'^{\rho(n)} + \mathfrak{m}^{\nu(i)}$ for $i = 1, 2, \ldots$ Since $\nu(i) \rightarrow +\infty$ and since R is a *semi-local* $\mathfrak{m}$-adic ring, we have $\bigcap_{i=1}^{\infty} (\mathfrak{m}'^{\rho(n)} + \mathfrak{m}^{\nu(i)}) = \mathfrak{m}'^{\rho(n)}$ [Theorem 4, b); § 6, (16)]. Hence $x_n \equiv 0$ (mod $\mathfrak{m}'^{\rho(n)}$), $\lim \rho(n) = +\infty$, and this shows that $\{x_n\}$ is a null sequence in the $\mathfrak{m}'$-adic ring R, and that therefore R'^* is a subring of R^*.

We now consider the ideal $\mathfrak{m}_1 = R'^* \mathfrak{m}$ and we observe that $\mathfrak{m}_1^n = R'^* \mathfrak{m}^n$ and hence $\mathfrak{m}_1^n \cap R = \mathfrak{m}^n$, since R is, by hypothesis, a semi-local $\mathfrak{m}'$-adic ring (7). Moreover, since $\mathfrak{m}$ is contained in the kernel $\mathfrak{K}$ of R (Theorem 5), the ideal $\mathfrak{m}_1$ is contained in the kernel

(7) The semi-local character of the $\mathfrak{m}'$-adic ring R is already a consequence of the assumption that R is a semi-local $\mathfrak{m}$-adic ring and that $\mathfrak{m}' \equiv 0$ (mod $\mathfrak{m}$) [see Theorem 5], provided we assume furthermore that $\mathfrak{m}' \neq (0)$

$R'^*\mathfrak{K}$ of R'^* (Corollary 2 to Theorem 6). From these remarks it follows that R'^* can be regarded as a (semi-local) $\mathfrak{m}_1$-adic ring and that if it is so regarded then R'^* contains the $\mathfrak{m}$-adic ring R as a subspace (§ 5, Lemma 2). Hence, by Theorem 1, the complete $\mathfrak{m}^*$-adic ring R^* ($\mathfrak{m}^*=R^*\mathfrak{m}$) is a subring and a subspace of the $\mathfrak{m}_1$-adic completion of R'^*. Let this last mentioned completed ring be denoted by R^*_1. We have then that:

a) the $\mathfrak{m}^*$-adic complete ring R^* is a subring and subspace of the complete $\mathfrak{m}^*_1$-adic ring R^*_1 ($\mathfrak{m}^*_1=R^*_1\,\mathfrak{m}_1$).

We have proved above that R'^* is a subring of R^*. We have $\mathfrak{m}_1=R'^*\mathfrak{m}$ and $\mathfrak{m}^*=R^*\mathfrak{m}$, hence

$$\mathfrak{m}_1\equiv 0 \pmod{\mathfrak{m}^*}. \tag{24}$$

On the other hand we have $\mathfrak{m}^{*n} \cap R'^*=R^*\mathfrak{m}^n \cap R'^*\subseteq R^*_1\,\mathfrak{m}^n\cap R'^*$ (by *a*)$=R^*_1.\ R'^*\,\mathfrak{m}^n\cap R'^*=R^*_1\,\mathfrak{m}_1{}^n\cap R'^*=\mathfrak{m}_1{}^n$, i. e. $\mathfrak{m}^{*n}\cap R'^*=\mathfrak{m}_1{}^n$, and consequently, in view of (24):

$$\mathfrak{m}^{*n}\cap R'^*=\mathfrak{m}_1{}^n. \tag{24'}$$

Relations (24) and (24′) show that the $\mathfrak{m}_1$-adic ring R'^* is (not only a subring but also) a subspace of the $\mathfrak{m}^*$-adic complete ring R^*. Consequently:

b) The $\mathfrak{m}^*_1$-adic complete ring R^*_1 (the $\mathfrak{m}_1$-adic completion of R'^*) is a subring and a subspace of the $\mathfrak{m}^*$-adic complete ring R^*.

From *a*) and *b*) we conclude that the two complete rings R^*_1 and R^* are identical as topological rings, and this completes the proof of the theorem.

COROLLARY. *If R is complete $\mathfrak{m}$-adic ring and if $\mathfrak{m}'\equiv 0\ (\mathfrak{m})$ then R is also a complete $\mathfrak{m}'$-adic ring.* (8).

If in the above theorem we take for $\mathfrak{m}$ the kernel $\mathfrak{K}$ of R, the condition $\mathfrak{m}'\equiv 0 \pmod{\mathfrak{m}}$ is automatically satisfied whenever $\mathfrak{m}'$ is an ideal in R such that R is a semi-local $\mathfrak{m}'$-adic ring (Theorem

(8) However, the $\mathfrak{m}$-adic topology of R and the $\mathfrak{m}'$-adic topology of R are in general distinct.

5). Hence the completion of a semi-local ring R with respect to its kernel $\mathfrak{K}$ is a universal ring for all $\mathfrak{m}$-adic completions of R, provided only such ideals $\mathfrak{m}$ are considered in R that satisfy the condition that R is a semi-local $\mathfrak{m}$-adic ring.

THEOREM 8. *Let $\bar{R}$ be a homomorphic image of an $\mathfrak{m}$-adic ring R and let $\bar{\mathfrak{m}}$ be the ideal in R which is the image of $\mathfrak{m}$.*

a) If the nucleus $\mathfrak{N}$ of the homomorphism is a closed ideal in R then $\bar{R}$ is an $\bar{\mathfrak{m}}$-adic ring, and conversely.

b) If $\mathfrak{m} \not\equiv 0 \pmod{\mathfrak{N}}$ and if every maximal prime ideal in R which divides $\mathfrak{N}$ also divides $\mathfrak{m}$, then $\bar{R}$ is a semi-local $\bar{\mathfrak{m}}$-adic ring, and conversely. In particular, if R is a semi-local $\mathfrak{m}$-adic ring then the condition $\mathfrak{m} \not\equiv 0 \pmod{\mathfrak{N}}$ is necessary and sufficient in order that $\bar{R}$ be a semi-local $\bar{\mathfrak{m}}$-adic ring.

c) If R is a complete $\mathfrak{m}$-adic ring, then $\bar{R}$ is a complete $\bar{\mathfrak{m}}$-adic ring.

Proof.

a) Let τ denote the given homomorphism $R \sim \bar{R}$. We have $\tau^{-1}\bar{\mathfrak{m}}^n = \mathfrak{N} + \mathfrak{m}^n$. Hence $\bigcap_{n=1} \bar{\mathfrak{m}}^n = (\bar{0})$ if and only if $\bigcap_{n=1}^{\infty} (\mathfrak{N} + \mathfrak{m}^n) = \mathfrak{N}$, i..e if and only if $\mathfrak{N}$ is a closed ideal in the $\mathfrak{m}$-adic ring R.

b) The ring $\bar{R}$ is an $\bar{\mathfrak{m}}$-adic semi-local ring if and only $\bar{\mathfrak{m}} \neq 0$ and $\bar{\mathfrak{m}}$ is contained in every maximal prime ideal of $\bar{R}$ (Theorem 5). Since the maximal prime ideals of $\bar{R}$ are the τ-images of the maximal prime ideals in R which divide $\mathfrak{N}$, the assertion follows.

c) We have only to show that every Cauchy sequence $\{\bar{x}_n\}$ in $\bar{R}$ has a limit. We may assume that $\bar{x}_n - \bar{x}_{n-1} \equiv \bar{0} \pmod{\bar{\mathfrak{m}}^n}$. That means that there exists an element v_n in $\mathfrak{m}^n$ such that $\tau v_n = \bar{x}_n - \bar{x}_{n-1}$. Let x_0 be some element in R such that $\tau x_0 = \bar{x}_0$. We set $x_n = x_0 + \sum_{i=n}^{n} v_i$. The sequence $\{x_n\}$ converges to some element x^* in R, since R is complete and since $v_n \equiv 0 \pmod{\mathfrak{m}^n}$. We have $\tau x_n = \tau x_{n-1} + \tau v_n = \bar{x}_n + \tau x_{n-1} - \bar{x}_{n-1} = \bar{x}_n + \tau x_{n-2} - \bar{x}_{n-2} = \ \ldots \ = \bar{x}_n + \tau \bar{x}_0 - \bar{x}_0 = \bar{x}_n$.

Hence $\lim \bar{x}_n = \tau x^*$.

THEOREM 9. *Let R^* be the $\mathfrak{m}$-adic completion of an $\mathfrak{m}$-adic ring R and let $\mathfrak{N}$ be a closed ideal in R. If $\bar{\mathfrak{m}}$ denotes the ideal $(\mathfrak{m}+\mathfrak{N})/\mathfrak{N}$ in the ring $\bar{R}=R/\mathfrak{N}$, then the ring $\bar{R}^*=R^*/R^*\mathfrak{N}$ is the $\bar{\mathfrak{m}}$-adic completion of the ring $\bar{R}$.*

Proof. Since $\mathfrak{N}$ is a closed ideal, we have $R^*\mathfrak{N} \cap R=\mathfrak{N}$, whence *$\bar{R}$ is a subring of $\bar{R}^*$*. By theorem 8, part c, *the ring $\bar{R}^*$ is a complete $\bar{\mathfrak{m}}^*$-adic ring*, where $\bar{\mathfrak{m}}^*=R^*(\mathfrak{m}+\mathfrak{N})/R^*\mathfrak{N}$. We have $\bar{\mathfrak{m}}^{*n} \cap \bar{R}=R^*(\mathfrak{m}^n+\mathfrak{N})/R^*\mathfrak{N} \cap R/\mathfrak{N}=[R^*(\mathfrak{m}^n+\mathfrak{N}) \cap R]/\mathfrak{N}=[\mathfrak{m}^n+\mathfrak{N}]/\mathfrak{N}$ (9) $=\bar{\mathfrak{m}}^n$. *Therefore $\bar{R}$ is a subspace of $\bar{R}^*$*. The same reasoning as the one employed in the proof of Theorem 8, part c, shows that *$\bar{R}$ is everywhere dense in $\bar{R}^*$*. This completes the proof of the theorem.

§ 8. Analytically irreducible semi-local rings.

Let R be a semi-local ring with kernel $\mathfrak{K}$.

DEFINITION 2. *R is analytically irreducible if its $\mathfrak{K}$-adic completion is an integral domain.*

This definition implies, of course, that if R is an analytically irreducible semi-local ring then R itself must be an integral domain.

DEFINITION. 3. *Let R be a Noetherian ring which has no zero divisors and let $\mathfrak{p}$ be a maximal prime ideal in R. The ring R is locally irreducible at $\mathfrak{p}$ if the $\mathfrak{p}$-adic completion of R is an integral domain.*

The above definitions are suggested by algebro-geometric considerations. Let V be an irreducible algebraic variety in an affine n-space, over a ground field k, and let $(\xi_1, \xi_2, \ldots, \xi_n)$ be the general point of V. We let $R=k[\xi_1, \xi_2, \ldots, \xi_n]$. If $\mathfrak{p}$ is a maximal ideal in R and P is the corresponding point of V, and if we regard V as an analytical variety V^* in the neighborhood of P, then the arithmetic meaning of the statement: "V is locally irreducible at P" (i.e. "V^* is irreducible") is exactly this: the $\mathfrak{p}$-adic completion of R is an integral domain (compare with Definition 3).

(9) Observe that $\mathfrak{m}^n + \mathfrak{N} + \mathfrak{m}^i = \mathfrak{m}^n + \mathfrak{N}$ if $i \geq n$ and that consequently $\bigcap_{i=1}^{\infty} (\mathfrak{m}^n + \mathfrak{N} + \mathfrak{m}^i) = \mathfrak{m}^n + \mathfrak{N}$.

Let W be an algebraic subvariety of V, not necessarily an irreducible one. Let $\mathfrak{m}$ be the ideal of W in R, i.e. the ideal consisting of those elements of R which vanish on every irreducible component of W. We point out explicitly that we are in an affine n-space (since $\xi_1, \xi_2, \ldots, \xi_n$ are non-homogeneous coordinates) and that therefore all points and subvarieties of V which are being considered are at finite distance. The ideal $\mathfrak{m}$ is, of course, its own radical, i.e. $\mathfrak{m}$ is a finite intersection of prime ideals.

We apply to R and $\mathfrak{m}$ the quotient ring construction given in the beginning of § 7. The ring R_S is a semi-local $\mathfrak{M}$-adic ring where $\mathfrak{M} = R_S.\ \mathfrak{m}$. The maximal prime. ideals of R_S arise (by extension) from those maximal prime ideals in R which divide $\mathfrak{m}$, i.e. which represent points of V which lie on W. Thus the structure of the ring R_S controls the *semi-local geometry of V along W*. If we apply Hilbert's Nullstellensatz to the irreducible components of W we see immediately that $\mathfrak{m}$ coincides with the intersection of the maxima prime ideals in R which divide $\mathfrak{m}$. *Hence the radical* $\mathfrak{R}$ *of* R_S *coincides with* $\mathfrak{M}$.

The variety V may very well be locally reducible at the *generic* point of W (i.e. at all points of W outside proper algebraic subvarieties of the components of W) and V may even be locally reducible at each point W, and still V may be at the same time *analytically irreducible along* W. That may happen because the analytical irreducible branches of V at the variable point P of W may undergo a transitive group of permutations as P varies on W. It is not difficult to see that the analytical irreducibility of V along W has, arithmetically speaking, just this meaning: *the* $\mathfrak{R}$*-adic completion of* R_S *an integral domain* (compare with Definition 2: the ring R of the definition is now the ring R_S).

DEFINITION 4. *The kernel* $\mathfrak{R}$ *of a semi-local ring R is connected if it is not possible to write* $\mathfrak{R}$ *in the form:* $\mathfrak{R} = \mathfrak{A} \cap \mathfrak{B}$, *where* $\mathfrak{A}$ *and* $\mathfrak{B}$ *are ideals in R satisfying the conditions;*

$$\mathfrak{A} + \mathfrak{B} = (1), \qquad \mathfrak{A} \neq (1), \qquad \mathfrak{B} \neq (1). \tag{25}$$

In the geometric set-up discussed above, the connectivity of the kernel $\mathfrak{K}$ of R_S means that W is not the sum of two algebraic varieties neither of which is empty and which have no points in common.

Suppose that R is a semi-local ring whose kernel $\mathfrak{K}$ is not connected: $\mathfrak{K}=\mathfrak{A}\cap\mathfrak{B}$, where the ideals $\mathfrak{A}$ and $\mathfrak{B}$ satisfy (25). In the construction of the quotient ring R_S given in § 7 we replace the ideal $\mathfrak{m}$ by $\mathfrak{A}$ and $\mathfrak{B}$ respectively, and we denote by R_a and R_b the corresponding quotient rings. Let $\mathfrak{K}_a=\mathfrak{A}.\ R_a$, $\mathfrak{K}_b=\mathfrak{B}.\ R_b$.

It is easily seen that $\mathfrak{K}_a$ and $\mathfrak{K}_b$ are the kernels of the rings R_a and R_b respectively. By a reasoning identical to that employed by Chevalley (loc. cit. note 2, Proposition 8, p. 700) it can be shown that the $\mathfrak{K}$-adic completion of R is isomorphic to the direct sum of the $\mathfrak{K}_a$-adic completion of R_a and the $\mathfrak{K}_b$-adic completion of R_b.

It follows that the kernel $\mathfrak{K}$ of an analytically irreducible semi-local ring must be connected. However, it is clear geometrically that the connectivity of the kernel does not by itself guarantee the analytical irreducibility of a semi-local ring. It is our main pur pose in this section to prove that *the connectivity of the kernel $\mathfrak{K}$ together with the local irreducibility of R at each of its maximal prime ideals $\mathfrak{p}$ are sufficient for the analytical irreducibility of the semi-local ring R.*

For the proof of this result we prove first several lemmas.

LEMMA 5. *Let R be a Noetherian integral domain and let $\mathfrak{p}$ and $\mathfrak{p}_1$ be two prime ideals in R such that $\mathfrak{p}$ is maximal in R and $\mathfrak{p}_1\equiv 0$* (mod $\mathfrak{p}$). *If R is locally irreducible at $\mathfrak{p}$, then the symbolic power* [10] *$\mathfrak{p}_1^{(i)}$ of $\mathfrak{p}_1$ is contained in a power $\mathfrak{p}^\nu$ of $\mathfrak{p}$ whose exponent $\nu=\nu(i)$ increases indefinitely with i.*

Proof. It $\tilde{R}$ denotes the quotient ring $R_{\mathfrak{p}}$ and if $\tilde{\mathfrak{p}}=\tilde{R}.\mathfrak{p}$ and $\tilde{\mathfrak{p}}_1=\tilde{R}.\mathfrak{p}_1$, then $\tilde{\mathfrak{p}}_1^{(i)}\cap R=\mathfrak{p}_1^{(i)}$ and $\tilde{\mathfrak{p}}^\nu\cap R=\mathfrak{p}^\nu$. This shows that it is

(10) In a normal decomposition of an ordinary power $\mathfrak{p}_1^i$ into primary components there occurs one primary component $\mathfrak{q}$ belonging to the prime ideal $\mathfrak{p}_1$. while all the remaining primary components belong to embedded prime ideals. The only isolated primary component $\mathfrak{q}$ of $\mathfrak{p}_1^i$ is the symbolic power $\mathfrak{p}_1^{(i)}$ of $\mathfrak{p}_1$.

sufficient to prove the lemma under the assumption $R=R_{\mathfrak{p}}$. We shall therefore assume that R is a local $\mathfrak{p}$-adic integral domain.

Let R^* be the $\mathfrak{p}$-adic completion of R and let $\mathfrak{p}^*=R^*\mathfrak{p}$. From the definition of the symbolic power $\mathfrak{p}_1^{(i)}$ it follows that there exists in R an element c such that:

$$c\ \mathfrak{p}_1^{(i)}\equiv 0 \pmod{\mathfrak{p}_1^{i}}, \qquad c\not\equiv 0\ (\mathfrak{p}_1). \tag{26}$$

We consider the ideal $\mathfrak{A}^*_i=R^*\mathfrak{p}_1^{(i)}$. By (26) we have $c\ \mathfrak{A}^*_i\equiv 0$ $(\mathrm{mod}\ \mathfrak{A}^{*i}_1)$, and hence if $\mathfrak{P}^*$ denotes an *isolated* prime ideal of $\mathfrak{A}^*_1$ then

$$c\ \mathfrak{A}^*_i\equiv 0 \pmod{\mathfrak{P}^{*(i)}}. \tag{27}$$

We show next that

$$\mathfrak{P}^*\cap R=\mathfrak{p}_1 \tag{28}$$

for *any* prime ideal $\mathfrak{P}^*$ of $\mathfrak{A}^*_1$ $(=R^*\mathfrak{p}_1)$. For that it is necessary to show that if a is an element of R not in $\mathfrak{p}_1$, then $R^*\,\mathfrak{p}_1\colon R^*.a= =R^*\,\mathfrak{p}_1$. Let x^* be an element of the quotient $R^*.\mathfrak{p}_1\colon R^*.a$, and let $x^*=\lim\ x_i$, where $x_i\ \epsilon\ R$ and $x^*\equiv x_i$ $(\mathrm{mod}\ \mathfrak{p}^{*i})$. We have $ax^*\ \epsilon\ R^*.\mathfrak{p}_1$, and also $ax^*\equiv ax_i$ $(\mathrm{mod}\ a\mathfrak{p}^{*i})$, whence $ax_i\ \epsilon\ R^*.(a\mathfrak{p}^i+\mathfrak{p}_1)\cap R=R.(a\mathfrak{p}^i+\mathfrak{p}_1)$. Therefore there exists an element ω_i in $\mathfrak{p}^i$ such that $a(x_i-\omega_i)\equiv 0$ $(\mathrm{mod}\ \mathfrak{p}_1)$. Since $a\not\equiv 0$ $(\mathrm{mod}\ \mathfrak{p}_1)$, we have $x_i\equiv\omega_i$ $(\mathrm{mod}\ \mathfrak{p}_1)$, $x_i\ \epsilon\ \mathfrak{p}_1+\mathfrak{p}^i$, and consequently $x^*\ \epsilon\ R^*.\ \mathfrak{p}_1+\mathfrak{p}^{*i}$, $i=1, 2, \ldots$, whence $x^*\ \epsilon\ R^*\ \mathfrak{p}_1$. Since x^* is any element of $R^*.\ \mathfrak{p}_1\colon R^*a$, it follows that a is prime to $R^*.\ \mathfrak{p}_1$, as was asserted.

If $\mathfrak{P}^*$ is an isolated prime ideal of $\mathfrak{A}^*_1$, then it follows from (27), (28), and from $c\not\equiv 0$ $(\mathfrak{p}_1)$ that $\mathfrak{A}^*_i\equiv 0$ $(\mathfrak{P}^{*(i)})$. Now we shall use our assumption that R is locally irreducible at $\mathfrak{p}$. By this assumption the ring R^* is an integral domain, and in Noetherian integral domain the intersection of all symbolic powers $\mathfrak{P}^{*(i)}$ of a prime ideal $\mathfrak{P}^*$ is the zero ideal. Hence, in view of the preceding congruence, we conclude that $\bigcap_{i=1}^{\infty}\mathfrak{A}^*_i=(0)$. By a theorem on local rings due to Chevalley (loc. cit. note 2, p. 695, Lemma 7) it follows then that $\mathfrak{A}^*_i \subseteq \mathfrak{p}^{*\nu(i)}$, Lim $\nu(i) = +\infty$. This completes the proof of the lemma since $\mathfrak{p}_1^{(i)} = \mathfrak{A}^*_i \cap R$ and $\mathfrak{p}^{\nu(i)} = \mathfrak{p}^{*\nu(i)} \cap R$.

Let $\mathfrak{m}$ be an ideal in a Noetherian integral domain and let $\mathfrak{p}$ be a maximal prime ideal in R which divides $\mathfrak{m}$. The $\mathfrak{p}$-adic completion $R^*_{\mathfrak{p}}$ of R contains a homomorphic image of the $\mathfrak{m}$-adic completion R^* of R (see the first few lines of the proof of Theorem 7, § 7). Let this homomorphism be denoted by $\tau_{\mathfrak{p}}$ If x^* is any element of R^*, its image $\tau_{\mathfrak{p}} \cdot x^*$ in $R^*_{\mathfrak{p}}$ shall be referred to as *the analytical element of* x^* *at* $\mathfrak{p}$.

LEMMA 6. *Let R^* be an $\mathfrak{m}$-adic completion of a Noetherian integral domain R, and let x^* be an element of R^*. If every analytical element of x^* vanishes (i. e. if $\tau_{\mathfrak{p}} x^* = 0$ for every maximal prime ideal $\mathfrak{p}$ in R which divides $\mathfrak{m}$) then $x^* = 0$.*

Proof. Let for a given integer n

$$\mathfrak{m}^n = [\mathfrak{q}_1, \mathfrak{q}_2, \ldots, \mathfrak{q}_g] \tag{29}$$

be a normal decomposition of $\mathfrak{m}^n$ into primary components, and let $x^* = \lim x_i$, where $x_i \in R$ and $x_i \equiv x_j \pmod{\mathfrak{m}^i}$ for all $j \geqq i$. We fix a maximal prime ideal $\mathfrak{p}_{os}$ which divides $\mathfrak{q}_s$, $s = 1, 2, \ldots, g$. The hypothesis $\tau_{\mathfrak{p}_{os}} x^* = 0$ implies that x_j belongs to a power of $\mathfrak{p}_{os}$ whose exponent increases indefinitely with j. Hence we have for *any* integer i: $x_i \in \bigcap_{\rho=1}^{\infty} (\mathfrak{m}^i + \mathfrak{p}_{os}^{\rho})$, and in particular $x_n \in \bigcap_{\rho=1}^{\infty} (\mathfrak{m}^n + \mathfrak{p}_{os}^{\rho})$, i.e. x_n belongs to the closure of the ideal $\mathfrak{m}^n$ in the $\mathfrak{p}_{os}$-adic ring R. By theorem 3, § 6, this implies that $x_n \in \mathfrak{q}_s$, $s = 1, 2, \ldots, g$, and hence, by (29), $x_n \in \mathfrak{m}^n$. This proves that $x^* = 0$, as asserted.

LEMMA 7. *Let R^* be an $\mathfrak{m}$-adic completion of a Noetherian integral domain R and let $\mathfrak{p}_1$ be an isolated prime ideal of the ideal $\mathfrak{m}$ in R. Let x^* be an element of R^*. If the analytical element of x^* vanishes at one maximal prime ideal $\mathfrak{p}_o$ which divides $\mathfrak{p}_1$, then it also vanishes at any other maximal prime ideal $\mathfrak{p}$ which divides $\mathfrak{p}_1$, provided R is locally irreducible at $\mathfrak{p}$.*

Proof. We use the notation of the proof of the preceding lemma. Since $\mathfrak{p}_1$ is an isolated prime ideal of $\mathfrak{m}$, one of the components $\mathfrak{q}_s$

in (29) is $\mathfrak{p}_1^{(n)}$, say $\mathfrak{q}_1 = \mathfrak{p}_1^{(n)}$. We take for $\mathfrak{p}_{o1}$ the given ideal $\mathfrak{p}_o$ and we therefore find that $x_n \in \mathfrak{p}_1^{(n)}$. If $\mathfrak{p}$ is any maximal prime in R which divides $\mathfrak{p}_1$ and if we assume that R is locally irreducible at $\mathfrak{p}$, then it follows from Lemma 5 that $x_n \in \mathfrak{p}^{\nu(n)}$, $\lim \nu(n) = +\infty$ This shows that $\tau_{\mathfrak{p}} x^* = 0$, as asserted.

We are now ready for the proof of our theorem which we shall here re-state:

Theorem 9. *Let R be a semi-local integral domain and let $\mathfrak{R}$ be the kernel of R. If $\mathfrak{R}$ is connected and if the ring R is locally irreducible at each of its maximal prime ideal, then R is analytically irreducible.*

Proof. Every maximal prime ideal in R divides the kernel $\mathfrak{R}$. We fix one maximal prime ideal $\mathfrak{p}_o$ in R. We have $\mathfrak{R} \subseteqq \mathfrak{p}_o$ and therefore we have our homomorphic mapping $\tau_{\mathfrak{p}_o}$ of the $\mathfrak{R}$-adic completion R^* of R into the $\mathfrak{p}_o$-adic completion $R^*_{\mathfrak{p}_o}$ of R. Our proof of the theorem will consist in showing that $\tau_{\mathfrak{p}_o}$ *is an isomorphism*, i.e. that R^* is a subring of $R^*_{\mathfrak{p}_o}$, for then it will follow that R^* is an integral domain, since, by our hypothesis, $R^*_{\mathfrak{p}_o}$ is an integral domain.

We have to show that if x^* is an element of R^* then $\tau_{\mathfrak{p}_o} x^* = 0$ implies $x^* = 0$. Let:

$$\mathfrak{R} = [\mathfrak{p}_1, \mathfrak{p}_2, \ldots, \mathfrak{p}_g]$$

be a normal representation of the kernel $\mathfrak{R}$ as a finite intersection of prime ideals. Since $\mathfrak{R}$ is connected, we can find a set of g-1 maximal prime ideals $\mathfrak{p}_{o1}, \mathfrak{p}_{o2}, \ldots, \mathfrak{p}_{o,g-1}$ in R such that

$$(30) \qquad [\mathfrak{p}_1, \mathfrak{p}_2, \ldots, \mathfrak{p}_i] + \mathfrak{p}_{i+1} \equiv 0 \ (\mathfrak{p}_{o,i}), \ i = 1, 2, \ldots, g\text{-}1.$$

We may assume without loss of generality that $\mathfrak{p}_1 \equiv 0$ $(\mathfrak{p}_o)$. From $\tau_{\mathfrak{p}_o} x^* = 0$ and from our assumption that R is everywhere locally irreducible, it follows, by Lemma 7, that $\tau_{\mathfrak{p}} x^* = 0$ for every maximal prime ideal $\mathfrak{p}$ which divides $\mathfrak{p}_1$ In particular, we have

$\tau_{\mathfrak{p}_{o1}} x^* = 0$. Since $\mathfrak{p}_{o1}$ is also a divisor of $\mathfrak{p}_2$, it follows in the same fashion that $\tau_{\mathfrak{p}} x^* = 0$ for every maximal prime ideal $\mathfrak{p}$ which divides $\mathfrak{p}_2$. In particular, we have $\tau_{\mathfrak{p}_{o2}} x^* = 0$, and since $\mathfrak{p}_{o2}$ is also a divisor of $\mathfrak{p}_3$ we conclude that $\tau_{\mathfrak{p}} x^* = 0$ for every maximal prime ideal $\mathfrak{p}$ which divides $\mathfrak{p}_3$. By using in this fashion successively the connecting bridges $\mathfrak{p}_{o1}, \mathfrak{p}_{o2}, \ldots, \mathfrak{p}_{o,g-1}$ and observing that every maximal prime ideal in R divides at least one of the ideals $\mathfrak{p}_1, \mathfrak{p}_2, \ldots, \mathfrak{p}_g$, we come to the conclusion that $\tau_{\mathfrak{p}} x^* = 0$ *for every maximal prime ideal in* R. But then, in view of Lemma 6, we must have $x^* = 0$. This completes the proof of the theorem.

FACULDADE DE FILOSOFIA, CIENCIAS E LETRAS,
UNIVERSIDADE DE S. PAULO, BRASIL
AND
UNIVERSITY OF ILLINOIS
URBANA, ILLINOIS, U. S. A.

ANNALS OF MATHEMATICS
Vol. 49, No. 2, April, 1948

ANALYTICAL IRREDUCIBILITY OF NORMAL VARIETIES

BY OSCAR ZARISKI

(Received August 11, 1947)

1. Introductory concepts

By a *local domain* we mean an integral domain which is at the same time a local ring in the sense of Krull [4]. If $\mathfrak{m}$ is the ideal of non-units in a local domain $\mathfrak{o}$ and if $\mathfrak{o}^*$ denotes the completion of $\mathfrak{o}$ (with respect to the powers of $\mathfrak{m}$), we say that $\mathfrak{o}$ is *analytically unramified* if the zero ideal in $\mathfrak{o}^*$ is an intersection of prime ideals. In other words: $\mathfrak{o}$ is analytically unramified if $\mathfrak{o}^*$ has no nilpotent elements.

If $\mathfrak{p}$ is a prime ideal in an arbitrary local ring $\mathfrak{o}$, we say that $\mathfrak{p}$ is *analytically unramified* if the local domain $\mathfrak{o}/\mathfrak{p}$ is analytically unramified. It is well known that if $\mathfrak{o}^*$ is the completion of $\mathfrak{o}$ then $\mathfrak{o}^*/\mathfrak{o}^*\mathfrak{p}$ is the completion of $\mathfrak{o}/\mathfrak{p}$ (Chevalley [1], Proposition 5). It follows that a prime ideal $\mathfrak{p}$ in a local ring $\mathfrak{o}$ is analytically unramified if and only if the extended ideal $\mathfrak{o}^*\mathfrak{p}$ in the completion $\mathfrak{o}^*$ of $\mathfrak{o}$ is an intersection of prime ideals.

The following theorem has been conjectured by the author and proved by Chevalley ([2], Lemma 9 on p. 9, last sentence, and Theorem 1 on p. 11): *The local ring of a point P of an irreducible algebraic variety V is analytically unramified.* It follows that any prime ideal $\mathfrak{p}$ in such a ring is also analytically unramified, because $\mathfrak{p}$ defines an irreducible subvariety W of V, and the residue class ring $\mathfrak{o}/\mathfrak{p}$ is the local ring of the point P, this point now being regarded as a point of W. Note the following special case: V is the affine n-space over k, and P is the origin. In this case the completion of the local ring of the point P is the ring $k\langle x\rangle$ of formal power series in n independent variables $x_1, x_2, \cdots, x_n$, with coefficients in k, and therefore it follows that *every prime ideal in the polynomial ring* $k[x]$ *splits into prime ideals in the power series ring* $k\langle x\rangle$. In informal geometric language this result signifies that an irreducible algebraic variety V can decompose in the neighborhood of a point P only into "simple" analytical branches (i.e., none of the branches has to be "counted" more than once). At any rate, it is true in the complex domain that the analytical reducibility of V in the neighborhood of a point can be no worse ideal-theoretically than it is set-theoretically.

We say that a local domain is *analytically irreducible* if its completion has no zero divisors, and that a variety V is *analytically irreducible* at a point P of V if the local ring of P is analytically irreducible. We recall that V is said to be *locally normal at P* if the local ring of P is integrally closed. The object of this paper is to prove the following theorem:

If an irreducible algebraic variety V is locally normal at a point P, then it is analytically irreducible at P.

In the course of the proof of this Theorem we shall arrive incidentally at another proof of Chevalley's result.

Our theorem is to be compared with another result concerning normal varieties and proved by the author elsewhere ([7], Definition 4, Theorem 8(A) and Theorem 10, pp. 512–514). We have shown namely that if V is locally normal at P and if a birational transformation of V into another variety V' sends P into a *finite* set of points of the variety V', then this set consists necessarily of a single point. We saw in this result a strong indication of the analytical irreducibility of normal varieties. For if a variety V consists of s branches in the neighborhood of a point P, then one would expect that these s branches could be separated by a suitable birational transformation. Such a transformation would then replace P by s distinct points.

2. Some auxiliary lemmas

The first two of the following lemmas refer to an arbitrary local ring $\mathfrak{o}$ and its completion $\mathfrak{o}^*$. The ideals of non-units in these two rings are denoted by $\mathfrak{m}$ and $\mathfrak{m}^*$ respectively.

Lemma 1. *If $\mathfrak{A}$ is an ideal in $\mathfrak{o}$ and b is any element of $\mathfrak{o}$, then $\mathfrak{o}^*\mathfrak{A} : \mathfrak{o}^*b = \mathfrak{o}^*(\mathfrak{A} : b)$.*

Proof. It is sufficient to prove the inclusion $\mathfrak{o}^*\mathfrak{A} : \mathfrak{o}^*b \subset \mathfrak{o}^*(\mathfrak{A} : b)$. Let u be any element of $\mathfrak{o}^*\mathfrak{A} : \mathfrak{o}^*b$, $u = \lim u_i$, $u_i \in \mathfrak{o}$. We have

$$u_i \in u + \mathfrak{m}^{*i+1}, \qquad u_i b \in ub + b\mathfrak{m}^{*i+1}, \quad \text{i.e.,} \quad u_i b \in \mathfrak{o}^*\mathfrak{A} + \mathfrak{o}^* b\mathfrak{m}^{i+1}.$$

Therefore $u_i b \in \mathfrak{A} + b\mathfrak{m}^{i+1}$,[1] $u_i \in \mathfrak{A} : b + \mathfrak{m}^{i+1}$, and hence $u \in \mathfrak{o}^*(\mathfrak{A} : b) + \mathfrak{m}^{*i+1}$, for all i. This implies that $u \in \mathfrak{o}^*(\mathfrak{A} : b)$, as asserted.

Lemma 2. *If $\mathfrak{A}$ and $\mathfrak{B}$ are ideals in $\mathfrak{o}$ and if $\mathfrak{A} : \mathfrak{B} = \mathfrak{A}$, then $\mathfrak{o}^*\mathfrak{A} : \mathfrak{o}^*\mathfrak{B} = \mathfrak{o}^*\mathfrak{A}$.*

Proof. Let $\mathfrak{p}_1, \mathfrak{p}_2, \cdots, \mathfrak{p}_g$ be those prime ideals of $\mathfrak{A}$ which are not contained in any other prime ideal of $\mathfrak{A}$ (i.e., they are maximal with respect to the property of being prime ideals of $\mathfrak{A}$), and let a_{ij} be an element of $\mathfrak{p}_i$ which is not in $\mathfrak{p}_j$ $(i, j = 1, 2, \cdots, g, i \neq j)$. From $\mathfrak{A} : \mathfrak{B} = \mathfrak{A}$ it follows that $\mathfrak{B}$ is not contained in any of the prime ideals of $\mathfrak{A}$. Hence we can find an element a_{ii} which is in $\mathfrak{B}$ but not in $\mathfrak{p}_i (i = 1, 2, \cdots, g)$. We set

$$b_j = a_{1j} a_{2j} \cdots a_{gj}, \; b = b_1 + b_2 + \cdots + b_g.$$

Then $b \in \mathfrak{B}$, $b \notin \mathfrak{p}_i$, $i = 1, 2, \cdots, g$, and therefore $\mathfrak{A} : \mathfrak{o}b = \mathfrak{A}$. It follows then from Lemma 1 that $\mathfrak{o}^*\mathfrak{A} : \mathfrak{o}^*b = \mathfrak{o}^*\mathfrak{A}$, and since $\mathfrak{o}^*\mathfrak{A} : \mathfrak{o}^*\mathfrak{B} \subset \mathfrak{o}^*\mathfrak{A} : \mathfrak{o}^*b$, the inclusion $\mathfrak{o}^*\mathfrak{A} : \mathfrak{o}^*\mathfrak{B} \subset \mathfrak{o}^*\mathfrak{A}$ follows, and that completes the proof of the lemma.

In the remaining lemmas we assume that $\mathfrak{o}$ *is a local domain which is integrally closed in its quotient field.* Lemmas 3–7 refer to a fixed *minimal* prime ideal $\mathfrak{p}$ in $\mathfrak{o}$, *where we assume that $\mathfrak{p}$ is analytically unramified*, whence

$$\mathfrak{o}^*\mathfrak{p} = \mathfrak{p}_1^* \cap \mathfrak{p}_2^* \cap \cdots \cap \mathfrak{p}_h^*, \tag{1}$$

each $\mathfrak{p}_i^*$ being a prime ideal in $\mathfrak{o}^*$. The lemmas 3–7 concern certain properties

[1] Here we are making use of the relation $\mathfrak{o}^*\mathfrak{B} \cap \mathfrak{o} = \mathfrak{B}$ which holds for any ideal $\mathfrak{B}$ in $\mathfrak{o}$ (Krull [4], Theorem 15; also Chevalley [1], Proposition 5).

of these h prime ideals $\mathfrak{p}_i^*$. Let $\mathfrak{p}^*$ be one these prime ideals, say $\mathfrak{p}^* = \mathfrak{p}_1^*$. Let $\mathfrak{P}^*$ be one of the prime ideals of the zero ideal in $\mathfrak{o}^*$ such that $\mathfrak{P}^* \subset \mathfrak{p}^*$. We set

$$(2) \qquad \Omega = \mathfrak{o}^*/\mathfrak{P}^*, \qquad \mathfrak{P} = \mathfrak{p}^*/\mathfrak{P}^*.$$

LEMMA 3. *The quotient ring $R = \Omega_{\mathfrak{P}}$ is a discrete valuation ring of rank 1. If ω is an element of $\mathfrak{o}$ such that $\omega \in \mathfrak{p}$, $\omega \notin \mathfrak{p}^{(2)}$ and if $\bar{\omega}$ is the $\mathfrak{P}^*$-residue of ω, then $R\bar{\omega}$ is the ideal of non-units in R.*

PROOF. Let a^* be an element of $\mathfrak{p}_2^* \cap \mathfrak{p}_3^* \cap \cdots \cap \mathfrak{p}_h^*$ which does not belong to $\mathfrak{p}_1^*$, and let b be an element of $\mathfrak{o}\omega : \mathfrak{p}$, not in $\mathfrak{p}$. If we set $c^* = a^*b$, then $c^* \notin \mathfrak{p}_1^*$ since b is not in $\mathfrak{p}_1^*$.[2] We have $\mathfrak{p}^*a^* \subset \mathfrak{o}^*\mathfrak{p}$, $\mathfrak{p}^*c^* \subset \mathfrak{o}^*\mathfrak{p}b \subset \mathfrak{o}^*\omega$. Hence if $\bar{c}$ denotes the $\mathfrak{P}^*$-residue of c^*, then $\mathfrak{P}\bar{c} \subset \Omega\bar{\omega}$, and therefore $R\mathfrak{P} \subset R\bar{\omega}$, since $\bar{c} \notin \mathfrak{P}$. This shows that $R\mathfrak{P}$, the ideal of non-units in R, is the principal ideal $R\bar{\omega}$, and this completes the proof of the lemma.

LEMMA 4. *For any integer n the ideal $\mathfrak{P}^*$ belongs to the symbolic power $\mathfrak{p}^{*(n)}$.*

PROOF. Since $\mathfrak{P}^* \subset \mathfrak{p}^*$, it follows from the preceding proof that $\mathfrak{P}^*c^* \subset \mathfrak{o}^*\omega$. Therefore if α^* is any element of $\mathfrak{P}^*$ then $\alpha^*c^* = \alpha_1^*\omega \in \mathfrak{P}^*$ where α_1^* is an element of $\mathfrak{o}^*$. Since ω is in $\mathfrak{o}$, it is not a zero divisor in $\mathfrak{o}^*$ (see footnote 2), and therefore $\omega \notin \mathfrak{P}^*$. Consequently $\alpha_1^* \in \mathfrak{P}^*$, $\alpha_1^* c \in \mathfrak{o}^*\omega$, and this shows that $\mathfrak{P}^* \cdot (c^*)^2 \subset \mathfrak{o}^*\omega^2$. In a similar fashion it can be shown that $\mathfrak{P}^* \cdot (c^*)^n \subset \mathfrak{o}^*\omega^n \subset \mathfrak{p}^{*(n)}$, and this completes the proof, since $c^* \notin \mathfrak{p}^*$.

LEMMA 5. *Every primary ideal of $\mathfrak{p}^*$ is a symbolic power of $\mathfrak{p}^*$.*

PROOF. By Lemma 4 every primary ideal of $\mathfrak{p}^*$ contains the ideal $\mathfrak{P}^*$. Hence there is (1, 1) correspondence between the primary ideals of $\mathfrak{p}^*$ in $\mathfrak{o}^*$ and the primary ideals of $\mathfrak{P}$ in Ω. But the latter primary ideals are all symbolic powers of $\mathfrak{P}$, since the quotient ring of $\mathfrak{P}$ is a discrete valuation ring of rank 1 (Lemma 3). This completes the proof.

LEMMA 6. *Each of the h prime ideals $\mathfrak{p}_i^*$ contains one and only one of the prime ideals of the zero ideal in $\mathfrak{o}^*$.*

PROOF. Since $\bigcap_{n=1}^{\infty} \mathfrak{P}^{(n)} = (0)$ and since, by Lemma 4, $\mathfrak{p}^{*(n)}$ is the full inverse image of $\mathfrak{P}^{(n)}$ under the homomorphism $\mathfrak{o}^* \sim \Omega$, it follows that $\bigcap_{n=1}^{\infty} \mathfrak{p}^{*(n)} = \mathfrak{p}^*$, and so $\mathfrak{P}^*$ is uniquely determined by $\mathfrak{p}^*$. Since $\mathfrak{p}^*$ can be any of the h ideals $\mathfrak{p}_i^*$, the lemma follows.

LEMMA 7. *The relation $\mathfrak{o}^*\mathfrak{p}^{(n)} = \bigcap_{i=1}^{h} \mathfrak{p}_i^{*(n)}$ holds for any interger n.*

PROOF. It is sufficient to prove the inclusion $\mathfrak{o}^*\mathfrak{p}^{(n)} \supset \bigcap_{i=1}^{h} \mathfrak{p}_i^{*(n)}$ Let α^* be any element of the ideal on the right. We have $\alpha^* \in \mathfrak{o}^*\mathfrak{p}$, and hence $\alpha^* b = \alpha_1\omega$, where b is the element which was introduced in the proof of Lemma 3. Since $\omega \notin \mathfrak{p}^{(2)}$,

[2] It has been proved by Chevalley ([1], Proposition 6) that if $\mathfrak{o}$ is a local ring and $\mathfrak{o}^*$ is the completion of $\mathfrak{o}$, then any prime ideal of the zero ideal of $\mathfrak{o}^*$ contracts in $\mathfrak{o}$ to a prime ideal of the zero ideal of $\mathfrak{o}$. If we apply this result to the local domain $\mathfrak{o}/\mathfrak{p}$, where $\mathfrak{p}$ is a prime ideal of $\mathfrak{o}$, and if we take into account that $\mathfrak{o}^*/\mathfrak{o}^*\mathfrak{p}$ is the completion of $\mathfrak{o}/\mathfrak{p}$, we conclude that any prime ideal of $\mathfrak{o}^*\mathfrak{p}$ contracts in $\mathfrak{o}$ to $\mathfrak{p}$. In particular, we have in our present case: $\mathfrak{p}_1^* \cap \mathfrak{o} = \mathfrak{p}$.

it follows from this same lemma that $\alpha_1 \in \bigcap_{i=1}^{h} \mathfrak{p}_i^{*(n-1)}$,[3] and hence $\alpha_1 b = \overset{*}{\alpha_1}\omega$, i.e. $\alpha^* b^2 = \overset{*}{\alpha_1}\omega^2$. In this fashion we find, after n steps, the relation $\alpha^* b^n = \overset{*}{\alpha_n}\omega^n$, where $\overset{*}{\alpha_n}$ is some element of $\mathfrak{o}^*$. Hence $\alpha^* \in \mathfrak{o}^*\omega^n : \mathfrak{o}^* b^n \subset \mathfrak{o}^*\mathfrak{p}^{(n)} : \mathfrak{o}^* b^n$. Since $b \notin \mathfrak{p}$, it follows from Lemma 2 that $\alpha^* \in \mathfrak{o}^*\mathfrak{p}^{(n)}$, and this completes the proof of the lemma.

LEMMA 8. *Let* $\mathfrak{A} = \bigcap_{i=1}^{n} \mathfrak{q}_i$ *be an ideal in* $\mathfrak{o}$, *where* $\mathfrak{q}_1, \mathfrak{q}_2, \cdots, \mathfrak{q}_n$ *are primary ideals belonging to distinct minimal prime ideals* $\mathfrak{p}_1, \mathfrak{p}_2, \cdots, \mathfrak{p}_n$ *in* $\mathfrak{o}$. *If each of the n ideals* $\mathfrak{p}_i$ *is analytically unramified, then* $\mathfrak{o}^*\mathfrak{A} = \bigcap_{i=1}^{n} \mathfrak{o}^*\mathfrak{q}_i$.

PROOF. It is sufficient to prove the inclusion $\mathfrak{o}^*\mathfrak{A} \supset \bigcap_{i=1}^{n} \mathfrak{o}^*\mathfrak{q}_i$. Let α^* be any element of the ideal on the right. For each i we fix an element ω_i which is in $\mathfrak{p}_i$ but not in $\mathfrak{p}_i^{(2)}$ and not in $\mathfrak{p}_j$, for $j \neq i$. Moreover, let b_i be an element of $\mathfrak{o}\omega_i : \mathfrak{p}_i$, not in $\mathfrak{p}_j$, $j = 1, 2, \cdots, n$. We know from the proof of Lemma 7 that if ρ_i is the exponent of $\mathfrak{q}_i$ (so that $\mathfrak{q}_i$ is then necessarily the symbolic power $\mathfrak{p}_i^{(\rho_i)}$), then $\alpha^* b_1^{\rho_1}$ is a multiple of $\omega_1^{\rho_1}$, say $\alpha^* b_1^{\rho_1} = \overset{*}{\alpha_1}\omega_1^{\rho_1}$, $\overset{*}{\alpha_1} \in \mathfrak{o}^*$. Since $\omega_1 \notin \mathfrak{p}_j$, $j \neq 1$, it follows that $\overset{*}{\alpha_1} \in \bigcap_{i=2}^{n} \mathfrak{o}^*\mathfrak{q}_i$, and hence, by a similar argument we have $\overset{*}{\alpha_1} b_2^{\rho_2} = \overset{*}{\alpha_2}\omega_2^{\rho_2}$, where $\overset{*}{\alpha_2} \in \bigcap_{i=3}^{n} \mathfrak{o}^*\mathfrak{q}_i$. Ultimately we find $\alpha^* b_1^{\rho_1} b_2^{\rho_2} \cdots b_n^{\rho_n} = \overset{*}{\alpha_n}\omega_1^{\rho_1}\omega_2^{\rho_2} \cdot \omega_n^{\rho_n} \in \mathfrak{o}^*\mathfrak{A}$. Hence if we set $b = b_1^{\rho_1} b_2^{\rho_2} \cdots b_n^{\rho_n}$, then $\alpha^* \in \mathfrak{o}^*\mathfrak{A} : \mathfrak{o}^* b$, whence, by Lemma 1, $\alpha^* \in \mathfrak{o}^*(\mathfrak{A} : b)$, i.e., $\alpha^* \in \mathfrak{o}^*\mathfrak{A}$, since $b \notin \mathfrak{p}_i$, $i = 1, 2, \cdots, n$. This completes the proof.

LEMMA 9. *If there exists an element* $t \neq 0$ *in* $\mathfrak{o}$ *such that all the prime ideals of the principal ideal* $\mathfrak{o} \cdot t$ *are analytically unramified, then* $\mathfrak{o}^*$ *has no nilpotent elements.*

PROOF. Let $\mathfrak{p}_1, \mathfrak{p}_2, \cdots, \mathfrak{p}_n$ be the prime ideals of $\mathfrak{o} \cdot t$ and let us assume that all these ideals are analytically unramified. They are also minimal primes in $\mathfrak{o}$ since $\mathfrak{o}$ is integrally closed. Let $\mathfrak{o}^*\mathfrak{p}_i = \overset{*}{\mathfrak{p}}_{i1} \cap \overset{*}{\mathfrak{p}}_{i2} \cap \cdots \cap \overset{*}{\mathfrak{p}}_{i\nu_i}$, $i = 1, 2, \cdots, n$, and let $\mathfrak{P}_1^*, \mathfrak{P}_2^*, \cdots, \mathfrak{P}_s^*$ be the prime ideals of the zero ideal in $\mathfrak{o}^*$ which contain at least one of the prime ideals $\overset{*}{\mathfrak{p}}_{ij}$ $(i = 1, 2, \cdots, n; j = 1, 2, \cdots, \nu_i)$. We know from Lemma 6 that each of the prime ideals $\overset{*}{\mathfrak{p}}_{ij}$ contains one and only one of the s ideals $\mathfrak{P}_\mu^*$, and from the proof of that lemma it follows that

$$\mathfrak{P}_1^* \cap \mathfrak{P}_2^* \cap \cdots \cap \mathfrak{P}_s^* = \bigcap_{i=1}^{n} \bigcap_{j=1}^{\infty} \{\mathfrak{p}_{i1}^{*(j)} \cap \mathfrak{p}_{i2}^{*(j)} \cap \cdots \cap \mathfrak{p}_{i\nu_i}^{*(j)}\},$$

or, in view of Lemma 7,

$$\mathfrak{P}_1^* \cap \mathfrak{P}_2^* \cap \cdots \cap \mathfrak{P}_s^* = \bigcap_{i=1}^{n} \bigcap_{j=1}^{\infty} \mathfrak{o}^* \mathfrak{p}_i^{(j)},$$

or finally, by Lemma 8,

$$\mathfrak{P}_1^* \cap \mathfrak{P}_2^* \cap \cdots \cap \mathfrak{P}_s^* = \bigcap_{j=1}^{\infty} \mathfrak{o}^* \{\mathfrak{p}_1^{(j)} \cap \mathfrak{p}_2^{(j)} \cap \cdots \cap \mathfrak{p}_n^{(j)}\} \subset \mathfrak{o}^* t^{\nu(j)},$$

where $\lim \nu(j) = +\infty$. Hence $\mathfrak{P}_1^* \cap \mathfrak{P}_2^* \cap \cdots \cap \mathfrak{P}_s^* \subset \mathfrak{m}^{*l}$, for all integers l. Consequently

$$\mathfrak{P}_1^* \cap \mathfrak{P}_2^* \cap \cdots \cap \mathfrak{P}_s^* = (0), \tag{3}$$ q.e.d.

[3] We have $\alpha_1\omega \in \mathfrak{p}_1^{*(n)}$, hence passing to the $\mathfrak{P}^*$-residues $\bar{\alpha}_1, \bar{\omega}$ we find $\bar{\alpha}_1 \bar{\omega} \in \mathfrak{P}^{(n)}$. By Lemma 3 it then follows that $\bar{\alpha}_1 \in \mathfrak{P}^{(n-1)}$, whence $\alpha_1 \in \mathfrak{p}_1^{*(n-1)}$. Similarly it is shown that $\alpha_1 \in \mathfrak{p}_i^{*(n-1)}, i = 1, 2, \cdots, h$.

3. Application to algebraic varieties

As a first application we shall show that from Lemma 9 it is possible to derive Chevalley's theorem stated in section 1 of this paper. We first observe that it is sufficient to prove that theorem under the assumption that V *is locally normal at* P. For suppose that the theorem has already been proved under this assumption and suppose that we are dealing with a variety V and a point P of V such that V is not locally normal at P. Then we pass to a derived normal model V' of V, and we consider the points $P_1', P_2', \cdots, P_s'$ which correspond on V' to the point P. Let $\mathfrak{o}$ denote the local ring of P and let $\bar{\mathfrak{o}}$ denote the intersection of the local rings of the points P_i'. Then $\bar{\mathfrak{o}}$ is a semilocal ring in the sense of Chevalley [1] and is a finite $\mathfrak{o}$-module ($\bar{\mathfrak{o}}$ is in fact the integral closure of $\mathfrak{o}$ in the function field $\mathcal{F}(V)$ of V; see [7], p. 511). The completion of $\bar{\mathfrak{o}}$ contains the completion $\mathfrak{o}^*$ of $\mathfrak{o}$ as a subring, in fact is a finite $\mathfrak{o}^*$-module ([1], Proposition 7, p. 699). To show that $\mathfrak{o}$ is analytically unramified, it is therefore sufficient to show that the completion of $\bar{\mathfrak{o}}$ has no nilpotent elements. Now the completion of $\bar{\mathfrak{o}}$ is the direct sum of the completions of the s local rings of the points P_i' ([1], Proposition 8, p. 700), and since V' is locally normal at each of the points P_i' it follows, from our hypothesis, that the local rings of points P_i' are all analytically unramified. Hence the completion of $\bar{\mathfrak{o}}$ has no nilpotent elements, since the rings of which it is a direct sum have no nilpotent elements.

We shall now proceed by induction with respect to the dimension r of V. If $r = 1$ and if the curve V is locally normal at P, then P is a simple point of V, the local ring of P is a regular ring ([8], p. 19), in fact a valuation ring, and the completion of this ring is itself a regular ring ([4]) which therefore has no zero divisors at all. Having shown this for normal curves, it follows from the preceding observation that Chevalley's theorem is true for algebraic curves. Now let us assume that this theorem is true for algebraic varieties of dimension less than r, and let V be an r-dimensional irreducible variety which is locally normal at a given point P. We have to show that the local ring $\mathfrak{o}$ of the point P is analytically unramified. By our induction hypothesis we have that every prime ideal in $\mathfrak{o}$ is analytically unramified (compare with section 1), and in particular every minimal prime ideal in $\mathfrak{o}$ is analytically unramified. Hence the assumption in Lemma 9 is automatically satisfied for *any* element t in $\mathfrak{o}$, $t \neq 0$, and since $\mathfrak{o}$ is integrally closed, it follows by that lemma that $\mathfrak{o}^*$ has no nilpotent elements.

Before proceeding to the proof of the analytical irreducibility of normal varieties, we shall make a few geometric comments about some of the lemmas proved in the preceding section. The local ring $\mathfrak{o}$ is now the local ring of the point P of an r-dimensional irreducible variety V, and V is locally normal at P. Let (3) represent the decomposition of the zero ideal of $\mathfrak{o}^*$. In that case the variety V consists, locally at P, of s analytical branches $M_1, M_2, \cdots, M_s$, each M_i being an analytical manifold. For each prime ideal $\mathfrak{P}_i^*$ of the zero ideal we define as in (2) the domain $\Omega_i = \mathfrak{o}^*/\mathfrak{P}_i^*$. Then Ω_i is the ring of holomorphic functions *on* the analytical manifold M_i. Now let $\mathfrak{p}$ be a minimal prime in $\mathfrak{o}$ and let (1) be its decomposition in $\mathfrak{o}^*$. Then $\mathfrak{p}$ represents an $(r-1)$-dimensional subvariety W of V which contains the point P and which decomposes, locally at P, into h analytical manifolds, $N_1, N_2, \cdots, N_h$. Lemma 6 signifies that *each of the h analytical* $(r-1)$*-dimensional branches* N_j *of W lies on* (one and) *only one of the analytical branches* M_i *of V*. This result was to be expected from an intuitive geometric viewpoint, since the intersection of two distinct analytical branches M_i and M_j of V is part of the singular manifold of V and since, on the other hand, the singular manifold of V cannot possess an $(r-1)$-dimensional component at the point P where V is locally normal. Naturally this entire geometric picture which we are painting has real significance only in the classical case, because in the abstract case an analytical manifold is not a point set at all (there is no conceivable incidence relation between M_i and points of the affine ambient space of V, except that we may say that the origin P of the branch M_i is on M_i). At any rate is is now becoming clear why the hypothesis that $\mathfrak{o}$ is integrally closed was *a priori* necessary in the proof of Lemma 6. As to Lemma 3, we shall only make the following observation: the fact that the quotient ring R is a valuation ring is to be interpreted geometrically in the sense that *if* N_j *lies on* M_i *then* N_j *is not singular for* M_i. This interpretation is again in connection with the fact that the singular manifold of V must be, locally at P, of dimension $< r$. The remarks just made disregard our final result (which we shall now proceed to prove) that actually V is analytically irreducible at P, whence $s = 1$, and $\mathfrak{o}^*$ is an integral domain. It is still an open question whether $\mathfrak{o}^*$ is integrally closed. In other words: *if V is locally normal at P, is V also normal as an analytical manifold?*

We now proceed to the proof of the theorem on the analytical irreducibility of normal varieties stated in section 1. Let (3) be the decomposition of the zero ideal of $\mathfrak{o}^*$ into prime ideals. *We shall prove in the next section the following relations:*

$$(4) \qquad (\mathfrak{P}_i^* + \mathfrak{P}_j^*) \cap \mathfrak{o} \neq (0), \quad \text{if} \quad i \neq j.$$

Assuming for the moment relation (4), we now show that the assumption that $s > 1$ leads to a contradiction. Let us then assume that $s > 1$ and let u_j be an element of $(\mathfrak{P}_1^* + \mathfrak{P}_j^*) \cap \mathfrak{o}$, $u_j \neq 0$, $j = 2, 3, \cdots, s$. If we set $u = u_2 u_3 \cdots u_s$, we can write u in the form

$$u = v^* + w^*, \text{ where } v^* \in \mathfrak{P}_1^* \text{ and } w^* \in \mathfrak{A}^* = \mathfrak{P}_2^* \cap \mathfrak{P}_3^* \cap \cdots \cap \mathfrak{P}_s^*.$$

Let $\mathfrak{o}u = \mathfrak{p}_1^{(\rho_1)} \cap \mathfrak{p}_2^{(\rho_2)} \cap \cdots \cap \mathfrak{p}_n^{(\rho_n)}$, where the $\mathfrak{p}_i$ are minimal prime ideals in $\mathfrak{o}$' and let $\mathfrak{o}^*\mathfrak{p}_i = \mathfrak{p}_{i1}^* \cap \mathfrak{p}_{i2}^* \cap \cdots \cap \mathfrak{p}_{i\nu_i}^*$. (Since we are dealing with the local ring of a point of an algebraic variety V, we know that every prime ideal of $\mathfrak{o}$ is analytically unramified.) All the lemmas of section 2 are applicable. In particular, we find by lemmas 7 and 8 that

$$\mathfrak{o}^* u = \bigcap_{i=1}^{n} \bigcap_{j=1}^{\nu_i} \mathfrak{p}_{ij}^{*(\rho_i)}.$$

Each of the ideals $\mathfrak{p}_{ij}^*$ contains one of the ideals $\mathfrak{P}_\mu^*$. If $\mathfrak{p}_{ij}^* \supset \mathfrak{P}_\mu^*$, then by Lemma

4 any symbolic power of $\overset{*}{\mathfrak{p}}_{ij}$ contains $\overset{*}{\mathfrak{P}}_\mu$, and in particular $\mathfrak{p}_{ij}^{*(\rho_i)} \supset \overset{*}{\mathfrak{P}}_\mu$. If $\mu = 1$ then $v^* \in \overset{*}{\mathfrak{P}}_\mu \subset \mathfrak{p}_{ij}^{*(\rho_i)}$, and since also $u \in \mathfrak{p}_{ij}^{*(\rho_i)}$ it follows that $w^* = u - v^* \in \mathfrak{p}_{ij}^{*(\rho_i)}$. If $\mu \neq 1$, then $w^* \in \mathfrak{p}_{ij}^{*(\rho_i)}$, and hence we have again $v^* \in \mathfrak{p}_{ij}^{*(\rho_i)}$. Therefore in either case we find that both v^* and w^* are in $\mathfrak{p}_{ij}^{*(\rho_i)}$, for all i and j. Hence both v^* and w^* belong to the ideal $\mathfrak{o}^*u$, say $v^* = \overset{*}{v}_1 u$ and $w^* = \overset{*}{w}_1 u$. From $\overset{*}{v}_1 + \overset{*}{w}_1 = 1$[4] it follows that $\overset{*}{v}_1$ and $\overset{*}{w}_2$ cannot both belong to $\mathfrak{m}^*$, and hence at least one of these two elements must be a unit in $\mathfrak{o}^*$. On the other hand we have $\overset{*}{v}_1 u \in \overset{*}{\mathfrak{P}}_1$, $\overset{*}{w}_1 u \in \mathfrak{A}^*$, and u does not belong to any of the prime ideals $\overset{*}{\mathfrak{P}}_\mu$, since $u \neq 0$ is an element of $\mathfrak{o}$ and is therefore not a zero divisor in $\mathfrak{o}^*$. Hence $\overset{*}{v}_1 \in \overset{*}{\mathfrak{P}}_1$ and $\overset{*}{w}_1 \in \mathfrak{A}^*$, so that neither $\overset{*}{v}_1$ nor $\overset{*}{w}_1$ can be a unit in $\mathfrak{o}^*$. Thus the assumption $s > 1$ leads to a contradiction, and this proves our theorem.

4. Proof of relation (4)

We first point out the geometric meaning of relation (4). The two prime ideals $\overset{*}{\mathfrak{P}}_i$ and $\overset{*}{\mathfrak{P}}_j$ represent two distinct analytical branches M_i and M_j of V through the point P. The sum $\overset{*}{\mathfrak{P}}_i + \overset{*}{\mathfrak{P}}_j$ represents the intersection L of these two branches. The left hand member of (4) is an ideal in the local ring of P whose zero manifold is the least algebraic subvariety W of V which contains the analytical manifold L. Therefore relation (4) is equivalent to the assertion that *W is not the entire variety V*. The geometric facts which lie behind this assertion are the following: a) as an intersection of two analytical branches the manifold L must belong to the singular manifold of V; b) therefore also W belongs to the singular manifold of V; c) the singular manifold of V is a proper subvariety of V. The formal proof shall now be given.

Let the ambient linear space S of V be of dimension n and let $\mathfrak{q}$ be the prime ideal of V in the polynomial ring $k[x]$ of the n variables $x_1, x_2, \cdots x_n$. Let R denote the local ring $Q(P/S)$ of the point P regarded as a point of S, and let R^* be the completion of R. The decomposition (3) of the zero ideal in $\mathfrak{o}^*$ implies a corresponding decomposition of the ideal $R^*\mathfrak{q}$ into s prime ideals:

$$R^*\mathfrak{q} = \overset{*}{\mathfrak{q}}_1 \cap \overset{*}{\mathfrak{q}}_2 \cap \cdots \cap \overset{*}{\mathfrak{q}}_s. \tag{5}$$

The ring R^* is a complete local domain of dimension n. Each of the prime ideals $\overset{*}{\mathfrak{q}}_i$ is of dimension r, the same as the dimension of the prime R-ideal $R \cdot \mathfrak{q}$ ([2], Theorem 1). Relation (4) is equivalent to the following relation:

$$(\overset{*}{\mathfrak{q}}_i + \overset{*}{\mathfrak{q}}_j) \cap R \neq R\mathfrak{q}. \tag{6}$$

Let $\mathfrak{Q}^*$ be any prime ideal of $\overset{*}{\mathfrak{q}}_i + \overset{*}{\mathfrak{q}}_j$. To prove (5) we have to show that for any such prime ideal $\mathfrak{Q}^*$ we have:

$$\mathfrak{Q}^* \cap R \neq R\mathfrak{q}. \tag{7}$$

[4] Note that u is not a zero divisor in $\mathfrak{o}^*$, since $u \in \mathfrak{o}$. Therefore the relation $(v_1^* + w_1^*)u = u$ implies $v_1^* + w_1^* = 1$.

Let ρ be the dimension of $\mathfrak{Q}^*$. Since $\mathfrak{Q}^*$ is a proper divisor of $\mathfrak{q}_i^*$ and $\mathfrak{q}_j^*$, it follows that $\rho < r$. We pass to the quotient ring $\mathfrak{O} = R^*_{\mathfrak{Q}^*}$. This ring is a local ring of dimension $n - \rho$, and since the ring R^* is a regular complete ring, it follows that also $\mathfrak{O}$ is a regular ring (Cohen [3], Theorem 20, p. 97). We consider the ideal $\mathfrak{O}\cdot\mathfrak{q}$. By well known properties of quotient rings, the decomposition of the ideal $\mathfrak{O}\cdot\mathfrak{q}$ is obtained from the decomposition (5) of $R^*\mathfrak{q}$ by replacing the prime ideals $\mathfrak{q}_i^*$ by their extensions $\mathfrak{O}\cdot\mathfrak{q}_i^*$. These extensions are prime ideals. Moreover $\mathfrak{O}\cdot\mathfrak{q}_i^*$ is the unit ideal in $\mathfrak{O}$ if and only if $\mathfrak{Q}^*$ does not contain $\mathfrak{q}_i^*$, and two distinct prime ideals which are contained in $\mathfrak{Q}^*$ give rise to distinct extensions. Since $\mathfrak{q}_i^*$ and $\mathfrak{q}_j^*$ are both contained in $\mathfrak{Q}^*$, it follows that $\mathfrak{O}\mathfrak{q}$ *is not a prime ideal.* Another property of the ideal $\mathfrak{O}\mathfrak{q}$ which we shall have to use is that *its prime ideals are all of dimension* $r - \rho$. This follows from the fact that the ideals $\mathfrak{q}_i^*$ are of dimension r and the ideal $\mathfrak{Q}^*$ of which $\mathfrak{O}$ is the quotient ring is of dimension ρ.

Now let $\{f_1(x), f_2(x), \cdots, f_N(x)\}$ be a basis of the prime polynomial ideal $\mathfrak{q}$. These N polynomials will also form a basis of the ideal $\mathfrak{O}\mathfrak{q}$. We denote by $\mathfrak{M}$ the ideal of non-units of the regular $(n - \rho)$-dimensional local ring $\mathfrak{O}$. The additive group $\mathfrak{M}/\mathfrak{M}^2$ can be regarded as a vector space $\mathcal{M}^*$, of dimension $n - \rho$, over the field $\mathfrak{O}/\mathfrak{M}$ ([8], p. 6). We now use the two properties of the ideal $\mathfrak{O}\mathfrak{q}$ which have been derived above. Since this ideal is pure $(r - \rho)$-dimensional, it follows that at most $n - r$ of the polynomials $f_i(x)$ map, modulo $\mathfrak{M}^2$, on independent vectors of $\mathcal{M}^*$. Moreover, since the ideal $\mathfrak{O}\mathfrak{q}$ is not prime, it follows that *the set of vectors of* $\mathcal{M}^*$ *which correspond to the polynomials* $f_i(x)$ *contains actually less than n-r independent vectors* ([3], corollary on p. 87). On the other hand, if we consider the local vector space $\mathcal{M} = \mathcal{M}(V/S)$ of V in S (see [8], p. 6) we find that *the set of vectors of* $\mathcal{M}$ *which correspond to the polynomials* $f_i(x)$ *contains exactly n-r independent vectors,* since any variety V in S is simple for S. We shall now show that relation (7), and hence also relation (4), follows from this discrepancy between the dimensionalities of the vector spaces spanned by the polynomials $f_i(x)$ in the vector spaces $\mathcal{M}$ and $\mathcal{M}^*$ respectively.

We first assume that the function field $\mathcal{F}(V)$ *is separably generated over* k. We use the results of our paper [8] concerning the vector space $\mathcal{M} = \mathcal{M}(V/S)$ and the vector space $\mathcal{D}(V)$ of local V-differentials in S ([8], p. 25). Let ξ_i be the $\mathfrak{q}$-residue of x_i and let ξ_i^* be the $\mathfrak{Q}$-residue of x_i. Since the function field of V is separably generated over k, it follows that the Jacobian matrix $\partial(f_1, f_2, \cdots, f_N)/\partial(x_1, x_2, \cdots, x_n)$ is of rank n-r at $x = \xi$ ([8], Theorem 7′, p. 31). On the other hand, since the polynomials $f_i(x)$ span in $\mathcal{M}^*$ a space of dimension less than n-r, it follows *a fortiori* that any n-r rows in the Jacobian matrix $\partial(f_1(x), f_2(x), \cdots, f_N(x))/\partial(x_1, x_2, \cdots, x_n)$ are linearly dependent over the field $\mathfrak{M}/\mathfrak{M}^2$.[5]

[5] A few words will suffice to explain this assertion. The partial derivations $\partial/\partial x_i$ have obviously the property of transforming into itself the quotient ring of any prime ideal in $k[x]$. In particular, these derivations transform R into itself. Moreover, if $\mathfrak{m}$ is the ideal of non-units in R, then the partial derivatives of any element of $\mathfrak{m}^\rho$ are elements of $\mathfrak{m}^{\rho-1}$.

Hence this matrix is of rank less than n-r. We conclude therefore that while all the (n-r)-rowed minors of the matrix $\partial(f_1(x), f_2(x), \cdots, f_N(x))/\partial(x_1, x_2, \cdots, x_n)$ belong to $\mathfrak{O}$, at least one of these minors is not in $\mathfrak{q}$. This establishes (7) and hence also (4).

If $\mathcal{F}(V)$ is not separably generated over k, the proof is the same except that instead of the ordinary Jacobian matrix we must use the mixed Jacobian matrix introduced in our paper [8] on p. 38.

In conclusion we observe that in our proof of the analytical irreducibility of normal varieties we have made use only of two special properties of local, integrally closed, domains which are true for local rings of points of normal varieties and which are not known to be true in general. These two properties are the following: 1) the local domain $\mathfrak{o}$, and every prime ideal in $\mathfrak{o}$, is analytically unramified; 2) if $\mathfrak{P}_i^*$ and $\mathfrak{P}_j^*$ are any two distinct prime ideals of the zero ideal in $\mathfrak{o}^*$, then relation (4) holds. The first property is certainly false for general local domains, if we drop the condition that $\mathfrak{o}$ is integrally closed ([6]; also [8], p. 24, where the ring $\mathfrak{o}$ defined in (8) is easily seen to be analytically ramified), but it is possible that it holds for all integrally closed local domains. Also the extent to which relation (4) is valid in the non-geometric case, is an unsolved question. On the answer to these questions depends the answer to the following general question: *if $\mathfrak{o}$ is an integrally closed local domain, is it true that the completion of $\mathfrak{o}$ is also an integral domain?* It is trivial that the answer is affirmative if $\mathfrak{o}$ is of dimension 1 (for $\mathfrak{o}$ is then a discrete rank valuation ring).

ADDED IN PROOF. A related question is the following: if $\mathfrak{o}$ is a local domain such that (a) its integral closure $\bar{\mathfrak{o}}$ is a finite $\mathfrak{o}$-module, is it true then that (b) $\mathfrak{o}$ is analytically unramified? It is known (Krull [5]) that if $\mathfrak{o}$ is of dimension 1 then (a) and (b) are equivalent. An affirmative answer to the first question would imply an affirmative answer to this second question, since it can be easily shown that under assumption (a) $\mathfrak{o}$ is analytically unramified if and only if $\bar{\mathfrak{o}}$ is analytically unramified.

HARVARD UNIVERSITY

REFERENCES

[1] C. CHEVALLEY, *On the theory of local rings*, Ann. of Math., vol. 44 (1943), pp. 690–708.
[2] C. CHEVALLEY, *Intersections of algebraic and algebroid varieties*, Trans., Amer. Math. Soc., vol. 57 (1945), pp. 1–85.

From this it follows that each of the derivations $\partial/\partial x_i$ has a unique extension in the complete ring R^*. This extension will be denoted by the same symbol $\partial/\partial x_i$. By the same argument, the extended derivation $\partial/\partial x_i$ in R^* can be further extended to the quotient ring $\mathfrak{O}$, and moreover, the partial derivatives of any element of $\mathfrak{M}^\rho$ are elements of $\mathfrak{M}^{\rho-1}$. Now consider any $n - r$ of the polynomials $f_i(x)$, say $f_1(x), f_2(x), \cdots, f_{n-r}(x)$. Since the corresponding vectors of $\mathfrak{M}/\mathfrak{M}^2$ are linearly dependent, there is a relation of the form: $A_1^* f_1(x) + A_2^* f_2(x) + \cdots + A_{n-r}^* f_{n-r}(x) \in \mathfrak{M}^2$, where the A_i^* are elements of $\mathfrak{O}$, not all in $\mathfrak{M}$. Applying the derivation $\partial/\partial x_i$ and observing that the polynomials $f_j(x)$ belong to $\mathfrak{M}$, we find that $A_1^* \partial f_1(x)/\partial x_i + A_2^* \partial f_2(x)/\partial x_i + \cdots + A_{n-r}^* \partial f_{n-r}(x)/\partial x_i$ is in $\mathfrak{M}$, for $i = 1, 2, \cdots, n$. Since the A_i^* are not all zero mod$\mathfrak{M}$, the assertion in the text follows.

[3] I. S. Cohen, *On the structure and ideal theory of complete local rings*, Trans., Amer. Math. Soc., vol. 59 (1946), pp. 54–106.

[4] W. Krull, *Dimensionstheorie in Stellenringen*, J. Reine Angew. Math., vol. 179 (1938), pp. 204–226.

[5] W. Krull, *Ein Satz über primäre Integritätsbereiche*, Math. Ann., vol. 103 (1930), pp. 540–565.

[6] F. K. Schmidt, *Über die Erhaltung der Kettensätze der Idealtheorie bei beliebigen endlichen Körpererweiterungen*, Math. Zeit., vol. 41 (1936), pp. 443–450.

[7] O. Zariski, *Foundations of a general theory of birational correspondences*, Trans., Amer. Math. Soc., vol. 53 (1943), pp. 490–542.

[8] A. Zariski, *The concept of a simple point of an abstract algebraic variety*, Trans., Amer. Math. Soc., vol. 61 (1947), pp. 1–52.

Reprinted from the Proceedings of the NATIONAL ACADEMY OF SCIENCES,
Vol. 35, No. 1, pp. 62–66. January, 1949

A SIMPLE ANALYTICAL PROOF OF A FUNDAMENTAL PROPERTY OF BIRATIONAL TRANSFORMATIONS

BY OSCAR ZARISKI

DEPARTMENT OF MATHEMATICS, HARVARD UNIVERSITY

Communicated November 5, 1948

1. *Valuations in Local Domains.*—Let $\mathfrak{o}$ be a local domain, i.e., a local ring which is at the same time an integral domain, and let $\mathfrak{m}$ be the maximal ideal in $\mathfrak{o}$. We consider a valuation v of the quotient field of $\mathfrak{o}$. Let R be the valuation ring of v and let $\mathfrak{P}$ be the maximal ideal of R. We say that v *has center* $\mathfrak{m}$ *in* $\mathfrak{o}$ if (1) $\mathfrak{o} \subset R$ and (2) $\mathfrak{m} = \mathfrak{P} \cap \mathfrak{o}$. If v has center $\mathfrak{m}$ in $\mathfrak{o}$, then the residue field $\mathfrak{o}/\mathfrak{m}$ of $\mathfrak{o}$ can be regarded as a subfield of the residue field $R/\mathfrak{P}$ of the valuation v. The transcendence degree of the latter field over the former will be called the $\mathfrak{o}$-*dimension* of v.

THEOREM 1. *If the local domain* $\mathfrak{o}$ *is of dimension* r *and if a valuation* v *of the quotient field of* $\mathfrak{o}$ *has center* $\mathfrak{m}$ *in* $\mathfrak{o}$ *and has rank 1, then the* $\mathfrak{o}$-*dimension of* v *is at most* $r - 1$.

Proof.—We shall first consider the case in which $\mathfrak{o}$ is complete. By a theorem due to I. S. Cohen[2] (Theorem 11, p. 79) $\mathfrak{o}$ contains, then, a coefficient ring Ω. In the equal-characteristic case (i.e., if $\mathfrak{o}$ and $\mathfrak{o}/\mathfrak{m}$ have the same characteristic) Ω is a *field* which is mapped isomorphically mod $\mathfrak{m}$ on the entire residue field $\mathfrak{o}/\mathfrak{m}$. It is this case that occurs in algebraic geometry. In the unequal-characteristic case (i.e., if $\mathfrak{o}$ is of characteristic 0 and $\mathfrak{o}/\mathfrak{m}$ is of characteristic $p \neq 0$) Ω is a complete discrete valuation ring of rank 1 in which the maximal ideal is generated by the prime number p and which is mapped mod $\mathfrak{m}$ (and therefore also mod p) on the entire residue field $\mathfrak{o}/\mathfrak{m}$.

Let $t_1, t_2, \ldots, t_r$ be a system of parameters in $\mathfrak{o}$, i.e., a set of r elements of $\mathfrak{m}$ such that the ideal $\mathfrak{o}(t_1, t_2, \ldots, t_r)$ is primary for $\mathfrak{m}$. Then $\mathfrak{o}$ contains the ring $\Omega < t_1, t_2, \ldots, t_r >$ of formal power series in the t's, with coefficients in Ω. We denote this ring by Ω^*. The ring Ω^* is a complete local domain of dimension r, has the same residue field as $\mathfrak{o}$, and $\mathfrak{o}$ is a finite module over Ω^* (see Chevalley,[1] Lemma 1 and Prop. 5, p. 702. For the general case, see I. S. Cohen,[2] Lemma 16, p. 90). If v' denotes the valuation of the quotient field of Ω^* induced by v and if $\mathfrak{m}'$ denotes the maximal ideal in Ω^*, then it is clear that v' has center $\mathfrak{m}'$ in Ω^* and that the residue field of v is an algebraic extension of the residue field of v' (since the quotient field of $\mathfrak{o}$ is an algebraic extension of the quotient field of Ω^*). It follows that it is sufficient to prove the theorem for Ω^* and v'. We therefore assume that $\mathfrak{o}$ itself is a power series ring $\Omega < t_1, t_2, \ldots, t_r >$ over the coefficient ring Ω.

Let $\xi = u/w$ be an arbitrary element in the valuation ring of v, where u and w are elements of $\mathfrak{o}$. Let u_i and w_i be the partial sums of the power series u and w, respectively. We have $u - u_i \epsilon \mathfrak{m}^i$, $w - w_i \epsilon \mathfrak{m}^i$. Since v has center $\mathfrak{m}$ in $\mathfrak{o}$ and since the rank of v is 1 (i.e., the value group of v consists of real numbers), it follows that $v(u) = v(u_i)$ and $v(w) = v(w_i)$ for all i sufficiently large. We have $\xi - u_i/w_i = [w_i(u - u_i) - u_i(w - w_i)]/ww_i$. As i becomes very large the value of the numerator (in the given valuation v) approaches $+\infty$, while the value of the denominator remains constantly equal to the value of w^2. Hence we have, for i very large, $v(\xi - u_i/w_i) > 0$, and therefore ξ and u_i/w_i have the same residue in the given valuation v. Since u_i and w_i are polynomials in $\Omega[t_1, t_2, \ldots, t_r]$, it follows that the residue field of v is the same as the residue field of the valuation v_1 induced by v in the ring S of rational functions $\Omega(t_1, t_2, \ldots, t_r)$, and from this our theorem follows at once. For in the equal-characteristic case Ω and s are fields, v_1 is trivial on Ω and is non-trivial on S. In the unequal-characteristic case we may assume that $t_1 = p$. In that case S has transcendence degree $r - 1$ over the quotient field of Ω, and it is clear that the transcendence degree, over $\Omega/\Omega p$, of the residue field of v_1 cannot exceed $r - 1$.

We now can easily prove the theorem for arbitrary (not necessarily complete) local domains $\mathfrak{o}$. Let $\mathfrak{o}^*$ and R^* be the $\mathfrak{m}$-adic and the $\mathfrak{P}$-adic completions of $\mathfrak{o}$ and of the valuation ring R, respectively. Since v is of rank 1, the completion of R, in the sense of the theory of local rings, coincides with the completion of R, in the sense of valuation theory. Hence, R^* is a complete valuation ring for an extension v^* of v, and v^* has the same value group and the same residue field as v. The assumption that v has center $\mathfrak{m}$ in $\mathfrak{o}$ implies that Cauchy sequences and zero sequences in $\mathfrak{o}$ are, respectively, also Cauchy sequences and zero sequences in R. We have therefore a natural homomorphism τ of $\mathfrak{o}^*$ into R^*. Let $\mathfrak{o}' = \tau\mathfrak{o}^*$.

Then $\mathfrak{o}'$ is a complete local domain, and we have $\mathfrak{o} \subset \mathfrak{o}'$, since τ reduces to the identity on $\mathfrak{o}$. If $\mathfrak{m}^*$ and $\mathfrak{m}'$ are the maximal ideals in $\mathfrak{o}^*$ and $\mathfrak{o}'$, respectively, then $\mathfrak{m}' = \tau\mathfrak{m}^*$, $\mathfrak{o}'/\mathfrak{m}' = \mathfrak{o}^*/\mathfrak{m}^*$, and since $\mathfrak{m}^* = \mathfrak{o}^*\mathfrak{m}$ and $\mathfrak{o}^*/\mathfrak{m}^* = \mathfrak{o}/\mathfrak{m}$, we conclude that $\mathfrak{m}' = \mathfrak{o}' \cdot \mathfrak{m}$, $\mathfrak{m}' \cap \mathfrak{o} = \mathfrak{m}$ and $\mathfrak{o}'/\mathfrak{m}' = \mathfrak{o}/\mathfrak{m}$. Finally we point out that $\mathfrak{o}'$ is of dimension $\leqq r$, since $\mathfrak{o}^*$ is exactly of dimension r. Now, let v' be the valuation of the quotient field of $\mathfrak{o}'$ induced by v^*. Since $\mathfrak{o}' \subset R^*$ and $\mathfrak{m}' = \mathfrak{o}'\mathfrak{m}$, it follows that the valuation v' (which is of rank 1) has center $\mathfrak{m}'$ in $\mathfrak{o}'$. Therefore by the previous case, the $\mathfrak{o}'$-dimension of v' is at most $r - 1$. Since v' is an extension of v and since the residue fields of $\mathfrak{o}$ and $\mathfrak{o}'$ are the same, it follows *a fortiori* that the $\mathfrak{o}$-dimension of v is at most $r - 1$. This establishes the theorem for arbitrary local domains.

The second part of the proof yields an important consequence in the following special case: *the local domain* $\mathfrak{o}$ *is analytically irreducible, i.e.,* $\mathfrak{o}^*$ *is an integral domain, and the valuation* v *(of rank* 1 *and center* $\mathfrak{m}$) *is of dimension* $r - 1$. In that case, also, v' must be exactly of dimension $r - 1$. In view of Theorem 1, this is possible only if in the inequality $\dim \mathfrak{o}' \leqq r$ the equality sign holds. Hence $\mathfrak{o}'$ is of dimension r. Since also the local *domain* $\mathfrak{o}^*$ is of dimension r, *it follows that the homomorphism* τ of $\mathfrak{o}^*$ *onto* $\mathfrak{o}'$ *must be an isomorphism.* We can therefore state the following theorem:

THEOREM 2. *Let* $\mathfrak{o}$ *be a local domain of dimension* r *and let* $\mathfrak{m}$ *be the maximal ideal of* $\mathfrak{o}$. *Let* v *be a rank 1 valuation of the quotient field of* $\mathfrak{o}$ *such that the* $\mathfrak{o}$*-dimension of* v *is* $r - 1$ *and such that* $\mathfrak{m}$ *is the center of* v *in* $\mathfrak{o}$. *If* $\mathfrak{o}$ *is analytically irreducible (i.e., if the completion* $\mathfrak{o}^*$ *of* $\mathfrak{o}$ *is an integral domain), then* $\mathfrak{o}$ *is a subspace of the valuation ring* R *of* v, *i.e., we have:* $\mathfrak{P}^i \cap \mathfrak{o} \subset \mathfrak{m}^{\nu(i)}$, $\nu(i) \rightarrow +\infty$, *as* $i \rightarrow +\infty$ *where* $\mathfrak{P}$ *is the maximal ideal of* R. *The valuation* v *can then be extended to a valuation of* $\mathfrak{o}^*$.

2. *An Application to Algebraic Varieties.*—Let V and V' be two algebraic varieties over a common ground field k, and assume that the function field Σ of V is contained in the function field Σ' of V' (whence V is a rational transform of V'). Let P and P' be two corresponding points of V and V'.

THEOREM 3. *If the local ring of* P' *contains the local ring of* P *and if* V *is locally irreducible at* P, *then the local ring of* P *is a subspace of the local ring of* P'.

Proof: Let $\mathfrak{o}$ and $\mathfrak{o}'$ denote the local rings of P and P', respectively. Let $\mathfrak{m}$ and $\mathfrak{m}'$ be the maximal ideals in these two rings. We have, by assumption, $\mathfrak{o} \subset \mathfrak{o}'$. Furthermore, the fact that P and P' are corresponding points signifies that $\mathfrak{m} \subset \mathfrak{m}'$ (and hence $\mathfrak{m}' \cap \mathfrak{o} = \mathfrak{m}$). Let r and r' be the dimensions of $\mathfrak{o}$ and $\mathfrak{o}'$, respectively. Let $s = \dim V$, $s' = \dim V'$. Note that the difference $s - r$ is merely the dimension of the point P over k, i.e., the dimension of the irreducible variety of which P is a general point. Similarly for s' and r'. We fix a divisor v' [i.e., an $(s' -$

1) dimensional valuation v'] of the function field Σ' which has center at the point P'. The induced valuation v in the function field Σ has then center P, and its dimension (the dimensions of v and v' are intended here with reference to k) is necessarily $s - 1$ (since dim v' − dim v cannot exceed $s' - s$). In the terminology of the preceding section, v has center $\mathfrak{m}$ in $\mathfrak{o}$, and its $\mathfrak{o}$-dimension is equal to $r - 1$. Similarly v' has center $\mathfrak{m}'$ in $\mathfrak{o}'$, and its $\mathfrak{o}'$-dimension is equal to $r' - 1$. Let $\mathfrak{P}$ and $\mathfrak{P}'$ be the prime ideals of the valuations v and v'. It is clear that high powers of $\mathfrak{P}'$ contract in the valuation ring of v to high powers of $\mathfrak{P}$. Since V is analytically irreducible at P, it follows from Theorem 2 that high powers of $\mathfrak{P}$ contract in $\mathfrak{o}$ to high powers of $\mathfrak{m}$. Therefore high powers of $\mathfrak{P}'$ contract in $\mathfrak{o}$ to high powers of $\mathfrak{m}$. But since $\mathfrak{m}' \subset \mathfrak{P}'$, it follows *a fortiori* that high powers of $\mathfrak{m}'$ contract in $\mathfrak{o}$ to high powers of $\mathfrak{m}$, i.e., the local ring $\mathfrak{o}$ is a subspace of the local ring $\mathfrak{o}'$, q. e. d.

From Theorem 3 it is possible to derive in a very simple manner the "main theorem" of our paper on birational transformations (Zariski,[3] p. 522). Let V and V' be birationally equivalent varieties and let T denote the birational transformation of V onto V'. We denote by $T\{P\}$ the set of points of V' which correspond to P. We say that a point P' of $T\{P\}$ is *isolated* if P' is not a specialization (over k) of any point Q' of $T\{P\}$ such that dim $Q'/k >$ dim P'/k.

THEOREM 4. *Under the assumptions of Theorem 3 and under the additional assumptions that P' is an isolated point of $T\{P\}$ and that dim P/k = dim P'/k, the set $T\{P\}$ consists of a finite number of points, any two of which are isomorphic over k.*

Proof: We use the notation of the proof of Theorem 3. Let $\mathfrak{o}^*$ and $\mathfrak{o}'^*$ be the completions of $\mathfrak{o}$ and $\mathfrak{o}'$, respectively. Since, by Theorem 3, $\mathfrak{o}$ is a subspace of $\mathfrak{o}'$, it follows that $\mathfrak{o}^*$ is also a subring and subspace of $\mathfrak{o}'^*$. The assumption that P' is an isolated point of $T\{P\}$ signifies that the ideal $\mathfrak{o}' \cdot \mathfrak{m}$ is primary for $\mathfrak{m}'$. Since, on the other hand, $\mathfrak{o}$ and $\mathfrak{o}'$ have, by assumption, the same dimension, it follows that if $\{t_1, t_2, \ldots, t_r\}$ is a system of parameters in $\mathfrak{o}$, then the t's form also a system of parameters in $\mathfrak{o}'$. Therefore $\mathfrak{o}'^*$ is a finite module over the power series ring $k < t_1, t_2, \ldots, t_r >$ (Chevalley[1]), and *a fortiori* $\mathfrak{o}'^*$ is a finite module over $\mathfrak{o}^*$. In particular, *every element of* $\mathfrak{o}'$ *is integrally dependent on* $\mathfrak{o}^*$.

Now let v be any divisor of the function field Σ, with center at P. By Theorem 2, v can be extended to a valuation v^* of the quotient field of the local *domain* $\mathfrak{o}^*$. Since every element of $\mathfrak{o}'$ is integrally dependent on $\mathfrak{o}^*$, it follows that $\mathfrak{o}'$ is contained in the valuation ring of v^* (since $\mathfrak{o}^*$ is contained in that valuation ring). Consequently $\mathfrak{o}'$ *is contained in the valuation ring of* v (since v^* is an extension of v). The center of the divisor v on the variety V' is therefore a point A' such that *the local ring of* A' *contains the local ring* $\mathfrak{o}'$ of P', i.e., a point A' such that P' is a specialization of A' over k. Since P and A' are centers of v, A' belongs to the set $T\{P\}$ Since P' is an isolated point of $T\{P\}$, we must have dim A'/k = dim P'/k. From this it follows that A' and P' are isomorphic points over k (since P' is a specialization of A' over k) and that consequently the local rings of A' and P' coincide. Since $\mathfrak{o} \subset \mathfrak{o}'$, the local field $k(A')$ contains the local field $k(P)$ of P, and hence A' is an algebraic point over $k(P)$ (since dim A'/k = dim P'/k = dim P/k). Since $T\{P\}$ is a variety over $k(P)$, we conclude that $T\{P\}$ is zero dimensional over $k(P)$ and therefore consists of a finite number of points. This completes the proof.

Our "Main Theorem" on birational transformations follows from the above theorem if one takes into account that if V is locally normal at P then (*a*) V is analytically irreducible at P (Zariski[4]) and (*b*) the *finite* set $T\{P\}$ necessarily consists of a single point (Zariski[3], theorem 8(A) and theorem 10, pp. 512–514).

[1] Chevalley, C., "On the Theory of Local Rings," *Ann. Math.*, **44**, 690–708 (1943).

[2] Cohen, I. S., "On the Structure and Ideal Theory of Complete Local Rings," *Trans. Am. Math. Soc.*, **59**, 54–106 (1946).

[3] Zariski, O., "Foundations of a General Theory of Birational Correspondences," *Ibid.*, **53**, 490–542 (1943).

[4] Zariski, O., "Analytical Irreducibility of Normal Varieties," *Ann. Math.*, **49**, 352–361 (1948).

A fundamental lemma from the theory of holomorphic functions on an algebraic variety.

Memoria di OSCAR ZARISKI (a Cambridge, Mass., USA).

1. The concept of a holomorphic function.

The contents of this article are in close connection with a theory of holomorphic functions in the large which we have developed for algebraic varieties over arbitrary ground fields ([1]). To give this article its proper background it will therefore be necessary to reproduce here and discuss briefly the concept of a holomorphic function in the large.

Let V be an irreducible algebraic variety over a given ground field k, and let Σ be a fixed function field of V/k (this field is only defined to within an arbitrary k-isomorphism). It P is a point of V, we denote by $\mathbf{o}(P/V)$ the local ring of V at P, by $\mathbf{m}(P/V)$ the ideal of non-units in $\mathbf{o}(P/V)$ and by $\mathbf{o}^*(P/V)$ the completion of $\mathbf{o}(P/V)$. Any element of the complete ring $\mathbf{o}^*(P/V)$ is called *a holomorphic function on* V, *defined at* P, or briefly: *a holomorphic function at P.*

Starting from this local, well known concept, we now consider an arbitrary set G of points on V and we denote by $\mathbf{O}^*(G/V)$ the direct product of the rings $\mathbf{o}^*(P/V)$, $P \varepsilon G$. If $\xi \varepsilon \mathbf{O}^*(G/V)$ and $P \varepsilon G$, we denote by $\xi[P]$ the P-component of ξ. Then $\xi[P]$ is a well defined holomorphic function at P, which we shall call *the analytical element of* ξ *at* P.

An infinite sequence $\{x_i\}$ of *rational* functions on $V (x_i \varepsilon \Sigma)$ converges at a point P if it is a CAUCHY sequence in the local ring of V at P. If $\{x_i\}$ converges at P, then its limit *at* P is a well defined holomorphic function x^* at P, and we write $x^* = \lim x_i$ *at* P.

([1]) A brief summary of this unpublished work has appeared in the *Abstracts of addresses given at the Conference on Algebraic Geometry and Algebraic Number Theory*, the University of Chicago, the Department of Mathematics, January, 1949. In the abstract of our address (*Theory and applications of holomorphic functions on algebraic varieties*), given at that conference, we also present the principal application of the theory of holomorphic functions, namely the derivation of the « principle of degeneration » in abstract algebraic geometry. Some of the ground work for this theory has been laid in the following papers of ours: (1) *Generalized semi-local rings*,, « Summa Brasiliensis Mathematicae », Vol. I, fasc. 8, 1946. (2) *Analytical irreducibility of normal varieties*, « Ann. of Math. », vol. 49, 1948. (3) *A simple analytical proof of a fundamental property of birational transformations*, « Proc. Mat. Acad. Sci. », vol. 35, 1949.

A sequence $\{x_i\}$ of rational functions on V *converges* on G if it converges et each point P of G. If $\{x_i\}$ converges on G, there is a well defined element ξ of $\mathbf{O}(G/V)$ such that $\xi[P] = \lim x_i$ at P, for all P in G. We say then that ξ is the limit of the sequence $\{x_i\}$ *on* G.

Uniform convergence is defined in the usual fashions: the sequence $\{x_i\}$ converges uniformly on G, if for any integer n there exists an integer $N(n)$ such that $x_i - x_j \varepsilon [\mathbf{m}(P/V)]^n$ for all $i, j > N(n)$ *and for all points* P *of* G.

Preliminary to the concept of a holomorphic function is the following definition of a *strongly holomorphic function:*

An element ξ *of* $\mathbf{O}^*(G/V)$ *is a strongly holomorphic function along* G *if it is the limit of a sequence of rational functions on* V *which converges uniformly on* G.

If G' is a subset of G, we denote by $\tau_{G,G'}$ the natural projection of $\mathbf{O}^*(G/V)$ onto $\mathbf{O}^*(G'/V)$. The projection into $\mathbf{O}^*(G'/V)$ of any element ξ of $\mathbf{O}^*(G/V)$ will be referred to as *the* G′-*component of* ξ and will be denoted by $\xi[G']$.

To define holomorphic functions on V, with G as domain of definition, or briefly: *holomorphic functions along* G, we use the topology of the variety V in which the closet sets are the algebraic subvarieties of V (2). There is then an induced topology in G, and the terms « open set », « open covering » in the following definition are in reference to this induced topology.

DEFINITION. - *An element* ξ *of* $\mathbf{O}^*(G/V)$ *is a holomorphic function along* G *if there exists a finite open covering* $\{G_\alpha\}$ *of* G *such that* $\xi[G_\alpha]$ *is strongly holomorphic on* G_α *(all* α*).*

It is not difficult to see that the holomorphic functions along a given set G form a subring of $\mathbf{O}^*(G/V)$. This subring will be denoted by $\mathbf{o}^*(G/V)$.

2. The main lemma and its application to holomorphic functions.

We now come to the connection between the concept of holomorphic functions and the question which we propose to treat in this paper.

We first point out that in all that precedes we use the term « point » in its widest possible sense: the coordinates of a point P are arbitrary quantities which are not necessarily algebraic over k. In the preceding definition the set G is arbitrary, but the case which is of interest is the one in which G is a subvariety of V, say W. Let W_0 be the set of algebraic points of W. Unless W is zero-dimensional, W_0 is a proper subset of W. We have, then, *a priori two* rings of holomorphic functions associated with W: $\mathbf{o}^*(W/V)$ and $\mathbf{o}^*(W_0/V)$. In applications one is primarily interested in algebraic points,

(2) See our paper *The compactness of the Riemann manifold of an abstract of algebraic functions,* « Bull. Amer. Math. Soc. », vol. 40, 1044.

and for that reason one would really wish to study the second of these two rings. On the other hand, it is much easier to study the ring $\mathbf{o}^*(W/V)$, since from the above definition we have many more data about the functions of this ring than of the ring $\mathbf{o}^*(W_0/V)$. The difference becomes very clear if we assume, for the sake of simplicity, that W is irreducible. Any non-empty open subset of W is the complement of a proper algebraic subvariety of W and hence contains all the general points of W. Hence when we study the ring $\mathbf{o}^*(W/V)$ we deal with sequences of rational functions on V *which necessarily converge at the general point of* W. This property of our sequences plays an essential role in the theory. It is not at all obvious that this same property belongs to all sequences which converge uniformly on the set of algebraic points of W or of some open subset of W (3). The main object of the present paper is to prove a general result (see main lemma below) from which it will follow that *the two rings* $\mathbf{o}^*(W/V)$ *and* $\mathbf{o}^*(W_0/V)$ *are isomorphic*, or, in less precise, but more descriptive terms: *every holomorphic function on* V, *which is defined along the set of all algebraic points of a given subvariety* W *of* V, *can be extended to a holomorphic function defined along the set of all points of* W. *and the extension is unique.* Hence one may safely replace the ring $\mathbf{o}^*(W_0/V)$ by the technically more manageable ring $\mathbf{o}^*(W/V)$, without fear of losing touch with function-theoretic realities.

For expository reasons we introduce the following terminology: if P is a point of V and z is a rational function on V, we shall say that z *vanishes to the order* ν *at* P if $z \in [\mathfrak{m}(P/V)]^\nu$, $z \notin [\mathfrak{m}(P/V)]^{\nu+1}$.

MAIN LEMMA. - *Let* W *be an irreducible subvariety of* V *and let* G *be a subset of* W *such that* W *is the closure of* G *(i.e.,* W *is the least algebraic variety containing* G*). If* z *is a rational function on* V *and* z *vanishes to an order* $\geq \nu$ *at each point* P *of* G, *then* z *also vanishes to an order* $\geq \nu$ *at each general point of* W.

We now apply this lemma to the question discussed above. For the sake of simplicity, we shall only consider the case of an irreducible subvariety W. The extension to holomorphic functions on V which are defined along a reducible variety W is straightforward.

Let ξ_0 be any element of $\mathbf{o}^*(W_0/V)$. There will exist then a finite open covering $\{G_{\alpha 0}\}$ of W_0 such that $\xi_0[G_{\alpha 0}]$ is strongly holomorphic on $G_{\alpha 0}$. For each $G_{\alpha 0}$ there is a subvariety W_α of W such that $G_{\alpha 0}$ is the set of all algebraic points of $W - W_\alpha$. Since the $G_{\alpha 0}$ cover W_0, it follows that the intesection of the W_α contains no algebraic points, and is therefore empty (HILBERT's Nullstellensatz). *Hence if we set* $G_\alpha = W - W_\alpha$, *then* $\{G_\alpha\}$ *is an open covering of* W.

(3) That the sequences in question *do* have this property, is proved below.

Let $\{x_{\alpha 1}, x_{\alpha 2}, \ldots\}$ be a sequence of rationai functions on V which converges uniformly on $G_{\alpha 0}$ to $\xi_0[G_{\alpha 0}]$. Then if n is a given integer, we will have that $x_{\alpha i} - x_{\alpha j}$ vanishes to an order $\geqq n$ at each point of $G_{\alpha 0}$, i.e., at each algebraic point of G_α, provided $i, j \geqq N(n)$. Now let P be an arbitrary point of G_α and let U be the irreducible algebraic variety whose general point is P. If U' is the closure of the set $U \cap G_{\alpha 0}$, then the variety $U' + W_\alpha$ contains all the algebraic points of U, since $U \subset W$. Hence $U \subset U' + W_\alpha$. Since U is irreducible and since $U \subset\!\!\!|\!= W_\alpha$, it follows that $U \subset U'$, and hence $U = U'$ since U' is the least variety containing the set $U \cup G_{\alpha 0}$. If in the main lemma we now replace W by $U \cap G_{\alpha 0}$, we conclude that $x_{\alpha i} - x_{\alpha j}$ vanishes to an order $\geqq n$ at P, for $i, j > N(n)$. *Hence the sequence* $\{x_{\alpha 1}, x_{\alpha 2}, \ldots\}$ *converges uniformly on* G_α. Let ξ_α be the limit of this sequence on G_α. It is clear that

$$\xi_\alpha[G_{\alpha 0}] = \xi_0[G_{\alpha 0}]. \tag{1}$$

If $\alpha \neq \beta$, then the sequence $\{x_{\alpha 1} - x_{\beta 1}, x_{\alpha 2} - x_{\beta 2}, \ldots\}$ converges uniformly to zero on $G_{\alpha 0} \cap G_{\beta 0}$ $(= W_0 - W_\alpha - W_\beta)$. It follows by the same argument as above that this sequence converges uniformly to zero also on $G_\alpha \cap G_\beta$. This signifies that any two of the functions ξ_α have the same analytical element at each common point of definition. Hence there exists a well defined element ξ iu $\mathbf{0}^*(W/V)$ such that

$$\xi_\alpha = \xi[G_\alpha], \quad all \quad \alpha. \tag{2}$$

Since each ξ_α is strongly holomorphic along G_α and since $\{G_\alpha\}$ is a finite open covering of W, it follows that ξ *is holomorphic along* W. From (1) and (2) we conclude that

$$\xi_0 = \xi[W_0]. \tag{3}$$

We have thus proved that *every holomorphic function along* W_0 *is the projection of at least one holomorphic function along* W. Hence the projection τ_{W, W_0} maps $\mathbf{o}^*(W/V)$ *onto* $\mathbf{o}^*(W_0/V)$. This mapping is clearly a ring homomorphism. On the other hand, the preceding proof shows also that if a holomorphic function along W, *different from zero,* is defined by certain uniformly convergent sequences $\{x_{\alpha 1}, x_{\alpha 2}, \ldots\}$, then these sequences could not possibly converge to zero at all points of W_0. It follows that the above homomorphism is actually an isomorphism, and this proves the essential identity of the two rings $\mathbf{o}^*(W/V)$ and $\mathbf{o}^*(W_0/V)$.

We now proceed to the proof of the main lemma.

3. Some properties of the local ring of a simple point.

Let P be a given point of the afine n-space S. We denote by $\mathbf{o}$ and $\mathbf{m}$ the local ring $\mathbf{o}(P/S)$ and the ideal $\mathbf{m}(P/S)$ respectively and by $k(P)$ the field $\mathbf{o}/\mathbf{m}$. This field is generated over k by the coordinates of the point P. For any integer ν, the additive group of the ring $\mathbf{m}^\nu/\mathbf{m}^{\nu+1}$ can be regarded as a vector space over $k(P)$ ([4]). This vector space will be denoted by $\mathfrak{M}_\nu(P)$.

If we denote by $n-r$ the dimension of the point P (i.e., the transcendence degree of $k(P)$ over k), then any minimal basis of $\mathbf{m}$ consists exactly of r elements (SP, p. 15, 4.1), and the elements of any such basis are called *local uniformizing parameters (l.u.p.) of* S *at* P. The dimension of the vector space $\mathfrak{M}_1(P)$ is equal to r. If $t_1, t_2, \ldots, t_r$ are l.u.p. of S at P, then the corresponding vectors $\bar{t}_1, \bar{t}_2, \ldots, \bar{t}_r$ in $\mathfrak{M}_1(P)$ form a basis of $\mathfrak{M}_1(P)$. If $u_1, u_2, \ldots, u_N$ denote the distinct power products of the t' s of a given degree ν, then the u' s form a *minimal* basis of the ideal $\mathbf{m}^\nu$ (in other words, the local ring $\mathbf{o}$ is a regular ring; see SP, p. 19, 5.1), and hence (SP, p. 12, 3.3) the corresponding vectors $\bar{u}_i$ form a basis of the vector space $\mathfrak{M}_\nu(P)$.

Let now W be an irreducible variety containing the point P and let A be a general point of W. We denote by $\mathbf{O}$ and $\mathbf{M}_1$ the ring $\mathbf{o}(A/S)$ and the ideal $\mathbf{m}(A/S)$ respectively ($\mathbf{O}$ and $\mathbf{M}_1$ depend only on W and not on the choice of the general point of W; $\mathbf{O}$ contains the ring $\mathbf{o}$). We denote by $n-s$ the dimension of W/k (this is also the dimension of the point A). If P is a *simple* point of W, then the following is known:

a) There exist l.u.p. $t_1, t_2, \ldots, t_r$ of S at P such that $t_1, t_2, \ldots, t_s$ are l.u.p. of S at A.

b) If $t_1, t_2, \ldots, t_r$ are chosen as in *a*), then

$$\mathbf{M}_1 \cap \mathbf{o} = \mathbf{o} \cdot (t_1, t_2, \ldots, t_s), \tag{4}$$

aud $t_1, t_2, \ldots, t_s$ form in fact a minimal basis of the ideal $\mathbf{M}_1 \cap \mathbf{o}$.

c) Conversely, everly minimal basis of $\mathbf{M}_1 \cap \mathbf{o}$ consists exactly of s elements $t_1, t_2, \ldots, t_s$, these elements are l.u.p. of S at A and are such that the set $\{t_1, t_2, \ldots . t_s\}$ can be extended to a set of l.u.p. of S at P.

All these assertions are either contained in, or are easy consequences of, SP, p. 13. Theorem 2. In that theorem the ideal $\mathbf{M}_1 \cap \mathbf{o}$ is referred to as *the local ideal of* W *at* P. We shall denote this ideal by $\mathbf{M}$:

$$\mathbf{M} = \mathbf{M}_1 \cap \mathbf{o}. \tag{5}$$

We shall now prove the following relations:

$$\mathbf{M}_1^\nu \cap \mathbf{o} = \mathbf{M}^\nu, \tag{6}$$

$$\mathbf{M}^\nu \cap \mathbf{m}^\mu = \mathbf{M}^\nu \mathbf{m}^{\mu-\nu}, \qquad \mu \geqq \nu, \qquad (\mathbf{m}^0 = \mathbf{o}). \tag{7}$$

We choose the l.u.p. of S at P as indicated in *a*), and we denote by $v_1, v_2, \ldots$ the various power products of $t_1, t_2, \ldots, t_s$, of degree ν. If x is an element of $\mathbf{M}^\nu$, then it follows from (4) and (5) that x can be expressed as a linear

([4]) See our paper *The concept of a simple point of an abstract algebraic variety*, « Trans. Amer. Math. Soc. », vol. 62, 1947, p. 12, section 3. 3. This paper will be referred to in the sequel as SP.

form in the v_i, with coefficients in $\mathfrak{o}$. If $x \,\overline{\in}\, \mathbf{M}^{\nu+1}$, then these coefficients are not all in $\mathbf{M}_1$, by (5), and hence $x \,\overline{\in}\, \mathbf{M}_1^{\nu+1}$, since the vectors in $\mathfrak{M}_\nu(A)$ which correspond to the v_i form a basis. What we have proved is that if $x \in \mathbf{M}^\nu$, $x \,\overline{\in}\, \mathbf{M}^{\nu+1}$, then $x \,\overline{\in}\, \mathbf{M}_1^{\nu+1}$. If we replace in this result the integer ν by any integer less than ν, we obtain (6).

To prove (4) it is sufficient to prove the inclusion $\mathbf{M}^\nu \cap \mathfrak{m}^\mu \subset \mathbf{M}\,\mathfrak{m}^{\mu-\nu}$, since $\mathbf{M} \subset \mathfrak{m}$. We may also assume that $\mu > \nu$, since for $\mu = \nu$ (7) is trivial. We therefore have to prove the following assertion: *if* $x \in \mathbf{M}^\nu \mathfrak{m}^\sigma$ *and* $x \,\overline{\in}\, \mathbf{M}^\nu \mathfrak{m}^{\sigma+1}$ $(\sigma \geqq 0)$, *then* $x \,\overline{\in}\, \mathfrak{m}^{\nu+\sigma+1}$. If $\sigma = 0$, the proof is as above, except that now we use the fact that the vectors in $\mathfrak{M}_\nu(P)$ which correspond to the elements v_i can be extended to a basis of $\mathfrak{M}_\nu(P)$ and hence are independent (we are now dealing with a linear form in the v' s, with coefficients in $\mathfrak{o}$ *and not all in* $\mathfrak{m}$). For $\sigma > 0$ we shall use induction from $\sigma - 1$ to σ. We can express x as a form of degree ν in $t_1, t_2, \ldots, t_s$, with coefficients in $\mathfrak{m}^\sigma$, and each of these coefficients can be expressed in its turn as a form of degree σ in $t_1, t_2, \ldots, t_r$, with coefficients in $\mathfrak{o}$. Hence $x = \varphi\,(t_1, t_2, \ldots, t_r)$, where φ is a form of degree $\nu + \sigma$, with coefficients in $\mathfrak{o}$, and each term of φ is of degree $\geqq \nu$ in $t_1, t_2, \ldots, t_s$. If at least one of the coefficients of φ is not in $\mathfrak{m}$, then $x \,\overline{\in}\, \mathfrak{m}^{\nu+\sigma+1}$, and our assertion is proved. We shall therefore assume that $\mathfrak{m}$ contains the coefficients of all the terms of φ which are exactly of degree ν in $t_1, t_2, \ldots, t_s$. If we denote by x_1 the sum of these latter terms, then we can write $x = x_1 + x_2$, where $x_1 \in \mathbf{M}^\nu \mathfrak{m}^{\sigma+1}$ and $x_2 \in \mathbf{M}^{\nu+1}\mathfrak{m}^{\sigma-1}$. Since $x \,\overline{\in}\, \mathbf{M}^\nu \mathfrak{m}^{\sigma+1}$, it follows that $x_2 \,\overline{\in}\, \mathbf{M}^\nu \mathfrak{m}^{\sigma+1}$, and hence *a fortiori* $x_2 \,\overline{\in}\, \mathbf{M}^{\nu+1}\mathfrak{m}^\sigma$. Since, on the other hand, $x_2 \in \mathbf{M}^{\nu+1}\mathfrak{m}^{\sigma-1}$, it follows, from the induction hypothesis, that $x_2 \,\overline{\in}\, \mathfrak{m}^{\nu+\sigma+1}$. Since $x \in \mathbf{M}^\nu \mathfrak{m}^{\sigma+1} \subset \mathfrak{m}^{\nu+\sigma+1}$, we conclude that $x = x_1 + x_2 \,\overline{\in}\, \mathfrak{m}^{\nu+\sigma+1}$. This proves our assertion and completes the proof of (7).

We consider again the various power products $v_1, v_2, \ldots$ of $t_1, t_2, \ldots, t_s$, of degree ν. It follows from (4) and (5) that the v_i form a basis of the ideal $\mathbf{M}^\nu$. We know that the vectors which correspond to the v_i in $\mathfrak{M}_\nu(A)$ or in $\mathfrak{M}_\nu(P)$ are independent (in $\mathfrak{M}_\nu(A)$ these vectors form even a basis). From either one of these two facts it follows that the v_i form a *minimal* basis of the ideal $\mathbf{M}^\nu$. Any other minimal basis of $\mathbf{M}^\nu$ is related to the basis $v_1, v_2, \ldots$ by a linear homogeneous transformation with coefficients in $\mathfrak{o}$ and with determinant not in $\mathfrak{m}$. We conclude therefore that if $u_1, u_2, \ldots$ is any minimal basis of $\mathbf{M}^\nu$, then the vectors which correspond to the u_i in $\mathfrak{M}_\nu(A)$ or in $\mathfrak{M}_\nu(P)$ are independent, i.e., the following two properties hold:

(8) $$\text{«} \textstyle\sum A_i u_i \in \mathbf{M}^{\nu+1},\ A_i \in \mathfrak{o} \text{»} \rightarrow \text{« all } A_i \text{ are in } \mathbf{M} \text{»}.$$

(9) $$\text{«} \textstyle\sum A_i u_i \in \mathfrak{m}^{\nu+1},\ A_i \in \mathfrak{o} \text{»} \rightarrow \text{« all } A_i \text{ are in } \mathfrak{m} \text{»}.$$

We point out the following consequences. By (7), we have

(10) $$\mathbf{M}^\nu \cap \mathfrak{m}^\mu = \textstyle\sum \mathfrak{m}^{\mu-\nu} u_i, \qquad \mu \geqq \nu, \qquad (\mathfrak{m}^0 = \mathfrak{o}),$$

and by (8)

(11) $$\text{«} \textstyle\sum A_i u_i \in \mathbf{M}^{\nu+1},\ A_i \in \mathfrak{m}^{\mu-\nu} \text{»} \rightarrow \text{«} A_i \in \mathbf{M} \cap \mathfrak{m}^{\mu-\nu} \text{»}.$$

Suppose now that we have a relation of the form $\Sigma A_i u_i \in \mathbf{m}^{\mu+1}$, $A_i \in \mathbf{m}^{\mu-\nu}$. Then by (10) (where μ is to be replaced by $\mu+1$) we can write: $\Sigma A_i u_i = \Sigma B_i u_i$, where the B_i are in $\mathbf{m}^{\mu-\nu+1}$. From $\Sigma(A_i - B_i)u_i = 0$ it follows, as a special case of (11), that $A_i - B_i \in \mathbf{M} \cap \mathbf{m}^{\mu-\nu}$. Hence we have proved that

$$(12) \qquad «\Sigma A_i u_i \in \mathbf{m}^{\mu+1},\ A_i \in \mathbf{m}^{\mu-\nu}» \rightarrow «A_i \in \mathbf{M} \cap \mathbf{m}^{\mu-\nu} + \mathbf{m}^{\mu-\nu+1}», \qquad (\mu \geqq \nu).$$

We shall denote by $g_{\nu\mu}$ and $G_{\nu\mu}$ the sets of vectors in $\mathfrak{M}_\nu(P)$ and $\mathfrak{M}_\nu(A)$ respectively which correspond to the elements of $\mathbf{M}^\nu \cap \mathbf{m}^\mu$, $\mu \geqq \nu$. Note that $g_{\nu\mu}$ is a subspace of $\mathfrak{M}_\mu(P)$, but that $G_{\nu\mu}$ is only a subgroup of the additive group of $\mathfrak{M}_\nu(A)$ (since $\mathbf{M}^\nu \cap \mathbf{m}^\mu$ is not an $\mathbf{O}$-module). These groups $g_{\nu\mu}$ and $G_{\nu\mu}$ will play an essensial role in the sequel. For the moment we make the following remarks in the special case $\mu = \nu$.

1) $G_{\nu\nu}$ *spans the entire space* $\mathfrak{M}_\nu(A)$. This follows from what has been said about minimal bases of the ideal $\mathbf{M}^\nu$.

2) *There is a natural homomorphism* τ *of* $G_{\nu\nu}$ *onto* $g_{\nu\nu}$, *defined as follows:* if $\bar{v}$ is any vector in $G_{\nu\nu}$ and if v is any element of $\mathbf{M}^\nu \cap \mathbf{m}^\mu$ to which $\bar{v}$ corresponds, then $\bar{v}\tau$ is in the vector in $g_{\nu\nu}$ which corresponds to v. That τ is single-valued (and hence a homomorphism) follows from the fact that $\mathbf{M}^{\nu+1} \subset \mathbf{m}^{\nu+1}$.

3) *Linearly dependent vectors in* $G_{\nu\nu}$ *are mapped by* τ *into linearly dependent vectors of* $g_{\nu\nu}$. For let $v_1, v_2, \ldots, v_q$ be elements of $\mathbf{M}^\nu$ such that the corresponding vectors in $\mathfrak{M}_\nu(P)$ are linearly independent. Then the set of q elements v_i can be extended to a minimal base of the ideal $\mathbf{M}^\nu$, and hence the vectors which correspond to the v_i in $\mathfrak{M}_\nu(A)$ are also independent.

4. (W, ν)-regular points of V.

We now consider a second irreducible variety V *such that* V *contains* W. We denote by $\mathbf{p}$ the local ideal of V at the point P and we set

$$(13) \qquad \mathbf{p}_{\nu\mu} = \mathbf{p} \cap \mathbf{M}^\nu \cap \mathbf{m}^\mu.$$

In each of the two groups $g_{\nu\mu}$ and $G_{\nu\mu}$ the ideal $\mathbf{p}_{\nu\mu}$ determines a subgroup. We shall denote these subgroups by $h_{\nu\mu}$ and $H_{\nu\mu}$ respectively We consider in particular the groups $h_{\nu\nu}$ and $H_{\nu\nu}$. It is clear that $h_{\nu\nu} = H_{\nu\nu}\tau$, where τ is the homomorphism defined above [remark 2)]. It follows by remark 3 of the preceding section that the dimension of $h_{\nu\nu}$ is not greater that the dimension of the space spanned by $H_{\nu\nu}$ in $\mathfrak{M}_\nu(A)$.

DEFINITION. - *The point* P *is said to be* (W, ν)*-regular for* V *if* P *is a simple point of* W *and if the dimension of the subspace* $h_{\nu\nu}$ *of* $\mathfrak{M}_\nu(P)$ *is the same as the dimension of the subspace of* $\mathfrak{M}_\nu(A)$ *spanned by* $H_{\nu\nu}$.

The concept of a (W, ν)-regular point of V is the key to our proof of the main lemma.

Theorem 1. - *If* P *is* (W, ν)*-regular for* V, *then*

$$\mathbf{p}_{\nu\mu} = \mathbf{p}_{\nu\nu}\mathbf{m}^{\mu-\nu} + \mathbf{p}_{\nu+1,\mu}, \qquad \mu \geq \nu. \tag{14}$$

Proof. Let $\rho = \dim h_{\nu\nu}$, $\sigma = \dim g_{\nu\nu}$ ($\rho \leq \sigma$) and let $u_1^*, u_2^*, \ldots, u_\sigma^*$ be a basis of $g_{\nu\nu}$ such that $u_1^*, u_2^*, \ldots, u_\rho^*$ is a basis of $h_{\nu\nu}$. For each $i = 1, 2, \ldots, \sigma$ we fix an element u_i in $\mathbf{M}^\nu$ whose $\mathbf{m}^{\nu+1}$-residue is u_i^* and which belongs to $\mathbf{p}$ if $1 \leq i \leq \rho$. Let x be any element of $\mathbf{p}_{\nu\mu}$. Since the σ elements u_i form a basis of the ideal $\mathbf{M}^\nu$ and since x belongs to the ideal $\mathbf{M}^\nu \cap \mathbf{m}^\mu$, we have by (10): $x = \Sigma A_i u_i$, $A_i \in \mathbf{m}^{\mu-\nu}$. Passing to the corresponding vectors $\bar{x}$, $\bar{A}_i$ and $\bar{u}_i$ in $\mathfrak{M}_\nu(A)$ [i.e., to the $\mathbf{M}^{\nu+1}$-residues; see (6)] we have

$$\bar{x} = \Sigma \bar{A}_i \bar{u}_i . \tag{15}$$

Here the σ vectors $\bar{u}_i$ form a basis of $\mathfrak{M}_\nu(A)$, since the σ elements u_i form a minimal basis of $\mathbf{M}^\nu$. Now let us assume that P is (W, ν)-regular for V. In that case ρ is also the dimension of the space spanned in $\mathfrak{M}_\nu(A)$ by $H_{\nu\nu}$. Since the first ρ elements u_i belong to $\mathbf{p}$, the ρ *independent* vectors $\bar{u}_1$, $\bar{u}_2, \ldots, \bar{u}_\rho$ belog to $H_{\nu\nu}$. *It follows that every vector in* $H_{\nu\nu}$ *is linearly dependent* on $\bar{u}_1, \bar{u}_2, \ldots, \bar{u}_\rho$. Since $\bar{x}$ belongs to $H_{\nu\nu}$, we conclude that the last $\sigma - \rho$ coefficients $\bar{A}_i$ in (15) must be zero. Hence $A_i \in \mathbf{M}$ for $\rho + 1 \leq i \leq \sigma$. If we now set $x_1 = \Sigma_{i=1}^{\rho} A_i u_i$, $x_2 = \Sigma_{i=\rho+1}^{\sigma} A_i u_i$, then $x_1 \in \mathbf{p}_{\nu\nu}\mathbf{m}^{\mu-\nu}$, $x_2 \in \mathbf{M}^{\nu+1} \cap \mathbf{m}^\mu$, and hence $x_2 \in \mathbf{p}_{\nu+1,\mu}$, since $x_2 = x - x_1$ and $x_1 \in \mathbf{p}_{\nu\mu}$. We have thus proved that $\mathbf{p}_{\nu\mu} \subset \mathbf{p}_{\nu\nu}\mathbf{m}^{\mu-\nu} + \mathbf{p}_{\nu+1,\mu}$, and since the opposite inclusion is obvious, the proof of the theorem is complete.

Te key result is the following

Theorem 2. - *If* P *is a* (W, ν)*-regular point of* V, *then*

$$h_{\nu\mu} \cap g_{\nu+1,\mu} = h_{\nu+1,\mu} . \tag{16}$$

Proof. We may assume that $\mu > \nu$, for if $\mu = \nu$ then $g_{\nu+1,\mu} = h_{\nu+1,\mu} = (0)$. We use the notation of the proof of the preceding theorem. It is suffficient to prove the inclusion $h_{\nu\mu} \cap g_{\nu+1,\mu} \subset h_{\nu+1,\mu}$. This is equivalent to proving that

$$\mathbf{p}_{\nu\mu} \cap [(\mathbf{M}^{\nu+1} \cap \mathbf{m}^\mu) + \mathbf{m}^{\mu+1}] \subset \mathbf{p}_{\nu+1,\mu} + \mathbf{m}^{\mu+1}.$$

In view of (14), we get an equivalent relation if we replace here $\mathbf{p}_{\nu\mu}$ by $\mathbf{p}_{\nu\nu}\mathbf{m}^{\mu-\nu}$. Let then x be any element of the ideal

$$\mathbf{p}_{\nu\nu}\mathbf{m}^{\mu-\nu} \cap [(\mathbf{M}^{\nu+1} \cap \mathbf{m}^\mu) + \mathbf{m}^{\mu+1})].$$

Since $x \in \mathbf{p}_{\nu\nu}\mathbf{m}^{\mu-\nu}$, we can write x in the form $\Sigma_{j=1}^{\rho} A_j u_j$, $A_j \in \mathbf{m}^{\mu-\nu}$, since, by construction, the elements $u_1, u_2, \ldots, u_\rho$ form a basis (in fact a minimal basis) of the ideal $\mathbf{p}_{\nu\nu}$. Since x also belongs to the ideal $(\mathbf{M}^{\nu+1} \cap \mathbf{m}^\mu) + \mathbf{m}^{\mu+1}$, it follows by (7) and (8) that there exist element $B_1, B_2, \ldots, B_\sigma$ in $\mathbf{M} \cap \mathbf{m}^{\mu-\nu}$ such that $x - \Sigma_{i=1}^{\sigma} B_i u_i \in \mathbf{m}^{\mu+1}$. We have, then, $\Sigma_{j=1}^{\rho} A_j u_j - \Sigma_{i=1}^{\sigma} B_i u_i \in \mathbf{m}^{\mu+1}$, and this implies, by (12), that $A_j \in \mathbf{M} \cap \mathbf{m}^{\mu-\nu} + \mathbf{m}^{\mu-\nu+1}$. Since $u_j \in \mathbf{p}_{\nu\nu} \subset \mathbf{M}^\nu \subset \mathbf{m}^\nu$ for $j = 1, 2, \ldots, \rho$, it follows that $x \in \mathbf{p}_{\nu+1,\mu} + \mathbf{m}^{\mu+1}$, and this completes the proof of the theorem.

We shall say that *almost all points* of an irreducible variety W have a given property α if the points of W which do not have property α lie on some *proper* algebraic subvariety of W (we do not mean to imply that the set of points of W which do not have property α is itself an algebraic variety).

THEOREM 3. - *For any given integer* ν *almost all points of* W *are* (W, ν)*-regular for* V.

Proof. Let $X_1, X_2, \ldots, X_n$ be non-homogeneous coördinates in our affine space S, and let $\mathbf{P}$ be the prime ideal of V in the polynomial ring $k[X_1, X_2, \ldots, X_n]$. The polynomials in $\mathbf{P} \cap \mathbf{M}_1^{\nu}$ (i.e., the polynomials which are zero on V and are zero to an order $\geqq \nu$ at the general point A of W) determine a set of vectors in $\mathfrak{M}_\nu(A)$. Let α be te dimension of the subspace of $\mathfrak{M}_\nu(A)$ spanned by that set of vectors, and let $\{ f_i(X), i = 1, 2, \ldots, \alpha \}$ be a set of polynomials in $\mathbf{P} \cap \mathbf{M}_1^{\nu}$ which determine independent vectors in $\mathfrak{M}_\nu(A)$. If σ is the dimension of $\mathfrak{M}_\nu(A)$, we choose other $\sigma - \alpha$ polynomials $g_j(X)$ such that the σ polynomials $f_i(X)$ and $g_j(X)$ determine together a basis of $\mathfrak{M}_\nu(A)$. These σ polynomials constitute then a minimal basis of the ideal $\mathbf{M}_1^{\nu}$ in the local ring $\mathbf{O}$ of the point A.

Let $\mathbf{A}$ denote the *polynomial ideal* generated by the above σ polynomials. It is clear that not only is $\mathbf{P}$ an isolated prime ideal of $\mathbf{A}$, but also that $\mathbf{P}$ will appear as a component in any normal decomposition of $\mathbf{A}$ into primary components. Let $\mathbf{P}_1, \mathbf{P}_2, \ldots$ be the other prime ideals of $\mathbf{A}$, both isolated or embedded. Let $L(f_i, g_j)$ denote the sum of the following *proper* subvarietes of W: 1) the variety of singular points of W; 2) the intersection of W with the variety of $\mathbf{P}_q$, $q = 1, 2, \ldots$ We *claim that any point* P *of* W, *which is not on* L(f_i, g_j) *is* (W, ν)*-regular for* V. The proof of this assertion will estabilish our theorem.

It is clear that the α polynomials $f_i\,(X)$ form a basis of the ideal $\mathfrak{p}_{\nu\nu}$. Hence $H_{\nu\nu}$ spans in $\mathfrak{M}_\nu(A)$ a subspace of dimension α (this is true for any point P of M). On the other hand, it follows from our choice of the point P that the σ polynomials $f_i(X)$ and $g_j(X)$ form a basis of the ideal $\mathbf{M}^{\nu}$, necessarily a *minimal* basis, since we know that any minimal basis of $\mathbf{M}^{\nu}$ must have exactily σ elements $[\sigma = \dim \mathfrak{M}_\nu(A)]$. But then the vectors which correspond in $\mathfrak{M}^{\nu}(P)$ to the polynomials f_i, g_j are also independent. Since the $f_i(X)$ are in $\mathfrak{p}_{\nu\nu}$, we conclude that $\dim h_{\nu\nu} \geqq \alpha$. It follows from remark 3) of section 3 that $\dim h_{\nu\nu} = \alpha$, i.e., P is (W, ν)-regular for V, as asserted.

COROLLARY 1. - *For any given integer* ν, *almost all point of* W *are* (W, λ)*-regular*, $\lambda = 1, 2, \ldots, \nu$.

COROLLARY 2. - *For any integer* ν *the set of points of* W *which are not* (W, ν)*-regular for* V *is an algebraic variety.* We refer to the minimal basis $\{ u_1, u_2, \ldots, u_\sigma \}$ of $\mathbf{M}_\nu$ introduced in the proof of Theorem 1. It is clear that we may assume that the u_i are polynomials. If we set $f_i(X) = u_i$, $i = 1, 2, \ldots, \rho$, $g_j(X) = u_{\rho+j}$, $j = 1, 2, \ldots, \sigma - \rho$, then the σ polinomials $f_i(X)$ and $g_j(X)$ are of the type used in the preceding proof, and furthermore the point P does not belong to $L(f_i, g_j)$. It follows that the set of points of W which are not (W, ν)-regular for V is given by the intersection of all the varieties $L(f_i, g_j)$ obtained by choosing the polynomials f_i and g_j in all possible ways. Since this intersection is an algebraic variety, our assertion follows.

5. Proof of the main lemma.

We have, by assumption, that W is the closure of the given set G. It follows from Corollary 1 of Theorem 3 that the set of points of G which are (W, λ)-regular for V, for $\lambda = 1, 2, \dots, \nu$, also has the property that its closure is W. *Hence we may assume that all the points of* G *are* (W, λ)*-regular for* V, $\lambda = 1, 2, \dots, \nu$.

The main lemma is obvious if $\nu = 1$. Hence we shall proceed by induction from ν to $\nu + 1$. Let, then, z be a rational function on V which vanishes to an order $\geqq \nu + 1$ at each point P of G. By induction hypothesis, z vanishes to an order $\geqq \nu$ at the general point A of W, i e., $z \varepsilon [\mathbf{m}(A/V)]^{\nu}$. We have to prove that $z \varepsilon [m(A/V)]^{\nu+1}$.

Let $x_1, x_2, \dots, x_n$ be the coördinates of the general point of V such that $k(x_1, x_2, \dots, x_n)$ is our fixed function field Σ of V. We go back to the independent variables $X_1, X_2, \dots, X_n$, to the local ring $\mathbf{O}$ of the affine space S at A and to the maximal ideal $\mathbf{M}_1$ of $\mathbf{O}$. Every element of $[\mathbf{m}(A/V)]^{\nu}$ can be written in the form $\varphi(x)/\psi(x)$, where $\varphi(X)$ and $\psi(X)$ are polynomials, $\psi(x) \neq 0$ on W, and $\varphi(X)/\psi(X) \varepsilon \mathbf{M}_1{}^{\nu}$. We fix one such repprésentation for our element z: $z = \varphi(x)/\psi(x)$, and we set $Z = \varphi(X)/\psi(X)$.

Since $\psi(X) \neq 0$ on W, the points of W where $\psi(X)$ is zero form a proper subvariety of W. Hence the set of points of G at which $\psi(X)$ does *not* vanish is still such that its closure is the entire variety W. We may therefore assume that $\psi(X) \neq 0$ at *each* point of G. Under this assumption, the element Z belongs to the local ring of S at P, *for any point* P *of* G.

We fix a point P in G and we use the notation of the preceding sections. We have, then, $Z \varepsilon \mathbf{M}^{\nu}$ [see (6), section 3]. Our assumption that z vanishes to an order $\geqq \nu + 1$ at P is equivalent to assuming that Z belongs to the ideal $\mathbf{p} + \mathbf{m}^{\nu+1}$, where $\mathbf{p}$ is the local ideal of V at P and $\mathbf{m}$ is the maximal ideal of the local ring $\mathbf{o}$ of S at P. We can therefore write: $Z = Z_1 + Y$, where $Z_1 \varepsilon \mathbf{m}^{\nu+1}$ and $Y \varepsilon \mathbf{p}$. Since $Z \varepsilon \mathbf{M}^{\nu} \subset \mathbf{m}^{\nu}$, Z defines a vector $\bar{Z}$ in $\mathfrak{M}_{\nu}(P)$. We have $\vec{Z} \varepsilon g_{\nu\nu}$, since $Z \varepsilon \mathbf{M}^{\nu}$. On the other hand, Z and Y determine the same vector in $\mathfrak{M}_{\nu}(P)$ since $Z - Y = Z_1 \varepsilon \mathbf{m}^{\nu+1}$. Since $Y \varepsilon \mathbf{p} \cap \mathbf{m}^{\nu}$ and $\mathbf{p} \subset \mathbf{M}$, in follows that $Y \varepsilon \mathbf{p}_{1\nu}$, and consequently $\bar{Z} \varepsilon h_{1\nu}$. Hence $\bar{Z} \varepsilon h_{1\nu} \cap g_{\nu\nu}$. Since $g_{2\nu} \supset g_{\nu\nu}$, we can write $\bar{Z} \varepsilon (h_{1\nu} \cap g_{2\nu} \cap g_{\nu\nu}$, and applying Theorem 2 we obtain $\bar{Z} \varepsilon h_{2\nu} \cap g_{\nu\nu}$ [since P is $(W, 1)$-regular for V]. This, again, we can write in the form: $\bar{Z} \varepsilon (h_{2\nu} \cap g_{3\nu}) \cap g_{\nu\nu}$ (since $g_{3\nu} \supset g_{\nu\nu}$), and applying again Theorem 2 we obtain: $Z \varepsilon h_{3\nu} \cap g_{\nu\nu}$ [since P is $(W, 2)$-regular for V]. Ultimately we find in this fashion that $\bar{Z} \varepsilon h_{\nu\nu} \cap g_{\nu\nu}$, i.e., $\bar{Z} \varepsilon h_{\nu\nu}$. *We conclude that* Z *belongs to the ideal* $\mathbf{P} \cap \mathbf{M}^{\nu} + \mathbf{m}^{\nu+1}$. Since $Z \varepsilon \mathbf{M}^{\nu}$ and since $\mathbf{M}^{\nu} \cap \mathbf{m}^{\nu+1} = \mathbf{M}^{\nu}\mathbf{m}$ [see (7), section 3], we can now assert that

$$(17) \qquad Z \varepsilon \mathbf{p} \cap \mathbf{M}^{\nu} + \mathbf{M}^{\nu}\mathbf{m}.$$

Relation (17) holds for each point P of G. Now let us fix a set of polynomials $f_i(X)$ and $g_j(X)$ as in the proof of Theorem 3. Since these polynomials form a base of the ideal $\mathbf{M}^{\nu}$ and since $Z \varepsilon \mathbf{M}^{\nu}$, we have

$$(18) \qquad Z = \Sigma A_i f_i + \Sigma B_j g_j,$$

where A_i, $B_j \varepsilon \mathbf{O}$. It is clear that for almost all points P of W it is true that the elements A_i and B_j belong to the local ring of S at P. Replacing, if necessary, the set G by a subset whose closure ist still the entire variety W, we may therefore assume that the A_i and B_i belong to $\mathbf{o}(P/S)$, for all points P in G. A similar argument shows that we may assume, without loss of generality, that no point of G belongs to the variety $L(f_i, g_j)$ of points which we had to avoid in the proof of Theorem 3. Because of this last assumption, we may assert that if P is any point of G, then the f_i form a basis of the ideal $\mathbf{P}_{\nu\nu}$, and that the polynomials f_i and g_j form a basis of $\mathbf{M}^{\nu}$. Hence, in view of (17), Z can be expressed locally, at P, in the following form: $z = \Sigma C_i f_i + \Sigma D_j g_j$, where the C_i and D_j are in $\mathbf{o}(P/S)$ *and the* D_j *are in* $\mathbf{m}(P/S)$. If we compare this local expression of Z with that given by (18) and if we recall that the polynomials f_i, g_j form a *minimal* basis of $\mathbf{M}^{\nu}_{\cdot}$ (see proof of Theorem 3), we conclude, by (9), section 3, *that all the* B_j *are zero at* P. Since this holds for each poin P of G and since W is the closure of G, it follows that *the* B_j *are zero on the entire variety* W, i.e., the B_j belong to the ideal $\mathbf{M}_1$. Therefore the sum $\Sigma B_j g_j$ belongs to $\mathbf{M}^{\nu+1}$, and since the sum $\Sigma A_i f_i$ is zero on V, we conclude that the original element z of the fun tion field of V belongs to $[\mathbf{m}(A/V)]^{\nu+1}$. This completes the proof of the main lemma.

6. Another application of the main lemma.

Let $R = k[x_1, x_2, \dots, x_n]$ be the non-homogeneous coördinate ring of V, where the x_i are the coördinates of the general point of V. Let $\mathbf{p}$ be a prime ideal in R and let W be the irreducible subvariety of V defined by $\mathbf{p}$ (we are dealing with varieties in the affine space). As an application of the main lemma, we shall prove that

$$(19) \qquad \bigcap_{P \varepsilon W} [\mathbf{m}(P\, V)]^{\nu} \cap R \subset \mathbf{p}^{\rho_\nu}.$$

were $\rho_\nu \to \infty$ *as* $\nu \to \infty$. In other words: *if an element* z *of* R *vanishes to a high order at each point* P *of* W, *then* z *belongs to a high power of the prime ideal* $\mathbf{p}$ *of* W.

Proof. For the proof it will be sufficient to show that given any integer ρ, there exists an integer ν such that the left-hand member of (19) is contained in $\mathfrak{p}^\rho$. Let $\mathfrak{p}^\rho = \mathfrak{p}^{(\rho)} \cap \mathfrak{q}_1 \cap \mathfrak{q}_2 \cap \dots \cap \mathfrak{q}_h$ be a normal decomposition of $\mathfrak{p}^\rho$ into primary components, where $\mathfrak{p}^{(\rho)}$ is the ρ-th symbolic power of p, and let $\mathfrak{p}_1$, $\mathfrak{p}_2, \dots, \mathfrak{p}_h$ be the prime ideals of the primary components $\mathfrak{q}_1, \mathfrak{q}_2, \dots, \mathfrak{q}_h$. It is known that $\mathfrak{p}_i \supset \mathfrak{p}$, $i = 1, 2, \dots, h$. Let ν_i be the exponent of $\mathfrak{q}_i$. Then $\mathfrak{p}_i^{(\nu_i)} \subset \mathfrak{q}^i$. Let $\nu = \max(\rho, \nu_1, \nu_2, \dots, \nu_h)$.

Suppose now that an element z of R vanishes to an order $\geqq \nu$ at each point P of W. The main lemma implies that if $\mathfrak{p}'$ is any prime ideal in R such that $\mathfrak{p}' \supset \mathfrak{p}$, then z belongs to the ν-th symbolic power of $\mathfrak{p}'$. We have, therefore, in particular: $z \in \mathfrak{p}'^{(\nu)} \subset \mathfrak{p}^{(\rho)}$ and also $z \in \mathfrak{p}_i^{(\nu)} \subset \mathfrak{p}_i^{(\nu_i)} \subset \mathfrak{q}_i$. Hene $z \in \mathfrak{p}^\rho$, and this establishes (19).

7. The case of a simple subvariety W of V.

The whole point of our proof of the main lemma is that it establishes the lemma for arbitrary subvarieties W of V, hence also for *singular* subvarieties W. In the case of a simple subvariety a much shorter proof can be given, as we shall now show.

Let A be a general point of W. By assumption, A is a simple point of V. Therefore the results of section 3 continue to hold if the affine space S is replaced by V. Accordingly, we shall now mean by $\mathbf{O}$ and $\mathbf{o}$ the rings $\mathbf{o}(A/V)$ and $\mathbf{o}(P/V)$ respectively. Actually we shall only make use of (6) and (9), section 3.

Let $t_1, t_2, \dots, t_\rho$ be l.u.p. of V at A, where $\rho = \dim V - \dim W$. We proceed, as in section 5, by induction from ν to $\nu + 1$. We have then $z \in \mathbf{M}_1^\nu$, where $\mathbf{M}_1$ is the maximal ideal of $\mathbf{O}$, and hence $z = \varphi(t_1, t_2, \dots, t_\rho)$, where φ is a form of degree ν, with coefficients in $\mathbf{O}$.

We may assume that the t_i's belong to the coördinate ring $k[x_1, x_2, \dots, x_n]$ of V. Let $\mathbf{A}$ be the ideal generated in this ring by the ρ elements t^i. Let $L(\mathbf{A})$ denote the sum of the following *proper* subvarieties of W: 1) the singular locus of W; 2) the intersection of W with the varieties (other than W) of the prime ideals of $\mathbf{A}$. It follows as in section 5 that it is permissible to assume that no point of the set G belongs to $L(\mathbf{A})$. Under this assumption, the l.u.p. $t_1, t_2, \dots t_\rho$ constitute a minimal base of the local ideal $\mathbf{M}$ of W at P, where P is any point of G [$\mathbf{M} = \mathbf{m}(A/V) \cap \mathbf{m}(P/V)$], and the power products of the t_i, of degres ν, constitute a minimal base of $\mathbf{M}^\nu$. Since $z \in [\mathbf{m}(P/V)]^{\nu+1}$, it follows, by (9), that the coefficients of the form φ belong to $\mathbf{m}(P/V)$ (as in section 5, we may assume here that these coefficients all belong to $\mathbf{o}(P/V)$, for all points P of G)]. Since this holds for any point of G and since W is the closure of G, it follows that the coefficients of the form φ belong to $\mathbf{M}_1$. Hence $z \in \mathbf{M}_1^{\nu+1}$, q.e.d.

QUELQUES QUESTIONS
CONCERNANT LA THÉORIE DES FONCTIONS HOLOMORPHES
SUR UNE VARIÉTÉ ALGÉBRIQUE

par M. O. ZARISKI.

1. PRINCIPE DE DÉGÉNÉRESCENCE.

Ce fut en recherchant une démonstration du principe de dégénérescence que je fus conduit à la théorie des fonctions holomorphes sur une variété algébrique abstraite. Pour énoncer ce principe de façon précise, il nous faut commencer par poser diverses définitions.

Soit V une variété algébrique projective, et k un corps de définition de V (c'est-à-dire que V peut être définie par un système d'équations homogènes à coefficients dans k); nous dirons que V est *connexe sur k* (ou que *V/k est connexe*) si V n'est pas la réunion de deux variétés définies sur k, non vides, et disjointes.

Ceci est une définition relative : une variété V peut être connexe sur un corps de définition k, et non connexe sur un autre. Nous dirons que V est *absolument connexe* si elle est connexe sur chacun de ses corps de définition. Si k est un corps de définition *algébriquement clos* de V, on voit aisément que V est absolument connexe si, et seulement si V/k est connexe.

Pour une caractéristique donnée p, je considère un espace projectif universel S_n de dimension donnée n (c'est-à-dire un espace projectif de dimension n sur un *domaine universel* à la Weil). Dans S_n je considère l'ensemble de toutes les variétés absolument irréductibles de dimension donnée r, et je prends ces variétés pour générateurs d'un groupe abélien libre (écrit additivement); les éléments de ce groupe sont appelés *cycles de dimension r*, ou *r-cycles;* les générateurs de ce groupe, c'est-à-dire les variétés absolument irréductibles ci-dessus, sont appelés des *cycles premiers.*

Si un r-cycle Γ est donné par

$$(1) \qquad \Gamma = \Sigma m_i \Gamma_i$$

où les Γ_i sont des cycles premiers et les m_i des entiers, nous dirons que Γ est *effectif* si $m_i \geq 0$ pour tout i. Nous ne nous occuperons que de cycles effectifs et non nuls. Tout cycle Γ (effectif et non nul) détermine de façon évidente une variété purement r-dimensionnelle que nous noterons $|\Gamma|$. Par abus de langage nous attribuerons au cycle Γ toutes les propriétés («irréductibilité absolue», «connexion absolue», «connexion sur k», par exemple) que pourrait posséder la *variété* $|\Gamma|$, Γ étant défini par (1), le degré de Γ sera par définition $\Sigma m_i \nu_i$ où ν_i est le degré de la variété absolument irréductible $|\Gamma_i|$.

Chow et Van der Waerden ont montré que l'ensemble des r-cycles de degré donné m de S_n possède une structure algébro-géométrique. Il est en effet possible de mettre ces r-cycles en correspondance biunivoque avec les points d'une variété définie sur le corps premier, le point représentant le cycle Γ ayant pour coordonnées homogènes les coefficients de la *forme associée* («Zugeordnete Form») à Γ. Ceci étant, soit k un corps et M un ensemble de r-cycles de degré m de S_n; nous dirons que M est un *système algébrique* et que k est un *corps de définition* de M, si l'ensemble M* des points représentatifs des cycles de M est une variété définie sur k. Nous pourrons donc appliquer aux systèmes algébriques de cycles la terminologie en usage pour les variétés; nous pourrons en particulier parler de «cycle générique d'un système algébrique irréductible M/k» et de «spécialisation d'un cycle Γ sur un corps k». Remarquons que l'ensemble de tous les points de tous les cycles du système M/k est lui-même une variété définie sur k et appelée le *porteur du système* M. Je puis maintenant énoncer le

Principe de dégénérescence.. — *Si un cycle générique* Γ *d'un système algébrique irréductible* M/k *est absolument irréductible, alors tout cycle du système est absolument connexe.*

Dans le langage de la géométrie algébrique classique nous dirions que «la limite d'une variété irréductible est nécessairement connexe» (mais peut être réductible). On voit aisément que le principe de dégénérescence reste vrai si le cycle générique Γ est absolument connexe au lieu d'absolument irréductible.

Le principe de dégénérescence peut être considér comme un cas particulier d'un théorème plus général que je vais énoncer dans un instant. Soit σ/k une correspondance algébrique irréductible entre M/k (identifié à la variété des points représentatifs des cycles de M) avec quelqu'autre variété irréductible V/k. Si Q est un point de V, l'ensemble $\sigma\{Q\}$ des cycles de M qui correspondent à Q est un système algébrique défini sur $k(Q)$.

Soit (P, Γ) une paire générique de σ/k.

Théorème de connexion. — Si les conditions suivantes sont réalisées :

a. *Le corps* $k(P)$ *est quasi algébriquement fermé dans* $k(P, \Gamma)$ [c'est-à-dire que tout élément de $k(P, \Gamma)$ qui est algébrique et séparable sur $k(P)$ est contenu dans $k(P)$];

b. *Un cycle générique de* M/k *est absolument connexe;*

c. *La variété* V/k *est analytiquement irréductible en* Q (c'est-à-dire que le complété de l'anneau local de V/k en Q est un domaine d'intégrité).

Alors le porteur du système $\sigma\{Q\}$ *est connexe sur* $k(Q)$.

On déduit le principe de dégénérescence du théorème de connexion de la manière suivante :

a. Pour un cycle donné Γ_0 de M/k, nous passons du corps de base k à la clôture algébrique k_1 de $k(\Gamma_0)$;

b. Nous remplaçons, dans le théorème de connexion, M/k par une composante irréductible M_1/k_1 de M/k_1 contenant Γ_0;

c. Nous prenons pour V/k un modèle normal dérivé de M_1/k_1, et nous utilisons le fait bien connu qu'une variété normale est analytiquement irréductible en chacun de ses points.

2. FONCTIONS HOLOMORPHES. LE THÉORÈME FONDAMENTAL.

Il peut sembler naturel qu'un premier progrès dans la démonstration du théorème de connexion soit le suivant : *étant donnée une sous-variété* W/k *de la variété irréductible* V/k, *trouver un critère arithmértique de connexion de* W/k, *critère s'exprimant en termes des propriétés arithmétiques du couple* (V, W). L'analogie avec la théorie locale d'une part, et avec la géométrie algébrique classique d'autre part, suggère la marche suivante. On devra d'abord définir, de façon naturelle, la notion de *fonction holomorphe sur* V/k, *définie le long de* W; ces fonctions devront former un anneau $o^*(W)$; on dira alors, par définition, que «V/k *est analytiquement irréductible le long de* W» si $o^*(W)$ est un anneau d'intégrité. Un arrêt ici pour vérifier que cette définition coïncide bien, dans le cas classique, avec celle de la théorie des fonctions de plusieurs variables complexes. Si nous avons cette chance, nous continuerons, tout revigorés, par la démonstration du critère de connexion suivant, critère qui est presque évident dans ce cas classique :

Critère de connexion. — (1) *Si* W/k *est non connexe,* V/k *est analytiquement réductible au voisinage de* W.

(2) *Si* W/k *est connexe et si* V/k *est analytiquement irréductible en chaque point de* W, *alors* V/k *est analytiquement irréductible au voisinage de* W.

C'est naturellement (2) la partie la plus intéressante (et la plus difficile à démontrer) du critère de connexion. Remarquons que la condition que V/k est analytiquement irréductible en chaque point de W est automatiquement satisfaite si V est une variété normale.

Quant à la recherche d'une bonne définition des hypothétiques fonctions holomorphes, c'est chose relativement simple, car les données de la question nous laissent une liberté de choix limitée. Si F est le corps des fonctions rationnelles sur V/k, la théorie locale nous dit déjà ce qu'est une suite convergente en un point Q de V/k de fonctions x_i de F. D'autre part, dans le cas abstrait, la variété W/k, qui est le domaine de définition de nos fonctions holomorphes, n'a qu'une topologie naturelle : c'est celle où les ensembles fermés sont les sous-variétés de W/k; elle satisfait à l'axiome de séparation T_1 de Fréchet, et à l'axiome de Borel-Lebesgue (espace «quasi-compact»). On doit alors commencer par considérer des suites (x_i), $(x_i \in F)$ qui convergent uniformément sur un ouvert de W. Ceci amène à considérer un recouvrement ouvert fini arbitraire (G_α) de W/k, et à définir une fonction holomorphe sur V/k le long de W/k comme un ensemble (x_{α_i}) de suites telles que :

(1) (x_{α_i}) converge uniformément sur G_α;

(2) Si Q est un point commun à G_α et G_β, alors les deux suites (x_{α_i}) et (x_{β_i}) sont équivalentes dans l'anneau local de Q.

D'un point de vue formaliste il est préférable de partir du produit direct infini

$$O^*(W) = \Pi_{Q \in W} o^*(Q)$$

où $o^*(Q)$ est l'anneau complété de l'anneau local de V/k en Q. Si ξ est un élément de $O^*(W)$ et Q un point de W, la projection de ξ sur $o^*(Q)$ sera notée $\xi[Q]$. Ceci étant, la définition formaliste des fonctions holomorphes sur V/k le long de W est la suivante :

Définition. — *Un élément* ξ *de* $O^*(W)$ *est une fonction holomorphe sur* V/k *le long de* W, *s'il existe un recouvrement ouvert fini* (G_α) *de* W/k *et un ensemble correspondant de suites* (x_{α_i}), $(x_{\alpha_i} \in F)$ *tels que :*

(1) *Chaque suite* (x_{α_i}) *converge uniformément sur* G_α;

(2) *Si* $Q \in G_\alpha$, *alors* $\lim (x_{\alpha_i}) = \xi[Q]$ *en* Q.

Il est aisé de montrer que l'ensemble $o^*(W)$ de toutes les fonctions holomorphes sur V/k le long de W est un sous-anneau de $O^*(W)$. Comme je l'ai dit plus haut, on peut maintenant prouver que le critère de connexion est valable en toute généralité.

Le point central de la théorie, qui contient le théorème de connexion comme cas très particulier, est le théorème suivant, relatif à *l'invariance des anneaux de fonctions holomorphes :*

Théorème 1. — *Soit* T *une correspondance algébrique irréductible entre deux variétés irréductibles* V/k *et* V′/k, *et soit* (P, P′) *une paire générique de* T/k; *soient* W/k *une sous-variété de* V/k *et* W′ = T{W} *la transformée totale de* W *sur* V′. *On suppose :*

a. *Le corps* k(P) *est algébriquement fermé dans* k(P, P′);

b. V/k *est localement normale en tout point de* W;

c. *La transformation* T^{-1} *est rationnelle, c'est-à-dire* k(P) ⊂ k(P′);

d. T^{-1} *est semi-régulière en tout point de* W′ (c'est-à-dire que, si Q′ est un point quelconque de W′ et Q un point quelconque de W correspondant à Q, alors l'anneau local de V′/*k* en Q′ contient l'anneau local de V/*k* en Q; ceci implique que Q est le seul point correspondant à Q′).

Dans ces conditions l'anneau o*(W) *des fonctions holomorphes sur* V *le long de* W *est* k*-isomorphe à l'anneau* o*(W′) *des fonctions holomorphes sur* V′ *le long de* W′.

La démonstration en est longue et difficile, et je n'essaierai même pas d'en donner une idée. A la place je vais indiquer brièvement comment ce théorème sert à démontrer le théorème de connexion. Remarquons d'abord le corollaire suivant :

Corollaire. — *Dans les conditions* a, b, c *et* d *du théorème, la connexion de* W/k *entraine celle de* W′/k.

En effet, si W/*k* est connexe, (*b*) et la partie (2) du critère de connexion montrent que *o**(W) est un anneau d'intégrité. Il en est alors de même de *o**(W′) d'après le théorème 1, et par conséquent W′/*k* est connexe d'après la partie (1) du critère de connexion.

Mais il est clair qu'une correspondance algébrique univoque en tous points transforme une variété connexe en une variété connexe (image continue d'un ensemble connexe!). Ainsi, on voit aussitôt, en passant au *graphe de* T/*k*, *que le corollaire reste vrai si les conditions* c *et* d *sont omises.*

Un raisonnement fort simple (et dont je ne parlerai pas) montre que le corollaire reste vrai si les conditions *a* et *b* sont remplacées respectivement par les conditions suivantes :

a′. *Le corps* k(P) *est quasi algébriquement fermé dans* k(P, P′);

b′. V/k *est analytiquement irréductible en tout point de* W

Soit maintenant Q un point quelconque de V tel que V/*k* soit analytiquement irréductible en Q. Si Q est un point rationnel sur *k*, Q est une variété définie sur *k*, et donc, si la condition *a′* est satisfaite, T{Q} est connexe sur *k*(Q) = *k*. Si Q n'est pas un point rationnel, nous posons $k_1 = k(Q)$ et nous étendons le corps de base de *k* en k_1. Or, il n'est pas difficile de démontrer le résultat suivant : *si* V/k *est analytiquement irréductible en* Q, V/k_1 *possède une seule composante irréductible contenant* Q, *soit* V_1, *et* V_1/k_1 *est analytiquement irréductible en* Q. D'autre part, il s'ensuit aisément de *a′* que les composantes irréductibles T_i de T/k_1 sont associées de façon biunivoque aux composantes V_i de V/k_1, chaque T_i étant une correspondance irréductible entre V_i/k_1 et quelque composante irréductible de V'/k_1. De plus, pour tout *i*, la correspondance T_i/k_1 vérifie une condition analogue à a′) Puisque $Q \notin V_i$ pour $i \neq 1$, on a $T\{Q\} = T_1\{Q\}$. Comme V_1/k_1 est analytiquement irréductible en Q, et comme Q est rationnel sur k_1, on déduit du cas précédent que T{Q} *est connexe sur* k(Q). Nous avons ainsi démontré le théorème suivant :

Théorème 2. — *Soit* T/k *une correspondance algébrique irréductible entre deux variétés irréductibles* V/k *et* V′/k, *et soit* (P, P′) *une paire générique de* T/k. *Si le corps* k(P) *est quasi algébriquement fermé dans* k(P, P′), *alors, à tout point* Q *de* V *tel que* V/k *soit analytiquement irréductible en* Q, *il correspond sur* V′ *une variété* T{Q} *qui est connexe sur le corps* k(Q).

Le théorème de connexion s'en déduit immédiatement. D'abord, dans le cas particulier où V/*k* est un modèle normal dérivé de M/*k*, le théorème 2 donne aussitôt le principe de dégénérescence. Dans le cas général, le théorème 2 nous indique que le système σ{Q} *est connexe sur* k(Q). D'après le principe de dégénérescence nous savons déjà que *tout cycle de* σ{Q} *est absolument connexe* (car σ{Q} ⊂ M). Il s'ensuit alors aisément que le poteur de σ{Q} est connexe sur *k*(Q).

3. DEUX PROBLÈMES NON RÉSOLUS.

Considérons les fonctions *rationnelles* sur V/*k* qui sont holomorphes le long de W. Elles forment un sous-anneau *o*(W.) de *o**(W). Il est clair que l'on a

$$o(W) = \bigcap_{P \in W} O(P).$$

Si W *est irréductible,* alors la trace sur W de tout élément de *o*(W) est une fonction rationnelle sur W qui est holomorphe en *tout* point de W, c'est-à-dire une *constante.* On en déduit que :

Un élément u *de* o(W) *n'est pas inversible dans* o(W) *si, et seulement si la trace de* u *sur* W *est nulle. Les éléments non inversibles de* o(W) *forment un idéal.*

Si W est réductible, mais connexe, tout élément de *o*(W) a la même trace sur chaque composante irréductible de W/*k*, et l'énoncé ci-dessus reste vrai. Dans le cas général, on en déduit que *o*(W) n'a qu'un nombre fini d'idéaux maximaux, et que ce nombre n'est pas supérieur à celui des composantes connexes de W/*k*. Ainsi, en tous cas, il serait possible d'affirmer que *o*(W) est un *anneau semi-local* (au sens de Chevalley; un anneau local si W/*k* est connexe), *à condition* que l'on sache que *o*(W) satisfait à la condition maximale (ou condition des chaînes

9·

ascendantes) pour ses idéaux. Mais ce problème est non résolu :

PROBLÈME A. — o(W) *est-il un anneau noethérien* ?

Un second problème non résolu concerne les relations entre les anneaux $o(W)$ et $o^*(W)$. Soit $m(W)$ l'intersection des idéaux maximaux de $o(W)$; que $o(W)$ soit ou non nœthérien, nous pouvons toujours considérer le complété de $o(W)$ par rapport à la topologie définie par les puissances de $m(W)$; il est clair que ce complété s'identifie. à un sous anneau de l'anneau $o^*(W)$ des fonctions holomorphes sur V/k le long de W.

PROBLÈME B. — *Si* W *est connexe,* $o^*(W)$ *est-il le complété de* o(W)?

Si W est non connexe, l'exemple suivant montre que $o^*(W)$ n'est pas toujours le complété de $o(W)$; si $o(W)$ ne contient que les constantes, il est son propre complété, et ce complété est un corps, tandis que $o^*(W)$ n'est pas un anneau d'intégrité (d'après le critère de connexion).

4. APPLICATION AUX SOUS-VARIÉTÉS EXCEPTIONNELLES.

DÉFINITION. — *Une sous-variété* W/k *d'une variété irréductible* V/k *est dite exceptionnelle s'il existe une transformation rationnelle* T *de* V/k *dans une autre variété* V'/k *telle que :* a. T *est semi-régulière en tout point de* W; b. T{W} *est une variété* W'_0/k *de dimension zéro;* c. $T^{-1}\{W'_0\} = W$.

Le terme «*courbe exceptionnelle de première espèce*», utilisé dans la théorie classique des surfaces algébriques, se rapporte au cas particulier où V/k est une surface, W/k une courbe, et où W'_0 est simple sur V'.

Nous supposerons que V est une variété normale. Si W est exceptionnelle, on voit aisément qu'il existe une transformation rationnelle T qui non seulement satisfait aux conditions de la définition, mais qui est aussi telle que la variété transformée V' satisfasse aux conditions : (1) V'/k est normale; (2) le corps des fonctions rationnelles sur V'/k est algébriquement fermé dans le corps des fonctions rationnelles sur V/k. Nous supposerons que ces conditions (1) et (2) sont vérifiées. Dans ces conditions, un raisonnement classique de valuations montre que l'on a : $o(W) = o(W'_0)$. Ceci résoud affirmativement le problème A dans le cas d'une variété exceptionnelle W, car, pour une sous-variété *de dimension zéro* W'_0 de V', l'anneau $o(W'_0)$ est nœthérien (c'est alors un anneau semi-local au sens de Chevalley). De plus, $o^*(W'_0)$ est le complété de $o(W'_0)$ (de nouveau à cause du fait que W'_0 est de dimension zéro), et, d'après le théorème fondamental, $o^*(W)$ s'identifie à $o^*(W'_0)$. Par conséquent, le problème B admet aussi une réponse affirmative dans le cas d'une sous-variété exceptionnelle W.

Comme T est univoque en tout point de W (condition *a*), toute composante connexe de W/k doit être transformée par T en une composante connexe (et donc irréductible) de W'_0/k. D'autre part, on déduit du théorème 2 que toute composante irréductible de W'_0/k doit être transformée par T^{-1} en une variété connexe sur k. Nous en concluons que les composantes connexes de W/k sont en correspondance biunivoque avec les composantes irréductibles de W'_0/k, et *sont donc elles-mêmes des sous-variétés exceptionnelles de* V/k. Nous pouvons donc nous limiter aux variétés exceptionnelles connexes.

Nous allons établir des conditions nécessaires et suffisantes pour qu'une sous-variété donnée W/k de V/k soit exceptionnelle. Pour ce faire nous établirons un résultat général, applicable à des sous-variétés quelconques W. Pour simplifier nous supposerons W/k connexe.

Il *peut* exister des points Q de V, non sur W, tels que toute fonction rationnelle sur V/k qui est holomorphe et *nulle* le long de W est nécessairement aussi holomorphe et nulle en Q. Soit H[W] l'ensemble de ces points (W compris), et soit $\overline{H[W]}$ la plus petite variété (définie sur k) contenant H[W].

Si u est une fonction rationnelle sur V/k, nous noterons Z_u la variété des zéros de u. Z_u est de dimension $r - 1$ (si V est de dimension r) et contient les points d'indétermination de u sur V; posons $Z = \bigcap_{u \in m(W)} Z_u$

LEMME. — *La variété* Z *admet la décomposition suivante :* $Z = \overline{H[W]} \cup U$, *où* U *est une variété sans points communs avec* H[W]. *De plus, si* o(W) *contient des fonctions non constantes, il existe une transformation rationnelle* T *de* V/k *dans une variété* V'/k *vérifiant les conditions :*

a. T *est semirégulière en tout point de* H[W];

b. T{H[W]} *est une variété irréductible* W'_0/k *de dimension zéro;*

c. $T^{-1}\{W'_0\} = \overline{H[W]}$

Démonstration (résumée). — Si Z_1 est une composante irréductible de Z/k qui rencontre H[W], tout point générique de Z_1/k appartient à H[W]; en effet, si u est un élément quelconque de $m(W)$, on a $Z_1 \subset Z \subset Z_u$, et Z_1 n'est pas contenue dans la variété d'indétermination de u. Donc, $Z_1 \subset \overline{H[W]}$ et ceci prouve la première partie du lemme.

Si l'anneau $o(W)$ est de degré de transcendance ≥ 1 sur k, il existe certainement des transformations rationnelles T vérifiant les conditions *a* et *b* du lemme. Choisissons-en une pour laquelle la variété $T^{-1}\{W'_0\}$ soit la plus petite possible. Pour une telle T, on voit aisément que $T^{-1}\{W'_0\} \subset Z$. Comme $T^{-1}\{W'_0\}$ est connexe (nous pouvons en effet supposer que V'/k est normale et que le corps des fonctions rationnelles sur V'/k est algébriquement fermé dans le corps des fonctions rationnelles sur V/k) et contient $\overline{H[W]}$, on déduit de la première partie du lemme que $T^{-1}\{W'_0\} = \overline{H[W]}$.

Comme conséquence immédiate du lemme nous avons

la caractérisation suivante des variétés exceptionnelles connexes :

Théorème 3. — *Pour qu'une sous-variété connexe* W/k *de* V/k *soit exceptionnelle, il faut et il suffit que les conditions suivantes soient satisfaisantes :*

I. *L'anneau* o(W) *contient des fonctions non constantes;*

II. H[W] = W.

La nécessité est conséquence facile de la définition des variétés exceptionnelles. La suffisance se déduit aussitôt du lemme.

5. Analyse du problème A.

Si le degré de transcendance de $o(W)/k$ est strictement inférieur à r ($r = dim\ V/k$), soit s, il est possible de trouver un modèle projectif V'/k du corps des quotients de $o(W)$ et une sous-variété W'/k de V'/k tels que $o(W) = o(W')$. Si l'on suppose que la réponse au problème A est affirmative pour toutes les variétés V de dimension inférieure à r, la réponse sera aussi affirmative si $dim V = r$ et $s < r$. Nous pouvons donc supposer $s = r$. En ce cas le corps des quotients de $o(W)$ coïncide avec le corps Σ des fonctions rationnelles sur V/k (car $o(W)$ est intégralement fermé dans Σ).

Si $\{C\}$ est un système linéaire de diviseurs de V/k, et si W/k est une sous-variété connexe de V/k qui n'est pas variété base de $\{C\}$, nous dirons que V est une *variété fondamentale* de $\{C\}$ si la condition suivante est vérifiée : *si* C_1 *et* C_2 *sont deux membres de* $\{C\}$ *tels que* $W \not\subset C_1 \cap C_2$, *et si* x *est une fonction de* Σ *telle que* $(x) = C_1 - C_2$, *alors la trace sur* W *de* x *est une constante.*

Soient $(u_1, u_2, \ldots, u_h)$ des éléments de $o(W)$ linéairement indépendants sur k et $\{C\}$ le système linéaire déterminé par le système linéaire des fonctions $(c_0 + c_1u_1 + \ldots + c_hu_h)$; le système $\{C\}$ a les deux propriétés :

(1) W *est variété fondamentale de* $\{C\}$;

(2) *Aucun point base de* $\{C\}$ *n'est sur* W.

Inversement, si un système linéaire $\{C\}$ de dimension h, a les propriétés (1) et (2), il provient de h fonctions (u_i) de $o(W)$ par le procédé ci-dessus.

On voit aussitôt que l'ensemble des conditions (1) et (2) équivaut à : *il existe une variété* C_0 *dans* $\{C\}$ telle que $C_0 \cap W = \emptyset$. On en déduit que, *si* C *satisfait aux conditions* (1) *et* (2), *il en est de même du système complet* $|nC|$ *pour tout entier* $n \geq 1$.

Pour résoudre affirmativement le problème A, il serait suffisant de montrer que H[W] *est une variété* (définie sur k), car on déduirait du lemme du paragraphe 4 que H[W] est une variété exceptionnelle, et, comme $o(H[W]) = o(W)$, notre affirmation se déduirait du paragraphe 4. Autrement dit, il suffirait de montrer que $H[W] = \overline{H[W]}$. Nous allons maintenant transformer la conjecture « $H[W] = \overline{H[W]}$ » en une conjecture relative aux systèmes linéaires.

Remarquons d'abord : *si* $\{C\}$ *est un système linéaire quelconque vérifiant les conditions* (1) *et* (2), *et si* Z *est une composante irréductible de* $\overline{H[W]}$, *alors* Z *est variété fondamentale de* $|nC|$ *pour tout* $n \geq 1$. En effet, si $(u_1, \ldots, u_3)$ est un ensemble de fonctions définissant $|nC|$ ($u_i \in o(W)$), et si P est un point générique de Z/k, nous savons alors que $P \in H[W]$ et que, par conséquent, les traces sur Z des u_i sont des constantes. Ceci implique que Z est fondamentale pour $|nC|$.

Pour montrer que $\overline{H[W]} = H[W]$, nous avons à montrer que toute composante irréductible Z de $\overline{H[W]}$ appartient à H[W]. Supposons le contraire, et soit Q un point de Z n'appartenant pas à H[W]. Il existerait alors une fonction u de $m(W)$ ayant Q pour point d'indétermination. Soit $\{C\}$ le faisceau défini par le système de fonctions $(c_0 + c_1u)$. Alors $\{C\}$ vérifie les conditions (1) et (2) et Q est point base de $\{C\}$. *J'affirme que* Q *est point base du système* $|nC|$ *pour tout* $n \geq 1$. En effet, dans le cas contraire, on en déduirait que toute variété C_n de $|nC|$ qui contient Q aurait à contenir Z (car, d'après la remarque précédente, Z est fondamentale pour $|nC|$); ceci est contradictoire, car, si C_0 est une variété de $\{C\}$ qui ne contient pas Z, alors $nC_0 \in |nC|$, $Q \in nC_0$ et $nC_0 \not\supset Z$.

Nous avons donc atteint la conclusion suivante : *si* H[W] *n'est pas une variété* (définie sur k), *il existe un système linéaire* $\{C\}$, *une variété* Z/k *sur* V, *et un point* Q *de* Z *tels que* Z *soit variété fondamentale et* Q *un point base de* $|nC|$ *pour tout* $n \geq 1$. Mais je conjecture que le résultat suivant est vrai :

A. *Si une variété irréductible* Z/k *sur* V *est fondamentale pour tous les multiples d'un système linéaire donné* $\{C\}$, *alors, pour* n *suffisamment grand, le système* $|nC|$ *n'a pas de point base sur* Z.

Si V/k n'a pas de singularités, je conjecture même que le résultat plus fort qui suit est vrai :

B. *Si* $|C|$ *est un système linéaire complet arbitraire sur une variété (non singulière)* V, *s'il est de dimension strictement positive et sans composantes fixes, alors le système complet* $|nC|$ *n'a pas de point base pour* n *suffisamment grand.*

Pour les surfaces algébriques, B est conséquence du théorème de Riemann-Roch généralisé. En réalité, il n'est pas besoin d'utiliser dans la démonstration toute la puissance du théorème de Riemann-Roch. Il suffit d'utiliser la formule qui donne la dimension de $|nC|$ pour n assez grand (une sorte de théorème de Riemann généralisé). Il semble possible d'obtenir une démonstration de B selon cette méthode, et valable pour les variétés non singulières supérieures, bien que cette démonstration ne se laisse pas présager facile, au moins dans le cas abstrait.

Pour les variétés normales en général, la conjecture A semble poser un problème fort difficile. En plus de la difficulté de généraliser le théorème de Riemann à une variét normale arbitraire, il y a la difficulté supplémentair e du à l'absence d'une théorie des intersectio[illegible] variétés singulières.

THEORY AND APPLICATIONS OF HOLOMORPHIC FUNCTIONS ON ALGEBRAIC VARIETIES OVER ARBITRARY GROUND FIELDS

By Oscar Zariski

CONTENTS

INTRODUCTION

1. The present memoir had its origin in the search of a method which could be used for extending the so-called "principle of degeneration" of classical algebraic geometry to algebraic varieties over arbitrary ground fields. Indeed, our proof (given in III, §22), which establishes this principle in all its generality, represents one of the main applications of the general theory of holomorphic functions developed in the first two parts of this memoir (§§1-19). It is noteworthy that while the proof of the principle of degeneration in the classical case (i.e., in the complex domain) is essentially a simple exercise in topology, our proof of this same principle in the abstract case can be given only after a long and difficult journey whose goal is a fundamental theorem of invariance of rings of holomorphic functions under rational transformations (II,§11 and §16). This theorem, which we first have to prove for birational transformations (§§12-15) before we can prove it for rational transformations (§§17-19), is the high point of this memoir. The significance of this theorem resides in the fact that it provides a birationally invariant basis for the theory of holomorphic functions in the large developed in Part I. In this theory, we develop the concept of an analytic function on a given variety V, defined and holomorphic along a given subvariety W of V. This theory can be regarded as an extension of the theory of generalized semi-local rings and their completions which we gave in our paper [7]. The extension is from varieties in the affine space to varieties in the projective space. There is a profound difference between the affine case and the projective case as far as the theory of holomorphic functions is concerned. The difference is due to the following fact: while holomorphic functions on an affine model are definable directly as elements of the completion of the coordinate ring of V (this ring being provided with a suitable $\mathfrak{m}$-adic topology depending on the given subvariety W of V), holomorphic functions on a projective model must be built up piecewise out of "analytical elements". As a consequence, the various rings of holomorphic functions on affine models are explicitly constructible as completions of coordinate rings, but no such explicit algebraic construction exists for rings of holomorphic functions on projective models. This difference accounts for the fact that while there is not a single problem left open for rings of holomorphic functions on affine models, some of the unsolved questions in the theory of holomorphic functions on projective models (see I, §4) undoubtedly are among the most

difficult problems in the theory of algebraic varieties.

2. We can best present the underlying ideas of this memoir if we retrace in reverse the order of exposition followed in the text and take therefore the principle of degeneration as starting point. We recall that in the classical case this principle (first formulated by Enriques) asserts that if an irreducible variety V varies in a continuous system of varieties and degenerates, in the limit, into a reducible variety V_o, then this limit variety V_o is connected. This principle is almost self-evident since V_o is a continuous image of V, and a continuous image of a connected set (V, being irreducible, is connected) is connected.* Now the principle of degeneration, as stated above, can be transformed into an equivalent statement in which no reference is made to continuity and limits and which makes sense also in the abstract case. It was shown namely by Chow and van der Waerden [2] that the algebraic varieties of given order and dimension, in a given projective space, form an algebraic system. It follows that it is permissible to assume that our variety V varies continuously in an irreducible algebraic system M. Using properties of the associated form of a variety (see [2]) it is easy to prove that if an irreducible system M does not consist entirely of reducible varieties, then the varieties in M which are reducible form an algebraic subsystem of M. In the latter case it follows that every reducible variety belonging to M is the limit of irreducible varieties in M. Let us agree to use the following terminology: we shall say that "almost all" points of an irreducible variety V (or "almost all" members of an irreducible algebraic system M) have a given property α if the points of V (or the members of M) which do not have property α belong to a proper algebraic subvariety of V (or to a proper algebraic subsystem of M).**

*We say "almost self-evident" because in order to assert that V_o is a continuous image of V it is necessary to show that the continuous variation of V can be accompanied by a continuous deformation of V onto V_o. The existence of such a deformation has always been taken more or less for granted.

** In the language of the Italian goemeters this would be expressed by saying that the "generic" point of V (or the "generic" member of M) has property α. We prefer not to use the term "generic" in order to avoid confusion with the term "general point". The concept of "general point" (in the sense of E. Noether-van der Waerden) is not and could not be identical with the concept of "generic point" (in the sense of the Italian school) for the simple reason that the Italian geometers never meant to populate an algebraic variety with points having coordinates in an algebraic function field. It is hardly necessary to add the linguistic information that the Italians have a perfectly good word for "general" and that this word is not "generico" but "generale".

We can then state the principle of degeneration as follows:

DP. I. If almost all varieties of an irreducible algebraic system M are irreducible, then every variety in M is connected.

In this form the principle of degeneration has a meaning also in the abstract case⁺. Only a few finishing touches are necessary in order to make this statement precise. In the first place, the members of an algebraic system must be thought of as cycles and not varieties, i.e., they are formal linear combinations of absolutely irreducible varieties with non-negative integral coefficients, and that is so whether we are dealing with the classical or the abstract case. We give in III, § 21, a brief exposition of the concept of an algebraic system of cycles, including certain specialization properties of cycles which we have not found explicitly stated or proved in the literature and which are essential for our purpose (see the lemma proved in III, § 21). In the second place, it is necessary to specify in which sense are the terms "irreducible" and "connected" used in DP. I, since these terms refer to relative properties of varieties (i.e., relative to a specified ground field). In the classical case everything is clear, especially if, following the Italian geometers, we include in M only cycles which are defined over the field of complex numbers, i.e., cycles whose associated forms have complex coefficients. That is equivalent to saying that on the representative variety M* of the system M (see III, § 21) we consider only points with complex coordinates. In that case the terms "irreducibility" and "connectedness" refer naturally to the field of complex numbers as ground field. Now suppose that we are dealing with an arbitrary ground field k and that M is an algebraic system defined and irreducible over k. If, still following the Italian geometers, we agree to include in M only cycles which are defined over the algebraic closure $\bar{k}$ of k, then we prove that DP. I remains true if the terms "irreducibility" and "connectedness" refer to the field $\bar{k}$, i.e., if we speak of "absolute irreducibility" and "absolute connectedness". Thus, the following is a precise formulation of the principle of degeneration in the abstract case:

DP. II. If almost all cycles of an irreducible algebraic system M/k are absolutely irreducible, then every cycle of M is absolutely connected.

However, it is customary in modern expositions (and is also metho-

⁺A variety V, defined over a field k, is said to be <u>connected over</u> k, if V is not the sum of two <u>proper</u> subvarieties defined over k and having no common points.

doligically more convenient) to include in an algebraic system cycles with "transcendental" coordinates (i.e., transcendental over k), and, in particular, the general cycles of the system. If we do that for the system M/k in DP. II, we strengthen both the assumption and the conclusion of that statement. However, it is easy to see that if the general cycle of an irreducible algebraic system is absolutely irreducible, then almost every cycle of the system is absolutely irreducible. Hence we can re-formulate DP. II as follows:

> DP. III. If the general cycle of an irreducible algebraic system M/k is absolutely irreducible, then every cycle of the system is absolutely connected.

It is under this final form that we prove the principle of degeneration.

3. We now analyze the principle of degeneration, in its form DP. III, from the standpoint of algebraic correspondences. Associated with the algebraic system M there is an algebraic correspondence T between the representative variety M* of M and the carrier variety V of M, i.e., the variety V which carries the cycles of M. In this correspondence T (the <u>incidence correspondence</u> of M) the total transform $T\{Q^*\}$ of a point Q* of M* is the variety $|\Delta|$ of the cycle Δ of M represented by the point Q*, i.e., the set of points which belong to the cycle Δ. Both T and V are defined and irreducible over k (III, §21). If Γ is a general cycle of M/k and P* is its representative point on M*, then P* is a general point of M*/k, and the variety $|\Gamma|$ is defined over k(P*). The assumption that Γ is absolutely irreducible signifies that the variety $|\Gamma|$ is absolutely irreducible. Hence if P is a general point of this variety over k(P*), then the field k(P*) is quasi-maximally algebraic in k(P*,P). On the other hand, (P*,P) is a general pair of the correspondence T/k. Hence in terms of the incidence correspondence T the hypothesis in DP. III can be formulated as follows:

> A. If (P*, P) is a general point pair of T/k ($P^* \in M^*$, $P \in V$), then the field k(P*) is quasi-maximally algebraic in k(P*,P).

The conclusion in DP. III is to the effect that if Q* is any point of M*, then the total transform $T\{Q^*\}$ is an absolutely connected subvariety of V. If T, M* and V were arbitrary, this conclusion would not necessarily follow from condition A. However, in the present case we have a special situation: M* is the representative variety of an algebraic system, V is the carrier of that system and T is the incidence correspondence. We consider a derived normal model $\overline{M}^*$ of M^* and we denote by φ the birational transformation of $\overline{M}^*$ onto M^*. This transformation is single-valued on $\overline{M}^*$ without exceptions. Let $\overline{P}^*$ be a point of $\overline{M}^*$ whose φ-transform is the above general point of M^*/k (there is only a finite number of such points and each of them is a general point of $\overline{M}^*/k$) and let $\overline{T}$ denote the irreducible algebraic correspondence (over k) between $\overline{M}^*$ and V having $(\overline{P}^*,P)$ as general point pair. Again, were T, M^* and V quite arbitrary, it could not be asserted that $\overline{T}$ coincides with the product φT at <u>every</u> point of $\overline{M}^*$. However, in the present special case we prove (III, §22) that we have indeed $\overline{T} = \varphi T$. Since φ is single-valued, it follows that the individual cycles of the system M are also total $\overline{T}$- transforms of points of the variety $\overline{M}^*$. So now we can say that the conclusion in DP. III is to the effect that if $\overline{Q}^*$ is any point of $\overline{M}^*$, then the total transform $\overline{T}\{\overline{Q}^*\}$ is an absolutely connected subvariety of V. The correspondence $\overline{T}$ shares with T property A, i.e., we have that the field $k(\overline{P}^*)$ is quasi-maximally algebraic in $k(\overline{P}^*,P)$ (since $k(\overline{P}^*) = k(P^*)$), but this time the variety $\overline{M}^*$ (which now plays the role of M^*) is <u>normal</u>. This being so, we obtain the principle of degeneration as a special case of the following general result on algebraic correspondences (III, §20):

THE CONNECTEDNESS THEOREM FOR ALGEBRAIC CORRESPONDENCES: If an irreducible algebraic correspondence T/k between two varieties M^* and V satisfies condition A and if Q^* is any point of M^* at which M^* is analytically irreducible (over k), then the total transform $T\{Q^*\}$ is connected over k^+.

Since normality implies analytical irreducibility (Zariski [8]), the above connectedness theorem holds also if M^* is locally normal at Q^*. This theorem does not yet yield directly the desired result, since in the principle of degeneration it is asserted that $T\{Q^*\}$ is absolutely connected. However, all that remains to do is to introduce the algebraic closure K of the field $k(Q^*)$ and consider an irreducible component M_1 of M/K which contains the cycle Δ represented by the point Q^*. If we then apply the connectedness theorem to the incidence correspondence of the system M_1, we conclude that the variety $|\Delta|$ is connected over K and hence is absolutely connected (since K is algebraically closed and $T\{Q^*\}$ is a variety defined over K).

+ The point set $T\{Q^*\}$ is a variety, but not necessarily over k. However, <u>connectedness over</u> k is defined for any point set on V in terms of the topology in which the closed sets are the subsets of V which are varieties defined over k.

The connectedness theorem for algebraic correspondences covers a good deal more of ground than does the principle of degeneration. The incidence correspondence of an algebraic system has the special property that it possesses no fundamental points (on M*): to every point of the representative variety M* of the system M there corresponds, on the carrier variety V, a variety of the same fixed dimension, equal to the dimension of the cycles of the system M. In the connectedness theorem, however, it is not excluded that the correspondence T has fundamental points on M*. Thus, the variety T {Q*} whose k-connectedness is being claimed may very well have dimension higher than that of the total transform of the general point of M*. How much more is gained thereby can be best illustrated in the case in which T is a birational transformation. If Q* is not a fundamental point of T and if we assume, for simplicity, that M* is locally normal at Q*, then it is well known that the set T {Q*} consists of a single point. In this case the conclusion of the connectedness theorem is trivial. Similarly, still assuming that Q* is not fundamental for T, if M* is analytically irreducible at Q* it is a simple matter to show that T {Q*} consists of a finite set of points which are isomorphic over k (I, §7, Theorem 8), and hence also in this case the k-connectedness of T {Q*} is obvious. But if Q* is a fundamental point of the birational transformation T, the connectedness of the variety T {Q*} (which now has positive dimension) is not at all trivial even in the classical case (and- to our knowledge - has never been proved in that case*).

4. We have said in the beginning of this introduction that the proof of the principle of degeneration represents the main application of the theory of holomorphic functions developed in this memoir. It would have been more correct to say that the principal application is precisely the proof of the general connectedness theorem for algebraic correspondences. However, before the holomorphic functions can be brought into play, a two-fold reduction of the connectedness theorem must be carried out. In

* Here is an indication of a possible topological proof in the classical case. Let us assume for simplicity that T is a birational transformation, that Q* is an isolated fundamental point and that T^{-1} is everywhere single valued on V. If N is an open neighborhood of Q* on M, then T{N} is an open neighborhood of the variety T {Q*} on V. If M* is analytically irreducible at Q*, then there exists an open neighborhood N of Q* (on M*) such that N - Q* is a connected set. Since Q* is an isolated fundamental point, N - Q* and T{N} - T{Q*} are homeomorphic sets (if N is sufficiently small). Hence the set T{N} - T{Q*} is connected, and this implies that T{Q*} is connected.

the first place we show that it is sufficient to prove the connectedness theorem under the assumption that the point Q^* is rational over k (III, §20). In the text, this special case of the connectedness theorem is incorporated in a more general result in which the point Q^* is replaced by an arbitrary subvariety W of V, defined over k. The condition that M^* is analytically irreducible at Q^* must then be replaced by the condition that M^* be analytically irreducible at each point of W. Furthermore we must naturally assume that W is connected over k. Our modified version of the connectedness theorem states, then, that under all these assumptions the total transform $T\{W\}$ is connected over k. A second reduction results from replacing V by the graph of the correspondence T and from replacing the field k by the field $k^{-p^{\infty}}$. When that is done, we have a reduction to the case in which T^{-1} is a rational transformation, semi-regular at each point of V; and furthermore, condition A is now replaced by the stronger condition that the field $k(P^*)$ be maximally algebraic in the field $k(P^*,P)$. After all these reductions, the connectedness theorem for algebraic correspondences takes the form of Theorem 14 in III, §20 (except for differences in notation). It is at this point that the fundamental theorem of invariance of rings of holomorphic functions comes into play. We shall now indicate briefly the background and the substance of this theorem. In conformity with the notation of the text (parts I and II) we shall designate the varieties M^* and V by V and V' respectively. It should be clearly understood that we are dealing with varieties in the projective space.

With each subvariety W/k of V/k we associate "functions" on V which are defined and holomorphic along W (see I, §3, where the definition of these functions is given not only for subvarieties W but also for arbitrary subsets of V). These functions form a ring which we denote by o_W^*. We prove (I, §6, Theorem 6) that

> if V is analytically irreducible at each point of W, then W is connected over k if and only if the ring o_W^* is an integral domain.

(Actually Theorem 6 is more general since it refers to arbitrary subsets of V, which need not be varieties over k). The really significant part of this result is the implication "o_W^* - an integral domain" ---> " W is k-connected", and its proof is based on a lemma which we gave in [7] (see §6, Lemma 1). The proof of the opposite implication is rather straightforward.

Let now T be an algebraic correspondence between V/k and some other variety V'/k, such that T^{-1} is a rational transformation, semi-regular at each point of V'. Let W' be the total transform $T\{W\}$ of W. We have the ring $o^*_{W'}$ of functions on V', defined and holomorphic along W'. We show that there always exists a natural isomorphism $H_{W,W'}$ of o^*_W into $o^*_{W'}$ (in the case of birational transformations T this result is contained in Theorem 13, II, §10; in the general case see II, §16). Let now (P,P') be a general point pair of T. The fundamental theorem asserts the following:

> if (a) $k(P)$ is maximally algebraic in $k(P,P')$ and (b) V is locally normal at each point of W, then $H_{W,W'}$ is an isomorphism of o^*_W onto $o^*_{W'}$.

We observe that condition (a) is automatically satisfied if T is a birational transformation, for in that case $k(P) = k(P')$.

The fundamental theorem, in conjunction with the connectedness criterion just stated above (I, §6, Theorem 6), gives at once the connectedness theorem for algebraic correspondences (as formulated, in the reduced form, in Theorem 16, III, §20) under the additional assumption that V is locally normal along W. For if we assume that W is k-connected, then by the connectedness criterion the ring o^*_W is an integral domain. Hence, by the fundamental theorem, also $o^*_{W'}$ is an integral domain, and therefore, again by the connectedness criterion, also W' (i.e., $T\{W\}$) is k-connected, The general case, in which V is only assumed to be analytically irreducible at the points of W, can be easily reduced to the case in which V is locally normal at the points of W (see II, §11, proof of the connectedness theorem for birational transformations).

5. The entire second part of this memoir is dedicated to the proof of the fundamental theorem. We have already mentioned earlier in this introduction that we first prove this theorem for birational transformations (§§10-15) and then for arbitrary rational transformations T^{-1} (§§16-19). This division of the proof into two parts is dictated by the very nature of the problem. The rationality and semi-regularity of T^{-1} imply that the dimension of $T\{W\}$ is not less than the dimension of W, and unless T is birational and regular (in which case the fundamental theorem follows trivially from the very definition of holomorphic functions) the dimension of $T\{W\}$ is greater than the dimension of W. The difference $\dim T\{W\} - \dim W$ is the sum of two non-negative integers, say s and h. Here s is equal to $\dim V' - \dim V$, and hence $s = 0$ if and only if T is birational (since $k(P)$ is, by assumption, maximally algebraic in $k(P,P')$).

As to h, it is positive if and only if W is fundamental for T. What we do, then, is to first prove the fundamental theorem in the case $s = 0$ and then in the case $h = 0$. The general case is easily reducible to these two special cases (see II, §17).

The basic ideas of the two parts of the proof are very similar, but there are some important technical differences. In both cases ($s = 0$ or $s > 0$, $h = 0$) the problem is reduced to a special case. In this special case the variety V' is obtained from V in two elementary steps. The first step is the projective analogue of what in the affine case would be a simple ring extension of the coordinate ring of V. This step leads to a variety which we denote by Vot. In the birational case ($s= 0$), t is an element of the function field of V. In the case $s > 0$, t is a transcendental over that field (and Vot is the product variety of V with the projective line). The second step is the transition to a derived normal model of Vot. Here the normalization is relative to the function field of Vot in the birational case, and to some algebraic extension of that function field in the case $s > 0$.

The transition from V to Vot requires subtle considerations in the birational case, but is relatively simple in the case $s > 0$. The transition from Vot to the normalized model is based, in the birational case, primarily on the results of our paper [8], while in the case $s > 0$ the proof depends- rather unexpectedly- on the consideration of the Dedekind- Weber normal basis in fields of algebraic functions of one variable.

We add a few concluding remarks to guide the reader through Part I and also to call again attention to the marked difference between the affine case and the projective case in the theory of holomorphic functions (see this Introduction,1). We begin this memoir by defining the concept of a function on V, defined and <u>strongly holomorphic</u> along a subset G of V (I, §1). A strongly holomorphic function is essentially an entity which is given by a single series which converges uniformly on the set G. We prove in I, § 2 (Theorem 1) that if G is an affine subvariety W of V (i.e., if G consists of the points at finite distance, belonging to a subvariety of V) , then the ring of strongly holomorphic functions along G coincides with the $\mathfrak{m}$-adic completion of the non-homogeneous coordinate ring of V, where $\mathfrak{m}$ is the ideal of W in that ring. The general holomorphic functions defined in I, §3 are built up piecewise out of strongly holomorphic functions. We prove in I, § 9 (Theorem 10) that <u>in the affine case every holomorphic function is strongly holomorphic</u>. It is this theorem that settles all possible questions for holomorphic functions on affine models (see I, §4). On the other hand, this theorem is also very useful for the theory of holomorphic functions on projective models. For instance, it plays an essential role in the proof of the fundamental theorem given in Part II.

In concluding this introduction we wish to thank C. Chevalley who has read the manuscript of this memoir and gave us a number of valuable suggestions which we have incorporated in this work. Our indebtedness to Chevalley goes beyond this acknowledgement, since without the theory of local rings, developed systematically by Krull and Chevalley, the present extension of the analytical methods in abstract algebraic geometry to problems in the large would not have been possible.

We also wish to express our thanks to Dr. Gino Turrin for his invaluable assistance in the preparation of the manuscript.

PART I
GENERAL THEORY OF HOLOMORPHIC FUNCTIONS.

§ 1. STRONGLY HOLOMORPHIC FUNCTIONS.

Let V be an irreducible variety over a given ground field k (which will remain fixed throughout this paper) in a projective n-space and let Σ be the field of rational functions on V[1]. If P is a point[2] of V we denote by o_P the local ring[3] of P and by o_P^* the ring of holomorphic functions at P on V, i.e. the completion of o_P with respect to the maximal ideal $\mathfrak{m}_P$ of o_P. We point out at once that if P and Q are k-isomorphic points of V, then $o_P = o_Q$ and hence $o_P^* = o_Q^*$.

Let G be a non-empty subset of V. By <u>a function on</u> V, <u>defined along</u> G, we shall mean a mapping ξ^* which associates with each point P of G a unique element ξ_P^* of o_P^*, <u>provided</u> $\xi_P^* = \xi_Q^*$ whenever P and Q are k-isomorphic points of G. The element $\xi_P^*(P \in G)$ will be called <u>the analytical element</u> of ξ^* <u>at</u> P. The set of all functions on V, defined along G, shall be denoted by o_G^*.

DEFINITION 1. A function ξ^* on V, defined along G, is strongly holomorphic along G, if there exists a sequence of elements ξ_i in the field Σ, such that the following conditions are satisfied:

a. For any point P of G the elements ξ_i belong to o_P and the sequence $\{\xi_i\}$ converges in o_P^* to ξ_P^*. In other words: the sequence $\{\xi_i\}$ converges on G to the function ξ^* (in symbols: $\xi^* = \lim \xi_i$ on G).

b. The convergence $\xi_i \to \xi^*$ is uniform on G, i.e., there exists a sequence $\{\nu(n)\}$ such that

$$\xi_n - \xi_P^* \in \mathfrak{m}_P^{*\,\nu(n)}, \ \lim \nu(n) = \infty,$$

for all points P in G. (Here $\mathfrak{m}_P^*$ denotes the maximal ideal of o_P^*.)

Given a sequence $\{\xi_i\}$ of rational functions on V ($\xi_i \in \Sigma$), such a sequence defines a function on V, strongly holomorphic along G, if the following conditions are satisfied: a'. $\xi_i \in o_P$, all i and all P in G; b'. There exists a sequence $\nu(n)$ such that $\xi_i - \xi_j \in \mathfrak{m}_P^{\nu(n)}$, $\lim \nu(n) = \infty$, for all P in G and for $i \geq n$, $j \geq n$. Under these conditions $\{\xi_i\}$ is a Cauchy sequence in o_P, $P \in G$, and therefore converges to an element ξ_P^* of o_P^*. It is clear that if P and Q are k-isomorphic points of G, then $\xi^*_P = \xi^*_Q$. Hence the mapping ξ^*: P --> ξ^*_P is a function on V, defined along G. Furthermore, the sequence $\{\xi_i\}$ converges uniformly to ξ^* on G, and ξ^* is therefore strongly holomorphic along G.

The functions ξ^* on V, defined along G, can be thought of as elements of the direct sum of the rings o^*_P, $P \in G$. They are completely arbitrary elements of this direct sum except for the condition imposed that $\xi^*_P = \xi^*_Q$, if P and Q are k-isomorphic points of G, and they form a ring O^*_G. It is clear that the strongly holomorphic functions along G form a subring of O^*_G.

If $c \in k$ then we can identify c with the limit of the sequence $\{c, c, \cdots\}$. Hence the elements of k are strongly holomorphic functions on V, defined along any subset G of V (in particular, along V itself).

Let k' be the algebraic closure of k in Σ (k' = the field of constant rational functions on V) , and let G be any subset of V. It is obvious that if an element c' of k' is contained in the local ring of each point P of G, then c' is a function on V, defined and strongly holomorphic along G. In particular, if V is locally normal at each point of G, then $k' \subset o_P$ for all P in G (since o_P is integrally closed), and hence every constant rational function on V is then strongly holomorphic along G.

It is well known that $k \subset \bigcap_{P \in V} o_P \subset k'$. Hence every function which is strongly holomorphic along the entire variety V is necessarily a constant rational function on V.

We emphasize that the ring of functions on V which are strongly holomorphic along a set G depends only on the set of local rings of the points of G and not, for instance, on the variety V which carries these points. Hence, if V' is another projective model of the field Σ and if G, G' are sets of corresponding points of V and V' such that the birational correspondence between V and V' is regular at each point of G and of G', then the ring of functions on V, defined along G, and the ring of strongly holomorphic functions along G coincide respectively with the ring of functions on V', defined along G', and the ring of strongly holomorphic functions along G'. For a similar reason, the ring of strongly holomorphic functions along G remains unaltered if we replace G by any other set G' of points of V, provided each point of G' is k-isomorphic to at least one point of G and each point of G is k-isomorphic to at least one point of G'. It follows that for the purposes of the theory of holomorphic functions it would have been perfectly legitimate to identify the notion of a point of a variety V with that of the local ring of V at that point- a suggestion which has been made orally to me by Chevalley. That would have meant identifying k-isomorphic points of one and the same variety and regularly corresponding points of birationally equivalent varieties. However such an identification would have caused some technical difficulties in the study of the behaviour of rings of holomorphic functions under arbitrary birational transformations, and for this reason we have preferred to deal directly in this paper with points-as given by their coordinates - rather than only with the local ring of a point.

If G_1 is a subset of G, then every element ξ^* of O_G^* determines a unique element ξ_1^* of $O_{G_1}^*$ such that $\xi_P^* = \xi_{1P}^*$ for all P in G_1. The element ξ_1^* will be called <u>the projection of</u> ξ^* into $O_{G_1}^*$. Every element of $O_{G_1}^*$ is the projection of at least one element of O_G^*. We have, therefore, a mapping $\xi^* \dashrightarrow \xi_1^*$ of O_G^* onto $O_{G_1}^*$. This mapping, which is obviously a homomorphism, will be called <u>the projection of</u> O_G^* into $O_{G_1}^*$ and will be denoted by τ_{G, G_1}. In particular, if G_1 consists of a single point P, then $\tau_{G,P}$ maps every element ξ^* of O_G^* into its analytical element ξ_P^* at P.

It is clear that if ξ^* is a strongly holomorphic function along G, then the projection of ξ^* into $O_{G_1}^*$ is a strongly holomorphic function along G_1. We shall indicate this fact by saying that ξ^* <u>is strongly holomorphic along</u> G_1, although what we actually mean is that $\tau_{G, G_1}(\xi)$ is strongly holomorphic along G_1. However, it is not necessarily true that every strongly holomorphic function along G_1 is the projection of a strongly holomorphic function along G. The projection of the ring of strongly holomorphic functions along G is, in general, a proper subring of the ring of strongly holomorphic functions along G_1. In particular, if P is a point of G, then the analytical elements ξ_P^* of the functions which are strongly holomorphic along G will form a subring of o_P^*, and--in general--this will be a proper subring of o_P^*.

Completely arbitrary subsets G of V will seldom occur in the sequel. As a rule we shall only deal with particular sets G of the form $W - \Gamma$, where W is some (non-empty) subvariety of V and Γ is a proper (possibly empty) subvariety of W. We proceed to discuss an important special case.

§ 2. STRONGLY HOLOMORPHIC FUNCTIONS ON AFFINE MODELS.

Let V lie in a projective n-space S_n and let W be a (non-empty) subvariety of V. We fix in S_n a hyperplane H not containing W and we take for G the set W-H. Then G is non empty. We proceed to find the strongly holomorphic functions along this set G.

We take H as hyperplane at infinity and we regard V and W as affine models in the affine space $S_n - H$. Let R be the corresponding non-homogeneous coordinate ring of $V^{(4)}$ and let $\mathfrak{m}$ be the ideal of W in R. We have $\mathfrak{m} \neq R$, since $W \not\subset H$. We consider the completion R^* of the $\mathfrak{m}$-adic ring R (For the definition and properties of $\mathfrak{m}$-adic rings see our paper [7] . Let $\{\xi_i\}$ be any Cauchy sequence in the $\mathfrak{m}$-adic ring R. For any integer n there exists then an integer $N = N(n)$ such that $\xi_i - \xi_j \in \mathfrak{m}^n$ for all $i, j > N$. If P is any point of G, then $\mathfrak{m} \subset \mathfrak{m}_P$. It follows that $\xi_i - \xi_j \in [\mathfrak{m}_P]^n$ for all $i, j > N$, i.e., the sequence $\{\xi_i\}$ converges uniformly on G and thus has a well-defined limit ξ^*, where ξ^* is a strongly holomorphic function along G. By the same argument it follows that equivalent Cauchy sequences in the $\mathfrak{m}$-adic ring R converge on G to one and the same strongly holomorphic function. We have then a natural mapping σ of R^* into the ring of strongly holomorphic functions along G. It is immediate that σ is a homomorphism.

THEOREM 1. The mapping σ is an <u>isomorphism</u> of R^* <u>onto</u> the ring of strongly holomorphic functions along $G (G = W - H)$.

PROOF. We first show that σ is an isomorphism. Let $\{\xi_i\}$ be a Cauchy sequence in the $\mathfrak{m}$-adic ring R such that $\lim \xi_i = 0$ on G. We have to show that $\{\xi_i\}$ is a null sequence in R. Let n be an arbitrary integer and let

$$\mathfrak{m}^n = \mathfrak{q}_1 \cap \mathfrak{q}_2 \cap \cdots \cap \mathfrak{q}_g$$

be a normal decomposition of the ideal $\mathfrak{m}^n$ into primary components. Let $\mathfrak{p}_i$ be the prime ideal of $\mathfrak{q}_i$, ν_i the exponent of $\mathfrak{q}_i$ and $\nu = \max(\nu_1, \nu_2, \cdots, \nu_g)$. (The prime ideals $\mathfrak{p}_i$ which are embedded as well as the integer g may depend on n). Since the sequence $\{\xi_i\}$ converges uniformly to zero on G, there exists an integer $N = N(\nu)$ such that $\xi_i \in \mathfrak{m}_P^\nu$ for all P in G and all $i > N$. For each $j = 1, 2, \cdots, g$ we fix a general point P_j of the irreducible subvariety W_j of V defined by the prime ideal $\mathfrak{p}_j$. Since $\mathfrak{p}_j \supset \mathfrak{m}$, W_j is contained in W. Since W_j is not contained in the hyperplane at infinity it follows that $P_j \in G\ (= W - H)$, $j = 1, 2, \cdots, g$. We have, then, for $i > N$ and $j = 1, 2, \cdots, g$: $\xi_i \in \mathfrak{m}_{P_j}^\nu$. Now ξ_i belongs to R, and since P_j is a general point of W_j we have $\mathfrak{m}_{P_j}^\nu \cap R = \mathfrak{p}_j^{(\nu)}$ = symbolic ν-th power of $\mathfrak{p}_j$. Hence

$$\xi_i \in \mathfrak{p}_j^{(\nu)} \subset \mathfrak{p}_j^{(\nu_j)} \subset \mathfrak{q}_j, \quad j = 1, 2, \cdots, g, \ i > N,$$

and therefore $\xi_i \in \mathfrak{m}^n$ for all $i > N$. This proves our assertion that $\{\xi_i\}$ is a null sequence in R.

We now complete the proof of the theorem by showing that σ maps R^* onto the ring of functions which are strongly holomorphic along G. Let ξ^* be an element of the latter ring: $\xi^* = \lim \xi_i$, where the sequence $\{\xi_i\}$ satisfies conditions a and b of Definition 1. For a given i, let $\mathfrak{A}_i$ denote the set of elements a of R such that $a \cdot \xi_i \in R$. Then $\mathfrak{A}_i$ is an ideal in R having the following property: a point P of the affine model V − H belongs to the variety $\mathcal{V}(\mathfrak{A}_i)$ of the ideal $\mathfrak{A}_i$ if and only if $\xi_i \notin \mathfrak{o}_P$. Since for any point P of the affine variety G(= W − H) it is true that all the ξ_i are in $\mathfrak{o}_P$, it follows that $G \cap \mathcal{V}(\mathfrak{A}_i) = \emptyset$. Hence $\mathfrak{A}_i + \mathfrak{m} = (1)$, for all i. Hence we have also $\mathfrak{A}_i + \mathfrak{m}^i = (1)$, all i. For each i we select an element α_i in $\mathfrak{A}_i$ and an element π_i in $\mathfrak{m}^i$ such that $\alpha_i + \pi_i = 1$. By the definition of $\mathfrak{A}_i$ we can write $\eta_i \xi_i = \zeta_i$, where $\zeta_i \in R$. We set $\omega_i = \xi_i - \zeta_i$. We have $\omega_i = \zeta_i \pi_i / \alpha_i \in \mathfrak{m}_P^i$ for all point P of G since $\pi_i \in \mathfrak{m}^i$ and since $\alpha_i (= 1 - \pi_i)$ is a unit in the local ring of each point P of G. Hence the sequence $\{\omega_i\}$ converges (uniformly) on G to the function 0, and therefore the sequence $\{\zeta_i\} = \{\xi_i - \omega_i\}$ converges uniformly on G and has the same limit ξ^* as that of $\{\xi_i\}$. We may therefore replace the sequence $\{\xi_i\}$ by the sequence $\{\zeta_i\}$, the elements of the latter sequence being now elements of R. We shall therefore assume that the elements of the original sequence $\{\xi_i\}$ belong to our coordinate ring R. We now proceed as in the first part of the present proof (in which we have established that σ is an isomorphism). Using the same notation, we start with a given integer n and we determine a corresponding integer N such that $\xi_\alpha - \xi_\beta \in \tilde{\mathfrak{m}}_{P_j}$ for $j = 1, 2, \ldots, g$ and all $\alpha, \beta > N$. We then conclude that $\xi_\alpha - \xi_\beta \in \mathfrak{m}^n$ for all $\alpha, \beta > N$. Hence $\{\xi_\alpha\}$ is a Cauchy sequence in the $\mathfrak{m}$-adic ring R, and the limit of that sequence in R^* is an element whose σ- image is the given strongly holomorphic function ξ^*. This completes the proof of the theorem.

In view of the theorem just proved, we shall identify R^* with the ring of functions on V which are strongly holomorphic along the set G of points of W which are at finite distance. We note the special case of Theorem 1 in which W = V. In that case our set G is the affine part of the variety V, i.e. G = V − H, the ideal $\mathfrak{m}$ is the zero ideal. Hence the $\mathfrak{m}$-adic completion of R is then the ring R itself and Theorem 1 asserts in this case that the only functions on V which are strongly holomorphic along the set of all points of V which are at finite distance are the elements of the (non-homogeneous) coordinate ring R of V. This result can also be obtained directly as follows. In the first place it is easy to see that the intersection of all the local rings o_P, $P \in V - H$, is the coordinate ring R[5]. From this it follows that the intersection of all the maximal ideals $\mathfrak{m}_P$, $P \in V - H$, contains only the element zero of Σ. If, then, $\{\xi_i\}$ is a sequence which converges uniformly on V - H, then all the ξ_i must belong to R and we must have $\xi_i = \xi_{i+1} = \cdots$ for i sufficiently large.

§ 3. THE GENERAL CONCEPT OF A HOLOMORPHIC FUNCTION.

Let G be any subset of V.

DEFINITION 2. A function ξ^* on V, defined along G is holomorphic along G, if there exists a finite set of subvarieties $\Gamma_1, \Gamma_2, \cdots, \Gamma_h$ of V such that:

1) $G \cap \Gamma_1 \cap \Gamma_2 \cap \cdots \cap \Gamma_h = \emptyset$;
2) $\tau_{G, G-\Gamma_j}(\xi^*)$ is strongly holomorphic along $G - \Gamma_j$, $j = 1,2,\cdots, h$.

The h sets $G - \Gamma_j$ cover G, in view of 1), and on each of these sets the function ξ^* (or rather, the projection of ξ^* into $O^*_{G - \Gamma_j}$) is the limit of a uniformly convergent sequence $\{\xi_{1j}, \xi_{2j}, \cdots\}$ of elements of the field Σ. At each common point P of two sets $G - \Gamma_j$ and $G - \Gamma_{j'}$, the two sequences $\{\xi_{ij}\}$ and $\{\xi_{ij'}\}$ must converge to the analytical element ξ^*_P of ξ^* at P, these two sequences being regarded as Cauchy sequences in the local ring of P.

If H is a point set on V which contains G and if η^* is an element of O^*_H, we shall say that η^* is holomorphic along G if the projection $\tau_{H,G}(\eta^*)$ is holomorphic.

We may express the above definition in topological terms by introducing the following natural topology on the abstract variety V: we define as closed sets the algebraic subvarieties of V. With this definition of closed sets, V becomes a compact topological space. The h sets $G - \Gamma_j$ of the above definition are open in G and cover G. We may therefore say that, according to that definition, a function ξ^* on V, defined along G, is holomorphic along G if there exists an open covering G_j of G such that ξ^* is strongly holomorphic along each G_j.

THEOREM 2. The holomorphic functions along G form a ring (a subring of O^*_G).

PROOF. Let ξ^* and η^* be two functions on V defined and holomorphic along G. We have then a set of subvarieties $\Gamma_1, \Gamma_2, \cdots, \Gamma_h$ of V, such that conditions 1) and 2) of Definition 2 are satisfied for ξ^*, and we have a similar set of subvarieties $\Delta_1, \Delta_2, \cdots, \Delta_g$ relative to η^*. We set $\Omega_{ij} = \Gamma_i \cap \Delta_j$, $i = 1, 2, \cdots, h$; $j = 1, 2, \cdots, g$. The intersection of the hg varieties Ω_{ij} and G is empty. Moreover, ξ^* and η^* are strongly holomorphic along $G - \Omega_{ij}$, since $G - \Omega_{ij} = (G - \Gamma_i) \cap (G - \Delta_j)$. Since the strongly holomorphic functions along $G - \Omega_{ij}$ form a ring we conclude that $\xi^* - \eta^*$ and $\xi^*\eta^*$ are strongly holomorphic along $G - \Omega_{ij}$. Hence $\xi^* - \eta^*$ and $\xi^*\eta^*$ are holomorphic functions along G in the sense of our definition, q.e.d.

The ring of holomorphic functions along G will be denoted by o^*_G.

It is obvious that if G_1 is a subset of G, then $\tau_{G, G_1}(o^*_G) \subset o^*_{G_1}$.

As in the case of strongly holomorphic functions, we emphasize also here that the ring of holomorphic functions along a given point set G depends only on the set of the local rings of the points of G and not on the variety V which carries these points. For suppose that V is another variety, birationally equivalent to V, and suppose that the birational correspondence T between V and V' is regular at all points of G. Let G' be the set of points of V' which correspond under T to the points of G' and let ξ^* be a function on V which is holomorphic along G. Clearly ξ^* is also a function on V', defined along G'. To show that ξ^* is holomorphic along G' we fix a finite set of subvarieties Γ_i of V such that the sets $G - \Gamma_i$ cover G and such that ξ^* (or rather the projection $\tau_{G, G-\Gamma_i}(\xi^*)$) is strongly holomorphic along each $G - \Gamma_i$. Let $\Gamma_i' = T\{\Gamma_i\}$ = total transform of Γ_i under T. By the regularity of T on G we see at once that $T\{G - \Gamma_i\} = G' - \Gamma_i'$. Hence the sets $G' - \Gamma_i'$ cover G'. Since Γ_i' is a subvariety of V' and since from the fact that ξ^* is strongly holomorphic along $G - \Gamma_i$ follows that ξ^* is also strongly holomorphic along $G' - \Gamma_i'$, we conclude that ξ^* is holomorphic along G'. - We may express all this by saying that rings of holomorphic functions are invariant under regular birational transformations.

Given a point set G on V and given a function ξ^* on V which is holomorphic on G, there will exist, in general, more than one open covering $\{G - \Gamma_i\}$ of G with the property that ξ^* is strongly holomorphic on each set $G - \Gamma_i$. We proceed to show that upon replacing V, if necessary, by a suitable regular birational transform we may always arrange matters so that the varieties Γ_i are hyperplane sections of V. More precisely, we shall prove the following

LEMMA. Given a subset G of V and given a function ξ^* on V which is holomorphic along G, there exists a regular birational transformation T of V into another variety V' and there exists a set $\{\pi'_\nu\}$ of independent hyperplanes in the ambient projective space of V' such that if G' is the T-transform of G and if $\Gamma'_\nu = V' \cap \pi'_\nu$, then the sets $G' - \Gamma'_\nu$ cover G' and ξ^* is strongly holomorphic along each set $G' - \Gamma'_\nu$.

PROOF. Let $(y_0, y_1, \cdots, y_n)$ be homogeneous coordinates of a general point of V and let $f_0(Y), f_1(Y), \cdots, f_N(Y)$ be an independent k-basis of set of all forms in $k[Y]$ (= $k[Y_0, Y_1, \cdots Y_N]$), of a given degree m (the Y's are indeterminates). We set $z_j = f_j(y)$ and we call V' the irreducible variety in S_N having $(z_0, z_1, \cdots, z_N)$ as general point. It is well known (and also immediately verifiable) that the variety V' is in regular birational correspondence with V (see, for instance, Zariski [6], Corollary of Lemma 8, p. 534). We shall show that the transformation T: V--> V' satisfies the desired conditions of the lemma provided the degree m of the forms $f_j(y)$ is sufficiently high.

Let us fix a finite set of subvarieties Γ_i of V, $i = 1, 2, \cdots, g$, such that the sets $G - \Gamma_i$ cover G and such that ξ^* is strongly holomorphic along each set $G - \Gamma_i$. We choose m so high as to satisfy the following condition: for each $i = 1, 2, \cdots, g$ the homogeneous ideal of Γ_i in $k[Y]$ possesses a basis which consists of forms of degree $\leq m$. Let $\{\varphi_{i1}, \varphi_{i2}, \cdots, \varphi_{is_i}\}$ be a maximal linearly independent set of forms in $k[Y]$, of degree m, which vanish on Γ_i. By our choice of m, <u>the forms φ_{ij}</u> (i = fixed) <u>do not vanish simultaneously at any point outside of</u> Γ_i. We extract from the above set of $s_1 + s_2 + \cdots + s_g$ forms φ_{ij} $(1 \leq i \leq g)$ a maximal set of linearly independent forms, say $\psi_1, \psi_2, \cdots, \psi_h$. Let $\psi_\nu(Y) = \sum_{j=0}^N c_{\nu j} f_j(Y)$ and let π'_ν be the hyperplane defined by the equation $\sum_{j=0}^N c_{\nu j} Z_j = 0$.

Since the forms $f_j(Y)$ are linearly independent over k and since also the forms $\psi_\nu(Y)$ are linearly independent over k, it follows that <u>the hyperplanes π'_ν are independent</u> (over k).

Each form $\psi_\nu(Y)$ vanishes on at least one of the varieties Γ_i. Since the section of V with the hypersurface $\psi_\nu(Y) = 0$ is transformed by T onto the section Γ'_ν of V' with the hyperplane π'_ν, it follows that each variety Γ'_ν contains the T-transform of at least one of the varieties

Γ_1. Hence ξ is strongly holomorphic along each of the sets $G' - \Gamma'_\nu$ (since ξ^* is strongly holomorphic along each of the sets $G' - T\{\Gamma_1\}$).

Let P' be any point of G' and let P be the corresponding point of G. Since the sets $G - \Gamma_i$ cover G, at least one of the varieties Γ_i does not contain P. Let, say, $P \notin \Gamma_1$. Then at least one of the forms $\varphi_{1j}(Y)$ is different from zero at P. Therefore also at least one of forms $\psi_\nu(Y)$ must be different from zero at P. This implies that at least one of the hyperplanes Π'_ν does not contain P', and hence P' belongs to at least one of the sets $G' - \Gamma'_\nu$. It follows that the sets $G' - \Gamma'_\nu$ cover G'. - The proof of the lemma is now complete.

As an application of the above lemma we prove the following

THEOREM 3. A function ξ^* on V which is holomorphic along the entire variety V is necessarily a constant rational function on V.

PROOF. We may assume, by the preceding lemma, that ξ^* is strongly holomorphic along sets $V - \Gamma_i$, $i = 1, 2, \cdots, g$, where the Γ_i are certain hyperplane sections of V. For any i, let R_i denote the non-homogeneous coordinate ring of the affine model $V - \Gamma_i$. If ξ_i^* denotes the $\tau_{V, V-\Gamma_i}$ - projection of ξ^*, then by Theorem 1 (in the special case W = V) it follows that $\xi_i^* \in R_i$. If $i \neq j$, then ξ_i^* and ξ_j^* are two elements of the function field Σ of V which belong to the local ring of each point P of $V - \Gamma_i - \Gamma_j$ and have the same analytical element at each such point. Since $V - \Gamma_i - \Gamma_j$ is non-empty it follows that $\xi_i^* = \xi_j^*$. We have, then, $\xi_1^* = \xi_2^* = \cdots = \xi_g^* \in \Sigma$. Hence ξ^* is a rational function on V and is therefore necessarily a constant since it belongs to the local ring of each point of V (see §1).

In the sequel we shall have occasion to encounter the following situation: two point sets G_1 and G_2 are given on V and it is known that a given function ξ^* on V, which is defined along $G_1 \cup G_2$, is holomorphic along each of the two sets G_1 and G_2. In such a case it does not yet follow necessarily that ξ^* is holomorphic along $G_1 \cup G_2$ [For instance, if G_1 and G_2 are arbitrary complementary subsets of V and if ξ^* is defined by the conditions: $\xi_P^* = 0$ if $P \in G_1$ and $\xi_P^* = 1$ if $P \in G_2$, then ξ^* is holomorphic (even strongly holomorphic) along G_1 and G_2, but, by Theorem 3, ξ^* is not holomorphic along $G_1 \cup G_2 (=V)$.] The following is a sufficient condition in order that ξ^* be holomorphic along $G_1 \cup G_2$: if $H = G_1 \cap G_2$, then

(1) $\overline{G_1 - H} \cap G_2 = \overline{G_2 - H} \cap G_1 = \emptyset$,

where $\overline{G_1 - H}$ denote the closure of $G_1 - H$ (in the natural topology of V; $\overline{G_1 - H}$ = least algebraic variety which contains $G_1 - H$). The proof is as follows:

Let $\{G_1 - \Gamma_i\}$ and $\{G_2 - \Delta_j\}$ be open coverings of G_1 and G_2 respectively such that ξ^* is strongly holomorphic along each of the sets $G_1 - \Gamma_i$ and $G_2 - \Delta_j$. Let $\Gamma = \overline{G_2 - H}$ and $\Delta = \overline{G_1 - H}$. It follows at once from condition (1) that $(G_1 \cup G_2) - \Gamma - \Gamma_i = G_1 - \Gamma_i$ and $(G_1 \cup G_2) - \Delta - \Delta_j = G_2 - \Delta_j$. Hence the sets $G_1 - \Gamma_i$ and $G_2 - \Delta_j$ are open also in $G_1 \cup G_2$, and therefore ξ^* is holomorphic along $G_1 \cup G_2$.

COROLLARY. Let ξ^* be a function on V which is defined along a subset G of V and let $\{G - \Gamma_i\}$ be an open covering of G. If ξ^* is holomorphic along each set $G - \Gamma_i$, then ξ^* is also holomorphic along G.

It is sufficient to give the proof in the case in which the covering of G consists of two sets $G_1 = G - \Gamma_1$, $G_2 = G - \Gamma_2$. We have $H = G_1 \cap G_2 = G - \Gamma_1 - \Gamma_2$, and hence $G_1 - H = G_1 \cap \Gamma_2 \subset \Gamma_2$. Since Γ_2 is closed it follows that $\overline{G_1 - H} \subset \Gamma_2$ and hence $\overline{G_1 - H} \cap G_2 = \emptyset$. Similarly we find that $\overline{G_2 - H} \cap G_1 = \emptyset$, and hence conditions (1) are satisfied.

§4. RATIONAL HOLOMORPHIC FUNCTIONS. SOME UNSOLVED PROBLEMS.

Let x be a rational function on V ($x \in \Sigma$) and let P be a point of V. The expression "x is holomorphic at P" can be intended <u>a priori</u> in two senses: (1) $x \in o_P$ or (2) $x \in o_P^*$, i.e., there exist elements y and z in o_P and an element ξ^* in o_P^* such that $y \neq 0$, $yx = z$ and $y\xi^* = z$. However, it is not difficult to see that if $x \in o_P^*$ then necessarily $x \in o_P$ (the converse is, of course, obvious). For the relation $y\xi^* = z$ signifies that y divides z in o_P^*. Hence, by a well-known theorem of Krull, y also divides z in o_P, which proves that $x \in o_P$. We also see that necessarily $x = \xi^*$, for we have $y(x - \xi^*) = 0$ and y cannot be a zero divisor in o_P^* (since o_P is a local domain and y is an element of o_P, different from zero; see Chevalley [1], Proposition 6).

Let now G be a subset of V. From what has been said above it follows that a rational function x on V is holomorphic at each point P of G if and only if x belongs to the local ring o_P of each point P of G. Such functions x form a ring which we shall denote by o_G:

$$o_G = \bigcap_{P \in G} o_P .$$

It is clear that each element x of o_G defines a function ξ^* on V which is holomorphic (even strongly holomorphic) along G: ξ^* is that element of o_G^* which is defined by the sequence $\{x, x, \cdots\}$ and whose analytical element at any point P of G is therefore the given rational function x.

It is also clear that the mapping x --> ξ^* is an isomorphism of o_G into o_G^*. We shall therefore identify each element x of o_G with the corresponding element ξ^* of o_G^* and we shall refer to x as a rational holomorphic function (defined along G).

We suppose now that G is a subvariety W of V and we analyze in this case the relationship between the two rings o_W and o_W^*. Since there is profound difference between the case of projective models V and that of affine models, we shall consider these two cases separately.

I. THE AFFINE CASE. Let R be the non-homogeneous coordinate ring of the affine variety V and let $\mathfrak{m}$ be the ideal of W in R. The elements of R which are $\equiv 1 \pmod{\mathfrak{m}}$ form a multiplicative set S (i.e., a set closed under multiplication). We show that the ring o_W coincides with the quotient ring R_S. The inclusion $R_S \subset o_W$ is obvious, since if P is any point of W then $R \subset o_P$, $\mathfrak{m} \subset \mathfrak{m}_P$ and hence every element of S is $\equiv 1 \pmod{\mathfrak{m}_P}$ and is therefore a unit in o_P. On the other hand, if x is any element of o_W, let $\mathfrak{A}$ denote the ideal in R consisting of the elements α of R such that $\alpha x \in R$. If $P \in W$ then x has an expression of the form ξ/η, where $\xi, \eta \in R$ and $\eta \neq 0$ at P. Since $\eta \in \mathfrak{A}$, it follows that P is not a zero of the ideal $\mathfrak{A}$. Since this holds for any point P of W, it follows that the two ideals $\mathfrak{A}$ and $\mathfrak{m}$ have no common zeros, and hence $\mathfrak{A} + \mathfrak{m} = (1)$. There exists then in $\mathfrak{A}$ an element s which belongs to S, and since $sx \in R$ it follows that $x \in R_S$. Hence $o_W \subset R_S$, $o_W = R_S$, as asserted.

Let $\mathfrak{M} = \mathfrak{m}R_S$. We have shown in [7] that R_S is a (generalized) semi-local $\mathfrak{M}$-adic ring, i.e.: every element of R_S which is $\equiv 1 \pmod{\mathfrak{M}}$ is a unit in R_S. The $\mathfrak{m}$-adic ring R and the $\mathfrak{M}$-adic ring R_S have the same completion, [7]. We have seen in §2 that the $\mathfrak{m}$-adic completion R^* of R coincides with the ring of functions on V which are strongly holomorphic along W (Theorem 1). We shall prove in the last section of Part I that every holomorphic function along W is in fact strongly holomorphic along W. Anticipating this result, we can therefore state the following.

THEOREM 4. If V is an affine variety and W is a subvariety of V, then o_W is a generalized semi-local $\mathfrak{M}$-adic ring, where $\mathfrak{M} = \mathfrak{m}\, o_W$ ($\mathfrak{m}$ = ideal of W in the non-homogeneous coordinate ring of V), and o_W^* is the $\mathfrak{M}$-adic completion of o_W.

Note that as a consequence of this theorem, the ring o_W^* is Noetherian (see [7]).

II. THE PROJECTIVE CASE. We shall begin by defining traces of a holomorphic function. We first consider the local case. Let P be a point of

V and let W be an irreducible subvariety of V (over k) such that $P \in W$. Let $\mathfrak{p}$ be the prime ideal of W in the local ring o_P of P on V. It is then well known that if $\bar{o}_P$ denotes the local ring of P on W, then there is a natural homomorphism σ of o_P^* onto $\bar{o}_P^*$ such that the kernel of σ is the ideal $o_P^* \cdot \mathfrak{p}$. If ξ^* is any holomorphic function on V defined at P, i.e., if ξ^* is any element of o_P^*, we shall call $\sigma(\xi^*)$ the trace of ξ^* on W; in symbols: $\sigma(\xi^*) = Tr_W \xi^*$. Hence $Tr_W \xi^*$ is a holomorphic function on W, defined at P, and every such function is the trace of some element ξ^* of o_P^*. Let now W be again an irreducible subvariety of V and let G be an arbitrary point set on V. We shall assume that the set $G_0 = G \cap W$ is non-empty. If ξ^* is any function on V, defined along G, we define a unique function $\bar{\xi}^*$ on W, defined along G_0, by the condition: $\bar{\xi}_P^* = Tr_W \xi_P^*$, for all P in G_0, and we call this function $\bar{\xi}^*$ the trace of ξ^* on W: $\bar{\xi}^* = Tr_W \xi^*$.

It is immediately seen that if ξ^* is strongly holomorphic along G, then $Tr_W \xi^*$ is strongly holomorphic along G_0 (this follows from the fact that in the above homomorphism of o_P onto $\bar{o}_P$ the maximal ideal $\mathfrak{m}_P$ of o_P is transformed into the maximal ideal $\bar{\mathfrak{m}}_P$ of $\bar{o}_P$). If we now apply our definition of holomorphic functions we see at once that if ξ^* is holomorphic along G, then also $Tr_W \xi^*$ is holomorphic along G_0.

Suppose now that we are dealing with a function ξ^* on V (V = a projective model) which is holomorphic along an irreducible variety W. Then it follows from Theorem 3 (§ 3) that $Tr_W \xi^*$ is a constant rational function on W. Let us suppose that $Tr_W \xi^* \neq 0$ and let $\{\xi_{1j}, \xi_{2j}, \cdots\}$, $1 \leq j \leq h$, be a set of convergent sequences which together define ξ^* along W (see Definition 2, § 3). Without loss of generality we may assume that $\xi_{nj} - \xi_{mj} \in \mathfrak{m}_P^n$ for all P in $W - \Gamma_j$ and for all $m \geq n$ $(n = 1, 2, \cdots)$. Since $Tr_W \xi^*$ is a constant, different from zero, it follows $Tr_W \xi_P^* \notin \mathfrak{m}_P^*$ for all P in W, and hence $\xi_{nj} \notin \mathfrak{m}_P$ for $P \in W - \Gamma_j$ and $n = 1, 2, \cdots$. Hence all the reciprocals $1/\xi_{nj}$ belong to $o_P (P \in W - \Gamma_j)$ and it is clear that the sequence $1/\xi_{nj}$ converges uniformly on $W - \Gamma_j$, and that these h sequences define together a function η^* on V which is holomorphic along W. Furthermore, we have $\xi^* \eta^* = 1$, and hence we have proved the following: if $\xi^* \in o_W^*$ and $Tr_W \xi^* \neq 0$, then ξ^* is a unit in o_W^*. Hence the non-units of o_W^* are those and only those elements of o_W^* whose W-trace is zero. The non-units of o_W^* form therefore an ideal.

A similar argument shows that also the non-units of o_W form an ideal, and the proof in this case is even simpler since it does not involve the consideration of infinite sequence but only the remark that in the natural homomorphism of o_P onto $\bar{o}_P$ the ideal $\mathfrak{m}_P$ is mapped onto the ideal $\bar{\mathfrak{m}}_P$. The ideal of non-units of o_W is given by $\bigcap_{P \in W} \mathfrak{m}_P$.

Were it known that the rings o_W and o_W^* are Noetherian it would then follow that o_W and o_W^* are local rings. This, then, is our first unsolved problem:

PROBLEM A. Are o_W and o_W^* Noetherian rings?
Regardless of whether o_W is or is not Noetherian we can consider the completion of o_W with respect to the powers of its ideal of non-units. It is immediately seen that the completion of o_W can be identified, in a natural fashion, with a subring of o_W^*, and that the elements of this subring are strongly holomorphic along W.

PROBLEM B. Is o_W^* the completion of o_W?

The following question is closely connected to Problem B:

PROBLEM C. If a function ξ^* on V is holomorphic along W, is ξ^* necessarily strongly holomorphic along W? (Compare with Theorem 4 for the affine case).

It is clear that an affirmative answer to Problem B implies an affirmative answer to Problem C. However, to prove the converse (and hence the equivalence of the two problems) it would still be necessary to prove the following statement: if $\mathfrak{M}$ is the maximal ideal of o_W, there exists a sequence $\vartheta(n)$ such that $\bigcap_{P \in W} \mathfrak{m}_P^n \subset \mathfrak{M}^{\vartheta(n)}$, where $\lim \vartheta(n) = \infty$. This statement is an easy consequence of the lemma given in the next section.

Until now we have assumed that W is irreducible. We shall now discuss briefly the case in which W is reducible (over k).

We say that W is connected (over k) if W is not the sum of two disjoint and non empty varieties (defined over k). If W is not connected, it is the sum $W_1 \cup W_2 \cup \cdots \cup W_h$ of connected varieties $(h > 1)$ such that no W_i is empty and $W_i \cap W_j = \emptyset$ for $i \neq j$. The varieties W_i are called the connected components of W and are uniquely determined by W.

Let Γ be an irreducible component of W and let ξ^* be any element of o_W^*. Since ξ^* is also holomorphic along Γ, we can consider the trace $\bar{\xi}^*$ of ξ^* on Γ. We know that this trace must be a constant, say c. If f(Z) is the irreducible polynomial in k[Z] such that f(c) = 0, [whence $Tr_\Gamma\ f(\xi^*) = 0$] and if P is any point of Γ, then $f(\xi^*)$ necessarily belongs to $\mathfrak{m}_P^*$. It is clear that at any given point P of Γ there can exist only one irreducible polynomial f(Z) in k[Z] with the property: $f(\xi^*) \in \mathfrak{m}_P^*$. It follows that if Γ_1 is another irreducible component of W and if Γ and Γ_1 have at least one point in common, then also $Tr_{\Gamma_1} f(\xi^*) = 0$. If W is connected, we conclude that $f(\xi^*)$ has zero trace on each irreducible component of W, i.e., ξ^* <u>has the same constant trace on each irreducible component of</u> W. From this it follows at once that the non-units of o_W^* still form an ideal (as in the case of an irreducible W). In a similar fashion it follows that also the non-units in o_W form an ideal. The problems A, B and C, formulated above, are intended to refer to connected as well as to irreducible varieties W.

If W is not connected and consists of h connected components W_i, it is easily seen that the elements of o_W^* (or of o_W) which have zero trace on a given component W_i form a maximal ideal, and that the h (not necessarily distinct) ideals thus obtained are the only maximal ideals of o_W^* (or of o_W). Were it known that o_W and o_W^* are Noetherian rings it would follow that o_W and o_W^* are semi-local rings in the sense of Chevalley [1].

As to Problems B and C, the answer is not necessarily in the affirmative if W is not connected. We prove in §6 that if W is not connected then o_W^* is not an integral domain. On the other hand, it may very well happen that o_W consists only of constants, and in that case the completion of o_W coincides of course with o_W. Hence in this case the completion of o_W does not coincide with o_W^*. However, the case of a disconnected variety W is of little interest in itself since it can be reduced to the case of connected varieties in virtue of the fact that o_W^* is always the direct sum of the rings $o_{W_i}^*$, when the W_i are the connected components of W (see §6).

§5. DIGRESSION ON ALGEBRAIC POINTS.

We have consistently used the term "point" in its widest possible sense: the coordinates of a point are arbitrary quantities which are not necessarily algebraic over k. However, in applications one may be interested primarily in the algebraic points of a variety. This is certainly so in the classical case where a variety is never intended to carry points other than those which have complex coordinates. Points with transcendental coordinates are after all only a convenient algebraic tool, not a geometric reality (except for the fact that each such point determines an algebraic irreducible variety of which it is a general point).

Now suppose that, we restrict ourselves to algebraic points of V (in the affine or in the projective space). In that case, given a subvariety W of V and denoting by W_0 the set of algebraic points of W, we would not be speaking of functions on V, holomorphic along W (this purely arithmetic concept could not even be defined in terms of classical function theory) but only of functions on V, defined along W_0. In other words, we would be dealing now with the ring $o^*_{W_0}$ rather than with o^*_W. The machinery used in the preceding sections is not immediately applicable to the ring $o^*_{W_0}$ since an essential ingredient of our proofs (especially of the proof of Theorem 1) was the fact that the sequences used to define the elements of o^*_W converge also at transcendental points of W, and- in particular- at each general point of each irreducible component of W.

We shall prove now that there is a natural isomorphism between the two rings $o^*_{W_0}$ and o^*_W. This isomorphism will be given by the projection τ_{W,W_0}; in other words: <u>every function in $o^*_{W_0}$ has an extension to a function in o^*_W and the extension is unique</u>. In view of this result, everything that will be proved in this paper for holomorphic functions is automatically applicable if the domains of definition of these functions are restricted so as contain only algebraic points. This result also signifies that even in the classical case one gains an extra powerful tool when one replaces the ring $o^*_{W_0}$ by the ring o^*_W and that this can be done without losing touch with function-theoretic realities.

Our proof that τ_{W,W_0} is an isomorphism is based on the following:

LEMMA. Let W be an irreducible subvariety of V (in the affine or in the projective space) and let G be a subset of W such that $W = \overline{G}$ (i.e., W is the least algebraic variety containing G). If x is a rational function on V such that $x \in \mathfrak{m}_Q^n$ for all Q in G, then $x \in \mathfrak{m}_P^n$ for any general point P of W.

The proof of this lemma is given in our paper [10]. Assuming this lemma, we proceed as follows (for the sake of simplicity we shall only consider the case of an irreducible W. The extension of the proof to a reducible W is straightforward.)

Let ξ_0^* be any element of $o^*_{W_0}$. There will exist then a finite open covering $\{G_{\alpha 0}\}$ of W_0 such that ξ_0^* is strongly holomorphic on $G_{\alpha 0}$. For each $G_{\alpha 0}$ there is subvariety W_α of W such that $G_{\alpha 0}$ is the set of all algebraic points of $W - W_\alpha$. Since the $G_{\alpha 0}$ cover W_0, it follows that the intersection of the W_α contains no algebraic point, and is therefore empty (Hilbert's Nullstellensatz). <u>Hence, if we set</u> $G_\alpha = W - W_\alpha$, <u>then</u> $\{G_\alpha\}$ <u>is an open covering of</u> W.

Let $\{x_{\alpha 1}, x_{\alpha 2}, \dots\}$ be a sequence of rational functions on V which converges uniformly on $G_{\alpha 0}$ to ξ_0^*. Then if n is a given integer, we will have that $x_{\alpha i} - x_{\alpha j} \in \mathfrak{m}_Q^n$ for each algebraic point Q of G_α, provided i,j exceed some integer N= N(n). Now let P be an arbitrary point of G_α and let U be the irreducible algebraic variety whose general point is P. If U' is the closure of the set $U \cap G_{\alpha 0}$, then the variety U' + W_α contains all the algebraic points of U, for any algebraic point of U either belongs to U' or does not belong to $G_{\alpha 0}$, and in the latter case it belongs to W_α (since $U \subset W$). Hence $U \subset U' + W_\alpha$. Since U is irreducible and since $U \not\subset W_\alpha$ (since $P \in G_\alpha$), it follows that $U \subset U'$, and hence U= U' since U' is the least variety containing the set $U \cap G_{\alpha 0}$. If in the main lemma we now replace W by U and G by $U \cap G_{\alpha 0}$, we conclude that $x_{\alpha i} - x_{\alpha j} \in \mathfrak{m}_P^n$ for i, $j \geq N(n)$. <u>Hence the sequence</u> $\{x_{\alpha 1}, x_{\alpha 2}, \cdots\}$ <u>converges uniformly on</u> G_α. If $\alpha \neq \beta$, then the sequence $\{x_{\alpha 1} - x_{\beta 1}, x_{\alpha 2} - x_{\beta 2}, \cdots\}$ converges uniformly to zero on $G_{\alpha 0} \cap G_{\beta 0}$ (=$W_0 - W_\alpha - W_\beta$). It follows by the same argument as above that this sequence converges uniformly to zero also on $G_\alpha \cap G_\beta$. This signifies that the two sequences $\{x_{\alpha 1}, x_{\alpha 2}, \cdots\}$ and $\{x_{\beta 1}, x_{\beta 2}, \cdots\}$ have the same limit at each common point of G_α and G_β. Hence the various sequences $\{x_{\alpha 1}, x_{\alpha 2}, \cdots\}$ define together an element ξ^* of o_W^*, and it is clear that $\xi_0^* = \tau_{W, W_0}(\xi^*)$. Hence the projection τ_{W,W_0} maps o_W^* onto $o_{W_0}^*$. This mapping is clearly a ring homomorphism. On the other hand, the preceding proof shows also that if a holomorphic function along W, different from zero, is defined by certain uniformly convergent sequences $\{x_{\alpha 1}, x_{\alpha 2}, \cdots\}$, then these sequences could not possibly converge to zero at all points of W_0. It follows that the above homomorphism is actually an isomorphism, and this proves the essential identity of the two rings o_W^* and $o_{W_0}^*$.

§6. HOLOMORPHIC FUNCTIONS, ANALYTICAL IRREDUCIBILITY, AND A CONNECTEDNESS CRITERION.

Let W be subvariety of V in the projective or in the affine n-space and let $W_1, W_2, \cdots, W_h$ be the connected components of W. It is clear that the ring O_W^* is the direct sum of the h rings $O_{W_i}^*$. It is not difficult to see that also <u>the ring of holomorphic functions</u> o_W^* <u>is the direct sum of the h rings</u> $o_{W_i}^*$. Namely, for any given value of i and for any element c in k, the function on V, whose analytical element at each point of W_i is equal to c is strongly holomorphic along W_i. Hence the function ϵ_i^* such that $\epsilon_{iP}^* = 1$ if $P \in W_i$ and $\epsilon_{iP}^* = 0$ if $P \in W_j$, $j \neq i$, has a unique analytical element at each point of W (since $W_i \cap W_j = \emptyset$, if $i \neq j$) and is a holomorphic function along W in the sense of Definition 2 (in the present case we may choose for $\Gamma_1, \Gamma_2, \cdots, \Gamma_h$ the components $W_1, W_2, \cdots, W_h$). We have $\varepsilon_i^{*2} = \varepsilon_i^*$, $\varepsilon_i^* \varepsilon_j^* = 0$ if $i \neq j$, $\varepsilon_1^* + \varepsilon_2^* + \cdots + \varepsilon_h^* = 1$, and it is clear that in the direct decomposition $o_W^* = \sum_{i=1}^h o_W^* \cdot \varepsilon_i^*$ each term $o_W^* \cdot \varepsilon_i^*$ is projected isomorphically by τ_{W,W_i} onto $o_{W_i}^*$.

We observe that in the affine case each of the h functions ε_i^* is strongly holomorphic along W. For let $\mathfrak{m}_i$ be the ideal of W_i in the non-homogeneous coordinate ring R of V. Since $W_i \cap W_j = \emptyset$ if $i \neq j$, it follows that $\mathfrak{m}_i + \mathfrak{m}_j = (1)$ and that, more generally, $\mathfrak{m}_i^\nu + \mathfrak{m}_j^\nu = (1)$ for any integer ν and for $i \neq j$. We can therefore find an element $\xi_{i\nu}$ of R such that $\varepsilon_{i\nu} - 1 \in \mathfrak{m}_i^\nu$, $\varepsilon_{i\nu} \in \mathfrak{m}_j^\nu$ if $j \neq i$. The sequence $\{\varepsilon_{i1}, \varepsilon_{i2}, \cdots\}$ converges then uniformly to ε_i^* on W. This remark, however, is included, as a very special case, in Theorem 10, §9.

In [8] we have defined analytical irreducibility of a variety V at a point P by the condition that the ring o_P^* be an integral domain. We now extend this definition to arbitrary subsets G of V:

DEFINITION 3. V is analytically irreducible along G if the ring o_G^* of holomorphic functions along G is an integral domain.

We point out explicitly that if G is an irreducible subvariety W of V (in the affine or in the projective space) the analytical irreducibility of V along W and the analytical irreducibility of V at a general point P of W mean different things altogether, and neither one implies the other. In the subsequent sections we shall obtain full and precise information about the geometric content of the concept of analytical irreducibility, both in the local case (analytical irreducibility at a point) and the semi-local case (analytical irreducibility along a subvariety W).

We have seen above that if W is not connected, then V is analytically reducible along W. This statement holds both in the projective and the affine case. It is perfectly clear, however, that the connectedness of W does not imply necessarily the analytical irreducibility of V along W[6]. In [7] we have proved the following result in the affine case: if W is connected and if V is analytically irreducible at each point of W, then the ring of strongly holomorphic functions along W is an integral domain. By Theorem 4, §4, we have therefore the following result:

If W is an affine subvariety of V and if V is analytically irreducible at each point of W, then V is analytically irreducible along W if and only if W is connected.

In this section we shall prove that this result remains true if W is replaced by an arbitrary point set G. No use will be made of Theorem 4, §4.

Since the quoted paper [7] may not be easily accessible to many readers, we reproduce below one essential ingredient of the proof given in that paper, namely the following lemma [7]:

LEMMA 1. Let R be a Noetherian domain and let $\mathfrak{p}$ and $\mathfrak{p}_1$ be prime ideals in R such that $\mathfrak{p}$ is maximal and $\mathfrak{p}_1 \subset \mathfrak{p}$. If R is analytically irreducible at $\mathfrak{p}$, i.e., if the $\mathfrak{p}$-adic completion of R is an integral domain, then high symbolic powers of $\mathfrak{p}_1$ are contained in high ordinary powers of $\mathfrak{p}$.

PROOF. If $\tilde{R}$ denotes the quotient ring $R_{\mathfrak{p}}$ and if $\tilde{\mathfrak{p}} = \tilde{R}.\mathfrak{p}$ and $\tilde{\mathfrak{p}}_1 = \tilde{R}.\mathfrak{p}_1$, then $\tilde{\mathfrak{p}}_1^{(i)} \cap R = \mathfrak{p}_1^{(i)}$ and $\tilde{\mathfrak{p}}^{\nu} \cap R = \mathfrak{p}^{\nu}$. This shows that it is sufficient to prove the lemma under the assumption $R = R_{\mathfrak{p}}$. We shall therefore assume that R is a local domain and that $\mathfrak{p}$ is the ideal of non units in R.

Let R* be the $\mathfrak{p}$-adic completion of R and let $\mathfrak{p}^* = R^*\mathfrak{p}$. From the definition of the symbolic power $\mathfrak{p}_1^{(i)}$ it follows that there exists in R an element c_1 such that:

(2) $$c_1 \mathfrak{p}_1^{(i)} \subset \mathfrak{p}_1^{i}, \quad c_1 \notin \mathfrak{p}_1.$$

We consider the ideal $\mathfrak{a}_i^* = R^* \mathfrak{p}_1^{(i)}$. By (2) we have $c_1 \mathfrak{a}_i^* \subset \mathfrak{a}_1^{*i}$, and hence if $\mathfrak{P}^*$ denotes a prime ideal of $\mathfrak{a}_1^*$, then

(3) $$c_1 \mathfrak{a}_i^* \subset \mathfrak{P}^{*(i)}.$$

It is well known (and is an easy consequence of Proposition 6 in Chevelley, [1]) that

(4) $$\mathfrak{P}^* \cap R = \mathfrak{p}_1$$

for <u>any</u> prime ideal $\mathfrak{P}^*$ of $\mathfrak{a}_1^* (= R^* \cdot \mathfrak{p}_1)$. It follows from (3), (4), and from $c_1 \notin \mathfrak{p}_1$, that $\mathfrak{a}_i^* \subset \mathfrak{P}^{*(i)}$. Now we shall use our assumption that R is analytically irreducible at $\mathfrak{p}$. By this assumption the ring R* is an integral domain, and in Noetherian integral domain the intersection of all symbolic powers $\mathfrak{P}^{*(i)}$ of a prime ideal $\mathfrak{P}^*$ is the zero ideal. We conclude that $\bigcap_{i=1}^{\infty} \mathfrak{a}_i^* = (0)$. By a theorem on local rings due to Chevalley ([1], p. 695, Lemma 7) it follows then that $\mathfrak{a}_i^* \subset \mathfrak{p}^{*\nu(i)}$, $\lim \nu(i) = +\infty$. This completes the proof of the lemma since $\mathfrak{p}_1^{(i)} = \mathfrak{a}_i^* \cap R$ and $\mathfrak{p}^{\nu(i)} = \mathfrak{p}^{*\nu(i)} \cap R$.

LEMMA 2. Let P and P_1 be two points of V such that P is a specialization of P_1 (over k), and let ξ^* be a function on V which is strongly holomorphic on the set (P, P_1). If $\xi^*_P = 0$, then also $\xi^*_{P_1} = 0$. Conversely, if $\xi^*_{P_1} = 0$, then also $\xi^*_P = 0$, provided V is analytically irreducible at P.

PROOF. Let o and o_1 denote the local rings of P and P_1, and let $\mathfrak{m}$

and $\mathfrak{m}_1$ be the maximal ideals in these rings. Since P is a specialization of P_1, we have $\mathfrak{o} \subset \mathfrak{o}_1$. We set $\mathfrak{m}_1 \cap \mathfrak{o} = \mathfrak{p}_1$. Let $\xi^* = \lim \xi_i$ on the point pair (P, P_1). Replacing, if necessary, the sequence $\{\xi_i\}$ by a subsequence, we may assume that $\xi_i - \xi_j \in \mathfrak{m}^i \cap \mathfrak{m}_1^i$ for all $j \geq i$. Since $\xi_i - \xi_j \in \mathfrak{o}$ and $\mathfrak{m}_1^i \cap \mathfrak{o} = \mathfrak{p}_1^{(i)}$, it follows that $\xi_i - \xi_j \in \mathfrak{p}_1^{(i)}$, $j \geq i$. Suppose that $\xi_P^* = 0$. Then we may assume that $\xi_j \in \mathfrak{m}^j$ for all j. Hence
$\xi_i \in \bigcap_{j=1}^{\infty} (\mathfrak{p}_1^{(i)} + \mathfrak{m}^j)$, i.e. $\xi_i \in \mathfrak{p}_1^{(i)}$.
Therefore $\xi_i \in \mathfrak{m}_1^i$, i.e. $\xi_{P_1}^* = 0$.

Conversely, assume that $\xi_{P_1}^* = 0$. Then we may suppose that $\xi_i \in \mathfrak{m}_1^i$, i.e. $\xi_i \in \mathfrak{p}_1^{(i)}$. Now if we assume furthermore that V is analytically irreducible at P, then, by Lemma 1, we have $\mathfrak{p}_1^{(i)} \subset \mathfrak{m}^{\nu(i)}$, where $\nu(i) \to +\infty$ as $i \to +\infty$. In that case the sequence $\{\xi_i\}$ is also a zero sequence at P, i.e., we have $\xi_P^* = 0$, as asserted.

THEOREM 5. Let W be an irreducible subvariety of V and let G be a subset of W which contains at least one general point of W/k. Let ξ^* be a function on V which is holomorphic along G. If the analytical element $\xi_{P_0}^*$ of ξ^* at one point P_0 of G is zero, then $\xi_P^* = 0$ for every point P of G such that V is analytically irreducible at P and also for every point P of G which is a general point of W/k.

PROOF. Let $\Gamma_1, \Gamma_2, \cdots, \Gamma_h$ be varieties such that ξ^* is strongly holomorphic along $G - \Gamma_j$, $j = 1, 2, \cdots, h$, and such that $\bigcup_{j=1}^{h} (G - \Gamma_j) = G$ (see Definition 2). We may assume, of course, that no Γ_j contains G. The point P_0 belongs to one of the sets $G - \Gamma_j$, say to $G - \Gamma_1$. Let P_1 be a general point of W belonging to G. Since $\Gamma_j \not\supset G$, Γ_j does not contain W, and therefore $P_1 \notin \Gamma_j$. The point P_1 belongs therefore to each of the h sets $G - \Gamma_j$, and each point of G is a specialization of P_1 (over k). Since $P_0, P_1 \in G - \Gamma_1$, and since ξ^* is strongly holomorphic along $G - \Gamma_1$, we have, by the first part of Lemma 2, $\xi_{P_1}^* = 0$. If now P is any point of G where V is analytically irreducible, and if say, $P \in G - \Gamma_j$, we can apply the second part of Lemma 2 to the points P, P_1, and we conclude that $\xi_P^* = 0$. This completes the proof.

The preceding theorem can be extended to arbitrary connected varieties W. If W is a subvariety of V and G is a subset of W, we say that W is G-<u>connected</u> if W is not the union of two proper subvarieties W' and W'' having no common points in G[8]. We can state then the following theorem:

THEOREM 5'. Let W be a subvariety of V and let G be a subset of W which contains at least one general point of each irreducible component of W. Let G_1 be the set of points of G at which V is analytically irreducible and let ξ^* be any function on V which is holomorphic along G. If W is G_1-connected, then the vanishing of the analytical element of ξ^* at one point of G implies the vanishing of the analytical element of ξ^* at each point of G_1.

To prove this theorem it is only necessary to use the points of G_1 which serve as connecting bridges between the various irreducible components of W and apply Theorem 5 to each of these irreducible components.

COROLLARY 1. The notation being that of Theorem 5', assume that G contains at least one general point of each irreducible component of W and that W is G_1- connected. Then the projection $\tau_{G,\,G-H}$ induces an isomorphism of o^*_G into o^*_{G-H}, where H is any subset of G_1 such that G - H is not empty.

For if ξ^* is an element of o^*_G such that $\xi^*\tau = 0$, then ξ^* has a vanishing analytical element at each point of G - H, and since G - H is not empty it follows from Theorem 5' that $\xi^* = 0$.

COROLLARY 2. Under the assumptions of Corollary 1 and under the additional assumption that $G = G_1$, V is analytically irreducible along G.

For under the above assumptions we can apply Corollary 1 to the set $H = G - \{P\}$, where P is any point of G, and we then conclude that o^*_G can be mapped isomorphically into o^*_P. Since this latter ring is an integral domain, it follows that also o^*_G is an integral domain.

We know from an earlier part of this section that if V is analytically irreducible along W, then W is connected. If we set G = W in Corollary 2 we conclude at once that _if_ V _is analytically irreducible at each point of_ W _then the connectedness of_ W _implies the analytical irreducibility of_ V _along_ W. Thus under the assumption that V is analytically irreducible at each point of a given subvariety W, we have an arithmetic criterion (" o^*_W is an integral domain") of the connectednessof W. This connectedness criterion will be used in the proof of the principle of degeneration which we shall give in Part III of this paper.

We shall now extend the connectedness criterion to arbitrary subsets of V. Let G be a set of points of V and let $\overline{G}$ be the least variety containing G. We say that G is _connected_ if $\overline{G}$ is G-connected (this definition refers to our fixed ground field k). If we use the topology on V in which closed subsets of V are the algebraic subvarieties of V, then it is easily seen that our definition of connected subsets of V coincides with the definition of connected sets based on that topology.

THEOREM 6. If G is a subset of V and if V is analytically irreducible at each point of G, then V is analytically irreducible along G if and only if G is connected.

PROOF. Assume that G is not connected. We have then $\overline{G} = W' \cup W''$ and $W' \cap W'' \cap G = \emptyset$, where W' and W" are proper subvarieties of $\overline{G}$. Since $\overline{G}$ is the least variety containing G, it follows that $W' \cap G$ and $W'' \cap G$ are both non empty sets. Since these two sets have no points in common, it follows that o^*_G contains the function ε^*_1 whose analytical element at each point P of G is either 1 or zero, according as P belongs to $W' \cap G$ or to $W'' \cap G$ (see Corollary at the end of § 3) and that this function ε^*_1 is not the zero element of o^*_G. In a similar fashion we define an element ε^*_2 of o^*_G by interchanging the roles of W' and of W". We have $\varepsilon^*_1 \neq 0$, $\varepsilon^*_2 \neq 0$, $\varepsilon^*_1 \varepsilon^*_2 = 0$, and hence o^*_G is not an integral domain and V is not analytically irreducible along G.

Assume now that G is connected and that V is analytically irreducible at each point of G. Let $W_1, W_2, \cdots, W_g$ be the irreducible components of $\overline{G}$ and let P_i be a general point of W_i. We consider the set $H = G \cup \{P_1, P_2, \cdots, P_g\}$. We identify in Theorem 5' the variety W and the set G with $\overline{G}$ and H respectively. The set G_1 of Theorem 5' contains in the present case our present set G, and since $\overline{G}$ is G-connected it is *a fortiori* G_1- connected, and therefore Theorem 5' is applicable. Let A be a point of G and let ξ^* be a function in o^*_H. It follows then from Theorem 5' that if $\xi^*_A = 0$, then $\xi^*_B = 0$ for all points B of G. Since $H \cap W_i$ contains the general point P_i of W_i and since the *non-empty* set $G \cap W_i$ (non-empty, because G is connected) is contained in $H \cap W_i$, it follows from Theorem 5 that also $\xi^*_{P_i} = 0$, $i = 1, 2, \cdots, g$. Hence ξ^* is the zero element of o^*_H. We have therefore proved that the projection $\tau_{H,A}$ induces an isomorphism of o^*_H into o^*_A. Since $A \in G$, V is analytically irreducible at A, o^*_A is an integral domain, and therefore also o^*_H is an integral domain. We shall now prove that also o^*_G is an integral domain by showing that the homomorphism of o^*_H into o^*_G induced by the projection $\tau_{H,G}$ is an *isomorphism onto* o^*_G. This will complete the proof of Theorem 6.

Let ξ^* be any element of o^*_G and let $\Gamma_1, \Gamma_2, \cdots, \Gamma_h$ be a set of subvarieties of V such that $G \cap \Gamma_1 \cap \Gamma_2 \cap \cdots \cap \Gamma_h = \emptyset$ and such that ξ^* is strongly holomorphic along each of the h sets $G - \Gamma_j$. Let, for a given j, $\{\xi_{j\nu}\}$ be a sequence of rational functions on V which converges uniformly to ξ^* on $G - \Gamma_j$. Since G is the union of the h sets $G - \Gamma_j$, $\overline{G}$ is the union of the h sets $\overline{G - \Gamma_j}$, where $\overline{G - \Gamma_j}$ is the least variety containing $G - \Gamma_j$. Hence each of the irreducible components W_i of $\overline{G}$ must be contained in at least one of the varieties $\overline{G - \Gamma_j}$. If $W_i \subset \overline{G - \Gamma_j}$, then W_i is an irreducible component of $\overline{G - \Gamma_j}$ (since $\overline{G - \Gamma_j} \subset \overline{G}$ and W_i is an irreducible component of $\overline{G}$) and is the closure of $W_i \cap (G - \Gamma_j)$. Hence it follows from the lemma of §5 that the sequence $\{\xi_{j\nu}\}$ also converges uniformly on the set consisting of $G - \Gamma_j$ and the point P_i and that if $\lim \xi_{j\nu} = 0$ on $G - \Gamma_j$, then also $\lim \xi_{j\nu} = 0$ at P_i. Now we observe that we have for all i and j: $(\overline{G - \Gamma_j}) \cup \Gamma_j \supset \overline{G} \supset W_i$. If $P_i \notin \Gamma_j$, then $W_i \not\subset \Gamma_j$ and hence $W_i \subset \overline{G - \Gamma_j}$. If $P_i \in \Gamma_j$, then $W_i \subset \Gamma_j$, $\overline{G - \Gamma_j} \subset \overline{G} - W_i$, and hence $W_i \not\subset \overline{G - \Gamma_j}$. We have thus shown that $W_i \ \overline{G - \Gamma_j}$ if and only if $P_i \notin \Gamma_j$, and from this we conclude at once that each sequence $\{\xi_{j\nu}\}$ converges uniformly on the set $H - \Gamma_j$ and that if the limit of that sequence is zero on $G - \Gamma_j$ it is also zero on $H - \Gamma_j$. This shows that $\tau_{H,G}$ is an isomorphism of o^*_H onto o^*_G, as asserted.

We point out that if the set G is a variety, $G = W$, then in the above connectedness criterion it is permissible to replace the assumption that V is analytically irreducible at each point of W by the assumption that V is analytically irreducible at each algebraic point of W. For on the one hand the set W_o of algebraic points of W is connected if and only if W is connected, and on the other hand we know that o^*_W and $o^*_{W_o}$ are isomorphic rings (§5).(9)

Lemma 1, which has played an essential role in the proof of Lemma 2 (and hence also in the proof of Theorem 6) can be generalized to the case in which the ideal $\mathfrak{p}$ is not maximal:

LEMMA 3. If $\mathfrak{p}$ and $\mathfrak{p}_1$, $\mathfrak{p} \supset \mathfrak{p}_1$, are **prime ideals in a** Noetherian integral domain R and if R is analytically irreducible at $\mathfrak{p}$, then high symbolic powers of $\mathfrak{p}_1$ are contained in high symbolic powers of $\mathfrak{p}$. If R is analytically irreducible at each prime ideal $\mathfrak{p}_o$ such that $\mathfrak{p}_o \supset \mathfrak{p}$, then high symbolic powers of $\mathfrak{p}_1$ are contained in high ordinary powers of $\mathfrak{p}$.

PROOF. Let R' denote the quotient ring of $\mathfrak{p}$ and let $\mathfrak{p}' = R'\mathfrak{p}$, $\mathfrak{p}_1' = R'\mathfrak{p}_1$. Then the assumptions of Lemma 1 are satisfied if R, $\mathfrak{p}$ and $\mathfrak{p}_1$

are replaced respectively by R', $\mathfrak{p}'$ and $\mathfrak{p}_1'$. Since we have $\mathfrak{p}'^{(i)} \cap R = \mathfrak{p}_1^{(i)}$ and $\mathfrak{p}'^{i} \cap R = \mathfrak{p}^{(i)}$, the first part of the lemma follows immediately.

For the proof of the second part of the lemma, we fix an integer n and we decompose $\mathfrak{p}^n$ into primary components: $\mathfrak{p}^n = \mathfrak{p}^{(n)} \cap \mathfrak{q}_{o1} \cap \mathfrak{q}_{o2} \cap \cdots \cap \mathfrak{q}_{oG}$. If $\mathfrak{p}_{oj}$ is the prime ideal of $\mathfrak{q}_{oj}$, then we have $\mathfrak{p} \subset \mathfrak{p}_{oj}$. Hence, by assumption, R is analytically irreducible at each $\mathfrak{p}_{oj}$, and also at $\mathfrak{p}$, and therefore, by the first part of the lemma, we can find an integer i_n such that $\mathfrak{p}_1^{(i_n)} \subset \mathfrak{p}_{oj}^{(\rho_j)} \subset \mathfrak{q}_{oj}$, where ρ_j is the exponent of $\mathfrak{q}_{oj}$, j= 1,2,...,g, and $\mathfrak{p}_1^{(i_n)} \subset \mathfrak{p}^{(n)}$. We will have then $\mathfrak{p}_1^{(i_n)} \subset \mathfrak{p}^n$, which completes the proof of the lemma.

COROLLARY. If R is analytically irreducible at each prime ideal $\mathfrak{p}_o$ such that $\mathfrak{p}_o \supset \mathfrak{p}$, $\mathfrak{p}_o \neq \mathfrak{p}$, then high symbolic powers of $\mathfrak{p}$ are contained in high ordinary powers of $\mathfrak{p}$.

Except for the inequality $\mathfrak{p}_o \neq \mathfrak{p}$, this corollary is a special case of the second part of Lemma 3 (let $\mathfrak{p}_1 = \mathfrak{p}$). That in the present case it is sufficient to require the analytical irreducibility of R at all proper prime ideal divisors $\mathfrak{p}_o$ of $\mathfrak{p}$ follows from the fact that we have $\mathfrak{p}_1^{(i_n)} \subset \mathfrak{p}^{(n)}$ for $i_n \geqq n$, if $\mathfrak{p}_1 = \mathfrak{p}$.

§ 7. ANALYTICAL IRREDUCIBILITY AND NORMALIZATION.

The applications which will be developed in this section are based on the theorem proved in our paper [8], to the effect that a normal variety is analytically irreducible at each point. It follows therefore from the connectedness criterion of the preceding section that in the case of a normal variety V, a necessary and sufficient condition in order that a subvariety W of V be connected is that the ring $\mathfrak{o}_W^*$ be an integral domain. This result holds for varieties in the affine space as well as varieties in the projective space.

If R is a non-homogeneous coordinate ring of a variety V, then R is an integrally closed domain if and only if V is normal in the affine space. Applying Lemma 3 of § 6 and the theorem of analytical irreducibility of normal varieties, we obtain at once the following theorem:

THEOREM 7. If R is an integrally closed finite integral domain (over a given ground field k), and if $\mathfrak{p}_1$ and $\mathfrak{p}$ are any two prime ideals in R, such that $\mathfrak{p}_1 \subset \mathfrak{p}$ (in particular, if $\mathfrak{p}_1 = \mathfrak{p}$), then high symbolic powers of $\mathfrak{p}_1$ are contained in high ordinary powers of $\mathfrak{p}$.

Now let V be an arbitrary variety and let $\overline{V}$ be a derived normal model of V. If P is any point of V and if $\overline{P}_1, \overline{P}_2, \cdots, \overline{P}_g$ are the points of $\overline{V}$ which correspond to V, then it is well known that $\overline{o}_P = \bigcap_{i=1}^{g} o_{\overline{P}_i}$, where $\overline{o}_P$ is the integral closure of o_P in the field Σ (= $\mathcal{F}(V)$), and that $\overline{o}_P$ is a finite o_P-module (Zariski, [6], p. 511), say $\overline{o}_P = \sum_{i=1}^{m} o_P \cdot \omega_i$. Let $\{\overline{P}_1, \overline{P}_2, \cdots, \overline{P}_h\}$ be a subset of $\{\overline{P}_1, \overline{P}_2, \cdots, \overline{P}_g\}$ such that no two of the points $\overline{P}_1, \overline{P}_2, \cdots, \overline{P}_h$ are isomorphic (over k) and such that each of the points $\overline{P}_1, \overline{P}_2, \cdots, \overline{P}_g$ is isomorphic to one of the points $\overline{P}_1, \overline{P}_2, \cdots, \overline{P}_h$. Since k-isomorphic points have the same local ring, it follows that $\overline{o}_P = \bigcap_{i=1}^{h} o_{\overline{P}_i}$. It is known that, on the one hand, the completion of the semilocal ring $\bigcap_{i=1}^{h} o_{\overline{P}_i}$ is isomorphic with the direct sum of h rings $o^*_{\overline{P}_i}$ (Chevalley [1], Proposition 2, p. 693) and that, on the other hand, the ring o^*_P can be regarded as a subring and a subspace of the completion $(\overline{o}_P)^*$ of the semilocal ring $\overline{o}_P$, and that, if it is so regarded, then $\sum_{i=1}^{m} o^*_P \cdot \omega_i$ coincides with $(\overline{o}_P)^*$ (Chevalley [1], Proposition 7, p. 699). Hence we can write:

$$(5) \qquad (\overline{o}_P)^* = o^*_{\overline{P}_1} \oplus o^*_{\overline{P}_2} \oplus \cdots \oplus o^*_{\overline{P}_h} = \sum_{i=1}^{m} (o_P)^* \cdot \omega_i .$$

Since o_P and $\overline{o}_P$ have the same quotient field Σ, there exists an element $c \neq 0$ in o_P such that $c \cdot \omega_i \in o_P$, $i = 1, 2, \cdots, m$. Then it follows from (5) that $c \cdot (\overline{o}_P)^* \subset o^*_P$, and since no element c of $\overline{o}_P$, different from zero, is a zero divisor in $(\overline{o}_P)^*$ (Chevalley [1], Proposition 6, p. 699), we conclude that o^*_P <u>is an integral domain if and only if</u> $(\overline{o}_P)^*$ <u>is an integral domain</u>. Now, by the analytical irreducibility of normal varieties, the rings $o_{\overline{P}_i}$ are integral domains. We therefore conclude that o^*_P is an integral domain, if and only if h = 1, i.e., we have the following theorem:

THEOREM 8. A variety V is analytically irreducible at a point P if and only if the points which correspond to P on a derived normal model of V are isomorphic over k.

The condition of Theorem 8 is purely of a local character. This condition can be also stated equivalently in either one of the following forms:

(a) The integral closure $\overline{o}_P$ of the local ring o_P is still a local ring;

(b) to the irreducible subvariety W of V whose general point is P there corresponds on a derived normal model $\overline{V}$ of V an irreducible variety $\overline{W}$.

Here, if T denotes the birational transformation of V into $\overline{V}$, we mean by the variety which corresponds to W on $\overline{V}$ the variety T[W] (see Zariski [6], p. 519), not the total transform T{W} of W. This latter variety T{W} may have irreducible components which correspond to proper subvarieties of W, and so may be reducible even if T [W] is irreducible.(10)

Assume now that V is analytically reducible at P. Then, by (b), T[W] has at least two distinct irreducible components, say $\overline{W}_1$ and $\overline{W}_2$. To every point P_o of W there will correspond at least one point P_{oi} on $\overline{W}_i$, i=1, 2. If P_{o1} and P_{o2} are k-isomorphic points then they belong to $\overline{W}_1 \cap \overline{W}_2$. Since to $\overline{W}_1 \cap \overline{W}_2$ there corresponds on V a proper subvariety of W, we conclude by Theorem 8 that if V is analytically reducible at a point P, then V is also analytically reducible at almost all points (and in particular- at almost all algebraic points) of the subvariety W of V whose general point is P. If then V is analytically irreducible at almost all algebraic points of W, then V is analytically irreducible at almost all points of W (in particular- at each general point of W), and if V is analytically irreducible at all algebraic points of W, then V is analytically irreducible at every point of W. This shows that in the case of finite integral domains the second part of Lemma 3 (§6) remains true if we only require that R be analytically irreducible at all the maximal prime ideals $\mathfrak{p}_o$ containing $\mathfrak{p}$. The following corresponding questions for arbitrary Noetherian domains R are left open: (1) If R is analytically irreducible at all maximal prime ideals $\mathfrak{p}_o$ which contain a given prime ideal $\mathfrak{p}$, is R also analytically irreducible at $\mathfrak{p}$? (2) Under the same assumption as in (1), is it true that if $\mathfrak{p}_1$ is a prime ideal containing $\mathfrak{p}$, then high symbolic powers of $\mathfrak{p}_1$ are contained in high ordinary powers of $\mathfrak{p}$? (If the answer to (1) is in the affirmative, then, of course, also the answer to (2) is in the affirmative, in view of Lemma 3, §6.

As a final application we shall now derive a necessary and sufficient condition in order that, given two prime ideals $\mathfrak{p}_1 \subset \mathfrak{p}$ in a finite integral domain R, high symbolic powers of $\mathfrak{p}_1$ be contained in high symbolic powers of $\mathfrak{p}$. We denote by $\overline{R}$ the integral closure of R in its quotient field.

THEOREM 9. A necessary and sufficient condition that high symbolic powers of $\mathfrak{p}_1$ be contained in high symbolic powers of $\mathfrak{p}$

is that every prime ideal $\bar{\mathfrak{p}}$ in $\bar{R}$ which lies over $\mathfrak{p}$ (i.e., which is such that $\bar{\mathfrak{p}} \cap R = \mathfrak{p}$) contain at least one prime ideal $\bar{\mathfrak{p}}_1$ in $\bar{R}$ which lies over $\mathfrak{p}_1$.[11]

PROOF. Let $\bar{\mathfrak{p}}_1, \bar{\mathfrak{p}}_2, \cdots, \bar{\mathfrak{p}}_g$ be the prime ideals of $\bar{R}$ which lie over $\mathfrak{p}_1$ (the number of the ideals $\bar{\mathfrak{p}}_i$ is finite, since $\bar{R}$ is a finite R-module). Let $\bar{\mathfrak{A}}_j = \bar{\mathfrak{p}}_1^{(j)} \cap \bar{\mathfrak{p}}_2^{(j)} \cap \cdots \cap \bar{\mathfrak{p}}_g^{(j)}$. The intersection O of the quotient rings of the ideals $\bar{\mathfrak{p}}_1, \bar{\mathfrak{p}}_2, \cdots, \bar{\mathfrak{p}}_g$ in $\bar{R}$ is a finite module over the quotient ring of $\mathfrak{p}_1$ in R, and if $\mathfrak{m}$ is the intersection of the maximal ideals in the semilocal ring O, then $\mathfrak{m}^j \cap \bar{R} = \bar{\mathfrak{A}}_j$. It follows therefore (Chevalley [1], Lemma 7, p. 695) that $\bar{\mathfrak{A}}_j \cap R \subset \mathfrak{p}_1^{(n_j)}$, where $n_j \to +\infty$ as $j \to +\infty$. Let $c \neq 0$ be an element of R such that $c\bar{R} \subset R$. Then

$$\text{(6)} \qquad c \cdot \bar{\mathfrak{A}}_j \subset \bar{\mathfrak{A}}_j \cap R \subset \mathfrak{p}_1^{(n_j)} .$$

Now let us assume that high symbolic powers of $\mathfrak{p}_1$ are contained in high symbolic powers of $\mathfrak{p}$, and let $\bar{\mathfrak{p}}$ be any prime ideal in $\bar{R}$ which lies over $\mathfrak{p}$ From (6) it follows then that $c \cdot \bar{\mathfrak{A}}_j \subset \mathfrak{p}^{(\nu_j)}$, where $\nu_j \to +\infty$ as $j \to +\infty$. From $\bar{\mathfrak{p}} \cap R = \mathfrak{p}$ follows directly that $\mathfrak{p}^{(i)} \subset \bar{\mathfrak{p}}^{(i)}$ for all i. Hence $c \cdot \bar{\mathfrak{A}}_j \subset \bar{\mathfrak{p}}^{(\nu_j)}$, and therefore by Lemma 9 of Chevalley [1] (p. 699), applied to the quotient ring of $\bar{\mathfrak{p}}$ in $\bar{R}$, it follows that $\bar{\mathfrak{A}}_j \subset \bar{\mathfrak{p}}^{(s_j)}$, where $s_j \to +\infty$ as $j \to +\infty$. We shall only use a very weak consequence of this result, namely that $\bar{\mathfrak{A}}_j \subset \bar{\mathfrak{p}}$ if j is sufficiently high. This implies at once that one of the prime ideals $\bar{\mathfrak{p}}_1, \bar{\mathfrak{p}}_2, \cdots, \bar{\mathfrak{p}}_g$ must be contained in $\bar{\mathfrak{p}}$, and proves that the condition stated in the theorem is necessary.

So far we have not used the assumption that $\bar{R}$ is the integral closure of R in the quotient field of R, but only that $\bar{R}$ is a finite R-module. We shall use the assumption that $\bar{R}$ is the integral closure of R in the proof that the condition is sufficient. Since $\bar{\mathfrak{p}}_i \cap R = \mathfrak{p}_1$ for $i = 1, 2, \cdots, g$, we have $\mathfrak{p}_1^{(n)} \subset \bar{\mathfrak{p}}_i^{(n)}$, $i = 1, 2, \cdots, g$, whence $\mathfrak{p}_1^{(n)} \subset \bar{\mathfrak{A}}_n$. Let $\bar{\mathfrak{p}}_{o1}, \bar{\mathfrak{p}}_{o2}, \cdots, \bar{\mathfrak{p}}_{oh}$ be the prime ideals in $\bar{R}$ which lie over $\mathfrak{p}$. Each $\bar{\mathfrak{p}}_{oj}$ contains, by assumption, at least one of the g ideals $\bar{\mathfrak{p}}_i$, and if $\bar{\mathfrak{p}}_{oj} \supset \bar{\mathfrak{p}}_i$ then, by Theorem 7, high symbolic powers of $\bar{\mathfrak{p}}_1$ are contained in high symbolic powers of $\bar{\mathfrak{p}}_{oj}$. From $\mathfrak{p}_1^{(n)} \subset \mathfrak{A}_n$ we conclude therefore that $\mathfrak{p}_1^{(n)} \subset \bigcap_{j=1}^{h} \bar{\mathfrak{p}}_{oj}^{(\nu_n)}$, where $\nu_n \to +\infty$ as $n \to +\infty$. If we denote by $\bar{\mathfrak{B}}_\nu$ the ideal $\bigcap_{j=1}^{h} \bar{\mathfrak{p}}_{oj}^{(\nu)}$, then $\bar{\mathfrak{B}}_\nu \cap R$ is contained in a high symbolic power of $\mathfrak{p}$ if ν is very large (see the argument used above in order to show that $\bar{\mathfrak{A}}_j \cap R$ is contained in a high symbolic power of $\mathfrak{p}_1$ if j is very large). Hence $\mathfrak{p}_1^{(n)}$ is contained in a high symbolic power of $\mathfrak{p}$ if n is very large. This completes the proof of the theorem.

§8. SOME LEMMAS ON $\mathfrak{m}$-ADIC RINGS.

In this section we shall extend to $\mathfrak{m}$-adic rings some results on local rings due to Chevalley [1]. In the next section these extended results will be applied to the theory of holomorphic functions on affine models. For the sake of simplicity we shall confine ourselves to $\mathfrak{m}$-adic domains. With slight modifications, indicated below, the proofs are valid for $\mathfrak{m}$-adic rings having proper zero divisors, but in the applications in which we are immediately interested only $\mathfrak{m}$-adic domains will occur.

Let R be a Noetherian domain and let $\mathfrak{m}$ be an arbitrary ideal[12] in R. Let R* denote the $\mathfrak{m}$-adic completion of R.

LEMMA 1. If $c \neq 0$ is any element of R[13], then $\mathfrak{m}^n : Rc \subset \mathfrak{m}^{\nu(n)}$, where $\nu(n) \to +\infty$ as $n \to +\infty$. Furthermore, c is not a zero divisor in R*.

PROOF. The lemma has been proved by Chevalley ([1], Lemma 9, p. 699) in the case in which R is a local ring and $\mathfrak{m}$ is the ideal of non-units in R. In the general case we first observe that if $\mathfrak{p}$ is any prime ideal in R, then by passing to the quotient ring $R_{\mathfrak{p}}$ [14] and using Chevalley's result it follows that $\mathfrak{p}^n : Rc$ is contained in a symbolic power of $\mathfrak{p}$ whose exponent increases indefinitely with n. Now let ν be a fixed integer and let $\mathfrak{m}^\nu = \mathfrak{q}_1 \cap \mathfrak{q}_2 \cap \cdots \cap \mathfrak{q}_h$ be a normal decomposition of $\mathfrak{m}^\nu$ into primary components. Let $\mathfrak{p}_i$ be the prime ideal of $\mathfrak{q}_i$ and let ρ_i be the exponent of $\mathfrak{q}_i$. Let n be a sufficiently high integer such that $\mathfrak{p}_i^n : Rc \subset \mathfrak{p}_i^{(\rho_i)}$, for $i = 1, 2, \cdots, h$. Since $\mathfrak{p}_i^{(\rho_i)} \subset \mathfrak{q}_i$ and $\mathfrak{m} \subset \mathfrak{p}_i$, we have $\mathfrak{m}^n : Rc \subset \mathfrak{p}_i^n : R.c \subset \mathfrak{q}_i$, and hence $\mathfrak{m}^n : R.c \subset \mathfrak{m}^\nu$. This completes the proof of the first part of the lemma.

Let $cu^* = 0$, $u^* \in R^*$. We can find a sequence $\{u_n\}$ in R such that $u^* - u_n \in \mathfrak{m}^{*n}$, where $\mathfrak{m}^* = R^*\mathfrak{m}$ We have, then, $cu_n = c(u_n - u^*) \in \mathfrak{m}^{*n} \cap R$, and $\mathfrak{m}^{*n} \cap R = \mathfrak{m}^n$ by Zariski [7], p. 9. Hence $cu_n \in \mathfrak{m}^n$, and thus, by the first part of the lemma, $\lim u_n = 0$, i.e. $u^* = 0$. This completes the proof of the lemma.

LEMMA 2. Let R and R' be Noetherian domains such that R is a subring of R' and R' is a finite R- module. Let $\mathfrak{m}$ be an ideal in R and let $\mathfrak{m}' = R'.\mathfrak{m}$. Then the $\mathfrak{m}$-adic ring R is a subspace of the $\mathfrak{m}'$-adic ring R', and the completion R* of the $\mathfrak{m}$-adic ring R is a subring and a subspace of the completion R'* of the $\mathfrak{m}'$-adic ring R'. Moreover, if $R' = \sum_{i=1}^m R.\,y_i$, then $R'^* = \sum_{i=1}^m R^*.\,y_i$.

The proof is the same as that of Proposition 7 in Chevalley ([1], p. 699). It is only necessary to point out that Chevalley's use of his Lemma 7 (which does not extend to arbitrary $\mathfrak{m}$-adic rings) is irrelevant for the proof of our lemma. For since R' is a finite R-module, we can find an element c in R, $c \neq 0$, and r elements $x_1, x_2, \cdots, x_r$ in R', $(x_1 = 1)$, which are linearly independent over R, such that $cR' \subset \sum_{i=1}^r Rx_i$ (see Chevalley [1], proof of Lemma 4 on p. 694). We will have then $c\,\mathfrak{m}'^n \subset \sum_{i=1}^r \mathfrak{m}^n x_i$, $c(\mathfrak{m}'^n \cap R) \subset \mathfrak{m}^n$, i.e., $\mathfrak{m}'^n \cap R \subset \mathfrak{m}^n : Rc \subset \mathfrak{m}^{\nu(n)}$, where $\nu(n) \longrightarrow \infty$. The rest of Chevalley's proof carries over without any changes whatsoever.

Our next- and final- lemma concerns arbitrary Noetherian rings.

LEMMA 3. Let $\mathfrak{A}$ and $\mathfrak{B}$ be two ideals in a Noetherian ring R and let S denote the set of all prime ideals in R which contain the ideal $\mathfrak{A} + \mathfrak{B}$. If $\mathfrak{p} \in S$ and i is an integer ≥ 1, let $\mathfrak{q}_i(\mathfrak{p})$ denote the primary component of $\mathfrak{A} + \mathfrak{p}^{(i)}$ which has $\mathfrak{p}$ as associated prime ideal . Then $\bigcap_{\mathfrak{p} \in S} \mathfrak{q}_i(\mathfrak{p}) \subset \mathfrak{A} + \mathfrak{B}^{\nu(i)}$, where $\nu(i) \longrightarrow \infty$ as $i \longrightarrow \infty$.

PROOF. We first consider the case $\mathfrak{A} = (0)$. In this case we have $\mathfrak{q}_i(\mathfrak{p}) = \mathfrak{p}^{(i)}$, and the lemma asserts that $\bigcap_{\mathfrak{p} \in S} \mathfrak{p}^{(i)} \subset \mathfrak{B}^{\nu(i)}$. For any given integer n, let $\mathfrak{q}_{1n} \cap \mathfrak{q}_{2n} \cap \cdots \cap \mathfrak{q}_{s_n n}$ be a normal decomposition of $\mathfrak{B}^n$ into primary components, let $\mathfrak{p}_{jn}$ be the prime ideal of $\mathfrak{q}_{jn}$ (whence $\mathfrak{p}_{jn} \in S$) and let ρ_{jn} be the exponent of $\mathfrak{q}_{jn}$. For a given i, let $\nu(i)$ be the greatest of all the integers n with the property: $\rho_{jn} \leq i$, $j = 1,2,\ldots,s_n$. It is evident that $\nu(i) \longrightarrow \infty$ as $i \longrightarrow \infty$. We have: $\mathfrak{p}_{jn}^{(i)} \subset \mathfrak{p}_{jn}^{(\rho_{jn})} \subset \mathfrak{q}_{jn} \subset \mathfrak{B}^n$, for all $n \leqq \nu(i)$, and this proves our assertion.

In the general case we pass to the residue class ring $\bar{R} = R/\mathfrak{a}$ and to the ideal $\bar{\mathfrak{B}} = (\mathfrak{a}+\mathfrak{B})/\mathfrak{a}$. If τ denotes the natural homomorphism of R onto $\bar{R}$ and if $\bar{S}$ is the set of all prime ideals in $\bar{R}$ which contain $\bar{B}$, there is a (1,1) correspondence between the ideals $\mathfrak{p}$ in S and the ideals $\bar{\mathfrak{p}}$ in $\bar{S}$, each ideal $\mathfrak{p}$ in S being the full inverse image $\bar{\mathfrak{p}}\tau^{-1}$ of the corresponding prime ideal $\bar{\mathfrak{p}}$. Furthermore, it is easily seen that $(\bar{\mathfrak{p}}^{(i)})\tau^{-1} = \mathfrak{q}_i(\mathfrak{p})$. Our lemma then follows at once from the first part of the proof.

COROLLARY 1. The notation being as in Lemma 3, let it be assumed that the Noetherian ring R is an integral domain. Then $\bigcap_{\mathfrak{p}\in S}\{(R_{\mathfrak{p}}\mathfrak{a} + \mathfrak{m}_{\mathfrak{p}}^i) \cap R\} \subset \mathfrak{a}+\mathfrak{B}^{\nu(i)}$, where $R_{\mathfrak{p}}$ is the quotient ring of R with respect to $\mathfrak{p}$ and $\mathfrak{m}_{\mathfrak{p}}$ is the maximal ideal in R .

For by known properties of quotient rings, we have $R_{\mathfrak{p}}\mathfrak{a} + \mathfrak{m}_{\mathfrak{p}}^i = R_{\mathfrak{p}}(\mathfrak{a} + \mathfrak{p}^{(i)})$, and hence the intersection of $R_{\mathfrak{p}}\mathfrak{a} + \mathfrak{m}_{\mathfrak{p}}^i$ with R is equal to the intersection of those primary components of $\mathfrak{a} + \mathfrak{p}^{(i)}$ whose associated prime ideals are contained in $\mathfrak{p}$. If $\mathfrak{p} \in S$ then $\mathfrak{p}$ is the only isolated prime ideal of $\mathfrak{a} + \mathfrak{p}^{(i)}$, and hence the above intersection is $\mathfrak{q}_i(\mathfrak{p})$.

§ 9. HOLOMORPHIC FUNCTIONS ON AFFINE MODELS.

Let V be an irreducible variety (over k) in an affine space and let R be the non-homogeneous coordinate ring of V. Let W be a subvariety of V and let $\mathfrak{m}$ be the ideal of W in R. Our object in this section is to prove the following theorem:

THEOREM 10. Every function on V which is holomorphic along W is necessarily strongly holomorphic along W. In other words (see Theorem 1, § 2), the ring $\mathfrak{o}_W^*$ can be identified with the $\mathfrak{m}$-adic completion R* of R.

The proof will be divided into several parts.

a). Let V be a variety in the affine or in the projective space and let $\bar{V}$ be a derived normal model of V. Let H be an arbitrary subset of V and let $\bar{H}$ be the total transform of H under the birational transformation T of V into $\bar{V}$. If $P \in H$ and if $\bar{P}_1, \bar{P}_2, \cdots, \bar{P}_g$ are the points of $\bar{V}$ which correspond to P under T, then we have that $\mathfrak{o}_P \subset \mathfrak{o}_{\bar{P}_i}$ and $\mathfrak{m}_P \subset \mathfrak{m}_{\bar{P}_i}$. Hence if a sequence $\{\xi_j\}$ of elements of the field Σ converges uniformly on a subset G of H (in the sense of Definition 1, § 1), it also converges uniformly on the corresponding subset $\bar{G}$ of $\bar{H}$, where $\bar{G} = T\{G\}$. Moreover, if we denote by ξ^* and by $\bar{\xi}^*$ the limit of $\{\xi_i\}$ on G and $\bar{G}$ respectively, then $\xi^*_P = 0$ implies $\bar{\xi}^*_{\bar{P}_i} = 0$, $i = 1, 2, \ldots g$. Conversely, if $\bar{\xi}^*_{\bar{P}} = 0$, for $i = 1, 2, \cdots, g$, then $\xi^*_P = 0$, since $\bigcap_{i=1}^g \mathfrak{m}^n_{\bar{P}_i} \cap o_P \subset \mathfrak{m}_P^{\nu_n}$ with $\nu_n \to +\infty$, as $n \to +\infty$. It follows that the ring o^*_H of holomorphic functions along H can be identified in a natural fashion with a subring of the ring $o^*_{\bar{H}}$ of holomorphic functions along $\bar{H}$. We shall therefore regard o^*_H as a subring of $o^*_{\bar{H}}$. In particular, we shall regard o^*_P as a subring of the ring $o^*_{\{\bar{P}_1, \bar{P}_2, \ldots, \bar{P}_g\}}$ (in this case H consists of the single point P). If ξ^* is any element of o^*_H and if the points $P \in H$, $\bar{P}_1, \bar{P}_2, \ldots \bar{P}_g$ are as indicated above, then ξ^*_P is to be regarded as a holomorphic function along the set $\{\bar{P}_1, \bar{P}_2, \ldots \bar{P}_g\}$. The analytical element of ξ^*_P at $\bar{P}_i$ ($i = 1, 2, \ldots g$) is then merely the analytical element $\xi^*_{\bar{P}_i}$ of ξ^*, when ξ^* is regarded as a holomorphic function along $\bar{H}$.

b). In this part of the proof we shall assume that V is a normal variety (in the affine space) and that W is irreducible. We proceed to establish our theorem under these assumptions.

Let ξ^* be any element of o^*_W. There exists then a set $\Gamma_1, \Gamma_2, \ldots, \Gamma_h$ of proper subvarieties of W such that $\bigcap_{\alpha=1}^h \Gamma_\alpha = \emptyset$ and such that ξ^* is strongly holomorphic along $W - \Gamma_\alpha$ for $\alpha = 1, 2, \ldots h$. Let then ξ^* [more precisely: $\tau_{W, W-\Gamma_\alpha}(\xi^*)$] be the limit of a sequence $\{\xi_{1\alpha}, \xi_{2\alpha}, \ldots\}$ which converges uniformly on $W - \Gamma_\alpha$. Replacing, if necessary, each of the h sequences $\{\xi_{i\alpha}\}$ by a suitable subsequence, we may assume that

(7) $\xi_{i\alpha} - \xi_{j\alpha} \in \mathfrak{m}_P^i$, for $j \geqq i$ and all P in $W - \Gamma_\alpha$;

(7') $\xi_{i\alpha} - \xi_{i\beta} \in \mathfrak{m}_P^i$, for $\alpha, \beta = 1, 2, \ldots, h$ and all P in $W - \Gamma_\alpha - \Gamma_\beta$.

We denote by $\mathfrak{m}$ the prime ideal of W in the non-homogeneous coordinate ring R of V. We set $\Omega = R_{\mathfrak{m}}$ = quotient ring of W; $\mathfrak{M} = \Omega \cdot \mathfrak{m}$ = ideal of non units in Ω. Any general point P of W belongs to each set $W - \Gamma_\alpha$, and for such a point P we have $\mathfrak{m}_P = \mathfrak{M}$. Hence we have by (7) and (7'):

(8) $\xi_{i\alpha} - \xi_{j\alpha} \in \mathfrak{M}^i$, for $j \geqq i$ ($\alpha = 1, 2, \ldots, h$);

(8') $\xi_{i\alpha} - \xi_{i\beta} \in \mathfrak{M}^i$, all i ($\alpha, \beta = 1, 2, \ldots, h$).

We fix one of the varieties Γ_α, say Γ_1. We proceed to show that there exists a sequence of elements $u_1, u_2, \ldots, u_i, \ldots$ in R such that

(9) $\xi_{i1} - u_i \in \mathfrak{M}^i$, all i.

Since ξ_{i1} belongs to the local ring of each point of $W - \Gamma_1$, ξ_{i1} certainly belongs to the quotient ring Ω of W. We may therefore write

(10) $\xi_{i1} = A_{i1}/B_{i1}$,

where $A_{i1}, B_{i1} \in R$ and

(11) $B_{i1} \notin \mathfrak{m}$.

Let P be any point of W. For some α we have $P \in W - \Gamma_\alpha$, $\xi_{i\alpha} \in o_P$. Since $\xi_{i1} - \xi_{i\alpha} \in \mathfrak{M}^i$ by (8'), it follows from (10) that $A_{i1} = B_{i1}\xi_{i\alpha} + \beta_{i\alpha}$, where $\beta_{i\alpha}$ belong to $\mathfrak{M}^i$ and to o_P and therefore also to $o_P \cdot \mathfrak{m}^{(i)}$. Hence

$$A_{i1} \in R \cap o_P (RB_{i1} + \mathfrak{m}^{(i)}).$$

Since this inclusion holds for all points P of W, we conclude [15] that $A_{i1} \in RB_{i1} + \mathfrak{m}^{(i)}$. We can therefore write $A_{i1} = u_i B_{i1} + \pi_i$ where $u_i \in R$ and $\pi_i \in \mathfrak{m}^{(i)}$. We have

(12) $\xi_{i1} - u_i = \pi_i / B_{i1}$, $\pi_i \in \mathfrak{m}^{(i)}$,

and hence, in view of (11), the sequence $\{u_i\}$ satisfies condition (9).

From (8), for $\alpha = 1$, and from (9) it follows that $u_i - u_j \in \mathfrak{M}^i$ if $j \geq i$, i.e., $u_i - u_j \in \mathfrak{m}^{(i)}$, since the u_i are in R and since $\mathfrak{M}^i \cap R = \mathfrak{m}^{(i)}$. Since R is integrally closed we have by Theorem 7, § 7 that high symbolic powers of the prime ideal $\mathfrak{m}$ are contained in high ordinary powers of $\mathfrak{m}$. Hence the sequence $\{u_i\}$ is a Cauchy sequence in the $\mathfrak{m}$-adic ring R and therefore converges uniformly to a u^* on V <u>which is strongly holomorphic along</u> W. The two functions ξ^* and u^* are both holomorphic along W, and by (9) these two functions have the same analytical element at some point of W (namely at each general point of W). Since W is irreducible (hence connected) and V is normal (hence analytically irreducible at each point), it follows from Theorem 6, § 6, that $\xi^* = u^*$, and this proves our theorem in the case now under consideration.

c) Our next step will be to prove that if our theorem is true for irreducible subvarieties W (whether V is normal or not), then it also is true for any reducible W. This, incidentally, will complete the proof of the theorem for normal varieties V.

We shall proceed by induction with respect to the number of irreducible components of W. Let W_1 be one of the irreducible components of W and let W_2 be the union of the remaining irreducible components. Let ξ^* be any element of o_W^* and let ξ_α^* be the projection of ξ^* into $o_{W_\alpha}^*$ (α=1, 2). Let $\mathfrak{m}_\alpha$ be the ideal of W_α in the coordinate ring R of the affine variety V. By our induction hypothesis, ξ_α^* is a strongly holomorphic function on W_α, <u>in fact an element of the completion of the $\mathfrak{m}_\alpha$-adic ring</u> R (see Theorem 1, § 2). Hence we can write $\xi_\alpha^* = \lim \xi_{i\alpha}$, where $\{\xi_{i\alpha}\}$ is a Cauchy sequence in the $\mathfrak{m}_\alpha$-adic ring R (α= 1,2). We may assume that

$$\left.\begin{array}{ll} (13) & \xi_{i1} - \xi_{j1} \in \mathfrak{m}_1^i \\ (13') & \xi_{i2} - \xi_{j2} \in \mathfrak{m}_2^i \end{array}\right\}, \text{ for } j \geqq i.$$

At each common point P of $W_1 \cap W_2$ (if such points exist) the holomorphic functions ξ_1^* and ξ_2^* must have the same analytical element, namely ξ_P^*. Hence, replacing, if necessary, each of the given sequences by suitable subsequences, we may assume, in view of (13) and (13'), that

$$(14) \qquad \xi_{j1} - \xi_{j2} \in \mathfrak{p}^{(j)},$$

<u>for all prime ideals $\mathfrak{p}$ which contain the ideal</u> $\mathfrak{m}_1 + \mathfrak{m}_2$ (if such prime ideals exist at all; i.e., if $\mathfrak{m}_1 + \mathfrak{m}_2$ is not the unit ideal.)

Combining (13), (13') and (14) we find, for any given prime ideal $\mathfrak{p}$ containing $\mathfrak{m}_1^i + \mathfrak{m}_2^i$, that

$$\xi_{i1} - \xi_{i2} \in \bigcap_{j=1}^{\infty} (\mathfrak{m}_1^i + \mathfrak{m}_2^i + \mathfrak{p}^{(j)}).$$

Passing to the quotient ring $R_\mathfrak{p}$ of $\mathfrak{p}$ and recalling that in the local ring $R_\mathfrak{p}$ we have for any ideal $\mathfrak{a}$: $\mathfrak{a} = \bigcap_{j=1}^{\infty}(\mathfrak{a} + \mathfrak{P}^j)$, where $\mathfrak{P} = R_\mathfrak{p}\cdot\mathfrak{p}$ we conclude that $\xi_{i1} - \xi_{i2} \in R_\mathfrak{p}\cdot(\mathfrak{m}_1^i + \mathfrak{m}_2^i)$. Since this holds for all prime ideals $\mathfrak{p}$ containing $\mathfrak{m}_1^i + \mathfrak{m}_2^i$ and since $\xi_{i1} - \xi_{i2} \in R$, it follows (see footnote 15) that $\xi_{i1} - \xi_{i2} \in \mathfrak{m}_1^i + \mathfrak{m}_2^i$. Let then $\xi_{i1} - \xi_{i2} = \pi_{i2} - \pi_{i1}$, where $\pi_{i\alpha} \in \mathfrak{m}_\alpha^i$, α= 1, 2, and let $u_i = \xi_{i1} + \pi_{i1} = \xi_{i2} + \pi_{i2}$. We have by (13) and (13'): $u_i - u_j = (\xi_{i1} - \xi_{j1}) + (\pi_{i1} - \pi_{j1}) \in \mathfrak{m}_1^i$ if $j \geqq i$. Similarly $u_i - u_j \in \mathfrak{m}_2^i$ if $j \geqq i$. Hence the sequence $\{u_i\}$ converges uniformly on $W_1 \cup W_2$, i.e., on W, to a strongly holomorphic function u^*. Since $u_i - \xi_{i1} = \pi_{i1} \in \mathfrak{m}_1^i$, we have $u_P^* = \xi_P^*$ for all point P of W_1. Similarly $u_P^* = \xi_P^*$ on W_2. Hence $u^* = \xi^*$, and this completes the proof of our assertion.

d). We shall now prove our theorem for an arbitrary affine variety V, and an arbitrary W. Let $\bar{V}$ be a derived normal affine model of V and let R be the non-homogeneous coordinate ring of V. The non-homogeneous coordinate ring of $\bar{V}$ is then the integral closure $\bar{R}$ of R, and $\bar{R}$ is a finite R-module. Let $\mathfrak{m}$ be the ideal of W in R and let $\bar{\mathfrak{m}} = \bar{R}\mathfrak{m}$. Then the variety of $\bar{\mathfrak{m}}$ is the total transform $\bar{W}$ of W under the birational transformation of V into $\bar{V}$, and since the radical of $\bar{\mathfrak{m}}$ is the ideal of $\bar{W}$, it follows that the ring of functions on $\bar{V}$ which are strongly holomorphic along $\bar{W}$ is the $\bar{\mathfrak{m}}$-adic completion $\bar{R}^*$ of $\bar{R}$. By parts b) and c) of the proof we conclude that $o^*_{\bar{W}} = \bar{R}^*$, since $\bar{V}$ is normal.

Let ξ^* be any element of o^*_W. By part a) of the proof we have $o^*_W \subset o^*_{\bar{W}}$, and hence, by part b), the element ξ^* belongs to $\bar{R}^*$. Let $c \neq 0$ be an element of R such that $c\bar{R} \subset R$. Then, by Lemma 5 (§ 8), we have $c.\bar{R}^* \subset R^*$, and, in particular, $c\xi^* \in R^*$. Let, therefore, $c\xi^* = \lim \omega_i$, where $\{\omega_i\}$ is a Cauchy sequence in $\mathfrak{m}$-adic ring R. We may assume without loss of generality

(15) $\omega_i - \omega_j \in \mathfrak{m}^i$, for $j \geq i$.

Let P be any point of W and let $\{\xi_{1P}, \xi_{2P}, \cdots\}$ be a Cauchy sequence in o_P whose limit is ξ^*_P. We may assume that $\xi_{iP} - \xi_{jP} \in \mathfrak{m}^i_P$, if $j \geq i$. Since the two sequences $\{\omega_1, \omega_2, \cdots\}$ and $\{c\xi_{1P}, c\xi_{2P}, \cdots\}$ are equivalent at P, it follows from (15) that $\omega_i - c\xi_{iP} \in \mathfrak{m}^i_P$. Hence $\omega_i \in o_P\, c + \mathfrak{m}^i_P$. Since this holds for any point P of W, we conclude, by Lemma 3, Corollary 1 (§ 8) that $\omega_i \in R.c + \mathfrak{m}^{\vartheta(i)}$, where $\vartheta(i) \to \infty$ as $i \to \infty$. Hence for each i = 1, 2, ... there exists an element u_i in R such that the sequence $\{cu_1, cu_2, \cdots\}$ is a Cauchy sequence in the $\mathfrak{m}$-adic ring R and is equivalent to the sequence $\{\omega_1, \omega_2, \cdots\}$. It follows at once from Lemma 2 (§8) that also $\{u_1, u_2, \cdots\}$ is a Cauchy sequence in the $\mathfrak{m}$-adic ring R, and its limit is therefore a function u^* which is strongly holomorphic along W. We have, by assumption, $\omega^* = c\xi^*$, and also $\omega^* = cu^*$. Hence $c(\xi^* - u^*) = 0$. At any point P of W the two holomorphic functions ξ^* and u^* have therefore the same analytical element, since c is not a zero divisor in o^*_P. Hence $\xi^* = u^*$, i.e., ξ^* is strongly holomorphic along W, q.e.d.

We point out the following application to varieties V in the projective space S_n. Let W be a subvariety of V and let $H_0, H_1, \ldots, H_s$ be s + 1 hyperplanes in S_n such that $W \cap H_0 \cap H_1 \cap \cdots \cap H_s = \emptyset$. (We can take, for instance, n + 1 independent hyperplanes in S_n) . Let R_i, i= 0,1,...,s, be a non-homogeneous coordinate ring of the affine variety $V - H_i$, and let $\mathfrak{m}_i$ be the ideal of $W - H_i$ in the ring R_i. Let R_i be the $\mathfrak{m}_i$-adic completion of R_i. If ξ^* is any function on V which is holomorphic along W, its projection ξ_i^* in $O^*_{W-H_i}$ is, by the theorem just proved, an element of R_i^*. Conversely, if an element ξ^* of O_W^* is such that the above projection ξ_i^* belongs to R_i^* for $i = 0, 1, \cdots, s$, then $\xi^* \in o_W^*$. We conclude that o_W^* consists of those and only those elements of O_W^* whose projections in $o^*_{W-H_0}, o^*_{W-H_1}, \cdots, o^*_{W-H_s}$ belong respectively to $R_0^*, R_1^*, \cdots, R_s^*$. In this sense one may think of o_W^* as the "intersection" of the s + 1 complete rings R_i^*. It is thus seen that the unspecified subvarieties Γ_α which appear in our definition of holomorphic functions (Definition 2, §3) can always be identified -without loss of generality - with suitable hyperplane sections of V, and that the various sequences $\{\xi_{1\alpha}, \xi_{2\alpha}, \cdots\}$ which together define a given holomorphic function along W can be assumed to be Cauchy sequences in the corresponding $\mathfrak{m}$-adic non-homogeneous coordinate rings R_α.

Part c) of the proof of the last theorem established in fact if W is a subvariety of an affine model V and if a function ξ^* on V, defined along W, is strongly holomorphic along each irreducible component of W, then ξ^* is also (strongly) holomorphic along W. We shall have to use in Part II this same result for holomorphic functions, i.e. the following theorem:

THEOREM 11. Let W be a subvariety of a (projective or affine) model V. If a function ξ^* on V, defined along W, is holomorphic along each irreducible component of W, then ξ^* is also holomorphic along W.

PROOF. In the affine case the theorem follows at once from Theorem 10 and from the result just stated above. In the projective case we merely observe that if H is any hyperplane, then the assumptions of the theorem imply that ξ^* is holomorphic along each irreducible component of the affine variety W- H. Hence, by the affine case of the theorem, ξ^* is holomorphic along W - H. Since this is so for every choice of the hyperplane H, the theorem follows from the corollary at the end of §3.

In the course of part d) of the above proof we came very close to proving the following theorem:

THEOREM 12. Let W be a subvariety of a projective or affine model V and let c be a rational function on V which is holomorphic along W ($c \neq 0$). If a function ω^* in o_W^* is such that c divides ω^* locally at each point P of W (i.e., if c divides ω_P^* in o_P^*), then c divides ω^* in o_W^*.

PROOF. We first consider the case of an affine model V. In this case we know, by Theorem 10, that ω^* is strongly holomorphic along W, but it happens that this knowledge does not contribute anything to the proof of the present theorem, and for this reason we shall not make use of Theorem 10. Instead we shall first <u>assume</u> that ω^* is strongly holomorphic along W. Then, by assumption, ω^* is the limit of a Cauchy sequence $\{\omega_1, \omega_2, \ldots\}$ in the $\mathfrak{m}$-adic ring R, where R is the non-homogeneous coordinate ring of V and $\mathfrak{m}$ is the R-ideal of W. At each point P of W we have, by assumption, a Cauchy sequence $\{\xi_{1P}, \xi_{2P}, \ldots\}$ in o_P^* such that the two sequences $\{\omega_i\}$ and $\{c\xi_{iP}\}$ are equivalent in o_P^*. It follows as in the part c) of the proof of Theorem 10, that $\omega_i \in o_P c + \mathfrak{m}_P^i$ for all points P of W, and that therefore $\omega_i \in R\cdot c + \mathfrak{m}^{\nu(i)}$, where $\nu(i) \to \infty$. We reach the same conclusion as in the quoted part of the proof of Theorem 10, namely that $\omega^* = cu^*$, where u^* is a function on V which is strongly holomorphic along W.

In the general case- and that includes the affine as well as the projective case- we proceed as follows. Let $W - \Gamma_1, W - \Gamma_2, \cdots, W - \Gamma_g$ be a finite open covering of W (i.e., a covering by complements of algebraic subvarieties Γ_α of W) such that ω^* is strongly holomorphic on each set $W - \Gamma_\alpha$. Without loss of generality we may assume that we are dealing with a projective model V, for in the affine case we can pass from V and W to the corresponding projective models without affecting the differences $W - \Gamma_\alpha$, provided we add to each Γ_α the section of V with the hyperplane at infinity. We now apply Theorem 3 (§ 3). According to this theorem we may assume that the Γ_α are hyperplane sections of W, whence $V - \Gamma_\alpha$ and $W - \Gamma_\alpha$ are affine models. Since ω^* is strongly holomorphic along each $W - \Gamma_\alpha$, it follows by the first part of this proof that c divides ω^* along $W - \Gamma_\alpha$: $\omega^* = cu_\alpha^*$, $u_\alpha^* \in o^*_{W-\Gamma_\alpha}$.

If P is a common point of $W - \Gamma_\alpha$ and $W - \Gamma_\beta$, then $cu^*_{\alpha P} = cu^*_{\beta P}$ $(= \omega_P^*)$, and hence $u^*_{\alpha P} = u^*_{\beta P}$ since c is not a zero divisor in o_P^*. It follows that the g functions u_i^* are the projections of one and the same function u^* defined along W. Since u^* is strongly holomorphic along each $W - \Gamma_\alpha$, u^* is holomorphic along W, and it is clear that we have $\omega^* = cu^*$. This completes the proof.

PART II

INVARIANCE OF RINGS OF HOLOMORPHIC FUNCTIONS UNDER RATIONAL TRANSFORMATIONS

§10. HOLOMORPHIC FUNCTIONS AND SEMI-REGULAR BIRATIONAL TRANSFORMATIONS.

Let V and V' be two birationally equivalent varieties, over a given ground field k, and let T be a given birational transformation of V into V'. We consider pairs (P,P') of corresponding points of V and V' such that T^{-1} is semi-regular at P'.[16] Let o and $\mathfrak{m}$ be respectively the local ring of the point P and the ideal of non-units in o. Let o' and $\mathfrak{m}'$ have the similar significance for the point P'. The above assumption of semi-regularity implies that $o \subset o'$ and $\mathfrak{m} \subset \mathfrak{m}'$. Therefore, Cauchy sequences in o are also Cauchy sequences in o', amd any zero Cauchy sequence in o is also a zero Cauchy sequence in o'. We thus have a natural homomorphism of the ring o^* of holomorphic functions at the point P ($o^* = \mathfrak{m}$-adic completion of o) into (but not necessarily onto) the ring o'^* of holomorphic functions at the point P'. We shall denote this homomorphism by $H_{P,P'}$.

Let T' be a birational transformation of V' into another variety V", and let P" be a point of V" which corresponds to P'. If we assume that T'^{-1} is semi- regular at P", then also $H_{P',P''}$ is defined, and it is evident that

(1) $$H_{P,P'} \cdot H_{P',P''} = H_{P,P''}.$$

The preceding definition of the homomorphism $H_{P,P'}$ can be extended now to the case in which, instead of a point pair (P,P'), we have a pair (P,G') consisting of a point P of V and a collection G' of points P' of V', such that (1) every point P' of G' corresponds to P and (2) T^{-1} is semi-regular at every point of G'. It is then evident that every Cauchy sequence in o converges uniformly on G' and therefore defines a function on V' which is holomorphic along the set G' I, §3, Definition 2; actually we are dealing here with a function which is even strongly holomorphic along G'; see I, §1, Definition 1). Zero sequences in o define the function zero, i.e., the element zero of the ring $o^*_{G'}$ of functions on V' which are holomorphic along G'. We therefore have again a natural homomorphism of o^* into (but not necessarily onto) $o^*_{G'}$. We denote this homomorphism by $H_{P,G'}$. Let $\tau_{G',P'}$ be the projection of $o^*_{G'}$ into $o^*_{P'}$, where P' is any point of G' (this projection associates with each function in $o^*_{G'}$ the analytical element of that function at the point P'.) From the definition of the homomorphisms $H_{P,G'}$ and $H_{P,P'}$ it follows at once that

(2) $$H_{P,G'}\,\tau_{G',P'} = H_{P,P'}.$$

If ξ^* is any element of o^*, relation (2) exhibits the analytical elements of $H_{P,G'}(\xi^*)$ at the various points P' of the set G', whence (2) constitutes in effect a definition of $H_{P,G'}$ in terms of the homomorphisms $H_{P,P'}(P' \in G')$.

Relation (1) has the following obvious generalization:

(3) $$H_{P,P'}\,H_{P',G''} = H_{P,G''}.$$

Here G" is a collection of points of V" such that (1) every point of G" corresponds to P' and (2)T'^{-1} is semi-regular at each point of G". Moreover, P and P' are correspondint points of V and V', and T^{-1} is semi-regular at P'.

In order to avoid verbose and repetitious explanations of our semi-regularity conditions, we introduce a notation which we shall use throughout this work. If G and G' are subsets of the varieties V and V' respectively , we shall write: $G < G'$, if the following conditions are satisfied: (1) T^{-1} is semi-regular at each point of G'; (2) $T^{-1}\{G'\} = G$.[17] Thus, the homomorphism $H_{P,G'}$ is defined only under the assumption that $P < G'$. Similarly, (3) holds under the assumptions: $P < P' < G''$.

Let now G and G' be subsets of V and V' such that $G < G'$. We consider the ring o_G^* of functions on V which are holomorphic along G. Let ξ^* be any element of o_G^*. If we observe that to every point P' of G' there corresponds a unique point P of V and that this point P belongs to G, we see that we can associate with ξ^* a unique element ξ'^* of the direct product $O_{G'}^* = \prod_{P' \in G} o_{P'}^*$. This element ξ'^* is defined as follows: if P' is any point of G' and if $P = T^{-1}(P')$, then the analytical element $\xi_{P'}'^*$ of ξ'^* is equal to $H_{P,P'}(\xi_P^*)$, where ξ_P^* is the analytical element of ξ^* at the point P. It is not difficult to see that $\xi'^* \in o^*_{G'}$. This is obvious if ξ^* is strongly holomorphic along G, for in that case ξ^* is the limit of a sequence of rational functions on V which converges uniformly on G and which therefore also converges uniformly on G' to the function ξ'^*. In the general case there will exist a set of subvarieties of V, say $\Gamma_1, \Gamma_2, \cdots, \Gamma_h$, such that $G \cap \Gamma_1 \cap \Gamma_2 \cap \ldots \cap \Gamma_h = \emptyset$ and such that ξ^* is strongly holomorphic along each set $G - \Gamma_i$. Let $\Gamma_i' = T\{\Gamma_i\}$. From $G < G'$ it follows that $G - \Gamma_i < G' - \Gamma_i'$ and that $G' \cap \Gamma_1' \cap \Gamma_2' \cap \cdots \cap \Gamma_h' = \emptyset$. Since ξ'^* is strongly holomorphic along each set $G' - \Gamma_i'$, we conclude that ξ'^* is holomorphic along G', as was asserted.

The mapping $\xi^* \dashrightarrow \xi'^*$ is, then, a homomorphism of o_G^* into $o_{G'}^*$. We denote this homomorphism by $H_{G,G'}$. The following relations, similar to (2) and (3), follow directly from the definition of $H_{G,G'}$:

$$(4)\qquad H_{G,G'}\,\tau_{G',P'} = \tau_{G,P}\,H_{P,P'}, \quad (P = T^{-1}\{P'\},\ P' \in G');$$

$$(5)\qquad H_{G,G'}\,H_{G',G''} = H_{G,G''}, \qquad (G < G' < G'').$$

We are now in position to prove the following preliminary result:

THEOREM 13. If $G < G'$ and if either (a) V is locally normal at each point of G or (b) $G' = T\{G\}$, then the homomorphism $H_{G,G'}$ is an isomorphism (into, not necessarily onto $o^*_{G'}$).

PROOF. We deal first with assumption (a). In the Proceedings note [9] we have proved that if V is locally normal at a point P and if P' is a point of V' such that $P < P'$, then the local ring o of P is a subspace of the local ring o' of P', or equivalently: high powers of $\mathfrak{m}'$ contract in o to high powers of $\mathfrak{m}$. It follows that a Cauchy sequence in o is a zero sequence in o' if and only if it is a zero sequence in o. Hence $H_{P,P'}$ is an isomorphism, and this is precisely what Theorem 13 asserts in the special case $G = P$ and $G' = P'$. In the general case we use relation (4). Let ξ^* be an element of o^*_G such that $H_{G,G'}(\xi^*) = 0$, and let $\xi^*_P = \tau_{G,P}(\xi^*)$, where ξ^*_P is, then, the analytical element of ξ^* at P (P - any point of G). Since $G < G'$, there exist in G' a point which corresponds to P. If P' is such a point, then it follows from (4) that $H_{P,P'}(\xi^*_P) = 0$, and hence $\xi^*_P = 0$, since $P < P'$ and since V is locally normal at P. Since P is any point of G, we conclude that $\xi^* = 0$, and hence $H_{G,G'}$ is indeed an isomorphism.

We now pass to assumption (b) and we first observe that in the case in which V' is a derived normal model of V (Zariski [5] and [6]) (in which case the relation $G < T\{G\}$ holds for any subset G of V) the theorem has already been proved in part a) of the proof of Theorem 10 (§9). In the general case we proceed as follows. Let V_1 and V'_1 be derived normal models of V and V' respectively, and let T_1 and T'_1 be the birational transformations respectively of V into V_1 and of V' into V'_1. We set $G_1 = T_1\{G\}$, $G'_1 = T'_1\{G'\}$, where $G' = T\{G\}$. We have $G < G' < G'_1$, whence by (5):

$$(6)\qquad H_{G,G'}\,H_{G',G'_1} = H_{G,G'_1}.$$

We now compare the two point sets G_1 and G'_1. The birational transformations T, T_1 and T'_1 determine in an obvious fashion a birational transformation S of V_1 into V_1'[18]. From $G < G_1'$ and $G_1 = T_1\{G\}$, it follows that $S^{-1}\{G_1'\} \subset G_1$. On the other hand, it follows from $G < G_1$ and $G' = T\{G\}$, $G_1' = T_1'\{G'\}$, that $S(G_1) \subset G_1'$. <u>Hence</u> $G_1 = S^{-1}\{G_1'\}$ (and $G_1' = S\{G_1\}$). Moreover, since $G < G_1'$ and since T_1 is finitely valued on V, it follows that S^{-1} is finitely valued on G_1'. <u>Therefore</u> S^{-1} <u>is semi-regular at each point of</u> G_1', since V_1' is a normal variety. We have thus shown that $G_1 < G_1'$, and hence we can write by (5):

$$H_{G,G_1} H_{G_1,G_1'} = H_{G,G_1'}. \tag{7}$$

From the preceding part of the proof we know that the homomorphisms which appear on the left-hand side of (7) are isomorphisms. Therefore $H_{G,G_1'}$ is also an isomorphism. But then it follows from (6) that $H_{G,G'}$ is an isomorphism, and this completes the proof of the theorem.

§ 11. ABSOLUTE BIRATIONAL INVARIANCE OF RINGS OF HOLOMORPHIC FUNCTIONS.

Our main object in this part of the paper is to establish the following

FUNDAMENTAL THEOREM. Let T be a birational transformation of an irreducible variety V into another irreducible variety V', and let W be an arbitrary subvariety of V (the variety W may be reducible). Let $W' = T\{W\}$. We assume that V is locally normal at each point of W and that $W < W'$. Under these assumptions, the rings of holomorphic functions o_W^* and $o_{W'}^*$ are isomorphic, and namely $H_{W,W'}$ is an isomorphic mapping of o_W^* <u>onto</u> $o_{W'}^*$.

We note that the assumptions of the theorem are to the effect that both conditions (a) and (b) of Theorem 13 are satisfied simultaneously, with G and G' replaced respectively by W and W'. Each one of these conditions, taken alone, implies, according to Theorem 13, that $H_{W,W'}$ is an isomorphism of o_W^* into $o_{W'}^*$; the fundamental theorem asserts that the two conditions, taken together, imply that $H_{W,W'}$ is an isomorphism of o_W^* <u>onto</u> $o_{W'}^*$.

In §§ 16-19 we shall extend the fundamental theorem to rational transformations.

The most important single consequence of the fundamental theorem is the following theorem:

CONNECTEDNESS THEOREM (THE BIRATIONAL CASE). Let T be a birational transformation of an irreducible variety V into another irreducible variety V', and let W be a subvariety of V. If V is analytically irreducible at each point of W (in particular, if V is locally normal at each point of W) and if W is connected, then also the total transform $T\{W\}$ is a connected variety.

PROOF. Let $W' = T\{W\}$. Assume first that V is locally normal at each point of W and that $W < W'$. Since W is connected, it follows that the ring o^*_W is an integral domain (I, § 6, Theorem 6). From the fundamental theorem it follows that o^*_W and $o^*_{W'}$ are isomorphic rings. Hence also $o^*_{W'}$ is an integral domain, and therefore (I, § 6) W' is connected.

We next assume only that V is locally normal at each point of W. We pass to the graph V* of the birational transformation T (V* = join of V and V'). Let W* be the total transform of W on V. We have $W < W^*$, and hence it follows from the preceding case that W* is connected. We also have $W' < W^*$, and from this it follows directly that also W' is connected.

We now drop the assumption of local normality and we maintain only the initial assumption that V is locally irreducible at each point of W. We pass to a derived normal V_1 of V. Let T_1 be the birational transformation of V into V_1. We know (I, § 7, Theorem 8) that if V is analytically irreducible at a point P, then $T_1\{P\}$ is a (finite) set of points which are isomorphic over k. It follows that for each point P of W the set $T_1\{P\}$ has the following property: if a subvariety Γ_1 of V_1 contains one of the points of $T_1\{P\}$, then Γ_1 contains the entire set $T_1\{P\}$. Therefore, if two subvarieties Γ_1 and Δ_1 of V_1 have no common points in $T_1\{W\}$ (in particular, if $\Gamma_1 \cap \Delta_1 = \emptyset$), the corresponding subvarieties $T_1^{-1}\{\Gamma_1\}$ and $T_1^{-1}\{\Delta_1\}$ of V will have no common points in W. From this we conclude at once that $T_1\{W\}$ is connected since W is connected. Now it is clear that, in the birational transformation between V_1 and V', the variety W' is the total transform of $T_1\{W\}$. Since we have just proved that the latter variety is connected and since V_1 is normal, it follows from the preceding case that W' is connected. This completes the proof of the theorem.

The fundamental theorem can be extended to the case of arbitrary birational transformations. In this extension the assumption of semiregularity is eliminated (we refer to the assumption $W < T\{W\}$). However, we must now require that both varieties V and V' be normal, or rather locally normal on W and on $T\{W\}$ respectively. It will be sufficient to state the result for normal varieties only.

THE FUNDAMENTAL THEOREM FOR NORMAL VARIETIES. Let T be a birational transformation of a normal variety V into a normal variety V', and let W, W' be subvarieties of V and V' respectively. If $W' = T\{W\}$ and $W = T^{-1}\{W'\}$, then the rings of holomorphic functions o^*_W and $o^*_{W'}$ are isomorphic.

For the proof we pass to the graph V_1 of the birational transformation T and we denote by W_1 the total transform of W on V_1. Under the assumptions on W and W' made in the theorem, it follows that W_1 is also the total transform of W' on V_1. We can therefore apply the fundamental theorem (as formulated in the beginning of this section) to the pair of varieties (V,V_1) and also to the pair of varieties (V', V_1), since $W < W_1$ and $W' < W_1$. We thus conclude that the rings o^*_W and $o^*_{W'}$ are isomorphic, since they both are isomorphic to the ring $o^*_{W_1}$.

We note that the isomorphism in question is given by $H_{W,W_1} \cdot H^{-1}_{W',W_1}$. If V_2 is another variety in the birational class of V, such that $V < V_2$, and $V' < V_2$, we obtain in a similar fashion an isomorphism of o^*_W onto $o^*_{W'}$, given by $H_{W,W_2} \cdot H^{-1}_{W',W_2}$, where W_2 is the total transform of W on V_2. However, it is immediately seen that we get the same isomorphism in both cases [consider the join V_3 of V_1 and V_2 and apply the composition formula (5) to each of the four triples (W, W_i, W_3), (W', W_i, W_3), $i = 1, 2$]. <u>We have therefore a natural isomorphism of o^*_W onto $o^*_{W'}$.</u>

§12. REDUCTION OF THE PROOF OF THE FUNDAMENTAL THEOREM TO A SPECIAL CASE.

We shall prepare the ground for the proof of the fundamental theorem by first reducing the whole problem to the case in which the birational transformation of V into V' is of a certain special type. It will be essentially the same reduction that we have used in [6] for the proof of the " main theorem" (loc. cit., p. 523), except that this time the reduction must be carried out carefully in the projective space, rather than in the affine space.

We observe in the first place that it is permissible to replace V by a derived normal model $\bar{V}$. For since V is locally normal at each point of W, the birational transformation of V into $\bar{V}$ is regular on W and therefore does not affect at all the ring o^*_W (I, §3). <u>We shall therefore assume that V is a normal variety.</u>

We next observe that it is permissible to replace V' by any birationally equivalent variety V_1 such that $V < V_1$ and $V' < V_1$. For if V_1 is such a variety and if W_1 denotes the total transform of W on V_1, we have, by (5); $H_{W,W_1} = H_{W,W'} \cdot H_{W',W_1}$. By Theorem 13, the three homomorphisms which appear in this relation are all isomorphisms. If, then, the fundamental theorem is true for the pair of varieties V and V_1, i.e., if H_{W,W_1} is an isomorphism of o^*_W onto $o^*_{W_1}$, then the above relation implies that also $H_{W,W'}$ is an isomorphism of o^*_W onto $o^*_{W'}$. We shall apply this observation in a moment.

We now introduce a notation. If $y_0, y_1, \cdots, y_n$ are homogeneous coordinates of the representative general point A of V and if t is an element of the function field Σ, $t \neq 0$, we denote by Vot the variety which lies in a projective space of dimension 2n + 1 and has the general point $(y_0, y_1, \cdots, y_n, ty_0, ty_1, \cdots ty_n)$. We also denote this general point by Aot. It is clear that Vot and V are birationally equivalent varieties. We point out the following property of the variety Vot: if P_1 is any point of Vot, then at least one of the two elements t, 1/t belongs to the local ring of P_1. For if $z_0, z_1, \cdots, z_{2n+1}$ are homogeneous coordinates of P_1 and if z_i is one of the coordinates which is $\neq 0$, then either t $(= ty_i/y_i = z_{i+n}/z_i)$ or 1/t $(= y_{i-n}/ty_{i-n} = z_{i-n}/z_i)$ is in o_{P_1}, according as $i \leq n$ or $i > n$. We shall make frequent use of this property. Since P_1 is a specialization of Aot, each one of the two sets $(z_0, z_1, \ldots, z_n)$ $(z_{n+1}, z_{n+2}, \ldots, z_{2n+1})$ either consists of zeros only or represents a point P of V, and if neither set consists of zeros only then both sets represent one and the same point P of V. It is clear that in either case this point P corresponds to P_1 in the birational transformation between V and Vot, and that it is the only point which corresponds to P_1. Moreover, it is obvious that $o_P \subset o_{P_1}$. Hence we have $V < \text{Vot}$.

Let $y'_0, y'_1, \ldots, y'_m$ be homogeneous coordinates of the general point A' of V'. We may assume, without loss of generality, that $y'_i \neq 0$, $i = 0, 1, \cdots, m$. Let $N = \binom{m+1}{2}$, and let $t_1, t_2, \cdots, t_N$ denote the quotients y'_i/y'_j, $i < j$, taken in some order. We now define N pairs of varieties $(V_\alpha, \overline{V}_\alpha)$ as follows: 1) $V_1 = \text{Vot}_1$, $\overline{V}_1$ = a derived normal model of V_1; 2) $V_\alpha = \overline{V}_{\alpha-1}\, ot_\alpha$, $\overline{V}_\alpha$ = a derived normal model of V_α ($\alpha = 2, 3, \cdots, N$).

We also set $V_0 = \overline{V}_0 = V$, $\overline{V}_N = V^*$. We have $V_\alpha < \overline{V}_\alpha < V_{\alpha+1}$, and hence $V < V^*$. It is not difficult to see that we also have $V' < V^*$. For let P^* be any point of V^*, and let P_α be the point of V_α which corresponds to P^* (there is only one such point P_α, since $V_\alpha < V^*$). We know that at least one of the two elements t_α and $1/t_\alpha$ belongs to the local ring of P_α. Since $P_\alpha < P^*$, we conclude that the local ring of P^* contains at least one element of each pair $(t_\alpha, 1/t_\alpha)$. In other words: given any two of the homogeneous coordinates of the general point A' of V', say y_i' and y_j', the local ring of P^* contains at least one of the quotients y_i'/y_j', y_j'/y_i'. From this it follows that it is possible to choose one of the coordinates y_i', say y_0', in such a fashion that all the m quotients y_j'/y_0' belong to the local ring of P^*. But then this local ring contains the non-homogeneous coordinate ring $k[y_1'/y_0', y_2'/y_0', \ldots, y_m'/y_0']$ of V', and that implies that the birational transformation of V^* into V' is semi-regular at P^*. Since P^* is any point of V^*, it follows that $V' < V^*$, as asserted.

We have, then, $V < V^*$ and $V' < V^*$. Hence by the observation made earlier, it is permissible to replace V' by V^*. Since the sequence of normal varieties $V, \overline{V}_1, \ldots, \overline{V}_{N-1}, V^*$ is monotone (with respect to the relation $<$), it follows that in order to prove the fundamental theorem for the pair (V,V^*) it is sufficient to prove the theorem for each pair $(\overline{V}_\alpha, \overline{V}_{\alpha+1})$ of consecutive varieties of the sequence [this follows from relation (5), § 10]. We have therefore reduced the problem to the following special case: <u>prove the fundamental theorem when</u> V' <u>is a derived normal model of</u> Vot, <u>where</u> V <u>is a normal variety and where</u> t <u>is an arbitrary non-zero element of the function field</u> Σ of V.

In view of the presence of the intermediate (not necessarily normal) variety Vot, the special problem divides naturally into two parts. Let W be the given subvariety of V, and let W_1 and W' be the total transforms of W on Vot and V' respectively. By Theorem 13(§ 10) the homomorphisms $H_{W,W_1}, H_{W_1,W'}, H_{W,W'}$ are isomorphisms. In order that $H_{W,W'}$ be an isomorphism "onto", it is necessary and sufficient that H_{W,W_1} and $H_{W_1,W'}$ be isomorphisms "onto". The assertion that H_{W,W_1} is an isomorphism of o_W^* onto $o_{W_1}^*$ is still a special case of the fundamental theorem, since V is normal. This special case will be considered in the next two sections. After that, we shall take up the case of the two varieties Vot and its derived normal model V', and we shall show that $H_{W_1,W'}$ is an isomorphism of $o^*_{W_1}$ onto $o^*_{W'}$. Note that in this second part of the proof of the fundamental theorem we shall not be dealing with a special case of the theorem itself, since Vot need not be a normal variety, nor need it be locally normal on W_1. However, W_1 is now a special subvariety of V_1, since it is the total transform of a subvariety W of V.

§13. THE BIRATIONAL TRANSFORMATION V --> Vot.

We shall discuss in this section some properties of the birational transformation of V into Vot which will be needed later on. We denote this transformation by T, and we denote the variety Vot by V'. Let $y_0, y_1, \cdots, y_n$ be homogeneous coordinates of the general point A of V. The homogeneous coordinates of the general point A' of V' are, then, $y_0, y_1, \cdots, y_n, y_0t, y_1t, \cdots, y_nt$. We denote these coordinates by $y'_0, y'_1, \cdots, y'_n, y'_{n+1}, y'_{n+2}, \cdots, y'_{2n+1}$. We introduce the n+1 non-homogeneous coordinate rings R_i of V, where $R_i = k[y_0/y_i, y_1/y_i, \cdots, y_n/y_i]$, $i = 0, 1, \cdots, n$. If $P(z_0, z_1, \cdots z_n)$ is a point of V, we shall say that P <u>is covered by the ring</u> R_i <u>if</u> $z_i \neq 0$. We say that a subset G of V is covered by the ring R_i if every point of G is covered by R_i. The set of points of V which are covered by R_i is an affine variety and shall be denoted by V_i. Every point of V is covered by at least one of the n+ 1 coordinate rings R_i.

Similarly, we have 2(n+ 1) non-homogeneous coordinate rings for V'. We shall denote these rings by R'_{i0} and $R'_{i\infty}$, where $R'_{i0} = k[y'_0/y'_i, y'_1/y'_i, \cdots, y'_{2n+1}/y'_i]$ and $R'_{i\infty} = k[y'_0/y'_{i+n+1}, y'_1/y'_{i+n+1}, \cdots, y'_{2n+1}/y'_{i+n+1}]$, $i = 0, 1, \cdots, n$. It is clear that

(8) $$R'_{i0} = R_i[t], \quad R'_{i\infty} = R_i[1/t].$$

A point $P(z'_0, z'_1, \cdots, z'_{2n+1})$ of V' is covered by the ring R'_{i0} or $R'_{i\infty}$, if respectively $z'_i \neq 0$ or $z'_{i+n+1} = 0$. We shall denote by $\underline{V'_{i0}}$ and $\underline{V'_i}$ the affine varieties consisting of the points of V' which are covered respectively by the rings R'_{i0} and $R'_{i\infty}$.

If P(z) and P'(z') are corresponding points of V and V', then, as has been explained in the preceding section, we have:

(9) $$(z') = (z, zu) = (z_0, z_1, \cdots, z_n, z_0u, z_1u, \cdots, z_nu), \text{ if } t \in o_{P'},$$

and

(9') $(z') = (zu,z) = (z_0u,z_1 u,\cdots,z_nu,z_0,z_1,\cdots,z_n)$, if $1/t \in o_{P'}$, where u is a suitable quantity. It follows that _if_ P _is covered by_ R_i, _then_ P' _is covered by at least one of the two rings_ R'_{i0} _and_ $R'_{i\infty}$; _and conversely_. More precisely: P' _is covered by_ R'_{i0} _if_ $t \in o_{P'}$, _and_ P' _is covered by_ $R'_{i\infty}$ _if_ $1/t \in o_{P'}$.

If G is any subset of V, we denote by G_i the set of points of G which are covered by the ring R_i (whence $G_i = G \cap V_i$). In a similar fashion we define G'_{i0} and $G'_{i\infty}$, if G' is a subset of V'. From what has been said a moment ago it follows that _if_ $G' = T\{G\}$ = _total transform of_ G, _then_

(10) $$G'_{i0} \cup G'_{i\infty} = T\{G_i\} .$$

Let, again, P and P' be corresponding points of V and V', and let us assume for a moment that either t or 1/t belongs to the local ring o_P. If, say, R_i covers P and if t belongs to o_P, then R'_{i0} covers P'. Since $R_i \subset o_P$, it follows by (8) that $R'_{i0} \subset o_P$, and hence $P' < P$. Since we also have $P < P'$, _we conclude that_ T _is regular at the point_ P. In a similar fashion we find that T is regular at P if 1/t belongs to the local ring of P. On the other hand, if neither t nor 1/t belongs to o_P, then o_P is a proper subring of $o_{P'}$, since this latter ring does contain at least one of the two elements t, 1/t. We conclude that _a point_ P _of_ V _is a fundamental point of the birational transformation_ T _if and only if neither_ t _nor_ 1/t _belongs to the local ring of_ P.

Let P(z) be a fundamental point of T. Equations (9) and (9') show that the total transform $T\{P\}$ of the point P is contained in the projective line $S_1/k(P)$, where k(P) is the local (residue) field of P. Now $T\{P\}$ is an algebraic variety over k(P), and, moreover, $T\{P\}$ is an infinite set, since P is a fundamental point. Hence $T\{P\} = S_1/k(P)$. Therefore $T\{P\}$ contains a point (z, zu), _where_ u _is a transcendental over_ k(P), and every point of $T\{P\}$ is, then, a specialization of (z, zu) over k(P). It follows that the least variety, over k, which contains the set $T\{P\}$, is irreducible, since it has the above point (z, zu) as general point. Now this least variety is T[W], where W is the irreducible subvariety of V which has P as general point (see §11). We have thus shown that _if_ W _is an irreducible subvariety of_ V _which is fundamental for_ T, _then_ T[W] _is irreducible_.

Let now P be an arbitrary point of V, and let W be the irreducible subvariety of V having P as general point. Let $W' = T\{W\}$. We fix among the rings R_i one that covers P; let it be R_j. For simplicity of notation we shall drop the index j and we shall therefore denote the coordinate ring R_j by R and the rings R'_{j0}, $R'_{j\infty}$ by R'_0 and R'_∞ respectively. Accordingly, we shall also denote V'_{j0}, W'_{j0} and $V'_{j\infty}$, $W'_{j\infty}$ by V'_0, W'_0 and V'_∞ and W'_∞ respectively. These are, then, affine varieties- the affine parts of V' and W' relative to the coordinate rings R'_0 and R'_∞ respectively. Let $\mathfrak{m}$ be the prime ideal of W in R and let $\mathfrak{m}'_0 = R'_0 \cdot \mathfrak{m}$, $\mathfrak{m}'_\infty = R'_\infty \cdot \mathfrak{m}$. Note that, by (9), $R'_0 = R[t] \supset R$, $R'_\infty = R[1/t] \supset R$.

LEMMA. The subvariety $V(\mathfrak{m}'_0)$ of V'_0 defined by the ideal $\mathfrak{m}'_0$ coincides with the variety W'_0. Moreover, the relation $\mathfrak{m}'_0 \cap R = \mathfrak{m}$ holds true, except in the following case: $t \notin o_P$, $1/t \in o_P$. Similarly, $W'_\infty = V(\mathfrak{m}'_\infty)$, and we have $\mathfrak{m}'_\infty \cap R = \mathfrak{m}$, except if $1/t \notin o_P$ and $t \in o_P$.

PROOF. It will be sufficient to prove the lemma for the ring R'_0 and the ideal $\mathfrak{m}'_0$. Let B' be any point of V'_0 and let B be the corresponding point of V. Then necessarily B is covered by R. Let $\mathfrak{p}'$ be the prime ideal of B' in the ring R'_0, and let similarly $\mathfrak{p}$ be the prime ideal of B in R. We have $\mathfrak{p}' \cap R = \mathfrak{p}$. If $B' \in W'_0$, then $\mathfrak{p} \supset \mathfrak{m}$ and consequently $\mathfrak{p}' \supset \mathfrak{m}'_0$. Conversely, if $\mathfrak{p}' \supset \mathfrak{m}'_0$, then $\mathfrak{p} \supset \mathfrak{m}$, and therefore $B \in W$, whence $B' \in W' \cap V'_0 = W'_0$. This proves the first part of the lemma.

Assume now that $t \notin o_P$, $1/t \in o_P$. Then t can be written in the form of a quotient v/w, where $w \in \mathfrak{m}$ and $v \in R$, $v \notin \mathfrak{m}$. We have $v = tw \in R'_0 \cdot \mathfrak{m} = \mathfrak{m}'_0$, i.e., $v \in \mathfrak{m}'_0 \cap R$, and since $v \notin \mathfrak{m}$ it follows that $\mathfrak{m}'_0 \cap R \neq \mathfrak{m}$. Conversely, assume that $\mathfrak{m}'_0 \cap R \neq \mathfrak{m}$, and let v be an element of $\mathfrak{m}'_0 \cap R$ which is not in $\mathfrak{m}$. Since $v \in \mathfrak{m}'_0$ we can write v in the form: $v = a_0 t^g + a_1 t^{g-1} + \cdots + a_g$, $a_j \in \mathfrak{m}$, $j = 0, 1, \ldots, g$. From this it follows in the first place that $t \notin o_P$, for otherwise we would have $v \in o_P \cdot \mathfrak{m} \cap R = \mathfrak{m}$ (this equality holds because o_P is the quotient ring of the prime ideal $\mathfrak{m}$ in R), contrary to our assumption. On the other hand, since $a_g \in \mathfrak{m}$ and $v \notin \mathfrak{m}$, the element $a_g - v$ is a unit in o_P, and therefore the above expression for v yields a relation of integral dependence for 1/t over o_P. Since o_P is integrally closed, we conclude that $1/t \in o_P$, and this completes the proof of the lemma.

REMARK 1. The case which will be of particular interest to us in the sequel is the one in which P is a fundamental point of T. We have seen above that in that case neither t nor 1/t belongs to o_P, and hence none of the exceptional cases referred to in the above lemma can arise if

P is fundamental. Hence if P is a fundamental point, then $\mathfrak{m}'_0 \cap R = \mathfrak{m}'_\infty \cap R = \mathfrak{m}$. We shall use this remark in the later sections.

REMARK 2. Another fact that we shall make use of later on is the following: <u>if</u> P <u>is a fundamental point of</u> T <u>then</u> W'_0 <u>and</u> W'_∞ <u>have points in common</u> (in other words: the total transform W' of W carries points which are covered by both rings R'_0 and R'_∞). The proof is immediate and is as follows:

We have seen that if P is fundamental for T, then T[W] is irreducible. We have also seen that if P = (z) and if u is a transcendental over k(P), then (z, zu) is a general point of T[W] (over k). Now since P is covered by $R(= R_j)$, we have $z_j \neq 0$, $z_j u \neq 0$, and hence the point (z, zu) is covered by both rings R'_0 and R'_∞ . Since $T[W] \subset W'$ (= T $\{W\}$), <u>we see that</u> W'_0 <u>and</u> W'_∞ <u>have in common the general point</u> (z, zu) <u>of</u> T[W] (and, in fact, must consequently have in common every general point of T[W]/k). This proves the above assertion.

REMARK 3. For the sake of completeness we shall explain briefly the geometric facts which lie behind the exceptional cases referred to in the above lemma. We know already that in either one of these cases we are dealing necessarily with a point P which is regular for our birational transformation. For the sake of concreteness, let us assume that we are dealing, for instance, with case $t \notin \mathfrak{o}_P$, $1/t \in \mathfrak{o}_P$. Let P' be <u>the</u> point of V' which corresponds to P. The variety T[W] is irreducible, has the same dimension as W and has P' as general point. Since T is regular at P, we have $\mathfrak{o}_P = \mathfrak{o}_{P'}$, whence $t \notin \mathfrak{o}_{P'}$. This implies that P' is not covered by the coordinate ring R'_0. Hence no point of T[W] is covered by R'_0 (or equivalently: with respect to the non-homogeneous coordinate ring R'_0 the variety T[W] lies entirely "at infinity"). It follows that if the total transform W' of W possesses irreducible components which do not lie entirely "at infinity" with respect to the ring R'_0 (or equivalently: if W'_0 is not empty), then these components cannot correspond to W itself but only to some proper subvarieties of W (which are, then, necessarily fundamental for T). The corresponding algebraic fact is the following: the ideal $\mathfrak{m}'_0 \cap R$ is in this case always a proper overideal of $\mathfrak{m}$ (or: the ideal $\mathfrak{m}$ is "lost" in R'_0). If $\mathfrak{m}'_0 \cap R$ is not the unit ideal then the variety of this ideal consists entirely of fundamental points of T (this variety is, of course, a subvariety of W).

§14. PROOF OF THE FUNDAMENTAL THEOREM IN THE CASE OF THE TRANSFORMATION V --> Vot.

The notation being the same as in the preceding section, we recall that we have assumed V to be a normal variety, and we consider an arbitrary subvariety W of V. Our object, then, is to prove the fundamental theorem for V, W and V', where V' = Vot. Our first step will be a reduction to the case of an irreducible W. We proceed to show that if the fundamental theorem is true in this latter case, it is also true for any reducible subvariety W of V.

Let $W_1, W_2, \cdots, W_g$ be the irreducible components of W, and let $W' = T\{W\}$, $W'_i = T\{W_i\}$. Our assumption is that for each i, i= 1, 2,$\cdots$, g, the isomorphism H_{W_i,W'_i} maps $o^*_{W_i}$ onto $o^*_{W'_i}$. We have to show that also $H_{W,W'}$ is an isomorphism of o^*_W <u>onto</u> $o^*_{W'}$. Let ξ'^* be an arbitrary element of $o^*_{W'}$ and let ξ'^*_i be its projection into $o^*_{W'_i}$: $\xi'^*_i = \tau_{W',W'_i}(\xi'^*)$. By assumption, there exists a (unique) element ξ^*_i in $o^*_{W_i}$ such that $H_{W_i,W'_i}(\xi^*_i) = \xi'^*_i$. Suppose that P is a point of W which belongs to two distinct components W_i and W_j of W, and let ξ^*_{iP} and ξ^*_{jP} be the analytical elements of ξ^*_i and ξ^*_j at the point P. We fix a point P' of $W'_i \cap W'_j$ which corresponds to P. If we set $G = W_i$ and $G' = W'_i$ in the formula (4) of §10, we obtain $\xi'^*_{iP'} = H_{P,P'}(\xi^*_{iP})$. Similarly, $\xi'^*_{jP'} = H_{P,P'}(\xi^*_{jP})$. Since $\xi'^*_{iP'} = \xi'^*_{jP'}$ $(= \xi'^*_{P'})$ and since $H_{P,P'}$ is an isomorphism, it follows that $\xi^*_{iP} = \xi^*_{jP}$. This shows that the set of g functions ξ^*_i determines a unique analytical element at each point P of W, and therefore this set defines an element ξ^* of O^*_W (such that $\xi^*_P = \xi^*_{iP}$, if $P \in W_i$). Since the projection of ξ^* into $O^*_{W_i}$ is the function ξ^*_i which is holomorphic along W_i, it follows (I, end of § 3, Corollary) that ξ^* is a holomorphic function along W, $\xi^* \in o^*_W$. Again it follows directly from (4), § 10 for G= W and G' = W', that ξ'^* and $H_{W,W'}(\xi)^*$ have the same analytical element at each point P' of W'. Hence $\xi'^* = H_{W,W'}(\xi^*)$, and this proves that $H_{W,W'}$ maps o^*_W <u>onto</u> $o^*_{W'}$.

We therefore assume now that W is irreducible. We consider the various non-homogeneous coordinate rings R_i, R'_{io} and $R'_{i\infty}$, introduced in the preceding section, and also the various affine models W_i, W'_{io} and $W'_{i\infty}$, where $W' = T\{W\}$. Interchanging, if necessary, the roles of t and 1/t, we may assume that the exceptional case -"$t \notin o_P$, $1/t \in o_P$" - of the lemma of § 13 does not arise (P= a general point of W). In other words, we assume that t is not the reciprocal of a non-unit in o_P. It is clear that in that case also t + 1 is not the reciprocal of a non-unit in o_P. Now replacing, if necessary, t by t + 1, we may also assume that the second exceptional case - " $1/t \notin o_P$, $t \in o_P$" - of Lemma 1 does not arise. Since Vo (t+ 1) is related to Vot by a non-singular projective transformation, we conclude that it is permissible to assume that neither one of the two exceptional cases of Lemma 1 arises for W. Moreover, we may also assume that each of the n+ 1 coordinate rings R_i covers the general point P of W. Having made all these assumptions, we see that the prime ideals $\mathfrak{m}_i$ are defined for i = 0,1,...,n, and that we have, by Lemma 1:

(11) $$\mathfrak{m}'_{io} \cap R_i = \mathfrak{m}'_{i\infty} \cap R_i = \mathfrak{m}_i; \quad i = 0,1,\dots, n.$$

Let now ξ'^* be any element of $o^*_{W'}$. We shall denote by ξ'^*_{io} and $\xi'^*_{i\infty}$ the projections of ξ'^* into $o^*_{W'_{io}}$ and $o^*_{W'_{i\infty}}$ respectively. We shall also denote by ξ'^*_i the projection of ξ'^* into $o^*_{T\{W_i\}} = o^*_{W'_{io} \cup W'_{i\infty}}$ [see (10), § 13]. The fundamental theorem demands that we show the existence of an element ξ^* of o^*_W such that $H_{W,W'}(\xi^*) = \xi'^*$. To show this, it is sufficient to show that for each i, i = 0,1,...,n, there exists an element ξ^*_i in $o^*_{W_i}$ such that $H_{W_i,W'_i}(\xi^*_i) = \xi'^*_i$, where W'_i stands for $T\{W_i\}$. For if such a set of n + 1 functions ξ^*_i exists, then, from the fact that their transforms ξ'^*_i on V' are the projections of one and the same element ξ'^* of $o^*_{W'}$, one can draw immediately the conclusion (see I, end of § 3, Corollary) that also the n + 1 functions ξ^*_i are the projections of a function ξ^* which is holomorphic along the entire variety W, since W is the union of the n + 1 affine models W_i. That function ξ^*, which is therefore defined by the relations $\xi^*_i = \tau_{W,W_i}(\xi^*)$, is clearly such that its transform under $H_{W,W'}$ is the given element ξ'^* of $o^*_{W'}$.

Our problem now is, therefore, to show that for each i the isomorphism H_{W_i,W'_i} maps $o^*_{W_i}$ onto $o^*_{W'_i}$. This problem presents mixed affine and projective features. On the one hand, the original variety V and its subvariety W have now been replaced by the affine varieties V_i and W_i. On the other hand, W'_i stands for the total transform of W_i on the projective model V' and is therefore not an affine variety (except when W is of dimension zero and is not fundamental for T); it is the union of the two affine varieties W'_{io} and $W'_{i\infty}$, the latter being subvarieties of the two affine varieties V'_{io} and $V'_{i\infty}$ respectively.

Since we shall be dealing from now on with a fixed value of the index i, it will be most convenient to omit the index altogether. Accordingly, in the remainder of this section the symbols $V, W, R, \mathfrak{m}$; V'_o, V'_∞, W', W'_o, W'_∞, R'_o, R'_∞, $\mathfrak{m}'_o$, $\mathfrak{m}'_\infty$ will stand for the entities which have been denoted heretofore respectively by V_i, W_i, R_i, $\mathfrak{m}_i$, V'_{io}, etc. However, there will be no change in the meaning of the symbol V'.

We shall need the following fact: <u>high powers of $\mathfrak{m}'_o$ (or of $\mathfrak{m}'_\infty$) contract in R to high powers of $\mathfrak{m}$</u>. To prove this, we observe that we have, by (11): $\mathfrak{m}'_o \cap R = \mathfrak{m}$. Since $\mathfrak{m}$ is a prime ideal, it follows that there exists a prime ideal $\mathfrak{p}'_o$ in R'_o such that $\mathfrak{p}'_o \supset \mathfrak{m}'_o$, $\mathfrak{m}'_o \cap R = \mathfrak{m}$. Let P' be a point of V' whose prime ideal is $\mathfrak{p}'_o$ and let P be the corresponding point of V. Then it is clear that $\mathfrak{m}$ is the prime ideal of P (whence P is a general point of W). Since $P < P'$ and since V is normal, it follows (Zariski [9]) that high powers of the maximal ideal of $o_{P'}$ contract in o_P to high powers of the maximal ideal in o_P. That is equivalent to saying that high symbolic powers of $\mathfrak{p}'_o$ contract in R to high symbolic powers of $\mathfrak{m}$. We have $\mathfrak{m}'_o \subset \mathfrak{p}'_o$, and hence $\mathfrak{m}'^{i}_o \subset \mathfrak{p}'^{(i)}_o$ for any integer i. On the other hand, since V is normal, we have, by a previous result (I, Theorem 7, §7), that high symbolic powers of the prime ideal $\mathfrak{m}$ are contained in high ordinary powers of $\mathfrak{m}$. This establishes the fact stated above.

For any integer i, $i \geq 0$, we denote by $\nu(i)$ the greatest of all the integers j for which $\mathfrak{m}'^{i}_o \cap R \subset \mathfrak{m}^j$ <u>and</u> $\mathfrak{m}'^{i}_\infty \cap R \subset \mathfrak{m}^j$. We have therefore

(12) $$\mathfrak{m}'^{i}_o \cap R \subset \mathfrak{m}^{\nu(i)};\quad \mathfrak{m}'^{i}_\infty \cap R \subset \mathfrak{m}^{\nu(i)},\ \lim \nu(i) = +\infty .$$

We shall also need a lemma on integrally closed Noetherian domains. Let R be such a domain and let t be an element of the quotient field of R. Let f(X) be a polynomial with coefficients in R, say $f(X) = a_o X^g + a_1 X^{g-1} + \dots + a_g$, $a_i \in R$, and let $f_j(X) = a_o X^j + a_1 X^{j-1} + \dots + a_j$, $0 \leq j \leq g$. It is not difficult to see that the following is true:

<u>If</u> $f(t) = 0$, <u>then</u> $f_j(t) \in R$, $j = 0, 1, \dots, g$.

The proof is immediate and is as follows. Let $\mathfrak{p}$ be any minimal prime ideal of R and let v be the valuation of the quotient field of R which is determined by $\mathfrak{p}$. If $v(t) \geq 0$, then also $v[f_j(t)] \geq 0$. On the other hand, we have $f_j(t) = -(a_{j+1}/t + a_{j+2}/t^2 + \dots + a_g/t^{g-j})$ since $f(t) = 0$, and from this we conclude that if $v(t) < 0$ then $v[f_j(t)] > 0$.

Hence in either case $f_j(t)$ has non-negative value, and since this is so for every minimal prime ideal of R, it follows that $f_j(t)$ belongs to R.

The lemma which we shall need is an immediate consequence of the above result and is as follows:

> LEMMA. Let R be an integrally closed Noetherian domain and let t, f(X) and $f_j(X)$ have the same meaning as above. Let $\mathfrak{m}$ be an ideal in R and let $\mathfrak{m}'$ be the extended ideal of $\mathfrak{m}$ in the ring R[t]. Then if $f(t) \in \mathfrak{m}'^{\rho}$, there exist elements $b_0, b_1, \dots, b_g$ in R such that $f_j(t) - b_j \in \mathfrak{m}'^{\rho}$, $j = 0, 1, \dots, g$.

PROOF. By assumption, there exists a polynomial F(X) with coefficients in $\mathfrak{m}^{\rho}$ such that $f(t) - F(t) = 0$. We regard the polynomials $\varphi(X) = f(X) - F(X)$ and F(X) as polynomials of some formal degree $h \geq g$ and we apply to $\varphi(X)$ the result proved above. We find then that $\varphi_s(t) \in R$, $s = 0, 1, \dots, h$. In particular, for $s = h - g + j$, $j = 0, 1, \dots, g$, we have $\varphi_{h-g+j}(t) = f_j(t) - F_{h-g+j}(t) = b_j \in R$. Hence $f_j(t) - b_j = F_{h-g+j}(t)$, and since the coefficients of $F_{h-g+j}(t)$ are in $\mathfrak{m}^{\rho}$, it follows that $f_j(t) - b_j \in \mathfrak{m}'^{\rho}$, q.e.d.

We now proceed to the proof that $H_{W,W'}$ maps o^*_W <u>onto</u> $o^*_{W'}$. Let ξ'^* be any element of $o^*_{W'}$, and let ξ'^*_0 and ξ'^*_∞ be the projections of ξ'^* into $o^*_{W'_0}$ and $o^*_{W'_\infty}$ respectively. Since V'_0 and W'_0 are affine varieties, it follows that ξ'^*_0 is strongly holomorphic along W'_0 (I, Theorem 10, §9). Since W'_0 is the variety of the ideal $\mathfrak{m}'_0$ (§13, Lemma), it follows that ξ'^*_0 belongs to the $\mathfrak{m}'_0$- adic completion of the ring R'_0 (I, Theorem 1, §2). We have therefore

(13) $$\xi'^*_0 = \lim f_i(t),$$

where $\{f_i(t)\}$ is a Cauchy sequence in the $\mathfrak{m}'_0$ - adic ring R'_0. Similarly we find that

(14) $$\xi'^*_\infty = \lim g_i(1/t),$$

where $\{g_i(1/t)\}$ is a Cauchy sequence in the $\mathfrak{m}'_\infty$ -adic ring R'_∞. Here $f_i(X)$ and $g_i(X)$ are polynomials with coefficients in R. Replacing, if necessary, each of these two sequences by a suitable subsequence, we may assume that

(15) $$f_i(t) - f_j(t) \in \mathfrak{m}'^{i}_0, \quad j \geq i,$$

and that similarly

(16)
$$g_i(1/t) - g_j(1/t) \in \mathfrak{m}'^{\,i}_\infty \quad , j \geq i.$$

Let P' be any common point of W'_0 and W'_∞ , and let $\mathfrak{m}_{P'}$ be the maximal ideal of the local ring $o_{P'}$ of P'. We have $\mathfrak{m}'_0 \subset \mathfrak{m}_{P'}$, $\mathfrak{m}'_\infty \subset \mathfrak{m}_{P'}$, and hence (15) and (16) continue to hold if $\mathfrak{m}'_0$ and $\mathfrak{m}'_\infty$ are replaced by $\mathfrak{m}_{P'}$. Moreover, in the local ring $o_{P'}$, the two sequences $\{f_i(t)\}$ and $\{g_i(1/t)\}$ are equivalent, since they both converge to the analytical element $\xi'^*_{P'}$ of the given function ξ'^*. Hence it follows from (15) and (16) that

(17)
$$f_i(t) - g_i(1/t) \in \mathfrak{m}^i_{P'}, \quad \underline{\text{for all}}\ P' \ \underline{\text{in}}\ W'_0 \cap W'_\infty \ .$$

Let s_i be the degree of the polynomial $g_i(X)$. We set $g_i(1/X) = \varphi_i(X)/X^{s_i}$, where $\varphi_i(X)$ is a polynomial of degree s_i, with coefficients in R. We may multiply (17) by t^{s_i} , since $t \in o_{P'}$ for all points P' of W'_0, and we thus obtain the following relations:

(18)
$$t^{s_i} f_i(t) - \varphi_i(t) \in \mathfrak{m}^i_{P'}, \quad \underline{\text{for all}}\ P' \ \underline{\text{in}}\ W'_0 \cap W'_\infty \ .$$

Relations (18) do not necessarily hold for those points P' of W'_0 which do not belong to W'_∞ . These are the points of W' which are covered by the ring R'_0 and are not covered by the ring R'_∞ . It has been pointed out in section 13 that any such point P' is characterized by the condition that t is a non-unit in the local ring $o_{P'}$. Hence if $P' \in W'_0$ and $P' \notin W'_\infty$, then $t \in \mathfrak{m}_{P'}$. Therefore it follows from (18) that

(19)
$$t^{s_i+i} f_i(t) - t^i \varphi_i(t) \in \mathfrak{m}^i_{P'}, \quad \underline{\text{for all}}\ P' \ \underline{\text{in}}\ W'_0.$$

Denote by $F_i(X)$ the polynomial $X^{s_i+i} f_i(X) - X^i \varphi_i(X)$. Relations (19) show that the sequence $\{F_i(t)\}$ converges uniformly to zero on W'_0. It follows (I, first part of the proof of Theorem 1, § 2), that $\{F_i(t)\}$ is a zero sequence in the $\mathfrak{m}'_0$ - adic ring R'_0.

We have therefore

(20)
$$F_i(t) \in \mathfrak{m}'^{\,\rho(i)}_0 \quad , \lim \rho(i) = +\infty \ .$$

We now apply the above lemma. As polynomial f(X) of the lemma we take the polynomial $F_i(X)$. The corresponding polynomials $f_j(X)$ of the lemma will now be denoted by $F_{ij}(X)$. If u_i denotes the leading coefficient of $\varphi_i(X)$ and if l_i denotes the degree of $f_i(X)$, we find at once that $F_{il_i}(X) = f_i(X) - u_i$. Hence by Lemma 2, there exists an element v_i in R such that $f_i(t) - u_i - v_i \in \mathfrak{m}_o'^{\rho(i)}$. We set $u_i + v_i = \xi_i$. Then ξ_i is an element of R, and we have:

$$f_i(t) - \xi_i \in \mathfrak{m}_o'^{\rho(i)}, \ \xi_i \in R.$$

The two sequences $\{f_i(t)\}$ and $\{\xi_i\}$ are, therefore, equivalent sequences in the $\mathfrak{m}_o'$- adic ring R_o', and hence we have by (13):

(21) $\xi_o'^* = \lim \xi_i$.

It has been pointed out earlier [see (12)] that high powers of $\mathfrak{m}_o'$ contract in R to high powers of $\mathfrak{m}$. It follows that the sequence $\{\xi_i\}$, <u>which consists of elements of</u> R, converges uniformly also on W. It therefore defines a function ξ^* on V which is holomorphic along W:

(22) $\lim \xi_i = \xi^*$ (along o_W^*).

<u>We have therefore shown the existence of an element</u> ξ^* <u>of</u> o_W^* <u>such that</u> $H_{W,W'}(\xi^*)$ <u>coincides with</u> ξ'^* <u>at every point of</u> W_o'. Interchanging the roles of W_o' and W_∞', we conclude that there also exists an element η^* of o_W^* such that $H_{W,W'}(\eta^*)$ coincides with ξ'^* at every point of W_∞'. If P is a general point of W, then from our assumption that neither t nor 1/t is the reciprocal of a non-unit in o_P it follows that the affine varieties W_0' and W_∞' have points in common; in fact they have in common all the general points of T[W], in particular, all the general points of T[W] which correspond to P (see Remark 2 at the end of §13). Let, then, P' be a common point of W_0' and W_∞' such that P and P' are corresponding points. Since $H_{W,W'}(\xi^*)$ and $H_{W,W'}(\eta^*)$ have the same analytical element at P', namely $\xi_{P'}'^*$, and since $H_{P,P'}$ is an isomorphism, it follows that ξ^* and η^* have the same analytical element at P, and therefore $\xi^* = \eta^*$, by I, Theorem 5, §6. Hence $H_{W,W'}(\xi^*)$ coincides with ξ'^*, since these two functions coincide locally at each point of the variety W'. This completes the proof of the fundamental theorem in the case of the birational transformation V--> Vot.

§15. LAST STEP OF THE PROOF OF THE FUNDAMENTAL THEOREM: TRANSITION TO THE DERIVED NORMAL MODEL $\overline{\text{Vot}}$.

Let $\overline{V}$ denote a derived normal model of Vot (= V'), and let $\overline{T}$ denote the birational transformation of V' into $\overline{V}$. It is well known that if R' is a non-homogeneous coordinate ring of V' and if $\overline{R}$ denotes the integral closure of R' (in the quotient field of R'). then $\overline{R}$ is a non-homogeneous coordinate ring of $\overline{V}$, and $\overline{R}$ covers the total transform, under $\overline{T}$, of any subset of V' which is covered by R'. Therefore, if $\overline{R}_0$ and $\overline{R}_\infty$ denote the integral closure of R'_0 and R'_∞ respectively and if $\overline{W}$, $\overline{W}_0$, $\overline{W}_\infty$ denote respectively the total $\overline{T}$- transform of W', W'_0, W'_∞, then $\overline{R}_0$ and $\overline{R}_\infty$ are non-homogeneous coordinate rings of $\overline{V}$, $\overline{R}_0$ covers $\overline{W}_0$, $\overline{R}_\infty$ covers $\overline{W}_\infty$, and hence $\overline{W}(= \overline{W}_0 \cup \overline{W}_\infty)$ is covered by the pair of rings $\overline{R}_0$, $\overline{R}_\infty$. In order to simplify our notation we shall undertake the identification of certain rings which our preceding proofs have shown to be isomorphic. In the first place, we identify the ring o^*_W with the ring $o^*_{W'}$. This is permissible since we have proved in the preceding section that $H_{W,W'}$ is an isomorphism "onto". We shall denote the ring o^*_W by o^*. We also denote the rings $o^*_{W'_0}$ and $o^*_{W'_\infty}$ by o'^*_0 and o'^*_∞ respectively. Let ξ^* be any element of o^* and let P' be any point of W'. If we apply relation (4) of § 10 to $G = W$ and $G' = W'$ and if we observe that in view of the above identification of the rings o^*_W and $o^*_{W'}$ the isomorphism $H_{W,W'}$ is now the identity, we see that if P is the point of W which corresponds to P', then the analytical element of ξ^* at P' is the $H_{P,P'}$- transform of the analytical element of ξ^* at P. If $\xi^* \neq 0$, then ξ^* has a non-zero analytical element at each point P of W (I, Theorem 5, § 6). Since $H_{P,P'}$ is an isomorphism, we conclude that every non-zero element of o^* has a non-zero analytical element at <u>each</u> point of W'. It follows <u>a fortiori</u> that every non- zero element of o^* has a non-zero projection into the ring o'^*_0, i.e., the projection τ_{W',W'_0} induces an isomorphism of o^* into o'^*_0. We shall therefore identify the ring o^* with the corresponding subring of o'^*_0, so that now each element of o^* is its own projection into o'^*_0. In a similar fashion we shall regard o^* as a subring of o'^*_∞. Finally, we denote by $\overline{o}^*_0$ and $\overline{o}^*_\infty$ the rings $o^*_{\overline{W}_0}$ and $o^*_{\overline{W}_\infty}$ respectively, and we identify o'^*_0 with a subring of $\overline{o}^*_0$ and o'^*_∞ with a subring of $\overline{o}^*_\infty$. This is possible since, by Theorem 13, $H_{W'_0,\overline{W}_0}$ and $H_{W'_\infty,\overline{W}_\infty}$ are isomorphisms.

Let C'_0 be the conductor of R'_0 with respect to $\overline{R}_0$, and let similarly C'_∞ be the conductor of R'_∞ with respect to $\overline{R}_\infty$. We have $C'_0 \cap R \neq (0)$, since $C'_0 \neq (0)$ and since R and R'_0 have the same quotient field. Similarly we have $C'_\infty \cap R \neq (0)$. Therefore $C'_0 \cap C'_\infty \cap R \neq (0)$. We fix an element c in R, $c \neq 0$, such that c belongs to both conductors C'_0 and C'_∞. We will have then $c.\overline{R}_0 \subset R'_0$ and $c.\overline{R} \subset R'_\infty$. Let $\overline{\mathfrak{m}}_0$ be the extended ideal of $\mathfrak{m}'_0$ in the ring $\overline{R}_0$ and let $\overline{R}^*_0$ be the $\overline{\mathfrak{m}}_0$ - adic completion of $\overline{R}_0$. From $c\overline{R}_0 \subset R'_0$ it follows, in view of Lemma 2 of I, § 8, that $c.\overline{R}^*_0 \subset R'^*_0$, where $R_o'^*$ is the $\mathfrak{m}_o'$ - adic completion of the ring R_o'. Now $R_o'^*$ is the ring of functions on V' which are strongly holomorphic along W_o' (Theorem 1, §2), and since W_o' is an affine variety this ring coincides with $o_o'^*$ (I, §9, Theorem 10). Similarly, $\bar{R}_o^* = \bar{o}_o^*$. This same argument applies to the rings $o_\infty'^*$ and $\bar{o}_\infty^*$, and hence we have

(23) $$c.\bar{o}_o^* \subset o_o'^* , \qquad c.\bar{o}_\infty \subset o_\infty'^*, \qquad (o_o'^* \subset \bar{o}_o , o_\infty'^* \subset \bar{o}_\infty^*).$$

We denote by $\bar{o}^*$ the ring $o_{\bar{W}}^*$ and we proceed to the proof of our main objective, which is to prove that $H_{W,\bar{W}}$ is an isomorphism of o^* onto $\bar{o}^*$. Since $H_{W,\bar{W}}$ is at any rate an isomorphism "into", we identify o^* with a subring of $\bar{o}^*$, and our object now is to prove that o^* is the entire ring $\bar{o}^*$

Let $\bar{\xi}^*$ be any element of $\bar{o}^*$. Let $\bar{\xi}_o^*$ and $\bar{\xi}_\infty^*$ be the projections of $\bar{\xi}^*$ into $\bar{o}_o^*$ and $\bar{o}_\infty^*$ respectively, and let $\eta_o'^* = c\cdot\bar{\xi}_o^*, \eta_\infty'^* = c\cdot\bar{\xi}_\infty^*$. By (23), the elements $\eta_o'^*$ and $\eta_\infty'^*$ belong to the rings $o_o'^*$ and $o_\infty'^*$ respectively. They are, therefore, holomorphic functions on V', defined along W_o' and W_∞' respectively. If P' is a point of W_o', and $\eta_{oP'}'^*$ is the analytical element of $\eta_o'^*$ at P', then it follows from relations (4) of §10 that the function $\bar{\xi}^*$ coincides on $\bar{T}\{P'\}$ with the holomorphic function $H_{P',\bar{T}\{P'\}}(\eta_{oP'}'^*)$. A similar remark applies to the function $\eta_\infty'^*$ and to any point of W_∞'. Since $H_{P',\bar{T}\{P'\}}$ is an isomorphism of $o_{P'}^*$, we conclude that <u>if P' is any common point of W_o' and W_∞' then $\eta_o'^*$ and $\eta_\infty'^*$ have the same analytical element at P'</u>. It follows that $\eta_o'^*$ and $\eta_\infty'^*$ coincide with one and the same element of the common subring o^* of $o_o'^*$ and $o_\infty'^*$ (this common subring o^* being the ring of functions on V' which are holomorphic along W'). Hence if we set $\eta_o'^* = \eta_\infty'^* = \eta^*$, then we have proved that

(24) $$c.\ \bar{\xi}^* = \eta^* \in o^*,$$

i.e. <u>we have proved that if $\bar{\xi}^*$ is any element of $\bar{o}^*$, then the product $c\cdot\bar{\xi}^*$ belongs to</u> o^*. If we now apply this result to any power of $\bar{\xi}^*$, we find $c\cdot\bar{\xi}^{*\nu} = \eta_\nu^*$, where η_ν^* is a suitable element of o^*. Comparing with (24), we have $\eta^{*\nu} = c^{\nu-1}\eta_\nu^*$, <u>and hence $c^{\nu-1}$ divides $\eta^{*\nu}$ in o^* (ν - any positive integer)</u>. We proceed to show that this result implies <u>that c divides η^* in o^*</u>. This will prove that $\bar{\xi}^* \in o^*$ (and that consequently $o^* = \bar{o}^*$) since c, being an element of the function field of $\bar{V}$, is not a zero divisor of the ring $\bar{o}^*$. The proof of the fundamental theorem will then be complete.

To show that c divides η^* in o^*, it is sufficient to show that c divides η^* locally at each point P of W, i.e., that c **divides** η_P^* in o_P^*, where η_P^* is the analytical element of η^* at P [see I, §9, Theorem 12]. What we know is that $c^{\nu-1}$ divides $\eta_P^{*\nu}$ in o_P^*, for any positive integer ν. Since V is normal, the local domain o_P is integrally closed. We shall now use certain results which we have proved in our paper [8], and which concern the minimal primes ideals in o_P and their extensions in o_P^*.

Let

$$(25) \qquad o_P \cdot c = \mathfrak{p}_1^{(\alpha_1)} \cap \mathfrak{p}_2^{(\alpha_2)} \cap \ldots \cap \mathfrak{p}_h^{(\alpha_h)}$$

be the decomposition of the principal ideal $o_P \cdot c$ into symbolic powers of minimal prime ideals $\mathfrak{p}_i$ in o_P. It is known that each prime ideal in o_P is analytically unramified (Chevalley [1], Lemma 9 and Theorem 1; also Zariski [8], §3). Let then

$$(26) \qquad o_P^* \cdot \mathfrak{p}_i = \mathfrak{p}_{i1}^* \cap \mathfrak{p}_{i2}^* \cap \ldots \cap \mathfrak{p}_{ig_i}^*, \quad i = 1, 2, \cdots, h,$$

where the $\mathfrak{p}_{ij}^*$ are prime ideals in o_P^*. From Lemmas 7 and 8 of our paper [5] it follows that

$$(27) \qquad o_P^* \cdot c = \bigcap_{i=1}^{h} \bigcap_{j=1}^{g_i} \mathfrak{p}_{ij}^{*(\alpha_i)}.$$

Using the main result ("V normal" implies " V is analytically irreducible at each point") and lemma 3 of our quoted paper [8], we have that the quotient ring of each ideal $\mathfrak{p}_{ij}^*$ in o_P^* is a discrete valuation ring. Let v_{ij} be the corresponding valuation, and let $v_{ij}(\eta_P^*) = n_{ij}$. Since $c^{\nu-1}$ divides $\eta_P^{*\nu}$ in o_P^*, and since we have, by (27), $v_{ij}(c) = \alpha_i$, it follows that $\nu n_{ij} \geq (\nu-1)\alpha_i$. Since this inequality must hold for any integer ν, we conclude that $n_{ij} \geq \alpha_i$, where $i = 1, 2, \cdots, h$ and where, for a given i, the index j takes the values $1, 2, \cdots, g_i$. From these inequalities it follows, in view of $\eta_P^* \in \mathfrak{p}_{ij}^{*(n_{ij})}$, that $\eta_P^* \in \mathfrak{p}_{ij}^{*(\alpha_i)}$ and hence, by (27), c divides η_P^* in o_P^*, as asserted.

This completes the proof of the fundamental theorem.

§16. EXTENSION OF THE FUNDAMENTAL THEOREM TO RATIONAL TRANSFORMATIONS

We consider an irreducible variety V' and a rational transformation T^{-1} of V' into another irreducible variety V. Let (A,A') be a general point pair of T (T being now regarded as an irreducible correspondence between V and V'). Then A is a general point of V, A' is a general point of V', and the field k(A) is contained in the field k(A'). We shall denote these fields by Σ and Σ' respectively. These are fields of rational functions on V and V' respectively.

We say that T^{-1} is <u>semi-regular</u> at a point P' of V' if to P' there corresponds a single point P on V and if $o_P \subset o_{P'}$. If that is so, we write $P < P'$. More generally, if G is any subset of V and G' is any subset of V', we shall write $G < G'$ if G is the total transform of G' under T^{-1} (in symbols: $G = T^{-1}\{G'\}$) and if T^{-1} is semi-regular at each point of G'.

All the results which have been established in § 10, where T was a birational transformation, continue to hold in the present case of a rational transformation. In the first place, if $P < P'$, then $\mathfrak{m}_P \subset \mathfrak{m}_{P'}$, and we have, then, a natural homomorphism $H_{P,P'}$ of o_P^* into $o_{P'}^*$. For arbitrary subsets G and G' of V and V', such that $G < G'$, we have again a natural homomorphism of o_G^* into $o_{G'}^*$. This homomorphism, which we shall denote again by $H_{G,G'}$, satisfies relations (4) and (5). Concerning (5), it must be understood that V' is a rational transform of V". As to Theorem 13, the proof remains essentially unaltered. To see this, a few words of explanation will suffice. The first part of the proof dealt with the case in which V is locally normal at each point of G [assumption (a) of the theorem]. Here we have used the fact that if $P < P'$ and V is locally normal at P, then o_P is a subspace of $o_{P'}$. In the quoted Proceedings note [9] we have established this fact not only for birational but also for rational transformations. Therefore, the first part of the proof of Theorem 13 is valid also for rational transformations. The second part of the proof dealt with the case (b) of Theorem 13, under the additional assumption that V' is a derived normal model of V. By the very nature of this assumption, this part of the proof is irrelevant when T is not a birational transformation and should therefore be omitted. In the last part of the proof, the transformation S^{-1} will now be a rational transformation of V_1' into V_1. The established fact that S^{-1} is finitely valued at each point of G_1' will imply also in the present case that S^{-1} is semi-regular at each point of G_1' (the proof of this property of any rational transformation S^{-1} of a normal variety V_1' is the same as for birational transformations). Hence we have again the relation $G_1 < G_1'$, and therefore (7) is still valid. Now H_{G,G_1} is an isomorphism, by the birational case of Theorem 13, and $H_{G_1, G_1'}$ is an isomorphism by case (a) of Theorem

13. From this it follows, in view of (6), that also $H_{G,G'}$ is an isomorphism.

We shall now state our generalization of the fundamental theorem.

THE GENERALIZED FUNDAMENTAL THEOREM. Let T^{-1} be a rational transformation of an irreducible variety V' into an irreducible variety V, and let W be a subvariety of V. If W' denotes the total transform $T\{W\}$ of W, then $H_{W,W'}$ is an isomorphism of o_W onto $o_{W'}$, provided the following conditions are satisfied: (1) V is locally normal at each point of W: (2) $W < W'$; (3) the function field Σ of V is maximally algebraic in the function field Σ' of V'.

Note condition (3) above. This condition, which is automatically satisfied in the case of a birational transformation T, will re-appear in its natural geometric setting when we shall discuss the principle of degeneration.

§ 17. REDUCTION OF THE PROOF TO A SPECIAL CASE.

Let s be the degree of transcendence of Σ' over Σ. If s= 0, then Σ' is an algebraic extension of Σ, and hence we must have $\Sigma = \Sigma'$, in view of condition (3). Hence if s= 0, then we are dealing with a birational transformation T, and in this case the theorem has already been proved. We shall therefore assume that $s > 0$.

As in the birational case, it is seen also in the present case that it is permissible to replace V by any derived normal model of V and that it is also permissible to replace V' by any birationally equivalent variety V_1' such that $V' < V_1'$. In particular, we may replace V' by a derived normal model of the graph of the transformation T. Hence we may assume that both V and V' are normal and that $V < V'$.

At this stage we must open a parenthesis in order to give an extension of the concept of a derived normal model. Let V be now an arbitrary irreducible variety in an n-dimensional projective space, and let $y_0, y_1, \cdots, y_n$ be homogeneous coordinates of a general point of V. Let Σ be the function field of V which is generated, over the ground field k, by the quotients of the quantities y_i. We shall assume that y_0 is a transcendental over Σ. Let K be a finite algebraic extension of Σ. We shall define the concept of a derived normal model of V in the field K.

We regard y_0 as a transcendental over K and we consider the field $K(y_0)$. The elements of this field which are of the form uy^i, $u \in K$, will be called <u>homogeneous</u> (of degree i). We consider the elements of $K(y_0)$ which are integral functions of the y's. It can be shown that any such function is a sum of integral functions which are homogeneous, of degree ≥ 0. Since it is known that the integral closure of the ring $k[y]$ ($=k[y_0, y_1, \cdots y_n]$) in $K(y_0)$ is a finite module over $k[y]$, it follows that the integral functions of the y's, which are homogeneous of a given degree $m \geq 0$, form a finite-dimensional vector space over k. Let, for a given $m > 0$, $\omega_0, \omega_1, \ldots, \omega_N$ be a basis of that space, and let $\overline{V}_m$ be the irreducible variety whose general point has the ω's as homogeneous coordinates. It is clear that $\overline{V}_m$ is a projective model of K. It can be shown that if m is sufficiently large, the varieties $\overline{V}_m$ are all normal and any two of them are in regular birational correspondence. Any of these normal varieties $\overline{V}_m$ is called <u>a derived normal model of</u> V <u>in the field</u> K. The proofs of the various statements made above are the same as in the birational case ($\Sigma = K$); see Zariski [5], [6].

We shall now show that the proof of the generalized fundamental theorem can be reduced to the following special case:

s= 1 <u>and</u> V' <u>is a derived normal model, in the field</u> Σ', <u>of a variety of the form</u> Vot, <u>where</u> t <u>is an element of</u> Σ' <u>which is transcendental with respect to</u> Σ. Here the symbol Vot has the same meaning as in § 12: if $(y_0, y_1, \cdots y_n)$ is the general point of V, then Vot is the variety whose general point is $(y_0, y_1, \ldots, y_n, y_0 t, y_1 t, \ldots, y_n t)$.

To see this, we proceed as in §12. Let $y_0', y_1', \cdots y_n'$ be homogeneous coordinates of the general point of V', and let $t_1, t_2, \cdots, t_N$ denote the quotients y_i'/y_j', $i < j$, taken in some order. We now introduce the following sequence of N + 1 fields Σ_α, $0 \leq \alpha \leq N$: (a) $\Sigma_0 = \Sigma$: (b) Σ_α is the relative algebraic closure of $\Sigma_{\alpha-1}(t_\alpha)$ in the field Σ' ($1 \leq \alpha \leq N$). We next introduce the following sequence of N + 1 pairs of varieties $(V_\alpha, \overline{V}_\alpha)$: (a) $V_0 = \overline{V}_0 = V$; (b) $V_\alpha = \overline{V}_{\alpha-1} o t_\alpha$, $\overline{V}_\alpha$ = derived normal model of V_α <u>in the field</u> Σ_α. We have $\Sigma_N = \Sigma'$, and hence $\overline{V}_N$ and V' are birationally equivalent varieties. Moreover $V' < \overline{V}_N$. It follows as in § 12 that it is sufficient to prove the fundamental theorem for each pair of varieties $\overline{V}_{\alpha-1}, \overline{V}_\alpha$. Now, if t_α happens to be algebraic over the field $\Sigma_{\alpha-1}$, then $t_\alpha \in \Sigma_{\alpha-1}$, for $\Sigma_{\alpha-1}$ is maximally algebraic in Σ_α. In this case we are dealing with a birational transformation, and the fundamental theorem is applicable to the pair $(\overline{V}_{\alpha-1}, \overline{V}_\alpha)$. If t_α is transcendental over $\Sigma_{\alpha-1}$, then we are dealing with the special case described above, since $\Sigma_{\alpha-1}$ is maximally algebraic in Σ_α.

We now proceed to deal with the above special case. As in the birational case, so also in the present case, we may replace V and W by their affine parts (relative to a fixed hyperplane at infinity). We therefore assume that V and W are affine varieties. However, V' remains a projective model, and W' is the total transform on V' of the affine variety W. Moreover, we may assume that W is irreducible (see §14).

§18. THE TRANSFORMATION V--> Vot.

In this section we are dealing with case in which V' = Vot (and hence $\Sigma' = \Sigma(t)$. Let $x_1, x_2, \cdots, x_n$ be the non-homogeneous coordinates of the general point of V, and let $R = k[x_1, x_2, \cdots, x_n] = k[x]$. Let $\mathfrak{m}$ be the prime ideal of W in the coordinate ring R. We set $R'_0 = R[t], R'_\infty = R[1/t]$. These two rings are non-homogeneous coordinate rings of Vot, and together they cover W'. Since t is a transcendental over the field Σ, it is clear that Vot is the direct product of V and a projective line (over k). Hence the total transform of any irreducible subvariety of V is irreducible. In particular, if we denote by W'_0 and W'_∞ the portions of W' which are covered respectively by the rings R'_0 and R'_∞ , then W'_0 and W'_∞ are the affine parts of one and the same irreducible subvariety of V'. We set $\mathfrak{m}'_0 = R'_0 \cdot \mathfrak{m}$, $\mathfrak{m}'_\infty = R'_\infty \cdot \mathfrak{m}$. Then $\mathfrak{m}'_0$ and $\mathfrak{m}'_\infty$ are prime ideals (since t is a transcendental over R), and their varieties are W'_0 and W'_∞ respectively.

The proof of the fundamental theorem in the present case is in all respects similar to the proof given in §14 for the birational case, and is in fact a good deal simpler, because this time t is a transcendental over the function field Σ of V. The simplification takes the following form: while several points of the proof developed in §14 depended essentially on some deeper properties of normal varieties, in the present case these same points refer to fairly trivial facts in which the normality of V plays no role whatsoever. In fact, the whole proof is valid for an arbitrary variety V. It will be sufficient to review the salient points of the proof.

(a) High powers of $\mathfrak{m}'_0$ (or of $\mathfrak{m}'_\infty$) contract in R to high powers of $\mathfrak{m}$. In the birational case, the proof of this statement was based on Theorem 7 (I, §7) and on properties of normal varieties proved in our Proceedings Note [9]. In the present case the above statement is trivial, and we can even assert that $\mathfrak{m}'^{i}_0 \cap R = \mathfrak{m}'^{i}_\infty \cap R = \mathfrak{m}^i$.

(b) The lemma of § 13 is also trivial, and the assumption that R is integrally closed is superfluous. As element $b_j (0 \leq j \leq g)$ we can take in the present case the "constant" term of the polynomial $f_j(t)$.

(c) There is no change in the rest of the proof, up to and including the point where we have shown the existence of two elements ξ^* and η^* in o^*_W such that (1) $H_{W,W'}(\xi^*)$ coincides with ξ'^* at every point of W'_0 and (2) $H_{W,W'}(\eta^*)$ coincides with ξ'^* at every point of W'_∞. We have shown in the birational case that $\xi^* = \eta^*$, by showing that ξ^* and η^* have the same analytical element at one point of W (namely at the general point of W) and by applying Theorem 5 of I, § 6. In the present case it is not necessary to use this theorem. If L denotes the set of points of the projective line other than the points t = 0 and t = ∞, then it is clear that $W'_0 \cap W'_\infty = W \times L$. Hence every point of W arises from some common point (in fact, from infinitely many common points) of $W'_0 \cap W'_\infty$, and from this it follows at once that ξ^* and η^* have the same analytical element at each point of W.

§ 19. THE TRANSFORMATION Vot --> $\overline{\text{Vot}}$.

We are now dealing with a field Σ' which is an algebraic extension of the field $\Sigma(t)$, where Σ is the function field of V. We denote by $\overline{V}$ a derived normal model $\overline{\text{Vot}}$ of Vot in the field Σ'. The variety Vot itself will be denoted by V'. We shall use the same notation and the same ring identifications as in § 15. For the sake of clarity we shall recall briefly the meaning of the various rings and varieties that will occur in the rest of the proof.

The variety V is an affine model, and R is a non-homogeneous coordinate ring of V. We are given on V an irreducible subvariety W. The ring of functions on V which are holomorphic along W is denoted by R*. This ring is the $\mathfrak{m}$-adic completion of R, where $\mathfrak{m}$ is the prime ideal of W in R.

For the variety V'= Vot we have to introduce two non-homogeneous coordinate rings: $R'_0 = R[t]$ and $R'_\infty = R[1/t]$. The subset of V' which is the total transform of the affine variety W is denoted by W'. The two rings R'_0 and R'_∞ cover together the entire set W'. The subsets of W' which are covered by R'_0 and R'_∞ respectively are denoted by W'_0 and W'_∞. These two subsets are affine varieties. The ring of functions on V' which are holomorphic along W'_0 is denoted by R'^*_0. This ring is the $\mathfrak{m}'_0$-adic completion of the coordinate ring R'_0, where $\mathfrak{m}'_0$ is the extended ideal of $\mathfrak{m}$ in R'_0. The ring denoted by R'^*_∞ has a similar significance for W'_∞ and R'_∞. Our ring identifications signify that R* is to be regarded as a subring of both rings R'^*_0 and R'^*_∞ and also as the ring of functions on V' which are holomorphic along W'. This last identification is permissible in view of the proof given in the preceding section. We have then:

(28) $$R \subset R^* \subset R_0'^* \cap R_\infty'^* ,$$

(28') $$R \subset R_0' \subset R_0'^* , \quad R \subset R_\infty' \subset R_\infty'^* .$$

For the variety $\bar{V}$ we also have to introduce two non-homogeneous coordinate rings: $\bar{R}_0$ = integral closure of R_0' in the field Σ' and $\bar{R}_\infty$ = integral closure of R_∞' in the same field Σ'. The total transform of W' on $\bar{V}$ is denoted by $\bar{W}$, and the subsets of $\bar{W}$ which are covered by the ring $\bar{R}_0$ or the ring $\bar{R}_\infty$ are denoted by $\bar{W}_0$ and $\bar{W}_\infty$ respectively. The sets $\bar{W}_0$ and $\bar{W}_\infty$ are affine varieties, and $\bar{W}$ is their set- theoretic sum. The ring of functions on $\bar{V}$ which are holomorphic along $\bar{W}_0$ is denoted by $\bar{R}_0^*$. This ring is the $\bar{\mathfrak{m}}_0$- adic completion of $\bar{R}_0$, where $\bar{\mathfrak{m}}_0$ is the extended ideal of $\mathfrak{m}_0'$ in $\bar{R}_0$. Since $\bar{R}_0$ is a finite module over R_0', the ring $R_0'^*$ can be regarded as a subring of $\bar{R}_0^*$, and $\bar{R}_0^*$ is a finite module over $R_0'^*$. The ring $\bar{R}_\infty^*$ has a similar meaning for $\bar{W}_\infty$ and $\bar{R}_\infty$. Furthermore, the ring of functions on $\bar{V}$ which are holomorphic along $\bar{W}$ is denoted by $\bar{R}^*$. The other ring identifications carried out in § 15 (and which can be carried out also in the present case) signify that R^* is to be thought as a subring of $\bar{R}^*$ and that the ring $\bar{R}^*$ itself is to be regarded as a subring of both rings $\bar{R}_0^*$ and $\bar{R}_\infty^*$. We have therefore the following inclusions:

(29) $$R_0'^* \subset \bar{R}_0^* , \quad R_\infty'^* \subset \bar{R}_\infty^* ,$$

(29') $$R^* \subset \bar{R}^* \subset \bar{R}_0 \cap \bar{R}_\infty^* .$$

Our objective is to prove the equality: $R^* = \bar{R}^*$.

We now regard Σ' as a field of algebraic functions of the one variable t, with Σ as ground field. Let m be the relative degree of Σ' over $\Sigma(t)$, and let $\{\omega_{1o}, \omega_{2o}, \cdots, \omega_{mo}\}$ and $\{\omega_{1\infty}, \omega_{2\infty}, \cdots, \omega_{m\infty}\}$ be complementary normal integral bases in Σ' (in the sense of Dedekind-Weber) for the rings of integral functions of t and 1/t respectively. We may assume that $\omega_{1o} = \omega_{1\infty} = 1$. We denote by r_i the exponent of ω_{io}. We will have then $r_1 = 0$ and

(30) $$\omega_{i\infty} = \omega_{io}/t^{r_i} .$$

The ω_{io} are integral over $\Sigma[t]$. Upon multiplication of the ω_{io}, $i > 1$, by suitable elements of the "ground" field Σ, we can arrange matters so that

(a) the ω_{io} be integral over $R[t]$.

This is possible, since Σ is the quotient field of R.

Let

(31) $$\omega_{io}\,\omega_{jo} = \sum_{\nu=1}^{m} u_{ij}^{(\nu)}\,\omega_{\nu o},\quad u_{ij}^{(\nu)} \in \Sigma[t]\ ,$$

be the multiplication table for the ω_{io}. We have $u_{ij}^{(\nu)} = v_{ij}^{(\nu)}/b$, where $v_{ij}^{(\nu)} \in R[t]$ and $b \in R$. If we set $\omega_{1o}' = \omega_{1o}$, $\omega_{io}' = b\,\omega_{io}$, $i = 2,3,\dots,m$, then the new normal integral basis $\{\omega_{io}'\}$ still consists of elements which are integral over R[t], but now the coefficients of the multiplication table are elements of R[t]: $\omega_{io}'\,\omega_{jo}' = \sum_{\nu=1}^{m} v_{ij}^{(\nu)}\,\omega_{\nu o}'$. We shall therefore assume that the original basis $\{\omega_{io}\}$ satisfies condition (a) <u>and</u> the further condition that the coefficients $u_{ij}^{(\nu)}$ in (31) all belong to R[t] (= R_o'). Under these conditions we have the following: <u>if we set</u>

(32) $$\Omega_o = \sum_{i=1}^{m} R_o' \cdot \omega_{io},$$

<u>then Ω_o is a ring, and moreover</u>

(32') $$R_o' \subset \Omega_o \subset \overline{R}_o\ .$$

Let

(33) $$\omega_{io}^{m} + c_{i1}\omega_{io}^{m-1} + \dots + c_{im} = 0,\ c_{ij} \in R[t],$$

be the field equation for ω_{io} over $\Sigma(t)$.[19] By well known properties of normal integral bases, we have that $c_{ij}/t^{jr_i} \in \Sigma[1/t]$. Since $c_{ij} \in R[t]$, it follows that $c_{ij}/t^{jr_i} \in R[1/t]$(= R_∞'). We conclude therefore at once that

(b) <u>the $\omega_{i\infty}$ are integral over</u> R[1/t].

From (30) and (31) we find that the coefficients of the multiplication table of the $\omega_{i\infty}$ are $u_{ij}^{(\nu)}/t^{r_i + r_j - r_\nu}$. Since these coefficients must belong to $\Sigma[1/t]$ and since the $u_{ij}^{(\nu)}$ are in R[t], it follows that these coefficients belong to R[1/t] (= R_∞'). Hence if we set

(34) $$\Omega_\infty = \sum_{i=1}^{m} R_\infty' . \omega_{i\infty}\ ,$$

<u>then Ω_∞ is a ring, and moreover</u>

(35) $$R_\infty' \subset \Omega_\infty \subset \overline{R}_\infty\ .$$

The elements of $\bar{R}_o$ are integral over R'_o (= R[t]) and hence are integral functions of t. Therefore $\bar{R}_o \subset \Sigma[t]\omega_{1o} + \Sigma[t]\omega_{2o} + \ldots + \Sigma[t]\omega_{mo}$. On the other hand, $\bar{R}_o$ is a finite R'_o- module. From these two facts it follows at once that there exists an element c_o in R, $c_o \neq 0$, such that $c_o \cdot \bar{R}_o \subset \Omega_o$. Similarly, there exists an element c_∞ in R, different from zero, such that $c_\infty \cdot R_\infty \subset \Omega_\infty$. We set $c_o \cdot c_\infty = c$, and we have then

(36) $$c.\bar{R}_o \subset \Omega_o, \quad c \cdot \bar{R}_\infty \subset \Omega_\infty .$$

The elements ω_{1o} belong to $\bar{R}_o$, hence also to $\bar{R}^*_o$. Similarly, the $\omega_{i\infty}$ belong to $\bar{R}^*_\infty$. Hence, in view of (29) and the fact that the coefficients $u_{ij}^{(\nu)}$ in (31) belong to R[t], the following sets

(37) $$\Omega^*_o = \sum_{i=1}^m R'^*_o \omega_{io}, \quad \Omega^*_\infty = \sum_{i=1}^m R'^*_\infty \omega_{i\infty}$$

are rings and are subrings of $\bar{R}^*_o$ and $\bar{R}^*_\infty$ respectively. Note that since the ω_{io} are linearly independent over R'_o, they are also linearly independent over R'^*_o (I, §8, Lemma 2). Similarly, the $\omega_{i\infty}$ are linearly independent over R'^*_∞. In view of the relations $c.\bar{R}_o \subset \Omega_o$, $c.\bar{R}_\infty \subset \Omega_\infty$, it follows that

(38) $$c\bar{R}^*_o \subset \Omega^*_o, \quad c\bar{R}^*_\infty \subset \Omega^*_\infty .$$

Now let $\bar{\xi}^*$ be any element of $\bar{R}^*$. By (29'), $\bar{\xi}^*$ belongs both to $\bar{R}^*_o$ and $\bar{R}^*_\infty$. From (37) and (38) we see that we have two expressions for $c\bar{\xi}^*$, according as this product is regarded as an element of Ω^*_o or Ω^*_∞:

(39) $$c\bar{\xi}^* = \sum_{i=1}^m \alpha^*_{io} \omega_{io}, \quad \alpha^*_{io} \in R'^*_o ;$$

(39') $$c\bar{\xi}^* = \sum_{i=1}^m \alpha^*_{i\infty} \omega_{i\infty}, \quad \alpha^*_{i\infty} \in R'^*_\infty .$$

Let P' be any common point of W'_o and W'_∞, and let $\bar{G}$ denote the (finite) set of points which correspond to P' on $\bar{V}$. The set $\bar{G}$ belongs to $W_o \cap W_\infty$, and therefore we can apply to both sides of (39) and (39') the projections $\tau_{W_o, \bar{G}}$ and $\tau_{W_\infty, \bar{G}}$ respectively. We then obtain the following relation:

(40) $$\sum_{i=1}^m \alpha^*_{io,\bar{G}} \omega_{io} = \sum_{i=1}^m \alpha^*_{i\infty,\bar{G}} \omega_{i\infty} ,$$

where we have set $\alpha^*_{io,\bar{G}} = \tau_{W_o,\bar{G}}(\alpha^*_{io})$, $\alpha^*_{i\infty,\bar{G}} = \tau_{W_\infty,\bar{G}}(\alpha^*_{i\infty})$.

Since $P' \in W'_o \cap W'_\infty$, t is a unit in o'_P. Since $o_{P'} \subset o^*_{P'} \subset o^*_{\bar{G}}$, t is also a unit of $o^*_{\bar{G}}$. We therefore obtain from (40) and (30) the following relation in $o^*_{\bar{G}}$:

$$(41) \qquad \sum_{i=1}^{m}(\alpha^*_{i0,\bar{G}} - \frac{\alpha^*_{i\infty,\bar{G}}}{t^{r_i}})\omega_{i0} = 0.$$

From known properties of completions of semi-local rings it follows that the ω_{i0} are linearly independent over $o^*_{\mathfrak{P}'}$ [or, more precisely: over $H_{P',\bar{G}}(o^*_{P'})$]. Hence we conclude from (41) that $\alpha^*_{i0,\bar{G}} = \alpha_{i\infty,\bar{G}}/t^{r_i}$. Since $H_{P',\bar{G}}$ is an isomorphism, it follows that α^*_{i0} and $\alpha^*_{i\infty}/t^{r_i}$ have the same analytical element at P'. Since this is so for every common point of W'_0 and W'_∞, it follows that the two functions α^*_{i0} and $\alpha^*_{i\infty}/t^{r_i}$ arise, by projection, from one and the same element of the ring R^* [$= o^*_{W'}$]. In view of our previous ring identifications [see, (28) and (28')], we conclude that

$$(42) \qquad \alpha^*_{i0} = \alpha^*_{i\infty}/t^{r_i} = \alpha^*_i \in R^*.$$

In a similar fashion it follows that

$$(42'). \qquad \alpha^*_{i\infty} = \alpha^*_{i0}\, t^{r_i} = \beta^*_i \in R^* .$$

We have here two elements α^*_i and β^*_i of the ring R^*, which are related as follows: $\beta^*_i = \alpha^*_i \cdot t^{r_i}$. However, it is obvious that t is a transcendental over R^*[20]. We conclude therefore that $\alpha^*_i = 0$, $i = 2, 3, \dots, m$, since Σ is maximally algebraic in Σ' and since therefore the exponents r_i are positive for $i > 1$. We have, then, by (39) and (42): $c\,\bar{\xi}^* = \alpha^*_{00} = \alpha^*_0 \in R^*$. We have thus shown that $c\bar{R}^* \subset R^*$. This is the analogue of (24) of §15. The reasoning used in §15 in order to deduce from (24) that $R^* = \bar{R}^*$ can be applied also to the present case, without any change whatsoever. Hence $R^* = \bar{R}^*$, and this completes the proof of the generalized fundamental theorem.

PART III.
THE PRINCIPLE OF DEGENERATION.

§20. A CONNECTEDNESS THEOREM FOR ALGEBRAIC CORRESPONDENCES.

The fundamental theorem, as proved for rational transformations in Part II, implies a connectedness theorem for algebraic correspondences similar to the connectedness theorem proved in II, §11 for birational transformations. This theorem is as follows:

THEOREM 14. Let T be an irreducible algebraic correspondence between two irreducible algebraic varieties V/k and V'/k, and let (P,P') be a general point pair of T/k. Let W/k be a subvariety of V and let $W' = T\{W\}$ be the total transform of W on V. It is assumed that the following conditions are satisfied:

(a) The field k(P) is maximally algebraic in the field k(P,P').

(b) The variety V/k is analytically irreducible at each point of W. Then if W/k is connected, also W'/k is connected.

The proof is the same as that of the connectedness theorem of II, §11. It is only necessary to observe the following: (1) The theorem is a direct consequence of the fundamental theorem if the following additional assumptions are made: T^{-1} is a rational transformation, V is locally normal at each point of W, and $W < W'$. (2) If V* denotes the graph of the algebraic correspondence T, then k(P,P') is the field of rational functions on V* and hence V is a rational transform of V*.

REMARK. The above theorem remains true if condition (a) is replaced by the following:

(a') <u>The field</u> k(P) <u>is quasi-maximally algebraic in</u> k(P,P').

To see this, let $(y_0, y_1, \cdots, y_m)$ be homogeneous coordinates of the point P' and let $\bar{P}$ be the point $(y_0^{p^h}, y_1^{p^h}, \cdots, y_m^{p^h})$, where h is an integer. If we denote by K the field k(P), then we have

$K(\bar{P}) = K.[K(P')]^{p^h}$. By assumption, every element of K(P') which is algebraic over K is pure inseparable over K. Since the relative algebraic closure of K in K(P') is a finite extension of K, it is clear that for a suitable integer h the field K will be maximally algebraic in $K(\bar{P})$. With this choice of h, Theorem 14 is applicable to the pair of varieties V and $\bar{V}$, where $\bar{V}$ is the variety over k whose general point if $\bar{P}$, and to the algebraic correspondence having $(P,\bar{P})$ as general point pair. On the other hand, the algebraic correspondencebetween $\bar{V}$ and V', defined by the general point pair $(\bar{P}, P')$, is (1,1) without exceptions and transforms therefore every connected subvariety of $\bar{V}$ into a connected subvariety of V'.

In Theorem 14 the subvariety W of V must be a variety defined over k. In particular, if W consists of only one point then the coordinates of that point must be pure inseparable quantities over k. In the proof of the principle of degeneration we shall require, however, the knowledge that Theorem 14 remains true if in the statement of that theorem the variety W/k is replaced by an <u>arbitrary</u> point of V. It is obvious that all we need for that purpose is the following theorem:

THEOREM 15. The fundamental theorem remains true if in the statement of that theorem the subvariety W of V is replaced by an arbitrary point Q of V.

PROOF. Let us assume first that Q is an algebraic point over k. Let W denote the set of conjugate points of Q over k. Then W is a variety defined over k. Since conjugate points have the same local ring, it follows that $o^*_Q = o^*_W$. Let $G' = T\{Q\}$, $W' = T\{W\}$. Then G' is a subset of W', and each point of W' has at least one conjugate point in the set G'. Therefore also $o^*_{G'}$ and $o^*_{W'}$ are identical rings, and our theorem is in this case a direct consequence of the fundamental theorem.

Let now Q be of dimension s over k, $s > 0$, and let again G' denote the set $T\{Q\}$. It is obvious that it is permissible to replace in the proof the point Q by any point which is k-isomorphic to Q. This being so, let (P,P') be a general point pair of T and let $z_1, z_2, \cdots, z_n$ be the non-homogeneous coordinates of Q; $x_1, x_2, \cdots, x_n$ the non-homogeneous coordinates of P. We may assume that $z_1, z_2, \cdots z_s$ are algebraically independent over k. Then necessarily also $x_1, x_2, \cdots, x_s$ are algebraically independent over k. Since Q can be replaced by any k-isomorphic point, we may assume that $z_i = x_i$, $i=1,2,\cdots,s$. We set $k_1 = k(x_1, x_2, \cdots, x_s)$ and we take k_1 as our new ground field. Let V_1 and V'_1 be the irreducible varieties <u>over</u> k_1 having respectively P and P' as general points. Since $k_1 \subset k(P) \subset k(P')$, we have $k_1(P) = k(P)$ and $k_1(P') = k(P')$, and hence the function fields of V and V' are not affected by our ground field extension. The varieties V_1 and V'_1 are subsets of the varieties V and V' respectively, since any specialization of P (or of P') over k_1 is also a specialization over the smaller field k. Since $x_i = z_i$, $i = 1,2,\cdots,s$, the point Q is a specialization of P not only over k but also over k_1. Hence Q belongs to V_1. Moreover, since $z_1, z_2,\cdots,z_s$ are algebraically independent over k, the local ring of Q (on V) contains the entire field k_1. From this it follows at once that the local ring of Q on V_1 coincides with the local ring of Q on V. We express this by writing: $o_{Q/V} = o_{Q/V_1}$, the notation being self-explanatory. We shall use a similar notation for the rings of holomorphic functions on V and on V_1. Thus we have at once: $o^*_{Q/V} = o^*_{Q/V_1}$. We also note that since the local ring of the point Q has not been affected, <u>the variety V_1 is locally normal at Q</u>.

We now consider the irreducible algebraic correspondence T_1 (over k_1) between V_1 and V'_1, defined by the general point pair (P,P'). <u>We assert that $T_1\{Q\} = T\{Q\}$</u>. For in the first place it is clear that $T_1\{Q\} \subset T\{Q\}$, since any specialization of (P,P') over k_1 is also one over k. On the other hand, since $z_i = x_i$, $i = 1,2,\cdots, s$, any specialization, over k_1, of (P,P') such that P is specialized to Q, is necessarily also a specialization over k, and this proves our assertion.

Let Q' be any point of G' $(=T\{Q\} = T_1\{Q\})$. We have, by assumption, $o_{Q/V} \subset o_{Q'/V'}$, and hence $k_1 \subset o_{Q'/V'}$, since $k_1 \subset o_{Q/V}$. From this it follows easily that $o_{Q'/V'} = o_{Q'/V'_1}$. Since this holds true for each point Q' of G', we conclude that $o^*_{G'/V'} = o^*_{G'/V'_1}$. Since Q is an algebraic point over k_1, we know from the first part of the proof that $H_{Q,G'}$ is an isomorphism of o^*_{Q/V_1} onto o^*_{G'/V'_1}. Hence in order to complete the proof of the theorem it is only necessary to show that $o^*_{G'/V'} = o^*_{G'/V'_1}$.

If Γ' is a variety over k, Γ' is also a variety over k_1, and hence from the definition of strongly holomorphic functions it follows that every function on V' which is strongly holomorphic along $G' - \Gamma'$ is also a function on V'_1, strongly holomorphic along $G' - \Gamma'$, since the local rings of the points of G' have not been affected by our ground field extension. We conclude that $o^*_{G'/V'} \subset o^*_{G'/V'_1}$.

Now let ξ'^* be any element of o^*_{G'/V'_1}. There exists then a set of varieties $\Gamma'_1, \Gamma'_2,\ldots, \Gamma'_h$, <u>over</u> k_1, such that $G' \cap \Gamma'_1 \cap \Gamma'_2 \cap\cdots\cap \Gamma'_h = \emptyset$ and such that ξ'^* is strongly holomorphic on each of the h sets $G' - \Gamma'_i$. However, the varieties Γ'_i are not <u>a priori</u> varieties over k, and therein lies an apparent difficulty, since in the construction of holomorphic functions on V'/k only varieties over k must be used. Nevertheless, this difficulty is eliminated at once if we recall (see §9) that -as a consequence of Theorem 10 - the unspecified varieties Γ'_j may always be taken to be the sections of V' with the hyperplanes $Y_i = 0$, and hence may be assumed to be varieties over k. From this the inclusion $o^*_{G'/V'} \supset o^*_{G'/V'_i}$ follows at once, and this completes the proof of the theorem.

§21. ALGEBRAIC SYSTEMS OF r-CYCLES.

By a <u>field of definition</u> of a variety V in the projective space we mean any field k such that V can be defined by a system of homogeneous equations with coefficients in k. We say that a variety V is <u>absolutely connected</u>, if V is connected over each of its fields of definition. If k is a given algebraically closed field of definition of V, it is easy to see that V is absolutely connected if and only if V/k is connected.

For a given characteristic p, we consider the universal projective space S_n of a given dimension n (i.e., the projective n-space over a universal domain; see A. Weil [4], p. 1). In this space S_n we consider the totality of absolutely irreducible varieties of a given dimension r and we regard these varieties as free generators of an abelian (additive) group. The elements of this group we call <u>r-cycles</u>. The generators of this group, i.e., the absolutely irreducible varieties regarded as cycles, will be called <u>prime cycles</u>.

If an r-cycle Γ is given by $\sum m_i \Gamma_i$, where the Γ_i are distinct prime cycles and the m_i are integers, we say that Γ is an effective cycle if the m_i are non-negative. We shall deal only with effective non-zero cycles.

Every (non-zero and effective) cycle Γ determines in an obvious fashion a set of points in S_n, which will be denoted by $|\Gamma|$ (and which is clearly a variety over suitable ground fields). We shall attribute to the cycle Γ such properties as "absolute irreducibility", "connectedness over k", "absolute connectedness", whenever these properties belong to the variety $|\Gamma|$.

If a cycle Γ is given by $\Gamma = \sum m_i \Gamma_i$, where the Γ_i are prime cycles, then we define the <u>order</u> ν of Γ by $\nu = \sum m_i \nu_i$, where ν_i is the order of the (absolutely irreducible) variety $|\Gamma_i|$ (ν_i = number of intersections of $|\Gamma_i|$ with a general (n- r) -dimensional linear subspace of S_n:, here the term "general" refers to some fixed (but arbitrary) field of definition of $|\Gamma_1|$).

Van der Waerden and Chow [2] (see also [3], p. 153) have shown that the set of all r-cycles of a given order ν in S_n has a natural algebro-geometric structure. Namely, it is possible to represent these r-cycles, in (1,1) fashion, by points of a certain variety defined over the prime field of the given characteristic p, the representative point of a cycle Γ being the point whose homogeneous coordinates are the coefficients of the <u>associated form</u> of Γ. This being so, let M be any set of r-cycles of order ν in S_n and let k be any field. We say that M is an <u>algebraic system</u> and that k is a <u>field of definition</u> of M if the set M* of representative points of the cycles belonging to M is a variety defined over k. With this definition, the terminology used for algebraic varieties can be carried over to algebraic systems of cycles. In particular, we can speak of an irreducible algebraic system M/k, of the general cycle of such a system, of the specializations of a given cycle Γ, over a given field k, etc.

We recall certain properties of associated forms.

Let $F(U^1, U^2, \dots, U^{r+1}, A)$ be the general multiply homogeneous form, with indeterminate coefficients $A_0, A_1, \dots$, in r+1 sets U^i of indeterminates $U^i_0, U^i_1, \dots, U^i_n$, of degree ν in the indeterminates of each set U^i. We introduce other indeterminates Y_μ, $s^i_{\mu\nu}$, where $s^i_{\mu\nu} = -s^i_{\nu\mu}$ $(\mu, \nu = 0, 1, \dots, n;\ i = 1, 2, \dots, r+1)$ and we denote by s^iY the set of n+1 expressions $\sum_{\mu=0}^{n} s^i_{\mu\nu} Y_\mu$, $\nu = 0, 1, \dots, n$. Let $F(s^1Y, s^2Y, \dots, s^{r+1}Y, A) = G(s, Y, A)$, where (s) denotes the set of indeterminates $s^i_{\mu\nu}$. Then G is homogeneous in the indeterminates of each of the sets s, Y, A (it is linear in the A's), and its coefficients belong to the prime field. If Γ is an r-dimensional cycle of order ν in S_n and if $F(U^1, U^2, \dots, U^{r+1}, a)$ is the associated form of Γ (whence P* (a) is the representative point of Γ), then a necessary and sufficient condition that a point (y) of S_n belong to the variety $|\Gamma|$ is that G(s,y,a) be identically zero in the $s^i_{\mu\nu}$'s. From this property of the associated form it follows that if $\{H_q(Y,A)\}$ is the set of coefficients of G when G is regarded as a form in the $s^i_{\mu\nu}$'s then the equations $H_q(Y,a) = 0$ form a set of defining equations of the variety $|\Gamma|$. <u>Hence if</u> k <u>is any field, then</u> $|\Gamma|$ <u>is defined over</u> k(P*), the field k(P*) being the field generated over k by the ratios of the quantities $a_0, a_1, \dots$.

Another consequence of the above property of the associated forms is the following: <u>if a cycle $\bar{\Gamma}$ is a specialization, over k, of a cycle Γ and if</u> $Q(z)$ <u>and</u> $\bar{Q}(\bar{z})$ <u>are points of</u> S_n <u>such that</u> $Q \in |\Gamma|$ <u>and</u> $(\bar{\Gamma}, \bar{Q})$ <u>is a specialization, over k, of</u> (Γ, Q), <u>then</u> $\bar{Q}$ <u>belongs to</u> $|\bar{\Gamma}|$. For let (a) and $(\bar{a})$ be the representative points of Γ and $\bar{\Gamma}$. The assumption that $Q \in |\Gamma|$ implies that all $H_i(z,a)$ are zero, and hence also the $H_i(\bar{z},\bar{a})$ are zero [since, by assumption, $(\bar{a},\bar{z})$ is a specialization, over k, of (a,z) and since the coefficients of the forms H_i are in prime field]; i.e., $\bar{Q} \in |\bar{\Gamma}|$.

We shall also need the following result: <u>if $\bar{\Gamma}$ is a specialization, over k, of Γ and if $\bar{Q}$ is any point of</u> $|\bar{\Gamma}|$, <u>then there exist points</u> Q <u>on</u> $|\Gamma|$ <u>such that</u> $(\bar{\Gamma}, \bar{Q})$ <u>is a specialization, over</u> k, <u>of</u> (Γ, Q). We shall need, in fact, the following stronger result:

LEMMA. Let Γ and $\bar{\Gamma}$ be cycles in S_n and let R and $\bar{R}$ be points in some projective space S_m. If $(\bar{\Gamma}, \bar{R})$ is a specialization, over k, of (Γ, R), then for any point $\bar{Q}$ on $|\bar{\Gamma}|$ there exist points Q on $|\Gamma|$ such that $(\bar{\Gamma}, \bar{Q}, \bar{R})$ is a specialization, over k, of (Γ, Q, R).

PROOF. We denote by K the field $k(R,\bar{R})$ and we introduce (n+1) r quantities u_j^i ($j = 0,1,\dots,n$; $i = 1,2,\dots,r$) which are algebraically independent over $K(\Gamma, \bar{\Gamma})$. We denote by u^i the set of quantities $u_0^i, u_1^i, \dots, u_n^i$, and by Π_i the hyperplane $u_0^iY_0 + u_1^iY_1 + \dots + u_n^iY_n = 0$. If (a) is the representative point of Γ, then, by known properties of associated forms, $F(u^1, u^2, \dots, u^r, U, a)$ factors completely into linear factors:

$$(1) \qquad F(u^1, u^2, \dots, u^r, U, a) = \prod_{s=1}^{\nu}(y_0^s U_0 + y_1^s U_1 + \dots + y_n^s U_n).$$

The ν points $P^s(y_0^s, y_1^s, \dots, y_n^s)$ have the following properties: (1) they are the intersections of the variety $|\Gamma|$ with the linear space $L = \Pi_1 \cap \Pi_2 \cap \dots \cap \Pi_r$; (2) each point P^s is a general point of an irreducible component of $|\Gamma|/K(\Gamma, \bar{\Gamma})$: (3) conversely, each irreducible component of $|\Gamma|/K(\Gamma, \bar{\Gamma})$ has at least one of the points P^s as general point.

Since the (n+1)r quantities u_j^i are algebraically independent over $k(\Gamma, R)$, $(\bar{\Gamma}, \bar{R})$ is a specialization of (Γ, R) also over the field $k(u^1, u^2, \dots, u^r)$. This latter specialization has at least one extension to a specialization of $(\Gamma, R, y^1, y^2, \dots, y^{\nu})$. Let $(\bar{\Gamma}, \bar{R}, \bar{y}^1, \bar{y}^2, \dots, \bar{y}^{\nu})$ be one such extended specialization [over the field $k(u^1, u^2, \dots, u^r)$]. Applying this specialization to (1) we find:

(2) $$F(u^1, u^2, \ldots, u^r, U, \bar{a}) = \prod_{s=1}^{\nu} (\bar{y}_o^s U_o + \bar{y}_1^s U_1 + \cdots + \bar{y}_n^s U_n),$$
where $(\bar{a})$ is the representative point of $\bar{\Gamma}$. In (2) we have a factorization similar to (1) but relative to the associated form of the cycle $\bar{\Gamma}$. Since the u_j^i are algebraically independent over $K(\Gamma, \bar{\Gamma})$, it follows that the points $\bar{P}^s(\bar{y}_o^s, \bar{y}_1^s, \cdots, \bar{y}_n^s)$ enjoy, in relation to the cycle Γ, properties similar to (1), (2) and (3) above. Now $(\bar{\Gamma}, \bar{R}, \bar{P}^s)$ is a specialization, of (Γ, R, P^s) over $k(u^1, u^2, \cdots, u^r)$, hence *a fortiori* over k. By the above property (3), as applied to the cycle $\bar{\Gamma}$ and the points $\bar{P}^s$, every point $\bar{Q}$ of $|\bar{\Gamma}|$ is a specialization, over $K(\Gamma, \bar{\Gamma})$ - hence also over $k(\bar{\Gamma}, \bar{R})$ - of one of the points $\bar{P}^s$. Hence, by the transitivity of specializations, $(\bar{\Gamma}, \bar{Q}, \bar{R})$ is a specialization, over k, of one of the triples (Γ, P^s, R). Since the points P^s belong to $|\Gamma|$, the lemma is proved.

Let M/k be an irreducible algebraic system of r-cycles in S_n and let M^*/k be the representative variety of the system. Let P^* be a general point of M^*/k and let Γ be the corresponding cycle of M/k. Let $V_1, V_2, \ldots, V_h$ be the irreducible components of the variety $|\Gamma|/k(P^*)$ and let P_i be a general point of $V_i/k(P^*)$. Let T_i/k be the irreducible correspondence between M^* and S_n, defined by the general point pair (P^*, P_i) and let T be the correspondence whose graph is the union of the graphs of the h correspondences T_i. This correspondence T is defined on the whole of M^* (since T is defined at the general point of M^*/k). The T-transform of M^* is a certain variety in S_n, defined over k. From the lemma just proved above, it follows at once that if $\bar{\Gamma}$ is any cycle in M and $\bar{Q}$ is any point of S_n, *then $\bar{\Gamma}$ and $\bar{Q}$ are corresponding elements of T* (more precisely: the representative point of $\bar{\Gamma}$ on M^* and the point Q are corresponding points under T) *if and only if $\bar{Q} \in |\bar{\Gamma}|$*. It follows that the T-*transform of* M^* *is the union of the varieties $|\bar{\Gamma}|$ of the cycles $\bar{\Gamma}$ belonging to* M. We shall call this variety the *carrier* of the system M, and we shall call T the *incidence correspondence* between the system M and its carrier. It is clear that both T and the carrier of M are independent of the choice of the field of definition of M.

§22. PROOF OF THE PRINCIPLE OF DEGENERATION.

By the principle of degeneration we mean the following theorem: If the general cycle of an irreducible algebraic system M/k is absolutely connected then every cycle of the system is absolutely connected.

PROOF. We first reduce the proof of the theorem to the case in which the general cycle of M/k is prime, i.e. absolutely irreducible. Assume that the theorem has already been proved in this case. Let $\Gamma = n_1Z_1 + n_2Z_2 + \cdots + n_hZ_h$ be a general cycle of M/k, where $Z_1, Z_2, \cdots, Z_h$ are distinct prime cycles. By assumption, the theorem is true if $h = 1$. We shall therefore proceed by induction with respect to h. Since Γ is absolutely connected, the following will be true for a suitable labeling of the cycles Z_i: if $\Gamma_1 = n_1Z_1 + n_2Z_2 + \cdots + n_{h-1}Z_{h-1}$ and $\Gamma_2 = n_hZ_h$, then Γ_1 is absolutely connected and the two varieties Γ_1 and Γ_2 have a point in common. Let Δ be any cycle in M. Then Δ is of the form $\Delta_1 + \Delta_2$, where the pair of cycles (Δ_1, Δ_2) is a specialization, over k, of the pair of cycles (Γ_1, Γ_2). The cycle Γ_2 is absolutely irreducible, while the number of absolutely irreducible components of the cycle Γ_1 is h-1. Hence, by the induction hypothesis, both cycles Δ_1 and Δ_2 are absolutely connected. Let Q be a common point of the varieties $|\Gamma_1|$ and $|\Gamma_2|$. We can extend the specialization $(\Gamma_1, \Gamma_2) \xrightarrow{k} (\Delta_1, \Delta_2)$ (the letter k above the arrow indicates that the specialization is over k) to a specialization $(\Gamma_1, \Gamma_2, Q) \xrightarrow{k} (\Delta_1, \Delta_2, R)$. It follows (see § 21) that the point R must belong to both varieties $|\Delta_1|$ and $|\Delta_2|$. Hence the cycle Δ is absolutely connected.

We now consider the case in which the cycle Γ is absolutely irreducible. Let M*/k be the representative variety of the system M/k. Let V/k be the carrier of the system M/k and let T be the incidence correspondence between M*/k and V/k. Since the variety $|\Gamma|$ is absolutely irreducible and is defined over k(P*), where P* is the representative point of Γ, the variety $|\Gamma|$/k(P*) is irreducible. Hence T/k is irreducible, and a general point pair of T/k is given by (P*,P), where P is a general point of $|\Gamma|$/k(P*).

We now consider a derived normal model $\bar{M}$/k of M*/k. Let $\bar{P}$ be a general point of $\bar{M}$/k such that P* corresponds to $\bar{P}$ in the birational correspondence between M*/k and $\bar{M}$/k. We denote by $\bar{T}$ the irreducible correspondence (over k) between $\bar{M}$ and V, which has $(\bar{P}, P)$ as general point pair. Since the variety $|\Gamma|$ is absolutely irreducible and since P is a general point of this variety over k(P*), it follows that the field k(P*) is a quasi-maximally algebraic in the field k(P*,P). Since $k(\bar{P})$ = k(P*), also $k(\bar{P})$ is maximally algebraic in $k(\bar{P}, P)$. Furthermore, since the variety $\bar{M}$/k is normal, it is analytically irreducible at each point.

Let $\overline{Q}$ be any point of $\overline{M}$. To the point $\overline{Q}$ there corresponds on M^* a unique point Q^* (since $M^*/k < \overline{M}/k$). Let Δ be the cycle in M whose representative point is Q^*. We shall show that $\overline{T}\{\overline{Q}\} = |\Delta|$, and that consequently Δ is absolutely connected. Since any point Q^* of M^* corresponds to at least one point $\overline{Q}$ of $\overline{M}$, this will prove that any cycle in M is absolutely connected. As was explained in the introduction, 3, the proof of the principle of degeneration will be complete if we show that $\overline{T}\{\overline{Q}\} = |\Delta|$.

Let Q be any point of $\overline{T}\{\overline{Q}\}$. Then $(\overline{Q},Q)$ is a specialization, over k, of $(\overline{P},P)$. Extend this to a specialization $(\overline{P},P,P^*) \xrightarrow{k} (\overline{Q},Q,Q_1^*)$. Then Q_1^* necessarily coincides with the above point Q^* since Q^* is the only point of M^*/k which corresponds to $\overline{Q}$. From $(P,P^*) \xrightarrow{k} (Q,Q^*)$ and from the fact that $P \in |\Gamma|$, it follows (§21) that $Q \in |\Delta|$. Hence $\overline{T}\{\overline{Q}\} \subset |\Delta|$.

Now let Q be any point of $|\Delta|$. We have the following specialization $(P^*, \overline{P}) \xrightarrow{k} (Q^*,\overline{Q})$, or - using the corresponding cycles - $(\Gamma, \overline{P}) \xrightarrow{k} (\Delta, \overline{Q})$. By the lemma of the preceding section, there exists a point P_1 on Γ such that $(\Delta, \overline{Q}, Q)$ is a specialization, over k, of $(\Gamma,\overline{P},P_1)$. Now P is a general point of $|\Gamma|/k(P^*)$ and $P_1 \in |\Gamma|$. Hence P_1 is a specialization, over $k(P^*)$, of P. Since $k(\Gamma) = k(P^*) = k(\overline{P})$, it follows that $(\Gamma, \overline{P}, P_1)$ is a specialization of $(\Gamma,\overline{P},P)$ over k. Hence, by the transitivity of specializations, we have that $(\Delta,\overline{Q},Q)$ is a specialization of $(\Gamma,\overline{P}, P)$ over k. Hence also $(\overline{Q}, Q)$ is a specialization of $(\overline{P}, P)$ over k, and this shows that $Q \in \overline{T}\{\overline{Q}\}$. We have thus shown that $|\Delta| \subset \overline{T}\{\overline{Q}\}$, and hence $|\Delta| = \overline{T}\{\overline{Q}\}$, as asserted.

We note the following result, that is both a consequence and a generalization of the principle of degeneration:

> Let N/k be an algebraic subsystem of M. If the general cycle of M/k is absolutely connected and if the system N/k is connected (i.e., if the representative variety N^* of N is connected over k), then the carrier of the system N is connected over k.

The proof is as follows:

Let W' be the carrier of the system N and let $W'_1 \cup W'_2$ be a decomposition of W' into two varieties defined over k and having no points in common. We set $N_i^* = N^* \cap T^{-1}\{W'_i\}$, $i = 1,2$, where T is the incidence correspondence of M. We now show that N_1^* and N_2^* have no points in common. For let Q^* be any point of N^*. The variety $T\{Q^*\}$ is a cycle of M and is therefore, by the principle of degeneration, absolutely connected. Since $T\{Q^*\} \subset W_1' \cup W_2'$ and since W_1' and W_2' have no points in common, it follows that $T\{Q^*\}$ is contained in one of the two varieties W_i', and if, say, $T\{Q^*\}$ is contained in W_1', then $T\{Q^*\}$ and W_2' have no points in common. In that case we have $Q^* \in N_1^*$, $Q^* \notin N_2^*$. We have thus shown that no point of N^* can belong simultaneously to N_1^* and N_2^*, as asserted. Since $N^* = N_1^* \cup N_2^*$ and since, by hypothesis, N^* is connected over k, one of the two varieties N_i^* must be empty, and therefore also one of the two varieties W_i' must be empty. This completes the proof.

Harvard University
Cambridge, Mass.

FOOTNOTES

(1) The field Σ is defined to within an arbitrary k-isomorphism. We agree to fix a general point $(x_1, x_2, \cdots, x_n)$ of V/k and we take for Σ the field $k(x_1, x_2, \cdots, x_n)$.

(2) We use the term "point" in a wide sense. We do not restrict ourselves to algebraic points, i.e., points with coordinates which are algebraic over k. The point coordinates may be arbitrary quantities taken from a universal domain; see A. Weil [4], p. 1.

(3) This local ring o_P is defined with reference to the fixed function field Σ of V/k and is therefore a subring of Σ.

(4) R is the ring $k[x_1, x_2, \cdots, x_n]$, where the x_i are the non-homogeneous coordinates of the fixed general point of V/k. See footnote (1).

(5) PROOF. Let $\xi/\eta \in \bigcap_{P \in G} o_P$, where ξ and η are elements of R, and let $\mathfrak{q}$ be a primary component of the principal ideal $R \cdot \eta$. Let $\mathfrak{p}$ be the associated prime ideal of $\mathfrak{q}$ and let A be a general point of the irreducible variety $(\mathfrak{p})$ of V. Since $A \in G$, we have $\xi/\eta \in o_A$, i.e., $\xi/\eta = \xi'/\eta'$, where ξ', $\eta' \in R$ and $\eta' \notin \mathfrak{p}$. From $\xi\eta' = \xi'\eta$ and $\eta' \notin \mathfrak{p}$ it follows that $\xi \in \mathfrak{q}$. Since $\mathfrak{q}$ is any of the primary components of $R \cdot \eta$, we conclude that $\xi \in R \cdot \eta$, $\xi/\eta \in R$, as asserted.

(6) Trivial example: W is the set of conjugate algebraic points P_i, and W is analytically irreducible at P_i.

(7) This is Lemma 5 of our paper [7].

(8) Note that if W is G-connected, W is connected.

(9) In the next section we shall prove that the analytical irreducibility of V at all the algebraic points of a subvariety W implies the analytical irreducibility of V at each point of W.

(10) A few words about the case in which V is analytically irreducible at P, whence T[W] is an irreducible variety $\overline{W}$. If $\overline{P}$ is one of the points which correspond to P, then the point pair $(P, \overline{P})$ is a general point pair of an irreducible algebraic correspondence T_0 whose inverse is a rational transformation of $\overline{W}$ onto W. This correspondence is $(1, \nu)$; it is single-valued, without exceptions, on $\overline{W}$, and is finitely valued, also without exceptions, on W. Moreover, we have $T\{Q\} = T_0\{Q\}$ for almost all points Q of W. If $\nu = 1$, i.e., if the function field of $\overline{W}$ is a pure inseparable extension of the function field of W (in particular, if T_0 is a birational transformation), then V is analytically irreducible at almost every point of

W (the points of W where V is analytically reducible are, in that case, the points Q such that $T\{Q\} \not\subset W$ and possibly the singular points of W.) If $\nu > 1$ and if k is algebraically closed, then V is analytically reducible at almost every algebraic point of W. If k is not algebraically closed and $\nu > 1$, then the various possibilities are less obvious, and we shall not attempt here to analyze them.

(11) In the terminology used by Seidenberg and Cohen in their paper "Prime ideals and integral dependence" [Bull. Amer. Math. Soc., v. 52 (1946)], the condition of Theorem 9 is that the "going up" theorem holds for R and $\bar{R}$.

(12) If R is not an integral domain, then it is necessary to impose the condition $\bigcap_{i=1}^{\infty} \mathfrak{m}^i = (0)$.

(13) If R is not an integral domain, it is necessary to require that c be not a zero divisor.

(14) If R has zero divisors, then $R_{\mathfrak{p}}$ should be defined as in Chevalley [1].

(15) We make use here of the following fact: if $\mathfrak{A}$ is an ideal in a Noetherian ring R and if M denotes the set of all prime ideals in R which contain A, then $\bigcap_{\mathfrak{p} \in M}(R \cap R_{\mathfrak{p}} \cdot \mathfrak{A}) = \mathfrak{A}$. This follows readily from the well known relations between ideals in R and ideals in quotient rings $R_{\mathfrak{p}}$ of R. The above statement remains true if M is replaced by the set M_0 of all maximal ideals containing $\mathfrak{A}$.

(16) T^{-1} is said to be semi-regular at P' if $o_P \subset o_{P'}$. This condition implies that P is the only point of V which corresponds to P' (see Zariski [6]).

(17) The following notation is used throughout this paper: if A is a subset of V, $T\{A\}$ denotes the total transform of A on V', i.e., the set of all points of V' which correspond to points of A. Similarly for T^{-1} and $T^{-1}\{A'\}$, where A' is any subset of V'.

(18) S can be defined by assigning its general point pair, as follows: let (P,P') be a general point pair of T and let P_1 and P_1' be the (general) points of V_1 and V_1' which correspond to P and P' under T_1 and T_1' respectively; then (P_1, P_1') is a general point pair of S. The transformation S does not necessarily coincide with the transformation $s = T_1^{-1} T T_1'$. However, we do have $S\{A_1\} \subset s\{A_1\}$ for any

point A_1 of V_1 at which T_1^{-1} is single-valued. Similarly, $S^{-1}\{A_1'\} \subset S^{-1}\{A_1'\}$ for any point A_1' of V_1' at which $T_1'^{-1}$ is single valued.

(19) The coefficients C_{ij} belong indeed to R[t] since the coefficients of the multiplication table of the elements ω_{i0} have been assumed to belong to R[t].

(20) Suppose that we have an algebraic relation for t over R^*, say: $a_0^*t^h + a_1^*t^{h-1} + \dots + a_h^* = 0$, where $a_i^* \in R^*$. For any integer n we can find elements a_{in} in R $(0 \leq i \leq h)$ such that $a_i^* - a_{in} \in \mathfrak{m}^{*n}$, where $\mathfrak{m}^* = R^*\mathfrak{m}$. We have, then, $a_{on}t^h + a_{1n}t^{h-1} + \dots + a_{hn} \in R_0^* \cdot \mathfrak{m}^{*n} \cap R = \mathfrak{m}^n$. Since t is a transcendental over R it follows that $a_{in} \in \mathfrak{m}^n$, and hence $a_i^* \in \mathfrak{m}^{*n}$. Since this holds for every integer n, we conclude that $a_i^* = 0$, $0 \leq i \leq h$.

REFERENCES

[1] C. CHEVALLEY, On the theory of local rings, Ann. of Math., vol. 44 (1943), pp. 690-708.

[2] W. L. CHOW and B. L. VAN DER WAERDEN, Über zugeordnete Formen und algebraische Systeme von algebraischen Manningfaltigkeiten, Zur algebraischen Geometrie IX, Math. Ann., vol. 113 (1937), pp.692-704.

[3] B. L. VAN DER WAERDEN, Einführung in die algebraische Geometrie, Dover Publications, New York (1945).

[4] A. WEIL, Foundations of algebraic geometry, Amer. Math. Soc, Colloquium Publications, vol. XXIX (1946).

[5] O. ZARISKI, Some results in the arithmetic theory of algebraic varieties, Amer. Journ. of Math., vol. 61 (1938), pp. 249-294.

[6] " " Foundations of a general theory of birational correspondences, Trans. Amer. Math. Soc., vol. 53 (1943), pp. 490-542.

[7] " " Generalized semi-local rings, Summa Brasiliensis Mathematicae, vol. 1, fasc. 8 (1946), pp. 1-27.

[8] " " Analytical irreducibility of normal varieties, Ann. of Math., vol. 49 (1948), pp. 352-361.

[9] " " A simple analytical proof of a fundamental property of birational transformations, Proc. Nat. Acad. Sc., vol. 35 (1949), pp. 62-66.

[10] " " A fundamental lemma from the theory of holomorphic functions on an algebraic variety, Annali di Matematica pura ed applicata, s. 4, vol. 29 (1949), pp. 187-198.

SUR LA NORMALITÉ ANALYTIQUE DES VARIÉTÉS NORMALES

par Oscar ZARISKI (Harvard-University).

Soit $\mathfrak{o}$ l'anneau local d'un point *normal* P d'une variété algébrique V définie sur un corps k ; l'hypothèse de normalité veut dire que $\mathfrak{o}$ est un anneau intégralement clos. Nous avons montré [5] que l'anneau local complété $\bar{\mathfrak{o}}$ de $\mathfrak{o}$ est un anneau d'intégrité, c'est-à-dire que V est *analytiquement irréductible* en P. Nous allons montrer ici que $\bar{\mathfrak{o}}$ est aussi un anneau *intégralement clos,* c'est-à-dire que V, considérée comme variété algébroïde, est normale en P. Nous apporterons aussi quelques simplifications à notre démonstration d'irréductibilité analytique donnée en [5] ; nous ne supposerons de [5] que les lemmes du § 2. Nous entendrons par *domaine local* un anneau local sans diviseurs de zéro.

Théorème 1. — *Soient $\mathfrak{o}$ un domaine local intégralement clos, $\bar{\mathfrak{o}}$ son complété, et $\bar{\mathfrak{o}}'$, la fermeture entière de $\bar{\mathfrak{o}}$ dans son anneau total de fractions* A. *Supposons qu'il existe un élément $d \neq 0$ de $\mathfrak{o}$ tel que*

1) $d\bar{\mathfrak{o}}' \subset \bar{\mathfrak{o}}$.

2) *les idéaux premiers $\mathfrak{p}_i$ de $\mathfrak{o}d$ sont analytiquement non ramifiés. Alors $\bar{\mathfrak{o}}$ est lui-même un domaine local intégralement clos, et on a $\bar{\mathfrak{o}}' = \bar{\mathfrak{o}}$.*

Puisque $d \in \mathfrak{o}$, tout idéal premier $\bar{\mathfrak{p}}$ de $\bar{\mathfrak{o}}d$ est minimal dans $\bar{\mathfrak{o}}$, et est tel que l'anneau de fractions $\bar{\mathfrak{o}}_{\bar{\mathfrak{p}}}$ est l'anneau d'une valuation discrète v du corps des fractions de $\bar{\mathfrak{o}}/\mathfrak{v}$ ($\mathfrak{v}$ désignant l'unique idéal premier de (O) de $\bar{\mathfrak{o}}$ contenu dans $\bar{\mathfrak{p}}$; cf. [5] lemmes 3, 4, 5, 6) ; nous noterons w l'application composée de l'homomorphisme canonique de $\bar{\mathfrak{o}}$ sur $\bar{\mathfrak{o}}/\mathfrak{v}$ et de v. Soit $w(d) = s$, c'est-à-dire que l'on a $d \in \bar{\mathfrak{p}}^{(s)}$ et $d \notin \bar{\mathfrak{p}}^{(s+1)}$. Pour tout élément $z \in \bar{\mathfrak{o}}'$, on a $dz = y \in \bar{\mathfrak{o}}$ d'après 1). Comme z est entier sur $\bar{\mathfrak{o}}$, on a $w(z) \geqslant 0$, d'où $w(y) \geqslant s$. Ceci étant

11

vrai pour tout idéal premier de $\bar{\mathfrak{o}}d$, d divise y dans $\bar{\mathfrak{o}}$, et on a $y = dz'$ avec $z' \in \bar{\mathfrak{o}}$. D'où $d(z - z') = 0$. Comme $\mathfrak{o}$ est un anneau d'intégrité, d n'est pas diviseur de zéro dans $\bar{\mathfrak{o}}$([1], prop. 6), ni donc dans l'anneau total des fractions de $\bar{\mathfrak{o}}$; par conséquent $z = z'$, et on a $\bar{\mathfrak{o}} = \bar{\mathfrak{o}}'$.

De cette dernière égalité nous allons déduire que $\bar{\mathfrak{o}}$ est un domaine local, ce qui démontrera le th. 1. Or, en vertu de la condition 2) et du lemme 9 de [5], $\mathfrak{o}$ est analytiquement non ramifié. La conclusion résultera alors du lemme suivant :

Lemme. — *Si un anneau local* R *est intégralement fermé dans son anneau total des fractions* S *et n'a pas d'élément nilpotent, alors* R *est un anneau d'intégrité.*

Il est clair que S est composé direct d'un nombre fini h de corps. Soit $1 = e_1 + \cdots + e_h$ la décomposition correspondante de 1 en idempotents orthogonaux. On a $e_1(1 - e_1) = 0$; donc, si $h \neq 1$, on a $e_1 \notin$ R, sinon $1 - e_1$ serait inversible dans R contrairement au fait que c'est un diviseur de zéro. Mais, comme $e_1^2 - e_1 = 0$, ceci est contraire au fait que R est intégralement fermé dans S.

Nous allons maintenant déduire du th. 1 *à la fois* l'irréductibilité et la normalité analytique des variétés localement normales :

Théorème 2. — *Si* $\mathfrak{o}$ *est l'anneau local d'un point* P *normal sur une variété algébrique* V, *le complété* $\bar{\mathfrak{o}}$ *de* $\mathfrak{o}$ *est un domaine local intégralement clos.*

Il suffit de montrer que les conditions du th. 1 sont satisfaites. Tout élément non nul d de $\mathfrak{o}$ satisfaisant à 2) ([2], lemme 9, p. 9), il s'agit de vérifier 1). Si $\mathfrak{m}$ désigne l'idéal maximal de $\mathfrak{o}$, $\mathfrak{o}$ contient un corps k, sur lequel $\mathfrak{o}/\mathfrak{m}$ est de degré fini, et sur lequel le corps des fractions F de $\mathfrak{o}$ est séparable (un « basic field » de [1]). Soit $(x_1, \ldots, x_r)$ un système de paramètres de $\mathfrak{o}$ qui soit aussi une base de transcendance séparante de F sur k. Alors l'anneau des fractions de l'idéal premier $(x_1, \ldots, x_r)$ de $k[x_1, \ldots, x_r]$ est un anneau local régulier $\mathfrak{r}$ sur lequel $\mathfrak{o}$ est un module de type fini ([1], prop. 5, p. 702). Le complété $\bar{\mathfrak{r}}$ de $\mathfrak{r}$ est l'anneau de séries formelles $k[[x_1, \ldots, x_r]]$; il s'identifie canoniquement à un sous-anneau de $\bar{\mathfrak{o}}$ sur lequel $\bar{\mathfrak{o}}$ est un module de type fini.

Soit a un élément primitif de F sur $k(x_1, \ldots, x_r)$, que nous pouvons supposer entier sur $k[x_1, \ldots, x_r]$. Si n désigne le degré de F sur $k(x_1, \ldots, x_r)$, il existe un élément non nul c de $\mathfrak{r}$ tel que

$$c\mathfrak{o} \subset \mathfrak{r} + \mathfrak{r}a + \cdots + \mathfrak{r}a^{n-1}.$$

On a alors $c\overline{\mathfrak{o}} \subset \overline{\mathfrak{r}} + \overline{\mathfrak{r}}a + \cdots + \overline{\mathfrak{r}}a^{n-1}$. Comme c n'est pas diviseur de zéro dans $\overline{\mathfrak{o}}$ ([1], prop. 6), il s'ensuit que, K désignant le corps des fractions de $\overline{\mathfrak{r}}$, l'anneau total des fractions A de $\overline{\mathfrak{o}}$ est de la forme $K[a]$.

Comme $\overline{\mathfrak{r}}$ est intégralement clos, l'équation unitaire de plus petit degré satisfaite par a sur K est à coefficients dans $\overline{\mathfrak{r}}$; soit $f_1(a) = 0$. Le polynome f_1 divise le polynome minimal f de a sur $k(x_1, \ldots, x_r)$, dont les coefficients sont dans $\mathfrak{r}$. Donc le discriminant d de f (qui n'est pas nul en vertu de l'hypothèse de séparabilité) est un multiple (dans $\overline{\mathfrak{r}}$) du discriminant d_1 de f_1. Il va donc nous suffire de montrer que l'on a $d_1\overline{\mathfrak{o}'} \subset \overline{\mathfrak{r}}[a]$. Or, comme $\overline{\mathfrak{o}'}$ est un anneau entier sur $\overline{\mathfrak{r}}$, on peut appliquer le raisonnement classique (cf. [6], § 101, p. 81), en opérant dans le produit des corps des racines des divers facteurs irréductibles de f_1 (au lieu du corps des racines de f_1 dans le cas où f_1 est irréductible), et en tenant compte du fait que $\overline{\mathfrak{r}}$ est intégralement clos et ne contient aucun diviseur de zéro de A.

Voici enfin une application du th. 2 aux variétés non nécessairement normales :

Théorème 3. — *Soient $\mathfrak{o}$ l'anneau local d'un point* P *d'une variété algébrique* V, *et $\mathfrak{o}'$ la clôture intégrale de $\mathfrak{o}$. Si $\overline{\mathfrak{o}}$ désigne le complété de $\mathfrak{o}$, alors la fermeture intégrale $\overline{\mathfrak{o}'}$ de $\overline{\mathfrak{o}}$ dans son anneau total de fractions* A *est identique au complété* $(\overline{\mathfrak{o}'})$ *de l'anneau semi-local* $\mathfrak{o}'$ (autrement dit les opérations « fermeture intégrale » et « complétion » commuttent).

Comme il existe c non nul dans $\mathfrak{o}$ tel que $c\mathfrak{o}' \subset \mathfrak{o}$ et donc $c(\overline{\mathfrak{o}'}) \subset \overline{\mathfrak{o}}$, $(\overline{\mathfrak{o}'})$ est un sous-anneau de A, et c'est un module de type fini sur $\overline{\mathfrak{o}}$ puisqu'il en est ainsi de $\mathfrak{o}'$ sur $\mathfrak{o}$. Il ne nous reste alors plus qu'à montrer que $(\overline{\mathfrak{o}'})$ est intégralement fermé dans A ; or ceci est clair puisque les anneaux *locaux* complets dont $(\overline{\mathfrak{o}'})$ est composé direct sont des domaines locaux intégralement clos d'après le cas des variétés normales (th. 2).

Remarque. — Ce qui précède reste valable pour une variété analytique complexe V : la normalité au sens des séries convergentes implique la normalité au sens des séries formelles.

Remarques sur le mémoire « Sur les variétés algébroïdes » de M. P. Samuel.

1) Le th. 1 de ce mémoire [4] peut aussi se déduire du lemme qui le précède par application du théorème de Bertini au système linéaire irréductible découpé par les hyperplans passant par l'origine sur la variété algébrique dont la variété algébroïde étudiée est une nappe.

2) La démonstration du th. 3 (§ 3) de [4] peut se simplifier en utilisant notre th. 3. Reprenons le raisonnement au point où il s'agit de montrer l'existence d'un idéal principal $\mathfrak{o}a$ ayant mêmes composantes isolées que $\mathfrak{r}$ (lignes avant la remarque traitant du cas d'une variété normale). Soit $\mathfrak{o}'$ la clôture intégrale de $\mathfrak{o}$, et $\bar{\mathfrak{o}}'$ la fermeture intégrale de $\bar{\mathfrak{o}}$ dans son anneau total de fractions. Alors $\bar{\mathfrak{o}}'\mathfrak{r}$ et $\bar{\mathfrak{o}}'x$ ne diffèrent que par des composantes immergées. Mais, d'après notre th. 3, $\bar{\mathfrak{o}}'$ est le complété de l'anneau semi-local $\mathfrak{o}'$. On montre alors, comme dans la remarque susdite, que, si $(a_1, \ldots, a_s)$ est une base de $\mathfrak{r}$, on a, par exemple, $\mathfrak{o}'\mathfrak{r} = \mathfrak{o}'a_1$. Ceci montre que $\mathfrak{o}a_1$ et $\mathfrak{r}$ ne diffèrent que par des composantes immergées.

BIBLIOGRAPHIE

[1] C. Chevalley, « On the theory of local rings », *Ann. of Math.*, vol. 44 (1943), pp. 690-708.

[2] C. Chevalley, « Intersections of algebraic and algebroid varieties », *Trans. Amer. Math. Soc.*, vol. 57 (1945), pp. 1-85.

[3] C. Chevalley, « On the notion of the ring of quotients of a prime ideal », *Bull. Amer. Math. Soc.*, vol. 50 (1944).

[4] P. Samuel, « Sur les variétés algébroïdes », *Ann. Institut Fourier*, vol. 2 (1950), pp. 147-160.

[5] O. Zariski, « Analytical irreducibility of normal varieties », *Ann. of Math.*, vol. 49 (1948), pp. 352-361.

[6] Van der Waerden, « Moderne Algebra », Springer (Berlin), 1940.

(Parvenu aux Annales en novembre 1950.)

The Connectedness Theorem for Birational Transformations†

Oscar Zariski

1. Introduction

In our memoir [4] on abstract holomorphic functions we have proved the so-called 'principle of degeneration' in abstract algebraic geometry, by establishing a general connectedness theorem for algebraic correspondences. It is desirable to establish the connectedness theorem without using the theory of holomorphic functions which we have developed in the above-cited memoir. In the classical case this theorem follows, for birational transformations, from very simple topological considerations, as was pointed out in our memoir ([4], footnote on p. 7). By Lefschetz's principle it follows that in the case of characteristic zero we have a proof of the connectedness theorem for birational transformations, independent of the theory of abstract holomorphic functions. In the present paper we propose to deal algebraically with the connectedness theorem for birational transformations, without making any use of our theory of abstract holomorphic functions. We shall prove the connectedness theorem, in the case of birational transformations, only for non-singular varieties (or, more precisely, for simple points, since the question is of a purely local character).

2. The connectedness theorem for birational transformations

Let V be a variety defined over a ground field k. We say that V/k *is connected*, or that V *is connected over* k, if V is not the union of two proper subvarieties which are defined over k and have no points in common.

Let V and V' be two irreducible varieties, of dimension r, defined and birationally equivalent over k, and let T be a birational trans-

† This work was supported by a research project at Harvard University, sponsored by the Office of Ordnance Research, U.S. Army, under Contract DA–020–ORD–3100.

formation of V/k into V'/k (we assume, of course, that T itself is defined over k, i.e., that the graph of T, on the direct product $V \times V'$, is a variety defined over k; this is indicated by our saying that T is a birational transformation of V/k into V'/k). The following is the

CONNECTEDNESS THEOREM FOR BIRATIONAL TRANSFORMATIONS. *If P is a point of V at which V/k is analytically irreducible, then the variety $T\{P\}$ is connected over $k(P)$.*

(By $T\{P\}$ we mean the set of points of V' which correspond to P, under T. It is known that this set of points is a variety defined over the field $k(P)$ which is generated over k by the non-homogeneous coordinates of P.)

We shall now prove this theorem in the case in which P is a simple point on V/k. In the proof we may assume that P is a fundamental point of P, for otherwise $T\{P\}$ would consist of a single point, and there would be nothing to prove. We shall now recall from our paper ([1], p. 529) the definition of *isolated* fundamental points, since in our proof we shall consider separately two cases, according as P is or is not an isolated fundamental point of T.

Let V^* be the join of V and V', i.e., let V^* be the graph of T on the direct product $V \times V'$. The projection T^* of V^* onto V is a birational transformation defined over k and semi-regular at each point of V^* (we use the terminology of our paper [1]). Let F be the fundamental locus of T. Then F is also the fundamental locus of T^{*-1}. An irreducible subvariety W of F/k is said to be an *isolated* fundamental variety of T if it corresponds to an irreducible component of $T^{*-1}\{F\}/k$, and a point P of F is said to be an *isolated fundamental point of* T if P is a general point, over k, of an isolated fundamental variety of T. It is clear that every irreducible component of F/k is an isolated fundamental variety; but the converse is not necessarily true.

It is known that if an isolated fundamental variety W of T is *simple* for V/k, then every irreducible component of $T^{*-1}\{F\}/k$ which corresponds to W has dimension $r-1$ ([1], p. 532). It follows that an irreducible simple subvariety W of V/k is an isolated fundamental variety of T if and only if the following two conditions are satisfied: (1) $\dim W < r-1$; (2) there exists an irreducible $(r-1)$-dimensional subvariety W^* of V^*/k such that W and W^* are corresponding subvarieties under T^* (i.e., such that W is the projection of W^*). Condition (2) can also be expressed as follows: if P is a general point of W/k, then there exists on V' a point P' which corresponds to P under T and which is such that $\dim (P \times P')\, k = r-1$. If we denote by s the dimension of P/k (i.e. the transcendence degree of $k(P)/k$; s is also the dimension of W), then the relation $\dim(P \times P')/k = r-1$ is equivalent to $\dim P'/k(P) = r-1-s$. Summarizing, we have the following characterization of *simple* isolated fundamental points of T, which makes no direct use of the join V^*: *A simple point P of V/k, of dimension s over k, is an isolated fundamental point of T if and only if the following two conditions are satisfied:*

$$(1)\ s < r-1; \quad (2)\ \dim T\{P\} = r-1-s.$$

(For non-isolated fundamental points P we always have

$$\dim T\{P\} < r-1-s.)$$

3. The case of a non-isolated fundamental point P

We assume now that the simple point P of V/k is a non-isolated fundamental point of T. We observe that $T\{P\}$ is the projection of $T^{*-1}\{P\}$ into V'. Since the projection is a single-valued transformation and since the V'-projection of any variety Z^* defined over some ground field K ($Z^* \subset V \times V'$) is again a variety Z' defined over K ($Z \subset V'$), it follows that $T\{P\}$ is connected over $k(P)$ if and only if $T^{*-1}\{P\}$ is connected over $k(P)$. We can therefore replace V' by V^* and T by T^{*-1}. *We may therefore assume that T^{-1} is semi-regular at each point of V'.*

We observe that the connectedness theorem is obvious in the case $r=1$ (always in the case of birational transformations), for in that case there are no fundamental points at all. We shall therefore use induction on r. We assume therefore that $r \geqq 2$ and that the connectedness theorem, for birational transformations, is true for simple points of varieties of dimension $< r$.

We assume that V lies in some affine space of dimension n and that P is at finite distance. Let $x_1, x_2, \ldots, x_n$ be non-homogeneous coordinates of a general point of V/k. We denote by $\mathfrak{o}$ the local ring of P on V/k ($\mathfrak{o}$ = set of all quotients $f(x)/g(x)$, where f and g are polynomials with coefficients in k and where $g(P) \neq 0$) and by $\mathfrak{m}$ the maximal ideal of $\mathfrak{o}$ ($f(x)g(x) \in \mathfrak{m}$ if and only if $f(P)=0$). Since the fundamental point P is not isolated, the isolated fundamental varieties of T which pass through P are all of dimension $> s$. Let $F'_1, F'_2, \ldots, F'_h$ be the $(r-1)$-dimensional subvarieties of V' which correspond to the isolated fundamental varieties passing through P and let F_i be the variety which corresponds to F'_i. Let $\mathfrak{p}_i$ be the prime ideal in $\mathfrak{o}$ which corresponds to F_i. Since $\dim F_i > s$, $\mathfrak{p}_i$ is a proper subideal of $\mathfrak{m}$ ($i = 1, 2, \ldots, h$).

We fix a polynomial $f(X)$ in $k[X]$ such that the following conditions are satisfied:

$$(1) \qquad f(x) \in \mathfrak{m}, \quad f(x) \notin \mathfrak{m}^2;$$

$$(2) \qquad f(x) \notin \mathfrak{p}_i \quad (i = 1, 2, \ldots, h).$$

We denote by H the hypersurface $f(X) = 0$. By (1), this hypersurface has a regular intersection with V at the point P. That means that the intersection $H \cap V$ has only one irreducible component passing through P (we shall call that component W), that P is a simple point of W/k and that the principal ideal generated by $f(x)$ in $\mathfrak{o}$ is the prime ideal of W. We also note that W has dimension $r-1$. Conditions (2) signify that $F_i \not\subset W (i = 1, 2, \ldots, h)$.

Since $\dim W = r-1$, W is not fundamental for T, and hence T is regular at any general point of W/k (since T^{-1} is semi-regular). Thus to W there corresponds a unique irreducible subvariety W' of V'/k; this variety W' is also of dimension $r-1$, and the restriction of T^{-1} to W' is a semi-regular birational transformation of W' into W. We denote the inverse of this induced birational transformation by T_1. Since P is a simple point of W/k, we have, by our induction hypothesis, that the variety $T_1\{P\}$ is connected over $k(P)$. We also have that $T_1\{P\} \subset T\{P\}$, since the graph of T_1 is contained in the graph of T. We shall now show that $T_1\{P\} = T\{P\}$, and this will complete the proof of the connectedness theorem in the present case.

Let P' be any point of $T\{P\}$ and let $\mathfrak{o}'$ be its local ring on V'/k. Since T^{-1} is semi-regular, we have $\mathfrak{o} \subset \mathfrak{o}'$, whence $f(x) \in \mathfrak{o}'$. Since P and P' are corresponding points under T, every non-unit of $\mathfrak{o}$ is also a non-unit of $\mathfrak{o}'$. Hence $f(x)$ is a non-unit of $\mathfrak{o}'$. We fix some isolated prime ideal $\mathfrak{p}'$ of the principal ideal generated by $f(x)$ in $\mathfrak{o}'$ and we denote by G' the irreducible subvariety of V'/k which contains P' and is defined by $\mathfrak{p}'$. Then G' is of dimension $r-1$, and the irreducible subvariety G of V/k which corresponds to G' under T passes through P and is defined in $\mathfrak{o}$ by the prime ideal $\mathfrak{p} = \mathfrak{o} \cap \mathfrak{p}'$. Since $f(x) \in \mathfrak{p}$, it follows from (2) that $\mathfrak{p} \neq \mathfrak{p}_i$ $(i = 1, 2, \ldots, h)$. Hence G is not an isolated fundamental variety of T. On the other hand, G' corresponds to G and has dimension $r-1$. Hence G is not a fundamental variety at all. It follows that G has dimension $r-1$, and $\mathfrak{p}$ is a minimal prime ideal of $\mathfrak{o}$. Since $f(x) \in \mathfrak{p}$, it follows that the variety W coincides with G. Hence $G' = W'$ and $P' \in W'$. The point of W which corresponds to P' under the *semi-regular* transformation T_1^{-1} must be the point P (since the graph of T_1 is contained in the graph of T and since P is the only point of V which corresponds to P' under T). Hence $P' \in T_1\{P\}$, as asserted.

4. The case of an isolated fundamental point P

We now consider the case in which the point P is an isolated fundamental point of T. We apply to V a locally quadratic transformation ϕ, defined over k and having center P (this is another way of saying that ϕ is a monoidal transformation of V/k whose center is the irreducible variety having P as general point over k). Let V_1 denote the ϕ-transform of V. By known properties of monoidal transformations we have that the variety $\phi\{P\}$ is defined and irreducible over $k(P)$, that it is non-singular over $k(P)$, has dimension $r-1-s$, where $s=\dim P/k$, and that all its points are simple for V_1/k. The given birational transformation T of V into V' and the birational transformation ϕ of V into V_1 define, in an obvious fashion, a birational transformation T_1 of V_1 into V'. It is clear that $T\{P\}$ and $\phi\{P\}$ are the total transforms of each other under T_1 and T_1^{-1} respectively. We now show that if for every point P_1 of $\phi\{P\}$ *it is true that* $T_1\{P_1\}$ *is connected over* $k(P_1)$, *then* $T\{P\}$ *is connected over* $k\{P\}$. Assume the contrary, and let $T\{P\}$ be the union of two varieties W_1' and W_2' having no points in common and neither of which is empty (it is assumed that both varieties are defined over $k(P)$). Let G_1 and G_2 be the total T_1^{-1}-transforms of W_1' and W_2' respectively. The two varieties G_1 and G_2 are defined over $k(P)$ and their union is $\phi\{P\}$. Since $\phi\{P\}$ is irreducible over $k(P)$, one of the varieties G_i contains the other. Let, say, $G_2 \subset G_1$. Then if P_1 is any point G_2, the variety $T_1\{P_1\}$ meets both varieties W_1' and W_2'. The variety $T_1\{P_1\}$ is defined over $k(P_1)$, and the field $k(P_1)$ contains the field $k(P)$ since ϕ^{-1} is semi-regular at any point of V_1. Hence the intersections of $T_1\{P_1\}$ with W_1' and W_2' are varieties defined over $k(P_1)$; these intersections are non-empty and have no points in common. Therefore, $T_1\{P_1\}$ is disconnected over $k(P_1)$, contrary to our assumption.

Let then P_1 be any point of $\phi\{P\}$. If P_1 is not an isolated fundamental point of T_1, then $T_1\{P_1\}$ is connected over $k(P_1)$, by the preceding case of the proof. Assume that P_1 is an isolated fundamental point of T_1. We now proceed with P_1 as we did with P, i.e., we apply to V_1 a quadratic transformation ϕ_1 with center P_1, we denote by V_2 the ϕ_1-transform of V_1 and by T_2 the birational transformation of V_2 into V' which is defined in an obvious fashion by the two .birational transformations ϕ_1 and T_1. If $\phi_1\{P_1\}$ carries no isolated fundamental points of T_2, the proof is complete. In the contrary case, we consider a point P_2 on $\phi_1\{P_1\}$ which is an isolated fundamental point of T_2 and we repeat the above procedure. The proof of the theorem now depends on showing that this process must come to an end after a finite number of steps. This we now proceed to prove by an indirect argument.

Suppose that the contrary is true. We will have then an infinite sequence of varieties $V, V_1, V_2, \ldots, V_i, \ldots$ with the following properties: (a) V_{i+1} is the transform of V_i by a quadratic transformation ϕ_i whose center is a point P_i of V_i ($V_0 = V, \phi_0 = \phi, P_0 = P$); (b) the point P_i ($i > 0$) belongs to $\phi_{i-1}\{P_{i-1}\}$; (c) if T_{i+1} is the birational transformation of V_{i+1} into V' which is determined in a natural fashion by the birational transformations ϕ_i and T_i, then P_{i+1} is an isolated fundamental point of T_{i+1}. If V is a surface ($r=2$), then every fundamental point is isolated and the situation which we have just described signifies that it is not possible to eliminate the fundamental point P of the birational transformation T by applying to V successive quadratic transformations. However, we have proved in earlier papers of ours that the contrary is true, i.e., that the fundamental point P will be automatically eliminated if we apply a sufficient number of quadratic transformations ([3], p. 681). Our proof for surfaces is such that its generalization to varieties of any dimension leads to the conclusion that any *isolated* fundamental point can be eliminated by consecutive quadratic transformations. We have actually carried out this generalization in the case of three-dimensional varieties ([2], p. 535), and it will be sufficient to indicate here the main steps of the proof.

We assume then the existence of an infinite sequence of varieties V_i having the above indicated properties and we denote by s_i the dimension of the point P_i, over k. We have $s_i < r-1$ (since P_i is a fundamental point of T_i) and $s_i \leqq s_{i+1}$ (since ϕ_i^{-1} is semi-regular at P_{i+1}, whence $k(P_i)$ is a subfield of $k(P_{i+1})$). Hence for i sufficiently large we will have $s_i = s_{i+1} = \ldots$, and we may assume without loss of generality that this happens already for $i=0$, i.e., that $s = s_1 = s_2 = \ldots$. By the characterization of simple isolated fundamental points given at the end of § 2, we have $\dim T_i\{P_i\} = r-1-s$. On the other hand, we have

$$T_i\{P_i\} \subset T_{i+1}\{P_{i+1}\},$$

in view of the semi-regularity of ϕ_i^{-1}, and hence $T_i\{P_i\} = T_{i+1}\{P_{i+1}\} = \ldots$ for all sufficiently high values of i. There exists therefore a point P' of V' which belongs to all the varieties $T_i\{P_i\}$ and which has dimension $r-1-s$ over $k(P)$ (that point will then have dimension $r-1-s$ over each of the fields $k(P_i)$, since these fields have the same transcendence degree s over k). We have $\dim P'/k = r-1$, and therefore there is only a finite number of valuations whose center is P'; all these valuations are of dimension $r-1$, i.e., they are prime divisors. Since P' and P_i are corresponding points (under T_i), at least one of these valuations must have center P_i. Let L_i be the set of those valuations of center P' on V' whose center on V_i is P_i. Then each L_i is a non-empty finite set, and we have $L \supset L_1 \supset L_2 \supset \ldots \supset L_i \ldots$. It follows that there exists at least one prime divisor $\mathfrak{P}$ which is contained in all L_i. This prime divisor $\mathfrak{P}$ is of second kind with respect to all varieties V_i, since the center P_i of $\mathfrak{P}$ on V_i has dimension $s < r-1$. If then m_i denotes the maximal ideal of the local ring of P_i on V_i and if $v_{\mathfrak{P}}(m_i)$ denotes the minimum of all $v_{\mathfrak{P}}(x)$ for x in m_i, then we obtain as in [2], p. 536, or p. 493, Lemma 9.1, (see also [3], p. 681) the absurd sequence of inequalities

$$v_{\mathfrak{P}}(m) > v_{\mathfrak{P}}(m_1) > \ldots > v_{\mathfrak{P}}(m_i) > \ldots > 0,$$

where all the $v_{\mathfrak{P}}(m_i)$ are integers.

This completes the proof of the connectedness theorem for birational transformations, in the case of simple points.

HARVARD UNIVERSITY

REFERENCES

[1] O. ZARISKI, *Foundations of a general theory of birational correspondences*, Trans. Amer. Math. Soc., 53 (1943), pp. 490–542.

[2] ——, *Reduction of the singularities of algebraic three-dimensional varieties*, Ann. of Math., 45 (1944), pp. 472–542.

[3] ——, *Reduction of the singularities of an algebraic surface*, Ann. of Math., 40 (1939), pp. 639–689.

[4] ——, *Theory and applications of holomorphic functions on algebraic varieties over arbitrary ground fields*, Memoirs of Amer. Math. Soc., no. 5 (1951), pp. 1–90.

Part II

Linear Systems, the Riemann-Roch Theorem and Applications

Introduction by D. Mumford

Zariski's papers on the general topic of linear systems form a rather coherent whole in which one can observe at least two major themes which he developed repeatedly. One is the Riemann-Roch problem: to compute the dimension of a general linear system $|D|$ on a complete normal variety X and especially to consider the behavior of $\dim|nD|$ as n grows. The other is to apply the theory of linear systems in the 2-dimensional case to obtain results on the birational geometry of surfaces and on the classification of surfaces. In relation to his previous work, this research was, I believe, something like a dessert. He had worked long setting up many new algebraic techniques and laying rigorous foundations for doing geometry—and linear systems, which are the heart of Italian geometry, could now be attacked. In particular, the Italian theory could be recast from an algebraic point of view, freeing it of all dependence on the complex ground field and extending it to characteristic p; and many of its results could be extended from surfaces and curves to higher dimensional varieties.

In [56] he announces results on the dependence of the arithmetic genus on birational transformations. These results are proved in detail in a joint paper [57] with H. T. Muhly. Here, if $X \subset \mathbf{P}^n$ is a projective variety, the arithmetic genus $p_a(X)$ is defined by

$$p_a(X) = (-1)^{\dim X}(P(0)-1),$$

where

$$P(n) = \text{Hilbert polynomial of } X.$$

Using the concept of a "proper" birational map $f\colon X_1 \to X_2$ (which means: for generic linear sections $X_1 \cdot H_1 \cdot \cdots \cdot H_k = Z^{(k)}$ of every dimension, $f[Z^{(k)}]$ is normal—no connection with the presently used concept of "proper" as in Grothendieck's elements), they prove that $p_a(X_1) = p_a(X_2)$ if (1) f is biregular and X_1 and X_2 are normal or if (2) f is birational and X_1 and X_2 are non-singular and of dimension 2 (their extension of (2) to dimension 3 misquotes Zariski's earlier paper [47] on resolution in dimension 3 and is therefore incomplete). Subsequently the problem has been clarified by two main developments: Serre's introduction[29] of coherent sheaf cohomology gave a new expression for the arithmetic genus. He showed that

$$P(n) = \sum_{i=0}^{\dim X} (-1)^i \dim H^i(X, \mathcal{O}_X(n)), \qquad \forall\, n \in \mathbf{Z};$$

hence if $n_0 = \dim X$,

$$p_a(X) = \dim H^{n_0}(\mathcal{O}_X) - \dim H^{n_0-1}(\mathcal{O}_X) + \cdots + (-1)^{n_0-1} \cdot \dim H^1(\mathcal{O}_X).$$

This shows immediately that $p_a(X)$ depends only on X as a scheme and not on the particular projective embedding. Moreover it suggests the conjecture: if X_1, X_2 are non-singular and birational, then

$$(1) \quad \dim H^i(\mathcal{O}_{X_1}) = \dim H^i(\mathcal{O}_{X_2}), \qquad 1 \leq i \leq \dim X_1.$$

The second development is due to Matsumura[19] who showed that if $f\colon X_1 \to X_2$ is a birational map of non-singular varieties which can be factored,

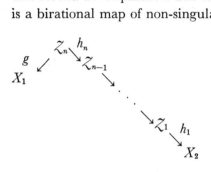

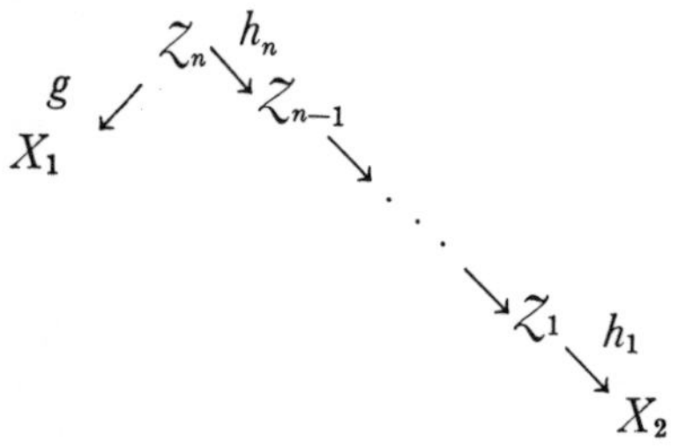

where g is a morphism and each h_i is a monoidal transformation with non-singular center, then

$$\dim H^i(\mathcal{O}_{X_1}) \leq \dim H^i(\mathcal{O}_{Z_n}) = \dim H^i(\mathcal{O}_{X_2}), \qquad 1 \leq i \leq \dim X_1.$$

It follows from the results of Hironaka[11] that in char. 0 this factorization is always possible, hence (1) is true in char. 0.

In [61], Zariski takes up the central point in the Italian Riemann-Roch theorem for surfaces, which he calls the lemma of Enriques-Severi and in a very thorough analysis, extends it to normal varieties of any dimension and characteristic and deduces from it a Riemann-Roch inequality for general normal surfaces. In Zariski's form, the lemma states that if X is a normal variety, D is any divisor on X, and C_m is a generic hypersurface section of degree m of X, then if m is large enough, $\mathrm{tr}_{C_m} |D|$ is a complete linear system on C_m. In the language of sheaves, this means that the map

$$H^0(\mathcal{O}_X(D)) \to H^0(\mathcal{O}_{C_m}(D.C_m))$$

is surjective if $m \gg 0$. Zariski's proof is notable in giving a very good bound on the m needed for this to be the case. He returned to this lemma in the notes [87] (taken from lectures delivered in 1957–1958) where he gave a second proof much closer in spirit to Severi's original proof. As in the case of the invariance of the arithmetic genus, the problem was greatly simplified by the introduction of higher cohomology groups. Using these, the problem is to show that

$$H^1(\mathcal{O}_X(D)(-m)) = (0) \quad \text{if} \quad m \gg 0.$$

This is proven by methods of homological algebra in Serre;[29] yet another proof is described in Zariski's report [67] which will be reproduced in volume III. The problem is also closely related to Kodaira's Vanishing Theorem: if X is non-singular over k of char. 0 and L is an ample line bundle, then

$$H^i(L^{-1}) = (0), \qquad 0 \leq i < \dim X.$$

(Cf. Kodaira;[15] also Akizuki-Nakano,[1] Mumford,[20] Ramanujam,[26] Grauert.[9])

The use of higher cohomology groups also shows why the Riemann-Roch theorem in its classical form has no simple generalization to dimensions $n \geq 3$: what one can readily compute is $\chi(\mathcal{O}_X(D))$ and this differs from the sought-after dim $H^0(\mathcal{O}_X(D))$ by too many terms. For the history of the Riemann-Roch theorem in higher dimensions see Hirzebruch,[13] appendix by R. Schwarzenberger; and Zariski [25], appendix to Ch. 4.

In [63], [77], and [79], Zariski looks in a new direction—at

(1) the function $n \to \dim |nD|$ and

(2) the rings $R_D = \bigcup_{n=0}^{\infty} \Gamma(\mathcal{O}_X(nD)) = \Gamma(X - \text{Support}(D), \mathcal{O}_X)$

and $R'_D = \bigoplus_{n=0}^{\infty} \Gamma(\mathcal{O}_X(nD))$

for arbitrary effective divisors D which need not be ample. He notes in [63] that the finite generation of these rings is a natural generalization of Hilbert's 14th problem. He shows in [63] that when $\dim X = 2$, and D is irreducible then R_D is finitely generated. Subsequently, Rees[27] showed that R_D may not be finitely generated when $\dim X = 3$; and Nagata[24] showed that Hilbert's 14th problem is false too. In [77] and [79], Zariski analyzed the rings R'_D when $\dim X = 2$ (these being isomorphic to particular rings of the type R_D when $\dim X = 3$) giving in particular a general structure theorem for the function dim $|nD|$.

The remaining papers in this section concern the application of linear systems to surfaces, especially to the study of exceptional curves and to the characterization of rational and ruled surfaces. Paper [26] (a joint paper with S. F. Barber) is an early work in which classical language is still used. They study a birational morphism $f: X_1 \to X_2$ between non-singular surfaces over a point $x \in X_2$ such that $\dim f^{-1}(x) = 1$. Among other things, they prove that f factors into a sequence of quadratic transformations, they study the intersection relations of the components E_i of $f^{-1}(x)$, and they characterize those configurations of curves $\{E_i\}$ on X_1 which arise from a morphism f in this way. Many of the same results were proved independently in a paper[7] by P. DuVal that appeared in the next volume of the American Journal of Mathematics. Similar questions are taken up by Zariski again in the monograph [69], but now in more modern language and in all characteristics. Zariski begins with a good deal of background including a new treatment of Bertini's theorem valid in char. p. In the second part he deals again with the factorization of birational morphism between non-singular surfaces and gives a general theory of exceptional curves. One important result is that a non-singular complete surface is always projective. Subsequently Nagata[22] showed that this was false in dimension 3 (see also Hironaka[12] for a large class of such examples). The third part is a summary of the papers [70] and [74] which prove that if K/k is a function field of transcendence degree 2 and K is not isomorphic to $K_0(t)$, where $\text{tr. d.}_{k} K_0 = 1$, (i.e. the models of K are not ruled surfaces), then K has a unique minimal non-singular model X dominated by all other non-singular models. Finally the results of these papers plus [71] together prove, as well, Castelnuovo's criterion: if X is a non-singular surface such that $p_a(X) = 0$ and $|2K_X| = \emptyset$ (K_X = canonical divisor of X), then X is rational. Most of these papers can be described as belonging to "synthetic algebraic geometry," that is to say, they depend on the Riemann-Roch theorem to construct specific linear systems giving one quite explicit grasp on the surface in question. Zariski in fact revived this whole line of investigation which, in the hands of Enriques,[8] had led to a general classification of algebraic surfaces. To be specific, the situation which Zariski quickly finds if he assumes that his function field K has no minimal models is that K has a non-singular model F with an irreducible rational but possibly singular curve E on it such that after at least one quadratic transformation of F, the proper transform of E can be "blown down" to a simple point. In this case, he finds

$$p_a(E) = \sum_{i=1}^{m} \frac{s_i(s_i - 1)}{2},$$

$$(E^2) = -1 + n + \sum_{i=1}^{m} s_i^2,$$

$$(K \cdot E) = -1 - n - \sum_{i=1}^{m} s_i,$$

where

s_i = multiplicities of singular points on E, infinitely near, possibly;
K = canonical divisor class on F;

$n + m > 0$

Now if for instance, $(E^2) = 0$, then E is non-singular, $(E \cdot K) = -2$. So, by Riemann-Roch,

$$\dim|nE| \geq \frac{(nE \cdot nE - K)}{2} + p_a(F) - [\text{index of specialty } i(nE)].$$

So if $n \gg 0$, dim $|nE| > 0$ and nE moves in a pencil; then apply the classical theorem of Noether-Enriques that if F has a non-singular rational pencil on it, then F is ruled. In the cases $(E^2) > 0$, he examines instead one of the linear systems $|E + nK|$ or $|-K|$. The hardest case turns out to be $(K^2) = 2$, char. $= 2$ in which he has to look also at $|-2K|$, and via this he constructs a birational map to a certain specific singular quartic hypersurface whose rationality is then checked directly.

This line of research has been followed by many others since then: (1) Criteria for the contractibility of finite sets of curves were studied by Artin;[3] (2) The theory of rational and ruled surfaces was pursued by Andreotti,[2] Nagata,[23] and Hartshorne[10] among others; (3) Kodaira gave at the same time another rigorous treatment of Castelnuovo's criterion and went on first to establish all of Enriques' classification and later to carry it further, especially to non-algebraic compact complex analytic surfaces in a beautiful series of papers[16,17]; Šafarevitch[28] gave a seminar in which this theory was also worked out as well as more detailed investigations of $K3$-surfaces (surfaces X such that $K_X \equiv 0$ and $p_a(X) = 1$); Mumford[21] extended much of Enriques' theory to char. p; (4) Van der Ven,[30] Kodaira,[18] and Bombieri[5] studied the surfaces X where K_X is ample or "nearly so"; (5) Šafarevitch and Pjatetskij-Shapiro[25] recently have worked out a deep theory of $K3$-surfaces; (6) finally one of the most important corollaries of Castelnuovo's criterion is Lüroth's theorem in dimension 2: if f: $\mathbf{P}^2 \rightarrow X$ is a rational separable map of surfaces, then X is rational. This has recently been disproven in dimension 3.[4,6,14]

References

1. Y. Akizuki and S. Nakano, *Note on Kodaira-Spencer's proof of Lefschetz's Theorem*, Proc. Japan Acad., vol. 30 (1954).

2. A. Andreotti, *On the complex structures of a class of simply connected manifolds*, in *Algebraic Geometry and Topology*, Princeton University Press, 1957.

3. M. Artin, *Some numerical criteria for contractibility of curves on an algebraic surface*, Amer. J. Math., vol. 84 (1962) pp. 485–496.

4. M. Artin and D. Mumford, *Some elementary examples of unirational varieties which are not rational*, Proc. London Math. Soc., vol. 25 (1972) pp. 75–95.

5. E. Bombieri, *Canonical models of surfaces of general type*, to appear.

6. H. Clemens and P. Griffiths, *The intermediate Jacobian of the cubic threefold*, Ann. of Math., vol. 95 (1972) pp. 281–356.

7. P. DuVal, *Reducible exceptional curves*, Amer. J. Math., vol. 58 (1936) p. 285.

8. F. Enriques, *Le Superficie Algebriche*, Zanichelli, Bologna, 1949.

9. H. Grauert and O. Riemenschneider, *Verschwindungssätze für analytische Kohomologiegruppen*, Invent. Math., vol. 11 (1970) pp. 263–292.

10. R. Hartshorne, *Curves with high self-intersection on algebraic surfaces*, Inst. Hautes Études Sci., Publ. Math., vol. 36 (1969), pp. 111–126.

11. H. Hironaka, *On resolution of singularities*, Proc. Internat. Congress in Stockholm, 1962, pp. 507–521.

12. H. Hironaka, *An example of a non-Kählerian deformation*, Ann. of Math., vol. 75 (1962) pp. 190–208.

13. F. Hirzebruch, *Topological methods in algebraic geometry*, third ed., Springer-Verlag, Berlin, 1966.

14. V. Iskovskikh and Y. Manin, *Three-dimensional quartics and counterexamples to the Lüroth problem*, Mat. Sbornik, vol. 86 (1971) pp. 140–166.

15. K. Kodaira, *On a differential-geometric method in the theory of analytic stacks*, Proc. Nat. Acad. Sci. U.S.A., vol. 39 (1953) pp. 1268–1273.

16. K. Kodaira, *On compact complex analytic surfaces*, I, II, and III, Ann. of Math., vol. 71 (1960) pp. 111–152; vol. 77 (1963) pp. 563–626; and vol. 78 (1963) pp. 1–40.

17. K. Kodaira, *On the structure of compact complex analytic surfaces*, I, II, III, IV, Amer. J. Math., vol. 86 (1964) pp. 751–798; vol. 88 (1966) pp. 682–721; vol. 89 (1967) pp. 55–83; vol. 90 (1968) pp. 1048–1065.

18. K. Kodaira, *Pluricanonical systems on algebraic surfaces of general type*, J. Math. Soc. Japan, vol. 20 (1968) pp. 170–192.

19. H. Matsumura, *Geometric structure of the cohomology rings in abstract algebraic geometry*, Mem. Coll. Sci. Kyoto, vol. 32 (1959) pp. 33–84.

20. D. Mumford, *Pathologies III*, Amer. J. Math., vol. 89 (1967) pp. 94–104.

21. D. Mumford, *Enriques' classification of surfaces I*, in *Global Analysis*, Princeton University Press, 1969, pp. 325–339.

22. M. Nagata, *Existence theorems for non-projective complete algebraic varieties*, Ill. J. Math., vol. 2 (1958) pp. 490–498.

23. M. Nagata, *On rational surfaces*, I and II, Mem. Coll. Sci. Kyoto, vol. 32 (1960) pp. 351–370; and vol. 33 (1960) pp. 271–293.

24. M. Nagata, *On the 14th problem of Hilbert*, Amer. J. Math., vol. 81 (1959) pp. 766–772.

25. I. Piatetskij-Sapiro and I. Šafarevich, *Torelli's theorem for algebraic surfaces of type K3*, Izvestia Akad. Nauk Ser. Mat., SSSR, vol. 35 (1971) pp. 530–572.

26. C. P. Ramanujam, *Remarks on the Kodaira Vanishing Theorem*, J. Indian Math. Soc., vol. 36 (1972) pp. 41–50.

27. D. Rees, *On a problem of Zariski*, Ill. J. Math., vol. 2 (1958) pp. 145–149.

28. I. Šafarevich (and others), *Algebraic surfaces*, Proc. Steklov Inst. Math., vol. 75 (1965).

29. J. -P. Serre, *Faisceaux algébriques cohérents*, Ann. of Math., vol. 61 (1955) pp. 197–278.

30. Van der Ven, *On the Chern numbers of certain complex manifolds*, Proc. Nat. Acad. Sci. U.S.A., vol. 55 (1966) pp. 1624–1627.

Reprints of Papers

REDUCIBLE EXCEPTIONAL CURVES OF THE FIRST KIND.*

By S. F. Barber † and Oscar Zariski

Introduction. In a birational transformation between two surfaces to certain curves of one of the surfaces may correspond *simple* points of the other. Such curves have been called *exceptional curves.* Exceptional curves have been subdivided into those of the first or second kind according as a point of the curve is not or is transformed into a curve of the other surface. The ones previously considered have either been irreducible or consisted of at most two components. A treatment of reducible exceptional curves with components all simple is found in ([4], Chap. II, 6). We shall consider exceptional curves of the first kind with s irreducible components, each counted a certain number of times. This case arises when the birational transformation possesses infinitely near fundamental points on one or several irreducible algebroid branches.

In the sequel we shall use some fundamental notions of linear systems of curves on an algebraic surface and the theory of singularities as developed by Enriques ([3], pp. 327-399).

Sections 1 and 2 of this paper are devoted to definitions and the derivation of the virtual degree and genus of an exceptional curve of the first kind; the presentation here is parallel to but more complete than that found in ([4], Chap. II, 6). In sections 3 and 4 we introduce s fundamental points of the birational transformation, mentioned above, on that surface where the given exceptional curve has been changed into a point, and limit ourselves to the case in which these fundamental points lie on a single irreducible algebroid branch. In 5 we consider the correspondence between the irreducible components of the given curve on the one surface and the immediate neighborhood of each of the fundamental points on the other surface. In 6 we determine the multiplicities of these fundamental points on branches of lowest order passing through the first α ($\alpha \leq s$) of them. In 7 we describe the intersections of the irreducible components of the given exceptional curve, using extensively the notion of proximate points introduced by Enriques in his theory of singularities ([3], p. 381); in 8 we study the classification of the s fundamental points as free points or satellites when the intersection numbers

* Presented to the Society, March 30, 1934.
† National Research Fellow.

of the components of the exceptional curve are known. In 9 we extend to reducible exceptional curves of the first kind the characterization of irreducible exceptional curves of the first kind as found in ([1]). Finally in 10 we consider the situation when the fundamental points lie on several irreducible algebroid branches.

1. *Definitions.* Let F be an algebraic surface in S_ρ, free from singularities, and let Σ be a linear system of curves C on F, of dimension r. We shall say for brevity that a curve L on F, irreducible or not, is a total fundamental curve of Σ, if L is a fixed component of a linear subsystem, ∞^{r-1}, of Σ and if the curves of this subsystem do not have other fixed components outside L. We shall adopt the following

Definition. ([4], Chap. II, 6). *A curve L on F is an exceptional curve if there exists on F a linear system Σ of curves C such that*

(i) *Σ is irreducible, simple, of dimension $r \geq 3$;*

(ii) *L is a total fundamental curve of Σ;*

(iii) *The ∞^{r-1} curves C_1, which together with L give total curves of Σ, form a linear system Σ_1 of effective degree one less than the effective degree of Σ.*

We recall that the effective degree of a linear system is the number of *variable* intersections of two curves of the system. If the system is of dimension r, its effective degree is $\geq r-1$. It follows that the system Σ_1 is necessarily irreducible, since by (ii) it is free from fixed components and by (i) ($r \geq 3$) and (iii) it is of positive effective degree and therefore cannot be composed of the curves of a pencil.

If it is possible to find the above system Σ in such a manner that it should have no base points on L, then L is called an exceptional curve of the *first kind.* If this is not possible, L is called an exceptional curve of the *second kind.*

If we refer the ∞^r curves C of Σ to the hyperplanes of an S_r, we obtain in S_r a surface $\bar{F}$ birationally equivalent to F, since Σ is simple. This surface is of order n, where n is the effective degree of Σ, and its hyperplane sections $\bar{C}$ correspond to the curves C of Σ. In this correspondence to the subsystem $\Sigma_1 + L$ of Σ there corresponds the system of hyperplane sections of $\bar{F}$ on a fixed point O of S_r. This point O is on $\bar{F}$, since the effective degree of Σ_1 is less than the effective degree of Σ, and is a *simple* point of $\bar{F}$, since by (iii) two hyperplane sections of $\bar{F}$ on O have $n-1$ variable intersections. Thus in the birational correspondence T between F and $\bar{F}$ the curve L (irreducible or not) is transformed into a *simple* point O of $\bar{F}$ (a fundamental point of T^{-1}) and this is in agreement with the usual definition of an exceptional curve.

It is one of the purposes of this paper to study the nature of the correspondence between L and the neighborhood of O when L is reducible.

The fundamental points of T (on F) are at the base points of Σ. Hence an exceptional curve L is of the first or of the second kind (see above definition) according as it is or is not possible to transform L into a simple point of a surface $\bar{F}$, birationally equivalent to F, in such a manner that no point of L is at the same time transformed into a curve of $\bar{F}$.

In the sequel we shall be concerned only with exceptional curves of the first kind. We observe, however, that the properties of exceptional curves of the second kind can be deduced from those of the first kind. In fact, it is always possible to transform birationally F into a surface $\bar{F}$ in such a manner that the transformation should possess no fundamental points on $\bar{F}$ and that to Σ there should correspond on $\bar{F}$ a linear system $\bar{\Sigma}$, free from base points. For this it is sufficient to refer the sections of F by hyperquadrics on a base point of Σ to the hyperplanes of a linear space and to apply this procedure successively to the transformed surfaces until all the base points of Σ disappear. To the exceptional curve L there will correspond on $\bar{F}$ an exceptional curve $\bar{L}$ of the *first kind.* However, in general $\bar{L}$ will consist of the transform proper of L and of the fundamental curves of the transformation on $\bar{F}$, arising from the base points of Σ.

2. *The virtual characters of an exceptional curve of the first kind.* *Theorem* 1 ([4], Chap. II, 6). *If L is an exceptional curve of the first kind, then* $(L^2) = -1$ *and* $[L] = 0$.*

Proof. Let H denote the set of effective base points of Σ. We have $(C \cdot L) = ((C_1 + L) \cdot L) = 0$, since L, a total fundamental curve of Σ, does not have variable intersections with a C and since no base point of Σ is on L. Consequently $(C_1 \cdot L) = -(L^2)$. Now $C_1 + L$ is the limit of an irreducible curve C and hence C_1 and L must have at least one point in common (Principle of degeneration of Enriques [2]). It follows that $(C_1 \cdot L) > 0$ and hence $(L^2) < 0$. We also have† $n = (C^2)_H = (C_1^2)_H + 2(C_1 \cdot L) + (L^2) = (C_1^2)_H - (L^2)$. From the hypothesis that the effective degree of Σ_1 is $n-1$, it follows that $(C_1^2)_H \geq n-1$,‡ and that consequently $(L^2) \geq -1$.

* (L^2) is the virtual degree, $[L]$ the virtual genus. In evaluating these characters of L, we consider L as *virtually free from base points.*

† $(C^2)_H$ denotes the virtual degree of C with respect to the assigned set H of base points, in this case the effective degree of Σ.

‡ The set of effective base points of Σ_1 certainly includes H, since $C_1 + L = C$. *A priori* we cannot exclude the possibility that Σ_1 may possess accidental base points, whence the necessity of using the inequality sign.

It follows that $(L^2) = -1$ and incidentally

$$(C_1^2)_H = n - 1; \tag{1}$$

$$(C_1 \cdot L) = 1. \tag{2}$$

We observe that from (1) it follows that H is also the set of effective base points of Σ_1 and hence $[C_1]_H$ gives the effective genus of C_1. Since the curves C_1 correspond to the hyperplane sections of $\bar{F}$ on a simple point O, it follows that $[C_1]_H = [C]_H$. Furthermore

$$[C]_H = [C_1]_H + [L] + (C_1 \cdot L) - 1 = [C_1]_H + [L].$$

Hence $[L] = 0$, q. e. d.

The exceptional curve L may be reducible and some of its irreducible components may occur in L to a certain multiplicity. Let then $L = k_1L_1 + k_2L_2 + \cdots + k_sL_s$, where $L_1, L_2, \cdots, L_s$ are the distinct irreducible components of L and where $k_1, k_2, \cdots, k_s$ are positive integers. From (2) it follows that C_1 intersects one and only one of the components L_i and that this must be a simple component of L. Choosing our notation properly we may assume that

$$(C_1 \cdot L_1) = 1, \quad (C_1 \cdot L_i) = 0 \qquad (i = 2, \cdots, s), \tag{3}$$

$$k_1 = 1. \tag{3'}$$

Since $(C \cdot L_i) = 0$ $(i = 1, \cdots, s)$, it follows that

$$(L \cdot L_1) = -1, \quad (L \cdot L_i) = 0 \qquad (i = 2, \cdots, s), \tag{4}$$

all curves L_i being considered as virtually free from base points. The relations (4) can be rewritten as follows:

$$\tag{5} \begin{aligned} &(L_1^2) + \sum_{i=2}^{s} k_i(L_1 \cdot L_i) = -1, \\ &(L_j \cdot L_1) + \sum_{i=2}^{s} k_i(L_j \cdot L_i) = 0 \qquad (j = 2, \cdots, s). \end{aligned}$$

We observe that the curve $L_1 + L_2 + \cdots + L_s$ must be *connected,* i. e. it cannot be represented as the sum of two curves having no points in common. This follows immediately from the fact that C_1 meets only the curve L_1 and that $C_1 + L$ is the limit of the irreducible curve C. Since $(L_i \cdot L_j) \geqq 0$, if $i \neq j$, and since, by the preceding remark, for a given value of i one at least of the intersection numbers $(L_i \cdot L_j)$, $i \neq j$, must be positive, it follows from (5) *that the virtual degree* (L_i^2) *of any component* L_i *of* L *is negative.*

Applying the well known formula for the virtual genus of a reducible curve, we find

$$[L]=\sum_{i=1}^{s}k_ip_i+\sum_{i=1}^{s}[k_i(k_i-1)/2](L_i)^2 + (1/2)\sum_{\substack{i=1\\ i\neq j}}^{s}\sum_{j=1}^{s}k_ik_j(L_i\cdot L_j)-\sum_{i=1}^{s}k_i+1,$$

where $p_i=[L_i]$. By means of (4) or the equivalent relations (5) this expression of $[L]$ can be simplified:

$$[L]=\sum_{i=1}^{s}k_ip_i-k_1/2-(1/2)\sum_{i=1}^{s}k_i(L_i^2)-\sum_{i=1}^{s}k_i+1.$$

Since by Theorem 1 we have $[L]=0$ and since $k_1=1$ we arrive at the following relation:

$$\sum_{i=1}^{s}k_i[2p_i-(L_i^2)-2]+1=0. \tag{6}$$

3. *The fundamental points in the first neighborhood of O.* We now proceed to study the system Σ_1, the residual of Σ with respect to L. By (3) the curves $L_2,\cdots,L_s$ are fundamental curves of Σ_1. On the contrary L_1 is not a fundamental curve of Σ_1, since $(C_1\cdot L_1)=1$, and since the intersection of C_1 and L_1 cannot be a fixed intersection, because by (1) the effective base points of Σ_1 coincide with those of Σ and are all outside L. Let us assume for the sake of generality that $L_2+\cdots+L_s$ consists of several connected curves $\bar{L}^{(1)}, \bar{L}^{(2)},\cdots \bar{L}^{(\sigma)}$, having two by two no points in common (maximal connected components of $L_2+\cdots+L_s$). Each $\bar{L}^{(i)}$ is then itself a fundamental curve of Σ_1. Let us consider for instance $\bar{L}^{(1)}$ and let us denote by Σ_2 the residual system of Σ_1 with respect to $\bar{L}^{(1)}$. Let C_2 denote a generic curve of Σ_2. What are the possible fixed components of Σ_2? L_1 is not a fixed component since it is not fundamental for Σ_1. Neither can any component of $\bar{L}^{(i)}$, $i>1$, be a fixed component of Σ_2. In fact we have $(C_1\cdot L_1)=((C_2+\bar{L}^{(1)})\cdot L_1)=1$, and since $(\bar{L}^{(1)}\cdot L_1)>0$ (the curve $L_1+\cdots+L_s$ is connected), and $(C_2\cdot L_1)$ cannot be negative, it follows that $(C_2\cdot L_1)=0$ and incidentally that

$$(\bar{L}^{(1)}\cdot L_1)=1. \tag{7}$$

If any component of $\bar{L}^{(i)}$, $i>1$, were a fixed component of Σ_2, then the whole curve $\bar{L}^{(i)}$ would be a fixed component of Σ_2, since it is a fundamental curve of Σ_1. We would have then $C_2=\bar{L}^{(i)}+\bar{C}_2$ and $(\bar{C}_2\cdot L_1)=-(\bar{L}^{(i)}\cdot L_1)<0$, and this is impossible since L_1 is not a fixed component of Σ_2.

Thus the only possible fixed components of Σ_2 are either the irreducible components of $\bar{L}^{(1)}$ or irreducible curves which are not components of L. We can, however, avoid this second possibility by replacing the system Σ_1 by the system $\Sigma^{(1)}$, which corresponds on F to the system cut out on $\bar{F}$ by the hyperquadrics (instead of by the hyperplanes) passing through the fundamental point O. This system $\Sigma^{(1)}$ contains all the reducible curves $C + C_1$, and it is obvious that the preceding conclusions apply as well to $\Sigma^{(1)}$ as to Σ_1. The curves of $\Sigma^{(1)}$ which contain $\bar{L}^{(1)}$ as a fixed component do not automatically possess other fixed components which are not components of L, since among these curves we find in particular the curves $L + 2C_1$. We conclude that the system $\Sigma^{(1)}$, just defined, possesses a *total* fundamental curve of the type

$$(8) \qquad L^{(1)} = k_2{}^{(1)}L_2 + k_3{}^{(1)}L_3 + \cdots + k_{s'}{}^{(1)}L_{s'} \qquad (s' \leq s),$$

where $L_2 + \cdots + L_{s'} = \bar{L}^{(1)}$ and where $k_2{}^{(1)}, \cdots, k_{s'}{}^{(1)}$ are positive integers. In a similar manner the remaining maximal connected components $\bar{L}^{(2)}, \cdots, \bar{L}^{(\sigma)}$ of $L_2 + \cdots + L_s$ give rise to total fundamental curves $L^{(2)}, \cdots, L^{(\sigma)}$ of $\Sigma^{(1)}$, where $L^{(i)}$ is made up of the irreducible components of $\bar{L}^{(i)}$, each counted to a proper positive multiplicity.

We have seen before that the curves C_1 of Σ_1 meet L_1 in one variable point. These curves correspond to the hyperplane sections of $\bar{F}$ on O. The curves C_1 passing through a fixed point of L_1 form a subsystem of Σ_1 of effective degree less than that of Σ_1, and hence correspond to the sections of $\bar{F}$ by hyperplanes through O and through another point O_1 of $\bar{F}$, at finite distance or infinitely near O. However, the point O_1 cannot be at a finite distance from O, since L does not carry fundamental points of the birational transformation between F and $\bar{F}$ and therefore to *every* point of the curve L there corresponds an unique point of $\bar{F}$. This point must coincide with(or be infinitely near) O, since the curve L is carried by the transformation into the point O. It follows that the curves C_1 on a fixed point of L_1 correspond to the hyperplane sections of $\bar{F}$ passing through O and touching there a fixed tangent line of $\bar{F}$. Since the curves C_1 meet L_1 in only *one* point, which varies as C_1 varies in Σ_1, it follows that there is a $(1, 1)$ correspondence between the points of L_1 and the tangential directions of $\bar{F}$ at O. The irreducible component L_1 of L thus corresponds to the first order neighborhood of the point O on $\bar{F}$. Incidentally this shows that *L_1 is a rational curve.*

We can be more precise and show that not only is L_1 a rational curve but that L_1 is free from multiple points, i. e. $p_1 = [L_1] = 0$. The system Σ_1 cuts out on L_1 a $g_1{}^1$. If L_1 had a point of multiplicity > 1, the subsystem of Σ_1 through this point would then contain L_1 as a fixed part, and this is impossible, since L_1 is not a fundamental curve of Σ_1. This property extends similarly to the curves L_i $(i = 2, \cdots, s)$. In fact, it will be shown in the next section that any curve L_{i+1} $(i = 1, 2, \cdots, s-1)$ plays the rôle of L_1 for a conveniently defined exceptional curve $L^{(i)}$ of the first kind [formula (16)]. Hence $p_i = 0$ $(i = 1, 2, \cdots, s)$, and (6) becomes

$$(6') \qquad \Sigma\, k_i[(L_i^2) + 2] = 1.$$

The above considerations apply to the system $\Sigma^{(1)}$ as well as to Σ_1. Let $C^{(1)}$ denote a generic curve of $\Sigma^{(1)}$. We have $C^{(1)} \equiv C + C_1$ and hence the following relations, similar to the relations (3), hold:

$$(9) \qquad (C^{(1)} \cdot L_1) = 1, \quad (C^{(1)} \cdot L_i) = 0 \qquad (i = 2, \cdots, s).$$

The curves $C^{(1)}$ meet L_1 in one variable point, and those which pass through a fixed point of L_1 correspond to the sections of $\bar{F}$ by hyperquadrics passing through O and touching there a fixed tangent line of $\bar{F}$. Let us now consider the curve $L^{(1)}$, given by (8). In view of (7), $L^{(1)}$ meets L_1 in one point, say $P^{(1)}$. Let $O_1{}^{(1)}$ be the point of $\bar{F}$, infinitely near O, which corresponds to $P^{(1)}$. The curves $C^{(1)}$ of $\Sigma^{(1)}$, which are constrained to pass through $P^{(1)}$, necessarily contain the whole curve $L^{(1)}$, since $L^{(1)}$ is a fundamental curve of $\Sigma^{(1)}$. The residual system, which we shall denote by $\Sigma_1{}^{(1)}$, is irreducible, since it does not possess fixed components ($L^{(1)}$ is a total fundamental curve of $\Sigma^{(1)}$) and is not composed of the curves of a pencil (the system $\Sigma_1{}^{(1)}$ contains partially the system Σ). Moreover $\Sigma_1{}^{(1)}$ is of effective degree one less than that of $\Sigma^{(1)}$, since the curves of $\Sigma_1{}^{(1)}$ correspond to the sections of $\bar{F}$ by the hyperquadrics through O and $O_1{}^{(1)}$. Hence $L^{(1)}$ *is itself an exceptional curve of the first kind.* In the birational correspondence between F and $\bar{F}$ to the curve $L^{(1)}$ corresponds the point $O_1{}^{(1)}$ infinitely near O. The exact nature of this correspondence is yet to be investigated.

In a similar manner it can be proved that the other total fundamental curves $L^{(2)}, L^{(3)}, \cdots, L^{(\sigma)}$ of $\Sigma^{(1)}$ are exceptional curves of the first kind and that they correspond to points $O_1{}^{(2)}, O_1{}^{(3)}, \cdots, O_1{}^{(\sigma)}$ on $\bar{F}$, infinitely near and in the first neighborhood of O.

The points $O_1{}^{(1)}, O_1{}^{(2)}, \cdots, O_1{}^{(\sigma)}$ *are distinct.* In fact, the point $O_1{}^{(\alpha)}$ corresponds, in the (1, 1) correspondence between the directions on $\bar{F}$ about O and the points of L_1, to the intersection $P^{(\alpha)}$ of $\bar{L}^{(\alpha)}$ or, what is the same, of $L^{(\alpha)}$ with L_1. These intersections $P^{(1)}, P^{(2)}, \cdots, P^{(\sigma)}$ are distinct, since the curves $\bar{L}^{(1)}, \bar{L}^{(2)}, \cdots, \bar{L}^{(\sigma)}$ have two by two no points in common.

We now prove that (7) can be replaced by the stronger relation:

$$(10) \qquad (L^{(1)} \cdot L_1) = 1.$$

In fact, if $C_1^{(1)}$ denotes a generic curve of $\Sigma_1^{(1)}$, so that $C^{(1)} \equiv C_1^{(1)} + L^{(1)}$, then it follows from (9) that $(L_1 \cdot C_1^{(1)}) = (L_1 \cdot C^{(1)}) - (L_1 \cdot L^{(1)}) = 1 - (L_1 \cdot L^{(1)})$. Since $(L_1 \cdot C_1^{(1)}) \geq 0$ and $(L_1 \cdot L^{(1)}) > 0$, (10) follows. We also have incidentally

$$(10') \qquad (C_1^{(1)} \cdot L_1) = 0.$$

Similar relations hold for $L^{(2)}, L^{(3)}, \cdots, L^{(\sigma)}$.

4. *The successive fundamental points $O, O_1, \cdots, O_{s-1}$ of T.* The points $O_1^{(1)}, O_1^{(2)}, \cdots, O_1^{(\sigma)}$, defined in the preceding section, are clearly to be considered as fundamental points, infinitely near O, of the birational transformation between F and $\bar{F}$. The considerations of the preceding section also show that the presence of two or more distinct fundamental points on $\bar{F}$ in the first neighborhood of O corresponds to the case in which the curve $L_2 + \cdots + L_s$ is not connected and consists of two or more maximal connected components. If then we wish to consider first the simplest case *in which there is only one fundamental point infinitely near and immediately following O,* we must assume that *$L_2 + \cdots + L_s$ is connected.* Let us make this assumption. Then the system $\Sigma^{(1)}$ possesses a total fundamental curve $L^{(1)}$ of the type:

$$(11) \qquad L^{(1)} = k_2^{(1)} L_2 + k_3^{(1)} L_3 + \cdots + k_s^{(1)} L_s,$$

where $k_2^{(1)}, k_3^{(1)}, \cdots, k_s^{(1)}$ are positive integers. The fundamental point on $\bar{F}$, infinitely near O, will be denoted by O_1. We may apply to the system $\Sigma^{(1)}$ and to the exceptional curve $L^{(1)}$ the results derived in section 2 for Σ and L. Thus, choosing properly our notation, we have the following relations, analogous to the relations (3), (3′) and (4):

$$(12) \qquad (C_1^{(1)} \cdot L_2) = 1, \qquad (C_1^{(1)} \cdot L_i) = 0, \qquad (i = 3, \cdots, s);$$

$$(12') \qquad k_2^{(1)} = 1;$$

$$(13) \qquad (L^{(1)} \cdot L_2) = -1; \quad (L^{(1)} \cdot L_i) = 0, \qquad (i = 3, \cdots, s).$$

The system $\Sigma^{(1)}$ corresponds to the sections of $\bar{F}$ by the hyperquadrics on O, and the system $\Sigma_1^{(1)}$ corresponds to the sections of $\bar{F}$ by the hyperquadrics on O and O_1. Let us transform birationally $\bar{F}$ into a surface $\bar{F}_1$ by referring the above hyperquadrics on O to the hyperplanes of a linear space. There arises a birational transformation T_1 between F and $\bar{F}_1$ in which to the system $\Sigma^{(1)}$ corresponds the system of hyperplane sections of $\bar{F}_1$. In the transformation between $\bar{F}$ and $\bar{F}_1$, which is locally quadratic at O, the first neighborhood of O is spread out into the points of a straight line; to the point (direction) O_1 there will correspond a point O'_1 of this line, and it is clear that the system $\Sigma_1^{(1)}$ goes into the system of hyperplane sections of $\bar{F}_1$ through O'_1. For the birational transformation T_1 between F and $\bar{F}_1$, the system $\Sigma^{(1)}$ and the exceptional curve $L^{(1)}$ play the same rôle as Σ and L played in the transformation T between F and $\bar{F}$. If $s > 2$, the transformation T_1 will necessarily possess on $\bar{F}_1$ fundamental points infinitely near O'_1 and hence the transformation T will possess on $\bar{F}$ fundamental points in the second order neighborhood of O. The number of these fundamental points equals the number of maximal connected components of the curve $L_3 + \cdots + L_s$. Again, if we wish to consider the simplest case in which T possesses on $\bar{F}$ only one fundamental point O_2 in the second order neighborhood of O, we must assume that $L_3 + \cdots + L_s$ is connected.

Considering the system $\Sigma^{(2)}$ which corresponds on F to the sections of $\bar{F}_1$ by the hyperquadrics on O'_1 (this system contains all reducible curves $C^{(1)} + C_1^{(1)}$) and assuming that $L_3 + \cdots + L_s$ is connected, we have that $\Sigma^{(2)}$ possesses a total fundamental curve of the type

$$L^{(2)} = k_3^{(2)} L_3 + \cdots + k_s^{(2)} L_s,$$

and that $L^{(2)}$ is an exceptional curve of the first kind. If $\Sigma_1^{(2)}$ denotes the residual system of $\Sigma^{(2)}$ with respect to $L^{(2)}$ and if $C^{(2)}$ and $C_1^{(2)}$ denote generic curves of $\Sigma^{(2)}$ and $\Sigma_1^{(2)}$ respectively, then, with a proper choice of notation, we have the following relations similar to (12), (12′) and (13):

$$(14) \qquad (C_1^{(2)} \cdot L_3) = 1, \qquad (C_1^{(2)} \cdot L_i) = 0 \qquad (i = 4, \cdots, s);$$

$$(14') \qquad k_3^{(2)} = 1;$$

$$(15) \qquad (L^{(2)} \cdot L_3) = -1, \quad (L^{(2)} \cdot L_i) = 0 \qquad (i = 4, \cdots, s).$$

If $s > 3$ the transformation T between F and $\bar{F}$ possesses on $\bar{F}$ fundamental points in the third order neighborhood of O (arising from fundamental points of T_1 on $\bar{F}_1$ in the second order neighborhood of O'_1) and there is but one such fundamental point if and only if $L_4 + \cdots + L_s$ is connected.

The general procedure is now straight-forward and serves to define the fundamental points of T on $\bar{F}$ in the 1st, 2nd, $\cdots$, $(s-1)$-th order neighborhoods of O. *We assume at present that each fundamental point O_i in the i-th neighborhood of O $(i < s - 1)$ is followed by just one fundamental point O_{i+1} in the neighborhood of order $i + 1$.* With these assumptions, T possesses on $\bar{F}$ in addition to O other $s - 1$ successive fundamental points $O_1, \cdots, O_{s-1}$, infinitely near O_1, *lying on one irreducible algebroid branch of origin O.*

We may define by induction the systems $\Sigma^{(i)}$, $\Sigma_1^{(i)}$, the exceptional curves

$L^{(i)}$ $(i=0,1,\cdots,s-1)$ and state their properties as follows. $L^{(i)}$ is a total fundamental curve of $\Sigma^{(i)}$ and we have

$$L^{(i)}=k_{i+1}^{(i)}L_{i+1}+k_{i+2}^{(i)}L_{i+2}+\cdots+k_s^{(i)}L_s, \tag{16}$$

where the k's are positive integers. $\Sigma_1^{(i)}$ is the residual system of $\Sigma^{(i)}$ (of dimension one less than $\Sigma^{(i)}$) with respect to $L^{(i)}$. We have the following relations, similar to the relations (14), (14′), (15):

(17) $(C_1^{(i)}\cdot L_{i+1})=1, \qquad (C_1^{(i)}\cdot L_j)=0 \qquad (j=i+2,\cdots,s);$

(17′) $k_{i+1}^{(i)}=1;$

(18) $(L^{(i)}\cdot L_{i+1})=-1; \quad (L^{(i)}\cdot L_j)=0 \qquad (j=i+2,\cdots,s).$

Here $C^{(i)}$ and $C_1^{(i)}$ denote total curves of $\Sigma^{(i)}$ and $\Sigma_1^{(i)}$ respectively. *The curve $L_{i+2}+\cdots+L_s$ is connected.* The system $\Sigma^{(i+1)}$ is the minimum linear system containing as total curves all reducible curves $C^{(i)}+C_1^{(i)}$.

Let $\bar{F}_i$ be the birational transform of the surface F obtained by referring the curves of $\Sigma^{(i)}$ to the hyperplanes of a linear space, and let T_i denote the birational transformation between F and $\bar{F}_i$ $(\bar{F}_0=\bar{F},\ \Sigma^{(0)}=\Sigma,\ T_0=T)$. The system $\Sigma_1^{(i)}$ corresponds by T_i to the system of hyperplane sections of $\bar{F}_i$ on a fixed point $O_i^{(i)}$, while the curves of $\Sigma^{(i+1)}$ correspond to the sections of $\bar{F}_i$ by the hyperquadrics through $O_i^{(i)}$. It is clear that $\bar{F}_{i+1}$ is a birational transform of $\bar{F}_i$, obtained by referring the sections of $\bar{F}_i$ by the hyperquadrics through $O_i^{(i)}$ to the hyperplane sections of $\bar{F}_{i+1}$. In this transformation, which is locally quadratic at $O_i^{(i)}$, the point $O_{i+1}^{(i+1)}$, corresponds to a direction about $O_i^{(i)}$. The points $O_0^{(0)}=O, O_1^{(1)}, O_2^{(2)},\cdots, O_{s-1}^{(s-1)}$ are the transforms of the s fundamental points $O, O_1,\cdots,O_{s-1}$ on $\bar{F}$ by the above successive locally quadratic transformations and serve to define the position of the fundamental points $O, O_1,\cdots,O_{s-1}$ on $\bar{F}$ on algebroid branches: every branch on $\bar{F}$ passing through $O, O_1,\cdots,O_i$ corresponds to a branch on $\bar{F}_i$ passing through $O_i^{(i)}$.

The following intersection formulas will be useful in the sequel and are obtained by applying the formulas (10) and (10′) to the exceptional curves $L^{(i)}$:

(19) $(L^{(i)}\cdot L_i)=1;$

(19′) $(C_1^{(i)}\cdot L_i)=0 \qquad (i=1,\cdots,s-1).$

Remark. Our notation implies a definite ordering of the irreducible components $L_1, L_2,\cdots,L_s$ of L. From (19) it follows that *each curve L_i*

intersects *one and only one of the curves* $L_{i+1}, \cdots, L_s$. If L_{i+a} $(\alpha > 0)$ is the curve which L_i intersects, then $(L_i \cdot L_{i+a}) = 1$ and L_{i+a} must be a simple component of $L^{(i)}$. *It does not, however, necessarily coincide with the simple component* L_{i+1}, *as we shall see in the sequel.* At any rate, all the intersection numbers $(L_i \cdot L_j)$, $i \neq j$, are either 0 or 1.

5. *The correspondence between* L_i *and the immediate neighborhood of* O_{i-1}. It has been proved in section 3 that, in the birational correspondence T between F and $\bar{F}$, the irreducible component L_1 of L corresponds to the first order neighborhood of the fundamental point O on $\bar{F}$. In a similar manner, in the birational correspondence T_{i-1} between F and $\bar{F}_{i-1}$ the irreducible component L_i of $L^{(i-1)}$ corresponds to the first order neighborhood of the fundamental point $O_{i-1}^{(i-1)}$ on $\bar{F}_{i-1}$. Hence, to a branch δ_{i-1} on $\bar{F}_{i-1}$ of origin $O_{i-1}^{(i-1)}$ and *possessing at* $O_{i-1}^{(i-1)}$ *a principal tangent line distinct from the singular direction at* $O_{i-1}^{(i-1)}$ (that direction to which there corresponds the point $O_i^{(i)}$ on $\bar{F}_i$), there corresponds a branch γ_{i-1} on F meeting L_i at point P_i. The branch δ_{i-1} arises, by the sequence of locally quadratic transformations applied to $\bar{F}, \bar{F}_1, \cdots$, from a branch $\bar{\gamma}_{i-1}$ passing through $O, O_1, \cdots, O_{i-1}$ *but not through* O_i. Hence we have the following

THEOREM 2. *To a branch* $\bar{\gamma}_{i-1}$ *on* $\bar{F}$, *passing through* $O, O_1, \cdots, O_{i-1}$ *but not through* O_i, *there corresponds a branch* γ_{i-1} *on* F *meeting* L_i. *As we approach on* $\bar{F}$ *the point* O *along such a branch* $\bar{\gamma}_{i-1}$ *the homologous point on* F *approaches a point* P_i *on* L_i. *This point* P_i *varies as the point on* $\bar{\gamma}_{i-1}$ *which immediately follows* O_{i-1} *varies.*

It will be noted that an apparent discontinuity arises as $\bar{\gamma}_{i-1}$ approaches a branch $\bar{\gamma}_i$ passing through $O, O_1, \cdots, O_{i-1}$ *and* O_i. The branch γ_i, which corresponds to $\bar{\gamma}_i$ on $\bar{F}$, is a branch meeting L_{i+1} but not necessarily L_i. The real situation is the following: as $\bar{\gamma}_{i-1}$ approaches a branch $\bar{\gamma}_i$, the corresponding branch γ_{i-1} on F approaches a locus which degenerates into the curve $L^{(i)}$ and into the branch γ_i.

Another paradoxical circumstance is the following. We have seen that there is a $(1,1)$ correspondence between the points of L_i and the points on $\bar{F}$ infinitely near and immediately following O_{i-1}. Among these points there is the fundamental point O_i, and it seems natural to expect that the corresponding point on L_i is a common point of L_i and L_{i+1}. However, this is not always the case, *as it may well happen that* L_i *and* L_{i+1} *do not intersect at all.* We shall deal with the general case in the next section. Here an example may suffice. Let $s = 3$ and let the fundamental points O, O_1, O_2 on $\bar{F}$ lie on a cuspidal branch. The general results of the following section show that in this case we have: $L = L_1 + L_2 + 2L_3$, $(L_1 \cdot L_2) = 0$, $(L_1 \cdot L_3) = 1$, $(L_2 \cdot L_3) = 1$ and incidentally $(L_1{}^2) = -3$, $(L_2{}^2) = -2$, $(L_3{}^3) = -1$. Thus the curve L_1 does not meet L_2, although there is a point on L_1 which corresponds formally to the direction $O\,O_1$. This point is the intersection of L_1 and L_3. If we consider a branch on F passing through this point and if we assume for simplicity that this branch is linear, we find that the corresponding branch on $\bar{F}$ passes *triply* through O and *simply* through O_1 and O_2.

6. *The geometric significance of the coefficients* $k_1{}^{(j)}$. A branch $\bar{\gamma}_{i-1}$ on $\bar{F}$ of *smallest order*, passing through $O, O_1, \cdots, O_{i-1}$, arises from a linear branch δ_{i-1} on $\bar{F}_{i-1}$, of origin $O^{(i-1)}_{i-1}$ and possessing at $O^{(i-1)}_{i-1}$ a generic tangent line. To such a branch there corresponds on F a *linear* branch γ_{i-1} of origin P_{i-1}, a generic point of L_i and therefore not on L_j, $j \neq i$.* It is well known from the theory of singularities that branches $\bar{\gamma}_{i-1}$ of lowest order are also characterized by the fact that such a branch $\bar{\gamma}_{i-1}$ passes *simply* through O_{i-1} and that the points which follow O_{i-1} on $\bar{\gamma}_{i-1}$ are all *free points*. ([3], pp. 365-366, 372).

The orders of multiplicity with which $\bar{\gamma}_{i-1}$ passes through the points $O, O_1, \cdots, O_{i-1}$ can be obtained as follows. The order of the multiple point O_{j-1} $(j \leq i)$ equals the multiplicity with which the branch δ, which corresponds on the surface $\bar{F}_{j-1}$ to $\bar{\gamma}_{i-1}$, passes through the point $O^{(j-1)}_{j-1}$, its origin. We may assume that δ is the branch of some irreducible algebraic curve $\bar{D}$ on $\bar{F}_{j-1}$ and that $\bar{D}$ does not possess other branches of origin $O^{(j-1)}_{j-1}$, distinct from δ. Then we may calculate the multiplicity of $\bar{D}$ (i. e. of δ) at $O^{(j-1)}_{j-1}$ as the difference between the number of intersections of a variable hyperplane section of $\bar{F}_{j-1}$ with $\bar{D}$ and the number of variable intersections with $\bar{D}$ of a hyperplane section through $O^{(j-1)}_{j-1}$. Going back to the surface F we see that the required multiplicity equals $(C^{(j-1)} \cdot D) - (C_1{}^{(j-1)} \cdot D)$, where D on F is the transform of $\bar{D}$ on $\bar{F}_{j-1}$. Now $C^{(j-1)} = C_1{}^{(j-1)} + L^{(j-1)}$ and hence $(C^{(j-1)} \cdot D) - (C_1{}^{(j-1)} \cdot D) = (L^{(j-1)} \cdot D) = k_i{}^{(j-1)}$, since the branch γ_{i-1}, and hence also D, has a simple intersection with the curve L_i, does not meet any other irreducible component of $L^{(j-1)}$, and since L_i is a $k_i{}^{(j-1)}$-fold component of $L^{(j-1)}$. We thus have the following

THEOREM 3. *The fundamental points* $O, O_1, \cdots, O_{i-1}$ $(i \leq s)$ *on* $\bar{F}$ *lie*

---

* The truth of this statement follows immediately if we recall that to the sections of $\bar{F}_{i-1}$ by hyperplanes through $O^{(i-1)}_{i-1}$ there correspond on F the curves $C_1{}^{(i-1)}$ of the system $\Sigma_1{}^{(i-1)}$ and that by (17) $(C_1{}^{(i-1)} \cdot L_i) = 1$.

on a branch $\bar{\gamma}_{i-1}$ *of lowest order, passing through these points with the multiplicities* $k_i, k_i^{(1)}, k_i^{(2)}, \cdots, k_i^{(i-1)} = 1$.

As a corollary we have the following inequalities:

$$k_i \geqq k_i^{(1)} \geqq k_i^{(2)} \geqq \cdots \geqq 1. \tag{20}$$

In particular the branch of lowest order passing through all the s fundamental points $O, O_1, \cdots, O_{s-1}$ has at these points the multiplicities $k_s, k_s^{(1)}, k_s^{(2)}, \cdots, k_s^{(s-1)} = 1$ respectively.

The branch $\bar{\gamma}_i$ passes through $O, O_1, \cdots, O_{i-1}$ and in addition through O_i. The multiplicity of any point O_j, $j \leqq i-1$, on this branch cannot be less than its multiplicity for the branch $\bar{\gamma}_{i-1}$ of lowest order passing through $O, O_1, \cdots, O_{i-1}$. Hence we also have the following inequalities:

$$k_j^{(j-1)} \leqq k_{j+1}^{(j-1)} \leqq k_{j+2}^{(j-1)} \leqq \cdots \leqq k_s^{(j-1)}. \tag{21}$$

7. *The determination of the intersection numbers* $(L_i \cdot L_j)$. In his theory of plane singularities, Enriques ([3], p. 381) has introduced the notion of *proximate points* on an algebroid branch. Let $O, O_1, \cdots, O_{s-1}$ be a sequence of infinitely near points on an algebroid branch γ, and let ν_i be the multiplicity of O_i on γ. We have for any i, $\nu_i \geqq \nu_{i+1}$. If $\nu_i = \nu_{i+1}$, then the set of proximate points of O_i on γ consists by definition of the single point O_{i+1}. If $\nu_i > \nu_{i+1}$ and if $\nu_i = h\nu_{i+1} + \nu'$ $(0 \leqq \nu' < \nu_{i+1})$, then it can be proved ([3], p. 381) that O_i is followed immediately on γ by h successive ν_{i+1}-fold points $O_{i+1}, O_{i+2}, \cdots, O_{i+h}$ and by one ν'-fold point O_{i+h+1} $(i+h+1 \leqq s-1)$, if $\nu' > 0$. These $h+1$ points constitute the set of proximate points of O_i. If $\nu' = 0$, the set of proximate points of O_i consists of h points $O_{i+1}, \cdots, O_{i+h}$. From the above definition one deduces the following fundamental property of the set of proximate points of a given point: *the multiplicity of* O_i *on* γ *equals the sum of the multiplicities of its proximate points.*

We shall have occasion to use the following property of proximate points:

THEOREM 4. *If* $O_{i+\alpha}$ *is in the set of proximate points of* O_i *on a given branch* γ, *then it also belongs to the set of proximate points of* O_i *on any other branch passing through* $O_{i+\alpha}$ *(and hence also through* O_i*).*

This theorem is implicitly contained in Enriques and follows from the construction of the branches of lowest order passing through a given set of successive infinitely near points $O, O_1, \cdots, O_i, O_{i+1}, \cdots, O_{s-1}$ on a given branch γ. If $O_{i+1}, \cdots, O_{i+\beta}$ is the set of proximate points of O_i on γ, it is found that on a branch of lowest order passing through $O, O_1, \cdots, O_{s-1}$ the set of proximate points of O_i consists of the set $O_{i+1}, \cdots, O_{i+\beta}$ if $s-1 > i+\beta$, and includes the set $O_{i+1}, \cdots, O_{s-1}$ if $s-1 \leq i+\beta$ ([4], Chap. I, 2). From this the above theorem follows immediately. We shall now prove the following

THEOREM 5. *Let γ_{s-1} be the branch of lowest order which passes through the infinitely near fundamental points $O, O_1, \cdots, O_{s-1}$ on $\bar{F}$ and let the set of proximate points of O_{i-1} on γ_{s-1} $(i < s)$ consist of α_i points $O_i, O_{i+1}, \cdots, O_{i+\alpha_i-1}$.* Then $(L_i \cdot L_{i+\alpha_i}) = 1$, $(L_i \cdot L_j) = 0$ if $j > i$ and $j \neq i + \alpha_i$. Moreover the virtual degree (L_i^2) of L_i equals $-(1+\alpha_i)$.*

For the proof we observe first of all that it is sufficient to prove the theorem for $i = 1$ $(O_{i-1} = O,\ L_i = L_1)$, for then we may apply the result to any exceptional curve $L^{(i-1)}$ and to the sequence of fundamental points $O^{(i-1)}_{i-1}, O_i^{(i-1)}, \cdots, O^{(i-1)}_{s-1}$ on $\bar{F}_{i-1}$ of the birational correspondence between F and $\bar{F}_{i-1}$. These points lie on a branch of lowest order, which is the transform of the branch γ_{s-1}, and the proximate points of $O^{(i-1)}_{i-1}$ are clearly the points $O_i^{(i-1)}, \cdots, O^{(i-1)}_{i+\alpha_i-1}$.

We first prove the theorem when $\alpha = 1$, i. e. when the set of proximate points of O consists of the single point O_1. By Theorem 3 we have then $k_s = k_s^{(1)}$. Applying Theorem 4, we see that the set of proximate points of O on any branch γ_{i-1} of lowest order passing through $O, O_1, \cdots, O_{i-1}$ will consist of the single point O_1, and hence, again by Theorem 3, we have $k_i^{(1)} = k_i$ $(i = 2, 3, \cdots, s)$, i. e. $L = L_1 + L^{(1)}$. By (18) we have $(L \cdot L_2) = 0$, $(L^{(1)} \cdot L_2) = -1$, hence $(L_1 \cdot L_2) = 1$ and this proves the first part of the theorem. We also have $(L \cdot L_1) = (L_1^2) + (L^{(1)} \cdot L_1)$ and by (18) and (19) $(L \cdot L_1) = -1$, $(L^{(1)} \cdot L_1) = 1$. Consequently $(L_1^2) = -2$, q. e. d.

By section 4, Remark, L_1 meets only one of the curves $L_2, \cdots, L_s$ and the corresponding intersection number is 1. To prove our theorem for α arbitrary, $\alpha > 1$, we shall assume that $(L_1 \cdot L_{\alpha+1}) = 1$ and we shall show that then the set of proximate points of O on the branch γ_{s-1} consists of the points $O_1, \cdots, O_\alpha$ and that $(L_1^2) = -(1+\alpha)$.

We have $(L^{(1)} \cdot L_1) = k^{(1)}_{\alpha+1}$ and hence by (19) $k^{(1)}_{\alpha+1} = 1$. Using the inequalities (20) we find $k^{(2)}_{\alpha+1} = k^{(3)}_{\alpha+1} = \cdots = k^{(\alpha)}_{\alpha+1} = 1$, and from this it follows, in view of the inequalities (21), that all the coefficients $k_i^{(j)}$ $(j = 1, 2, \cdots, \alpha,\ i = j+1, \cdots, \alpha+1)$, are 1. For the sake of clearness we write the simplified expression of the curves $L, L^{(1)}, \cdots, L^{(\alpha)}$:

* Here necessarily $i + \alpha_i - 1 \leq s - 1$ because on a branch of lowest order through a given sequence of infinitely near points $O, O_1, \cdots, O_{s-1}$, the set of proximate points of any point of the sequence except the last point O_{s-1} is contained in the given sequence.

$$
\begin{array}{lllllll}
L = L_1 + L_2 + k_3L_3 + \cdots + k_\alpha L_\alpha + k_{\alpha+1}L_{\alpha+1} + k_{\alpha+2}\, L_{\alpha+2} + \cdots \\
L^{(1)} = \quad L_2 + \quad L_3 + \cdots + \quad L_\alpha + \quad L_{\alpha+1} + k^{(1)}_{\alpha+2}\, L_{\alpha+2} + \cdots \\
\cdot \quad \cdot \quad \cdot \quad \cdot \quad \cdot \quad \cdot \quad \cdot \quad \cdot \quad \cdot \quad \cdot \quad \cdot \quad \cdot \quad \cdot \quad \cdot \quad \cdot \\
L^{(\alpha-1)} = \qquad\qquad\qquad L_\alpha + \quad L_{\alpha+1} + k^{(\alpha-1)}_{\alpha+2} L_{\alpha+2} + \cdots \\
L^{(\alpha)} = \qquad\qquad\qquad\qquad L_{\alpha+1} + k^{(\alpha)}_{\alpha+2}\, L_{\alpha+2} + \cdots.
\end{array}
$$

We have replaced k_2 by 1, since k_2 and $k_2^{(1)}$ $(=1)$ give the multiplicities of O and O_1 on the branch of lowest order passing through O and O_1, and hence clearly $k_2 = 1$. From Theorem 3 it follows that the branch γ_α of lowest order passing through $O, O_1, \cdots, O_\alpha$ has at these points the multiplicities $k_{\alpha+1}, 1, \cdots, 1$. Consequently on this branch the set of proximate points of O_i, $1 \leq i \leq \alpha - 1$, consists of the single point O_{i+1}. It follows by Theorem 4 that also on γ_{s-1} the set of proximate points of O_i, where now $1 \leq i \leq \alpha - 2$, consists of the single point O_{i+1}. Since we have already proved our theorem for the case $\alpha = 1$, we conclude that the following relations hold:

$$(22) \qquad (L_i \cdot L_{i+1}) = 1$$

$$(22') \qquad (L_i^2) = -2 \qquad (i = 2, 3, \cdots, \alpha - 1).$$

By (18) we have $(L \cdot L_1) = -1$ and $(L \cdot L_i) = 0$, $i > 1$. Using the relations (22), (22′) and recalling that by section 4, Remark, each curve L_i intersects only one of the components L_j, $j > i$, we find

$$(23) \qquad (L_1^2) + k_{\alpha+1} = -1;$$

$$(23') \qquad -2k_2 + k_3 = 0;$$

$$(23'') \qquad k_{i-1} - 2k_i + k_{i+1} = 0 \qquad (i = 3, \cdots, \alpha - 1).$$

Since $k_1 = k_2 = 1$ the relations (23′) and (23″) give the following values for $k_3, k_4, \cdots, k_\alpha$:

$$(24) \qquad k_3 = 2, k_4 = 3, \cdots, k_\alpha = \alpha - 1.$$

The curve L_α intersects one (and only one) of the curves $L_{\alpha+1}, L_{\alpha+2}, \cdots$. Let $(L_\alpha \cdot L_{\alpha+j}) = 1$, $j \geq 1$. From $(L^{(i)} \cdot L_\alpha) = 0$ $(i = 1, 2, \cdots, \alpha - 2)$, $(L^{(\alpha-1)} \cdot L_\alpha) = -1$ and $(L^{(\alpha)} \cdot L_\alpha) = 1$ we derive the following relations:

$$(25) \qquad 1 + (L_\alpha^2) + k^{(i)}_{\alpha+j} = 0 \qquad (i = 1, 2, \cdots, \alpha - 1);$$

$$(25') \qquad k^{(\alpha)}_{\alpha+j} = 1.$$

From these relations it follows that $k^{(1)}_{\alpha+j} = k^{(2)}_{\alpha+j} = \cdots = k^{(\alpha-1)}_{\alpha+j}$ and $k^{(\alpha)}_{\alpha+j} = k^{(\alpha+1)}_{\alpha+j} = \cdots = k^{(\alpha+j-1)}_{\alpha+j} = 1$ by (20). Hence the branch $\gamma_{\alpha+j-1}$ of lowest order passing through the points $O, O_1, \cdots, O_{\alpha+j-1}$ has at these points the following multiplicities: $O^{k_{\alpha+j}}, O_1^{k}, O_2^{k}, \cdots, O^{k}_{\alpha-1}, O_\alpha^{1}, \cdots, O^{1}_{\alpha+j-1}$, where we have put $k = k^{(1)}_{\alpha+j}$. We also have $(L \cdot L_\alpha) = 0$ and hence from (24)

$$(\alpha-2)+(\alpha-1)(L_\alpha^2)+k_{\alpha+j}=0.$$

Substituting for (L_α^2) the value $-k-1$ from (25), we find

$$k_{\alpha+j}=(\alpha-1)k+1.$$

This value of $k_{\alpha+j}$ shows that on the above branch $\gamma_{\alpha+j-1}$ the set of proximate points of O consists of the points $O_1, O_2, \cdots, O_\alpha$. By Theorem 4 it follows then that also on the branch γ_{s-1} the points $O_1, \cdots, O_\alpha$ constitute the set of proximate points of O, and this proves the first part of the theorem.

We have already observed that the branch γ_α of lowest order passing through the points $O, O_1, \cdots, O_\alpha$ possesses at these points the following multiplicities: $O^{k_{\alpha+1}}, O_1^1, \cdots, O_\alpha^1$. Also on this branch the points $O_1, \cdots, O_\alpha$ must constitute the set of proximate points of O. Hence $k_{\alpha+1}=\alpha$ and by (23), $(L_1^2)=-(1+\alpha)$, q. e. d.

Remark 1. By Theorem 1, the curve L_s is of virtual degree -1 because $L_s=L^{(s)}$, i. e. L_s is itself an exceptional curve of the first kind.

Remark 2. The preceding theorem shows that the intersection numbers $(L_i \cdot L_j)$ are completely determined by the characters of the branch γ_{s-1} of lowest order passing through the infinitely near fundamental points $O, \cdots, O_{s-1}$. Conversely, it is not difficult to show that *the intersection numbers* $(L_i \cdot L_j)$ *completely determine the characters of the branch* γ_{s-1}, i. e. *the multiplicities of the points* $O, O_1, \cdots, O_{s-1}$ *on this branch.* In fact, $-(L_i^2)-1$ gives the number of proximate points of O_i $(i<s-1)$, and hence the multiplicity of O_i on γ_{s-1} can be found if the multiplicities of the points $O_{i+1}, O_{i+2}, \cdots, O_{s-1}$ are known, because the multiplicity of O_i equals the sum of the multiplicities of the proximate points. Since the multiplicity of O_{s-1} is 1, the multiplicities of $O_{s-2}, O_{s-3}, \cdots$ can be determined step by step.

In a similar manner the multiplicities of the points $O, O_1, \cdots, O_{i-1}$ on the branch γ_{i-1} of lowest order containing them can be determined. Hence, by Theorem 3, it follows that *the intersection numbers* $(L_i \cdot L_j)$ *determine uniquely all the integers* $k_i^{(j)}$.

Remark 3. *The determinant* $\Delta_i=|(L_\alpha \cdot L_\beta)|$ $(\alpha,\beta=i,i+1,\cdots,s)$, *equals* $(-1)^{s-i+1}$. In fact, if we multiply in Δ_1 the second column by k_2, the third column by $k_3, \cdots$, the s-th column by k_s and add to the first column, we find, in view of equations (5), $\Delta_1=-\Delta_2$. In a similar manner we find $\Delta_2=-\Delta_3=\Delta_4=\cdots=\pm\Delta_s$. Since $\Delta_s=(L_s^2)=-1$, the statement follows. The equations (5) determine the integers k_i in terms of the intersection numbers $(L_i \cdot L_j)$. This is in agreement with the preceding remark.

8. *Free points and satellites.* Our analysis of reducible exceptional curves runs parallel to the analysis of plane singularities due to Enriques. It illustrates very concretely, by means of the intersection properties of the components of the exceptional curve, the notions and conventions used in this theory, as for instance the notion of proximate points. Another concept, also due to Enriques ([3], pp. 365-366, 372), is that of *free points and satellites* in a sequence of infinitely near points on an algebroid branch. We shall now show that the knowledge of the intersection numbers $(L_i \cdot L_j)$ allows us to decide whether the fundamental point O_i on γ_{s-1}, which corresponds to a given component L_{i+1} of L, is a free point or a satellite. For this we quote a few properties relative to the classification of the points $O, O_1, \cdots, O_{s-1}$ into free points and satellites ([4], Chap. I, 2).

Let $O_{k+1}, O_{k+2}, \cdots, O_{k+l}$ be a sequence of free points such that O_k (if $O_{k+1} \neq O$) and O_{k+l+1} (if $k + l < s - 1$) are satellites, and let ν_i denote the multiplicity of O_i on γ_{s-1}. Such a sequence enjoys the following characteristic properties:

(α) $\nu_{k+1} = \nu_{k+2} = \cdots = \nu_{k+l-1}$; $\nu_{k+l} < \nu_{k+l-1}$ except when $k + l = s - 1$, in which case $\nu_{k+l-1} = \nu_{k+l} = 1$.

(β) If O_{k+1} is distinct from O, then $k \geq 2$ and $\nu_{k-1} = \nu_k$. Moreover if $\nu_k = \nu_{k-1} = \cdots = \nu_{k-i+1}$ and $\nu_k \neq \nu_{k-i}$, then $\nu_{k-i} = i\nu_k$.

Recalling the properties of proximate points quoted in section 7, we conclude from (α) and (β) that

(1) Every satellite is the last proximate point of at least one point preceding it.

(2) If O_k is a satellite followed by a free point O_{k+1}, then O_k is the last proximate point of two points O_{k-1} and O_{k-i} preceding it, and conversely.

(3) The last proximate point of any point O_i is never a free point followed by a satellite.

From (1), (2) and (3) there follows

THEOREM 6. *If L_{i+1} does not meet any curve L_j, $j < i + 1$, then O_i is a free point followed by a satellite O_{i+1}, and conversely. If L_{i+1} meets two of the curves L_j, $j < i + 1$, then O_i is a satellite followed by a free point O_{i+1}, and conversely.*

This theorem characterizes the *last* point of a sequence of satellites followed by free points and the *last* point of a sequence of free points followed by satellites and hence enables us to find out the division of the set $O, O_1, \cdots, O_{s-1}$ into free points and satellites from the intersection properties of the curves $L_1, L_2, \cdots, L_s$.

We add a few remarks concerning the $(1,1)$ correspondence between the points of L_{i+1} and the points of the immediate neighborhood of O_i (points in the neighborhood of order $i+1$ of O). It is not difficult to show ([4], Chap. I, 2) that the simple infinity of points immediately following O_i contains *one* satellite, if O_i is a free point, and *two* satellites if O_i is itself a satellite. What point or points of L_{i+1} correspond to this satellite or these satellites? To answer this question it is necessary to observe that if $\bar{O}_{i+1}$ is a point in the immediate neighborhood of O_i, then the branch of lowest order passing through $O, O_1, \cdots, O_i, \bar{O}_{i+1}$ is of the same order as the branch γ_i of lowest order passing through $O, O_1, \cdots, O_i$ if $\bar{O}_{i+1}$ is a free point, and is of higher order than γ_i if $\bar{O}_{i+1}$ is a satellite. A branch γ_i corresponds to a *linear* branch on F, meeting L_{i+1} in a generic point. We obtain on $\bar{F}$ a branch of higher order than γ_i if and only if that linear branch intersects L_{i+1} in a point where L_{i+1} intersects some further component of L. Recalling that the fundamental point O_{i+1} corresponds formally to the intersection of L_{i+1} with a curve L_j, $j > i+1$ (section 5), we conclude as follows: if O_i is a free point and O_{i+1} is a satellite, then this satellite corresponds to the intersection of L_{i+1} with an L_j, $j > i+1$ (it should be noticed that in this case, by Theorem 6, L_{i+1} does not meet any curve L_j, $j < i+1$). If O_i and O_{i+1} are both free points, then, by Theorem 6, L_{i+1} necessarily intersects one and only one curve L_j, $j < i+1$; to this point of intersection corresponds *the* satellite which immediately follows O_i. If O_i is a satellite and O_{i+1} is a free point, then, by Theorem 6, L_{i+1} meets two curves L_j, $j < i+1$, and the two points of intersection correspond to the two satellites in the immediate neighborhood of O_i. Finally if O_i and O_{i+1} are both satellites, then by Theorem 6 L_{i+1} meets only one curve L_j, $j < i+1$, and the point of intersection corresponds to the second satellite, distinct from O_{i+1}, which immediately follows O_i.

9. *Characterization of reducible exceptional curves of the first kind.* It is known ([1]) that the conditions which characterize an *irreducible* exceptional curve L_1 of the first kind, on a surface free from singularities, are the following: $(L_1^2) = -1$, $[L_1] = 0$, these characters being evaluated on the assumption that L_1 is virtually free from base points. The curve L_1 is rational and *free from singularities*, because, if L_1 had singularities, the virtual genus $[L_1]$ would be positive.

We show that the properties of the intersection numbers $(L_i \cdot L_j)$, derived in the preceding section, together with the relations $[L_i] = 0$, proved in section 3, characterize these reducible exceptional curves of the first kind, which arise from birational transformations possessing a sequence of infinitely near fundamental points lying on an irreducible algebroid branch. In exact terms we prove the following

THEOREM 7. *Let $L_1, L_2, \cdots, L_s$ be a set of irreducible curves, virtually free from base points and of virtual genus zero, on a surface F free from singularities, and let there exist an arrangement of these curves, for instance the one written above, such that the following conditions are satisfied:* (i) *each curve L_i meets one and only one of the curves $L_{i+1}, \cdots, L_s$ which follow L_i;* (ii) *if L_i meets L_{i+a} $(i < s)$, then $(L_i \cdot L_{i+a}) = 1$, $(L_i^2) = -(1+\alpha)$ and $(L_{i+j} \cdot L_{i+j+1}) = 1$ $(j = 1, 2, \cdots, \alpha - 2)$;* (iii) $(L_s^2) = -1$. *Under these conditions a convenient combination $k_1L_1 + k_2L_2 + \cdots + k_sL_s$ $(k_1 = k_2 = 1)$ with positive integral coefficients is an exceptional curve of the first kind. The assumed arrangement $L_1, \cdots, L_s$ is uniquely determined and so also are the integers k_i.*

Proof. Since the theorem is true for $s = 1$ ([1]), we shall prove it by induction, assuming it to be true for $s - 1$. Since, by hypothesis, $[L_s] = 0$ and $(L_s^2) = -1$, L_s is an exceptional curve of the first kind. There exists therefore a birational transformation of F into a surface $\bar{F}$, in which to L_s there corresponds a simple point O of $\bar{F}$ and which does not possess fundamental points on L_s. It can also be assumed that the transformation possesses no fundamental points on F, that L_s is the only fundamental curve of the transformation on F and that $\bar{F}$ is free from singularities.* Let $\bar{L}_1, \bar{L}_2, \cdots, \bar{L}_{s-1}$ be the transforms of $L_1, \cdots, L_{s-1}$ on F. The virtual degree of a curve is invariant under birational transformations. However, if a given L_i meets L_s, then L_i passes through the point O *simply*, since $(L_i \cdot L_s) = 1$, and this point must be considered as an assigned base point of $\bar{L}_i$ ([4], Chap. II, 7). If we understand by $(\bar{L}_i^2)$ the virtual degree of $\bar{L}_i$ *virtually free from base points*, then $(\bar{L}_i^2) = (L_i^2)$ if $(L_i \cdot L_s) = 0$ and $(\bar{L}_i^2) = (L_i^2) + 1$ if $(L_i \cdot L_s) = 1$.

With these preliminaries we proceed to prove that the ordered sequence of curves $\bar{L}_1, \bar{L}_2, \cdots, \bar{L}_{s-1}$ on $\bar{F}$ satisfies the conditions for an exceptional curve stated in the above theorem. From (i) we deduce that $(L_{s-1} \cdot L_s) = 1$ and by (ii) $(L^2_{s-1}) = -2$. Hence $(\bar{L}^2_{s-1}) = (L^2_{s-1}) + 1 = -1$ and this shows that the condition (iii) holds. To show that the conditions (i) and (ii) are satisfied, we make the following self-evident remarks:

* The defining linear system Σ on F associated with the transformation, i. e. the system which goes into the system of hyperplane sections of $\bar{F}$, may be assumed to be free from base points and non-singular ([1; 4], Chap. IV, 4). If the transformation possesses on F fundamental curves other than L_s and hence fundamental points on $\bar{F}$ distinct and necessarily at a finite distance from O, these fundamental points can be transformed back into curves by means of a sufficiently general linear system on $\bar{F}$, free from fundamental curves and free from base points outside these fundamental points.

(α) If $(L_i \cdot L_s)$ and $(L_j \cdot L_s)$ are not both 1 $(i, j < s)$, then $(\bar{L}_i \cdot \bar{L}_j) = (L_i \cdot L_j)$.

(β) If $(L_i \cdot L_s) = (L_j \cdot L_s) = 1$, then $(\bar{L}_i \cdot \bar{L}_j) = (L_i \cdot L_j) + 1$.

We consider any curve L_i, $i < s - 1$. Let $(L_i \cdot L_{i+\alpha}) = 1$, $(L_i \cdot L_j) = 0$, $j > i$ and $j \neq i + \alpha$. If $i + \alpha \neq s$, then by (α) $(\bar{L}_i \cdot \bar{L}_{i+\alpha}) = 1$ and $(\bar{L}_i \cdot \bar{L}_j) = 0$ if $j > i$ and $\neq i + \alpha$. Moreover by (α) applied to the case $i = j$, it follows that $(\bar{L}_i^2) = (L_i^2) = -1 - \alpha$. If $i + \alpha = s$, then by (β) (where now $j = s - 1$) we have $(\bar{L}_i \cdot \bar{L}_{s-1}) = 1$, since $(L_i \cdot L_{s-1}) = 0$. Moreover by (ii) we have $(L_j \cdot L_s) = 0$ $(j = i + 1, \cdots, s - 2)$, and hence by ($\alpha$) $(\bar{L}_i \cdot \bar{L}_j) = 0$ $(j = i + 1, \cdots, s - 2)$, since $(L_i \cdot L_j) = 0$. We also have by (β), when we put $i = j$, $(\bar{L}_i^2) = (L_i^2) + 1 = -(s - i) - 1 + 1 = -(s - i - 1) - 1$. We have thus proved that every curve $\bar{L}_i$ $(i = 1, 2, \cdots, s - 2)$, meets one and only one of the curves $\bar{L}_j$, $j > i$, and that if $(\bar{L}_i \cdot \bar{L}_{i+\alpha}) = 1$, $\alpha > 0$, then $(\bar{L}_i^2) = -\alpha - 1$. Finally if $(L_i \cdot L_{i+\alpha}) = 1$, then by (ii) $(L_j \cdot L_{j+1}) = 1$ $(j = i + 1, \cdots, i + \alpha - 2)$, and consequently $(\bar{L}_j \cdot \bar{L}_{j+1}) = 1$ for the same values of j. Hence all the conditions (i), (ii), (iii) are satisfied by the ordered sequence $\bar{L}_1, \cdots, \bar{L}_{s-1}$.

From the hypothesis that our theorem is true for $s - 1$, it follows that a proper combination $\bar{L} = \bar{k}_1 \bar{L}_1 + \cdots + \bar{k}_{s-1} \bar{L}_{s-1}$ of the curves $\bar{L}_i$ is an exceptional curve of the first kind. There exists then a linear system $\bar{\Sigma}$ on $\bar{F}$ possessing $\bar{L}$ as a total fundamental curve, satisfying the conditions (i), (ii), and (iii) of the definition given in section 1 and *having no base points on* $\bar{L}$. It is seen immediately that the curves $L_1, \cdots, L_s$ are fundamental curves of the system Σ on F which corresponds to $\bar{\Sigma}$, that a proper combination $k_1 L_1 + \cdots + k_s L_s$ of these curves is a total fundamental curve of Σ (since by the conditions (i), (ii), (iii) of the preceding theorem the curve $L_1 + \cdots + L_s$ is connected) and that Σ satisfies the conditions (i), (ii), (iii) of the definition given in section 1. It has been proved already (section 7, Remark 2) that the integers k_i are uniquely determined. It remains to prove that the arrangement $L_1, L_2, \cdots, L_s$ is uniquely determined. Let us assume that $L_{i_1}, L_{i_2}, \cdots, L_{i_s}$ is another arrangement satisfying the conditions (i), (ii) and (iii), where $i_1, i_2, \cdots, i_s$ is a permutation of the indices $1, 2, \cdots, s$. Since by (iii) $(L_{i_s}^2) = -1$ and since by (ii) L_s is the only curve whose virtual degree is -1, it follows that L_{i_s} coincides with L_s. We shall therefore assume that $i_j = j$ for $j = \alpha, \alpha + 1, \cdots, s$ $(1 < \alpha \leqq s)$, and we shall prove that $i_{\alpha-1} = \alpha - 1$. Let $i_\beta = \alpha - 1$, and let $(L_{\alpha-1} \cdot L_{\alpha+\gamma}) = 1$, $\gamma \geqq 0$, by (ii). We have, by (ii), $(L^2_{\alpha-1}) = -(2 + \gamma)$, and by the same condition (ii), applied to the arrangement $L_{i_1}, L_{i_2}, \cdots, L_{i_{\alpha-1}}, L_\alpha, L_{\alpha+1}, \cdots, L_s$, we have $(L^2_{\alpha-1}) = (L^2_{i_\beta}) = -(\alpha + \gamma - \beta + 1)$. Hence $2 + \gamma = \alpha + \gamma - \beta + 1$, or $\beta = \alpha - 1$, q. e. d.

10. *Fundamental points on several irreducible branches.* In the previous sections we have treated the case of a birational transformation with a succession of infinitely near fundamental points on a single irreducible algebroid branch. We found it necessary to assume that the curves $L_i + \cdots + L_s$ $(i = 1, \cdots, s)$ were connected. When these curves were not connected we saw in section 3 that the aggregate of fundamental points would not lie all on one branch. It is the purpose of this section to consider this latter case.

The curve $L_1 + \cdots + L_s$ is connected (see 2), but we shall suppose that $L_2 + \cdots + L_s$ breaks up into ρ (≥ 1) maximal connected components $\bar{L}^{(1)}, \cdots, \bar{L}^{(\rho)}$, having two by two no points in common (see 3). Then on the surface $\bar{F}$, in the first neighborhood of the point O, we have the ρ distinct fundamental points $O_{1,1}, \cdots, O_{\rho,1}$ of the transformation T^{-1}. We shall fix our attention on one of these, say $O_{1,1}$. In the immediate neighborhood of $O_{1,1}$ T^{-1} may have several fundamental points, among which there may or may not be *the* * proximate point of O following $O_{1,1}$. We shall suppose that T^{-1} has a fundamental point $O_{1,2}$, immediately following $O_{1,1}$, which is a proximate point of O. Then in the immediate neighborhood of $O_{1,2}$, T^{-1} may have several fundamental points, among which there may or may not be the proximate point of O following $O_{1,2}$. We shall suppose that T^{-1} has a fundamental point $O_{1,3}$ immediately following $O_{1,2}$, which is a proximate point of O. We continue in this way to define the sequence of points $O, O_{1,1}, \cdots, O_{1,\sigma}$ such that each point is fundamental for T^{-1} and that $O_{1,i}$ is the proximate point of O immediately following $O_{1,i-1}$ $(i = 2, \cdots, \sigma)$.

We shall suppose that the fundamental points of T^{-1} which follow $O_{1,\sigma}$ include no proximate point of O. From the preceding sections it follows that the immediate neighborhood of each point of the sequence $O, O_{1,1}, \cdots, O_{1,\sigma}$ is represented in $(1,1)$ fashion by the points of irreducible components of L, namely $L_1, L_{1,1}, \cdots, L_{1,\sigma}$ respectively. Consider now a linear system of curves Φ on $\bar{F}$, cut out by a system of forms of a sufficiently high order, having base points only at the points $O, O_{1,1}, \cdots, O_{1,\sigma}$ and free from fundamental curves on $\bar{F}$. By relating the system Φ to the hyperplanes of a linear space, we induce a transformation T' and transform $\bar{F}$ into a surface F'; from the considerations in sections 1-9, it is clear that F' has an exceptional curve of the type treated in those sections. F' is the transform of F by the product

* We say "*the* proximate point of O," because it can be easily shown that there is only one such point in the immediate neighborhood of $O_{1,1}$. More generally, it is not difficult to prove that if $O_i, O_{i+1}, \cdots, O_{i+\beta}$ $(\beta > 0)$ is a sequence of successive points in the neighborhoods of order $i, i+1, \cdots, i+\beta$ of a point O, such that $O_{i+\beta}$ is a proximate point of O_i, then there is one and only one point $O_{i+\beta+1}$ immediately following $O_{i+\beta}$, which is a proximate point of O_i.

$T \cdot T'$, and this transformation is free from fundamental points on F. The exceptional curve on F' consists of the curves $L'_1, L'_{1,1}, \cdots, L'_{1,\sigma}$, the transforms of $L_1, \cdots, L_{1,\sigma}$ on F; the remaining components of L on F are replaced by points of F'.

Now in the exceptional curve on F', L'_1 must intersect $L'_{1,\sigma}$ by Theorem 5, since on $\bar{F}$ $O_{1,\sigma}$ is the last proximate point of O. Does this imply that the corresponding curves L_1 and $L_{1,\sigma}$ on F intersect? Let us assume that the intersection of L'_1 and $L'_{1,\sigma}$ arises from the intersection of L_1 and $L_{1,\sigma}$ on F with a curve L_β which has been replaced by a point on F' and let O_β be the fundamental point on $\bar{F}$ whose immediate neighborhood is in $(1,1)$ correspondence with the points of L_β. This point is not in the sequence $O, O_{1,1}, \cdots, O_{1,\sigma}$, since O_β is transformed by T' into a point of F'. On the other hand the fact that $L_{1,\sigma}$ and L_β intersect implies that if $L^{(j)}$ $(j = 0, 1, \cdots, \sigma - 1)$ denotes the exceptional curve on F which corresponds to the point $O_{1,j}$ $(O_{1,j} \equiv O,\ L^{(0)} \equiv L)$, then $L_{1,\sigma}$ and L_β belong to one and the same maximal connected component of $L^{(j)} - L_{1,j}$, i.e. to $L^{(j+1)}$. Hence there exists a branch passing through $O, \cdots, O_{1,\sigma}, O_\gamma, \cdots, O_\beta$. Set up a transformation T'', as T' was determined, possessing only $O, \cdots, O_\beta$ as fundamental points and transform $\bar{F}$ into F''; F'' has an exceptional curve of the type considered in sections 4-9. In this exceptional curve denote by L_j'' the transform of L_j on F. Then $(L''_1 \cdot L''_{1,\sigma}) = 1$ by Theorem 5 and $(L_1'' \cdot L''_\beta) = 1$ since $(L_1 \cdot L_\beta) = 1$ on F and since the transformation TT'' is free from fundamental points on F. But this is impossible, since L_1'' cannot meet two distinct irreducible components of the exceptional curve on F''.

Hence we must have on F, $(L_1 \cdot L_{1,\sigma}) = 1$. This constitutes an extension of Theorem 5 in whatever way the maximal connected components of $L_i + \cdots + L_s$ $(i = 2, \cdots, s)$ may decompose. In all cases we have shown that L_1 *intersects that curve which is the map of the immediate neighborhood of the fundamental point, which is the last proximate point of* O.

In Theorem 5 we were also able to derive the virtual degree of L_1 and hence the virtual degree of any component of L; we found $(L_1^2) = -(1 + \text{number of proximate points of } O)$; we wish to prove this for the present case when the fundamental points lie on several irreducible branches.*

The curve $\bar{L} = L_1 + \cdots + L_s$ is connected, but $L_2 + \cdots + L_s$ breaks up into $\rho \geq 1$ maximal connected components $\bar{L}^{(1)}, \cdots, \bar{L}^{(\rho)}$; L_1 meets $\bar{L}^{(\beta)}$ $(\beta = 1, \cdots, \rho)$ in one point each, and each $\bar{L}^{(\beta)}$ gives rise to an exceptional curve $L^{(\beta)}$ and to a fundamental point $O_{\beta,1}$ in the first neighborhood of O.

* The set of proximate points of a given point O, when several branches pass through O, is by definition the set of all the proximate points of O on the different branches through O ([4], Chap. I, 2).

We shall fix our attention on one of these points, say $O_{1,1}$, and on the corresponding curve $\bar{L}^{(1)} = L_2 + \cdots + L_s$. The intersection $(L_1 \cdot \bar{L}^{(1)}) = 1$ arises from the intersection of L_1 with some curve L_{a_α} of $\bar{L}^{(1)}$ such that $(L_1 \cdot L_{a_\alpha}) = 1$. Now $\bar{L}^{(1)} - L_2$ decomposes into several maximal connected components, one of which contains L_{a_α}; let this component be $\bar{L}^{(2)}_{a_2} = L_{a_2} + \cdots + L_{a_\alpha} + \cdots$; we have $(L_2 \cdot \bar{L}^{(2)}_{a_2}) = 1$ [as $(L_1 \cdot \bar{L}^{(1)}) = 1$] and $L^{(2)}_{a_2}$ gives rise to a fundamental point $O_{1,2}$ in the immediate neighborhood of $O_{1,1}$. Then $\bar{L}^{(2)}_{a_2} - L_{a_2}$ breaks up into several maximal connected components, one of which contains L_{a_α}; let this component be $\bar{L}^{(3)}_{a_3} = L_{a_3} + \cdots + L_{a_\alpha} + \cdots$. As before, there is a fundamental point $O_{1,3}$ in the immediate neighborhood of $O_{1,2}$. We continue in this manner and come finally to a fundamental curve $\bar{L}^{(\alpha)}_{a_\alpha} = L_{a_\alpha} + \cdots$ and this gives rise to a fundamental point $O_{1,\alpha}$ in the neighborhood of order α of O. From the proof in the earlier part of this section it follows that $O_{1,\alpha}$ is the last proximate point of O after $O_{1,1}$ among the fundamental points of T^{-1}.

The proof of section 5 extends directly to show that the immediate neighborhood of $O_{1,\alpha}$ is in $(1, 1)$ correspondence with the points of the curve L_{a_α}; then the proof of section 6 follows and shows that the branch of minimum order through $O, O_{1,1}, \cdots, O_{1,\alpha}$ has at these points the multiplicities $k_{a_\alpha}, k^{(1)}_{a_\alpha}, \cdots, k^{(\alpha)}_{a_\alpha} = 1$, where these values are the coefficients of L_{a_α} in the exceptional curves $L, L^{(1)}, L^{(2)}_{a_2}, \cdots, L^{(\alpha)}_{a_\alpha}$, derived from the fundamental curves $\bar{L}, \bar{L}^{(1)}, \cdots, \bar{L}^{(\alpha)}_{a_\alpha}$. But $k^{(1)}_{a_\alpha} = 1$ by (10) of section 3. Therefore $k^{(2)}_{a_\alpha} = \cdots = k^{(\alpha)}_{a_\alpha} = 1$. Since the multiplicity of a point is equal to the sum of the multiplicities of its proximate points, we have at once $k_{a_\alpha} = \alpha$. We apply this procedure to the curves $L^{(\beta)}$ $(\beta = 1, \cdots, \rho)$, and from the relation $(L \cdot L_1) = -1$ we derive in all cases the desired result: $(L_1{}^2) = -(1 + \text{number of proximate points of } O)$. As in Theorem 5, the proof extends to the other irreducible components of L.

REFERENCES.

[1] G. Castelnuovo and F. Enriques, "Sopra alcune questioni fondamentali nella teoria dellè superficie algebriche," *Annali di Matematica pura ed applicata*, Ser. 3, vol. 6 (1901), pp. 165-225.

[2] F. Enriques, "Sulla proprietà caratteristica delle superficie algebriche irregolari," *Rendiconti della Reale Accademia delle scienze di Bologna*, nuova serie, vol. 9 (1904-05).

[3] F. Enriques and O. Chisini, *Lezioni sulla teoria geometrica delle equazioni e delle funzioni algebriche*, vol. 2.

[4] O. Zariski, *Algebraic Surfaces*, Ergebnisse der Mathematik und ihrer Grenzgebiete, vol. III, 5, Berlin, Springer (1934).

THE JOHNS HOPKINS UNIVERSITY.

POSTULATION ET GENRE ARITHMÉTIQUE

par M. O. ZARISKI.

Soit F une surface normale de l'espace projectif S_n; considérons sa formule de postulation (nombre de conditions linéaires à imposer à une forme de degré s pour qu'elle contienne F)

$$\pi(F, s) = a_0\binom{s}{2} + a_1\binom{s}{1} + a_2.$$

a_0 est le degré de $F[a_0 = (F, S_{n-2})]$; a_1 est un caractère projectif des sections hyperplanes de F. L'entier $P_a(F) = a_2 - 1$ sera appelé le *genre arithmétique virtuel* de F; en général ce n'est pas encore l'invariant birationnel p_a.

Faisons varier F dans sa classe M d'équivalence birationnelle (F restant normale). Si F_1 et $F_2 \in M$, on notera $F_1 \leqslant F_2$ si la correspondance birationnelle entre ces deux surfaces (bien déterminée si l'on convient que tout élément de M se compose d'une surface G et d'une correspondance birationnelle entre F et G) n'a pas de point fondamental sur F_2 ("F_1 a moins de points que F_2"); ceci définit une structure d'ordre sur M; et la fonction $P_a(F)$ est *monotone décroissante*. Nous appellerons sa borne inférieure (si elle existe) le *genre arithmétique* de F.

Pour montrer l'existence de cette borne, nous utiliserons (en caractéristique 0) la résolution des singularités et le fait que, dans M, les surfaces non singulières sont cofinales et ont toutes même genre arithmétique virtuel (si F_1 et F_2 sont non singulières, on peut trouver F non singulière telle que $F_1 \leqslant F$ et $F_2 \leqslant F$; or on peut passer, par des transformations quadratiques locales, de F_1 à une surface G qui est en correspondance birégulière avec F; comme les transformations quadratiques conservent le genre virtuel, on a $P_a(F_1) = P_a(F) = P_a(F_2)$.

Quant à la généralisation aux dimensions supérieures, la démonstration de la monotonie dépend du fait que presque toute section hyperplane de F est normale, ce qui a été généralisé tout récemment aux variétés normales par A. SEIDENBERG. Les faits relatifs aux transformations quadratiques restent vrais en dimension 3, celles-ci étant remplacées par les transformations monoïdales. Mais il serait fort utile d'avoir une démonstration de l'existence de la borne inférieure indépendante d'une hypothétique résolution des singularités; par exemple, en caractéristique $p \neq 0$, une surface pour laquelle $P_a(F)$ atteint son minimum a des singularités d'un type assez simple pour que la résolution en soit facile; son existence faciliterait donc la résolution des singularités.

Les résultats ci-dessus seront démontrés dans les Transactions dans un mémoire du conférencier en collaboration avec H. T. Muhly.

A ce propos M. B. SEGRE a annoncé que DEURING aurait obtenu une méthode de résolution des singularités dans le cas général.

HILBERT'S CHARACTERISTIC FUNCTION AND THE ARITHMETIC GENUS OF AN ALGEBRAIC VARIETY

BY

H. T. MUHLY AND O. ZARISKI(1)

1. **Introduction.** The properties of the characteristic function of Hilbert associated with a doubly homogeneous ideal $\mathfrak{a}$ in a ring $R=k[X, Y]$ of polynomials in two sets of indeterminates (X) and (Y) were studied by van der Waerden in [1](2). Since such an ideal can be regarded as defining an algebraic correspondence between the varieties $U=\mathcal{V}(\mathfrak{a}\cap k[X])$ and $V=\mathcal{V}(\mathfrak{a}\cap k[Y])$, van der Waerden's results can be looked upon as belonging to the general theory of algebraic correspondences. In case U and V are birationally equivalent normal varieties(3), further results can be given. It is the object of this note to study the properties of the Hilbert characteristic function which is associated with such a pair of normal varieties. A new proof of the invariance of the arithmetic genus of two- and three-dimensional varieties as well as a proof of the Riemann-Roch theorem for a large class of linear systems on an algebraic surface are among the results obtained.

2. **Enumerative functions.** Let U and V be models of a field Σ of degree of transcendency r over a ground field k. The field k is assumed to be algebraically closed but otherwise arbitrary. Let $(\bar{x}_1, \bar{x}_2, \cdots, \bar{x}_g)$ and $(\bar{y}_1, \bar{y}_2, \cdots, \bar{y}_h)$ be nonhomogeneous coordinates of the generic points of U and V respectively, and let x_0 and y_0 be quantities which are algebraically independent over Σ. Let $x_i=x_0\bar{x}_i$, $i=1, 2, \cdots, g$; $y_j=y_0\bar{y}_j$, $j=1, 2, \cdots, h$, and form the ring $\mathfrak{o}=k[x_0, x_1, \cdots, x_g; y_0, y_1, \cdots, y_h]$. This ring is an integral domain with quotient field $\Sigma(x_0, y_0)$. If $R=k[X_0, X_1, \cdots, X_g; Y_0, Y_1, \cdots, Y_h]$ (where (X) and (Y) are indeterminates) and if τ is the homomorphic mapping of R onto $\mathfrak{o}$ which sends X_i into x_i, Y_j into y_j, then the kernel $\mathfrak{a}$ of τ is a doubly homogeneous prime ideal in R. The characteristic function $\chi(m, n; \mathfrak{a})$ defined in [1] as the number of doubly homogeneous forms of degree m in (X) and of degree n in (Y) which are linearly independent (over k) modulo $\mathfrak{a}$ is an enumerative function determined uniquely by the pair of models (U, V) of the field Σ. Van der Waerden's main result as-

Presented to the Society, September 10, 1948; received by the editors July 8, 1948.

(1) Research paper done under Office of Naval Research contract N6ori-71, Task Order XVIII at University of Illinois and Harvard University. The ideas expressed in this paper represent the personal views of the authors and are not necessarily those of the Office of Naval Research.

(2) Numbers in brackets refer to the bibliography.

(3) Throughout this paper the term "normal variety" will mean "locally normal variety" in the sense of [5, Definition 3].

serts that there exist integers a_{ij}, M, N such that

$$\chi(m, n; \mathfrak{a}) = \sum_r a_{ij} C_{m,i} C_{n,j}, \qquad \text{when } m \geqq M,\ n \geqq N, \tag{2.1}$$

where $\sum_r$ denotes the sum over all integers i, j such that $i+j \leqq r$. We shall denote the polynomial $\sum_r a_{ij} C_{m,i} C_{n,j}$ associated with the pair (U, V) by $\rho(m, n)$.

Let $\{A_m\}$ (4) ($\{B_n\}$) denote the system of $(r-1)$-dimensional subvarieties cut out on U (V) by the hypersurfaces of order m (n) in its ambient space S_g (S_h). We regard each of these systems as lying on the join W of U and V, and there we consider the minimal sum $\{A_m+B_n\}$ of $\{A_m\}$ and $\{B_n\}$ together with the complete sum $|A_m+B_n|$. Two enumerative functions $r(m, n)$ and $s(m, n)$ associated with the pair (U, V) are defined as follows:

$$r(m, n) = 1 + \dim |A_m + B_n|,$$
$$s(m, n) = 1 + \dim \{A_m + B_n\}.$$

Since it is clear from the above definition of the ideal $\mathfrak{a}$ that $\chi(m, n; \mathfrak{a}) = s(m, n)$, equation (2.1) can be written in the form

$$s(m, n) = \rho(m, n), \qquad \text{when } m \geqq M,\ n \geqq N. \tag{2.2}$$

Our main interest is in the function $r(m, n)$. We shall show that under suitable restrictions on the models U and V and on the birational correspondence between them, the minimal sum $\{A_m+B_n\}$ is complete for all m when n is large or for all n when m is large. Moreover, for such values of m and n, the equations $r(m, n) = s(m, n) = \rho(m, n)$ hold.

3. **Varieties of dimension one.** As a starting point for induction proofs to be undertaken later we consider the special case in which Σ is of degree of transcendency one over k. We recall a well known lemma of Castelnuovo which gives a sufficient condition for the minimal sum of two linear series on an algebraic curve to be complete.

CASTELNUOVO'S LEMMA. *If g_μ is a linear series without fixed points on a curve Γ, and if g_ν is a complete non-special series on Γ which partially contains g_μ and is such that the residual series $g_\nu - g_\mu$ is non-special, then the minimal sum of g_ν and g_μ is complete.*

This lemma leads immediately to the following one.

LEMMA 1. *If the variety V (of the pair (U, V)) is a normal curve, there exists an integer n_0 such that the minimal sum $\{A_m+B_n\}$ is complete if $n \geqq n_0$ and m is arbitrary.*

Proof. Let π be the genus of Σ and let μ and ν be the orders of U and V respectively. The system $\{A_m\}$ is then a $g_{m\mu}$ while $\{B_n\}$ is a $g_{n\nu}$. Since V is

(4) The notation $\{C\}$ is used to denote a specified linear system of generic member C, and the notation $|C|$ is reserved for the complete system determined by C.

normal, $\{B_n\}$ is complete if n is sufficiently large. We fix n_0 so that (a) $n\nu-\mu>2\pi-2$, (b) $g_{n\nu}$ partially contains g_μ, (c) $g_{n\nu}$ is complete, when $n\geqq n_0$. We proceed by induction on m since for $m=0$ the lemma is trivially true in view of (c). We therefore assume that the minimal sum of $g_{n\nu}$ and $g_{m\mu}$ is a complete series g_N $(N=m\mu+n\nu)$. We observe that g_N is non-special and that it partially contains g_μ. By condition (a), $N-\mu>2\pi-2$ so that g_N-g_μ is also non-special. By Castelnuovo's lemma the minimal sum of g_N and g_μ is complete, and since this minimal sum coincides with the minimal sum of $g_{(m+1)\mu}$ and $g_{n\nu}$, the lemma is proved.

LEMMA 2. *Under the hypotheses of Lemma* 1, *the ρ-function associated with the pair* (U, V) *is given by the formula*

$$\rho(m, n) = m\mu + n\nu - \pi + 1. \tag{3.1}$$

Moreover, $r(m, n)=\rho(m, n)$ *whenever* $m\mu+n\nu>2\pi-2$.

Proof. If $n\geqq n_0$, then $\{A_m+B_n\}$ is complete so that $r(m, n)=s(m, n)$ and $r(m, n)=\rho(m, n)$ for large values of m and n. Since the complete system $|A_m+B_n|$ is non-special when m and n are large, the Riemann-Roch theorem for curves yields $\rho(m, n)=r(m, n)=m\mu+n\nu-\pi+1$. The expression $m\mu+n\nu-\pi+1$ must then give the value of $\rho(m, n)$ for all m and n since ρ is a polynomial. It gives the value of $r(m, n)$ whenever $m\mu+n\nu>2\pi-2$, q.e.d.

4. **The virtual arithmetic genus.** A class of varieties which we shall call *sectionally normal* is defined inductively as follows.

DEFINITION 1. *An irreducible* 1-*dimensional variety is said to be sectionally normal if it is nonsingular. An irreducible r-dimensional variety is sectionally normal if almost all*(5) *of its sections by the hypersurfaces of its ambient space are sectionally normal varieties of dimension* $r-1$.

We next define a class of birational transformations.

DEFINITION 2. *A birational transformation T defined on an irreducible variety $W\subset S_n$ will be called a proper transformation if $T(W\cap S_t)$ is normal for almost all linear subspaces S_t of S_n* $(t=n-r+1), \cdots, n)$. *In particular, $T(W)$ is required to be normal.*

We point out that any normal curve or surface is sectionally normal as is any nonsingular variety of arbitrary dimension(6); moreover, if T is a regular birational transformation defined on a sectionally normal variety W, then T is proper. We shall have occasion to point out in the next section that there are proper transformations which are not regular.

(5) The term "almost all" here means all with the possible exception of a proper algebraic subsystem.

(6) It has recently been proved by A. Seidenberg (but not yet published) that almost all hyperplane sections of a normal variety V are normal varieties. This implies that every normal variety is sectionally normal.

THEOREM 1. *If the transformation T: $U \to V$ is proper then there exists an integer n_0 such that the minimal sum of $\{A_m\}$ and $\{B_n\}$ is complete when $n \geqq n_0$, $m \geqq 0$.*

Proof. The proof is by induction with respect to the dimension r of the field Σ of rational functions on U and V. By Lemma 1 the theorem is true if Σ is of dimension one over k and we assume that it is true if Σ is of dimension $r-1$ over k. The passage from $r-1$ to r is effected by induction with respect to m. Since T is proper, the variety V is normal so that the minimal sum $\{A_m+B_n\}$ is complete if $m=0$ and n is sufficiently large. We assume that the minimal sum $\{A_{m-1}+B_n\}$ is complete if n is large.

Let A be an irreducible hyperplane section of U such that $A'=T(A)$ is normal, and the transformation T': $A \to A'$ induced by T is proper. (Such hyperplane sections exist in view of the theorem of Bertini [4] and the fact that T is proper(7).) If $\{\bar{A}_m\}$ and $\{\bar{B}_n\}$ are the linear systems cut out on A and A' by the hypersurfaces of their respective ambient spaces, then these systems are cut out on A and A' by the systems $\{A_m\}$ and $\{B_n\}$. By the induction assumption the minimal sum $\{\bar{A}_m+\bar{B}_n\}$ is complete if $n \geqq \bar{n}_0$. We denote its dimension (increased by one) by $\bar{r}(m, n)$, and we fix n_0 ($\geqq \bar{n}_0$) so that $\{B_n\}$ is complete when $n \geqq n_0$. If G is the minimal sum of $\{A_m\}$ and $\{B_n\}$, then G cuts out the complete system $|\bar{A}_m+\bar{B}_n|$ on A, the dimension of which is $\bar{r}(m, n)-1$. The residual system of G with respect to A is the minimal sum $\{A_{m-1}+B_n\}$, which by the induction assumption is complete. Since the dimension of $|A_{m-1}+B_n|$ is given by $r(m-1, n)-1$ we conclude that

$$1 + \dim G = r(m-1, n) + \bar{r}(m, n).$$

(7) The theorem of Bertini is proved in [4] under the hypothesis that k is of characteristic zero. In the special case of the hyperplane sections of an irreducible variety the following considerations remove this restriction on the characteristic. Let V/k be an irreducible variety in S_n (k is algebraically closed) and let p be the prime homogeneous ideal of V in $k[Y]$ $(=k[Y_0, Y_1, \cdots, Y_n])$. Let $(u)=(u_0, u_1, \cdots, u_n)$ be a set of $n+1$ indeterminates (algebraically independent quantities over k) and let K denote the field generated over k by the quotients of the u's. If L denotes the linear form $\sum u_i Y_i$, then it is easily seen that the ideal (p, L) in $k(u)[Y]$ is prime. This implies that the intersection W of V with the hyperplane $L=0$ is irreducible over K. The variety W is of dimension $r-1$. If m is the order of V, any linear S_{n-r} which is general over k meets V in m distinct points. Hence any linear S_{n-r+1} which is general over K meets W in m distinct points. It follows that also W is of order m.

Let P be a general point of W/K. Then it is clear that P is also a general point of V/k and that K is a pure transcendental extension of $k(P)$ (of transcendence degree $n-1$). Hence K is maximally algebraic in $K(P)$ and therefore W is an absolutely irreducible variety. Also the associated form F of W (in the sense of Chow-van der Waerden) is therefore absolutely irreducible (the coefficients of F may be assumed to be forms in $k[u]$). Hence F remains irreducible for almost all specializations $(u) \to (\bar{u})$, $\bar{u}_i \in k$. Since in any such specialization, F is specialized to the associated form $\bar{F}$ of the corresponding hyperplane section $\bar{W}$ of V (here $\bar{W}$ is to be thought of as a cycle, not a variety), it follows at once that almost all hyperplane sections of V are irreducible varieties of order m.

However, if H is the complete system $|A_m+B_n|$ determined by G, the same reasoning applied to H would lead to the conclusion that

$$1 + \dim H = r(m - 1, n) + \bar{r}(m, n).$$

It follows that $G=H$, q.e.d.

COROLLARY 1. *Under the hypothesis of Theorem* 1, *the function* $r(m, n)$ *satisfies the addition formula*

$$r(m, n) = r(m - 1, n) + \bar{r}(m, n) \tag{4.1}$$

when $n \geqq n_0$.

COROLLARY 2. *The* r- *and* ρ-*functions associated with the pair* (U, V) *are equal when* $n \geqq n_0$ *and* m *is arbitrary.*

Proof. Let $\bar{\rho}(m, n)$ be the ρ-function associated with the pair (A, A') introduced above. Since the statement is true for varieties of dimension one by Lemma 2, we assume for purposes of induction that $\bar{r}(m, n) = \bar{\rho}(m, n)$ if $n \geqq \bar{n}_0$ and $m \geqq 0$. Since the minimal sum $\{A_m+B_n\}$ is complete, $r(m, n) = s(m, n)$, so that if both m and n are large, $r(m, n) = \rho(m, n)$. It follows from (4.1) that

$$\rho(m, n) = \rho(m - 1, n) + \bar{\rho}(m, n) \tag{4.2}$$

if both m and n are large, and since the ρ-functions are polynomials, (4.2) is valid for all values of m and n. Equations (4.1) and (4.2) together imply that if $r(m, n) = \rho(m, n)$ then also $r(m-1, n) = \rho(m-1, n)$. This completes the proof.

Let W be a normal variety defined by a homogeneous prime ideal $\mathfrak{p}$ in the ring $S = k[X_0, X_1, \cdots, X_h]$, and let $\{C_n\}$ be the linear system cut out on W by the hypersurfaces of order n in the ambient space S_h of W. If $r(n) - 1$ is the dimension of the complete system $|C_n|$ and if $\chi(n; \mathfrak{p})$ is the number of forms of degree n in S which are linearly independent modulo $\mathfrak{p}$, then $r(n) = \chi(n; \mathfrak{p})$ if n is sufficiently large, since W is normal. The function $\chi(n; \mathfrak{p})$ is a polynomial in n if n is large;

$$\chi(n; \mathfrak{p}) = \rho(n) = \sum_{i=0}^{r} a_i C_{n,i},$$

where the coefficients, $a_0, a_1, \cdots, a_r$ are integers uniquely determined by W, and r is the dimension of W. We call $\rho(n) - 1$ the *virtual dimension* of $|C_n|$, and note that for large values of n the *effective dimension* $r(n) - 1$ equals the virtual dimension $\rho(n) - 1$.

DEFINITION 3. *The virtual arithmetic genus* $p_a(W)$ *of the normal r-dimensional variety* W *is the integer* $(-1)^r(a_0 - 1)$.

THEOREM 2. *If* U *and* V *are normal models of* Σ *and if* $T: U \rightarrow V$ *and*

$T^{-1}: V \to U$ are both proper transformations, then $p_a(U) = p_a(V)$.

Proof. Let $r_1(m)$, $\rho_1(m)$; $r_2(n)$, $\rho_2(n)$ be the effective and virtual dimensions (increased by one) of $|A_m|$ and $|B_n|$ respectively, and let $r(m, n)$, $\rho(m, n)$ be the r- and ρ-functions associated with the pair (U, V). Let $\rho_1(m) = \sum c_{1i} C_{m,i}$, $\rho_2(n) = \sum c_{2j} C_{n,j}$. By definition, $r_1(m) = r(m, 0)$, $r_2(n) = r(0, n)$. Since both T and T^{-1} are proper, it follows by Corollary 2 of Theorem 1 that there exist integers m_0 and n_0 such that

$$r(m, 0) = \rho(m, 0), \qquad \text{if } m \geqq m_0,$$

$$r(0, n) = \rho(0, n), \qquad \text{if } n \geqq n_0.$$

It follows that

$$\rho_1(m) = \rho(m, 0), \qquad \rho_2(n) = \rho(0, n),$$

for all values of m and n. If a_{00} is the constant term in $\rho(m, n)$, then $c_{10} = a_{00} = c_{20}$. It follows that $p_a(U) = p_a(V)$, q.e.d.

COROLLARY 1. *If U and V are sectionally normal models in regular birational correspondence, then $p_a(U) = p_a(V)$. In other words, the virtual arithmetic genus of a sectionally normal model is a relative birational invariant.*

Proof. We have pointed out above that a regular birational transformation defined on a sectionally normal model is proper. Since the inverse of a regular transformation is also regular, the corollary follows.

5. **Varieties of dimension two and three.** In this article we confine our attention to the cases in which Σ is of dimension two or three over k and we restrict our remarks to *nonsingular* models of such fields. We assume throughout the remainder of the text that k is of characteristic zero.

If U and V are nonsingular models of Σ and if V dominates U, $U < V$(8), then it is immediately clear that $T: U \to V$ is proper. In fact, the system $T(\{A_1\})$ on V has no base points (since $U < V$) and since V has no singularities, the general member of $T(\{A_1\})$ has no singularities in view of Bertini's theorem [6]. If Σ is of dimension two this remark suffices to prove that T is proper. If Σ is of dimension three we observe in addition that a generic A_1 and its transform $T(A_1)$ are nonsingular and that $A_1 < T(A_1)$. Hence T induces a proper transformation on a generic A_1 so that it is itself proper.

We consider *quadratic* and *monoidal* transformations in the next two lemmas. These terms are used in the sense of [5, article 11].

LEMMA 3. *If U and V are nonsingular, and if V is obtained from U by a*

(8) The statement "V dominates U" implies that the birational transformation $T^{-1}: V \to U$ has no fundamental points on V; or equivalently, the local ring $Q(P')$ contains the local ring $Q(P)$, when $P(\subset U)$ and $P'(\subset V)$ are a pair of corresponding points in the birational correspondence T. This relationship is denoted by the symbols $U < V$.

quadratic transformation with center at a point P of U, then both $T: U \rightarrow V$ *and* $T^{-1}: V \rightarrow U$ *are proper transformations.*

Proof. In this case $U < V$ so that T is proper. Let $(\eta_0, \eta_1, \cdots, \eta_g)$ be homogeneous coordinates of the general point of U and let $\mathfrak{p}$ be the prime ideal of P in the ring $\mathfrak{o} = k[\eta]$. If $\phi_0, \phi_1, \cdots, \phi_h$ form a linear basis for the forms of degree ν in $\mathfrak{p}$, and if ν is sufficiently large, then the ideal $\mathfrak{a} = \sum \mathfrak{o}\phi_i$ differs from $\mathfrak{p}$ by at most an irrelevant component and $(\phi_0, \phi_1, \cdots, \phi_h)$ can be regarded as the general point of V. Hence the linear system defined by $\phi(\lambda) = 0$, where $\phi(\lambda) = \lambda_0\phi_0 + \lambda_1\phi_1 + \cdots + \lambda_h\phi_h$, $\lambda_i \in k$, is the inverse transform of the system $\{B_1\}$ of hyperplane sections of V. The point P is the only base point of this system, and since U is nonsingular, the generic member of this system has no singular points except possibly at P itself. If we assume that P is not on the hyperplane section $\eta_0 = 0$, and if we put $\psi_i = \phi_i/\eta_0^\nu$, then the quantities $\psi_0, \psi_1, \cdots, \psi_h$ will form a basis for the prime ideal of P in the quotient ring $Q(P/U)$. It follows that among the ψ_i there is at least one whose leading form at P is linear so that almost all members of $\{\phi(\lambda)\}$ have a simple point at P (see [6, p. 137]). The generic member of $\{\phi(\lambda)\}$ is therefore nonsingular and hence normal. This proves that T^{-1} is proper if U and V are surfaces.

If Σ is of dimension three, it is necessary to prove in addition that the general characteristic curve of the system $\{\phi(\lambda)\}$ is nonsingular, since such a curve is the transform by T^{-1} of a section $V \cap S_{h-2}$, where S_h is the ambient space of V. However, it is proved in [5] that a quadratic transformation with center at a point P of U induces a quadratic transformation on any subvariety of U through P. If we apply this remark to a generic member $\bar{\phi}$ of the system $\{\phi(\lambda)\}$ we conclude as above that the generic characteristic curve $\bar{\phi} \cap \phi(\lambda)$ has no singularities, q.e.d.

If U and V are three-dimensional varieties, and if T is a monoidal transformation with center along an irreducible curve $\Delta \subset U$, then since U is nonsingular, every point of Δ is a simple point of U. If in addition every point of Δ is a simple point of Δ, then we shall call T a *nonsingular* monoidal transformation.

LEMMA 4. *If U and V are nonsingular and if* $T: U \rightarrow V$ *is a nonsingular monoidal transformation, then T and* T^{-1} *are proper.*

Proof. Since $U < V$, T is proper. The proof that T^{-1} is proper is similar to the proof given in Lemma 3 for quadratic transformations. We use the same notations as in Lemma 3, except that now $\mathfrak{p}$ is the prime ideal of Δ in $\mathfrak{o}$. The general member of the system $\{\phi(\lambda)\}$ has no singularities except possibly at points of Δ. If P is a point of Δ not on the surface $\eta_0 = 0$, then $\psi_0, \psi_1, \cdots, \psi_h$ will form a basis for the prime ideal of Δ in $\mathfrak{Q} = Q(P/U)$. Then, since P is simple both for Δ and for U, it follows that among the ψ_i there will be at least one whose leading form at P is linear (see [5, footnote 34]). Hence P cannot be a singular base point of $\{\phi(\lambda)\}$. It follows that the generic member of $\{\phi(\lambda)\}$ is nonsingular.

To complete the proof we must show that the generic characteristic curve of $\{\phi(\lambda)\}$ is nonsingular, and for this it suffices to show, in view of Bertini's theorem, that the system cut by $\{\phi(\lambda)\}$ on a generic member ϕ (outside the fixed curve Δ) has no base points on Δ. The total transform $T(\Delta)$ of Δ is an irreducible surface F on V which is "ruled" by a pencil $\{f\}$ of rational curves. The curves f are in 1:1 correspondence with the points of Δ and each f is the total transform of the point of Δ to which it corresponds [7, article 5]. We choose a hyperplane section B_1 of V such that the section $F\cap B_1$ is an irreducible curve Γ which is not a component of any member of the pencil $\{f\}$. We further assume that $T^{-1}(B_1)$ is a nonsingular irreducible member $\bar{\phi}$ of $\{\phi(\lambda)\}$. If $\phi(\lambda)\in\{\phi(\lambda)\}$, we write $\phi(\lambda)\cap\bar{\phi}=\Delta+R(\lambda)$. Let us assume that $P\in\Delta$ is a base point of the residual system $\{R(\lambda)\}$. It then follows that every hyperplane section B_1' of V meets the section B_1 in a point of the variety $T(P)\cap\Gamma$. Since Γ is not a component of any member of $\{f\}$, it follows that $T(P)\cap\Gamma$ consists of a finite set of points, so that the conclusion that every hyperplane B_1' meets Γ in a point of $T(P)\cap\Gamma$ is absurd, q.e.d.

These lemmas yield the following further corollary to Theorem 2.

COROLLARY 2. *The virtual arithmetic genus of a nonsingular model of a field Σ of dimension two or three over k is invariant under quadratic and nonsingular monoidal transformations.*

The considerations of part IV of [7] show in particular that if U and V are nonsingular models of a three-dimensional field Σ, then there exist models U_1 and V_1 in regular birational correspondence such that U_1 (V_1) is obtained from U (V) by a sequence of quadratic and nonsingular monoidal transformations. It follows that $p_a(U)=p_a(U_1)=p_a(V_1)=p_a(V)$. Since a similar statement is true for nonsingular surfaces (see [3]) we can assert the following theorem.

THEOREM 3. *If Σ is a field of dimension two or three over k, then any two nonsingular models of Σ have the same virtual arithmetic genus.*

The virtual arithmetic genus of a nonsingular model can therefore be regarded as a character of the field rather than of a particular model. This character is called the *arithmetic genus* of Σ and is denoted by $p_a(\Sigma)$.

6. **Normal surfaces.** Our objective in this section is to show that if U is any normal model of a field Σ of dimension two over k, then the virtual arithmetic genus of U is not less than the arithmetic genus of Σ, $p_a(U)\geqq p_a(\Sigma)$. Since it is known that there exists a nonsingular model V of Σ such that $V>U$ [2], the following theorem will yield the desired result.

THEOREM 4. *If U and V are normal models of Σ (Σ of dimension two), and if $U < V$, then $p_a(U) \geqq p_a(V)$.*

The major portion of the proof of this theorem is contained in the following two lemmas.

LEMMA 5. *Let Γ be an irreducible algebraic curve of order ν in S_n, and let $g_{m\nu}$ be the series cut out on Γ by the hypersurfaces of order m in S_n. If g_μ is a linear series on Γ without fixed points, then there exists an integer M such that if $m \geqq M$, the deficiency of the minimal sum of $g_{m\nu}$ and g_μ is not greater than the deficiency of $g_{m\nu}$, in symbols, $\delta(g_{m\nu}+g_\mu) \leqq \delta(g_{m\nu})$.*

Proof. Let G_μ be a set of the series g_μ and choose nonhomogeneous coordinates on Γ so that no point of G_μ is at infinity. If $\mathfrak{a}$ is the ideal in the ring $\mathfrak{o} = k[x_1, x_2, \cdots, x_n]$ of nonhomogeneous coordinates on Γ which is determined by G_μ, then the length of $\mathfrak{a}$ is μ, and there exist μ elements $\phi_1, \phi_2, \cdots, \phi_\mu$ in $\mathfrak{o}$ which form an independent k-basis for $\mathfrak{o}$ modulo $\mathfrak{a}$. If m exceeds the degree of each of the polynomials ϕ_i, then the set G_μ will impose μ independent conditions on the series $g_{m\nu}$. That is, if $\bar{g}_{m\nu}$ is the subsystem of $g_{m\nu}$ consisting of those sets of $g_{m\nu}$ which contain G_μ, then dim $\bar{g}_{m\nu} =$ dim $g_{m\nu} - \mu$.

Fix M so large that the above condition holds when $m \geqq M$. Increase M if necessary so that $g_{m\nu} \supset g_\mu$, and the residual series $g_{m\nu} - g_\mu$ is non-special when $m \geqq M$. Let G'_μ be another fixed set of g_μ which has no point in common with G_μ. Consider the two series $g_{m\nu}+G_\mu$ and $g_{m\nu}+G'_\mu$ obtained by adding the fixed sets G_μ and G'_μ to the series $g_{m\nu}$. The union of these two series is contained in the minimal sum of $g_{m\nu}$ and g_μ. The common part of these series is the series $\bar{g}_{m\nu}$. Hence

$$m\nu + \mu - \pi - \delta(g_{m\nu} + g_\mu) \geqq 2(m\nu - \pi - \delta(g_{m\nu})) - (m\nu - \pi - \delta(g_{m\nu}) - \mu),$$

and $\delta(g_{m\nu}+g_\mu) \leqq \delta(g_{m\nu})$, q.e.d.

LEMMA 6. *If U and V are normal models of Σ such that $U < V$, then there exist integers m_0, d such that*

$$\rho(m, n) + d \geqq r(m, n) \geqq \rho(m, n) \tag{6.1}$$

when $m \geqq m_0$ and $n \geqq 0$.

Proof. Since $U < V$, and since U and V are normal, it follows that $T: U \rightarrow V$ is proper, so that by Corollary 2 of Theorem 1 there exists an integer n_0 such that $r(m, n) = \rho(m, n)$ when $n \geqq n_0$, $m \geqq 0$. Let B_1 be an irreducible nonsingular hyperplane section of V, and let $\bar{B}_1 = T^{-1}(B_1)$. If $r^*(m, n)$ and $\rho^*(m, n)$ are the r- and ρ-functions associated with the pair $(\bar{B}_1, B_1)$, then, by Lemma 2, $r^*(m, n) = \rho^*(m, n)$ for all m if n is large or for all n if m is large. The image of B_1 on the join of U and V is a normal curve, so that it follows as in the proof of Theorem 1 that $|A_m + B_n|$ will cut a complete series on B_1 if n is large. Since the residual system of $|A_m + B_n|$ with respect to B_1 is the complete system $|A_m+B_{n-1}|$, we conclude that $r(m, n)=r(m, n-1)+r^*(m, n)$ for large values of n. It follows that $\rho(m, n)-\rho(m, n-1)=\rho^*(m, n)$ for all values of m, n since the ρ-functions are polynomials.

Let $g_{m\nu}$ be the series cut out on $\bar{B}_1$ by the system $\{A_m\}$. If we apply Lemma 5 to the case in which the g_μ of that lemma is the series g_ν, we find $\delta(g_{(m+1)\nu}) \leqq \delta(g_{m\nu})$ for large values of m. It follows that $\delta(g_{m\nu})$ remains constant when m is large, say $\delta(g_{m\nu})=\delta$ if $m \geqq M_0$. Let $g_{s\mu}$ be the series cut on B_1 by $\{B_s\}$, $s=1, 2, \cdots, n_0$, consider it as a series on $\bar{B}_1$, and apply Lemma 5 to each of the series $g_{s\mu}$ in turn. This yields a set of integers $M_1, M_2, \cdots, M_{n_0}$. Let $m_0=\max(M_0, M_1, \cdots, M_{n_0})$. We can then assert that if $m \geqq m_0$, then $\delta(g_{m\nu}+g_{s\mu}) \leqq \delta$, $s=1, 2, \cdots, n_0$.

The system cut by $|A_m+B_s|$ on B_1 ($m \geqq m_0$, $s=1, 2, \cdots, n_0$) contains the minimal sum $g_{m\nu}+g_{s\mu}$ and is totally contained in the complete series determined by $g_{m\nu}+g_{s\mu}$. Since the residual system of $|A_m+B_s|$ with respect to B_1 is the complete system $|A_m+B_{s-1}|$, the following inequalities are valid:

$$r(m, s-1) + r^*(m, s) \geqq r(m, s) \geqq r(m, s-1) + r^*(m, s) - \delta, \qquad s = 1, 2, \cdots, n_0.$$

On combining these inequalities with the equalities $r(m, n)=\rho(m, n)$, $n \geqq n_0$, $m \geqq 0$; $r^*(m, n)=\rho^*(m, n)$, $m \geqq m_0$, $n \geqq 0$, and using the addition formula $\rho(m, n)-\rho(m, n-1)=\rho^*(m, n)$, we find

$$\rho(m, s) + (n_0 - s)\delta \geqq r(m, s) \geqq \rho(m, s), \qquad s = 0, 1, \cdots, n_0; m \geqq m_0.$$

If we put $d=\delta n_0$, the lemma follows, q.e.d.

Theorem 4 now follows quite readily. As in the proof of Theorem 2, let $r_1(m)$, $\rho_1(m)$; $r_2(n)$, $\rho_2(n)$ be the effective and virtual dimensions (increased by one) of $|A_m|$ and $|B_n|$, and let

$$\rho_1(m) = \sum_{i=0}^{2} c_{1i}C_{m,i}, \qquad \rho_2(n) = \sum_{j=0}^{2} c_{2j}C_{n,j}, \qquad \rho(m, n) = \sum_{2} a_{ij}C_{m,i}C_{n,j}.$$

Since $r_2(n)=r(0, n)=\rho(0, n)$ if $n \geqq n_0$, it follows that $c_{20}=a_{00}$. On the other hand, $\rho_1(m)=r_1(m)=r(m, 0)$ for large values of m so that, by Lemma 6, $\rho(m, 0)+d \geqq \rho_1(m) \geqq \rho(m, 0)$. It follows that the polynomials $\rho(m, 0)$ and $\rho_1(m)$ differ only by a constant and that $c_{10} \geqq a_{00}$. Hence $p_a(U) \geqq p_a(V)$, q.e.d.

If W is an arbitrary normal model of a two-dimensional field Σ, if the system $\{C\}$ of hyperplane sections of W is of degree ν and genus π, and if $p_a(W)=p_a(\Sigma)+s$, where $s \geqq 0$, then by a straightforward computation it is found that the function $\rho(n)$ associated with W is given by the expression

$$\rho(n) = \nu C_{n,2} + (\nu - \pi + 1)n + p_a(\Sigma) + s + 1.$$

It follows that

$$\dim |nC| = \nu C_{n,2} + (\nu - \pi + 1)n + p_a(\Sigma) + s$$

if n is sufficiently large. This is the Riemann-Roch theorem for the complete system $|nC|$.

Bibliography

1. B. L. van der Waerden, *On Hilbert's function, series of composition of ideals, and a generalization of a theorem of Bezout*, Proceedings of the Koninklijke Akademie van Wetenschappen, Amsterdam vol. 31 (1928).

2. O. Zariski, *The reduction of the singularities of an algebraic surface*, Ann. of Math. vol. 40 (1939).

3. ———, *A simplified proof of the resolution of singularities of an algebraic surface*, Ann. of Math. vol. 43 (1942).

4. ———, *Pencils on an algebraic variety and a new proof of a theorem of Bertini*, Trans. Amer. Math. Soc. vol. 50 (1941).

5. ———, *Foundations of a general theory of birational correspondences*, Trans. Amer. Math. Soc. vol. 53 (1943).

6. ———, *The theorem of Bertini on the variable singular points of a linear system of varieties*, Trans. Amer. Math. Soc. vol. 56 (1944).

7. ———, *Reduction of the singularities of algebraic three dimensional varieties*, Ann. of Math. vol. 45 (1944).

University of Illinois,
Urbana, Ill.
Harvard University,
Cambridge, Mass.

ANNALS OF MATHEMATICS
Vol. 55, No. 3, May, 1952
Printed in U.S.A.

COMPLETE LINEAR SYSTEMS ON NORMAL VARIETIES AND A GENERALIZATION OF A LEMMA OF ENRIQUES-SEVERI

BY OSCAR ZARISKI

(Received July 23, 1951)

1. Introduction

One of the central points of the proof of the Riemann-Roch theorem for algebraic surfaces, given by the Italian geometers, is a key lemma of Severi on adjoint surfaces. Let F be a surface is S_3 having only ordinary singularities and let n be the order of F. Let $| F_i |$ be the system of adjoint surfaces of F, of a given order i. Then the lemma of Severi states the following: *if C_m denotes the intersection of F with a generic surface of order m, then the system $| F_i |$ cuts out a complete linear series on C_m if m is sufficiently large* (see Severi [4b], pp. 372–382; also our monograph [9], p. 67).[1] The surface F is to be thought of as a "generic" projection of a non-singular surface in some projective space of high dimension, and in Severi's proof the realization of the surface as a model in S_3 having only ordinary singularities is essential. We will find it, however, more convenient to deal directly with the non-singular surface and to avoid projection into S_3. We also recall that the system $| F_i |$ of adjoint surfaces cuts out on F the complete linear system which is the sum of the canonical system $| K |$ and the system $| C_{i-n+4} |$. We therefore re-state the lemma of Severi as follows:

Let F be a non-singular surface is some projective space S and let C_m denote the intersection of F with a generic hypersurface of order m. Let $| K |$ denote the canonical system on F. Then for any integer i there exists an integer $m(i)$ such that the complete linear system $| K + C_i |$ cuts out a complete linear series on a generic C_m if $m \geqq m(i)$.

In his important memoir [4a], Severi studies extensively the geometry of a three-dimensional variety ([4a], pp. 33–82) and also briefly surveys possible generalizations to varieties of any dimension ([4a], pp. 83–87). In this memoir, Severi outlines a proof of his lemma for the case of an arbitrary variety, of dimension r, with ordinary singularities in S_{r+1} [see also Severi's letter to Bertini, *Osservazioni sul Restsatz per una curva iperspaziale*, Atti della R. Accademia delle Scienze di Torino, vol. 43 (1907–1908), pp. 852–855].

One of the objects of this paper is to prove the lemma of Severi for normal varieties of any dimension, over a field of arbitrary characteristic. Our method of proof is very different from that of Severi's approach to this question.

Throughout this paper we shall be concerned only with absolutely irreducible

[1] Actually the lemma of Severi refers not to the adjoint surfaces F_i of F but to the adjoint surfaces G_i of the curve C_m, i.e., to the surfaces passing through the intersections of C_m with the double curve of F. However, if i is sufficiently large with respect to m, the G_i are adjoint surfaces of F, and it is this consequence of Severi's lemma that plays a role in the proof of the Riemann-Roch theorem.

varieties. Furthermore, we shall assume that our varieties are absolutely normal, i.e., normal with respect to an algebraically closed field of definition (and hence normal with respect to any field of definition). Our result could be easily extended to absolutely irreducible varieties which are normal with respect to some field of definition without being absolutely normal.

We must point out at once that the canonical system $|K|$ will not appear in our generalization of Severi's lemma. The theorem which we shall prove refers not to the canonical system but to an arbitrary complete linear system $|D|$ and is therefore as follows:

THEOREM. *Let V be an irreducible absolutely normal variety and let C_m denote the intersection of V with a general hypersurface of order m (general with respect to some given algebraically closed field of definition of V). If $|D|$ is an arbitrary complete linear system on V, there exists an integer $m(D)$ (depending on D) such that the system $|D|$ cuts out a complete linear system on C_m if $m \geqq m(D)$.*[2]

The generalized lemma of Severi follows from this theorem by taking for $|D|$ the system $|K + C_i|$, i an arbitrary integer.

A very simple consideration will show that *if the above theorem holds for each of the complete linear systems $|D| = |C_i|$ $(i = 1, 2, \cdots)$, then the theorem is true for every complete linear system $|D|$* (see section 7). This fact shows very clearly that the ultimate reason for the validity of Severi's lemma and of our generalization of the lemma can have nothing to do with any particular properties of the canonical system. We put the burden of our proof on this special case $|D| = |C_i|$ (see Theorem 4′, section 8). In this sense, our approach to the problem has points of contact with an attempt made by Enriques. Enriques has attempted to prove, for algebraic surfaces, the theorem stated above.[3] However, the proof of Enriques has a serious gap and is therefore quite incomplete.[4] We have found, nevertheless, some of the ideas of the proof of Enriques extremely useful.

This paper is divided into three parts. In Part I we develop the necessary preliminary material on complete linear systems on absolutely normal varieties. Some of this material is familiar in the classical case, but the precise formulation in the abstract case is not available in the literature. In Part II we give the proof

[2] For a more precise formulation of this result see Theorem 4, section 7.

[3] Enriques considers two complete linear systems $|C|$ and $|D|$ on a non-singular surface F, such that $|D|$ is "normally large" (*normalmente grande*) with respect $|C|$. His fundamental lemma is then to the effect that $|C|$ cuts out a complete linear series on a generic member of $|D|$ (see [1], p. 129). The condition that $|D|$ be normally large with respect to $|C|$ seems to be satisfied if and only if $|D|$ is a sufficiently high multiple of the system of hyperplane sections of some birational transform of F.

[4] The proof of Enriques begins with two assumptions (see properties 1 and 2 in [1], p. 129), the second of which refers to the existence of an irreducible curve D_0 in $|D|$ satisfying certain conditions. In the presence of these two assumptions the lemma of Enriques follows readily. However, the rest of the proof is devoted exclusively to the justification of the first assumption, while the question of validity of the second assumption is not discussed (compare with footnote 10).

of our generalization of the lemma of Enriques-Severi. In Part III we develop the concept of the virtual arithmetic genus p_a of an arbitrary $(r-1)$-cycle on an r-dimensional absolutely irreducible and absolutely normal variety V. While the concept of the virtual arithmetic genus of a reducible *variety* is not new, (see Severi [4a]), our extension of this concept to *cycles* (where we can have prime components of arbitrary, positive or negative, multiplicity) is new.[4a] We apply this concept toward the derivation of a very simple and compact expression of the dimension of complete linear systems of the form $|D + C_m|$, where D is an arbitrary $(r-1)$-cycle on V and $|C_m|$ is the complete linear system cut out on V by the hypersurfaces of a sufficiently high order m (see Theorem 5, section 12). It will clearly appear from the contents of Part III that the notion of the virtual arithmetic genus of a cycle is a useful and natural tool in the theory of complete linear systems on higher varieties, whether these be non-singular or only normal. For normal varieties which have singularities this notion is even indispensible, since in the absence of an intersection theory there is no such thing as, say, the virtual degree of an arbitrary $(r-1)$-cycle. Toward the end of section 13 we discuss some significant points of contact between Severi's memoir [4a] on one hand and the present paper as well as our joint paper [2] with Muhly on the other hand. In this connection, we also discuss, from a different point of view, the question (already treated by Severi) of the equality $p_a(V) = P_a(V)$, where $P_a(V)$ is defined in (58) in terms of $p_a(V)$ and $p_a(-K)$. [*Added in proofs*. Meanwhile this equality has been proved rigorously, in the classical case, by K. Kodaira. An outline of this unpublished proof has been communicated by Professor Kodaira in a recent letter to the author.]

In a very short last section (section 14) we derive the theorem of Riemann-Roch for normal surfaces. This application to normal surfaces is of interest because it shows that the proof of the Riemann-Roch theorem for surfaces can be made entirely independent of the assumption that the surface V under consideration has no singularities. The normality of V is sufficient. The presence of singularities has only the following effect: in the formulation of the Riemann-Roch theorem on V there will appear the virtual arithmetic genus $p_a(V)$ of V [see formula (58)], which is not necessarily a birational invariant and may be greater than the arithmetic genus p_a of the function field of V. This is not surprising, since the Riemann-Roch theorem on a non-singular surface gives only a

[4a] *Added in proofs*. Chevalley has called my attention to a paper by Lefschetz, entitled "The arithmetic genus of an algebraic manifold immersed in another" (Annals of Mathematics, vol. 17, 1916, pp. 197–212), where the concept of the arithmetic genus of a cycle seems to have been introduced for the first time. Lefschetz makes extensive use of algebraic equivalence of cycles on a non-singular variety (in the classical case) and apparently postulates that equivalent cycles have the same virtual arithmetic genus. Since particular cycles may very well belong to more than one complete continuous system, it would seem that Lefshetz's definition is applicable to classes of cycles but not always to individual cycles. For instance, a line in S_3, counted 4 times, has virtual arithmetic genus 1 or 0 according as the corresponding cycle is regarded as the limit of a space quartic of the first or of the second kind.

lower bound for the dimension of a complete linear system. If V is transformed into a non-singular surface V' and if $\{C'\}$ denotes the linear system on V' which corresponds to the system of hyperplane sections of V, then one finds (see section 14) that the difference $p_a(V) - p_a$ is precisely the superabundance of the complete system $| mC' |$, for m large. We may add to all this that in the case of non-zero characteristic, when the existence of a non-singular model (and hence the very existence of the birational invariant p_a) is still an open question, a proof of the Riemann-Roch theorem that *precedes* the theorem of resolution of singularities is certainly an improvement on the present state of the theory of algebraic surfaces.

Part I. Linear Systems

2. Linear systems on absolutely normal varieties

Let V be an absolutely irreducible and absolutely normal variety of dimension r. A *prime s-cycle* on V is an s-dimensional absolutely irreducible subvariety of V. An *s-cycle* on V is a linear combination Z, with integral coefficients, of prime s-cycles; Z is a *positive* (*non-negative*) cycle if all the coefficients are positive (or non-negative). We shall be concerned exclusively with $(r-1)$-cycles, and from now the term "cycle" will mean "$(r-1)$-cycle."

Assuming that V lies in a projective n-space S_n, let Ω be the universal domain and let $\mathfrak{P}$ be the homogeneous prime ideal of V in the polynomial ring $\Omega[Y_0, Y_1, \cdots, Y_n]$. Let $y_0, y_1, \cdots, y_n$ be the $\mathfrak{P}$-residues of the Y's. We shall call (y) the *canonical general point of* V. By *a function on* V we mean any homogeneous element, of degree zero, of the quotient field of the residue class ring $\Omega[Y]/\mathfrak{P}$, i.e., any quotient of two forms in the y's, of like degree, with coefficients in Ω. The functions on V form a field which we shall denote by $\Omega(V)$. If k is a field of definition of V (whence $k \subset \Omega$ and Ω has infinite transcendence degree over k), and ξ is a function on V, we say that ξ *is defined over* k if $\xi \in k(y)$. If ξ is defined over k, then ξ can be expressed as the quotient of two forms in the y's, of like degree, with coefficients in k. The functions on V which are defined over k form a subfield of $\Omega(V)$, denoted by $k(V)$. It is clear that if P is any general point of V/k, then the fields $k(V)$ and $k(P)$ are k-isomorphic [$k(P)$ is the field generated over k by the non-homogeneous coördinates of P].

Let k be a field and let Q be a point in S_n. If $z_0, z_1, \cdots, z_n$ are homogeneous coördinates of Q, we shall say that they are *strictly homogeneous* coördinates of the point Q, *over* k, *or of* Q/k, if the ideal of their relations in $k[Y]$ is homogeneous. It is easily seen that the following is a necessary and sufficient condition for the strict homogeneity, over k, of a given set of homogeneous coördinates z_i of Q: *if for a given* j, $0 \leqq j \leqq n$, *we have* $z_j \neq 0$, *then* z_j *is a transcendental over* $k(Q)$.

In particular, if k is a field of definition of V and the z's are strictly homogeneous coördinates of a *general* point Q of V/k, then the ideal of the relations of the z's in $k[Y]$ is the prime homogeneous ideal of V/k in $k[Y]$. The ring $k[z]$ is then referred to as a *homogeneous coördinate ring of* V/k. It follows that there exists a k-isomorphism of $k[z]$ onto $k[y]$ which sends z_i into y_i, $i = 0, 1, 2, \cdots, n$.

We shall regard the ring $k[y]$ itself as a homogeneous coördinate ring of V/k, in fact we shall refer to it as the *canonical* homogeneous coördinate ring of V/k. The ring $\Omega[y]$ will be called *the homogeneous coördinate ring of* V (here, therefore, no ground field is specified).

Now let Γ be a prime $(r-1)$-cycle on V and let k be a common field of definition of Γ and V. Then Γ defines and is defined by a prime homogeneous ideal $\mathfrak{p}_k$ in $k[Y]$, and the residue class ring $k[Y]/\mathfrak{p}_k$ is precisely the canonical homogeneous coördinate ring of Γ/k. The local ring of Γ on V/k consists of all the functions ξ on V which are defined over k and which are of the form $f(y)/g(y)$, where $f(y)$ and $g(y)$ are in $k[y]$ and $g(Y)$ is not in $\mathfrak{p}_k$. Now suppose that k is algebraically closed. Then V/k is normal, and the above local ring is a discrete valuation ring in the field $k(V)$. The corresponding valuation, which we shall denote by $v_{\Gamma,k}$, is of dimension $r-1$, and its residue field is the field $k(\Gamma)$.

In a similar fashion, we have a local ring of Γ on V, *relative to the universal domain* Ω: this ring consists of all functions on V which are of the form $f(y)/g(y)$, where $f(y)$ and $g(y)$ belong to $\Omega[y]$ and $g(Y)$ does not belong to the prime homogeneous ideal of Γ in $\Omega[Y]$. From the fact that the local ring of Γ on V/k is integrally closed for every choice of the common algebraically closed field of definition k of V and Γ, it follows at once that also the local ring of Γ on V/Ω is integrally closed. Since Γ is of dimension $r-1$, this local ring is a valuation ring in $\Omega(V)$. The corresponding valuation of $\Omega(V)/\Omega$, which we shall denote by v_Γ, is discrete, of rank 1 and dimension $r-1$. Its residue field is precisely the field of functions on Γ.

Let $\mathfrak{p}$ denote the prime ideal of Γ in $\Omega[Y]$. It is clear that $\mathfrak{p}_k = \mathfrak{p} \cap k[Y]$. *Hence the valuation* v_Γ *is an extension of* $v_{\Gamma,k}$. Now if k is algebraically closed, then it is well known that every prime ideal in $k[Y]$ remains prime under any extension of k. We have therefore that $\mathfrak{p}$ is the extension of $\mathfrak{p}_k$ in $\Omega[Y]$. From this it follows that the prime ideal of v_Γ is the extension of the prime ideal of $v_{\Gamma,k}$. In other words: v_Γ *is an unramified extension of* $v_{\Gamma,k}$.

Our application of this result is the following:

Let Γ be any prime $(r-1)$-cycle on V and let ξ be any function on V, $\xi \neq 0$. Let k be an algebraically closed field which is a common field of definition of V, Γ and ξ. *Then* $v_{\Gamma,k}(\xi) = v_\Gamma(\xi)$, *and hence* $v_{\Gamma,k}(\xi)$ *depends only on* Γ *and* ξ (and not on k). We call $v_\Gamma(\xi)$ *the order* of the function ξ, *on* Γ. The cycle Γ is a *nullcycle* or a *polar* cycle of ξ according as the order of ξ on Γ is positive or negative.

A well known argument (which we need not repeat here) shows that any function ξ on V, $\xi \neq 0$, has only a finite number of prime nullcycles and prime polar cycles. As a consequence, the following sum

$$\textstyle\sum_\Gamma v_\Gamma(\xi)\cdot\Gamma,$$

extended to all prime cycles on V, is actually a finite sum and therefore defines an $(r-1)$-cycle on V. We call it *the cycle of the function* ξ and we denote it by (ξ). It is well known that if ξ is not a constant, i.e., if $\xi \notin \Omega$, then (ξ) is not the zero cycle and neither (ξ) nor $-(\xi)$ are positive.

Equivalence (or—more precisely—linear equivalence) of cycles is defined in the usual fashion: if Z is a cycle, then $Z \equiv 0$ if Z is the cycle of a function; if Z_1 and Z_2 are cycles then $Z_1 \equiv Z_2$ if $Z_1 - Z_2 \equiv 0$. That this is indeed a relation of equivalence follows from the relations $(\xi_1\xi_2) = (\xi_1) + (\xi_2)$, $(1/\xi) = -(\xi)$.

We now consider in $\Omega(V)$ any finite Ω-module $\mathfrak{L}$($\mathfrak{L}$, a finite dimensional vector space over Ω, $\mathfrak{L} \subset \Omega(V)$). There certainly exist cycles Δ with the following property: if ξ is any function in $\mathfrak{L}$, $\xi \neq 0$, then $(\xi) + \Delta$ is a non-negative cycle. This follows from the finiteness of the dimension of $\mathfrak{L}/\Omega$ and from the relations $v_\Gamma(\xi_1 + \xi_2) \geqq \min\{v_\Gamma(\xi_1), v_\Gamma(\xi_2)\}$, $v_\Gamma(c\xi) = v_\Gamma(\xi)$, which hold for any non-zero functions ξ_1, ξ_2, ξ on V, any prime cycle Γ on V and any non-zero element c of Ω.

Definition. *A non-empty set L of non-negative cycles on V is called a linear system if there exists a finite Ω-module $\mathfrak{L}$ in $\Omega(V)$ and a cycle Δ such that L coincides with the set of cycles $(\xi) + \Delta$, $\xi \in \mathfrak{L}$, $\xi \neq 0$. We say then that $\mathfrak{L}$ is a defining module of the linear system L.*

We note that any two cycles of a linear system are equivalent to each other. It follows that if a linear system contains one positive cycle, then all its cycles are positive. We note also that according to our definition, any non-negative cycle Z constitutes by itself a linear system (set $\mathfrak{L} = \Omega$, $\Delta = Z$).

Let $\mathfrak{L}$, L and Δ be as in the above definition. Consider the mapping $\xi \to (\xi) + \Delta$ of the set of non-zero elements of $\mathfrak{L}$ onto L. Two elements ξ_1 and ξ_2 are mapped into one and the same cycle of L if and only if $(\xi_1) = (\xi_2)$, i.e., if and only if $\xi_2 = c\xi_1$, $c \in \Omega$. It follows that the above mapping defines a projective structure in the linear system L. If, then, the module $\mathfrak{L}$ has dimension $s + 1$ over Ω ($s \geqq 0$), then the linear system L has the structure of a projective space over Ω, of dimension s.

Given two cycles Z_1 and Z_2 on V, we write $Z_1 > Z_2$ if $Z_1 - Z_2$ is a positive cycle; $Z_1 \geqq Z_2$, if $Z_1 - Z_2$ is a non-negative cycle.

If Z_0 is a positive cycle, Z_0 is said to be a *fixed component of* L if for every cycle Y in L we have $Y \geqq Z_0$. If Z_0 is a fixed component of L, then upon replacing the cycle Δ by the cycle $\Delta - Z_0$ we see that the set of cycles $Y - Z_0 (Y \in L)$ is also a linear system, of the same dimension as L. It is clear that if L has at all fixed components, then it has also a greatest fixed component D, i.e., a fixed component D such that for any other fixed component Z_0 of L we have $Z_0 \leqq D$. Upon subtracting D from each cycle Y of L we obtain a linear system L' of cycles $Y - D$, *which is free from fixed components.* The same module $\mathfrak{L}$ of functions ξ which defines the linear system L defines also this new linear system L'. The system L' is related to the module $\mathfrak{L}$ as follows: in the set of all cycles Δ such that $(\xi) + \Delta \geqq 0$ for all ξ in $\mathfrak{L}$, $\xi \neq 0$, there exists a unique smallest cycle Δ', and L' is precisely the set of all cycles $(\xi) + \Delta'$, $\xi \in \mathfrak{L}$, $\xi \neq 0$. We write Δ' in the form $\Delta_1' - \Delta_2'$ where Δ_1' and Δ_2' are non-negative cycles, free from common components, and we call Δ_1' *the polar cycle of the module* $\mathfrak{L}$. If the functions ξ in $\mathfrak{L}$ have no common prime nullcycle (in particular, if $\Omega \subset \mathfrak{L}$), then $\Delta_2' = 0$, and conversely. In that case, Δ' is the polar cycle of $\mathfrak{L}$, and in fact Δ' must then be a positive cycle, unless $\mathfrak{L} = \Omega$ (in the latter case L' consists only of the zero cycle).

Let $\mathfrak{L}$, L and Δ be as above and let α be any function on V, $\alpha \neq 0$. If $\mathfrak{L}^*$ denotes the set of all functions ξ^* of the form $\alpha\xi$, $\xi \in \mathfrak{L}$, then it is clear that our linear system L can also be defined by the module $\mathfrak{L}^*$: L coincides with the set of all cycles $(\xi^*) + \Delta^*$, $\xi^* \in \mathfrak{L}^*$, $\xi^* \neq 0$, where $\Delta^* = \Delta - (\alpha)$. Conversely, it is immediately seen that every Ω-module in $\Omega(V)$ which defines the given linear system L can be obtained from the module $\mathfrak{L}$ in the above indicated fashion. It follows that the projective structure of the linear system L is independent of the choice of the defining module of functions. We can therefore speak of linear subspaces of L, linearly independent cycles of L, etc. It is immediate that every linear subspace of L represents a linear system on V and that, conversely, any linear system M such that $M \subset L$ is represented by a linear subspace of L.

Of the various modules of functions which define the given linear system L, of particular importance are those which contain the universal domain Ω. Let $\mathfrak{L}$ be an arbitrary module of functions, which defines our linear system L, so that L is the set of cycles $(\xi) + \Delta$, $\xi \in \mathfrak{L}$, $\xi \neq 0$, and let Y_0 be a particular cycle of L: $Y_0 = (\xi_0) + \Delta$. In the above construction of the module $\mathfrak{L}^*$ we take for α the function $1/\xi_0$. Then $\mathfrak{L}^*$ contains Ω, and the auxiliary cycle Δ^* is now the cycle Y_0. The module $\mathfrak{L}^*$ consists (in addition to the function zero) of all functions ξ^* on V such that

$$(\xi^*) = Y - Y_0, \qquad Y \in L. \tag{1}$$

It follows that the module $\mathfrak{L}^*$ is uniquely determined by L and Y_0. Conversely, it is immediately seen that every defining module $\mathfrak{L}^*$ of L which is such that $\mathfrak{L}^* \supset \Omega$ is associated with some particular cycle Y_0 of L in the above indicated fashion. We shall denote by $L(Y_0)$ the module of functions associated with a given cycle Y_0 of L, i.e., the module of functions ξ^* satisfying (1).

Let L and M be two linear systems on V and let us assume that the two systems belong to one and the same linear equivalence class, i.e., that any cycle in L is equivalent to any cycle in M (for that it is sufficient that at least one cycle in L be equivalent to some cycle in M). We fix a cycle Y_0 in L and we consider the module $\mathfrak{M}$ consisting (in addition to zero) of the functions η such that $(\eta) = Z - Y_0$, $Z \in M$. It is clear that $\mathfrak{M}$ is a defining module of the linear system M. Let $\mathfrak{N}$ be the least Ω-module in $\Omega(V)$ spanned by the two modules $\mathfrak{M}$ and $L(Y_0)$, i.e., $\mathfrak{N}$ is the sum of the modules $\mathfrak{M}$ and $L(Y_0)$. Then $\mathfrak{N}$ has finite dimension over Ω. It is immediately seen that $(\beta) + Y_0 \geqq 0$ for any β in $\mathfrak{N}$, $\beta \neq 0$. Hence these cycles $(\beta) + Y_0$ form a linear system, say N. It is clear that L and M are subspaces of N. It is easy to see that N *is the least linear system containing both* L *and* M. For if N' is any linear system containing L and M, then the function module $N'(Y_0)$ obviously contains both modules $\mathfrak{M}$ and $L(Y_0)$ and hence must contain $\mathfrak{N}$. Therefore $N' \supset N$, as asserted.

It follows similarly that if we have any finite number of linear systems L_i, all belonging to the same linear equivalence class, there exists a unique least linear system N containing all the L_i.

Now let L and M be two arbitrary linear systems (not belonging necessarily to the same equivalence class). Let $\{Y_i\}$ and $\{Z_j\}$ be maximal sets of linearly independent cycles of L and M respectively ($i = 0, 1, \cdots, s; j = 0, 1, \cdots, t$). The $(s + 1)(t + 1)$ cycles $Y_i + Z_j$ are equivalent to each other, and there exists therefore a least linear system, say N, which contains them. For a fixed j, let N_j denote the least linear system containing the $s + 1$ cycles $Y_i + Z_j$. Then N_j is a subspace of N, and on the other hand N_j must also be contained in the linear system obtained from L by adding to each cycle of L the fixed cycle Z_j. It follows that Z_j is a fixed component of N_j. Let L' be the linear system obtained from N_j by deleting this fixed component Z_j. Then $L' \subset L$, and on the other hand L' contains the $s + 1$ independent cycles Y_i of L, and therefore L' *must contain* L (since L is obviously the least linear system containing the $s + 1$ cycles Y_i). It follows $L' = L$ and that N_j, and hence also N, contain all the cycles of the form $Y + Z_j$, $Y \in L$. By a similar argument, it follows now that N *contains all the cycles of the form* $Y + Z$, $Y \in L$, $Z \in M$, and, by construction, it is clear that N is the least linear system with this property. This linear system N is called *the minimal sum of the linear systems* L *and* M.

If $\mathfrak{L}$ and $\mathfrak{M}$ are function modules which define respectively the linear systems L and M and if $\mathfrak{N}$ denotes the Ω-module spanned by the products $\xi\eta$, $\xi \in L$, $\eta \in \mathfrak{M}$, then $\mathfrak{N}$ is a defining function module of the minimal sum N of L and M. Namely, if Δ and Δ' are fixed cycles such that L is the set of cycles $(\xi) + \Delta$ and M is the set of cycles $(\eta) + \Delta'(\xi \in \mathfrak{L}, \eta \in \mathfrak{M}, \xi\eta \neq 0)$, then N is the set of cycles $(\beta) + \Delta + \Delta'$, $\beta \in \mathfrak{N}$, $\beta \neq 0$.

In a similar fashion one defines the minimal sum of any finite number of linear systems L_i : it is the least linear system which contains all the cycles of the form $Y_1 + Y_2 + \cdots$, $Y_i \in L_i$. In particular, given any linear system L and any integer $m \geqq 1$, there exists a linear system which the minimal m-fold of L. This system will be denoted by mL.

A linear system L is *complete* if it is not contained in a larger linear system. It is clear from the preceding considerations that if L is a complete linear system and Y_0 is a cycle of L, then L coincides with the totality of non-negative cycles which are equivalent to Y_0. Hence a complete linear system is necessarily a full equivalence class of non-negative cycles. It follows that each linear system L is contained in *at most one* complete linear system. The proof that each linear system is always contained in a complete linear system depends on establishing the following assertion: if Y_0 is any cycle and L is the Ω-module in $\Omega(V)$ which consists (in addition to the function zero) of the functions ξ such that $(\xi) - Y_0 \geqq 0$, then L has finite dimension over Ω. The usual proof of this assertion, which proceeds by induction with respect to the dimension r of V, becomes somewhat elaborate when made rigorously. We shall give a non-inductive proof of this statement in Section 4.

We shall use the standard notation $|C|$ for the complete linear system which contains a given cycle C. We recall the well-known *residue theorem for linear system*: if L is a linear system and C_0 is any cycle, then the non-negative cycles D such that $C_0 + D \in L$ (the so-called *residues* of C_0 with respect to L) form a linear system (*denoted in the sequel by* $L - C_0$), and this system is complete if L is complete. The proof is straightforward.

We close this section with the proof of a result concerning linear systems on V, defined over a given algebraically closed ground field k.

A cycle Z is *rational* over k if all the prime components of Z are varieties defined over k. It is clear that if a function ξ on V is defined over k, then the cycle (ξ) is rational over k (provided, of course, that k is a field of definition of the variety V). The converse is also true and has been proved (for arbitrary ground fields) by A. Weil (see [5], Theorem 10, Corollary 1, pp. 240–241). For convenience of the reader we shall give here a proof of the converse, i.e., of the following lemma:

LEMMA 1. *Let Z be the cycle of a function on V and let k be an algebraically closed field of definition of V. Then if Z is rational over k, Z is the cycle of a function in $k(V)$.*

PROOF. We first observe that *any k-automorphism τ of Ω can be extended to a $k(V)$-automorphism of $\Omega(V)$, and the extension is unique.* The uniqueness of the extension is obvious. The existence of an extension is proved as follows. There certainly exists an extension of τ to a $k(Y)$-automorphism T of $\Omega(Y)$, where $(Y) = (Y_0, Y_1, \cdots, Y_n)$ is a set of indeterminates. If $\mathfrak{P}$ is the homogeneous prime ideal of V in the polynomial ring $\Omega[Y]$ and $\mathfrak{P}_k$ is the homogeneous prime ideal of V/k in $k[Y]$, then $\mathfrak{P}$ is the extended ideal of $\mathfrak{P}_k$ since k is algebraically closed. Hence T leaves invariant the ideal $\mathfrak{P}$ and therefore determines a $k(y)$-automorphism of $\Omega(y)$, where the y's are the $\mathfrak{P}$-residues of the Y's. It is clear that this automorphism is a $k(V)$-automorphism of $\Omega(V)$ which is an extension of τ. We shall denote by the same letter τ the $k(V)$-automorphism of $\Omega(V)$ which is an extension of τ.

By assumption, we have $Z = (\xi)$, where $\xi \in \Omega(V)$. We shall use τ as a superscript to denote τ-transforms of elements of $\Omega(V)$. Since Z is rational over k, it is clear $Z = (\xi^\tau)$ for any k-automorphism τ of Ω. Therefore $\xi^\tau = \rho_\tau \xi$, where $\rho_\tau \in \Omega$, $\rho_\tau \neq 0$. We can express ξ (in many ways) in the form

$$\xi = (\textstyle\sum_\alpha u_\alpha \xi_\alpha)/(\sum_\alpha v_\alpha \xi_\alpha), \tag{2}$$

where $u_\alpha, v_\alpha \in \Omega$, $\xi_\alpha \in k(V)$ and where *the ξ_α are linearly independent over k.* We assert that *among the various expressions* (2) *of ξ there is one in which the denominator belongs to $k(V)$.* To prove this, we fix among the expressions (2) of ξ one in which the set of v_α's has the least number h of non-zeros, $h \geqq 1$. Let, say, $v_1, v_2, \cdots, v_h \neq 0$, $v_{h+1} = v_{h+2} = \cdots = 0$. We may divide numerator and denominator by v_h and hence we may assume that $v_h = 1$. We have then

$$\xi(v_1\xi_1 + v_2\xi_2 + \cdots + \xi_h) = \textstyle\sum_\alpha u_\alpha \xi_\alpha . \tag{3}$$

Applying an arbitrary k-automorphism τ of Ω we find

$$\rho_\tau \xi(v_1^\tau \xi_1 + v_2^\tau \xi_2 + \cdots + \xi_h) = \textstyle\sum_\alpha u_\alpha^\tau \xi_\alpha . \tag{4}$$

Multiplying (3) by ρ_τ and subtracting from (4) we find

$$\rho_\tau \xi[(v_1 - v_1^\tau)\xi_1 + (v_2 - v_2^\tau)\xi_2 + \cdots + (v_{h-1} - v_{h-1}^\tau)\xi_{h-1}] = \textstyle\sum_\alpha w_\alpha \xi_\alpha , \tag{5}$$

where the w_α are elements of Ω. Since $\rho_\tau \neq 0$, the sum in the square brackets on the left hand side must be zero, for otherwise (5) would yield an expression for ξ, of the form (2), in which the denominator has less than h terms. It follows that $v_i = v_i^\tau$, $i = 1, 2, \cdots, h-1$, since the ξ_α, being linearly independent over k, are also linearly independent over Ω (the two fields $k(V)$ and Ω are linearly disjoint). Since τ is an arbitrary k-automorphism of Ω, we conclude that the v_α belong to k, and hence the denominator in (2) belongs to $k(V)$, as asserted.

Our result implies that ξ can be put in the form: $\xi = \sum_{\alpha=1}^{g} u_\alpha \eta_\alpha$, where $u_\alpha \in \Omega$, $\eta_\alpha \in k(V)$, and where we may assume that the η_α are linearly independent over k (and hence also over Ω). Applying any of our automorphisms τ, we find the relation

$$\sum_{\alpha=1}^{g} (\rho_\tau u_\alpha - u_\alpha^\tau)\eta_\alpha = 0,$$

and hence $\rho_\tau u_\alpha - u_\alpha^\tau = 0$, all α. Hence $u_\alpha/u_\beta = u_\alpha^\tau/u_\beta^\tau$, for all $\alpha, \beta = 1, 2, \cdots, g$ and for all k-automorphisms τ of Ω, and consequently the quotients u_α/u_β belong to k. Therefore ξ has the form $\rho\eta$, where $\rho \in \Omega$ and $\eta \in k(V)$, and this proves the lemma, since $Z = (\eta)$.

We say that a linear system L *is defined over* k if there exists a defining function module $\mathfrak{L}$ of L such that $\mathfrak{L}$ has an Ω-basis in $k(V)$. In view of Lemma 1, we have at once the following corollary:

COROLLARY. *Let L be a linear system, of dimension s, and let D_0 be the greatest fixed component of L. A necessary and sufficient condition that L be defined over a given algebraically closed field of definition k of V is that L contain $s+1$ independent cycles Z_i such that the cycles $Z_i - D_0$ are rational over k.*

3. The linear systems cut out on V by the hypersurfaces of a given order

We go back to the homogeneous coördinate ring $\Omega[y]$ of V, which we shall denote by R. For any integer $m \geq 1$ we denote by R_m the Ω-submodule of R consisting of those elements of R which are forms of degree m in the y's. We fix an element $f_0(y)$ in R_m, different from zero, and we denote by $\mathfrak{L}_m(f_0)$ the module of functions $f(y)/f_0(y)$, $f(y) \in R_m$. Then $\mathfrak{L}_m(f_0)$ contains Ω and is a finite-dimensional vector space over Ω. Let $Z(f_0)$ be the polar cycle of $\mathfrak{L}_m(f_0)$ and let L_m the linear system of cycles $(\xi) + Z(f_0)$, $\xi \in \mathfrak{L}_m(f_0)$, $\xi \neq 0$. We show now that the system L_m is independent of the choice of the element $f_0(y)$ in R_m and hence depends only on m.

Let $f_1(y)$ be any other element in R_m, different from zero, and let $\alpha = f_0(y)/f_1(y)$. Then $\mathfrak{L}_m(f_1)$ is the set of all products $\alpha\xi$, $\xi \in \mathfrak{L}_m(f_0)$. Since $(\xi) + Z(f_0) = (\alpha\xi) + Z(f_0) - (\alpha)$, we have only to prove that $Z(f_1) = Z(f_0) - (\alpha)$. By definition of $Z(f_1)$ we have $Z(f_1) \leq Z(f_0) - (\alpha)$. Suppose for a moment that we have $Z(f_1) < Z(f_0) - (\alpha)$. From the expression $(\xi^*) + Z(f_0) - (\alpha)$ of any cycle in L_m ($\xi^* \in \mathfrak{L}_m(f_1)$, $\xi^* \neq 0$) it would then follow that the positive cycle

$$Z(f_0) - (\alpha) - Z(f_1)$$

is a fixed component of L_m. This is impossible, since the module $\mathfrak{L}_m(f_0)$ contains the $n + 1$ functions $y_i^m/f_0(y)$ which certainly do not have any common null-cycle. This contradiction proves our assertion.

We have thus associated a non-negative cycle $Z(f)$ with every non-zero element f of R_m, and we have proved the relation

$$(6) \qquad Z(f_1) - Z(f_2) = (f_1/f_2),$$

where f_1 and f_2 are any two non-zero elements of R_m. It follows from this relation that if $\xi = f/f_0$ is any non-zero element of $L_m(f_0)$ then $(\xi) + Z(f_0) = Z(f)$. *Hence the linear system L_m coincides with the totality of cycles $Z(f)$, $f \in R_m$, $f \neq 0$.* We note that L_m has no fixed components.

Let $f(Y)$ be a form, of degree m, in the polynomial ring $\Omega[Y_0, Y_1, \cdots, Y_n]$ and let H_f be the hypersurface, or $(n - 1)$-cycle, in S_n, defined by the equation $f(Y) = 0$. Assume that H_f does not contain the variety V, or—equivalently—that $f(y) \neq 0$. Then the intersection of H_f with V is—set-theoretically—a pure $(r - 1)$-dimensional variety. Let Γ be any absolutely irreducible component of this variety. Fix an index i, $0 \leqq i \leqq n$, such that Γ does not lie in the hyperplane $Y_i = 0$. Then it is clear that $v_\Gamma(f(y)/y_i^m) > 0$. Since we have, by (6), $Z(f) = Z(y_i^m) + (f/y_i^m)$ (where f stands for $f(y)$), we see that the prime cycle Γ occurs in the cycle $Z(f)$ with a positive coefficient. This coefficient is the "intersection multiplicity" at Γ of H_f and V. Conversely, it is easily seen that each prime component of the cycle $Z(f)$ is an absolutely irreducible component of the variety intersection of H_f and V. The cycle $Z(f)$ now appears as the intersection cycle of H_f and V. We therefore refer to the linear system L_m as *the system cut out on V by the hypersurfaces of order m.* In particular, L_1 is the system of hyperplane sections of V.

If m and m' are integers $\geqq 1$, then $R_{m+m'}$, regarded as an Ω-module, is spanned by the products ff', $f \in R_m$, $f' \in R_{m'}$. It follows at once that $L_{m+m'}$ is the minimal sum of L_m and $L_{m'}$. It follows also that $L_m = mL_1$.

4. The deficiency of the linear system L_m

If k is a field of definition of V, we denote by R^k the canonical homogeneous coördinate ring $k[y]$ of V/k and by R_m^k the k-module of the forms of degree m in R^k. We denote by I and I^k the integral closure of R and R^k in $\Omega(y)$ and $k(y)$ respectively. Finally, we denote by I_m and I_m^k the set of elements of I and I^k respectively which are homogeneous of degree m (see our paper [6], p. 284). It is well known that I is a finite R-module and that similarly I^k is a finite R^k-module. It follows that I_m and I_m^k are finite modules over Ω and k respectively. Furthermore, R_m and R_m^k are submodules of I_m and I_m^k respectively. We point out that the quotient of any two elements of I_m is a function on V. Similarly, the quotient of any two elements of I_m^k is a function on V, defined over k.

THEOREM 1. *If D is any cycle which is linearly equivalent to the cycles of the linear system L_m, then there exists an element φ in I_m such that $D - Z(f) = (\varphi/f)$, where f is any element of R_m. Furthermore, if k is an algebraically closed field of defini-*

tion of V such that D is rational over k, then there already exists such an element φ in I_m^k.

PROOF. If f is any element of R_m, then, by our assumption, the cycle $D - Z(f)$ is the cycle of a function on V. Furthermore, if a field k satisfies the conditions stated in the theorem, then $D - Z(f)$ is the cycle of a function defined over k, provided $f \in R_m^k$ (see Lemma 1). We have therefore in particular: $D - Z(y_i^m) = (\xi_i)$, $i = 0, 1, \cdots, n$, where each of the $n + 1$ functions ξ_i is defined over k. Here it is permissible to assume that

$$\xi_i y_i^m = \xi_0 y_0^m. \tag{7}$$

Let us consider, in particular, the function ξ_0. This function has no polar cycles outside of the section of V with the hyperplane $Y_0 = 0$. Hence if we set $x_i = y_i/y_0$, $i = 1, 2, \cdots, n$, then ξ_0 (which is an element of $k(x)$) is integrally dependent on the ring $k[x]$. Since V/k is normal, the ring $k[x]$ is integrally closed, and hence ξ_0 belongs to $k[x]$. There exists therefore an integer s_0 such that

$$\xi_0 y_0^{s_0} \in R_{s_0}^k.$$

Similarly, there exists an integer s_i such that $\xi_i y_i^{s_i} \in R_{s_i}^k$. Let s be an integer greater than max $\{s_0, s_1, \cdots, s_n, m\}$. If we set $\varphi = \xi_0 y_0^m$, then $D - Z(y_0^m) = (\varphi/y_0^m)$, and hence, by (6): $D - Z(f) = (\varphi/f)$ for any element f in R_m, $f \neq 0$. On the other hand, from (7) and from the fact that the products $\xi_i y_i^s$ belong to R_s^k it follows that $y_i^{s-m}\varphi \in R_s^k \subset k[y]$, $i = 0, 1, \cdots, n$. This implies that φ is integrally dependent on $k[y]$, and hence $\varphi \in I_m^k$ since φ is homogeneous of degree m. This completes the proof of the theorem.

COROLLARY 1. *The linear system L_m is contained in a complete linear system $|L_m|$. If f_0 is any fixed element of R_m, then the system $|L_m|$ is defined by the module of functions φ/f_0, $\varphi \in I_m$, and consists of all the cycles $(\varphi/f_0) + Z(f_0)$, $\varphi \neq 0$.*

If a linear system L is contained in a complete linear system $|L|$, then the *deficiency* of L is defined as the difference between the dimensions of $|L|$ and L.

COROLLARY 2. *The deficiency of L_m is equal to the difference between the ranks of the two Ω-modules I_m and R_m. This deficiency is zero, and hence L_m is complete, if m is sufficiently large.*

The first part of the corollary follows directly from Corollary 1. The second part follows from the well known fact that for a normal variety V we have $I_m = R_m$ if m is sufficiently large.

As an application of the second part of Corollary 2, we shall now prove that every linear system is contained in a complete linear system. What we have to show is that given a cycle C_0, the set of all non-negative cycles which are linearly equivalent to C_0 is a linear system. In the proof of this assertion, we may assume that C_0 itself is a non-negative cycle.

If H_f is a hypersurface, defined by a homogeneous equation $f(Y) = 0$, and if H_f does not contain V, we say that H_f *passes through the cycle* C_0 if we have $Z(f) \geqq C_0$, where $Z(f)$ is the intersection cycle $H_f \cap V$. Since

$$H_{ff'} \cap V = H_f \cap V + H_{f'} \cap V$$

and since there exist hypersurfaces which pass through any given prime cycle on V (without containing V itself), it follows that for m sufficiently large there exist hypersurfaces of order m passing through the cycle C_0. We fix a hypersurface H_{f_0} of order m which passes through C_0 and we assume furthermore that m is sufficiently large so that the linear system L_m is complete. Let $Z(f_0) - C_0 = D_0$, where D_0 is, then, a non-negative cycle. The hypersurfaces of order m passing through the cycle D_0 cut out on V, outside of D_0, a linear system which contains C_0 and which is complete, by the residue theorem. This proves our assertion.

As another application of the preceding results we shall now prove the following theorem:

THEOREM 2. *Let k be an algebraically closed field of definition of V and let C_0 be a cycle on V. If C_0 is rational over k, then the complete linear system $| C_0 |$ is defined over k.*

PROOF. We write C_0 as a difference $C_0' - C_0''$ of two non-negative cycles, rational over k, and we fix a hypersurface H defined over k and of a sufficiently high order m, such that H passes through C_0' and that the linear system L_m is complete. Let $H \cap V = C_0' + D_0'$, where D_0' is, then, a non-negative cycle, rational over k. If we set $D_0 = D_0' + C_0''$, then D_0 is rational over k, and the complete linear system $| C_0 |$ is cut out on V by the hypersurfaces of order m passing through the non-negative cycle D_0. The proof of the theorem depends therefore on proving the following: *if $R_m(D_0)$ is the set of elements f of R_m such that $Z(f) \geqq D_0$* (we include 0 in this set), *then the Ω-module $R_m(D_0)$ has a basis in R_m^k.*

Let f be any element of $R_m(D_0)$. We can write f in the form $\sum u_\alpha f_\alpha$, where the f_α are in R_m^k and the u_α are linearly independent over k. The above assertion will be proved if we show that also the f_α necessarily belong to $R_m(D_0)$.

Let Γ be any prime component of D_0 and let s be the multiplicity to which Γ occurs in D_0. We may assume that Γ does not lie in the hyperplane $Y_0 = 0$, and we set $\xi = f/y_o^m$, $\xi_\alpha = f_\alpha/y_o^m$, whence $v_\Gamma(\xi) \geqq s$. Since Γ is rational over k, there exists in $k(V)$ a uniformizing parameter t of V at Γ, i.e., an element t of $k(V)$ such that $v_\Gamma(t) = 1$. Let $s' = \min \{v_\Gamma(\xi_\alpha)\}$. Then $v_\Gamma(\xi) \geqq s'$, since

$$\xi = \sum u_\alpha \xi_\alpha, \tag{8}$$

and hence the quotients $\xi/t^{s'}$, $\xi_\alpha/t^{s'}$ have finite residues in the valuation v_Γ. We denote these residues by $\bar{\xi}$ and $\bar{\xi}_\alpha$ respectively; they are elements of the function field $\Omega(\Gamma)$, and, in particular, the $\bar{\xi}_\alpha$ belong to $k(\Gamma)$ (since the f_α belong to R_m^k). Furthermore, *the $\bar{\xi}_\alpha$ are not all zero*, by the definition of s'. By (8), we have $\bar{\xi} = \sum u_\alpha \bar{\xi}_\alpha$. Since k is algebraically closed and the u_α are linearly independent over k, they are also linearly independent over $k(\Gamma)$. It follows that $\bar{\xi} \neq 0$. This implies that $v_\Gamma(\xi) = s'$. We have thus proved that $v_\Gamma(\xi_\alpha) \geqq s$, i.e., $Z(f_\alpha) \geqq s\Gamma$. Since this conclusion holds for every prime component Γ of D_0, it follows that $Z(f_\alpha) \geqq D_0$, i.e., the f_α belong to $R_m(D_0)$. This completes the proof of the theorem.

COROLLARY 1. *If a linear system L free from fixed components, is defined over a field of definition k of V (k, algebraically closed), then the complete linear system containing L is also defined over k.*

For, by Lemma 1, Corollary (section 2), L contains cycles which are rational over k.

COROLLARY 2. *The complete linear systems* $| L_m |$ *are defined over* k. *The deficiency of* L_m *is equal to the difference between the ranks of the* k*-modules* I_m^k *and* R_m^k.

The first part of the corollary follows directly from Corollary 1, since each linear system L_m is defined over k. Since L_m has no fixed components, also $| L_m |$ is free from fixed components. It follows by the Corollary of Lemma 1 that if ρ_m is the dimension of $| L_m |$, then there exist $\rho_m + 1$ independent cycles in $| L_m |$ which are rational over k. Consequently, by the second part of Theorem 1, I_m has an Ω-basis in I_m^k. This implies that the rank of the Ω-module I_m is the same as the rank of the k-module I_m^k. Since also the modules R_m and R_m^k have the same rank (over Ω and k respectively), the corollary follows from Theorem 1, Corollary 2.

5. Some properties of the general cycle of L_m/k

We shall denote cycles of L_m by C_m, and we shall attach superscripts or dashes to C_m if several cycles of L_m are dealt with simultaneously. *From now on we assume that the dimension* r *of* V *is at least* 2.

We fix an algebraically closed field of definition k of V. If a cycle C_m is cut out on V by a hypersurface $f(Y) = 0$, of order m, and if $a_0, a_1, a_2, \cdots$ are the coefficients of f, then it is clear that the cycle C_m is rational over the algebraic closure of the field $k(a)$. Since L_m is defined over k, it has the structure of a linear space over k, and we can speak of general cycles of L_m/k. If C_m is a general cycle of L_m/k, then C_m can be cut out by a hypersurface $f(Y) = 0$ such that the coefficients of f are algebraically independent over k.

The system L_m has no fixed components. On the other hand, L_m is not composite with a pencil.[5] Hence by the generalized theorem of Bertini (see [7], and also T. Matsusaka, The theorem of Bertini on linear systems in modular fields, Mem. Coll. Sci. Univ. Kyoto, Ser. A. Math. 26, pp. 51–62, 1950), any general cycle of L_m is prime, i.e., it represents an absolutely irreducible $(r - 1)$-dimensional subvariety of V (counted once). This conclusion does not depend on the assumption that V is absolutely normal.

THEOREM 3. *Any general cycle of* L_m/k *is an absolutely normal variety.*

PROOF.[6] Let C_m be a general cycle of L_m/k. As has been pointed out above, C_m is cut out on V by a hypersurface $f(Y) = 0$ such that the coefficients $u_0, u_1, \cdots$, of f are algebraically independent over k. Then C_m (regarded as a vari-

[5] For the purposes of this paper it will be sufficient to use the following purely algebraic criterion as a *definition*: if L is a linear system, of dimension ≥ 1, and $\mathfrak{L}$ is a defining function module of L such that $\mathfrak{L} \supset \Omega$, then L is composite with a pencil if either (a) L is of dimension 1 and the field generated by $\mathfrak{L}$ over Ω is not maximally algebraic in $\Omega(V)$ or (b) L is of dimension >1 and $\mathfrak{L}$ is contained in subfield of $\Omega(V)$ which has transcendence degree 1 over Ω.

[6] This theorem is of course included in the stronger theorem due to Seidenberg [3], to the effect that almost all members of L_m/k are normal varieties. However, we only need this result for the general cycle of L_m/k, and since the proof of this weaker result is very short we give it in the text.

ety) is defined and irreducible over $k(u)$. Let P be a general point of $C_m/k(u)$ and let $x_1, x_2, \cdots, x_n$ be non-homogeneous coördinates of P (we take as hyperplane at infinity any of the $n+1$ hyperplanes $Y_i = 0$). We have dim $P/k(u) = r - 1$, and hence if N denotes the number of coefficients u we have dim $k(P, u)/k = N + r - 1 = \dim k(P, u)/k(P) + \dim k(P)/k$. Now dim $k(P, u)/k(P) \leqq N - 1$ since the u's satisfy the linear relation $f(P) = 0$, and on the other hand dim $k(P)/k \leqq r$ since $P \in V$. Hence dim $k(P, u)/k(P) = N - 1$ and dim $k(P)/k = r$. We can therefore make the following two conclusions: (a) $N - 1$ of the quantities u_i, say $u_1, u_2, \cdots, u_{N-1}$, are algebraically independent over $k(P)$, while $u_0 \in k(P, u_1, u_2, \cdots, u_{N-1})$, in view of the relation $f(P) = 0$; (b) P is a general point of V/k. Since V/k is normal, it follows from (b) that the ring $k[x]$ is integrally closed. It follows then from (a) that also the ring $k[x, u_1, u_2, \cdots, u_{N-1}]$ is integrally closed. Since we can assume that u_0 is the coefficient of Y_i^m (where $Y_i = 0$ is the chosen hyperplane at infinity), it is permissible to assume that $u_0 \in k[x, u_1, u_2, \cdots]$, whence the ring $k[x, u]$ $(= k[x, u_0, u_1, \cdots])$ is integrally closed. It follows that also the non-homogeneous coördinate ring $k(u)[x]$ of $C_m/k(u)$ is integrally closed, since this ring is the quotient ring of $k[x, u]$ with respect to the multiplicative set $k[u]$. Since this conclusion holds for any choice of the hyperplane at infinity among the $n + 1$ planes $Y_i = 0$, we conclude that $C_m/k(u)$ is a normal variety.

To prove the absolute normality of C_m, it remains to prove, by a theorem due to A. Weil, that any point Q of C_m, of dimension $r - 2$ over $k(u)$, is absolutely simple for C_m (see [5], Appendix II, Proposition 4, p. 270). An easy dimensionality argument (similar to the one used above in the case of the general point P of $C_m/k(u)$) shows that dim $Q/k \geqq r - 1$. Since V is absolutely simple, it follows that Q is an absolutely simple point of V/k. Therefore, by the Jacobian criterion for absolutely simple points ([8], Theorem 13, p. 42) there exist $n - r$ polynomials $\varphi_i(X)$ in the non-homogeneous variables $X_1, X_2, \cdots, X_n$, with coefficients in k, such that the Jacobian matrix $\partial(\varphi_1, \varphi_2, \cdots, \varphi_{n-r})/\partial(X_1, X_2, \cdots X_n)$ has rank $n - r$ at Q. We may assume that

$$\partial(\varphi_1, \varphi_2, \cdots, \varphi_{n-r})/\partial(X_{r+1}, X_{r+2}, \cdots, X_n) \neq 0 \text{ at } Q. \tag{9}$$

At this stage we may achieve a simplification of the proof if we assume that $m = 1$. This assumption is permissible, for if $z_0, z_1, \cdots, z_n$ are homogeneous coördinates of a general point of V/k, we replace V by the birationally equivalent variety V'/k having as general point the point whose homogeneous coördinates are the various monomials in the z's, of degree m. The birational correspondence between V and V' is then regular (over k), and in this correspondence the variety C_m is transformed birationally and regularly into a general hyperplane section of V'/k. To prove the absolute normality of C_m it will then be sufficient to prove the absolute normality of a general hyperplane section of V'/k.

We assume therefore that $m = 1$, so that $f(Y) = u_0Y_0 + u_1Y_1 + \cdots + u_nY_n$. Let $F(X)$ be the corresponding dehomogenized polynomials:

$$F(X) = u_0 + u_1X_1 + \cdots + u_nX_n$$

(we assume that the point Q does not lie in the hyperplane $Y_0 = 0$ and we take this hyperplane as hyperplane at infinity). The $n - r + 1$ polynomials $\varphi_i(X)$, $F(X)$ are zero on C_1. To prove that Q is absolutely simple for C_1 it will be sufficient to prove that one of the r Jacobian determinants

(10) $\partial(\varphi_1, \varphi_2, \cdots, \varphi_{n-r}, F)/\partial(X_i, X_{r+1}, X_{r+2}, \cdots, X_n), \quad i = 1, 2, \cdots, r,$

is different from zero at Q. To prove this, we shall use again a dimension-theoretic argument. We have $\dim Q/k(u) = r - 2$, whence

$$\dim k(u, Q)/k = n + r - 1.$$

Since $\dim Q/k \leqq r$, we have $\dim k(u)/k(Q) \geqq n - 1$. Since $F(X)$ is zero at Q, u_0 belongs to the field $k(u_1, u_2, \cdots, u_n, Q)$. Therefore at least $n - 1$ of the quantities $u_1, u_2, \cdots, u_n$ must be algebraically independent over $k(Q)$. Since $r \geqq 2$, at least one of the r quantities $u_1, u_2, \cdots, u_r$ must be therefore transcendental over the field $k(Q, u_{r+1}, u_{r+2}, \cdots, u_n)$. If, say, u_1 is transcendental over this field, then it follows at once from (9) that the first of the determinants (10) ($i = 1$) is different from zero at Q. This completes the proof of the theorem.[7]

6. The C_m-trace of a linear system on V

As in the preceding section, let C_m be a general cycle of L_m/k, cut out on V by a hypersurface $H: f(Y) = 0$. Let Z be any $(r - 1)$-cycle on V no prime component of which coincides with C_m. Our variety V may have singular points. Nevertheless we are in position to speak of the *intersection cycle* $C_m \cdot Z$, this being a well defined $(r - 2)$-cycle on C_m. This is so for the following reasons:

The set-theoretic intersection of C_m with Z is the intersection of Z with the hypersurface H and hence is a pure $(r - 2)$-dimensional variety. The singular locus of V is a variety of dimension $\leqq r-2$. Since this singular locus is a variety *defined over* k and since H is a general hypersurface, over k, it follows that H meets the singular locus of V in a variety of dimension $< r-2$. Hence every absolutely irreducible subvariety of C_m, of dimension $r-2$, is a simple subvariety of V. This shows that if Γ is any prime $(r-2)$-cycle on C_m, the intersection multiplicity of Z and C_m at Γ (or along Γ) is a well defined non-negative integer $i(C_m, Z; \Gamma)$ which is positive if and only if Γ belongs to the variety of the cycle Z. We set $C_m \cdot Z = \sum i(C_m, Z; \Gamma)\, \Gamma$, the summation being extended to all the prime $(r-2)$-cycles Γ of C_m.

The machinery of the general intersection theory is not needed in the present case, since we are only concerned with $(r-1)$-cycles on V. The intersection theory in this case is just as straightforward as it is in the case of curves on an algebraic surface. We shall therefore produce here the definition of the intersection multiplicity $i(C_m, Z; \Gamma)$ under a form best suited for our present purposes.

We first consider the case in which Z is a prime cycle ($Z \neq C_m$). Since Γ is a

[7] This proof also establishes the following: *if Q is any point of C_m which is absolutely simple for V, then Q is also absolutely simple for C_m*. For we have at any rate $\dim Q/k(u) \geqq 0$, whence $\dim k(u)/k(Q) \geqq n - r + 1$, and also this implies that one of the quantities u_1, $u_2, \cdots, u_r$ is transcendental over $k(Q, u_{r+1}, u_{r+2}, \cdots, u_n)$.

simple subvariety of V, the prime cycle Z is locally, at Γ, complete intersection of V with a hypersurface H_1. (We assume, of course, that Γ lies on the variety of Z.) That means that Z is a simple component of the cycle $H_1 \cdot V$ and is the only component containing Γ. Then H_1 does not contain C_m (since Γ lies on C_m and $C_m \neq Z$), and hence $H_1 \cdot C_m$ is defined as an $(r-2)$-cycle on C_m. The cycle Γ is certainly a component of this $(r-2)$-cycle. *We define* $i(C_m, Z; \Gamma)$ *as the coefficient of* Γ *in* $H_1 \cdot C_m$.

We now show that this definition is independent of the choice of the hypersurface H_1. Let H_2 be another hypersurface such that Z is locally, at Γ, complete intersection of V with H_2. Without loss of generality, we may assume that H_1 and H_2 have the same order, since we may add to either hypersurface H_i any multiple of a hyperplane which does not contain the cycle Γ, without affecting the coefficient of Γ in $H_i \cdot C_m$. Let $f_i(Y) = 0$ be the equation of H_i. We consider the function ξ on V represented by the quotient $f_1(y)/f_2(y)$. The cycle (ξ) of this function is given by $Z(f_1) - Z(f_2)$, *and therefore no component of this cycle passes through* Γ. By the normality of V it follows that ξ *is a unit in the local ring of* Γ *on* V/k', where k' is any common field of definition of V, Γ and ξ. Since Γ lies on C_m, C_m is not a component of (ξ). In particular, C_m is not a polar cycle of ξ. Therefore ξ reduces module C_m to a well defined function $\bar{\xi}$ on C_m. We shall denote this function $\bar{\xi}$ by $\mathrm{Tr}_{C_m}\xi$ (*trace of* ξ *on* C_m). It is clear that $\bar{\xi}$ is a unit in the local ring of Γ on C_m/k'. *Therefore* Γ *is not a component of the cycle* $(\bar{\xi})$. If we denote by $\bar{y}_i$ the C_m-residue of y_i (whence $\Omega[\bar{y}]$ is the homogeneous coördinate ring of C_m), then $\bar{\xi} = f_1(\bar{y})/f_2(\bar{y}) = \bar{f}_1/\bar{f}_2$, where $\bar{f}_i = f_i(\bar{y})$. It follows that $(\bar{\xi}) = Z(\bar{f}_1) - Z(\bar{f}_2) = H_1 \cdot C_m - H_2 \cdot C_m$. Since Γ is not a component of $(\bar{\xi})$, we conclude that Γ occurs in $H_1 \cdot C_m$ and in $H_2 \cdot C_m$ with the same coefficient, and this proves that $i(C_m, Z; \Gamma)$ is independent of the choice of H_1.

Now if Z is any $(r-1)$-cycle on V, $Z = \sum a_\alpha Z_\alpha$, where the Z_α are prime cycles, and if no Z_α coincides with C_m, we define $i(C_m, Z; \Gamma)$ by linearity: $i(C_m, Z; \Gamma) = \sum_\alpha a_\alpha i(C_m, Z_\alpha; \Gamma)$.

From the preceding proof we can also derive very easily the following associative property of the intersection cycles under consideration: *If* G *is a hypersurface not containing* C_m, *then*

$$G \cdot C_m = (G \cdot V) \cdot C_m. \tag{11}$$

For let $G \cdot V = \sum a_\alpha Z_\alpha$, where the Z_α are prime $(r-1)$-cycles on V. Since G does not contain C_m, no Z_α coincides with C_m. Let Γ be an arbitrary prime $(r-2)$-cycle on C_m. For each α such that Γ belongs to the variety of Z_α let H_α be a hypersurface such that Z_α is complete intersection, locally at Γ, of V with H_α. If Γ does not belong to Z_α, we take for H_α the zero $(n-1)$-cycle in S_n. We set $H' = \sum a_\alpha H_\alpha$. It is clear that *the two cycles* $G \cdot V$ *and* $H' \cdot V$ *coincide locally at* Γ, i.e., they differ only in prime components which do not pass through Γ. Therefore, the reasoning developed in the above proof, in the case of the two cycles $H_1 \cdot C_m$ and $H_2 \cdot C_m$, is applicable also in the present case to the cycles

$G \cdot C_m$ and $H' \cdot C_m$, and leads to a similar conclusion, namely that Γ occurs with the same coefficient in these two cycles. We have $H' \cdot C_m = \sum a_\alpha (H_\alpha \cdot C_m)$, and hence the coefficient of Γ in $H' \cdot C_m$ is equal to $\sum a_\alpha i(C_m, Z_\alpha; \Gamma)$. By the distributive law, this integer is also equal to the coefficient of Γ in the cycle $(G \cdot V) \cdot C_m$. We have therefore proved that each prime $(r-2)$-cycle Γ on C_m occurs with the same coefficient in $G \cdot C_m$ and $(G \cdot V) \cdot C_m$, and this proves (11).

If η is a function on V such that C_m is not a component of the cycle (η), then the trace of η on C_m is defined and is different from zero. As an easy consequence of (11) we can now prove the following

LEMMA 2. *If η is a function on V such that C_m is not a component of the cycle (η), then*

$$C_m \cdot (\eta) = (\mathrm{Tr}_{C_m} \eta). \tag{12}$$

PROOF. Let $\eta = f_1(y)/f_2(y)$, where $f_1(Y)$ and $f_2(Y)$ are forms of like degree, and let H_i be the hypersurface $f_i(Y) = 0$, $i = 1, 2$. Then $(\eta) = H_1 \cdot V - H_2 \cdot V$. On the other hand, we have $\mathrm{Tr}_{C_m} \eta = f_1(\bar{y})/f_2(\bar{y})$, and hence $(\mathrm{Tr}_{C_m} \eta) = H_1 \cdot C_m - H_2 \cdot C_m$. The lemma now follows directly from (11).

COROLLARY. *Let Z_1 and Z_2 be $(r-1)$-cycles on V not containing C_m as component. Then if Z_1 and Z_2 are linearly equivalent on V, the cycles $C_m \cdot Z_1$ and $C_m \cdot Z_2$ are linearly equivalent on C_m.* Obvious.

Let now N be a linear system on V such that C_m is not a fixed component of N. We denote by $\mathrm{Tr}_{C_m} N$ (*trace of N on C_m*) the set $\bar{N}$ of cycles $Z \cdot C_m$, where Z is any cycle in N such that C_m is not a component of Z. Clearly, $\bar{N}$ is non-empty. We fix a cycle Z_0 in N such that C_m is not a component of Z_0 and we denote by $\mathfrak{N}$ the Ω-module of functions ξ on V such that $(\xi) + Z_0 \in N$, whence $\mathfrak{N}$ is a defining function module of the linear system N. Since C_m is not a component of Z_0, no function ξ in $\mathfrak{N}$ has C_m as polar cycle, and hence $\mathrm{Tr}_{C_m} \xi$ is defined for all ξ in $\mathfrak{N}$. It is clear that the set of all functions $\bar{\xi}$ in $\Omega(C_m)$ which are traces of elements ξ of $\mathfrak{N}$ form a finite Ω-module. We denote this module by $\bar{\mathfrak{N}}$ and we refer to $\bar{\mathfrak{N}}$ as *the trace of* $\mathfrak{N}$ *on* C_m : $\bar{\mathfrak{N}} = \mathrm{Tr}_{C_m} \mathfrak{N}$. Let ξ be any element of $\mathfrak{N}$ and let $\bar{\xi} = \mathrm{Tr}_{C_m} \xi$. We have $(\xi) = Z - Z_0$, $Z \in N$. If C_m is not a component of Z, then $\bar{\xi} \neq 0$, and by Lemma 2, we have $(\bar{\xi}) = C_m \cdot Z - C_m \cdot Z_0$. If C_m is a component of Z, then $\bar{\xi} = 0$. *It follows that* $\mathrm{Tr}_{C_m} N$ *is a linear system on* C_m *and that* $\mathrm{Tr}_{C_m} \mathfrak{N}$ *is a defining function module of this system.* Furthermore, if we denote by $\mathfrak{N}_0$ the submodule of $\mathfrak{N}$ consisting of the functions ξ having zero trace on C_m, then $\dim \bar{\mathfrak{N}} = \dim \mathfrak{N} - \dim \mathfrak{N}_0$. The linear subsystem of N which is defined by the submodule $\mathfrak{N}_0$ of $\mathfrak{N}$ is the set of cycles Z of N such that $Z \geqq C_m$. The dimension of this subsystem of N is the same as the dimension of $N - C_m$. We have

$$\dim \bar{\mathfrak{N}} = 1 + \dim \bar{N}, \quad \dim \mathfrak{N} = 1 + \dim N, \quad \dim \mathfrak{N}_0 = 1 + \dim (N - C_m).$$

Consequently

$$\dim \mathrm{Tr}_{C_m} N = \dim N - \dim (N - C_m) - 1. \tag{13}$$

PART II

THE GENERALIZED LEMMA OF ENRIQUES-SEVERI

7. A preliminary step of the proof

Using the notation introduced in Part I, we shall now re-state in precise form our generalization of the lemma of Enriques-Severi:

THEOREM 4. *Let N be a complete linear system $| D_0 |$ on V and let k be an algebraically closed field which is a common field of definition of V and D_0. There exists an integer $m(D_0)$, depending on D_0, such that for all $m \geqq m(D_0)$ the trace of N on a general member C_m of L_m/k coincides with the complete system $| C_m \cdot D_0 |$.*

It is important to point out that in this theorem we include the case in which N is an empty set, and in that case, it is to be understood that we mean by the trace of N the empty set. Our theorem includes therefore in part the following existence theorem: *if D_0 is an $(r - 1)$-cycle on V (not necessarily non-negative), then the existence of non-negative cycles on C_m which are linearly equivalent to $C_m \cdot D_0$ $(m \geqq m(D_0))$ implies the existence of non-negative cycles on V which are linearly equivalent to D_0.*

In this section we wish to show that if the above theorem holds for $N = L_i$, i sufficiently large, then the theorem holds for every cycle D_0. We assume therefore that for all large i there exists an integer $m(i)$ such that L_i (which we know to be complete if i is large) cuts out a complete linear system on a general member C_m of L_m/k if $m \geqq m(i)$. Here k can be an arbitrary algebraically closed field of definition of V.

Let, then, D_0 be an arbitrary $(r - 1)$-cycle on V, rational over k. There exists an integer s and a non-negative cycle E_0 such that the cycle $D_0 + E_0$ is non-negative and belongs to L_s. It follows that the linear system $L_i - D_0$ is non-empty if $i \geqq s$. If $i > s$ the system $L_i - D_0$ is not composite with a pencil, since it partially contains[8] the system L_1. *We show next that if i is sufficiently large then $L_i - D_0$ has no fixed components.* Let Δ_i be the greatest fixed component of $L_i - D_0$. It is clear that $\Delta_i \geqq \Delta_{i+1}$. Hence in order to prove that $\Delta_i = 0$ if i is sufficiently large, we have only to show the following: *if Γ is any prime $(r - 1)$-cycle on V then Γ is not a component of Δ_i if i is sufficiently large.* Now the system $L_s - D_0$ can be cut out on V by hypersurfaces, of a certain order h, passing through a certain non-negative cycle Z_0, and then $L_i - D_0$ is cut out by the hypersurfaces of order $h + i - s$ passing through Z_0 (we assume here that h is sufficiently large, so that L_m is complete if $m \geqq h$). If Γ is not a component of Z_0, there exist hypersurfaces which pass through Z_0 and do not pass through Γ. Hence in this case Γ is certainly not a component of Δ_i if i

[8] A complete linear system M is said to *partially contain* another linear system N if the (complete) linear system $M - N$ $(= | A - B |, A \in M, B \in N)$ exists. In that case M contains the minimal sum of N and $M - N$, and the manner in which a defining function module of this minimal sum can be obtained from defining function modules of N and $M - N$ (see section 2) shows at once that if N is not composite with a pencil then also M is not composite with a pencil (see footnote 5).

is sufficiently large. On the other hand, there exist hypersurfaces H passing through Z_0 such that no prime component of Z_0 is a component of $H \cdot V - Z_0$. This shows that no prime component of Z_0 can be a component of Δ_i if i is sufficiently large. This completes the proof of our assertion.

It follows that if i is sufficiently large then the linear system $L_i - D_0$ is irreducible. Now, k is a field of definition of this system (since D_0 is rational over k), and k is algebraically closed. Hence if Y is a general cycle of $L_i - D_0$, (over k), then Y is a prime cycle on V.

We fix an integer j such that $L_j - D_0$ is non-empty and irreducible and such that Theorem 4 holds for $N = L_i$, $i \geqq j$. We set $m(D_0) = \max \{m(j), j + 1\}$. Let m be any integer $\geqq m(D_0)$, and let C_m be a general cycle of L_m/k. *We shall prove that* $\mathrm{Tr}_{C_m} D_0$ *coincides with the complete linear system* $| C_m \cdot D_0 |$ *on* C_m.

Let C_m be cut out on V by the hypersurface H: $f(Y) = 0$, where we may assume that the coefficients $u_0, u_1, \cdots$ of the form f are algebraically independent over k. If we regard $L_j - D_0$ as a projective space over k, we can attach to each cycle in this linear system a set of homogeneous coördinates. We fix a cycle Y in $L_j - D_0$ whose homogeneous coördinates $v_0, v_1, \cdots$ are algebraically independent over $k(u_0, u_1, \cdots)$. Then Y is a general cycle of $L_j - D_0$, hence a prime cycle, and Y is defined over $k(v)$. Furthermore, our general cycle C_m of L_m/k is also a general cycle of L_m over $k(v)$.

We denote by L_{jm} the trace of L_j on C_m. Since L_j is complete and $m \geqq m(j)$, the linear system L_{jm} is complete. Since every cycle in L_j is equivalent to the cycle $Y + D_0$, the cycles of L_{jm} are equivalent to $C_m \cdot Y + C_m \cdot D_0$. Hence the complete linear system $| C_m \cdot D_0 |$ coincides with the system $L_{jm} - C_m \cdot Y$ (the system of residues of $C_m \cdot Y$ with respect to L_{jm}). If the latter system is empty, then there is nothing to prove. Suppose that $| C_m \cdot D_0 |$ is not empty, and let $\bar{D}$ be a cycle belonging to this system. *We have to prove that there exists a cycle* D *in* $| D_0 |$ *such that* $C_m \cdot D = \bar{D}$.

We have $\bar{D} + C_m \cdot Y \in L_{jm}$. Since L_{jm} is the trace of L_j on C_m, there exists a cycle C_j' in L_j such that $C_m \cdot C_j' = \bar{D} + C_m \cdot Y$. We have $C_j' = H' \cdot V$, where H' is a hypersurface of order j. Hence, by formula (11), section 6, we have

$$C_m \cdot H' = \bar{D} + C_m \cdot Y.$$

We now analyze this equality from a set-theoretic point of view. The prime cycle Y is an absolutely irreducible variety, defined over $k(v)$. The variety of the cycle $C_m \cdot Y$ is the intersection of Y with the hypersurfaces $f(Y) = 0$, and since the coefficients u_α of $f(Y)$ are algebraically independent over $k(v)$, it follows that the intersection is an absolutely irreducible variety W, of dimension $r - 2$ and order $m\nu$, where ν is the order of the variety Y. Now the above equality shows that W belongs to the hypersurface H', i.e., we have that W belongs to the intersection of H' with Y. Since H' is of order j *less than* m, we conclude that Y *is a subvariety of* H' (for otherwise the intersection of Y with H' would be a variety of dimension $r - 2$ and of order $\leqq j\nu < m\nu$). Therefore $H' \cdot V = Y + D$, where D is a non-negative cycle, and we have $D \in | D_0 |$, since $Y \in L_j - D_0$

and $Y + D \in L_j$. On the other hand, we also have $\bar{D} + C_m \cdot Y = C_m \cdot H' = C_m \cdot (D + Y) = C_m \cdot D + C_m \cdot Y$. Hence $\bar{D} = C_m \cdot D$. This completes the proof of the relation $\mathrm{Tr}_{C_m} \mid D_0 \mid = \mid C_m \cdot D_0 \mid$, for all $m \geqq m(D_0)$.

8. Proof of Theorem 4 in the case $N = L_s$

For any pair of integers i and m, the system cut out on C_m by L_i is also the system cut out on C_m by the hypersurfaces of order i. We shall denote this system by $L_i(C_m)$. According to the results of the preceding section, the proof of Theorem 4 is now reduced to the special case $N = \mid L_s \mid$, s a large integer. *We shall take s sufficiently large so that L_i is complete* for all $i \geqq s$. For the sake of clarity, we shall now re-state this special case of Theorem 4:

THEOREM 4′. *For any integer $s \geqq 1$, such that L_i is complete for all $i \geqq s$ there exists an integer $m(s)$ such that the linear system $L_s(C_m)$ is complete if $m \geqq m(s)$. Here C_m is a general cycle of L_m/k, and k is an arbitrary algebraically closed field of definition of V.*

Throughout this section the following will be fixed once and for always: (a) the integer s; (b) an integer m *greater than* s; (c) a general cycle C_m of L_m/k; (d) a cycle $\bar{D}$ of the complete system $\mid L_s(C_m) \mid$ containing $L_s(C_m)$.

For any integer $i \geqq 1$, the linear system $L_{s+i}(C_m) - \bar{D}$ (the system of residues of $\bar{D}$ with respect to $L_{s+1}(C_m)$) is contained in the complete system $\mid L_i(C_m) \mid$. We set

$$L_i^*(C_m) = L_i(C_m) \cap [L_{s+i}(C_m) - \bar{D}], \tag{14}$$

so that $L_i^*(C_m)$ is a linear subsystem of $L_i(C_m)$.

The cycle C_m is cut out on V by a hypersurface H_m, of order m:

$$H_m : f(Y) = 0, \tag{15}$$

and we may assume that the coefficients $u_0, u_1, \cdots$ of the form f are algebraically independent over k. We fix a hyperplane H_1:

$$H_1 : v_0 Y_0 + v_1 Y_1 + \cdots + v_n Y_n = 0 \tag{16}$$

such that $v_0, v_1, \cdots, v_n$ are algebraically independent over $k(u)$, and we denote by C_1 the cycle cut out on V by H_1. Then C_1 is a general cycle of $L_1/k(u)$, and C_m is also a general cycle of $L_m/k(v)$. For any $i \geqq 1$, we denote by $\delta_i(C_1)$ the deficiency of the linear system $L_i(C_1)$. Since C_1 is absolutely normal (Theorem 3, section 5), we have $\delta_i(C_1) = 0$ if i is large. Let m_0 be the least integer such that $\delta_i(C_1) = 0$ for all $i \geqq m_0$.

LEMMA 3. *For any integer $i \geqq m - s$ the following inequality holds:*

$$\begin{aligned} &\dim L_i(C_m) - \dim L_i^*(C_m) \\ &\qquad \leqq \dim L_{i+1}(C_m) - \dim L_{i+1}^*(C_m) + \delta_{s+i+1-m}(C_1). \end{aligned} \tag{17}$$

PROOF. We denote by $C_{m,1}$ the intersection of C_m with C_1, this intersection to be regarded an $(r - 2)$-cycle on the variety C_m. Since C_m is defined over $k(u)$ and C_1 is a general cycle of $L_1/k(u)$, $C_{m,1}$ is a prime cycle. We set

$$L_i'(C_m) = L_i(C_m) \cap [L_{i+1}^*(C_m) - C_{m,1}]. \tag{18}$$

Then $L'_i(C_m)$ is a linear subsystem of $L_i(C_m)$ since the linear system in the square brackets is contained in the complete linear system $| L_i(C_m) |$. By the definition of the system $L^*_{i+1}(C_m)$ [see (14)], the system $L^*_{i+1}(C_m) - C_{m,1}$ is the intersection of the two linear systems $L_{i+1}(C_m) - C_{m,1}$, $L_{s+i+1}(C_m) - C_{m,1} - \bar{D}$. The first of these two linear systems contains the system $L_i(C_m)$, since $C_{m,1}$ is the intersection of C_m with the hyperplane H_1. For a similar reason, the second contains the system $L_{s+i}(C_m) - \bar{D}$. Hence, by (14) the system $L^*_{i+1}(C_m) - C_{m,1}$ contains $L^*_i(C_m)$, and therefore, by (18):

$$(19) \qquad L_i(C_m) \supset L'_i(C_m) \supset L^*_i(C_m).$$

We denote by d the difference $\dim L_{i+1}(C_m) - \dim L^*_{i+1}(C_m)$. Any $d + 1$ cycles of $L_{i+1}(C_m)$ are then linearly dependent modulo $L^*_{i+1}(C_m)$. Now let

$$\{C^{\alpha}_{m,i}, \alpha = 0, 1, \cdots, d\}$$

be an arbitrary set of $d + 1$ cycles in $L_i(C_m)$. Then the cycles $C^{\alpha}_{m,i} + C_{m,1}$ belong to $L_{i+1}(C_m)$ and are therefore linearly dependent modulo $L^*_{i+1}(C_m)$. Since these cycles have $C_{m,1}$ as a common component, any cycle which is linearly dependent on them must also have $C_{m,1}$ as a component. There exists therefore a cycle in $L^*_{i+1}(C_m)$ which is of the form $C_{m,i} + C_{m,1}$, $C_{m,i} \geqq 0$, and such that $C_{m,i}$ is linearly dependent on the $d + 1$ cycles $C^{\alpha}_{m,i}$. We have $C_{m,i} \in L_i(C_m)$. On the other hand, we have $C_{m,i} \in L^*_{i+1}(C_m) - C_{m,1}$. Hence by (18), $C_{m,i} \in L'_i(C_m)$. The $d + 1$ cycles $C^{\alpha}_{m,i}$ are therefore linearly dependent modulo $L'_i(C_m)$. We have thus proved that $\dim L_i(C_m) - \dim L'_i(C_m) \leqq d$. Hence, by (19), the proof of the lemma will be complete if we prove the inequality

$$(20) \qquad \dim L'_i(C_m) - \dim L^*_i(C_m) \leqq \delta_{s+i+1-m}(C_1).$$

We set $\delta' = \delta_{s+i+1-m}(C_1)$ and we consider $\delta' + 1$ cycles $C^{\alpha}_{m,i}$ in $L'_i(C_m)$. For each α we fix a hypersurface H^{α}_i, of order i, such that $C^{\alpha}_{m,i} = H^{\alpha}_i \cap C_m$. By (18), each of the cycles $C^{\alpha}_{m,i} + C_{m,1}$ belongs to $L^*_{i+1}(C_m)$, and hence, by (14), each of the cycles

$$(21) \qquad C^{\alpha}_{m,s+i+1} = C^{\alpha}_{m,i} + C_{m,1} + \bar{D}, \qquad \alpha = 0, 1, \cdots, \delta'$$

belongs to $L_{s+i+1}(C_m)$. For each α we fix a hypersurface H^{α}_{s+i+1}, of order $s + i + 1$, such that $C^{\alpha}_{m,s+i+1} = H^{\alpha}_{s+i+1} \cap C_m$. By (21), each of these hypersurfaces contains the absolutely irreducible variety $C_{m,1}$. *We now regard* $C_{m,1}$ *as a cycle on* C_1. Set-theoretically, the variety of $C_{m,1}$ coincides with the intersection of C_1 with the hypersurface H_m given by (15). Since C_1 is defined over $k(v)$ and H_m is a general hypersurface, of order m, over $k(v)$, it follows that also the *cycles* $C_{m,1}$ and $H_m \cdot C_1$ (on C_1) coincide. We can therefore assert that each of the following $\delta' + 1$ cycles on C_1:

$$(22) \qquad (H^{\alpha}_{s+i+1} - H_m) \cdot C_1$$

is non-negative and therefore belongs to $| L_{s+i+1-m}(C_1) |$. Since δ' is the deficiency of $L_{s+i+1-m}(C_1)$, the cycles (22) must be linearly dependent modulo $L_{s+i+1-m}(C_1)$. There exists then a hypersurface $H_{s+i+1-m}$, or order $s + i + 1 -$

m, such that the cycle $H_{s+i+1-m} \cdot C_1$ (on C_1) is linearly dependent on the cycles (22). That is the same as saying that the cycle $(H_{s+i+1-m} + H_m) \cdot C_1$ is linearly dependent on the $\delta' + 1$ cycles $H^{\alpha}_{s+i+1} \cdot C_1$. This implies that there exists a hypersurface H_{s+i+1}, of order $s + i + 1$, such that

(a) H_{s+i+1} is linearly dependent on the $\delta' + 1$ hypersurfaces H^{α}_{s+i+1};

(b) the pencil determined by the two hypersurfaces H_{s+i+1} and $H_{s+i+1-m} + H_m$ contains a hypersurface H'_{s+i+1} passing through C_1.

Since H_m contains the variety of C_m, and since H'_{s+i+1} belongs to the pencil determined by H_{s+i+1} and $H_{s+i+1-m} + H_m$, it follows that $H'_{s+i+1} \cdot C_m = H_{s+i+1} \cdot C_m$. Hence, by (a), we can assert that

(c) *the cycle* $H'_{s+i+1} \cdot C_m$ *is linearly dependent on the cycles* $C^{\alpha}_{m,s+i+1}$.

On the other hand, we have by (b), that H'_{s+i+1} passes through C_1. Therefore the cycle $H'_{s+i+1} \cdot V - C_1$ is a member of the complete system $| L_{s+i} |$ on V. By our choice of the integer s, this complete system coincides with L_{s+i}, and therefore there exists a hypersurface H'_{s+i}, of order $s + i$, such that $H'_{s+i+1} \cdot V = H'_{s+i} \cdot V + C_1 = (H'_{s+i} + H_1) \cdot V$. Intersecting with C_m and using formula (11), section (6), we find: $H'_{s+i+1} \cdot C_m = H'_{s+i} \cdot C_m + C_{m,1}$. Hence, by (c) and (21), the cycle $H'_{s+i} \cdot C_m - \bar{D}$ is non-negative and *is linearly dependent on the cycles* $C^{\alpha}_{m,i}$. It therefore belongs to $L_i(C_m)$ [since $C^{\alpha}_{m,i} \in L'_i(C_m) \subset L_i(C_m)$]; on the other hand, it also belongs to the system $L_{s+i}(C_m) - \bar{D}$. *Hence, by* (14), *our cycle* $H'_{s+i} \cdot C_m - \bar{D}$ belongs to $L^*_i(C_m)$. We have thus proved that $\delta' + 1$ arbitrary cycles $C^{\alpha}_{m,i}$ in $L'_i(C_m)$ are linearly dependent modulo $L^*_i(C_m)$. This establishes the inequality (20) and completes the proof of the lemma.

Of fundamental importance for our proof of the generalized lemma of Enriques-Severi will be the following consequence of Lemma 3:

$$\dim L_{m-s}(C_m) - \dim L^*_{m-s}(C_m) \leqq \sum_{\nu=1}^{m_0-1} \delta_\nu(C_1). \tag{23}$$

This follows from Lemma 3 in view of the following facts: (a) $\delta_\nu(C_m) = 0$ if $\nu \geqq m_0$; (b) if i is sufficiently large, then $L_i(C_m)$ is complete, the system $L_{s+i}(C_m) - \bar{D}$ coincides with $L_i(C_m)$, *and hence* $L^*_i(C_m) = L_i(C_m)$.

9. Continuation of the proof

The system $L_{m-s}(C_m)$ is the trace, on C_m, of the linear system L_{m-s} on V. The subsystem $L^*_{m-s}(C_m)$ of $L_{m-s}(C_m)$ is then the trace of a subsystem L^*_{m-s} of L_{m-s}. Since no cycle in L_{m-s} contains C_m as a comopnent, we have by (13), section 6:

$$\dim L_{m-s}(C_m) = \dim L_{m-s}, \ \dim L^*_{m-s}(C_m) = \dim L^*_{m-s}.$$

Hence, by (23):

$$\dim L_{m-s} - \dim L^*_{m-s} \leqq \sum_{\nu=1}^{m_0-1} \delta_\nu(C_1). \tag{24}$$

Let us suppose for a moment that the linear system L^*_{m-s} *is irreducible.* We then fix a prime cycle C_{m-s} in L^*_{m-s} and a hypersurface H_{m-s}, of order $m - s$, such that $H_{m-s}\cdot V = C_{m-s}$. By the definition of $L^*_{m-s}(C_m)$, there exists a hypersurface H'_m, of order m, such that $H'_m\cdot C_m = C_{m-s}\cdot C_m + \bar{D}$. Let $f_{m-s}(Y) = 0$ and $f'_m(Y) = 0$ be equations of H_{m-s} and H'_m respectively. We denote by (y), $(\bar{y})$ and (z) the canonical general point of V, C_m and C_{m-s} respectively, whence $\Omega[y]$, $\Omega[\bar{y}]$ and $\Omega[z]$ are respectively the homogeneous coördinate rings of these three varieties.

Since C_m is absolutely normal and since $H'_m\cdot C_m \geqq H_{m-s}\cdot C_m$, an argument similar to the one used in the proof of Theorem 1, section 4, shows that if h is a sufficiently high integer then $f'_m(\bar{y})\bar{y}^h_i/f_{m-s}(\bar{y}) \in \Omega[\bar{y}]$, $i = 0, 1, \cdots, n$. Hence there exist forms $A_i(Y)$, of degree $s + h$, such that each of the $n + 1$ hypersurfaces $f'_m(Y)Y^h_i - A_i(Y)f_{m-s}(Y) = 0$ passes through C_m. The residual intersection (outside of C_m) of each of these hypersurfaces with V belongs to $| L_h |$, and if we take h sufficiently large these residual intersections belong to L_h and are therefore intersections of V with certain hypersurfaces $\varphi_i(Y) = 0$, of order h $(i = 0, 1, \cdots, n)$. It follows then that the $n + 1$ quotients

$$\frac{f'_m(y)y^h_i - A_i(y)f_{m-s}(y)}{f_m(y)\varphi_i(y)}$$

are constant functions on V. Since C_{m-s} is on V and since $f_{m-s}(z) = 0$, it follows that also the $n + 1$ quotients $f'_m(z)z^h_i/f_m(z)\varphi_i(z)$ are constant functions on C_{m-s}. Hence the products

$$\frac{f'_m(z)}{f_m(z)}\cdot z^h_i, \qquad i = 0, 1, \cdots, n,$$

belong to $\Omega[z]$, and this implies that the quotient $f'_m(z)/f_m(z)$ is integrally dependent on $\Omega[z]$. Since this quotient is a function on C_{m-s} [a homogeneous element of $\Omega(z)$, of degree zero] and since C_{m-s} is an absolutely irreducible variety [whence Ω is maximally algebraic in $\Omega(z)$], it follows that the quotient $f'_m(z)/f_m(z)$ is a constant function on C_{m-s}. There exists therefore an element c of Ω such that the hypersurface

$$H''_m: f'_m(Y) - cf_m(Y) = 0$$

passes through C_{m-s}. What we have proved is that *the pencil determined by H_m and H'_m contains a hypersurface H'' passing through C_{m-s}.*[9]

Since H_m contains the variety C_m, we have $H'_m\cdot C_m = H''_m\cdot C_m$, i.e.,

$$H''_m\cdot C_m = C_{m-s}\cdot C_m + \bar{D}. \tag{25}$$

[9] We wish to make a few comments about the above proof, and for the sake of simplicity we shall consider the case $r = 2$. We have two absolutely irreducible curves C_m and C_{m-s} on our surface V, intersections of V with hypersurfaces H_m and H_{m-s}, of order m and $m - s$ respectively. We also have another hypersurface H'_m, of order m, such that $H'_m\cdot C_m \geqq C_{m-s}\cdot C_m$ $(= H_{m-s}\cdot C_m)$. Or, in other words: the hypersurface H'_m is such that at every common point P of H_{m-s} and C_m the intersection multiplicity $i(H'_m, C_m; P)$ is not less than the intersection multiplicity $i(H_{m-s}, C_m; P)$ [= intersection multiplicity of C_{m-s} and C_m on V at P]. From this, by considering the various branches of C_{m-s} at P, it would not be difficult to derive the inequality $i(H'_m, C_{m-s}; P) \geqq i(H_m, C_{m-s}; P)$. Since the total number of intersections of H_m with C_{m-s} is the same as that of H'_m with C_{m-s} (theorem of Bezout), the above inequality

On the other hand, since H''_m passes through the cycle C_{m-s} on V, we have $H''_m \cdot V = C_{m-s} + C_s$, where C_s is a non-negative cycle and therefore belongs to L_s $(= | L_s |)$. Intersecting with C_m and comparing with (25) we find $\bar{D} = C_s \cdot C_m$, i.e., $\bar{D} \in L_s(C_m)$.

Our conclusion is therefore the following: *if for a given cycle* $\bar{D}$ *in* $| L_s(C_m) |$ *and for a given integer* $m > s$ *it happens that the linear system* L^*_{m-s} *is irreducible, then* $\bar{D} \in L_s(C_m)$. We shall now prove, however, that *there exists an integer* $m(s)$ *such that if* $m \geqq m(s)$ *and* $\bar{D}$ *is any cycle in* $L_s(C_m)$, *then the corresponding linear system* L^*_{m-s} *is irreducible.* The proof of this assertion will complete the proof of Theorem 4′ and hence also of Theorem 4. What plays the main role in the proof that follows is the fact that the right-hand side of the inequality (24) is a *constant* (an integer which depends only on V and is independent of m and s).[10]

We first observe that *if* Γ *is any prime* $(r - 1)$*-cycle in* S_n $(r - 1 < n)$, *then* Γ *imposes at least* $m(r - 1) + 1$ *independent linear conditions on the hypersurfaces of order* m *constrained to contain* Γ. For let k be a field such that the variety Γ is defined over k and let $\{P_i, 0 \leqq i \leqq m(r - 1)\}$ be a set of $m(r - 1) + 1$ k-independent general points of Γ/k. We shall show that given any of the points P_i, there exists a hypersurface of order m which does not pass through the given point and passes through the remaining $m(r - 1)$ points P_i.

We fix $r - 1$ of the points P_i, say $P_0, P_r, \cdots, P_{r-2}$, and we consider the linear system M of hyperplanes passing through these $r - 1$ points. Since $r - 1 < n$, M is non-empty (in fact, dim $M \geqq 1$). Since the base space of M is a linear space of dimension $\leqq r - 2$, it follows that Γ does not belong to the base space of M. Furthermore, the linear system M is defined over

$$k(P_0, P_1, \cdots, P_{r-2}).$$

Hence a general hyperplane of $M/k(P_0, P_1, \cdots, P_{r-2})$ does not contain Γ (since Γ is defined over k). We fix a general hyperplane π of

$$M/k(P_0, P_1, \cdots, P_{m(r-1)}).$$

shows that the two groups of points (or zero-cycles) $H_m \cdot C_{m-s}$ and $H'_m \cdot C_{m-s}$ coincide. If we now consider in the pencil determined by H_m and H'_m a hypersurface H''_m which passes through an additional point of C_{m-s}, *that hypersurface* H'' *will have to contain* C_{m-s}. This is essentially the method used by Enriques. It is clear that this method of intersection multiplicities is not extendable to our case of higher varieties, for C_{m-s} may have singularities and may even not be normal. Our proof, which is methodologically simpler than the proof of Enriques already in case $r = 2$, avoids these difficulties by a direct function-theoretic argument. Only the normality of C_m is essential in the proof. As to C_{m-s}, the proof makes use only of the absolute irreducibility of C_{m-s}.

[10] In the case of surfaces $(r = 2)$, Enriques uses the Riemann-Roch theorem on an algebraic curve and the postulation formula for surfaces, and in this fashion succeeds only in proving that the difference dim L_{m-s} − dim L^*_{m-s} has, for s fixed, at most the order of magnitude of a polynomial in m, of degree 1. Since dim L_{m-s} increases as a polynomial of degree 2 in m, it follows that L^*_{m-s} has *positive* dimension if m is large. However, this estimate does not suffice to insure the existence of *irreducible* curves in L^*_{m-s}. More generally, if we had used the method of Enriques in the general case, we would have only found that dim L_{m-s} − dim L^*_{m-s} increases at most as a polynomial of degree $r - 1$ in m.

In view of the k-independence of the points P_i, it is easily seen that there exists a field K of definition of π such that $K \supset k(P_0, P_1, \cdots, P_{r-2})$ and such that K and $k(P_0, P_1, \cdots, P_{m(r-1)})$ are free over $k(P_0, P_1, \cdots, P_{r-2})$. It follows that each of the points P_i, $i > r-2$, is a general point of Γ/K *and therefore none of these points belongs to* π (since $\Gamma \not\subset \pi$). If we divide the given set of points P_i into m disjoint sets of $r-1$ points each, leaving out one of the points P_i, and apply the above argument to each of these disjoint sets, we find a set of m hyperplanes π_ν such that the hypersurface $\cup\pi_\nu$, of order m, passes through all but one of the points P_i.

It follows that the points P_i impose $m(r-1)+1$ independent linear conditions on the hypersurfaces of order m. *A fortiori*, then, also the cycle Γ will impose at least that number of linear independent conditions on the hypersurfaces of order m.

From the above result we can draw at once the following conclusion: *if we set*

$$\delta = = \sum_{\nu=1}^{m_0-1} \delta_\nu(C_1), \tag{26}$$

then for any integer m such that

$$m \geqq s + \frac{\delta}{r-1} \tag{27}$$

and for any cycle $\bar{D}$ in $|L_s(C_m)|$, the system L^*_{m-s} has no fixed components. For were Γ a fixed component of L^*_{m-s}, where Γ is a prime $(r-1)$-cycle on V, then Γ would be a base variety of the linear system of hypersurfaces of order $m-s$ which intersect V in the cycles of L^*_{m-s}. The difference between the dimension of the linear system of all hypersurfaces of order $m-s$ and the dimension of the linear system of hypersurfaces of order $m-s$ which contain Γ is not less that $(m-s)(r-1)+1$, hence $\geqq \delta + 1$, by (27). It would then follow that $\dim L_{m-s} - \dim L^*_{m-s} \geqq \delta + 1$, in contradiction with (24).

We now fix an integer h such that

$$\dim L_h \geqq \delta. \tag{28}$$

We assert that if m is any integer such that

$$m \geqq s + h + 1 \tag{29}$$

and $\bar{D}$ is any cycle in $|L_s(C_m)|$, then the corresponding linear system L^*_{m-s} is not composite with a pencil. For let C_{m-s-h} be any cycle in L_{m-s-h} [note that $m-s-h \geqq 1$, by (29)]. The linear subsystem M of L_{m-s} obtained by adding to the cycles of L_h the fixed cycle C_{m-s-h} has dimension $\geqq \delta$, by (28). It follows then from (24) that M and L^*_{m-s} have a non-empty intersection. Since this is true for every cycle C_{m-s-h} in L_{m-s-h}, we see that the linear system L^*_{m-s} partially contains the linear system cut out on V by the hypersurfaces of (positive) order $m-s-h$. This implies that L^*_{m-s} cannot be composite with a pencil, as asserted.

We conclude now that if m satisfies both inequalities (27) and (29) [where h

is such that (28) holds], then for any cycle $\bar{D}$ in $|L_s(C_m)|$ the corresponding linear system L^*_{m-s} is irreducible. This completes the proof of the generalized lemma of Enriques-Severi. We note that the inequalities (27) and (29) assign a fixed lower bound to the difference $m - s$.

Part III

The Virtual Arithmetic Genus of a Cycle and Applications to Complete Linear Systems

10. A new expression for the characteristic function of Hilbert

Let $W = W^r$ be a pure r-dimensional variety in S_n, i.e., a variety all absolutely irreducible components of which are of dimension r. The Hilbert characteristic function of W is a polynomial $\chi(W, m)$ in m, with integral coefficients and of degree r, which for m sufficiently large represents the maximum number of linearly independent forms in $Y_0, Y_1, \cdots, Y_n$, with coefficients in Ω and of degree m, no linear combination of which is zero on W. Let $\mathfrak{A}$ be the homogeneous ideal of W in the polynomial ring $\Omega[Y_0, Y_r, \cdots, Y_n]$ and let R denote the residue class ring $\Omega[y_0, y_1, \cdots, y_n]$ of $\mathfrak{A}$, where y_i is the $\mathfrak{A}$-residue of Y_i. Then for m sufficiently large, $\chi(W, m)$ is equal to the dimension of the Ω-module R_m of the homogeneous elements in R, of degree m. In particular, if W is absolutely irreducible and absolutely normal, then the expression $\chi(W, m) - 1$ gives, for m sufficiently large, the dimension of the linear system L_m cut out on W by the hypersurfaces of order m.

The Hilbert characteristic function $\chi(W, m)$ has the form

$$\chi(W, m) = a_0\binom{m}{r} + a_1\binom{m}{r-1} + \cdots + a_{r-1}\binom{m}{1} + a_r, \tag{30}$$

where the a_i are integers. By analogy with the case of absolutely irreducible varieties studied in [2], we define the *virtual arithmetic genus* $p_a(W)$ *of* W as follows:

$$p_a(W) = (-1)^r(a_r - 1). \tag{31}$$

We note that in this definition we are referring directly to the universal domain, without specifying any particular field of definition of W. However, if k is any algebraically closed field of definition of W, then it is well known that the ideal $\mathfrak{A}$ is the extension of the homogeneous ideal of W in $k[Y]$. Therefore neither $\chi(W, m)$ nor $p_a(W)$ are affected if the polynomial ring $\Omega[Y]$ is replaced in the above definition by the ring $k[Y]$.

Let k be as above and let H be a general hyperplane over k. We set $W^{r-1} = W \cap H$. More generally, we shall denote by W^{r-i} the intersection of W with a general linear $(n - i)$-space in S_n/k. Then W^{r-i} is a pure $(r - i)$-dimensional variety, and W^{r-i-1} is the intersection of W^{r-i} with a hyperplane which is general with respect to a suitable algebraically closed field of definition of W^{r-i}. The following relation is well known (and is in fact used in the derivation of (30)):

$$\chi(W^r, m) - \chi(W^r, m - 1) = \chi(W^{r-1}, m). \tag{32}$$

If we replace in (32) the integer m successively by $m - 1, m - 2, \cdots, 1$ and note that $\chi(W^r, 0) = 1 + (-1)^r p_a(W)$, we find in a straightforward manner, using induction with respect to r, the following expression for $\chi(W^r, m)$ in terms of the virtual arithmetic genera $p_a(W^i)$, $i = 0, 1, \cdots, r$:

$$\chi(W^r, m) = \binom{m+r}{r} + \binom{m+r-1}{r} p_a(W^0) - \binom{m+r-2}{r-1} \cdot p_a(W^1) + \cdots + (-1)^{r-1}\binom{m}{1} p_a(W^{r-1}) + (-1)^r p_a(W^r). \tag{33}$$

For the proof it is only necessary to use the identity

$$\sum_{i=s}^{h} \binom{i}{s} = \binom{h+1}{s+1}$$

and to note that W^0 is a finite set of points, $\chi(W^0, m)$ is the constant $1 + p_a(W^0)$ (= number of points in W^0) and that consequently (33) holds if $r = 0$.

We shall now carry out a slight generalization of (33). Let s be an integer ≥ 1 and let $H_s^1, H_s^2, \cdots, H_s^r$ be r general and k-independent hypersurfaces in S_n/k, of order s. We denote by W_s^{r-i} the intersection $W \cap H_s^1 \cap H_s^2 \cap \cdots \cap H_s^i$. Then W_s^{r-i} is a pure $(r - i)$-dimensional variety, H_s^{i+1} is a hypersurface of order s which is general with respect to a suitable algebraically closed field of definition of W_s^{r-i}, and we have $W_s^{r-i-1} = W_s^{r-i} \cap H_s^{i+1}$. We have a relation similar to (32) if we replace $m - 1$ and W^{r-1} by $m - s$ and W_s^{r-1} respectively. In particular, if we replace in this new relation m by ms, we find

$$\chi(W^r, ms) - \chi(W^r, ms - s) = \chi(W_s^{r-1}, ms),$$

and from this one derives the following formula, similar to (33):

$$\chi(W^r, ms) = \binom{m+r}{r} + \binom{m+r-1}{r} p_a(W_s^0) - \binom{m+r-2}{r-1} \cdot p_a(W_s^1) + \cdots + (-1)^{r-1}\binom{m}{1} p_a(W_s^{r-1}) + (-1)^r p_a(W_s^r), \tag{34}$$

where $W_s^r = W^r = W$. In (34) we set $m = 1$ and then replace s by m. We then find the following expression of $\chi(W, m)$:

$$\chi(W, m) = r + 1 + p_a(W_m^0) - p_a(W_m^1) + \cdots + (-1)^{r-1} p_a(W_m^{r-1}) + (-1)^r p_a(W). \tag{35}$$

If W is our absolutely irreducible and absolutely normal variety V studied in Parts I and II, then $\chi(V, m) = 1 + \dim L_m$ for large m. Furthermore, V_m^{r-1} is now a general cycle C_m of L_m/k. We shall denote by C_m^i the intersection of i general and k-independent cycles of L_m/k. Then we have by (35):

$$\dim L_m = r + p_a(C_m^r) - p_a(C_m^{r-1}) + \cdots + (-1)^{r-1} p_a(C_m) + (-1)^r p_a(V). \tag{36}$$

Here for $i < r$, C_m^i is an absolutely irreducible and absolutely normal variety of dimension $r - i$, while C_m^r is a set of gm^r distinct points, where g is the order of V [and hence $p_a(C_m^r) = gm^r - 1$].

11. The virtual arithmetic genus of an $(r - 1)$-cycle on V

We shall now deal with $(r - 1)$-cycles on our absolutely irreducible and absolutely normal r-dimensional variety V.

LEMMA 4. *Let Z and Z' be positive $(r - 1)$-cycles on V having no multiple components. If Z and Z' are linearly equivalent and if we also denote by the same letters Z, Z' the varieties associated with these two cycles, then $\chi(Z, m) = \chi(Z', m)$ and hence $p_a(Z) = p_a(Z')$.*

PROOF. For large m, the linear system $L_m - Z$ exists and is cut out on V by the hypersurfaces of order m which pass through the *variety* Z (since Z has no multiple components). It follows at once that

$$\dim (L_m - Z) = \dim L_m - \chi(Z, m). \tag{37}$$

If m is sufficiently large, L_m is a complete linear system, and so is $L_m - Z$. Since Z and Z' are linearly equivalent cycles, we have $L_m - Z = L_m - Z'$ for m large, and hence, by (37), $\chi(Z, m) = \chi(Z', m)$ for large values of m. Since both sides of this equality are polynomials in m, the equality holds for all values of m, and the lemma is proved.

Let Z be a positive $(r - 1)$-cycle on V, free from multiple components, and let k be an algebraically closed field which is a common field of definition of V and of the prime cycles of Z. If C_m is a general cycle of L_m/k, then the intersection $Z \cdot C_m$ is a positive $(r - 2)$-cycle on C_m, free from multiple components, and hence $p_a(Z \cdot C_m)$ is well defined, this being the virtual arithmetic genus of the *variety* associated with the cycle $Z \cdot C_m$.

LEMMA 5. *Let Z and Z' be positive $(r - 1)$-cycles on V, free from multiple components, and let k be an algebraically closed field which is a common field of definition of V and of Z. Let C_m be a general cycle of L_m/k. Then if Z' and $Z + C_m$ are linearly equivalent, we have*

$$p_a(Z') = p_a(Z) + p_a(C_m) + p_a(Z \cdot C_m). \tag{38}$$

PROOF. Since C_m is not a component of Z, the cycle $Z + C_m$ is positive and free from multiple components. Since $Z' \equiv Z + C_m$, we have, by Lemma 4: $\chi(Z', i) = \chi(Z + C_m, i)$, for all i. Let $\mathfrak{p}$ and $\mathfrak{p}'$ be respectively the prime homogeneous ideals of Z and C_m in $\Omega[Y]$. The following relation (valid for any two homogeneous ideals $\mathfrak{p}$ and $\mathfrak{p}'$) is well known (and is in fact used in the derivation of the Hilbert characteristic function of a homogeneous ideal $\mathfrak{A}$):

$$\chi(\mathfrak{p}, i) + \chi(\mathfrak{p}', i) = \chi(\mathfrak{p} + \mathfrak{p}', i) + \chi(\mathfrak{p} \cap \mathfrak{p}', i). \tag{39}$$

We have $\chi(\mathfrak{p}, i) = \chi(Z, i)$, $\chi(\mathfrak{p}', i) = \chi(C_m, i)$, $\chi(\mathfrak{p} \cap \mathfrak{p}', i) = \chi(Z + C_m, i)$. On the other hand, by our choice of C_m it follows easily that $\mathfrak{p} + \mathfrak{p}'$ differs from the homogeneous ideal of the variety $Z \cdot C_m$ in $\Omega[Y]$ only by an irrelevant component. Hence we also have $\chi(\mathfrak{p} + \mathfrak{p}', i) = \chi(Z \cdot C_m, i)$. Substituting into (39) we find: $\chi(Z', i) = \chi(Z, i) + \chi(C_m, i) - \chi(Z \cdot C_m, i)$ [we have also used here the relation $\chi(Z', i) = \chi(Z + C_m, i)$]. Setting $i = 0$ and noting that Z, Z' and C_m are of dimension $r - 1$ while $Z \cdot C_m$ is of dimension $r - 2$, we find (38) in view of the definition (31) of the virtual arithmetic genus.

We shall now proceed to define the virtual arithmetic genus of any $(r - 1)$-cycle on V.

If $r = 1$, so that V is an algebraic curve, then if D is any 0-cycle on V, $D = \sum \alpha_i P_i$, where the α_i are integers and the P_i are points of V, we set $p_a(D) = \sum \alpha_i - 1$. We note that if the α_i are all equal to $+1$ then this definition coincides with the definition of the virtual arithmetic genus of a zero-dimensional variety. We also note that according to our definition, the virtual arithmetic genus of the null cycle on an algebraic curve is equal to -1.

We shall now proceed by induction with respect to r. We assume that for any integer s, $1 \leqq s < r$, and for any absolutely irreducible and absolutely normal variety V, of dimension s, we have already a definition of the virtual arithmetic genus $p_a(D)$ of any $(s - 1)$-cycle on V. We assume furthermore that the following conditions are satisfied:

I_s. If $D \equiv D'$, then $p_a(D) = p_a(D')$.

II_s. The virtual arithmetic genus of the null cycle (of dimension $s - 1$) is $(-1)^s$.

III_s. If k is an algebraically closed field which is a common field of definition of V and of the prime components of D and if C_m is a general cycle of L_m/k, then

$$p_a(D + C_m) = p_a(D) + p_a(C_m) + p_a(D \cdot C_m), \tag{40}$$

where $(D \cdot C_m)$ is to be regarded as a cycle on C_m.

IV_s. If V is transformed into an isomorphic variety V^τ by an automorphism τ of the universal domain Ω and if D^τ is the τ-transform of an $(s - 1)$-cycle D on V, then $p_a(D) = p_a(D^\tau)$.

These conditions are satisfied if $s = 1$. This is obvious in the case of the conditions I, II, and IV. As to III, it is clear that we have $p_a(D + C_m) = p_a(D) + p_a(C_m) + 1$, so that (40) is satisfied if we define $p_a(D \cdot C_m)$ to be $+1$. This is formally in agreement with II in the case $s = 0$, for C_m is now of dimension 0 and $D \cdot C_m$ can be regarded as a null cycle of dimension -1.

Now, let V be of dimension r and let D be any $(r - 1)$-cycle on V. Let $j = j(D)$ be the least non-negative integer such that the complete linear system $| D + L_j |$ exists and contains positive cycles without multiple components (see section 7). If $j = 0$ then we define $p_a(D)$ as follows: $p_a(D) = p_a(Z)$, where Z is any positive cycle in $| D |$, free from multiple components, and where $p_a(Z)$ is the virtual arithmetic genus of the *variety* Z. By Lemma 4, this definition is independent of the choice of the above cycle Z in $| D |$. If $j(D) > 0$, we proceed by induction with respect to $j(D)$. Assume that $p_a(D)$ has already been defined for all $(r - 1)$-cycles D such that $j(D)$ is less than a given positive integer h, and let D be an $(r - 1)$-cycle on V such that $j(D) = h$. Let k be an algebraically closed field which is a common field of definition of V and of the prime components of D and let C be a general cycle of L_1/k. Since $h > 0$ we have $j(D + C) = j(D) - 1 = h - 1$, and hence $p_a(D + C)$ is already defined. We now define $p_a(D)$ as follows:

$$p_a(D) = p_a(D + C) - p_a(C) - p_a(D \cdot C), \tag{41}$$

where $D \cdot C$ is to be regarded as an $(r - 2)$-cycle on the (absolutely irreducible and absolutely normal) variety C.

We shall now show that our definition of $p_a(D)$ is independent of the choice of k and C (if $h > 0$) and that conditions I_r — IV_r are satisfied. The proof will be by induction with respect to h (except, of course, for the proof of II_r).

We first consider the case $h = 0$. We have already observed above that in this case $p_a(D)$ is uniquely determined in view of the definition and Lemma 4, and by the same token also I_r is satisfied. Also IV_r is self-evident in this case. As to III_r, we first observe that since D is rational over k the linear system $| D |$ is defined over k (Theorem 2, section 4). Since $| D |$ contains positive cycles free from multiple components, the cycles in $| D |$ which have multiple components are contained in a proper algebraic subsystem of $| D |/k$.[11] Therefore, $| D |$ contains positive cycles which are defined over k and have no multiple components. Let Z be such a cycle. Then also $Z + C_m$ is a positive cycle free from multiple components (since C_m is not defined over k, whence C_m is not a component of Z). We have therefore: $p_a(D) = p_a(Z)$, $p_a(D + C_m) = p_a(Z + C_m)$, and now (40) follows from Lemma 5 and from I_{r-1}, since $D \cdot C_m \equiv Z \cdot C_m$.

Now let $j(D) = h > 0$. We first show the independence of $p_a(D)$ on the choice of k and C. We keep temporarily k fixed and we consider another general cycle C' of L_1/k. There exists then a k-automorphism τ of the universal domain Ω such that $C' = C^\tau$. This automorphism leaves D invariant (since D is rational over k) and therefore transforms $D \cdot C$ into $D \cdot C'$. Hence, by IV_{r-1}, we have $p_a(D \cdot C) = p_a(D \cdot C')$. On the other hand, since $j(D + C) = j(D + C') = h - 1$, it follows from our induction hypothesis, that $p_a(D + C) = p_a(D + C')$. (since $D + C \equiv D + C'$). Since $p_a(C) = p_a(C')$, we conclude that for a fixed choice of k the virtual arithmetic genus $p_a(D)$ is independent of the choice of C. If now k' is another algebraically closed field which is a common field of definition of V and of the prime components of D, the proof is completed by choosing C to be a general cycle of L_1/k *and* of L_1/k'.

If $D \equiv D'$, we take for our algebraically closed field k a field which is a common field of definition of V, D *and* of D'. Then in (41) we can use the same cycle C for both D and D'. We have by $I_{r-1} : p_a(D \cdot C) = p_a(D' \cdot C)$ (since $D \cdot C \equiv D' \cdot C$), and by our induction hypotheses, as applied to I_r, we have $p_a(D + C) = p_a(D' + C)$. Hence $p_a(D) = p_a(D')$, which establishes I_r.

[11] Perhaps the simplest way of proving this is to consider the associated form of any cycle in $| D |$ and to observe that a cycle D has multiple components if and only if its associated form has multiple factors (over Ω).

We next prove III_r. The cycle C_m is defined over the field $k(u_0, u_1, \cdots)$, where $u_0, u_1, \cdots$ are certain indeterminates. We denote by k' the algebraic closure of the field $k(u_0, u_1, \cdots)$ and we choose a general cycle C of L_1/k'. Then C is cut out on V by a hyperplane $v_0Y_0 + v_1Y_1 + \cdots + v_nY_n = 0$, where the v's are algebraically independent over k'. We have that the variety $C_m \cdot C$ is a general hyperplane section of C_m/k', and this same variety is also the section of C with a hypersurface of order m which is general with respect to the algebraic closure of the field $k(v_0, v_1, \cdots, v_n)$. The variety $C_m \cdot C$ is absolutely irreducible and absolutely normal.

We set $D' = D + C$. Then D' is rational over $k(v)$ and we have $j(D') = h - 1$. Hence by our induction hypothesis we have:

$$p_a(D' + C_m) = p_a(D') + p_a(C_m) + p_a(D' \cdot C_m). \tag{42}$$

On the other hand, the cycle $D + C_m$ is rational over $k(u)$ and we have $D' + C_m = (D + C_m) + C$. Hence upon replacing D by $D + C_m$ in (41) we find:

$$p_a(D' + C_m) = p_a(D + C_m) + p_a(C) + p_a(D \cdot C + C_m \cdot C). \tag{43}$$

Equating the two expressions (42) and (43) of $p_a(D' + C_m)$ we find:

$$p_a(D + C_m) = p_a(D + C) - p_a(C) + p_a(C_m) + p_a(D \cdot C_m + C \cdot C_m) - p_a(D \cdot C + C_m \cdot C),$$

or, by (41):

$$p_a(D + C_m) = p_a(D) + p_a(D \cdot C) + p_a(C_m) + p_a(D \cdot C_m + C \cdot C_m) - p_a(D \cdot C + C_m \cdot C). \tag{44}$$

Applying III_{r-1} to the varieties C_m and C respectively, we have:

$$p_a(D \cdot C_m + C \cdot C_m) = p_a(D \cdot C_m) + p_a(C \cdot C_m) + p_a(D \cdot C \cdot C_m),$$

$$p_a(D \cdot C + C_m \cdot C) = p_a(D \cdot C) + p_a(C_m \cdot C) + p_a(D \cdot C_m \cdot C).$$

In the first of these two equations, $C \cdot C_m$ is to be regarded as a cycle on C_m, while in the second equation $C_m \cdot C$ is a cycle on C. However, since both these cycles are prime, their virtual arithmetic genera coincide with the virtual arithmetic genus of the variety $C_m \cap C$. Hence $p_a(C \cdot C_m) = p_a(C_m \cdot C)$. The cycle $D \cdot C \cdot C_m$ in the first equation has the following meaning: it is the intersection, *on* C_m, of the two cycles $D \cdot C_m$ and $C \cdot C_m$, and it is a cycle *on the* (absolutely irreducible and absolutely normal) *variety* $C \cap C_m$. By the associative property of the intersection symbol [see (11) section 6, where V and C_m have now to be replaced by C_m and $C_m \cdot C$ respectively] we have therefore: $D \cdot C \cdot C_m = G \cdot V'$, where G is a hypersurface which cuts out on V the cycle D and where V' denotes the *variety* $C_m \cap C$. In a similar fashion we find that $D \cdot C_m \cdot C = G \cdot V'$, whence $p_a(D \cdot C \cdot C_m) = p_a(D \cdot C_m \cdot C)$. It follows that $p_a(D \cdot C_m + C \cdot C_m) - p_a(D \cdot C + C_m \cdot C) = p_a(D \cdot C_m) - p_a(D \cdot C)$. Substituting into (44) we find (40).

The proof of IV_r is straightforward.

To Prove II_r we apply III_r to the cycle $D = 0$. Then also $D \cdot C_m$ is the null cycle on C_m, and we find that II_r follows directly from II_{r-1}.

12. The dimension of $|D + C_m|$ for large values of m

If we apply (40) in the case of a cycle D which is linearly equivalent to $-C_m$ then $p_a(D + C_m) = (-1)^r$ and $p_a(D \cdot C_m) = p_a(-C_m^2)$, and hence (40) yields the following expression of $p_a(-C_m)$:

(45) $$p_a(-C_m) = (-1)^r - p_a(C_m) - p_a(-C_m^2),$$

where C_m^2 denotes the intersection of two general and k-independent cycles of L_m/k. More generally, if C_m^i denotes the intersection of i general and k-independent cycles of L_m/k, then applying (45) to the absolutely normal variety C_m^{i-1} instead of to V, we find:

(45_i) $$p_a(-C_m^i) = (-1)^{r-i+1} - p_a(C_m^i) - p_a(-C_m^{i-1}),$$

where i can take the values $1, 2, \cdots, r$. [Note that C_m^{r+1} is the null cycle (of dimension -1) on the zero dimensional variety C_m^r, and as previously agreed we have $p_a(-C_m^{r+1}) = +1$]. Multiplying (45_i) by $(-1)^{i-1}$ and adding for $i = 1, 2, \cdots, r$, we find

(46) $$p_a(-C_m) = (-1)^r(r + 1) + \sum_{i=1}^r (-1)^i p_a(C_m^i).$$

This expression of $p_a(-C_m)$, when substituted into (36), allows us to put the expression for dim L_m, m large, in the following compact form:

(47) $$\dim L_m = (-1)^r\{p_a(V) + p_a(-C_m)\} - 1.$$

We shall now derive the following generalization of (47):

THEOREM 5. *If D is any $(r - 1)$-cycle on V then for all sufficiently large values of m the dimension of the complete linear system $|D + C_m|$ is given by the following formula*:

(48) $$\dim |D + C_m| = (-1)^r\{p_a(V) + p_a(-D - C_m)\} - 1.$$

PROOF. We fix an integer s such that the linear system $|C_s - D|$ exists and is irreducible, and we fix a prime cycle E in this system. We have then $|D + C_m| = |C_{m+s} - E|$, and hence in order to prove the theorem we have only to prove that

(49) $$\dim |C_m - E| = (-1)^r\{p_a(V) + p_a(E - C_m)\} - 1$$

for m sufficiently large. Now for large m, the complete linear system $|C_m - E|$ is cut out on V by the hypersurfaces of order m passing through the prime cycle E. Hence we have for large m:

$$\dim |C_m - E| = \chi(V, m) - \chi(E, m) - 1,$$

or by (35):

(50) $$\dim |C_m - E| = (-1)^r p_a(V) + \sum_{i=1}^r (-1)^{r-i}[p_a(C_m^i) - p_a(E \cdot C_m^{i-1})].$$

(In this formula, the C_m^i must be chosen with reference to an algebraically closed ground field k which is a common field of definition of V and E). By $E \cdot C_m^0$ we intend the cycle E itself.

We now apply (40) under the following conditions: (a) V is the variety C_m^{i-1}, $i = 1, 2, \cdots, r$, where $C_m^0 = V$; (b) D is the cycle $E \cdot C_m^{i-1} - C_m^i$. Naturally, the cycle C_m in (40) must then be replaced by C_m^i, and we find

$$(51_i) \qquad p_a(E \cdot C_m^{i-1} - C_m^i) = p_a(E \cdot C_m^{i-1}) - p_a(C_m^i) - p_a(E \cdot C_m^i - C_m^{i+1}), \qquad i = 1, 2, \cdots, r$$

This formula is similar to (45_i), and also here we must set $p_a(E \cdot C_m^r - C_m^{r+1}) = +1$. Multiplying (51_i) by $(-1)^{i-1}$ and adding the resulting r relations, we find:

$$(52) \qquad p_a(E - C_m) = \sum_{i=1}^r (-1)^i [p_a(C_m^i) - p_a(E \cdot C_m^{i-1})],$$

and (49) now follows from (50) and (52).

13. The canonical system and adjoint systems

In the function field $\Omega(V)$ we have the familiar concept of an r-fold differential $Ad(x)$, where A is a function on V, i.e., $A \in \Omega(V)$, and $d(x)$ stands for

$$dx_1 dx_2 \cdots dx_r ,$$

the r quantities $x_1, x_2, \cdots, x_r$ being functions on V which form a separating transcendence basis of $\Omega(V)/\Omega$. If (x') is another separating basis of $\Omega(V)/\Omega$, then $Ad(x) = AJd(x')$, where J is the Jacobian determinant $\partial(x)/\partial(x')$.

Let Γ be a prime $(r - 1)$-cycle on V and let ω be an r-fold differential on V, different from zero. Since V is absolutely normal, Γ is absolutely simple for V. Hence there exist uniformizing coördinates $x_1, x_2, \cdots, x_r$ of Γ such that $\Omega(\Gamma)'$ is a separable algebraic extension of the field generated over Ω by the Γ-traces of the x_i (see [8], Theorem 14, p. 46). [We shall describe this condition by referring to $(x_1, x_2, \cdots, x_r)$ as a set of *separating uniformizing coördinates of* Γ.] Then the x_i form necessarily a separating transcendence basis of $\Omega(V)/\Omega$ (see the proof of Theorem 14, loc. cit.), and we can therefore write ω in the form $Ad(x)$, $A \in \Omega(V)$. From the above separability properties of the chosen uniformizing coördinates x_i it follows easily that if ξ is any function on V which has finite trace on Γ, then also the partial derivatives $\partial\xi/\partial x_i$ have finite traces on Γ. Now let (x') be any other set of separating uniformizing coördinates of Γ, and let $\omega = A'd(x')$. By the remark just made, both Jacobian determinants $J = \partial(x)/\partial(x')$ and $J^{-1} = \partial(x')/\partial(x)$ have finite traces on Γ. Consequently $v_\Gamma(J) = 0$ and $v_\Gamma(A) = v_\Gamma(A')$. The integer $v_\Gamma(A)$ is therefore independent of the choice of the separating uniformizing coördinates x_i. This integer is called *the order of* ω *at* Γ, and we shall denote it by $v_\Gamma(\omega)$. The cycle Γ is a *null cycle* or a *polar cycle* of ω according as $v_\Gamma(\omega)$ is positive or negative.

For later purposes we shall need the following lemma:

LEMMA 6. *Let* Δ *be an absolutely irreducible simple subvariety of* V, *of dimen-*

sion s $(0 \leqq s \leqq r - 1)$, and let $z_1, z_2, \cdots, z_r$ be separating uniformizing coördinates of Δ on V. Then the z's are also separating uniformizing coördinates of any subvariety Γ of V containing Δ.

PROOF. This lemma is implicitly contained in the proof of Theorem 14 of the quoted paper [8]. Let $(x) = (x_1, x_0, \cdots, x_n)$ be the non-homogeneous canonical general point of V. From the first part of the proof of Theorem 14 ([8], p. 46) it follows that there exist n polynomials F_i in $\Omega[X, Z]$ $(= \Omega[X_1, X_2, \cdots, X_n, Z_1, Z_2, \cdots, Z_r]$ such that $F_i(x, z) = 0$ and such that the Jacobian determinant $\partial(F_1, F_2, \cdots, F_n)/\partial(X_1, X_2, \cdots, X_n)$ is different from zero at Δ. This determinant is then also different from zero at Γ (since $\Gamma \supset \Delta$), and hence it follows from the second part of the proof of Theorem 14 ([8], p. 47) that the z's are also separating uniformizing coördinates of Γ on V.

COROLLARY. *The notation being the same as in Lemma 6, let $\omega = Ad(z)$ be an r-fold differential on V. Then for any prime $(r - 1)$-cycle Γ on V such that $\Gamma \supset \Delta$ we have $v_\Gamma(\omega) = v_\Gamma(A)$.*

Obvious.

Let $x_1, x_2, \cdots, x_n$ be, as above, the non-homogeneous coördinates of the canonical general point of V and let $f_i(X)$, $1 \leqq i \leqq N$ be a basis of the prime ideal of V in the polynomial ring $\Omega[X_1, X_2, \cdots, X_n]$. Without loss of generality we may assume that $x_1, x_2, \cdots, x_r$ form a separating basis of $\Omega(V)/\Omega$. Then the Jacobian matrix $\partial(f_1, f_2, \cdots, f_N)/\partial(X_1, X_2, \cdots, X_r)$ is of rank $n - r$ on V, and hence there is only a finite number of prime $(r - 1)$-cycles on V (at finite distance) at which the above Jacobian matrix is of rank $<n - r$. It follows then by known results (see [8], proof of Theorem 14, p. 47) that outside of a finite set of prime $(r - 1)$-cycles on V (including possibly the cycles at infinity) every prime $(r - 1)$-cycle Γ on V will have $x_1, x_2, \cdots, x_r$ as separating uniformizing coördinates. If then $\omega \neq 0$ is any r-fold differential of $\Omega(V)$ and if write ω in the form $Adx_1dx_2 \cdots dx_r$, then for all but a finite number of prime $(r - 1)$-cycles Γ we will have $v_\Gamma(\omega) = v_\Gamma(A)$. It follows that ω has only a finite number of null cycles and polar cycles and that consequently the sum $\Sigma v_\Gamma(\omega)\Gamma$, extended over the set of all prime $(r - 1)$-cycles of V, is actually a finite sum and represents therefore an$(r - 1)$-cycle on V. This cycle is *the cycle* (or more precisely: *the V-cycle*) *of the differential* ω and is denoted by (ω). Since the r-fold differentials of $\Omega(V)$ form a vector space over Ω, of dimension 1, the cycles of any two such differentials are linearly equivalent, and any $(r - 1)$-cycle on V which is linearly equivalent to a cycle of an r-fold differential is itself the cycle of an r-fold differential, the latter being uniquely determined to within an arbitrary non-zero constant factor. The linear equivalence class (on V) of cycles of r-fold differentials of $\Omega(V)$ is called the *canonical class of* V, and any cycle in that class is called a *canonical cycle*. The *canonical system* is the complete linear system consisting of the non-negative canonical cycles and will be denoted by $|K|$, where K, then, is any canonical cycle on V.

An r-fold differential ω is *of the first kind with respect to* V if $\omega = 0$ of if $\omega \neq 0$

and the cycle (ω) on V is non-negative. The number of r-fold differentials of the first kind with respect to V which are linearly independent over Ω will be called the *virtual geometric genus of* V and will be denoted by $p_g(V)$. The dimension of the canonical system $|K|$ is therefore $p_g(V) - 1$.

If V' is an absolutely normal variety birationally equivalent of V and such that the birational transformation between V and V' has no fundamental points on V', we write $V \leqq V'$. If $V \leqq V'$, then it is immediately seen that any r-fold differential of $\Omega(V)$ $[= \Omega(V')]$ which is of the first kind with respect to V' is also of the first kind with respect to V. Hence if $V \leqq V'$ then $p_g(V) \geqq p_g(V')$. Since $p_g(V)$ is always a non-negative integer, it follows that there exist varieties V' in the birational class of V for which $p_g(V')$ is minimum. This minimum, which is a non-negative integer, is called the *geometric genus of the field* $\Omega(V)$ and is denoted by p_g; it is also *the effective geometric genus* of the variety V. If V' is any absolutely normal variety in the birational class of V such that $p_g(V') = p_g$, then any r-fold differential of the common function field of V and V' which is of the first kind with respect to V' is also of the first kind with respect to any other variety in the birational class of V. It is not difficult to see that *the virtual geometric genus of a non-singular variety is equal to its effective geometric genus.*[12]

Let ω be an r-fold differential on V and let C_m be a general cycle of L_m/k, where k is an algebraically closed field of definition of V. Let t be a uniformizing parameter of C_m, i.e., t is a function on V having order $+1$ on C_m. We can then find other $r - 1$ elements $z_1, z_2, \cdots, z_{r-1}$ in $\Omega(V)$ which together with t form a separating set of uniformizing coordinates of C_m. Let

$$\omega = A dt dz_1 dz_2 \cdots dz_{r-1}$$

and let us assume that C_m is neither a polar cycle nor a null cycle of ω. In that case A has a finite non-zero trace on C_m. Let $\bar{A}$ and $\bar{z}_i$ be the C_m-traces of A and z_i respectively. Since the $r - 1$ functions $\bar{z}_i$ form a separating transcendence basis of $\Omega(C_m)/\Omega$, the expression $\bar{A}d(\bar{z})$ represents an $(r-1)$-fold differential on C_m, and this differential is different from zero since $\bar{A} \neq 0$. It is easily seen that the differential $\bar{A}d(\bar{z})$ is independent of the choice of the functions z_i and hence depends only of ω and t. We call this differential *the C_m-trace of ω relative to t* and we denote it by $\mathrm{Tr}^t_{C_m}\omega$.

We observe that the cycle $C_m - (t)$ does not have C_m as a component and that similarly C_m is not a component of (ω). Hence the intersection cycle

$$[(\omega) + C_m - (t)] \cdot C_m$$

is well defined as an $(r-2)$-cycle on C_m. Our object is to prove the following relation:

$$(\mathrm{Tr}^t_{C_m}\omega) = [(\omega) + C_m - (t)] \cdot C_m . \tag{53}$$

[12] See the paper "On the differential forms of the first kind on algebraic varieties" by Shoji Koizumi (J. of the Math. Soc. of Japan, vol. 1, No. 3, pp. 273–280, 1949), where a corresponding result is proved for differentials of the first kind, of any degree.

Let Δ be any prime $(r - 2)$-cycle on C_m. Since Δ is simple for V (see section 6, Part I), C_m is locally at Δ complete intersection of V with a hypersurface, i.e., there exists a function u on V such that no prime component of $(u) - C_m$ passes through Δ. Since Δ is also simple for C_m, we can choose $r - 1$ functions z_i on V in such a manner that $\bar{z}_1$ be a uniformizing parameter of Δ on C_m and that $\bar{z}_1, \bar{z}_2, \cdots, \bar{z}_{r-1}$ form a set of separating uniformizing coördinates of Δ on C_m (here $\bar{z}_i$ is the C_m-trace of z_i). Then the $\bar{z}_i$ form necessarily a separating transcendence base of $\Omega(C_m)/\Omega$ and hence the z_i and t as well as the z_i and u are separating uniformizing coördinates of C_m on V. Let then

$$\omega = Adtd(z) = Bdud(z). \tag{54}$$

We have $\mathrm{Tr}^t_{C_m}\omega = \bar{A}d(\bar{z})$, where $\bar{A}$ is the C_m-trace of A. Therefore the multiplicity to which Δ occurs as a component of the cycle $(\mathrm{Tr}^t_{C_m}\omega)$ is equal to $\bar{v}_\Delta(\bar{A})$, where $\bar{v}_\Delta$ is the valuation of $\Omega(C_m)/\Omega$ determined by the prime cycle Δ (since the $\bar{z}_i$ are separating uniformizing coördinates of Δ on C_m). Hence in order to prove (53) we have only to show that $\bar{v}_\Delta(\bar{A})$ is also the multiplicity of Δ as a component of the cycle on the right hand side of (53).

It is clear that the cycle of the function u/t differs from the cycle $C_m - (t)$ only in prime components which do not pass through Δ. Hence for the purpose of the proof it is permissible to replace the right hand side of (53) by the cycle $[(\omega) + (u/t)]\cdot C_m$. Since $\omega = Bdud(z)$ and since $u, z_1, z_2, \cdots, z_{r-1}$ are separating uniformizing coördinates of Δ on V (in view of the fact that u and z_1 are uniformizing parameters of Δ on V, while the Δ-traces of $z_1, z_2, \cdots, z_{r-1}$ form a separating transcendence basis of $\Omega(\Delta)/\Omega$), it follows from Lemma 6, Corollary, that the cycle (ω) coincides locally at Δ with the cycle (B). By Lemma 2 (section 6), the cycle $(Bu/t)\cdot C_m$ coincides with the cycle $(\mathrm{Tr}_{C_m}Bu/t)$. Hence in order to prove (53) we have only to prove the following relation:

$$\bar{v}_\Delta(\mathrm{tr}_{C_m}Bu/t) = \bar{v}_\Delta(\bar{A}). \tag{55}$$

By (54) we have $B = A\partial t/\partial u$, where partial derivatives are taken with respect to the r independent variables $u, z_1, z_2, \cdots, z_{r-1}$. If we set $t/u = \xi$, then ξ has finite C_m-trace and we have $\partial t/\partial u = \xi + u\partial\xi/\partial u$, whence

$$Bu/t = A + (u^2/t)\partial\xi/\partial u.$$

Since $\partial\xi/\partial u$ has finite C_m-trace and since the C_m-trace of u^2/t is zero, it follows that $\mathrm{Tr}_{C_m}Bu/t = \mathrm{Tr}_{C_m}A = \bar{A}$, and this establishes (55).

The significance of (53) is well known and is as follows:

The trace of L_m on C_m is called *the characteristic linear system on* C_m, and any cycle of this system is called a *characteristic cycle on* C_m. Let G be a characteristic cycle on C_m, so that $G = C'_m\cdot C_m$, where $C'_m \in L_m$. We have $C_m - C'_m = (t)$, where t is a function on V. If we use this particular function t, then the right hand side of (53) is of the form $(K + C'_m)\cdot C_m$, where $K = (\omega)$. On the other hand, the left hand side of (53) is a canonical cycle on C_m. We have therefore the following well known result: *if K is a canonical cycle on V such that C_m is not* *a component of K and if G is a characteristic cycle on C_m, then the cycle $G + K \cdot C_m$ is a canonical cycle on C_m.*

If $|D|$ is any complete linear system on V, then the system $|D + K|$ is called *the adjoint system of* $|D|$ and is denoted by $|D'|$. We can express the above result as follows: *the trace of $|C'_m|$ on C_m consists of canonical cycles of C_m.* It is clear that this result continues to hold true if we replace C_m by any irreducible complete linear system $|D|$ which serves to define a regular birational transformation of V. In particular, if V is non-singular and E is any $(r - 1)$-cycle on V, then the above result holds for the system $|D| = |E + C_m|$ if m is sufficiently large.

We shall now find an expression for $\dim|C'_m|$ when m is sufficiently large. We observe that $-K = -C'_m + C_m$ and hence

$$p_a(-K) = p_a(-C'_m) + p_a(C_m) + p_a(-C'_m \cdot C_m).$$

Since $C'_m \cdot C_m$ is a canonical cycle on C_m, it follows that

$$p_a(-C'_m) = p_a(-K) - [p_a(C_m) + p_a(-K(C_m))], \tag{56}$$

where we have denoted by $K(C_m)$ a canonical cycle on C_m. By Theorem 5 we have therefore for m sufficiently large:

$$1 + \dim|C'_m| = (-1)^r\{[p_a(V) + p_a(-K(V))] - [p_a(C_m) + p_a(-K(C_m))]\}. \tag{57}$$

Let us introduce the following numerical character of V:

$$P_a(V) = (-1)^r[p_a(V) + p_a(-K(V)) - 1]. \tag{58}$$

Then we can re-write (57) as follows:

$$1 + \dim|C'_m| = P_a(V) + P_a(C_m). \tag{59}$$

We can now also find an expression for the deficiency $\delta(C_m)$ of the linear system $\mathrm{Tr}_{C_m}|C'_m|$. The corresponding complete linear system on C_m is the canonical system, and hence $\delta(C_m) = p_g(C_m) - 1 - \dim \mathrm{Tr}_{C_m}|C'_m|$. On the other hand, $\dim \mathrm{Tr}_{C_m}|C'_m| = \dim|C'_m| - \dim|K(V)| - 1 = \dim|C'_m| - p_g(V)$. Hence by (59), we have for m large:

$$\delta(C_m) = [p_g(V) - P_a(V)] + [p_g(C_m) - P_a(C_m)]. \tag{60}$$

Concerning the character $P_a(V)$, we observe that if V is a non-singular curve of genus $p[= p_a(V)]$, then $-K$ is a divisor of degree $2 - 2p$ and hence $p_a(-K) = 1 - 2p$, so that (58) yields $P_a(V) = p = p_a(V)$. If V is a non-singular surface, we consider the complete linear system $|D| = |-K + C_m|$, where we take m sufficiently large so that $|D|$ be an irreducible linear system which defines a regular birational transformation of V (into a surface whose system of hyperplane sections corresponds to the system $|D|$). If, then, D is a general member of $|D|$ (with respect to an algebraically closed field of definition of V), then D is non-singular, and $\mathrm{Tr}_D|C_m|$ consists of canonical cycles of the curve D (since $|C_m| = |D'|$). Hence $p_a(-C_m \cdot D) = 1 - 2p_a(D)$, and this yields, after some straightforward computations [taking into account (40) and the fact that $D \equiv - K + C_m$]: $p_a(-K) = 1$. Hence by (58), for $r = 2$, we find again that $P_a(V) = p_a(V)$.

It may be conjectured that the equality $P_a(V) = p_a(V)$ holds for any non-singular variety V of dimension r, or—what is the same thing—that

(61) $$p_a(-K) = 1, \qquad \text{if } r \text{ is even,}$$

and

(61′) $$p_a(-K) = 1 - 2p_a(V), \qquad \text{if } r \text{ is odd.}$$

It is immediately seen that the very simple method which we have used above in order to prove (61) in the case $r = 2$ is applicable to any even r and leads to the conclusion that (61) holds for a given even integer $r = s$ provided (61′) holds for the dimension $r = s - 1$. So the conjecture has only to be proved for non-singular varieties of odd dimension, but for these varieties the problem is non-trivial. In his memoir [4a], Severi establishes the equality $P_a(V) = p_a(V)$ for $r = 3$,[13] but in the proof he uses the theorem of the completeness of the characteristic system of a complete continuous system of surfaces on a V_3 (see [4a], p. 57, Theorem VII; this theorem is then applied on p. 59 to the proof of the crucial inequality $p_g(V) \geqq p_a(V)$ for three-dimensional varieties V having "surface irregularity" zero). We know now that the algebro-geometric proof of the completeness of the characteristic system is not valid already in the case $r = 2$. On the other hand, we do not know of any transcendental proof of this theorem when $r > 2$. So it would seem that the equality $P_a(V) = p_a(V)$ has still to be proved for $r = 3$. However, the results which Severi has obtained in his quoted memoir, by means of an ingenious method of "complete intersections," represent a considerable advance in this problem, for he has reduced the proof of the equality $P_a(V) = p_a(V)$ to the problem of showing that if V is a three-dimensional non-singular variety, of surface irregularity zero, then $p_g(V) \geqq p_a(V)$.

As long as the equality $P_a(V) = p_a(V)$ remains an open question in the case $r = 3$, it must be clearly understood that "the birational invariance of the virtual arithmetic genus of a non-singular V_3" has one meaning in the memoir of Severi and has a different meaning in our joint paper [2] with Muhly. Severi proves the birational invariance of $P_a(V)$.[14] We prove the birational invariance

[13] Severi's definition of $P_a(V)$ (in the case $r = 3$) is different from our formal definition (58). He projects the non-singular V into a V' in S_4 having only ordinary singularities and defines $P_a(V)$ as the virtual dimension of the linear system of hypersurfaces in S_4, of order $n - 5$ (n = order of V'), which are adjoint to V' (i.e., pass through the double surface of V'). Since he proves later on that $1 + \dim |C'_m| = P_a(V) + p_a(C_m)$ and since for the non-singular *surface* C_m we have $p_a(C_m) = P_a(C_m)$, (59) shows that Severi's definition is equivalent to ours.

[14] Severi's proof is based on the fact that the surfaces F belonging to V all have (under certain conditions on F) the same irregularity $p_g(F) - p_a(F)$ (called the "surface irregularity" of V). The proof of this result depends on topological and transcendental considera-

of $p_a(V)$. Combined together, these two results imply, in view of (58), that *also the virtual arithmetic genus of the negative of a canonical cycle of a non-singular* V_3 *is a birational invariant.* This latter result is not trivial, for the canonical cycles are not absolutely invariant under birational transformations.

[*Added in proofs.* In regard to the above discussion and with reference to the classical case, see our reference in the "Introduction" to the unpublished proof of the equality $P_a(V) = p_a(V)$, due to Kodaira.]

14. The Riemann-Roch theorem for normal surfaces

We shall now briefly consider the case $r = 2$. Let then V be an absolutely normal surface and let D be any 1-cycle on V. We have

$$\dim |D| = \dim |D + C_m| - d - 1,$$

where $d = \dim \mathrm{Tr}_{C_m}|D + C_m|$. Hence if m is sufficiently large we have by Theorem 5:

$$(62) \qquad \dim |D| = p_a(V) + p_a(-D - C_m) - d - 2.$$

If j_m is the index of specialty of the linear series $\mathrm{Tr}_{C_m}|D + C_m|$, then by the Riemann-Roch theorem on C_m we have $d \leqq -p_a(-E_m \cdot C_m) - p_a(C_m) + j_m - 1$, where for simplicity of notation we have denoted by E_m the cycle $D + C_m$ and where the equality holds if and only if the linear series $\mathrm{Tr}_{C_m}|D + C_m|$ is complete. By substitution into (62) we find:

$$\dim |D| \geqq p_a(V) + p_a(-E_m) + p_a(C_m) + p_a(-E_m \cdot C_m) - j_m - 1,$$

or by (40):

$$(63) \qquad \dim D \geqq p_a(V) + p_a(-D) - j_m - 1.$$

By the preceding result (section 13) concerning the characteristic series on C_m, we have $j_m - 1 = \dim (|\mathrm{Tr}_{C_m}|K - D||)$. By Theorem 4 (Part II),

$$\mathrm{Tr}_{C_m}|K - D|$$

is a complete series if m is sufficiently large, and for large m we also have

$$\dim |K - D| = \dim \mathrm{Tr}_{C_m}|K - D|.$$

Hence if $i(D)$ denotes the dimension of the system $|K - D|$ increased by 1 [$i(D)$ = index of specialty of $|D|$], then it follows that if m is sufficiently large we have $j_m = i(D)$. Hence we may write, by (63):

$$(64) \qquad \dim D \geqq p_a(V) + p_a(-D) - i(D) - 1,$$

tions. Hence, at present, Severi's proof of the invariance of $P_a(V)$ is not immediately applicable to the abstract case. On the other hand, our proof of the invariance of $p_a(V)$ is purely algebraic. It should not be difficult to show, by a direct analysis, that $p_a(-K)$ is invariant under quadratic and monoidal transformations. This would show that $p_a(-K)$ is an absolute invariant (see Part IV of our paper "Reduction of the singularities of algebraic three-dimensional varieties", Annals of Mathematics, vol. 45, 1944) and, by (58), would establish the absolute invariance of $P_a(V)$ also in the abstract case.

and this inequality expresses the Riemann-Roch theorem on our normal surface V.

If V is non-singular then it is easy to see that formula (46) generalizes to any 1-cycle D, i.e., we have $p_a(-D) = 3 + p_a(D^2) - p_a(D)$, where now $p_a(D^2)$ is the virtual degree $\nu(D)$ of the cycle D, diminished by 1. Hence on a non-singular surface V we find the well known inequality:

$$\dim |D| \geqq \nu(D) - p_a(D) + p_a + 1 - i(D),$$

where $p_a = p_a(V)$ = arithmetic genus of the function field $\Omega(V)$.

We make one final remark. Let V' be a non-singular birational transform of V such that $V \leqq V'$, and let $L'_m = |C'_m|$ be the transform on V' of the complete linear system L_m (it is clear that L'_m is a complete linear system on V'). The Riemann-Roch theorem on V, as applied to L_m, tells us that if m is sufficiently large then $\dim L_m = \nu(C_m) - p_a(C_m) + p_a(V) + 1$ [see (47)]. On the other hand, the Riemann-Roch theorem on V' yields the following: $\dim L'_m = \nu(C'_m) - p_a(C'_m) + p_a + 1 + s(C'_m)$ (if m is large), where $s(C'_m) \geqq 0$ is the superabundance of L'_m. Since $\dim L_m = \dim L'_m$, $\nu(C_m) = \nu(C'_m)$ and $p_a(C_m) = p_a(C'_m)$, we deduce that if m is sufficiently large then $s(C'_m) = p_a(V) - p_a$.

HARVARD UNIVERSITY

REFERENCES

[1]. F. ENRIQUES, Le superficie algebriche, Bologna, Zanichelli, 1949.

[2]. H. T. MUHLY and O. ZARISKI, *Hilbert's characteristic function and the arithmetic genus of an algebraic variety*, Trans. Amer. Math. Soc., vol. 69 (1950), pp. 78–88.

[3]. A. SEIDENBERG, *The hyperplane sections of normal varieties*, Trans. Amer. Math. Soc., vol. 69 (1950), pp. 357–386.

[4a]. F. SEVERI, *Fondamenti per la geometria sulle varietà algebriche*, Rend. Circ. Mat. Palermo, vol. 28 (1909), pp. 33–87.

[4b]. F. SEVERI, Serie, sistemi d'equivalenza e corrispondenze algebriche sulle varietà algebriche (a cura di F. Conforto e di E. Martinelli), Roma, 1942.

[5]. A. WEIL, Foundations of algebraic geometry, Amer. Math. Soc. Colloquium Publications, vol. 29, New York, 1946.

[6]. O. ZARISKI, *Some results in the arithmetic theory of algebraic varieties*, Amer. J. Math., vol. 61 (1939), pp. 249–294.

[7]. O. ZARISKI, *Pencils on an algebraic variety and a new proof of a theorem of Bertini*, Trans. Amer. Math. Soc., v. 50, (1941), pp. 48–70.

[8]. O. ZARISKI, *The concept of a simple point of an abstract algebraic variety*, Trans. Amer. Math. Soc., vol. 62 (1947), pp. 1–52.

[9]. O. ZARISKI, Algebraic surfaces, Ergebnisse der Mathematik und ihrer Grenzgebiete, vol. III (5), Berlin, Springer, 1935.

INTERPRÉTATIONS ALGÉBRICO-GÉOMÉTRIQUES DU QUATORZIÈME PROBLÈME DE HILBERT (1);

PAR M. O. ZARISKI.

1. Dans sa célèbre conférence *Sur les problèmes futurs des Mathématiques*, au deuxième Congrès international de Mathématiques (Paris, 1900), Hilbert a proposé le problème suivant :

Soit $R' = k[X_1, X_2, \ldots, X_n]$ un anneau de polynomes sur un corps k, et soit F un corps contenant k et contenu dans le corps des quotients F' de R'. L'intersection $R' \cap F$ est-elle toujours un domaine (2) *fini* sur k?

C'est le quatorzième problème de Hilbert. On sait que Hilbert fut amené à poser ce problème par l'étude d'un cas particulier qui se présente dans la théorie des invariants des groupes algébriques de transformations linéaires sur $X_1, X_2, \ldots, X_n$, à coefficients dans k.

L'un des attraits de ce problème difficile est la simplicité de sa formulation, et l'on peut bien dire, en citant les propres paroles de Hilbert, que cette formulation « a été rendue tellement claire

(1) Cet article est le résumé d'une conférence qui devait avoir lieu à Paris en mai 1953.

(2) Nous disons pour simplifier, ici et par la suite, domaine pour domaine d'intégrité.

qu'on puisse la faire comprendre au premier individu rencontré dans la rue ».

Je me propose de discuter le quatorzième problème de Hilbert du point de vue de la géométrie algébrique.

Je transformerai d'abord ce problème en un problème concernant les systèmes linéaires sur une variété algébrique. La variété qui interviendra dans le problème algébrico-géométrique est un modèle projectif convenable du corps F; sa dimension r est donc égale au degré de transcendance de F/k. Dans le cas $r = 1$, la solution du problème algébrico-géométrique (et par conséquent celle du problème de Hilbert) est une conséquence immédiate du théorème de Riemann-Roch. Dans le cas général ($r > 1$), nous trouverons un résultat très curieux; nous verrons, en effet, que la difficulté du problème de Hilbert est due entièrement au fait qu'un système linéaire complet sur une variété peut posséder des points de base *accidentels*, c'est-à-dire des points de base qui n'ont pas été imposés au système. L'analyse de cette difficulté nous conduira à un autre énoncé du problème algébrico-géométrique en termes de la surface de Riemann du corps F. Dans ce nouvel énoncé interviennent, non pas les points du modèle V de F, mais les places du corps F. En ce qui concerne le point de vue de la théorie des places (ou théorie de la valuation), notre principal résultat consiste à montrer que la solution du problème revient à la démonstration de la propriété suivante : un certain sous-ensemble de la surface de Riemann de F est un ensemble fermé. En utilisant ce critère et le théorème de l'uniformisation locale, nous obtenons la solution du problème de Hilbert dans le cas des surfaces ($r = 2$) sur un corps k de caractéristique nulle.

Ces différents aspects algébrico-géométriques du quatorzième problème de Hilbert nous conduisent raisonnablement à penser qu'il n'est pas possible de donner une solution générale du problème de Hilbert sans avoir auparavant résolu quelques questions fondamentales de la théorie des variétés de dimension quelconque.

2. Nous commencerons par généraliser, dans une certaine mesure, le quatorzième problème de Hilbert, en partant de cette remarque : le fait que R′ est un anneau de polynomes ne joue aucun rôle dans notre étude. Ce qui importe, c'est que R′ est un *domaine* *fini intégralement fermé*. Nous supposerons donc que R′ satisfait à cette condition, mais, à part cela, R′ peut être un anneau arbitraire. Nous désignerons par F′ le corps des quotients de R′ et par R l'anneau $R' \cap F$. Puisque R′ est intégralement fermé dans F′, R est aussi intégralement fermé dans F. On peut supposer que F *est le corps des quotients* de R car, si F_1 est le corps des quotients de R, on a $R' \cap F = R' \cap F_1$ et l'on peut remplacer F par F_1.

Soient r' et r les degrés de transcendance de F'/k et F/k respectivement, $r' \geqq r \geqq 1$. Il semblerait, à première vue, que *deux* corps interviennent dans notre problème, à savoir F et F′. En fait, c'est seulement le plus petit corps F qui présente un intérêt. La valeur de r' n'a rien à voir avec le degré de difficulté du problème. Je vais illustrer immédiatement cette remarque en m'occupant d'abord du cas $r = 1$.

Soit $\{x_1, x_2, \ldots, x_n\}$ un système de générateurs de R'/k et soit V′ la variété irréductible sur k dont le point général est le point $(x) = (x_1, x_2, \ldots, x_n)$ (nous utilisons des coordonnées non homogènes). Fixons un modèle projectif normal C du corps F/k (C = courbe sans singularités). L'inclusion $F \subset F'$ signifie que nous avons une transformation rationnelle T de V′ en C. Puisque dans la définition de V′ nous avons utilisé des coordonnées non homogènes, il est permis de parler des points de V′ qui sont à l'infini. Des considérations élémentaires empruntées à la théorie des correspondances algébriques montrent qu'il y a au plus un nombre fini de points Q sur C tels que *tout* point de V′ qui correspond à Q (par T^{-1}) soit à l'infini. Soit M l'ensemble de ces points Q de C.

Je dis que l'anneau R *est constitué par les fonctions de* F *qui n'ont aucun pôle en dehors de l'ensemble* M.

Démonstration. — Soit $f \in R$, $P \in C$, $P \notin M$. Il existe un point P′ sur V′ qui correspond à P et qui est à distance finie (puisque $P \notin M$). D'après la définition des correspondances algébriques basée sur la théorie de la valuation, il existe une place p' de F'/k telle que :

(a) P′ est le centre de p' sur V′;

(b) si p est la restriction de p' à F, P est le centre de p sur C.

Puisque P est à distance finie, on a $x_i p' \neq \infty$, $i = 1, 2, \ldots, n$. Puisque $f \in R \subset R'$, il en résulte que $fp \neq \infty$. Donc P n'est pas un pôle de f.

Réciproquement, soit $f \in F$ tel que M contienne tous les pôles de f. Soit p' une place quelconque de F'/k telle que $x_i p' \neq \infty$, $i = 1, 2, \ldots, n$, et soit p la restriction de p' à F. Soient P′ et P les centres de p' et p sur V′ et C respectivement. Le point P′ est le point $(x_1 p', x_2 p', \ldots, x_n p')$ et il est par conséquent à distance finie, tandis que P et P′ sont des points correspondants par T. Donc, $P \notin M$; P n'est pas un pôle de f et par conséquent $fp' (= fp) \neq \infty$. Nous avons montré que $fp' \neq \infty$ pour toute place p' de F'/k telle que $x_i p' \neq \infty$, $i = 1, 2, \ldots, n$. Puisque $k[x_1, x_2, \ldots, x_n] (= R')$ est intégralement fermé, il en résulte que $f \in R$.

C. Q. F. D.

Si M est vide, R ne comprend que les constantes, c'est-à-dire R est la fermeture algébrique de k dans F et est donc un domaine d'intégrité fini sur k (ce cas peut être exclu puisque nous avons supposé que F est le corps des quotients de R).

Si M n'est pas vide, nous déterminons une fonction z dans F dont l'ensemble des pôles coïncide avec l'ensemble M. Il en résulte alors, d'après ce qui vient d'être démontré, que R est la fermeture intégrale de $k[z]$ dans F. Par suite, R est encore un domaine intégralement fermé fini sur k.

3. Nous passons maintenant au cas général. Considérons encore la variété V′ sur k, dont le point général est $(x_1, x_2, \ldots, x_n)$. Le sous-corps donné F de F′ détermine sur V′ un système algébrique involutif S de cycles de dimension $r' - r$, c'est-à-dire un système algébrique irréductible de dimension r de cycles de dimension $r' - r$ tel que par un point général de V′ sur k passe un cycle et un seul de S. Le système S est défini comme suit :

Soit Z la variété irréductible *sur* F qui admet $(x_1, x_2, \ldots, x_n)$ comme point général. Alors Z est évidemment une sous-variété de V′ et a la dimension $r' - r$. Considérons Z comme un cycle, en comptant chaque composant absolument irréductible de la *variété* Z comme un composant de multiplicité 1 du *cycle* Z. Alors S est le système algébrique irréductible, *sur* k, qui a Z comme cycle général.

Au moyen de la forme associée de Chow-van der Waerden, nous représentons Z par un point P dans un espace projectif convenable, et le système S comme une variété irréductible V sur k ayant P comme point général.

On peut montrer que si F′ est séparablement engendré sur F, on a $k(P) \subset F$ et F est une extension purement inséparable de $k(P)$. On peut également montrer que si F′ est séparablement engendré sur les deux corps F et k, on a $k(P) = F$. Il est clair que, dans le problème de Hilbert, il est permis de remplacer k par une extension purement inséparable de k, et qu'on ne perd aucune généralité en supposant F′ séparablement engendré sur F. On peut donc supposer que $k(P) = F$. *Notre variété représentative* V *du système algébrique* S *est donc un modèle projectif du corps* F/k.

Ce modèle V de F/k joue maintenant le rôle de la courbe C considérée dans le cas $r = 1$; c'est une transformée rationnelle de V′. Désignons par T la transformation rationnelle qui fait passer de V′ à V. Cette transformation a la propriété que *son inverse* T^{-1} *ne possède aucun point fondamental sur* V; pour tout point Q de V, la transformée totale $T^{-1}\{Q\}$ est la variété du cycle de S qui est représenté par le point Q, et cette variété est donc de dimension $r' - r$. Notons que cette propriété de T^{-1} est conservée si nous remplaçons V par un modèle normal de V. Nous supposerons donc que V est une variété normale.

Si ξ est un élément arbitraire de F nous écrivons

$$(\xi) = Z_0(\xi) - Z_\infty(\xi)$$

où $Z_0(\xi)$ est le cycle des zéros de ξ sur V et $Z_\infty(\xi)$ est le cycle des pôles de ξ sur V. Nous allons caractériser maintenant les éléments de R par leur cycle polaire $Z_\infty(\xi)$.

Soit Γ un cycle premier quelconque de dimension $r - 1$ sur V, défini sur k, et soit v_Γ le diviseur premier de F/k défini par Γ. Nous dirons que v_Γ *est à l'infini par rapport à* V′ si, pour toute extension v' de v_Γ à F′, le centre de v' sur V′ est à l'infini. Dans le cas contraire, nous dirons que v_Γ *est à distance finie par rapport à* V′.

Il est clair que, si $\xi \in R$, on a $v_\Gamma(\xi) \geqq 0$ pour tout v_Γ qui est à distance finie. La réciproque est vraie aussi parce que T^{-1} n'a aucun point fondamental sur V. En effet, pour démontrer que $\xi \in R (= R' \cap F)$, il suffit de démontrer que $v'(\xi) \geqq 0$ pour tout diviseur premier v' de F'/k qui est défini par un idéal premier minimal de R'. Le centre W' de v' sur V' est à distance finie et a la dimension $r' - 1$. Soit v la restriction de v' à F et soit Γ le centre de v sur V. W' et Γ sont alors des variétés correspondantes par T. Puisque T^{-1} n'a aucun point fondamental sur V, on a : dimension $W' \leqq \dim \Gamma + (r' - r)$, c'est-à-dire : $\dim \Gamma \geqq r - 1$. Si $\dim \Gamma = r$, v est trivial sur F et l'on a $v(\xi) = 0$. Si $\dim \Gamma = r - 1$, alors $v = v_\Gamma$ et v_Γ est à distance finie par rapport à V' (puisque v' est une extension de v et puisque W' est à distance finie). Donc, par hypothèse, $v(\xi) \geqq 0$, c'est-à-dire $v'(\xi) \geqq 0$.

On peut montrer facilement qu'il n'existe qu'un nombre fini de cycles premiers Γ sur V de dimension $r - 1$, tels que v_Γ soit à l'infini par rapport à V'. Soient $\Gamma_1, \Gamma_2, \ldots, \Gamma_s$, ces cycles premiers. Nous pouvons maintenant énoncer notre résultat de la façon suivante : R *est l'ensemble des fonctions* ξ *dans* F *telles que tout composant premier du cycle polaire* $Z_\infty(\xi)$ *de* ξ *appartienne à l'ensemble* $\{\Gamma_1, \Gamma_2, \ldots, \Gamma_s\}$.

D'autres énoncés équivalents de ce résultat sont les suivants :

(a) $\xi \in R$ si et seulement s'il existe des entiers $i_1, i_2, \ldots, i_s$ tels que $(\xi) + i_1\Gamma_1 + i_2\Gamma_2 + \ldots + i_s\Gamma_s \geqq 0$ (« $\geqq 0$ » signifie : « est un cycle effectif »).

(b) Soit $D = \Gamma_1 + \Gamma_2 + \ldots, + \Gamma_s$. Alors $\xi \in R$ si et seulement s'il existe un entier i tel que

$$(\xi) + iD \geqq 0. \tag{1}$$

Nous sommes ainsi conduit à la conjecture algébrico-géométrique suivantes :

Soit D *un cycle effectif de dimension* $(r - 1)$ *sur une variété* V/k *normale de dimension* r. *Nous supposons que* D *est défini sur* k, *et nous désignons par* R[D] *l'anneau des fonctions* ξ *sur* V/k *qui vérifient* (1) *pour un entier convenable* i. *Alors* R[D] *est un domaine fini sur* k.

Si cette conjecture est vraie, celle qui est énoncée dans le quatorzième problème de Hilbert est vraie également. Je n'ai aucune démonstration de la réciproque; donc, *a priori*, notre conjecture algébrico-géométrique semble plus forte que la conjecture de Hilbert. Par contre, du point de vue géométrique, la conjecture algébrico-géométrique paraît plus plausible que la conjecture originale de Hilbert.

Remarquons que les fonctions ξ de F qui vérifient (1) pour un entier *fixé* $i \geqq 1$ forment un espace vectoriel Ω_i de dimension finie sur k. La dimension de Ω_i, diminuée de 1, est égale à la dimension du système linéaire complet $|iD|$. On a

$$\Omega_1 \subset \Omega_2 \subset \ldots \subset \Omega_i \subset \ldots, \qquad R[D] = \bigcup_{i=1}^{\infty} \Omega_i.$$

Notre conjecture est équivalente à la suivante :

Il existe un entier m tel que si $\{z_0, z_1, z_2, \ldots, z_h\}$ *est une base de* Ω_m/k, *on a* $R[D] = k[z_0, z_1, z_2, \ldots, z_h]$.

Nous pouvons exprimer notre conjecture sous une forme plus géométrique. Désignons par L_i le système complet $|iD|$ et par jL_i le sous-système linéaire minimum de L_{ij} qui contient tous les cycles de la forme $Z_1 + Z_2 + \ldots, + Z_j$, où $Z_\nu \in L_i$, $\nu = 1, 2, \ldots, j$. Le système linéaire jL_i n'est pas nécessairement complet; c'est le *j-multiple minimum de* L_i. Si Z est un cycle quelconque de jL_m et si $z_0, z_1, \ldots, z_h$ sont les éléments de base de Ω_m/k, $Z - jmD$ est le cycle d'un polynome $f(z_0, z_1, \ldots, z_h)$ de degré inférieur ou égal à j, à coefficients dans k. Réciproquement, le cycle d'un tel polynome est de la forme $Z - jmD$, $Z \in jL_m$. Si notre conjecture est vraie la suivante sera vraie également : si i est un entier quelconque supérieur ou égal à 1 et si Y est un cycle quelconque donné dans L_i, il existe un entier j dépendant de Y et un cycle Z de jL_m tels que $Y - iD = Z - jmD$, c'est-à-dire :

$$Y + (jm - i)D \in jL_m. \tag{2}$$

La validité de (2) pour tout Y de L_i et pour un j convenable dépendant de Y implique aussi celle de (2) pour tout Y de L_i et pour un j convenable qui est indépendant de Y (mais qui dépend seulement de i). Nous pouvons donc exprimer notre conjecture comme suit :

Si m est suffisamment grand, pour tout entier i supérieur ou égal à 1, *il existe un entier* j_i *tel que le système linéaire* $j_i L_m - (j_i m - i) D$ [*système des résiduels du cycle* $(j_i m - i) D$ *par rapport à* $j_i L_m$] *coïncide avec le système complet* $|iD|$.

La difficulté de la démonstration de cette conjecture est due au fait que les multiples minimum jL_m d'un système linéaire complet L_m ne sont pas, en général, complets, et par conséquent, si j est quelconque, il n'est pas possible d'affirmer que $jL_m - (jm - i)D$ est complet. Notre conjecture dit précisément que ce système *est* complet si j est suffisamment grand (par rapport à i).

4. Nous commencerons l'étude de notre problème algébrico-géométrique en considérant un cas particulier pour lequel la solution est immédiate. Soit $D = \Gamma_1 + \Gamma_2 + \ldots + \Gamma_s$, où les Γ_i sont des cycles premiers, et supposons qu'il existe des entiers *positifs* $m_1, m_2, \ldots, m_s$ tels que le *système complet* $|m_1\Gamma_1 + m_2\Gamma_2 + \ldots + m_s\Gamma_s|$ *n'ait pas de points base*. Soit ρ la dimension de ce système et soient $\Delta_0, \Delta_1, \ldots, \Delta_\rho$, $\rho + 1$ cycles indépendants du système. Déterminons une fonction z_ν telle que

$$(3) \qquad (z_\nu) = \Delta_\nu - (m_1\Gamma_1 + \ldots + m_s\Gamma_s) \qquad (\nu = 0, 1, \ldots, \rho).$$

Les $\rho + 1$ fonctions z_ν appartiennent à l'anneau R. Posons :

$$(4) \qquad E_\nu = D - (D \cap \Delta_\nu),$$

l'intersection $D \cap \Delta_\nu$ étant prise au sens de la théorie des ensembles. Le système linéaire $|m_1\Gamma_1 + m_2\Gamma_2 + \ldots + m_s\Gamma_s|$ n'ayant pas de point base et étant engendré par $\rho + 1$ cycles Δ_ν, on a :

$$(5) \qquad D = \bigcup_{\nu=0}^{\rho} E_\nu.$$

Soit p une place quelconque de F/k telle que $z_\nu p \neq \infty$, $\nu = 0, 1, \ldots, \rho$, et soit P le centre de p sur V. *Je dis que* P *n'appartient pas à* D. Car, en supposant le contraire, p devrait appartenir, d'après (5), à l'un des ensembles E_ν, par exemple $P \in E_1$. Puisque les m_i sont positifs, il résulte de (4) que P appartient au cycle polaire de z_1 et n'appartient pas au cycle des zéros Δ_1 de z_1.

Mais alors, on doit avoir $z_1 p = \infty$ puisque P est le centre de p, ce qui est une contradiction.

Puisque $P \notin D$, on a *$zp \neq \infty$ pour tout z dans* R[D]. Nous avons montré que $zp \neq \infty$ lorsque $z_0 p, z_1 p, \ldots, z_\rho p$ sont tous $\neq \infty$. D'après un théorème connu sur les domaines intégralement fermés, on en déduit que R[D] est la fermeture intégrale de $k[z_0, z_1, \ldots, z_\rho]$ dans F. Par suite, R[D] *est un domaine d'intégrité fini sur* k.

Notons le corollaire suivant :

Soit V *une variété sans singularités et soit* Γ *un cycle effectif quelconque sur* V. *Si* C *désigne une section hyperplane de* V, *l'anneau* R[C + Γ] *est un domaine d'intégrité fini.*

En effet, il est bien connu que, si V est sans singularités, le système linéaire $|iC + \Gamma|$ n'a pas de points base lorsque i est suffisamment grand. La démonstration de cette propriété repose exclusivement sur la propriété suivante d'une variété V sans singularités : *l'anneau local de chaque point de* V *est un domaine à factorisation unique.* Notre corollaire vaut donc également pour toute variété V qui a la propriété locale qui vient d'être mentionnée.

A propos du corollaire ci-dessus, il est nécessaire de dire explicitement que ce corollaire ne *nous permet pas de conclure* que, si D est un cycle équivalent à C + Γ, R[D] est aussi un domaine fini. D étant un cycle premier équivalent à C + Γ, il peut très bien arriver que le système complet $|D|$ et tous ses multiples $|iD|$ aient nécessairement des points de base. Dans cet ordre d'idées on peut faire la remarque suivante à titre de mise en garde : *dans l'énoncé de notre problème fondamental algébrico-géométrique, il n'est pas permis de remplacer le cycle donné* D *par un cycle linéairement équivalent.* Cependant, le résultat suivant peut être démontré :

Soit $D_0, D_1, \ldots, D_\rho$ *un ensemble maximal de cycles linéairement indépendants dans* $|D|$. *Si chacun des* $\rho + 1$ *anneaux* $R[D_0], R[D_1], \ldots, R[D_\rho]$ *est un domaine fini, alors pour tout cycle* D′ *dans* [D] *l'anneau* R[D′] *est un domaine fini.*

5. D'après le résultat démontré précédemment, notre pro-

blème ne présente pas de difficultés si le cycle $D = \Gamma_1 + \Gamma_2 + \ldots \Gamma_s$ est tel que, pour des entiers *positifs* convenables $m_1, m_2, \ldots, m_s$, le système $|m_1\Gamma_1 + m_2\Gamma_2 + \ldots + m_s\Gamma_s|$ n'ait pas de points base.

En particulier, si D lui-même est un cycle premier, le problème est résolu lorsque $|D|$ ou l'un des multiples $|iD|$ de $|D|$ n'a pas de points base. Malheureusement, il peut très bien arriver que tous les systèmes complets $|iD|$ aient des points base. Ceux-ci sont alors des points base *accidentels* puisque, dans la définition d'un système complet, aucun point base n'a été imposé *a priori*. Il est évident que, si V a des points singuliers, ceux-ci peuvent apparaître comme points base accidentels. Mais il est également possible de montrer par des exemples que des points base accidentels peuvent se présenter dans tous les systèmes $|iD|$ même lorsque V n'a pas de points singuliers. *Du point de vue algébrico-géométrique, la difficulté essentielle du quatorzième problème de Hilbert est représentée par l'existence possible de points base accidentels dans chacun des systèmes complets* $|iD|$, $i = 1, 2, \ldots$.

Il est raisonnable de regarder la complication provenant des points base accidentels comme liée au modèle projectif V qui a été l'objet des considérations précédentes. On peut très bien espérer que cette complication n'a pas de signification invariante pour le corps F lui-même et qu'il est peut-être possible de faire disparaître cette complication en choisissant un autre modèle projectif de F/k. On doit donc essayer d'aborder tout le problème d'une façon invariante en faisant intervenir la surface de Riemann du corps F. Nous allons étudier maintenant notre problème du point de vue de la théorie des places (ou théorie des valuations).

6. Si nous remplaçons V par un autre modèle projectif normal V^* de F/k tel que $V < V^*$ (c'est-à-dire tel que la transformation birationnelle $V^* \to V$ soit régulière sur V^*), il est facile de voir qu'il existe sur V^* un cycle D^* de dimension $r-1$ vérifiant $R[D^*] = R[D]$. Les composants premiers de D^* sont 1° les transformés $\Gamma_1^*, \Gamma_2^*, \ldots, \Gamma_s^*$ des composants premiers $\Gamma_1, \Gamma_2, \ldots, \Gamma_s$ de D et 2°) éventuellement des cycles exceptionnels Γ_{s+1}^*; $\Gamma_{s+2}^*, \ldots, \Gamma_t^*$, de dimension $r-1$, qui correspondent aux variétés fondamentales sur V.

Soit Q^* un point de V^*. Nous dirons que Q^* est un *point* *d'indétermination* pour l'anneau R[D] si, pour des entiers *positifs quelconques* $i_1, i_2, \ldots, i_t$, le point $Q^\star$ est toujours un point de base du système linéaire $|i_1\Gamma_1^\star + i_2\Gamma_2^\star + \ldots + i_t\Gamma_t^\star|$.

Pour que $Q^\star$ soit un point d'indétermination de R[D], il *n'est pas* nécessaire que *toutes* les fonctions (non constantes) de R[D] soient indéterminées en $Q^\star$; il est seulement nécessaire (et aussi suffisant) que $Q^\star$ soit un point d'indétermination pour ~~au moins une~~ toutes les fonction ξ de R[D] telle que le cycle polaire de ξ sur $V^\star$ contienne *tous* les composants premiers de $D^\star$.

Si p est une place du corps F/k, nous dirons que p est une *place de base* de l'anneau R[D] lorsque, pour tout modèle projectif $V^\star > V$, le centre $Q^\star$ de p sur $V^\star$ est un point d'indétermination de l'anneau R[D].

Il est très facile de voir *qu'un domaine fini* R[D] *n'a pas de places de base.* En effet, si $R[D] = k[z_1, z_2, \ldots, z_m]$, les m fonctions z_i de F définissent une transformation rationnelle de V dans un espace projectif de dimension m. Si nous prenons pour $V^\star$ le graphe de cette transformation nous trouvons immédiatement que R[D] n'a pas de point d'indétermination sur $V^\star$.

Nous allons maintenant démontrer la réciproque :

Théorème 1. — *Si* R[D] *n'a pas de places de base,* R[D] *est un domaine fini.*

Démonstration. — Désignons par M(D) l'ensemble de toutes les places p de F/k telles qu'il existe au moins une fonction ξ de R[D] qui est infinie en p. L'ensemble M(D) va jouer un rôle important.

Par hypothèse, pour toute place p donnée il existe un modèle $V^\star > V$ (fonction de p) tel que le centre de p sur $V^\star$ ne soit pas un point d'indétermination pour l'anneau R]D]. Déterminons un tel modèle $V^\star$ pour toute place p et désignons-le par V^p. Désignons par D^p le plus petit cycle de dimension $r-1$ sur V^p tel que $R[D] = R[D^p]$. Soit $N(D^p)$ l'ensemble des places q telles que le centre de q sur V^p appartienne à D^p. L'ensemble $N(D^p)$ est un sous-ensemble fermé de la surface de Riemann $\mathfrak{M}$ de F/k.

Il est clair que, si $q \in M(D)$, on a $q \in N(D^p)$ pour toutes les places p. Donc,

$$M(D) \subset \bigcap_{p \in \mathfrak{M}} N(D^p).$$

D'autre part, si $q \notin M(D)$, soit Q le centre de q sur V^q. D'après la définition de V^q, il existe une fonction ξ^q de R[D] qui n'est pas indéterminée en Q et dont le cycle polaire sur V^q contient tous les composants premiers de D^q. Puisque $q \notin M(D)$, on a $\xi^q q \neq \infty$, et ceci implique $Q \notin D^q$. Par suite, $q \notin N(D^q)$. Nous en déduisons donc

$$M(D) = \bigcap_{p \in \mathfrak{M}} N(D^p),$$

et par suite, M(D) *est un sous-ensemble fermé* de $\mathfrak{M}$.

Pour toute place q appartenant à M(D), déterminons une fonction ξ^q de R[D] comme ci-dessus, c'est-à-dire : le cycle polaire de ξ^q sur V^q contient tous les composants premiers de D^q, et le centre de q sur V^q n'est pas un point d'indétermination de ξ^q. Puisque le centre de q doit appartenir à D^q, on a $\xi^q q = \infty$. Désignons par N^q l'ensemble de toutes les places p dont les centres sur V^q *ne sont pas* des points d'indétermination de ξ^q (c'est-à-dire que ces centres n'appartiennent pas simultanément à D^q au cycle des zéros de ξ^q). L'ensemble N^q a les propriétés suivantes : *a*. N^q est un sous-ensemble ouvert de $\mathfrak{M}$; *b*. $q \in N^q$; *c*. si $p \in N^q \cap M(D)$, on a $\xi^q p = \infty$ (parce que le centre de p appartient à D^q et n'est pas un point d'indétermination de ξ^q).

D'après *a* et *b*, l'ensemble $\{N^q;\ q \in M(D)\}$ est un recouvrement ouvert de l'ensemble *fermé* M(D). Soit $\{N^{q_i};\ i = 1, 2, \dots, m\}$ un sous-recouvrement fini de M(D), et soit $z_i = \xi^{q_i}$. En utilisant la propriété *c* nous trouvons que les m fonctions z_i de R[D] ont la propriété suivante :

Si p est une place quelconque de M(D) *l'une au moins des fonctions z_i est infinie en p.*

En d'autres termes : si p est une place quelconque de F/k telle que $z_i p \neq \infty$ pour $i = 1, 2, \dots, m$, on a $\xi p \neq \infty$ pour tout ξ de R[D]. Ceci implique que R[D] est la fermeture intégrale de $k[z_1, z_2, \dots, z_m]$ dans F et par conséquent R[D] est un domaine fini.
C. Q. F. D.

Au cours de la démonstration, on a montré indirectement que, *si* R[D] *est un domaine fini*, M(D) *est un sous-ensemble fermé de la surface de Riemann* $\mathfrak{M}$ *de* F/k. Ceci peut aussi se voir directement, car si $R(D) = k[z_1, z_2, \ldots, z_n]$ et si N_i désigne l'ensemble des places p telles que $z_i p = \infty$, N_i est un ensemble fermé et l'on a $M(D) = \bigcap_{i=1}^{n} N_i$. Il est possible de démontrer aussi la réciproque : *si* $M(D)$ *est un ensemble fermé*, $R[D]$ *est un domaine intégralement fermé fini*. Nous le démontrons en établissant que, si $M(D)$ est un ensemble fermé, il n'existe aucune place p qui soit une place de base de $R(D)$. La démonstration n'est pas facile ; elle est basée en partie sur une induction portant sur le rang de la valuation v associée à la place p.

7. Nous allons appliquer maintenant les résultats précédents au cas des surfaces. Cette application repose sur un lemme concernant les séries linéaires sur une courbe algébrique. Soit C une courbe algébrique sans singularités et soient L et L' deux séries linéaires sur C, sans points fixes. Soit $\{L_{ij}\}$ une suite de séries linéaires sur C satisfaisant aux conditions suivantes :

(*a*) $$L_{10} = L, \qquad L_{01} = L';$$

(*b*) $$L_{ij} \subset |iL + jL'|;$$

(*c*) $$L_{i_1+i_2,\, j_1+j_2} \supset L_{i_1, j_1} + L_{i_2, j_2}$$

(+ désigne la somme minimum).

Lemme. — *Il existe un entier* N *tel que :*

(6) $$L_{i+1,j} = L_{i,j} + L, \quad \text{pour tout } i \geqq N \text{ et tout } j \geqq 0;$$

(7) $$L_{i,j+1} = L_{i,j} + L', \quad \text{pour tout } i \geqq 0 \text{ et tout } i \geqq N :$$

De ce lemme nous déduisons le théorème suivant :

Soit $|D|$ *un système linéaire irréductible sur une surface normale* V. *Si* Q *est un point simple de* V, Q *n'est pas un point base de* $|iD|$ *lorsque* i *est suffisamment grand.*

Démonstration. — Désignons par M le système des sections hyperplanes de V, et soit $M' = |D|$. Posons $M_{i,j} = |iM + jD|$. Soit $\overline{C}$ une section hyperplane générique de V, et posons $L_{i,j} = \mathrm{Tr}\, M_{i,j}$ sur $\overline{C}$. La suite de séries linéaires $L_{i,j}$ satisfait aux conditions du lemme. Il résulte de (7) que $M_{i,j+1}$ et $M_{i.j} + M'$ ont la même trace sur $\overline{C}$, si $i \geqq 0$, $j \geqq N$.

Donc $M_{i,j+1}$ est engendré par les deux sous-systèmes de $M_{i,j+1}$: $M_{i,j}+M'$ et $(M_{i,j+1}-\overline{C})+\overline{C}$. Le deuxième de ces deux sous-systèmes est contenu dans $M_{i-1,j+1}+M$. Par suite,

(8) $M_{i,j+1}$ est engendré par $M_{i,j}+M'$ et $M_{i-1,j+1}+M$ $\quad (i \geqq 1,\ j \geqq N)$.

Si Q n'est pas un point base de $M'(=|D|)$ il n'y a rien à démontrer. Supposons que Q soit un point base de M'. Il résulte de (8) que si Q est un point base de $M_{i-1,j+1}$, Q est aussi un point base de $M_{i,j+1}(i \geqq 1, j \geqq N)$. Par applications répétées de ce résultat, et en partant de $i=1$, nous trouvons que, si Q est un point base de $|(j+1)D|\ (j \geqq N)$, Q est aussi un point base de $|iM+(j+1)D|$, pour tout i. Ceci est impossible puisque, j étant donné, le système $|iM+(j+1)D|$ n'a pas de point base au point *simple* Q si i est suffisamment grand. Donc Q n'est pas un point base de $|(j+1)D|$ si $j \geqq N$. C. Q. F. D.

* Si la caractéristique du corps de base k est zéro, nous pouvons déduire du théorème ci-dessus que l'anneau R[D] est un domaine fini. Il suffit de montrer qu'aucune place p de F/k n'est une place de base de R[D]. D'après le théorème d'uniformisation locale il existe un modèle projectif $V^* > V$ de F/k tel que le centre Q^* de p sur V^* est un point simple. Le théorème précédent implique alors que Q^* n'est pas un point d'indétermination de R[D] et par suite p n'est pas une place de base de R[D].

(Extrait du *Bulletin des Sciences mathématiques*, 2e série, t. LXXVIII, juillet-août 1954.)

*. Proof incomplete. Complete proof as follows.

Let $D_1, \ldots, D_q$ be the prime components of D.

<u>Case $q=1$</u>. In this case it is clear that <u>either</u> $\dim|nD|=0$ for all n, (in which case $R[D]=k$), or there exists an integer $n_0>0$ s.t. $|n_0 D|$ has no fixed components; and in this case, there also exists an integer m_0 s.t. $|n_0 m_0 D|$ has no base points. So $R[D] = R[n_0 m_0 D]$ is finitely generated.

<u>Case $q>1$, by induction on q</u>. If there exist <u>positive</u> integers $i_1, i_2, \ldots, i_q$ s.t. $|i_1 D_1 + \cdots + i_q D_q|$ has no fixed components then $R[D] = R[i_1 D_1 + \cdots + i_q D_q]$, finitely generated.

In the contrary case, let B_n be the fixed component of $|nD_1 + \cdots + nD_q|$, $B_n = \nu_1 D_1 + \cdots + \nu_q D_q$, where the ν_i are functions of n. $\nu_i = \nu_i(n)$

Then we must have, for each n: $\nu_i(n) = n$ for at least one value of $i = 1, 2, \ldots, q$ (this value depends on n). There must exist one some value of i, say $i=1$, s.t. $\nu_1(n) = n$ for infinitely many values of n. It is obvious however, that if $n' < n$ then $\nu_1(n') + (n - n') \geq \nu_1(n)$. So if $\nu_1(n) = n$, also $\nu_1(n') = n'$ for $n' < n$. Thus $\nu_1(n) = n$ for all n. That means that $R[D_1 + \cdots + D_q] = R[D_2 + \cdots + D_q]$. Now use induction hypothesis

INTRODUCTION TO THE PROBLEM OF MINIMAL MODELS IN THE THEORY OF ALGEBRAIC SURFACES

By

OSCAR ZARISKI
Professor of Mathematics
Harvard University

THE MATHEMATICAL SOCIETY OF JAPAN
1 9 5 8

PREFACE

The present monograph is based on a series of lectures given by me at the University of Tokyo and the University of Kyoto in September-October, 1956, during my visit as guest lecturer of the Science Council of Japan. The difference between the contents of those lectures and the contents of this monograph is the following:

(1) The underlying general theory of rational transformations and linear systems was basically taken for granted in the lectures, while in the monograph this background material (especially in its geometric aspects) is developed systematically and in considerable detail; it constitutes Part I of the monograph.

(2) The actual solution of the problem of minimal models for algebraic surfaces is only outlined in the monograph (Part III), while in the lectures (especially in Kyoto) the proofs were given in full detail. The complete proofs will be found in a forthcoming paper which will be published in the American Journal of Mathematics.

As to the intermediate stage of the problem of minimal models—the theory of exceptional curves of the first and of the second kind on an algebraic surface—the monograph (Part II) follows rather faithfully the treatment given in the lectures.

I wish to express my thanks to the Science Council of Japan for having offered me a greatly appreciated opportunity to address mathematical audiences in Japan. I also wish to acknowledge gratefully the many constructive criticisms and suggestions which I have constantly received from my audiences in Tokyo and Kyoto during my course of lectures. Finally, thanks are due to the Mathematical Society of Japan for the publication of this monograph and to Prof. Shoji KOIZUMI who has taken notes of my lecture and whose preliminary manuscript which he has sent me was very helpful to me during the preparation of the monograph.

Oscar ZARISKI

This research was supported by the United States Air Force through the Air Office of Scientific Research of the Air Research and Development Command under Contract No AF18 (600)-1503.

TABLE OF CONTENTS

PART I. RATIONAL TRANSFORMATIONS OF ALGEBRAIC VARIETIES

PART II. THEORY OF EXCEPTIONAL CURVES ON AN ALGEBRAIC SURFACE

PART III. THE PROBLEM OF MINIMAL MODELS FOR ALGEBRAIC SURFACES

PART I

RATIONAL TRANSFORMATIONS OF ALGEBRAIC VARIETIES

I. 1. Generalities.

While this monograph is devoted primarily to algebraic surfaces, we shall deal in this part (and only in this part) of the monograph with varieties of any dimension.

Following a—by now—universal procedure, we operate in a given universal domain, with ground fields taken from that domain. All our varieties will be tacitly assumed to be either projective or affine. When dealing with projective varieties we shall, as a rule, omit the term "projective." We shall say that a variety V (affine or projective) is *defined* over a ground field K if it can be defined by equations with coefficients in K, and we say that V is *regularly defined* over K if V/K is irreducible and the function field $K(V)$ of V/K is a regular extension of K.

We fix once and for always a ground field k. *Unless otherwise specified, the ground field k will be assumed to be algebraically closed*, and when we speak of a variety (without mentioning a field of definition) we mean a variety which is defined over k.

If a field K (a subfield of the universal domain) is finitely generated over k (*k—an arbitrary ground field*), we mean by a *projective* model of K/k a pair (V, P) consisting of an irreducible variety V and a given general point P of V/k such that $k(P)=K$. Actually we shall often refer also to projective models as *varieties* and shall write V instead (V, P), without fear of confusion.

Given two projective models (V, P) and (V', P') (of two finitely generated extensions K/k and K'/k respectively) there is a well-defined irreducible algebraic correspondence T between the two models: it is the correspondence (defined over k) whose general point pair is (P, P'). The correspondence T is a *rational* transformation of V (onto V') if $k(P')$ is a subfield of $k(P)$; *birational* if $k(P)=k(P')$. We shall say that T is a *quasi-rational* transformation of V (onto V') if the coordinates of P' are purely inseparable over $k(P)$; and that T is *quasi-birational* if both T and T^{-1} are quasi-rational.

If P is a point in the projective n-space (with coordinates in the universal domain) and if $y_0, y_1, \ldots, y_n$ are homogeneous coordinates of P, we say that these coordinates are *strictly homogeneous* if the following condition is satisfied: if for a given index i we have $y_i \neq 0$ then y_i is a transcendental over the field $k(P)$ generated over k by the ratios $y_0/y_i, y_1/y_i, \ldots, y_n/y_i$. An equivalent condition is that the field $k(y_0, y_1, \ldots, y_n)/k$ have transcendence degree $1+\dim P/k$. Every point P has (infinitely many) sets of strictly homogeneous coordinates. If $y_0, y_1, \ldots, y_n$ are strictly homogeneous coordinates of P then the polynomials f in the polynomial ring $k[Y_0, Y_1, \ldots, Y_n]$ such that $f(y)=0$ form a homogeneous ideal; and conversely.

If (V, P) is a projective model and if $y_0, y_1, \ldots, y_n$ are strictly homogeneous coordinates of the general point P of V/k, then we call the ring $k[y]$ the *homogeneous coordinate ring* of the model. This ring is uniquely determined by the model, up to an arbitrary K-isomorphism ($K=k(P)$) into the universal domain. Since the ideal of the relations between the y's is homogeneous, the coordinate ring $k[y]$ is itself a graded ring, and we can speak without ambiguity of homogeneous elements of $k[y]$, of the degree of such elements and of the homogeneous components of an element of $k[y]$.

We shall need a well-defined projective embedding of the product of two varieties V and V' (defined over an arbitrary ground field). This embedding is given by the C. SEGRE variety of the pairs of points of V and V', and is as follows:

If P is a point in the projective n-space, having homogeneous coordinates $y_0, y_1, \ldots, y_n$, and $P' \equiv (y'_0, \ldots, y'_m)$ is a point in the projective m-space, then we denote by $P \times P'$ the point which lies in a projective space of $(n+1)(m+1)-1$ dimensions and whose homogeneous coordinates are the products $\{y_{ij}\}=\{y_0y'_0, y_0y'_1, \ldots, y_0y'_m, y_1y'_0, \ldots, y_ny'_m\}$ (in this order). If, now, V and V' lie in projective spaces of dimensions n and m respectively, then $V \times V'$ is the set of all points $P \times P'$, with P in V and P' in V'.

It is seen at once that this yields indeed a projective embedding of the direct product $V \times V'$ in the usual sense, and that the variety $V \times V'$ is defined over every common field of definition of V and V'. It is clear that $V \times V'$ and $V' \times V$ are projectively equivalent varieties.

If P and P' are points in affine spaces of dimension n and m respectively, and if $x_1, x_2, \ldots, x_n$ and $x'_1, x'_2, \ldots, x'_m$ are the (non-homogeneous) coordinates of P and P' respectively, then we mean by $P \times P'$ the point $(x_1, x_2, \ldots, x_n, x'_1, x'_2, \ldots, x'_m)$ in the affine space of $n+m$ dimensions; and if V, V' are affine varieties contained in affine spaces of dimension n and m respectively, then $V \times V'$ is the set of all points $P \times P'$, with P in V and P' in V'.

Let V be a projective variety in a projective n-space. Then V has a covering by $n+1$ affine varieties U_i, where U_i is the set of points of V which do not belong to the hyperplane $Y_i=0$. Similarly, if V' is a projective variety in a projective m-space, then V' has a covering by $m+1$ affine varieties U'_j. In the ambient projective space of the product variety $V \times V'$ we have a similar covering of $V \times V'$ by $(n+1)(m+1)$ affine varieties U_{ij}, where U_{ij} is the set of points of $V \times V'$ which do not belong to the hyperplane $Y_{ij}=0$. It is clear that U_{ij} is a biregular embedding of $U_i \times U'_j$. We shall refer to the covering $\{U_{ij}\}$ as the *canonical affine covering* of $V \times V'$. This covering depends only on the choice of the system of homogeneous coordinates in the ambient spaces of V and V' respectively.

From now on, when we speak of the graph of an algebraic correspondence T between two varieties V and V' we mean the corresponding subvariety of the SEGRE variety $V \times V'$.

I. 2. Fundamental varieties of algebraic correspondences and exceptional varieties of rational transformations.

In this section k denotes an arbitrary ground field. Let $T: V \rightarrow V'$ be an irreducible algebraic correspondence between a projective model $V\,(=(V, P))$ and a projective model $V'\,(=(V', P'))$, with (P, P') as general point pair. We recall the terminology which we have used in [28] for birational transformations and which we now extend to algebraic correspondences. If G is any subset of V we denote by $T\{G\}$ the set of all points of V' which correspond to points of G under T, and we call $T\{G\}$ the *total T-transform* of G. If k_1 is a ground field containing k and if G is a variety defined over k_1 then also $T\{G\}$ is a variety defined over k_1.

Two irreducible subvarieties W and W' of V and V' respectively are said to correspond to each other under T if the graph of T contains a point $Q \times Q'$ such that Q is a general point of W/k and Q' is a general point of W'/k. If W and W' correspond to each other then $W' \subset T\{W\}$ (but not conversely). We denote by $T[W]$ the union of those irreducible components of $T\{W\}/k$ which correspond to W under T and we call $T[W]$ the *proper T-transform* of W (Cf. p. 85, (1)). The variety $T[W]$ is non-empty (but may be a proper subvariety of $T\{W\}$); it can also be defined as the smallest variety defined over k which contains the set $T\{Q\}$, where Q is any (preassigned) general point of W/k.[1)]

If Q is a point of the projective model (V, P), the *local ring of Q on V* is defined as the set of all quantities $f(P)$ where f is any rational function of V which is regular at Q. Thus the local ring of Q on V is a subring of the function field $K = k(P)$ of the model. Similarly for V'. If Q and Q' are corresponding points of V and V', under T, then T is said to be *regular at Q* if the local ring of Q on V contains the local ring of Q' on

1) The definition of $T[W]$ can be extended to the case in which neither T nor W are assumed to be irreducible over k, and the definition is such that $T[W]$ is independent of the choice of the common ground field of definition k of T and W. This is done as follows.

Let f be the restriction of T to W, i.e., let $f = T \cap (W \times V')$, where V' is the range of T and where we denote by T also the graph of T. Then f is an algebraic correspondence defined over k, with domain contained in W. Let $f_1, f_2, \ldots, f_h$ be the irreducible components of f/k and let W_i be the domain of f_i. Then W_i is an irreducible variety defined over k and contained in W. We assume that for $i = 1, 2, \ldots, s$ (and only for these values of i) the variety W_i is an irreducible component of W/k. Then we set $T_k[W] = \bigcup_{i=1}^{s} f_i[W_i]$. If W/k has d irreducible components and if on each of these components we fix a general point Q_ν of that component over k, then it is clear that $T_k[W]$ is the least variety over k which contains the set $\bigcup_{\nu=1}^{d} T\{Q_\nu\}$.

V'. If T is regular at Q then T is necessarily a rational transformation and is single-valued at Q, i.e., the set $T\{Q\}$ consists of a single point; if V is normal at Q and T is rational then also the converse is true. If T is regular at Q and T^{-1} is regular at the point $Q'=T\{Q\}$, then T is said to be *biregular at* Q; T is then necessarily a birational transformation. A rational (resp., birational) transformation of V which is regular (resp., biregular) at each point of V is said to be *regular* (resp., *biregular*).

We recall the concept of *fundamental varieties* of an *arbitrary* irreducible algebraic correspondence $T: V \to V'$. Let r and r' be the dimension of V and V' respectively and let s be the dimension of the graph G of T. Let f be the projection of G onto V and let g be the inverse of f. It is well known that if Q is any point of V then the total transform $g\{Q\}$ is a variety defined over $k(Q)$ and has dimension $\geqq s-r$. We say that Q is a *fundamental point of* T if $\dim g\{Q\} > s-r$, and we say that an irreducible subvariety W/k of V is a *fundamental variety* of T if a general point of W/k is fundamental for T. By the *fundamental locus* of T we mean the set of all fundamental points of T. The fundamental locus of T is a variety defined over k, of dimension $\leqq r-2$, and an irreducible subvariety W of V is a fundamental variety of T if and only if W is contained in the fundamental locus of T. It is clear that W is a fundamental variety of T if and only if the dimension of its proper transform $g[W]$ on G is

---

We now show that $T_k[W]$ is independent of the choice of the common field of definition k of T and W. It will be sufficient to show that if k' is a ground field containing k then $T_{k'}[W]=T_k[W]$. Let $f_{i1}, f_{i2}, \ldots, f_{it_i}$ be the irreducible components of f_i/k'. Then it is immediately seen that the $t_1+t_2+\cdots+t_h$ correspondences f_{ij_i} $(i=1,2,\ldots,h;\ j_i=1,2,\ldots,t_i)$ are distinct and are the irreducible components of f/k' (a general point of any f_{ij}/k' is also a general point of f_i/k, and hence $f_{ij} \not\subset f_{i'j'}$ if $i \neq i'$). For a given f_{ij} let W'_{ij} be the domain of f_{ij} and let $Q \times Q'$ be a general point of f_{ij}/k'. We have then that Q is a general point of W'_{ij} over k'. On the other hand, $Q \times Q'$ is also a general point of f_i/k with the property that $\dim Q \times Q'/k' = \dim Q \times Q'/k$. Hence $\dim Q/k' = \dim Q/k$, i.e., $\dim W'_{ij} = \dim W_i$, showing that W'_{ij} is an irreducible component of W_i/k'. It follows that W'_{ij} is an irreducible component of W/k' if and only if W_i is an irreducible component of W/k, i.e., if and only if $i=1,2,\ldots,s$. Hence

$$T_{k'}[W] = \bigcup_{i=1}^{s} \bigcup_{j_i=1}^{t_i} f_{ij_i}[W'_{ij_i}] = \bigcup_{i=1}^{s} f_i[W_i] = T_k[W].$$

greater than $s-r+\dim W$.

If T is a rational transformation we have $s=r$, and the above definitions imply that a point Q of V is a fundamental point of T if and only if $T\{Q\}$ has positive dimension. If V is normal at the point Q then Q is a fundamental point of T if and only if T is not regular at Q. An irreducible subvariety W of V will be a fundamental variety of T if and only if its proper transform $g[W]$ on the graph G of T has greater dimension than W. On the other hand, a point Q' of W' is fundamental for T^{-1} if the dimension of its total transform on G is greater than $r-r'$, and an irreducible subvariety W' of V' is fundamental for T^{-1} if the dimension of the proper transform of W' on G is greater than $r-r'+\dim W'$.

Again if T is a rational transformation (and only under this assumption) we define *exceptional varieties* of T, as follows:

An irreducible subvariety W of V is said to be an *exceptional variety* of a rational transformation $T:V\rightarrow V'$ of V onto a variety V' if there exists an irreducible subvariety W' of V', of dimension *less* than $\dim W-r+r'$, such that W and W' are corresponding varieties in T. We list below various consequences of this definition.

(1) *If W is an exceptional variety of T then* $\dim W>r-r'$ (*and hence has positive dimension*).

(2) *If the rational transformation T is regular at a general point of W/k then W is an exceptional variety of T if and only if the proper transform $T[W]$ of W has dimension less than* $\dim W-r+r'$ (*less than* $\dim W$, *if T is a birational transformation*). In fact, $T[W]$ is then the only irreducible subvariety of V' which corresponds to W in T.

(3) *If W is an exceptional variety of T then there exists a fundamental variety W' of T^{-1} such that W and W' are corresponding varieties.* This is obvious since if W' is an irreducible subvariety of V' which corresponds to W then W and W' are projections of one and the same irreducible subvariety W^* of the graph G of T. If then $\dim W'<\dim W-r+r'$ then *a fortiori* $\dim W'<\dim W^*-r+r'$, and hence W' is fundamental for T^{-1}.

The converse of (3) is not true, even if W happens to be an irreducible component of $T^{-1}[W']$. *A priori* it does not even seem possible to exclude the following possibility: W' is a fundamental variety of T^{-1}, and none of the irreducible components of $T^{-1}[W']$ is an exceptional variety of T.

(see footnote 2)). We shall prove, however, the following

PROPOSITION I. 2. 1. *If* $T: V \to V'$ *is a rational transformation and* W' *is a fundamental variety of* T^{-1}, *then* $T^{-1}[W']$ *contains an exceptional variety* W *which corresponds to* W'. *If, furthermore,* V' *is normal at a general point of* W'/k *then every irreducible component of* $T^{-1}[W']/k$ *is an exceptional variety of* T.

PROOF. We shall have to use the following lemma (the proof of which will be given below):

LEMMA I. 2. 2. *If* $T: V \to V'$ *is a rational transformation, if* G *is the graph of* T, *and* f, f' *are the projections of* G *onto* V *and* V' *respectively, then the exceptional varieties of* T *are the* f*-projections of the exceptional varieties of* f'.

Assuming the lemma we first observe that the first part of the proposition is trivial if T is a regular transformation, and in that case it is even true that at least one of the irreducible components of $T^{-1}[W']$ is an exceptional variety of T. This follows from the definition of fundamental varieties and from the fact that if T is regular then the projection f is biregular. Again, if T is regular, the second part of the proposition follows from the " main theorem " on birational transformations (ZARISKI [28], p. 522).

In the general case we observe that, by the regular case, $f'^{-1}[W']$ contains an exceptional variety of f', and since $T^{-1}[W']$ is the f-projection of $f'^{-1}[W']$ the first part of the proposition follows from the lemma. The second part of the proposition follows from the fact that each irreducible component of $T^{-1}[W']$ is the f-projection of an irreducible component of $f'^{-1}[W']$.[2)]

2) If V' is not normal at the general point of W'/k it may happen that one of the irreducible components of $f'^{-1}[W']$ has the same dimension as W', and it does not seem possible to exclude the possibility that the f-projection of that particular component contain the projections of all the other irreducible components of $f'^{-1}[W']$.

We take this occasion to warn the reader against a possible misunderstanding in regard to Proposition I. 2. 1. While we assert that $T^{-1}[W']$ contains an exceptional variety W *which corresponds to* W' we do *not* mean to assert that $\dim W > \dim W' + r - r'$. We only assert that there exists another irreducible subvariety W'' of V' such that also W and W'' are corresponding varieties and such that $\dim W > \dim W'' + r - r'$.

PROOF OF THE LEMMA.

Let W be an exceptional variety of T, let W' be an irreducible subvariety of V', of dimension $<\dim W-r+r'$, which corresponds to W in T, and let W^* be an irreducible subvariety of G which has W and W' as projections. Then $\dim W^* \geqq \dim W > \dim W'+r-r'$, and hence W^* is an exceptional variety of f'.

Conversely, let W^* be an exceptional variety of f' and let W and W' be the projections of W^* into V and V' respectively. Let $s=\dim W^*$, $t=\dim W$, $t'=\dim W'$, let $Q\times Q'$ be a general point of W^*/k and let X' $(=T\{Q\})$ be the locus of Q' over $k(Q)$. We can find, in the ambient projective m-space of V', a linear space L, of dimension $m-s+t$ and defined over k, such that $\dim L\cap W'=t'-s+t$. (Note that $s\geqq t$ and $m-s+t\geqq m-t'$, since $s\leqq t+t'$). Then each irreducible component of $L\cap W'$ has dimension $t'-s+t$. On the other hand, the intersection of L with X' is non-empty since $\dim X'=s-t$. Since $X'\subset W'$, it follows that at least one irreducible component of $L\cap W'$ has a non-empty intersection with X'. Let Y' be such a component, let R' be a common point of Y' and X' $(=T\{Q\})$, and let Z' be the locus of R' over k. Since R' belongs to $T\{Q\}$, and Q is a general point of W/k, *the varieties W and Z' correspond to each other in T.* Since R' belongs to Y', and Y' is of dimension $t'-s+t$, the dimension of Z' is at most $t'-s+t$. Since W^* is exceptional for f' and the dimension of G is r, we have $s>t'+r-r'$. Hence $\dim Z'<t-r+r'$, showing that W is an exceptional variety of T. Q.E.D.

COROLLARY I. 2. 3. *There exists a subvariety E of V (called the exceptional locus of T) having the following properties:* a) *each irreducible component of E is an exceptional variety of T;* b) *every exceptional variety of T is contained in E.*

By the above lemma it is sufficient to prove the corollary for regular transformations T. We assume therefore that T is regular. We also assume that the corollary is true for rational transformations of varieties of dimension less than r (in the case $r=1$ the exceptional locus E is empty). Let F' be the fundamental locus of T^{-1}, let E_1 be the union of those irreducible components of $T^{-1}\{F'\}$ which are exceptional varieties of T and let $W_1, W_2, \ldots$ be the remaining irreducible components of $T^{-1}\{F'\}$. Let $W'=T[W_i]$ $(=T\{W_i\})$. The restriction of T to

W_i is a rational (in fact, regular) transformation T_i of W_i onto W_i'. Since $\dim W_i - \dim W_i' = \dim V - \dim V'$, every exceptional variety of T_i is also an exceptional variety of T, and every exceptional variety of T which belongs to W_i is an exceptional variety of T_i. If then E_2 denotes the union of the exceptional loci of $T_1, T_2, \ldots$, then E_2 is a variety (by our induction hypothesis), and the union of E_1 and E_2 is the desired exceptional locus of T.

Corollary I. 2. 4. *If T is a rational transformation of V onto a normal variety V' then the exceptional locus of T is the total T^{-1}-transform of the fundamental locus F' of T^{-1}.*

This follows from the second part of Proposition I. 2. 1.

If T is a regular transformation, the irreducible components of the exceptional locus of T are called *isolated exceptional varieties of* T, and their T-transforms on V' are called *isolated* fundamental varieties of T^{-1}. If T is an arbitrary rational transformation and f, f' are the projections of the graph G of T onto V and V', then the isolated exceptional varieties of T and the isolated fundamental varieties of T^{-1} are defined as being respectively the f-projections of the isolated exceptional varieties of f', and the isolated fundamental varieties of f'^{-1}. Every irreducible component of the exceptional locus E of T is an isolated exceptional variety of T, but if T is not regular (and only in that case) there may exist isolated exceptional varieties of T which are properly contained in irreducible components of E. Similarly, every irreducible component of the fundamental locus F' is an isolated fundamental variety of T^{-1}, but this time there may exist other isolated fundamental varieties of T^{-1} (properly contained in irreducible components of F') even if T is regular.

I. 3. Linear systems and rational transformations.

Let $T: V \rightarrow V'$ be a rational transformation of a *normal* variety V onto a variety V' of *positive* dimension. Here both V and V' are to be thought of as projective models, and if $V=(V, P)$ then $V'=(V', P')$, where $P'=T\{P\}$. If $y_0, y_1, \ldots, y_n$ are strictly homogeneous coordinates of P/k then there exist forms $\varphi_0(y), \varphi_1(y), \ldots, \varphi_m(y)$ in $k[y]$, of like degree, such that if we set $y_i' = \varphi_i(y)$, then $y_0', y_1', \ldots, y_m'$ are strictly homogeneous coordinates of P'/k. We assume that V' does not lie in any proper linear subspace of the projective m-space. In that case the $m+1$ forms $\varphi_i(y)$

are linearly independent over k, and the module of homogeneous *functions* $\varphi = \sum a_i \varphi_i$ generated by the functions φ_i over the universal domain (the a_i being arbitrary quantities) determines on V a unique linear system L of dimension m, free from fixed components. If φ and φ' are any two forms in the module and if D and D' are the corresponding divisors in L, then $D-D'$ is the divisor of the rational function φ/φ' on V. If $\psi_0(y), \psi_1(y), \ldots, \psi_m(y)$ is another set of $m+1$ forms in $k[y]$, of like degree, such that the $\varphi_i(y)$ are strictly homogeneous coordinates of P', then the $\psi_i(y)$ are proportional to the $\varphi_i(y)$, and it follows that the linear system L depends only on the rational transformation T, or—equivalently—on the projective model (V', P'), the model (V, P) being fixed. We shall denote this linear system L by $\mathscr{L}(V')$.

Conversely, given on V a linear system L, of positive dimension m and defined over k, L can be defined by a module of homogeneous functions, generated over the universal domain by $m+1$ forms φ_i of like degree, with coefficients in k, and the projective model (V', P'), where P' is the point $(\varphi_0(y), \varphi_1(y), \ldots, \varphi_m(y))$ in the projective m-space, is a rational transform of V and is of positive dimension. It is clear that the model (V', P') is uniquely determined by the linear system L up to projective equivalence over k. We agree to identify projectively equivalent projective models and we denote the model (V', P') by $\mathscr{V}(L)$.

It is clear that if V' is a projective model which is a rational transform of V then $\mathscr{V}(\mathscr{L}(V'))=V'$, and if L is a linear system on V satisfying the above stated conditions, then the linear system $\mathscr{L}(\mathscr{V}(L))$ is obtained from L by deleting all fixed components of L. We have thus a (1, 1) correspondence between the linear systems on V, defined over k and free from fixed components, and the projective models which are rational transforms of V. Note that the condition that the linear system has no fixed components automatically implies that the system has positive dimension.

In the sequel we shall refer to the system $\mathscr{L}(V')$ as the *defining linear system* of the rational transformation $V \rightarrow V'$ (and also of the rational transform V' of V). For reasons which will become clear later on we shall refer to the operation of passing from V to the rational transform $\mathscr{V}(L)$ as that of *cutting L by hyperplanes*.

PROPOSITION I. 3. 1. *Let L be a linear system on V, defined over k*

and free from fixed components, and let T be the rational transformation $V \to V' = \mathscr{V}(L)$. Let P be the general point of the projective model V $(=(V, P))$ and let $P' = T\{P\}$. Let $D_0(P)$ denote the set of points of V which are common to all the divisors D in L which pass through P. Then $D_0(P)$ is a variety defined over $k(P')$. If $D(P)$ denotes the union of those irreducible components of $D_0(P)/k(P')$ which are not defined over k, then $D(P) = T^{-1}\{P'\}$ (whence $D(P)/k(P')$ is irreducible and has dimension $r - r'$, where $r = \dim V$ and $r' = \dim V'$). The remaining irreducible components of $D_0(P)/k(P')$ are contained in the base locus of L.

Proof. With the same notations as above, T is given by equations of the form $y'_i = \varphi_i(y)$ $(i = 0, 1, \ldots, m)$. Let D be any member of L and let D correspond to the form $\sum a_i\varphi_i$. Then D contains P if and only if $\sum a_i\varphi_i(y) = 0$. Let f be any $k(P')$-automorphism of the universal domain and let $Q = Pf$. Then also $Q \in T^{-1}\{P'\}$, i.e., $P' = T\{Q\}$. This shows that the quantities $\varphi_i(z)$ are proportional to the quantities $\varphi_i(y)$, if $z_0, z_1, \ldots, z_n$ are homogeneous coordinates of Q; therefore if D contains P then the divisor Df, which corresponds to the form $\sum a_i f . \varphi_i$, also passes through P. We have thus shown that the variety $D_0(P)$ is invariant under all $k(P')$-automorphisms of the universal domain. Hence $D_0(P)$ is defined over $k(P')$.

Since $P \in D_0(P)$ and P is a general point of the irreducible variety $T^{-1}\{P'\}/k(P')$, it follows that $T^{-1}\{P'\}$ is contained in $D_0(P)$. Also the base locus B of the linear system L is contained in $D_0(P)$. On the other hand, if $Q \equiv (z)$ is a point of V which is not a base point of L, then T is regular at Q. We can therefore find a set of $m+1$ forms $\psi_i(y)$ in $k[y]$, *not all zero at* Q, which are proportional to the $\varphi_i(y)$. If, furthermore, $Q \in D_0(P)$ then we must have $\sum a_i\psi_i(z) = 0$ for any set of $m+1$ quantities a_i such that $\sum a_i\psi_i(y) = 0$. This implies that the $\psi_i(z)$ are proportional to the $\psi_i(y)$, i.e., to the y'_i, showing that $Q \in T^{-1}\{P'\}$. We have therefore proved that

(1) $$D_0(P) = T^{-1}\{P'\} \cup B.$$

Since $P \notin B$ we have $T^{-1}\{P'\} \not\subset B$, and thus (1) shows that $T^{-1}\{P'\}$ is an irreducible component of $D_0(P)/k(P')$. Relation (1) also shows that any other irreducible component of $D_0(P)/k(P')$ must be contained in B and must be in fact an irreducible component of $B/k(P')$. Since k is algebraically closed and since B is defined over k, the irreducible components of B/k are absolutely irreducible and are therefore also the irreducible components of $B/k(P')$. On the other hand, the variety $T^{-1}\{P'\}$ is certainly not defined over k, since it is a proper subvariety of V and contains a general point of V/k. Consequently $T^{-1}\{P'\}=D(P')$, and this completes the proof of the proposition.

COROLLARY I. 3. 2. *A necessary condition that a rational transformation* $T: V \rightarrow V'=\mathscr{V}(L)$ *be birational is that the members* D *of the linear system* L *which pass through a general point* P *of* V/k *have, outside of the base locus of* L, *only the point* P *in common. On the other hand, if this condition is satisfied then the function field* $k(P)$ *of* V/k *is a purely inseparable extension of the function field* $k(P')$ *of* V'/k.

For the variety $T^{-1}\{P'\}$ consists of the single point P if and only if the coordinates of P are purely inseparable quantities over $k(P')$.

We shall use the following notation: if Z is a cycle on a variety V we denote by $\langle Z\rangle$ the variety of the cycle Z, i.e., the union of the prime components of Z (prime cycle=absolutely irreducible variety).

PROPOSITION I. 3. 3. *Let* D *be a general cycle of* L/k, *corresponding to a form* $\sum u_i\varphi_i$, *where* $u_0, u_1, \ldots, u_m$ *are algebraically independent quantities over* k, *and let* k^* *denote the field* $k(u_0, u_1, \ldots, u_m)$. *Then the variety* $\langle D\rangle$ *is defined and is irreducible over* k^*, *and assuming that* $\varphi_0(y)\neq 0$ *then a point* Q *of* $\langle D\rangle$ *is a general point of* $\langle D\rangle/k^*$ *if and only if* Q *is a general point of* $V/k(u_1, u_2, \ldots, u_m)$. *Furthermore, if* C' *denotes the cycle cut out on* V' *by the hyperplane* $\sum u_iY'_i=0$ (C'=*a general hyperplane section of* V') *then*

$$\langle D\rangle=T^{-1}[\langle C'\rangle]=T^{-1}\{\langle C'\rangle\}, \tag{2}$$

$$\langle C'\rangle=T[\langle D\rangle]. \tag{2'}$$

PROOF. The intersection of V with the hypersurface $\sum u_i\varphi_i(Y)=0$ is a pure $(r-1)$-dimensional variety defined over k^*, and $\langle D\rangle$ is the union of those irreducible components of that variety (over k^*) which do not belong simultaneously to all the $m+1$ hypersurfaces $\varphi_i(Y)=0$. Hence $\langle D\rangle$ is defined over k^*.

Let $Q\equiv(z)$ be a point of D. Then

$$u_0\varphi_0(z)+u_1\varphi_1(z)+\cdots+u_m\varphi_m(z)=0. \tag{3}$$

The point Q is a general point of an irreducible component of

$\langle D\rangle/k^*$ if and only if $\dim Q/k^*=r-1$, or equivalently, if and only if tr.d.$k^*(Q)/k=r+m$ (since tr.d. $k^*/k=m+1$). Now, again if Q is a general point of an irreducible component of $\langle D\rangle/k^*$ then not all $\varphi_i(z)$ are zero, and hence, by (3), we have tr.d. $k^*(Q)/k(Q)\leqq m$. Since tr.d. $k^*(Q)/k=r+m$, it follows that $\dim Q/k\geqq r$, whence $\dim Q/k=r$ and Q is a general point of V/k. Since we have assumed that $\varphi_0(y)\neq 0$, it follows now that also $\varphi_0(z)\neq 0$. Hence, by (3), we have $k^*(Q)=k(u_1, u_2, \ldots, u_m)(Q)$, and since tr.d. $k^*(Q)/k=r+m$ it follows that $\dim Q/k(u_1, u_2, \ldots, u_m)=r$, i.e., Q is a general point of $V/k(u_1, u_2, \ldots, u_m)$. Conversely, if Q satisfies this last mentioned condition and if Q belongs to $\langle D\rangle$, then using (3) and the assumption $\varphi_0(y)\neq 0$ (which implies $\varphi_0(z)\neq 0$, since also Q is a general point of V/k) we find that tr.d. $k^*(Q)/k=r+m$ and that consequently Q is a general point of an irreducible component of $\langle D\rangle/k^*$.

Now, let Q and Q' be two points of $\langle D\rangle$ which both are general points of $V/k(u_1, u_2, \ldots, u_m)$. Then there exists a $k(u_1, u_2, \ldots, u_m)$-isomorphism between the two fields $k(u_1, u_2, \ldots, u_m)(Q)$ and $k(u_1, u_2, \ldots, u_m)(Q')$ which sends Q into Q'. Relation (3) and the analogous relation for the point Q' show that u_0 belongs to both those fields and is left invariant by the isomorphism in question. Hence Q and Q' are also k^*-isomorphic points. This shows that $\langle D\rangle$ has only one irreducible component over k^*, i.e., $\langle D\rangle/k^*$ is irreducible.

We now proceed to the proof of the last part of the theorem (relations (2) and (2')).

By the preceding proof, applied to the linear system of hyperplane sections of V', the variety $\langle C'\rangle$ is defined and is irreducible over k^*. Let $Q\equiv(z)$ be a general point of $\langle D\rangle/k^*$. Since Q is a general point of V/k, T is regular at Q. Let Q' be the corresponding point $T\{Q\}$. Then $\varphi_0(z), \varphi_1(z), \ldots, \varphi_m(z)$ are homogeneous coordinates $z_0', z_1', \ldots, z_m'$ of Q', and by (3), $Q'\in\langle C'\rangle$. Furthermore, $k(Q')\subset k(Q)$. We know that $k^*(Q)=k(u_1, u_2, \ldots, u_m)(Q)$, tr.d. $k^*(Q)/k=r+m$, and also that $k^*(Q')=k(u_1, u_2, \ldots, u_m)(Q')$ since $u_0z_0'+u_1z_1'+\cdots+u_mz_m'=0$ and $z_0'=\varphi_0(z)\neq 0$. We can write tr.d. $k^*(Q)/k^*(Q')+$tr.d. $k^*(Q')/k=r+m$. Now, if r' is the dimension of V' then $\dim Q/k(Q')=r-r'$, whence $\dim Q/k^*(Q')\leqq r-r'$, and the preceding equality yields therefore the inequality tr.d. $k^*(Q')/k\geqq r'+m$. Therefore $\dim Q'/k^*\geqq r'-1$, and since

$Q' \in \langle C' \rangle$, it follows that $\dim Q'/k^* = r' - 1$, and Q' is therefore a general point of $\langle C' \rangle / k^*$. This establishes (2′).

From (2′) it follows that $\langle D \rangle \subset T^{-1}\{\langle C' \rangle\}$. Now let (R, R') be any point pair of T such that $R' \in \langle C' \rangle$ and let $z_0, z_1, \ldots, z_n$ be homogeneous coordinates of R. If R is not a base point of the linear system L then we may assume that the $\varphi_i(z)$ are not all zero (replace, if necessary, the forms $\varphi_i(y)$ by suitable proportional forms $\psi_i(y)$). Then the $\varphi_i(z)$ are homogeneous coordinates of R', (3) is satisfied since $R' \in \langle C' \rangle$, and hence $R \in \langle D \rangle$. If, on the other hand, R is a base point of L then R again belongs to $\langle D \rangle$. We have therefore shown that $\langle D \rangle = T^{-1}\{\langle C' \rangle\}$, and this completes the proof.

The relations (2) and (2′) between the general cycle D of the linear system L and the corresponding general hyperplane section C' of V' justify the classical geometric language whereby the rational transformation $T: V \to V' = \mathscr{V}(L)$ is said to have the effect of "*referring the system L to the system of hyperplane sections of $\mathscr{V}(L)$*", or—as we agreed to say—of "*cutting L by hyperplanes*". However, in the relations (2) and (2′) only the varieties of the cycles D and C' occur explicitly, not the cycles themselves. We shall now study the *cycles* D and C' themselves. This study will lead us, among other things, to another proof of the theorem of BERTINI on reducible linear system.[3)]

I. 4. Digression on algebraic systems and involutions.

Let M be an algebraic system of (positive) cycles (in a projective space) and let F be a field of definition of M. The union V of all the varieties $\langle Z \rangle, Z \in M$, is called the *carrier* of the system M; it is a variety defined over F. The correspondence between V and M (more precisely: between V and the CHOW variety of M) consisting of all pairs (Q, Z) such that the point Q belongs to $\langle Z \rangle$ is algebraic and is defined over F; it is called the *incidence correspondence* of the algebraic system M. We denote the incidence correspondence $Q \to Z$ by I and the correspondence $Z \to Q$ by I'. The domain of I is the carrier V, and its range is the system M (more precisely: the CHOW variety of M). We set $T = I \cdot I'$. Then

3) This theorem was proved by us in [26] in the case of characteristic zero. Our proof was extended by MATSUSAKA in [9] to the case of characteristic $p \neq 0$.

T is an algebraic correspondence defined over F, of which V is both the domain and range. A point pair (Q, Q') belongs to T if and only if there exists a cycle Z in M such that Q and Q' belong to $\langle Z \rangle$.

From now on we shall only consider cycles all prime components of which have the same dimension d (briefly: d-cycles). Given a variety W, defined and irreducible over a field F, let $\Gamma_1, \Gamma_2, \ldots, \Gamma_n$ be the absolutely irreducible components of W and let p^e be the order of inseparability of the function field $F(W)/F$. Then we denote by $\mathcal{Z}(W/F)$ the cycle $p^e(\Gamma_1+\Gamma_2+\cdots+\Gamma_n)$ and we call $\mathcal{Z}(W/F)$ *the cycle of* W/F. We extend this definition, by linearity, to reducible varieties W/F, provided all the irreducible components of W/F have the same dimension d (since we have agreed to deal only with d-cycles).

If the incidence correspondence I is irreducible over F (resp., absolutely irreducible) then also V and M are irreducible over F (resp., absolutely irreducible); but not conversely.

If Z is a general cycle of an irreducible algebraic system M then the total transform $I'\{Z\}$ of Z on the carrier V of M is the variety $\langle Z \rangle$ of the cycle Z; this variety $\langle Z \rangle$ is therefore defined over the field $F(Z)$ generated over F by the non-homogeneous coordinates of the Chow point of the cycle Z.

A necessary and sufficient condition that the incidence correspondence I of M be irreducible over F is that the variety $\langle Z \rangle$ be irreducible over $F(Z)$. It is obvious that the condition is necessary. On the other hand, assume that this condition is satisfied and let P be a general point of $\langle Z \rangle/F(Z)$. Let (Z_0, P_0) be any pair of I'. Since $P_0 \in \langle Z_0 \rangle$ and since Z_0 is a specialization of Z over F, there exists a point Q of $\langle Z \rangle$ such that (Z_0, P_0) is a specialization of (Z, Q) over F. Since P is a general point of $\langle Z \rangle/F(Z)$, (Z, Q) is a specialization of (Z, P) over F. Hence (Z_0, P_0) is a specialization of (Z, P) over F, showing that (Z, P) is a general pair of I'/F, whence I is irreducible over F.

By a similar argument one proves the following generalization: *if $\langle Z \rangle$ has exactly h irreducible components over $F(Z)$ and if $P_1, P_2, \ldots, P_h$ are general points of these components over $F(Z)$, then the incidence correspondence I has exactly h irreducible components $I_1, I_2, \ldots, I_h$ over F, where I_q is the locus of (P_q, Z) over F.*

Definition I. 4. 1. *An algebraic system M of d-cycles is an involution if its incidence correspondence I is absolutely irreducible and if, F being a field of definition of I, a general point of the carrier V/F of I belongs to only one cycle of M.*

It is clear that if M is an involution then both M and the carrier V of M are absolutely irreducible and are defined over every field of definition F of I. It is also clear that if the condition stated in the definition is satisfied for one field of definition of the incidence correspondence I, the condition is also satisfied for any field of definition of I, for if F and F' are any two fields of definition of I (and hence also of V) there exists a point which is a general point of V with respect to both fields F and F'. We observe that the incidence correspondence of an involution M of d-cycles is a quasi-rational transformation of its carrier V onto M (and that consequently $\dim V = d + \dim M$). Conversely, if the carrier V of an algebraic system M is an absolutely irreducible variety and if the incidence correspondence of M is a quasi-rational transformation of V, then M is an involution (it is sufficient to note that the graph of a quasi-rational transformation of an absolutely irreducible variety is itself absolutely irreducible).

Definition I. 4. 2. *Let M/F and M'/F be two irreducible algebraic systems of d-cycles and d'-cycles respectively, in one and the same projective space, and let Z be a general cycle of M/F. An irreducible correspondence T/F, with domain M and range M', is called a composition of M with M' if $\langle Z \rangle$ is the carrier of the subsystem $T\{Z\}$ of M'. The system M is said to be composite with M' if there exists a composition of M with M'.*

We can easily show that if T/F is a composition of M/F with M'/F then for *any* cycle X of M it is true that the variety $\langle X \rangle$ is the carrier of the system $T\{X\}$. For let (Z, Z') be a general pair of T/F and let X' be any cycle in $T\{X\}$. Then (X, X') is a specialization of (Z, Z') over F, and since $\langle Z' \rangle$ is contained in $\langle Z \rangle$ it follows that $\langle X' \rangle$ is contained in $\langle X \rangle$. This shows that $\langle X \rangle$ contains the carrier of $T\{X\}$. On the other hand, let Q be any point of $\langle X \rangle$. Since X is a specialization of Z over F it follows, by well-known properties of specializations of cycles, that there exists a point P of $\langle Z \rangle$ such that (X, Q) is a specialization of (Z, P) over F. Since $\langle Z \rangle$ is the carrier of $T\{Z\}$ there exists a cycle Z' in $T\{Z\}$ such that $P \in \langle Z' \rangle$. We extend the specialization $(Z, P) \overset{F}{\rightarrow} (X, Q)$ to a specialization $(Z, P, Z') \overset{F}{\rightarrow} (X, Q, X')$. Then $X' \in T\{X\}$ and furthermore $Q \in \langle X' \rangle$ since $P \in \langle Z' \rangle$. This shows that $\langle X \rangle$ is contained in the carrier of $T\{X\}$, and our assertion that $\langle X \rangle$ is the carrier of $T\{X\}$ is proved. Note that our result also shows that M and M' have the same carrier.

Given M/F and M'/F, there may exist, in general, more than one composition of M with M' (if there exist at all compositions of M with M'). *However, there is one case in which the composition, if it exists, is uniquely determined: it is the case in which M' is an involution.* For assume that M' is an involution and let T be a composition of M with M'. Let V be the common carrier of M and M'. Since M' is an involution, V is irreducible over F. Let P be a general point of V/F and let Z' be *the* cycle of M' which passes through P (whence Z' is a general member of M'/F). Let M_P denote the set of cycles of M which pass through P. It is clear that if Z is any cycle of M_P then $Z' \in T\{Z\}$, since $\langle Z \rangle$ is the carrier of the system $T\{Z\}$ and since therefore this system must contain a cycle which passes through P. Conversely, if Z is any member of M such that $Z' \in T\{Z\}$, then $\langle Z' \rangle \subset \langle Z \rangle$ and hence $Z \in M_P$. Consequently $M_P = T^{-1}\{Z'\}$, and this shows that T is uniquely determined.

We say that an involution is *prime* if its general cycle is prime. The following theorem gives some relevant information about arbitrary involutions and their relation to prime involutions.

THEOREM I. 4. 3. *Let M be an involution, let F be a field of definition of the incidence correspondence I of M and let $Z = m_1\Gamma_1 + m_2\Gamma_2 + \cdots + m_h\Gamma_h$ be a general cycle of M/F, where the Γ_i are the distinct prime components of Z. Then:*

(a) *the h absolutely irreducible varieties $\langle \Gamma_i \rangle$ are defined over the algebraic closure of $F(Z)$ and form a complete set of conjugate varieties over $F(Z)$.*

(b) *The prime cycles Γ_i have the same locus N over F (i.e., they are general cycles of one and the same irreducible algebraic system N/F), N is a prime involution and M is composite with N. Furthermore $m_1 = m_2 = \cdots = m_h$.*

(c) *If for any cycle $X = m(\Delta_1 + \Delta_2 + \cdots + \Delta_h)$ of M [where m denotes the common value of the m_i and where $(\Delta_1, \Delta_2, \ldots, \Delta_h)$ is a zero-cycle in N which is a specialization of the zero-cycle $(\Gamma_1, \Gamma_2, \ldots, \Gamma_h)$, over F] we denote by*

$\mathcal{Z}(X)$ the zero cycle $(\Delta_1, \Delta_2, \ldots, \Delta_h)$ in N, then the totality of cycles $\mathcal{Z}(X)$, $X \in M$, is an involution L on N (of zero-cycles of degree h).

(d) *Conversely, given any positive integer m, given any prime involution N of d-cycles and given an involution L of zero-cycles in N such that N is the carrier of L, the set of all d-cycles of the form $m(\Delta_1+\Delta_2+\cdots+\Delta_h)$ such that $(\Delta_1, \Delta_2, \ldots, \Delta_h) \in L$ is an involution M composite with N.*

PROOF. Since the incidence correspondence I of M is irreducible over F we know that the variety $\langle Z\rangle$ $(=I'\{Z\})$ is irreducible over $F(Z)$, and since the $\langle\Gamma_i\rangle$ are the absolutely irreducible components of $\langle Z\rangle$, assertion (a) is proved.

The Γ_i being conjugate over $F(Z)$, they are *a fortiori* isomorphic over F and therefore have the same locus N over F. Let J be the incidence correspondence of N. If P is a general point of $\langle Z\rangle/F(Z)$ and if, say, $P \in \langle\Gamma_1\rangle$, then (P, Γ_1) is a pair of J. Let (Q, Δ) be any pair of J, so that $\Delta \in N$ and $Q \in \langle\Delta\rangle$. Since Γ_1 is a general cycle of N/F, Δ is a specialization of Γ_1 over F, and thus there exists a point P_1 of $\langle\Gamma_1\rangle$ such that $(P_1, \Gamma_1) \xrightarrow{F} (Q, \Delta)$. We now observe that Z is the only specialization of Z over $F(\Gamma_1)$, for any specialization of Z over $F(\Gamma_1)$ is a cycle Z' of M of which Γ_1 is one of the prime components and which therefore must coincide with Z since $P \in \langle Z'\rangle$ and since P is a general point of V/F. From this observation it follows that Z (i.e., the CHOW point of Z) is purely inseparable over $F(\Gamma_1)$. Consequently $\dim P/F(\Gamma_1) = \dim P/F(Z) = \dim\langle Z\rangle = \dim\langle\Gamma_1\rangle$, showing that P is a general point of $\langle\Gamma_1\rangle/F(\Gamma_1)$. Consequently we have the specialization $(P, \Gamma_1) \rightarrow (P_1, \Gamma_1)$, and combined with the above specialization $(P_1, \Gamma_1) \xrightarrow{F} (Q, \Delta)$ this yields the specialization $(P, \Gamma_1) \xrightarrow{F} (Q, \Delta)$. We have therefore shown that J/F is irreducible and has (P, Γ_1) as general pair. Since I is absolutely irreducible, with (P, Z) as general pair over F, F is quasi-maximally algebraic in $F(P \times Z)$, and therefore also in $F(P \times \Gamma_1)$, since—as we have just pointed out—the field $F(P \times Z)$ is a purely inseparable extension of $F(P \times \Gamma_1)$. This shows that *J is absolutely irreducible* (Cf. p. 85, (2)). It is clear that the carrier V of M is also the carrier of N since the general point P of V/F belongs to $\langle\Gamma_1\rangle$. Let Δ be any cycle of N containing the point P. We extend the specialization $\Gamma_1 \xrightarrow{F} \Delta$ to a specialization $(\Gamma_1, Z) \xrightarrow{F} (\Delta, Z')$. Then Z' is a cycle of M, of the form $Z' = m\Delta + \cdots$ (since $Z = m\Gamma_1 + \cdots$). Thus $\langle Z'\rangle$ contains the point P and therefore $Z'=Z$. It follows that Δ must coincide with one of the cycles Γ_i. Now, P is a general point of $\langle\Gamma_1\rangle$ over the algebraic closure of $F(Z)$ (since we have shown above that $\dim P/F(Z)=\dim\langle\Gamma_1\rangle$) and hence if $i\neq 1$ then $P\notin\langle\Gamma_i\rangle$ (for all the varieties $\langle\Gamma_i\rangle$ are defined and irreducible over the algebraic closure of $F(Z)$ and since $\Gamma_i\not\supset\Gamma_1$ if $i\neq 1$). Since $P\in\langle\Delta\rangle$ we conclude that $\Delta\neq\Gamma_i$ if $i\neq 1$. Hence $\Delta=\Gamma_1$. We have therefore shown that Γ_1 *is the only cycle of* N *which contains the general point* P *of* V/F, and this, combined with the absolute irreducibility of the incidence correspondence J of N, implies that N *is an involution.* Let f denote the locus of the pair (Z,Γ_1) over F. Then f/F is an irreducible algebraic correspondence, with domain M and range N, and it is clear that $f\{Z\}$ consists of the h cycles Γ_i (since these cycles form a complete set of conjugates of $F(Z)$). Hence $\langle Z\rangle$ is the carrier of $f\{Z\}$, showing that M *is composite with* N. As the equalities $m_1=m_2=\cdots=m_h$ are obvious from the preceding considerations, part (b) of the theorem is proved.

If X is any cycle of M then X is of the form $m(\Delta_1+\Delta_2+\cdots+\Delta_h)$, where the h-tuple of cycles Δ_i is a specialization of the h-tuple $(\Gamma_1,\Gamma_2,\ldots,\Gamma_h)$ over F. Hence the totality of cycles $\mathscr{Z}(X)$ is an algebraic system L of zero-cycles (on N), defined and irreducible over F. It is clear that N is the carrier of L, and that the incidence correspondence of L is irreducible over F since any of the pairs $(\Gamma_i,\mathscr{Z}(Z))$ is a general pair of that correspondence, over F. Since the cycles Γ_i form a complete set of conjugates over $F(Z)$, the coordinates of the Chow point of $\mathscr{Z}(Z)$ are purely inseparable over $F(Z)$, whence also over $F(\Gamma_1)$. Since F is quasi-maximally algebraic in $F(\Gamma_1)$, it follows that F is quasi-maximally algebraic in $F(\Gamma_1\times\mathscr{Z}(Z))$, showing that the incidence correspondence of L is absolutely irreducible. Finally, the general cycle Γ_1 of N/F belongs to only one cycle of L, namely to $\mathscr{Z}(Z)$, and thus (c) is proved. The proof of (d) is now straightforward and may be left to the reader.

I. 5. Linear systems and rational transformations (continuation).

We now go back to our linear system L on V studied in I. 3 and to the rational transformation $T: V\rightarrow V'=\mathscr{V}(L)$ obtained by cutting L by hyperplanes. Let P be the general point of the projective model $V=(V,P)$

and let $P'=T\{P\}$ be the corresponding general point of the projective model $V'=(V', P')$. Since the variety $T^{-1}\{P'\}$ is defined and irreducible over $k(P')$ we can consider the cycle $\mathcal{Z}(T^{-1}\{P'\}/k(P'))$ (see I.4). We denote this cycle by Z. Let M/k be the irreducible algebraic system of d-cycles ($d=r-r'$, where $r=\dim V$ and $r'=\dim V'$) which is the locus of Z over k. The variety V is the carrier of M since $P\in\langle Z\rangle$. We denote by K and K' the function fields $k(P)$ $(=k(V))$ and $k(P')$ $(=k(V'))$ respectively. We have $Z=p^e(\Gamma_1+\Gamma_2+\cdots+\Gamma_s)$, where the Γ_i are the distinct absolutely irreducible components of $\langle Z\rangle$. Since $\langle Z\rangle$ is irreducible over $k(P')$, the prime components Γ_i of Z form a complete set of conjugates over K'. Hence the cycle Z (or the CHOW point of Z) is purely inseparable over K'. Consequently, if f denotes the locus of $Z\times P'$ over k then f^{-1} is a quasi-rational transformation of V' onto M. On the other hand, if Q' is any point of $f\{Z\}$ then there exists a point Q of $\langle Z\rangle$ such that the specialization $(Z,P')\overset{k}{\to}(Z,Q')$ can be extended to the specialization $(Z,P',Q)\overset{k}{\to}(Z,Q',P)$ (since $P\in\langle Z\rangle$). Since (Q,P') is a pair of T, it follows that also (P,Q') is a pair of T, and hence $Q'=P'$ since T is a rational transformation. Thus $f\{Z\}$ consists of the single point P', showing that also f is a quasi-rational transformation. *Hence f is a quasi-birational transformation of M onto V'.*

Let I be the incidence correspondence of M. Since f is quasi-birational, each of the two fields $k(P')$ and $k(Z)$ is contained in a purely inseparable extension of the other. Hence $\langle Z\rangle$ is also defined and is irreducible over $k(Z)$, and P is a general point of $\langle Z\rangle$ over $k(Z)$ (as well as over $k(P')$). Consequently, I/k is irreducible and has (P,Z) as general pair. Since k is algebraically closed, I is absolutely irreducible. Since Z is purely inseparable over $k(P')$ and since $k(P')$ is a subfield of $k(P)$, Z is also purely inseparable over $k(P)$. Hence I is a quasi-rational transformation of V onto M, *showing that M is an involution.*

We now use the notations and assumptions of Proposition I.3.3. We choose the algebraically independent quantities $u_0, u_1, \ldots, u_m$ over k as follows: we take for $u_1, u_2, \ldots, u_m$ a set of algebraically independent quantities over $k(P)$ and we take for u_0 the linear form $-\sum_{i=1}^{m} u_i\varphi_i(y)/\varphi_0(y)$. The coefficients $\varphi_i(y)/\varphi_0(y)$ of this linear form generate the function field K' of V'/k over k, and since tr.d.$K'/k\geqq 1$ the condition that $u_0, u_1, \ldots, u_m$ be algebraically independent over k is satisfied. With this choice of the u_i our general cycle D of L/k passes through P, and furthermore P is a general point of $V/k(u_1, u_2, \ldots, u_m)$. Hence, by Proposition I.3.3., P is a general point of $\langle D\rangle/k^*$. We now consider the locus h of the point $D\times Z$ over k; this is an irreducible correspondence, over k, having as domain the linear system L and as range the involution M. Since $\langle Z\rangle = T^{-1}\{P'\}$, we have, by Proposition I.3.1, that $\langle Z\rangle$ is contained in $\langle D\rangle$. Hence $\langle D\rangle$ contains the variety of any cycle which is a specialization of Zov er $k(D)$, i.e., $\langle D\rangle$ contains the carrier of $h\{D\}$. On the other hand, the cycle D is purely inseparable over the field k^* (since D is the only specialization of D over k^*), while the point $u\equiv(u_0, u_1, \ldots, u_m)$ in the projective m-space is purely inseparable over the field $k(D)$ (since, in view of the linear independence of the $\varphi_i(y)$ over k, the point u is its only own specialization over $k(D)$). It follows that the fields k^* and $k(D, u_0)$ are contained in purely inseparable extensions of each other. Therefore P is a general point of $\langle D\rangle$ (not only over k^* but also) over $k(D)$. Since the carrier of $h\{D\}$ is a variety defined over $k(D)$ and since P belongs to that carrier (P belongs to $\langle Z\rangle$), we conclude that $\langle D\rangle$ is the carrier of $h\{D\}$. We have therefore shown that the correspondence h is a composition of L with M, and thus *the linear system L is composite with the involution M.*

The general cycle Z of M has the form $p^e(\Gamma_1+\Gamma_2+\cdots+\Gamma_s)$. Let us denote by X the cycle $\Gamma_1+\Gamma_2+\cdots+\Gamma_s$ and let N be the locus of X over k. It is clear that M is merely the set of all cycles p^eX, $X\in N$. We express this fact by the notation $M=p^eN$. It is also clear that N is an involution and that L is also composite with N. Let us adopt the following terminology: a cycle is *unramified* if all its prime components occur in it with multiplicity 1; an irreducible algebraic system is *unramified* if its general member is an unramified cycle. Thus N is an unramified involution. We shall now derive a maximality property of N which will characterize N as the only unramified involution which has that property and with which the linear system L is composite.

Let N' be any other unramified involution with which L is composite. Let X' be the general member of N' which passes through P, let φ be the locus of $X\times X'$ over k and let ψ be the composition of L with N'. If D_0

is any member of L which passes through P then $\langle D_0 \rangle$ must be the carrier of $\psi\{D_0\}$ (see I.4). Since $P \in \langle D_0 \rangle$ and since X' is the only member of N' which passes through P, it follows that $X' \in \psi\{D_0\}$ and $\langle X' \rangle \subset \langle D_0 \rangle$. Since this inclusion holds for all cycles D_0 of L which pass through P it follows from Proposition I.3.1 that $\langle X' \rangle \subset T^{-1}\{P'\} = \langle Z \rangle = \langle X \rangle$. Hence the carrier of $\varphi\langle X \rangle$ is contained in $\langle X \rangle$. On the other hand, this carrier is a variety defined over $k(X)$ and contains the point P which is a general point of $\langle X \rangle / k(X)$ (since $k(X)$ is a purely inseparable extension of $k(Z)$ and since we have shown earlier in this section that P is a general point of $\langle Z \rangle / k(Z)$). Consequently $\langle X \rangle$ is the carrier of X, whence *φ is a composition of N with N'*. Let us say that an algebraic system S is *maximally composite* with an unramified involution N if S is composite with N and if there exists no unramified involution H, different from N, such that S is composite with H and H is composite with N. It is immediately seen for any involution N with which S is composite there exist unramified involutions H with which S is maximally composite and which are composite with N. The above involution N with which our linear system L is composite is therefore the only unramified involution with which L is maximally composite.

We can now summarize our preceding results in the following theorem:

THEOREM I.5.1. *A linear system L on V, defined over k and free from fixed components, is maximally composite with one and only one unramified involution N. If r is the dimension of V and d is the dimension of the cycles of N, then the rational transform $V' = \mathscr{V}(L)$ of V obtained by cutting L with hyperplanes has dimension $r-d$. A necessary and suffcient condition that the rational transformation $T: V \to V'$ be quasi-birational is that N reduce to the involution of the individual points of V. There exists a quasi-birational transformation f of N/k onto V'/k such that if X is a general cycle of N/k and P' is the corresponding general point of V'/k then $\langle X \rangle = T^{-1}\{P'\}$.*

An involution N of d-cycles whose carrier V has dimension $r = d+1$ is called a *pencil* (on V). The case in which the unramified involution N with which our linear system is maximally composite is a pencil is of special importance. In this case the rational transform V' of V is an irreducible curve. If D is a general cycle of L/k then the algebraic subsystem of N

of which $\langle D\rangle$ is the carrier must be of dimension 0 (since $\langle D\rangle$ and the cycles of N have the same dimension $r-1$; see I.4). Therefore D—and also every other cycle of L—is the sum of a certain number g of (not necessarily distinct) cycles in N. We shall see later on that there exists an integer $e \geqq 0$ such that $g = p^e g_0$ and such that every cycle of L is a sum of g_0 cycles of M, where $M = p^e N$. We shall also prove later on that the *general* cycle D of L/k is a sum of g_0 *distinct* cycles of M. Anticipating these results [see theorem I.6.3 (theorem of BERTINI)], it follows that *the curve V' is of order g_0*, since $\langle D\rangle$ is the total T^{-1}-transform of the general hyperplane section $\langle C'\rangle$ of V' (Proposition I.3.3) and since the total T^{-1}-transform of each point of C' is the variety of a cycle of M (the number of points of C' is the order of V', and each point of $\langle C'\rangle$ is a general point of V'/k). We observe that if L is composite with a pencil N then L is a *reducible linear system*, except when the following conditions are satisfied: $e=0$, $s=1$ (i.e., N is a prime involution) and $g_0=1$ (whence V' is a line). In that case, the general cycle D of L/k coincides with the general cycle X of N/k and hence L *coincides with* N, i.e., L *is an irreducible* (*necessarily linear*) *pencil.*

The preceding theorem enables one to throw some light on the behaviour of the linear system L at exceptional varieties of the rational transformation $T: V \rightarrow V' = \mathscr{V}(L)$ (see I.2). We shall make only the following remarks in this connection, *and we shall only deal with exceptional varieties which are not at the same time fundamental varieties of* T.

Let W be an exceptional variety (not fundamental for T) and let $W' = T[W]$. Then the restriction of T to W has one (and only one) irreducible component T_1 with the property that T_1 is defined at the general point of W/k; T_1 is then a rational transformation of W onto W'. Since W is not fundamental, W is not a base variety of L, and thus L has a trace L_1 on W, this trace being a linear system on W. It is clear that L_1 is the defining linear system of the rational transformation T_1. If W' is a point, then L_1 has dimension zero and consists of a single cycle. That implies and is implied by the following property of W: *the linear subsystem of L consisting of the cycles D_0 such that $\langle D_0\rangle \supset W$ has dimension $m-1$* ($m = \dim L$). If W' has positive dimension, then L_1, after deletion of its fixed components, must yield a linear system which is composite with an involution of

δ-cycles, *where* δ *is an integer greater than* d, since $\dim W - \dim W' > r - r' = d$. *Hence an irreducible subvariety* W *of* V *which is not fundamental for* T *is an exceptional variety of* T *if and only if either* (a) *the trace* L_1 *of* L *on* W *is of dimension zero and* $\dim W > r - r'$, *or* (b) L_1 *is of positive dimension and is composite with an involution of* δ*-cycles*, $\delta > r - r'$. Note that in case (a) the intersection of W with any cycle in L which does not contain W is entirely contained in the base locus of L (and is empty if T has no fundamental points).

REMARK. If V and V' are both surfaces ($r = r' = 2$) and W is a curve we see that W is not an exceptional curve of T if and only if $\dim L_1 > 0$.

I. 6. The theorem of BERTINI.

We shall derive here the classical theorem of BERTINI on reducible linear systems as a direct consequence of the two propositions given below.

PROPOSITION I.6.1. *Let* K/F *be a field of algebraic functions, of transcendence degree* $r \geqq 2$, *let* $z_1, z_2, \ldots, z_m$ *be elements of* K *such that the field* $F(z)$ ($= F(z_1, z_2, \ldots, z_m)$) *has transcendence degree* $s \geqq 2$ *over* F, *let* $u_1, u_2, \ldots, u_m$ *be algebraically independent elements over* K *and let* z_u *denote the linear form* $u_1 z_1 + u_2 z_2 + \cdots + u_m z_m$. *If* F *is quasi-maximally algebraic in* K (*abbrev.*: *q.m.a.*), *then the field* $F(z_u, u)$ ($= F(z_u, u_1, u_2, \ldots, u_m)$) *is q.m.a. in* $K(u)$.

PROOF.

(a) We first achieve a reduction to the case in which $s = r = 2$ and F is m.a. in K. We fix a transcendence basis $\{x_{r-s+1}, x_{r-s+2}, \ldots, x_r\}$ of $F(z)/F$ and we extend it to a transcendence basis $\{x_1, x_2, \ldots, x_r\}$ of K/F. Let F' be the algebraic closure of $F(x_1, x_2, \ldots, x_{r-2})$ in K. Then $\text{tr.d.}\, K/F' = \text{tr.d.}\, F'(z)/F' = 2$. Assuming the proposition in the case $s = r = 2$ we have that the field $F'(z_u, u)$ is q.m.a. in $K(u)$. Now, since at least one of the quantities z_i is a transcendental over F' it is easily seen by a specialization argument applied to the " indeterminates " u_i (and is also well known) that $u_1, u_2, \ldots, u_m$ and z_u are algebraically independent over F'. Hence, if F is q.m.a. in K (therefore also in F') then it follows by a known result (see ZARISKI [26], p. 61) that $F(z_u, u)$ is q.m.a. in $F'(z_u, u)$, and therefore also in $K(u)$.

(b) We assume now that $s = r = 2$ and that F is m.a. in K. We next achieve a reduction to the case $m = 2$. We introduce $2m + 2$ quantities

v_i, w_i, t_1, t_2 $(i=1, 2, \ldots, m)$ which are algebraically independent over K and we set

(1) $$u_i = t_1 v_i + t_2 w_i, \qquad i = 1, 2, \ldots, m;$$

(2) $$z_v = v_1 z_1 + v_2 z_2 + \cdots + v_m z_m;$$

(2') $$z_w = w_1 z_1 + w_2 z_2 + \cdots + w_m z_m;$$

(3) $$z_u = t_1 z_v + t_2 z_w = u_1 z_1 + u_2 z_2 + \cdots + u_m z_m.$$

We have by (1)

(4) $$K(t_1, t_2, v, w) = K(t_1, t_2, v, u),$$

and hence also the $2m+2$ quantities t_1, t_2, v_i, u_i are algebraically independent over K. In particular, the u_i are algebraically independent over K, as required by the conditions stated in the proposition.

We now set

$$F' = F(v, w), \qquad K' = K(v, w).$$

We have tr.d.$K'/F' = 2$ and, by the above cited result, F' is m.a. in K'. Again, by a specialization argument applied to the " indeterminates " v_i and w_i (see, for instance, ZARISKI [24], p. 287), it follows that z_v and z_w are algebraically independent over F'. Since t_1 and t_2 are algebraically independent over K' and since $z_u = t_1 z_v + t_2 z_w$ we conclude (assuming the truth of the proposition in the case $s=m=r=2$) that $F'(z_u, t_1, t_2)$ is q.m.a. in $K'(t_1, t_2)$. Now, the field $F'(z_u, t_1, t_2)$ is generated over $F(z_u, u)$ by the $m+2$ quantities t_1, t_2, v_i, as follows from (1), and by (4) these quantities are algebraically independent over $K(u)$, hence also over $F(z_u, u)$. Therefore $F(z_u, u)$ is m.a. in $F'(z_u, t_1, t_2)$. Consequently, $F(z_u, u)$ is q.m.a. in $K'(t_1, t_2)$, and therefore also in $K(u)$, since $K(u)$ is contained in the field $K'(t_1, t_2)$.

(c) We now assume that $s=m=r=2$ and we next achieve a reduction to the case in which K is a Galoisian extension of $F(z_1, z_2)$. The field K is an algebraic extension of $F(z_1, z_2)$. Let K_0 be the maximal separable extension of $F(z_1, z_2)$ in K. Then $F(z_u, u)$ is q.m.a. in $K(u)$ if it is q.m.a. in $K_0(u)$. Hence we may replace in the proof the field K by the field K_0. We may therefore assume that K is a separable algebraic extension of $F(z_1, z_2)$. Now let K' be the least normal extension of $F(z_1, z_2)$ containing K. Then the u_i are also algebraically independent over K', and hence if we denote by F' the algebraic closure of F in K' and if we assume the truth of the proposition in the case in which K is a Galoisian extension of $F(z_1, z_2)$, then $F'(z_u, u)$ is q.m.a. in $K'(u)$. We shall show now that

$F'(z_u,u)\cap K(u)=F(z_u,u)$, and this will prove that $F(z_u,u)$ is q.m.a. in $K(u)$. Let t be a common element of the two fields $F'(z_u,u)$ and $K(u)$. The elements u_1, u_2 and z_u $(=u_1z_1+u_2z_2)$ are algebraically independent over F, hence also over F'. Consequently the expression of t as a quotient of two relatively prime polynomials in u_1, u_2 and z_u, with coefficients in F', is uniquely determined provided we normalize one of these polynomials by imposing the condition that a preassigned non-zero coefficient of that polynomial be equal to 1. Now, if f is any automorphism of K'/K then we must have $f(t)=t$ since $t\in K(u)$, and therefore the coefficients of the two polynomials must remain invariant under f. Consequently, these coefficients must belong to K, and since they are algebraic over F they must belong to F. Hence $t\in F(z_u,u)$, and this proves our assertion.

(d) We now assume therefore that K is a Galoisian extension of $F(z_1,z_2)$. We introduce a second linear form $z_v=v_1z_1+v_2z_2$ with indeterminate coefficients v_1, v_2 which are assumed to be algebraically independent over $K(u)$. It is clear that $K(u,v)$ is a Galoisian extension of $F(u,v,z_1,z_2)$ since each automorphism of $K/F(z_1,z_2)$ has a unique extension to $K(u,v)/F(u,v,z_1,z_2)$ and since $[K(u,v):F(u,v,z_1,z_2)]=[K:F(z_1,z_2)]$. The Galois groups of $K(u,v)/F(u,v,z_1,z_2)$ and of $K/F(z_1,z_2)$ are isomorphic, and we shall identify these two groups by identifying each automorphism of $K/F(z_1,z_2)$ with its extension to $K(u,v)/F(u,v,z_1,z_2)$. We denote by H_u the algebraic closure of $F(z_u,u)$ in $K(u)$ and by H_v the algebraic closure of $F(z_v,v)$ in $K(v)$. We shall prove that

$$H_u(z_v,v)=H_v(z_u,u). \tag{5}$$

Since both fields in (5) contain the field $F(u,v,z_1,z_2)$ it will be sufficient to show that the Galois group of $K(u,v)/H_u(z_v,v)$ coincides with the Galois group of $K(u,v)/H_v(z_u,u)$. Let G_1 and G_2 denote these two Galois groups. The field $K(u,v)$ admits an automorphism f which interchanges u_i and v_i $(i=1,2)$ and which reduces to the identity on K. It is clear that $G_2=f^{-1}G_1f$. On the other hand, f commutes with each element of the Galois group of $K(u,v)/F(u,v,z_1,z_2)$ since the automorphisms in this group are extensions of automorphisms of $K/F(z_1,z_2)$. Therefore $G_1=G_2$, as asserted.

Since F is m.a. in K, F is also m.a. in $K(v)$ and therefore also in H_v. Now, z_u, u_1 and u_2 are algebraically independent over $F(z_v,v)$ since z_u and

z_v are algebraically independent over $F(u,v)$. Hence z_u, u_1, u_2 are also algebraically independent over H_v. Therefore $F(z_u,u)$ is maximally algebraic in $H_v(z_u,u)$, and this implies by (5) that $F(z_u,u)=H_u$. This completes the proof of Proposition I.6.1.

PROPOSITION I.6.2. *Let F, F' and K be subfields of the universal domain such that F' and K are extensions of F, free over F and K is a finitely generated and regular extension of F. Let $\mathfrak{P}$ be a prime divisor of K/F, with residue field L, let $K'=KF'$ be the compositum of K and F' and let $\mathfrak{P}'$ be an extension of $\mathfrak{P}$ to K'/F' such that the fields F' and L, regarded as subfields of the residue field L' of $\mathfrak{P}'$, be free over F.*[4] *If we set $p^e=[L:F]_i/[L':F']_i$, then p^e is the reduced ramification index of $\mathfrak{P}'$ with respect to $\mathfrak{P}$*, i.e., $\mathfrak{P}=\mathfrak{P}'^{p^e}$ *(we set $p^e=1$ if $p=0$).*[5]

PROOF. We observe that K' is a regular extension of F'. Hence if F'' is an extension of F' such that K' and F'' are free over F' and if the proposition holds for F, F', K and F', F'', K', then it also holds for F, F'', K. Hence it is sufficient to prove the proposition in the following two cases: (1) F' is a purely transcendental extension of F; (2) F' is an algebraic extension of F. In case (1) it is easily seen that there is only one divisorial extension $\mathfrak{P}'$ of $\mathfrak{P}$ satisfying our assumptions and that if $F'=F(\{u\})$, where $\{u\}$ is a transcendence set in F'/F, then L' can be identified with $L(\{u\})$ in such a way that $\{u\}$ is still a transcendence set in L'/L. Hence

4) The geometric meaning of our assumption concerning $\mathfrak{P}'$ can be explained as follows

Let $r=\text{tr.d.}\,K/F$, whence $\text{tr.d.}\,L/F=r-1$. Let V/F be a projective model of K/F such that the center W of $\mathfrak{P}$ on V is of dimension $r-1$. Let $x_1, x_2, \dots, x_n$ be the non-homogeneous coordinates of the general point of V/F and let $z_1, z_2, \dots, z_n$ the non-homogeneous coordinates of the general point of W/F, where $z_i=x_i\mathfrak{P}$. Our assumption that F' and L are free over F is equivalent with the assumption that $\text{tr.d.}\,F'(z)/F'=\text{tr.d.}\,F(z)/F$, i.e., $\text{tr.d.}\,F'(z)/F'=r-1$. This signifies that (z) is a general point of an irreducible component of W/F', or equivalently, that $\mathfrak{P}'$ is a prime divisor of K'/F' and that the center of $\mathfrak{P}'$ on projective model V/F' is an irreducible component of W/F' (note that V/F' is irreducible since V is an absolutely irreducible variety, in view of the regularity of the extension K/F). Thus the number of extensions $\mathfrak{P}'$ of $\mathfrak{P}$ satisfying the condition of the theorem is finite.

5) The present proposition can be established for arbitrary places $\mathfrak{p}$ of K/F (and not only for prime divisors, as in the text).

$[L:F]_i=[L':F']_i$, i.e., $p^e=1$. On the other hand it is also immediately seen that $\mathfrak{P}'$ is unramified in the present case. The details of the proof are straightforward and may be left to the reader. Hence we have only to consider the case in which F' is an algebraic extension of F.

We assume now that F' is an algebraic extension of F. Let F_1 be an intermediate field such that F_1 is a finite algebraic extension of F, let $\mathfrak{P}_1$ be the restriction of $\mathfrak{P}'$ to $K_1=KF_1$ and let L_1 be the residue field of $\mathfrak{P}_1$. We can find F_1 in such a way that $\mathfrak{P}'$ is unramified with respect to $\mathfrak{P}_1$. It is easily seen that we can also find F_1 in such a way as to have $[L':F']_i=[L_1:F_1]_i$. Hence we can find F_1 in such a way that both these conditions are satisfied simultaneously. For such a choice of F_1 the proposition is valid for the fields F_1, F', K_1 and the prime divisor $\mathfrak{P}_1$, since $\mathfrak{P}'$ is unramified over $\mathfrak{P}_1$ and since we have also $[L_1:F_1]_i/[L_1:F']_i=1$. Hence it is only necessary to prove the proposition for F, F_1 and K. In other words: we may assume that F' is a finite algebraic extension of F. We have therefore only two cases to consider: (a) F' is a simple separable algebraic extension of F; (b) F' is a purely inseparable extension of F of degree p.

CASE (a).

Let $F'=F(\alpha)$ and let $g(X)$ be the minimal polynomial of α over F. Since F is maximally algebraic in K, $g(X)$ is also irreducible over K. Let R be the valuation ring of the prime divisor $\mathfrak{P}$. We have $K'=K(\alpha)=K[\alpha]$. If n is the degree of $g(X)$ and if $t=a_0+a_1\alpha+\cdots+a_{n-1}\alpha^{n-1}$ $(a_i\in K)$ is any element of K' which is integral over R, then taking the conjugates of t over K and using the fact that $g(X)$ is irreducible over K we find that $a_i\sqrt{D}\in R$, where D is the discriminant of $g(X)$ ($D\neq 0$ since $g(X)$ is a separable polynomial). Since $D\in F$ it follows that the a_i are integral over R, whence $t\in R[\alpha]$. Thus $R[\alpha]$ is the integral closure of R in K'. By known results it follows then that $\mathfrak{P}'$ is determined by a maximal ideal $\mathfrak{p}'$ of $R[\alpha]$ in such a fashion that the valuation ring R' of $\mathfrak{P}'$ is the quotient ring of $R[\alpha]$ with respect to $\mathfrak{p}'$. Let $g(X)=g_1(X)g_2(X)\cdots g_m(X)$ be the factorization of $g(X)$ into irreducible factors in $L(X)$, where we assume that each $g_i(X)$ is a monic polynomial. For each polynomial $g_i(X)$ we fix a monic polynomial $G_i(X)$ with coefficients in R and of the same degree as $g_i(X)$, such that the coefficients of $g_i(X)$ are the $\mathfrak{P}$-residues of the corresponding coefficients of $G_i(X)$. The $\mathfrak{P}$-residues of the coefficients of the polynomial

$G_1(X)G_2(X)\cdots G_m(X)-g(X)$ are then zero, and consequently these coefficients belong to the maximal ideal $\mathfrak{p}$ of R. If we now replace X be α we conclude that

$$G_1(\alpha)G_2(\alpha)\cdots G_m(\alpha)\in R[\alpha]\cdot\mathfrak{p}. \tag{6}$$

Hence the $\mathfrak{P}'$-residue of one of the quantities $G_i(\alpha)$ must be zero. Let, say, the $\mathfrak{P}'$-residue of $G_1(\alpha)$ be zero, whence $G_i(\alpha)\in\mathfrak{p}'$. Let β be the $\mathfrak{P}'$-residue of α. Thus $g_1(\beta)=0$, consequently $g_i(\beta)\neq 0$ if $i\neq 1$. This shows that the product $G_2(\alpha)G_3(\alpha)\cdots G_m(\alpha)$ does not belong to $\mathfrak{p}'$. Consequently, by (6):

$$G_1(\alpha)\in R'\cdot\mathfrak{p}. \tag{7}$$

Now, let $t=f(\alpha)$ be any element of $\mathfrak{p}'$, where $f(X)$ is a polynomial with coefficients in R. Upon division by $G_1(X)$ we have $f(X)=A(X)G_1(X)+B(X)$, where $A(X)$ and $B(X)$ are polynomials with coefficients in R and where the degree of $B(X)$ is less than the degree of $G_1(X)$. If we denote by $b(X)$ the polynomial in $L[X]$ whose coefficients are the $\mathfrak{P}$-residues of the coefficients of $B(X)$, then substituting α for X in the above identity and taking $\mathfrak{P}'$-residues on both sides we find $b(\beta)=0$. Since the degree of $b(X)$ is less than the degree of $g_1(X)$ and since $g_1(X)$ is the minimal polynomial of β in $L[X]$, all the coefficients of $b(X)$ must be zero, i.e., all the coefficients of $B(X)$ must lie in $\mathfrak{p}$. Therefore $t-A(\alpha)G_1(\alpha)\in R[\alpha]\cdot\mathfrak{p}$, whence, by (7), $t\in R'\cdot\mathfrak{p}$. Since the maximal ideal of R' is generated by the elements of $\mathfrak{p}'$, we conclude that $R'\cdot\mathfrak{p}$ is the maximal ideal of R', and this shows that the prime divisor $\mathfrak{P}'$ is unramified over $\mathfrak{P}$. On the other hand, the residue field L' of $\mathfrak{P}'$ is obviously equal to $L(\beta)$ and is therefore a separable extension of L. Consequently, $[L':F']_i=[L(\alpha):F(\alpha)]_i=[L:F]_i$ (we have here identified β with α), i.e., $e=0$, and the proposition is proved in this case.

CASE (b).

Let F' be a purely inseparable extension of F, of degree p. In this case $\mathfrak{P}'$ is the only extension of $\mathfrak{P}$ to K' and we have, by a known result,[6] that $p=p^{\rho}[L':L]$, where p^{ρ} is the reduced ramification index of $\mathfrak{P}'$ over $\mathfrak{P}$. It is obvious that in the present case L' is a purely inseparable extension of L, whence $[L':L]_i=[L':L]$. We have $p^e=[L:F]_i/[L':F']_i=$

6) See my joint paper [6] with I. S. COHEN.

$[F':F]_i/[L':L]_i = p^\rho$, and this completes the proof of Proposition I.6.2.

In the following statement of the theorem of BERTINI we use the notations of I.3.

THEOREM I.6.3 (*theorem of* BERTINI). *Let V be an absolutely irreducible variety, defined and normal over an algebraically closed field k, and let L be a linear system on V, of positive dimension m, defined over k and free from fixed components. Let D be a general cycle of L, corresponding to indeterminate values $u_0, u_1, \ldots, u_m$ of the homogeneous parameters of the system L, and let k^* denote the field $k(u_0, u_1, \ldots, u_m)$. Then D coincides with the cycle $\mathscr{Z}(\langle D\rangle/k^*)$,* i.e., *$D$ is of the form $p^e\Delta$, where p^e is the order of inseparability of the irreducible variety $\langle D\rangle/k^*$ and Δ is the sum of the absolutely irreducible components of $\langle D\rangle$* (see I.4). *Furthermore, if L is not composite with a pencil* (*or equivalently, if the dimension of the rational transform $\mathscr{V}(L)$ is greater than* 1) *then Δ is a prime cycle* (*and hence $\langle D\rangle$ is an absolutely irreducible variety*).

PROOF. The $(r-1)$-dimensional variety $\langle D\rangle$ is defined and is irreducible over k^* (Proposition I.3.3). Hence $\langle D\rangle$ defines a unique prime divisor $\mathfrak{P}$ of $k^*(V)/k^*$ whose center is $\langle D\rangle$. If $y_0, y_1, \ldots, y_n$ are homogeneous coordinates of a general point of V/k we set $x_j = y_j/y_0, yx'_i = \varphi_i(y)/\varphi_0(y)$ (whence (x') is a general point of V'/k, where $\mathscr{V}' = \mathscr{V}(L)$). If $u_0, u_1, \ldots, u_m$ are algebraically independent quantities over $k(x)$ and if we set $t = u_0 + u_1x'_1 + \cdots + u_mx'_m$, then we may assume that D is the null divisor of the function t. If v denotes the valuation of $k^*(V)$ $(= k^*(x)\,;\ k^* = k(u))$ defined by $\mathfrak{P}$ then it is well known (and the proof is straightforward) that $v(t) = 1$. Let $\bar{k}^*$ be the algebraic closure of k^* in the universal domain, let $\Delta_1, \Delta_2, \ldots, \Delta_g$ be the absolutely irreducible components of $\langle D\rangle$ and let $\mathfrak{P}'_i$ be the prime divisor of $\bar{k}^*(x)/\bar{k}^*$ whose center on $V/\bar{k}^*$ is Δ_i. Let v'_i be the valuation of $\bar{k}^*(x)$ corresponding to the divisor $\mathfrak{P}'_i$. By the definition of the divisor of a rational function on V it follows that the null divisor D of the function t is of the following form: $D = m_1\Delta_1 + m_2\Delta_2 + \cdots + m_g\Delta_g$, where $m_i = v'_i(t)$. Since $v(t) = 1$, m_i is the reduced ramification index of $\mathfrak{P}'_i$ with respect to $\mathfrak{P}$. All the assumptions of Proposition I.6.2. are satisfied if we replace in that proposition F, F', K and $\mathfrak{P}'$ by $k^*, \bar{k}^*, k^*(x)$ and $\mathfrak{P}'_i$ respectively. Hence each m_i is a power of p. Since the Δ_i are conjugate prime cycles over k^* and since the rational function t is defined over k^*, it follows that $m_1 = m_2 = \cdots = m_g$. This shows that the cycle D is indeed of the form

$p^e \Delta$, where Δ is the unramified cycle $\Delta_1 + \Delta_2 + \cdots + \Delta_g$. Let $\mathfrak{P}'$ stand for any of the prime divisors $\mathfrak{P}'_i$. The residue field L of $\mathfrak{P}$ is a function field of $\langle D \rangle / k^*$. If L' is the residue field of $\mathfrak{P}'$ then $[L' : \bar{k}^*]_i = 1$ since $\bar{k}^*$ is an algebraically closed field. Therefore, by Proposition I.6.2, we have $p^e = [L : k(u_0, u_1, \ldots, u_m)]_i$, and this proves the first part of the theorem.

Let $\bar{x} = (\bar{x}_1, \bar{x}_2, \ldots, \bar{x}_n)$ be a general point of $\langle D \rangle / k^*$ and let $\bar{x}' = (\bar{x}'_1, \bar{x}'_2, \ldots, \bar{x}'_m)$ be the corresponding general point of $\langle D' \rangle / k^*$, where $\bar{x}'_i = \varphi_i(\bar{x}) / \varphi_0(\bar{x})$ (see I.3.). If L is not composite with a pencil then $\dim V' \geqq 2$, whence the field $k(\bar{x}'_1, \bar{x}'_2, \ldots, \bar{x}'_m)/k$ has transcendence degree $\geqq 2$. Since $u_0 = -(u_1 \bar{x}'_1 + u_2 \bar{x}'_2 + \cdots + u_m \bar{x}'_m)$ and $u_1, u_2, \ldots, u_m$ are algebraically independent over $k(\bar{x})$, it follows from Proposition I.6.1 that k^* $(= k(u_0, u_1, \ldots, u_m))$ is quasi-maximally algebraic in the function field L of $\langle D \rangle / k^*$ (note that $L = k^*(\bar{x}) = k(\bar{x}, u_1, u_2, \ldots, u_m)$ and that k is algebraically closed). Hence $\langle D \rangle$ is absolutely irreducible, $g = 1$ and Δ is a prime cycle. This completes the proof of the theorem of Bertini.

We observe that if L is the system of hyperplane sections of V then $V = V'$ and $p^e = 1$. *Hence the general hyperplane section of V/k is a prime cycle* (if $\dim V > 1$).

We shall now give other characterizations of the multiplicity p^e with which the absolutely irreducible components of the variety $\langle D \rangle$ occur in the general cycle D of the linear system L/k.

Proposition I.6.4. *Let $V' = \mathscr{V}(L)$ be the rational transform of V obtained by cutting L with hyperplanes and let p^e be the multiplicity with which the absolutely irreducible components of $\langle D \rangle$ occur in the general cycle D of L/k. Then $k(V') \subset (k(V))^{p^e}$ and $k(V') \not\subset (k(V))^{p^{e+1}}$.*

Proof. We shall use the notations of the proof of the theorem of Bertini. We first observe that if $k(V') \subset (k(V))^p$ and if we set $x'_i = z_i{}^p (z_i \in k(V))$, then D is the p-fold of the null-cycle of the function $v_0 + v_1 z_1 + v_2 z_2 + \cdots + v_m z_m$, where $v_i = \sqrt[p]{u_i}$, and L consists of the p-folds of the cycles of the linear system L_1 defined by the functions $1, z_1, z_2, \ldots, z_m$. We can then replace in the proof the system L by the system L_1. It follows that in order to prove the proposition we have only to prove the following: *if $e \geqq 1$ then $k(V') \subset (k(V))^p$.*

Suppose then that $e \geqq 1$. Let $(\bar{x}) = (\bar{x}_1, \bar{x}_2, \ldots, \bar{x}_n)$ be a general point of $\langle D \rangle / k^*$ and let $(\bar{x}') = (\bar{x}'_1, \bar{x}'_2, \ldots, \bar{x}'_m)$ be the corresponding general point of the general hyperplane section $\langle D'\rangle$ of V'/k $(x_i'=\varphi_i(\bar{x})/\varphi_0(\bar{x}))$. By BERTINI's theorem the field $k^*(\bar{x})$ is an inseparable extension of k^*. Since tr.d.$k^*(\bar{x})/k^*=r-1$, the space of derivations of $k^*(\bar{x})/k^*$ has dimension $\geqq r$. We may therefore assume that the generators $\bar{x}_i$ of $k^*(\bar{x})/k^*$ have been so numbered that there exist derivations $D_1, D_2, \ldots, D_r$ of $k^*(\bar{x})/k^*$ such that $D_i\bar{x}_i=1$ and $D_i\bar{x}_j=0$ if $i\neq j$, for $i,j=1,2,\ldots,r$. Since $(\bar{x})$ is also a general point of V/k (Proposition I.3.3) and since therefore tr.d.$k(\bar{x})/k=r$, it follows that $\{\bar{x}_1, \bar{x}_2, \ldots, \bar{x}_r\}$ is a separating transcendence basis of $k(\bar{x})/k$, that the restrictions of the D_i to $k(\bar{x})$ form a basis of the space of derivations of $k(\bar{x})/k$ and that finally each derivation D_i maps $k(\bar{x})$ into itself.

Now, we have $u_0+u_1\bar{x}_1'+\cdots+u_m\bar{x}_m'=0$, and hence

$$u_1D_i\bar{x}_1'+u_2D_i\bar{x}_2'+\cdots+u_mD_i\bar{x}_m'=0.$$

Since $D_i\bar{x}_q'\in k(\bar{x})$ $(i=1,2,\ldots,r;\ q=1,2,\ldots,m)$ and since $u_1, u_2, \ldots, u_m$ are algebraically independent over $k(\bar{x})$ (Proposition I.3.3) it follows that $D_i\bar{x}_q'=0$ for $i=1,2,\ldots,r$ and $q=1,2,\ldots,m$. Hence $D\bar{x}_q'=0$ for $q=1,2,\ldots,m$ and for every derivation of $k(\bar{x})/k$. This implies that the quantities $\bar{x}_1', \bar{x}_2', \ldots, \bar{x}_q'$ belong to $(k(\bar{x}))^p$. Since $k(\bar{x})$ is an isomorphic copy of the field $k(x)$ it follows that $k(x_1', x_2', \ldots, x_m')\subset(k(x_1, x_2, \ldots, x_n))^p$, i.e., $k(V')\subset(k(V))^p$, as asserted.

COROLLARY I.6.5. *If the general cycle D of a linear system L/k is the p-fold of a cycle D_1 (p=characteristic of k) then L consists of the p-folds of the cycles of another linear system L_1/k.*

Obvious.

COROLLARY I.6.6. *The section of V/k with a general hyperplane $u_0+u_1X_1+u_2X_2+\cdots+u_nX_n=0$ is an unramified cycle, and the variety of that cycle has order of inseparability 1 over $k(u_0, u_1, \ldots, u_n)$.*

In fact, we have in this case $V'=V$.

PROPOSITION I.6.7. *In the notations of the theorem of* BERTINI, *let D' be the hyperplane section of V' which corresponds to D (and which is therefore cut out by the hyperplane $u_0+u_1X_1'+u_2X_2'+\cdots+u_mX_m'=0$). Let W be an absolutely irreducible component of $\langle D\rangle$ and let W' be the corresponding irreducible component of $\langle D'\rangle$ (here $W=\langle D\rangle$ and $W'=\langle D'\rangle$ if L is not composite with a pencil). Then*

$$p^e=[k(V):k(V')]_i/[\bar{k}^*(W):\bar{k}^*(W')]_i, \tag{8}$$

where $\bar{k}^*$ denotes the algebraic closure of the field $k^*=k(u_0, u_1, \ldots, u_m)$.

PROOF. We shall use the following well-known formula concerning orders of inseparability, due to WEIL ([21], p. 23, Proposition 27): if L is a finitely generated extension of a field k and K is a finitely generated extension of L (where k, L and K are subfields of the universal domain) then

$$(9) \qquad [K:k]_i = [K:L]_i[L:k]_i/[\bar{k}K:\bar{k}L]_i,$$

where $\bar{k}$ is the algebraic closure of k. By the theorem of BERTINI and in the notations of the proof of that theorem we have

$$p^e = [k(\bar{x},u):k(u)]_i,$$

where $\bar{x}$ stands for $(\bar{x}_1, \bar{x}_2, \ldots, \bar{x}_n)$ and u stands for $(u_0, u_1, \ldots, u_m)$. If, in (9), we replace k, L and K by $k(u)$, $k(\bar{x}',u)$ and $k(\bar{x},u)$ respectively (where $\bar{x}'_i = \varphi_i(\bar{x})/\varphi_0(\bar{x})$) we find

$$(10) \qquad p^e = [k(\bar{x},u):k(\bar{x}',u)]_i[k(\bar{x}',u):k(u)]_i/[\overline{k(u)}(\bar{x}):\overline{k(u)}(\bar{x}')]_i,$$

where $\overline{k(u)}$ is the algebraic closure of $k(u)$. Since

$$u_0 + u_1\bar{x}'_1 + u_2\bar{x}'_2 + \cdots + u_m\bar{x}'_m = 0,$$

we have $k(\bar{x}, u) = k(\bar{x}, u_1, u_2, \ldots, u_m)$ and $k(\bar{x}', u) = k(\bar{x}, u_1, u_2, \ldots, u_m)$. Since $u_1, u_2, \ldots, u_m$ are algebraically independent over $k(\bar{x})$ it follows that

$$(11) \qquad [k(\bar{x}, u):k(\bar{x}', u)]_i = [k(\bar{x}):k(\bar{x}')]_i = [k(V):k(V')]_i.$$

The integer $[k(\bar{x}', u):k(u)]_i$ is the order of inseparability of the general hyperplane section D' of V'/k, with respect to the field $k(u)$, and is equal to 1 by Corollary I.6.6. Finally, the integer $[\overline{k(u)}(\bar{x}):\overline{k(u)}(\bar{x}')]_i$ is clearly equal to the order of inseparability $[\bar{k}^*(W):\bar{k}^*(W')]_i$. Our proposition now follows from (10) and (11).

REMARK. The equality

$$p^e = [k^*(\langle D \rangle):k^*]_i$$

given in the theorem of BERTINI (Theorem I.6.3) is a special case of a result due to WEIL ([21], p. 161, Theorem 11) which gives a determination of the multiplicity of the absolutely irreducible components of the general member of an irreducible algebraic system of cycles.

I. 7. Antiregular birational transformations.

DEFINITION I.7.1. *A birational transformation $T: V \to V'$ is said to be antiregular at a point Q of V if T^{-1} is regular at each point of $T\{Q\}$. The birational transformation T is said to be antiregular if it is antiregular at each point of V.*

If V and V' are birationally equivalent projective models and if T is the birational transformation of V onto V' we shall say that V' *dominates* V and we shall write $V < V'$ if T is antiregular (equivalently: if T^{-1} is regular). If $V < V'$ and $V' < V$, i.e., if T is biregular, then V and V' are biregularly equivalent models, and we write $V \equiv V'$.

If V, V' and V'' are birationally equivalent projective models and if T, T' and T'' are the birational transformations $V \to V'$, $V' \to V''$ and $V \to V''$ respectively, then we write $T'' = T \circ T'$. We do not always have $T \circ T' = TT'$, where TT' denotes the ordinary product of the two transformations T and T'. However, we note the following fact:

LEMMA I.7.2. *If for a given subset G of V we have that either T^{-1} or T' is single-valued at each point of $T\{G\}$ (in particular, if either T is antiregular at each point of G or T' is regular at each point of $T\{G\}$) then $T \circ T' = TT'$ on G.*

PROOF. Obvious.

DEFINITION I.7.3. *Two birational transformations $T_1 : V \to V'$ and $T_2 : V \to V''$ of a projective model V are said to be biregularly equivalent at a point Q of V if the birational transformation $T_1^{-1} \circ T_2 : V' \to V''$ is biregular at each point of $T_1\{Q\}$. The two birational transformations T_1 and T_2 are said to be biregularly equivalent if they are biregularly equivalent at each point of V.*

It is obvious that the above definition is symmetric in T_1 and T_2 and is also reflexive and transitive; it thus establishes indeed an equivalence relation (local or global) between the various birational transformations of a given projective model V. If T_1 and T_2 are biregularly equivalent at a given point Q then we have locally at Q: $T_2 = T_1 T$, where T $(= T_1^{-1} \circ T_2)$ is a birational transformation of V' which is biregular at each point of $T_1\{Q\}$; and conversely.

With each *rational* transformation $T : V \to V'$ we can associate an infinite sequence $\{T_1, T_2, \ldots\}$ of biregularly equivalent antiregular birational transformations $T_i : V \to V_i$ of V which, if T itself is an antiregular birational transformation, are biregularly equivalent to T. We define the T_i by induction on i as follows: $T_0 = T$, $V_0 = V'$; V_i is the graph of T_{i-1} (see I.1), and T_i is the inverse of the projection of V_i onto V. Since the projection onto V of the graph of any rational transformation of V is always a regular birational transformation, each T_i is an antiregular birational transformation of V onto $V_i (i \geqq 1)$. We have $T_i = T_{i-1} \circ T'_{i-1}$, where T'_{i-1} is the inverse of the projection of the graph of T_{i-1} onto V_{i-1} $(i \geqq 1)$. Since for $i \geqq 2$ the birational transformation T_{i-1} is antiregular, T'_{i-1} is biregular, and thus T_i and T_{i-1} are biregularly equivalent birational transformations if $i \geqq 2$, and this remains true also for $i=1$ provided T itself is birational and antiregular. We shall call T_i *the i-th augmentation of* T.

Let $P \equiv (y)$ be the general point of the projective model $V(=(V,P))$ (over k) and let $P' \equiv (y') = (\varphi_0(y), \varphi_1(y), \ldots, \varphi_m(y)) = T\{P\}$ be general point of the projective model $V'(=(V', P'))$. We use here the notations of I.3. The variety V_i is the projective model (V_i, P_i), where $P_i = T_i\{P\}$. The homogeneous coordinates of P_1 are the products $y_i \varphi_j(y)$ taken in a suitable order (see I.1). If L is the defining linear system of T (see I.3) then it is clear that the defining linear system of T_1, i.e., the linear system defined on V by the module of forms generated by the products $y_i \varphi_j(y)$, is the smallest linear system on V which contains all cycles of the form $X + Y$, where X is any cycle in L and Y is any hyperplane section of V. We refer to this linear system as the *minimal sum* of L and the system H of hyperplane sections of V and we denote it by $L+H$, or also by L_1. Similarly, we find that the defining linear system L_i of T_i is the minimal sum $L_{i-1} + H$, or also $L + iH$, where iH denotes the minimal sum $H + H + \cdots + H$ (i times), i.e., the linear system cut out on V by the hypersurfaces of order i.

Let $\mathfrak{A}$ be the homogeneous ideal in $k[y]$ generated by the $m+1$ forms $\varphi_j(y)$ and let $\mathfrak{A}(s)$ denote the k-module of homogeneous elements of $\mathfrak{A}$ of degree s. If we denote by d the common degree of the forms $\varphi_j(y)$ then the transformations T_i can also be described ideal theoretically in terms of the ideal $\mathfrak{A}$, as follows: the elements of a suitable basis of the k-module $\mathfrak{A}(d+i)$ form a set of strictly homogeneous coordinates of the point P_i, and (P, P_i) is a general pair of T_i/k. Since, as a rule, we shall identify projectively equivalent varieties, any k-basis of $\mathfrak{A}(d+i)$ can be used to describe the transformation T_i.

The ideal $\mathfrak{A}$ which occurs above is not the most general homogeneous ideal in $k[y]$ since it has a basis of forms of like degree. However, also for an arbitrary homogeneous ideal $\mathfrak{A}$ in $k[y]$ (different from the zero ideal) we can introduce a sequence of biregularly equivalent antiregular birational transformations T_i of $V (i \geqq 0)$: using the above notations, we need only take for d the smallest integer such that $\mathfrak{A}$ has a basis of forms of degree $\leqq d$. The rational transformation T_0 defined (to within projective equivalence) by the k-module $\mathfrak{A}(d)$ plays then the role of the transformation T. It is immediately seen that the ideal generated in $k[y]$ by the forms belonging to $\mathfrak{A}(d)$ coincides with $\mathfrak{A}$ in all the forms of degree $\geqq d$ and hence differs from $\mathfrak{A}$ only by an irrelevant primary component (i.e., by a primary component whose prime ideal is the irrelevant prime ideal $(y_0, y_1, \ldots, y_n)$). If we wish to indicate explicitly the ideal $\mathfrak{A}$ by means of which the transformations T_i $(i \geqq 1)$ can be defined we shall denote T_i by $T(d+i, \mathfrak{A})$, or also by $T(\mathfrak{A})$ if we are not concerned with any particular choice of the degree $d+i$ of the forms in the ideal $\mathfrak{A}$. Similarly, we shall denote by $V(d+i, \mathfrak{A})$, or by $V(\mathfrak{A})$, the transforms V_i of V obtained by applying to V the transformations T_i $(i \geqq 1)$.

If $\mathfrak{A}$ is a homogeneous ideal in $k[y]$ (different from the zero ideal) and Q is a point of V we define the *local component* $\mathfrak{A}_Q$ *of* $\mathfrak{A}$ *at* Q as the ideal in the local ring of Q on V consisting of all the quotients $f(y)/g(y)$ of forms in $k[y]$, of like degree, such that $f(y) \in \mathfrak{A}$ and $g(y) \neq 0$ at Q.

LEMMA I.7.4. *If Q is a point of V and $\mathfrak{A}$, $\mathfrak{B}$ are homogeneous ideals in the homogeneous coordinate ring $k[y]$ of V/k such that $\mathfrak{A}_Q = \mathfrak{B}_Q$, then the birational transformations $T(\mathfrak{A})$ and $T(\mathfrak{B})$ are biregularly equivalent at Q.*

PROOF. If $\mathfrak{A}_Q = \mathfrak{B}_Q$ then $(\mathfrak{A}+\mathfrak{B})_Q = \mathfrak{A}_Q + \mathfrak{B}_Q = \mathfrak{A}_Q = \mathfrak{B}_Q$, and hence it will be sufficient to show that $T(\mathfrak{A}+\mathfrak{B})$ and $T(\mathfrak{A})$ are biregularly equivalent at Q. We may therefore replace $\mathfrak{B}$ by $\mathfrak{A}+\mathfrak{B}$ and thus we may assume that $\mathfrak{B} \supset \mathfrak{A}$. Let d and d' be the integers analogous to the integer d introduced above and relative to the ideals $\mathfrak{A}$ and $\mathfrak{B}$ respectively. We fix an integer i such that $i \geqq 1 + \max\{d, d'\}$ and we take for $T(\mathfrak{A})$ and $T(\mathfrak{B})$ the transformations $T(i, \mathfrak{A})$ and $T(i, \mathfrak{B})$ respectively. We fix a k-basis $\{f_0(y), f_1(y), \ldots, f_s(y)\}$ of $\mathfrak{A}(i)$ and we complete it to a k-basis $\{f_0(y), f_1(y), \ldots, f_t(y)\}$ of $\mathfrak{B}(i)$. We set $z'_\alpha = f_\alpha(y)$, $\alpha = 0, \ldots, s$; $z''_\beta = f_\beta(y)$, $\beta = 0, 1, \ldots, t$. Then $z'_0, z'_1, \ldots, z'_s$ are strictly homogeneous coordinates of the general point $R' = T(\mathfrak{A})\{P\}$ of $V(\mathfrak{A})/k$, and $z''_0, z''_1, \ldots, z''_t$ are strictly homogeneous coordinates of the general point $R'' = T(\mathfrak{B})\{P\}$ of $V(\mathfrak{B})/k$. The equality $\mathfrak{A}_Q = \mathfrak{B}_Q$ implies the existence of relations of the form

$$(1) \qquad a(y) z''_\beta = \sum_{\alpha=0}^{s} a_{\alpha\beta}(y) z'_\alpha, \quad \beta = 0, 1, \ldots, t,$$

where $a(y)$ and the $a_{\alpha\beta}(y)$ are forms of like degree, with coefficients in k, and $a(y)\neq 0$ at Q. Now, let Q' be any point of $T(\mathfrak{A})\{Q\}$. Without loss of generality we may assume that the quotients z'_α/z'_0 belong to the local ring $\mathfrak{o}_{Q'}$ of Q' on $V(\mathfrak{A})$ (i.e., that $z'_0\neq 0$ at Q'). Since $\mathfrak{o}_Q\subset\mathfrak{o}_{Q'}$ ($T(\mathfrak{A})$ being antiregular) and since therefore all the quotients $a_{\alpha\beta}(y)/a(y)$ belong to $\mathfrak{o}_{Q'}$, it follows from (1) that the quotients $z''_\beta/z''_0(=z''_\beta/z'_0)$ belong to $\mathfrak{o}_{Q'}$. In other words: a suitable non-homogeneous coordinate ring of $V(\mathfrak{B})$ is contained in $\mathfrak{o}_{Q'}$. This implies that the birational transformation $\bar{T}=T(\mathfrak{A})^{-1}\circ T(\mathfrak{B})$ is regular at the point Q'. If, Q'' is the point $T\{Q'\}$ of $V(\mathfrak{B})$ then the quotients z''_β/z''_0 must be finite at Q'' (since their Q''-residues are precisely their Q'-residues). Therefore the quotients z''_k/z''_0 belong to the local ring $\mathfrak{o}_{Q''}$ of Q'' on $V(\mathfrak{B})$. For $\beta=0,1,\ldots,s$ we find then that the quotients z'_α/z'_0 belong to $\mathfrak{o}_{Q''}$, whence $\bar{T}^{-1}$ is regular at Q''. Consequently $\bar{T}$ is biregular at each point Q' of $T(\mathfrak{A})\{Q\}$, and since $T(\mathfrak{B})=T(\mathfrak{A})\circ\bar{T}$, the lemma is proved.

I. 8. Monoidal dilatations and quadratic transformations.

DEFINITION I.8.1. *If W is an irreducible subvariety of V/k, different from V, and $\mathfrak{p}$ is the homogeneous prime ideal of W in the homogeneous coordinate ring $k[y]$ of V/k, then any of the antiregular birational transformations $T(\mathfrak{p})$ of V is called a monoidal dilatation of V, with center W. In the special case in which Q is a point (algebraic over k) the transformation $T(\mathfrak{p})$ is called a quadratic transformation with center Q. Any birational transformation of V which is biregularly equivalent to a quadratic transformation with center Q is said to be a locally quadratic transformation with center Q.*

PROPOSITION I.8.2. *If all the points of an irreducible subvariety W of V are simple for both V and W then the monoidal transformation $T(\mathfrak{p})$ with center W ($\mathfrak{p}$=prime homogeneous ideal of W in the homogeneous coordinate ring $k[y]$ of V/k) has the following properties:*

(a) *For any point Q of W the variety $T(\mathfrak{p})\{Q\}/k(Q)$ is irreducible, biregularly equivalent (over $k(Q)$) to a projective space of dimension $r-1-s$, where s is the dimension of W, and all its points are simple for the variety $T(\mathfrak{p})\{V\}$.*

(b) *The variety $T(\mathfrak{p})\{W\}/k$ is irreducible, non-singular, of dimension $r-1$, and is birationally equivalent to the direct product of W and the projective space of dimension $r-1-s$. The totality of prime cycles $T(\mathfrak{p})\{Q\}$ ($Q\in W$) is* *an involution, and its carrier is* $T(\mathfrak{p})\{W\}$. *Through each point of* $T(\mathfrak{p})\{W\}$ *(and not only through each general point of* $T(\mathfrak{p})\{W\}/k$*) there passes only one cycle of the involution.*[7)]

PROOF. Let $d=\dim Q/k$ and let $t_1, t_2, \ldots, t_{r-s}, u_1, u_2, \ldots, u_{s-d}$ be local (i.e., regular) parameters at Q on V such that $\mathrm{Tr}_W t_i=0$ and $\{\mathrm{Tr}_W u_j\}$ is a set of local parameters at Q on W. If $\mathfrak{o}_W$ denotes the local ring of V along W (i.e., the local ring of a general point of W/k on V) and $\mathfrak{p}_W$ denotes the maximal ideal of $\mathfrak{o}_W$ then $\mathfrak{p}_W=\mathfrak{o}_W\cdot(t_1, t_2, \ldots, t_{r-s})$, and it is clear that $\mathfrak{p}_W$ is the local component of $\mathfrak{p}$ along W (i.e., at a general point of W/k). Hence $\mathfrak{p}_Q=\mathfrak{o}_Q\cdot(t_1, t_2, \ldots, t_{r-s})$. We set

$$t_i=\varphi_i(y)/\varphi_0(y), \qquad i=1, 2, \ldots, r-s,$$

where $\varphi_0(y)$ and the $\varphi_i(y)$ are forms in $k[y]$, of like degree, and where $\varphi_0(Q)\neq 0$. If we denote by $\mathfrak{A}$ the homogeneous ideal in $k[y]$ generated by $\varphi_1(y), \varphi_2(y), \ldots, \varphi_{r-s}(y)$, then we have $\mathfrak{p}_Q=\mathfrak{A}_Q$. Hence, by Lemma I.7.4, we can consider $\mathfrak{A}$ instead of $\mathfrak{p}$ in order to study the local behaviour at Q of the monoidal dilatation $T(\mathfrak{p})$. We therefore replace $T(\mathfrak{p})$ by $T(\mathfrak{A})$. Let then $P'\equiv(\varphi_1(y), \varphi_2(y), \ldots, \varphi_{r-s}(y))$, let V' be the locus of P' over k and let $T: V\to V'$ be the rational transformation with general point pair (P, P') over k, where $P=(y_0, y_1, \ldots, y_n)$. Since the $\varphi_i(y)$ are of like degree we can take for $V(\mathfrak{A})$ (I.7) the graph of T; then $T(\mathfrak{A})$ is the inverse of the projection of $V(\mathfrak{A})$ onto V. Let $\{U^*_{\alpha i};\ \alpha=0, 1, \ldots, n;\ i=1, 2, \ldots, r-s\}$ be the canonical affine covering of $V(\mathfrak{A})$ (induced by the canonical affine covering of $V\times V'$; see I.1) and let $\{U_\alpha;\ \alpha=0, 1, \ldots, n\}$ be the canonical affine covering of V. Let U^* denote any of the affine varieties $U^*_{\alpha i}$, say let $U^*=U^*_{01}$. We have to show that those assertions in (a) and (b) which are of local character remain true if we replace $T(\mathfrak{p})\{Q\}$, $T(\mathfrak{p})\{V\}$ and $T(\mathfrak{p})\{W\}$ by $T(\mathfrak{A})\{Q\}\cap U^*$, $V(\mathfrak{A})\{V\}\cap U^*$ and $T(\mathfrak{A})\{W\}\cap U^*$ respectively, Q being any point of U_0.

The non-homogeneous coordinates of the general point of U^*/k (more precisely: of the biregularly equivalent affine variety in $U_0\times U'_1/k$, where $\{U'_1, U'_2, \ldots, U'_{r-s}\}$ is the canonical affine covering of V') are

7) The assertions stated in (b) amount to saying that the variety $T(\mathfrak{p})\{W\}$, together with its rational map onto W (defined by the restriction to $T(\mathfrak{p})\{W\}$ of the *regular* transformation $T(\mathfrak{p})^{-1}$), is a fibre bundle whose base space is W and whose fibres are projective spaces of dimension $r-s-1$. The above proposition was first proved in our paper [28].

(1) $$x_1, x_2, \ldots, x_n, x'_2, x'_3, \ldots, x'_{r-s}$$

where

$$x_\alpha = y_\alpha / y_0 \quad \text{and} \quad x'_i = t_i / t_1.$$

Let $z_1, z_2, \ldots, z_n$ be the non-homogeneous coordinates of the point Q. The affine representative $T(\mathfrak{A})\{Q\} \cap U^*$ of the variety $T(\mathfrak{A})\{Q\}$ is the set of all points $(z_1, z_2, \ldots, z_n, z'_2, z'_3, \ldots, z'_{r-s})$ which are zeros of the ideal of algebraic relations satisfied, over k, by the $n+r-s-1$ quantities (1). Now, if $f(x_1, x_2, \ldots, x_n, x'_2, x'_3, \ldots, x'_{r-s}) = 0$ is any such relation, then upon the substitution $x'_i = t_i/t_1$ this relation takes the form $F(x_1, x_2, \ldots, x_n, t_1, t_2, \ldots, t_{r-s}) = 0$, where F is a polynomial with coefficients in k, homogeneous in $t_1, t_2, \ldots, t_{r-s}$. Since the x_α belong to $\mathfrak{o}_Q$ and since the t_i belong to a set of regular parameters of the regular local ring $\mathfrak{o}_Q$ it follows, by well-known properties of regular parameters, that if F is regarded as a form in the t_i, with coefficients $g_\nu(x)$ in $k[x]$, then all these coefficients must be zero at Q; i.e., we must have $g_\nu(z) = 0$. This signifies that the variety $T(\mathfrak{A})\{Q\} \cap U^*$ is the set of all points $(z_1, z_2, \ldots, z_n, z'_2, z'_3, \ldots, z'_{r-s})$, where $z'_2, z'_3, \ldots, z'_{r-s}$ are *arbitrary* quantities, and is thus irreducible over $k(z)$ $(=k(Q))$, has dimension $r-s-1$ and is in fact biregularly equivalent to the $(r-s-1)$-dimensional affine space over $k(z)$. If we repeat this reasoning for each of the affine varieties $U^*_{\alpha i}$ such that $Q \in U_\alpha$ we conclude that the total transform $T(\mathfrak{A})\{Q\}$ is (to within biregular equivalence) the set of all points whose homogeneous coordinates are the products $z_\alpha u_i$ $(=0, 1, \ldots, n; i = 1, 2, \ldots, r-s)$, where $z_0 = 1$ and where the u_i are arbitrary quantities. Hence $T(\mathfrak{A})\{Q\}$ is indeed irreducible over $k(Q)$, is of dimension $r-s-1$, and is biregularly equivalent, over $k(Q)$, to the projective space of dimension $r-s-1$.

For the given point Q we can find a "sufficiently small" affine representative $\bar{U}_0$ of V/k such that $Q \in \bar{U}_0 \subset U_0$ and such that $t_1, t_2, \ldots, t_{r-s}$ belong to a set of local parameters, on V, of *any* point R of $W \cap \bar{U}_0$. We will have then $\mathfrak{p}_R = \mathfrak{A}_R$ for any point R of $W \cap \bar{U}_0$ and hence we can still use the ideal $\mathfrak{A}$ instead of $\mathfrak{p}$ for the study of the (semi-local) behaviour, along $W \cap \bar{U}_0$, of the monoidal dilatation $T(\mathfrak{p})$. If we now consider the affine variety $\bar{U}^* = \bar{U}_0 \times U'_1 \cap V(\mathfrak{A})$ we find that the affine representative $T(\mathfrak{A})\{W\} \cap \bar{U}^*$ of $T(\mathfrak{A})\{W\}$ is the set of all points $(z_1, z_2, \ldots, z_n, z'_2, z'_3, \ldots, z'_{r-s})$, where $(z_1, z_2, \ldots, z_n)$ is any point of $W \cap \bar{U}_0$ and where $z'_2, z'_3, \ldots, z'_{r-s}$ are arbitrary quantities. Therefore $T(\mathfrak{A})\{W\} \cap \bar{U}^*$ is irreducible over k and is biregularly equivalent to the direct product of the affine variety $W \cap \bar{U}_0$ and the affine space of dimension $r-s-1$. If we repeat this argument for each of the affine varieties of a sufficiently fine covering of $V(\mathfrak{A})$ we obtain the first assertion made in (b).

Let f denote the restriction to $T(\mathfrak{p})\{W\}$ of the *regular* transformation $T(\mathfrak{p})^{-1}$. Then f is a rational regular transformation of $T(\mathfrak{p})\{W\}$ onto W. This rational transformation defines an involution M of $(r-s-1)$-cycles on $T(\mathfrak{p})\{W\}$ whose general element is the cycle $Z=\mathscr{Z}(f^{-1}\{Q\}/k(Q))$ (see I. 5) and whose carrier is $T(\mathfrak{p})\{W\}$; here Q denotes a general point of W/k. Now $f^{-1}\{Q\}$ is the variety $T(\mathfrak{p})\{Q\}$. This variety is absolutely irreducible since, as we have just proved, it is biregularly equivalent, over $k(Q)$, to a projective space. Hence Z is of the form p^eX, where X is a prime cycle and p^e is the order of inseparability of $k(Q')/k(Q)$; here Q' denotes a general point of $T(\mathfrak{p})\{Q\}/k(Q)$. In the present case we have $p^e=1$ since the preceding part of the proof shows that the function field of $T(\mathfrak{p})\{Q\}$ is a purely transcendental extension of $k(Q)$. Hence Z is a prime cycle. If Q_0' is any particular point of $T(\mathfrak{p})\{W\}$ and if $Q_0=T^{-1}\{Q_0'\}$ is the corresponding point of W, then also $\mathscr{Z}(T(\mathfrak{p})\{Q_0\}/k(Q_0))$ is a prime cycle X_0 of dimension $r-s-1$, and some multiple hX_0 of X_0 will be a specialization Z_0 of Z over k. Since Q_0 is the only point of W which corresponds to Q_0' under f, it follows that Z_0 is the only cycle in M which passes through Q_0'. This establishes part (b) of the proposition, except for the assertion that $h=1$. The first part of the proof shows, however, that there exists a birational transformation T_1 of V which is biregularly equivalent to T at the points Q_0 and Q and such that both total transforms $T_1\{Q_0\}$ and $T_1\{Q\}$ are projective spaces, hence prime cycles of order 1. (Note that Q is a *general* point of W/k). Since $T^{-1}T_1$ is biregular at all point of $T(\mathfrak{p})\{Q_0\}$ and $T(\mathfrak{p})\{Q\}$, it follows that the cycles Z and X_0 also must have the same order, and hence $Z_0=X_0$, $h=1$.

To complete the proof of the proposition it remains only to show that all points of $T(\mathfrak{p})\{W\}$ are simple for $T(\mathfrak{p})\{V\}$, and it will be sufficient to show this for the algebraic points of $T(\mathfrak{p})\{W\}$, over k. Using the notations of the first part of the proof, let $Q'\equiv(z_1, z_2, \ldots, z_n, z_2', z_3', \ldots, z_{r-s}')$ be an algebraic point of $T(\mathfrak{A})\{Q\}\cap U^*$. Then also Q is algebraic over k, and so $d=0$. To show that Q' is a simple point of $T(\mathfrak{A})\{V\}$ it will be sufficient to show that the elements

(2) $$t_1, x_2'-z_2', x_3'-z_3', \ldots, x_{r-s}'-z_{r-s}', u_1, u_2, \ldots, u_s$$

form a basis of the maximal ideal $\mathfrak{m}'$ of the local ring $\mathfrak{o}_{Q'}$ of Q' on U^*. We know that $\mathfrak{m}'$ has the following basis:

$$x_1-z_1, x_2-z_2, \ldots, x_n-z_n, x_2'-z_2', x_3'-z_3', \ldots, x_{r-s}'-z_{r-s}'.$$

On the other hand, the first n elements of this basis generate the maximal ideal $\mathfrak{m}$ of $\mathfrak{o}_Q$. Since also $t_1, t_2, \ldots, t_{r-s}, u_1, u_2, \ldots, u_s$ is a basis of $\mathfrak{m}$, it follows that the following elements form a basis of $\mathfrak{m}'$:

$$\{t_i, i=1, 2, \ldots, t-s\};\ \{u_j, j=1, 2, \ldots, s\};\ \{x_q'-z_q', q=2, 3, \ldots, r-s\}.$$

Now, we have $t_i=x_i't_1$, $i=2, 3, \ldots, r-s$, and hence $t_2, t_3, \ldots, t_{r-s}$ are redundant in the above basis. This completes the proof of Proposition I. 8. 2.

REMARK. It is immediately seen that $t_1=0$ is a local equation of $T(\mathfrak{p})\{W\}$ at Q' and that $t_1=u_1=u_2=\cdots=u_s=0$ are local equations of $T(\mathfrak{p})\{Q\}$ at Q'.

EXAMPLE 1. Let V be a three-dimensional projective space S_3 and let W be an irreducible non-singular curve in S_3. Let T be the monoidal dilatation with center W. Then $T\{W\}$ is a non-singular surface, and $T\{S_3\}$ is a non-singular three-dimensional irreducible variety. Consider a point Q on the curve W. Let l be the tangent line of W at Q. From above proof one can easily derive the following: as a point P of S_3 approaches Q along an arc which is not tangent to l then the point $T\{P\}$ approaches a point Q' of $T\{Q\}$ which depends only on the plane determined by l and the tangent line of the arc. The points of $T\{Q\}$ are thus in (1, 1) correspondence with the planes through l. This is in accordance with the established fact that $T\{Q\}$ is in the present case biregularly equivalent with a projective line.

If however, the arc of approach is tangent to l, then the limiting position of the variable point $T\{P\}$ is again indeterminate and may fall at any point of $T\{Q\}$ (depending on the arc).

EXAMPLE 2. If T is a quadratic transformation with center at a point Q (rational over k), then Q is the only fundamental point of T, and T is biregular at every point of V different from Q. The exceptional variety $T\{Q\}$ of T^{-1} is now biregularly equivalent to a projective $(r-1)$-space over k. The proof of Proposition I. 8. 2 shows without difficulty that the points of that variety are in (1, 1) correspondence with the tangential directions $t_1 : t_2 : \cdots : t_r = u_1 : u_2 : \cdots : u_r$ of V at Q. If we choose a coordinate system such that the homogeneous coordinates of Q are $z_0 = 1, z_i = 0$ $(i = 1, 2, \ldots, n)$, then the homogeneous prime ideal of Q is the ideal $(y_1, y_2, \ldots, y_n)$. Then the quadratic transformation T is given by the equations

$$y'_{ij} = y_i y_j, \quad (i = 1, 2, \ldots, n;\ j = 0, 1, \ldots, n)$$

which are quadratic in the y's. This is the reason for the term " quadratic " transformation. Note that the defining linear system of T is the one cut out on V by the system of hyperquadrics passing through Q.

EXAMPLE 3. Assume that the dimension of the center W of the monoidal dilatation T is $r-1$. In that case we have for any point Q of W that $\dim T\{Q\} = (r-1) - s = 0$, i.e., $T\{Q\}$ is a single point. Since Q is a simple, hence a normal, point of V, T is regular at Q. *Therefore T is a biregular transformation* (since T^{-1} is regular). This conclusion remains valid if we drop the assumption that W itself is a non-singular variety, but the condition that each point of W be simple for V is essential. For instance, if V is the quadric cone $z^2 - xy = 0$ and T is the monoidal dilatation whose center is the line W: $x = z = 0$ then it is easy to see that the total T-transform of the origin $x = y = z = 0$ is a curve (irreducible and non-singular), and thus the origin is a fundamental point of T. The total T-transform of the line W consists of two irreducible curves: the proper transform $T[W]$ of W and the total transform of the origin. However, T is biregular at each point of W other than the origin.

REMARK. If in Proposition I. 8. 2 we drop the assumption that every point of W be simple for V but assume only that the general points of W/k are simple for V, that it can be shown (ZARISKI [28], p. 536) that the *proper* transform $T[W]$ is irreducible, of dimension $r-1$ and is simple for V. However, the exceptional locus of T^{-1} may very well have irreducible components other than $T[W]$; these will correspond then to isolated fundamental varieties of T, embedded in W (see I. 2.). See in this connection our paper [28], p. 536.

PART II

THEORY OF EXCEPTIONAL CURVES ON AN ALGEBRAIC SURFACE

From now on we shall deal only with algebraic surfaces. We fix a field Σ of algebraic functions of two independent variables, over an algebraically closed field k, and we study the projective models of Σ/k. We denote by B the birational class of all *non-singular* projective models of Σ/k. All our surfaces are tacitly assumed to be members of the class B, unless the contrary is specified. Also, unless the contrary is specified, the term " point " will always refer to points which are rational over k. All the conventions made in I. 1 remain in force.

II. 1. The factorization of antiregular transformations into quadratic transformations.

Let F and F' be two members of the class B and let $T: F \to F'$ be the birational transformation of F onto F'. Let (Q, Q') be a point pair of T. We write $Q < Q'$ if T is regular at Q' and $Q \equiv Q'$ if T is biregular at Q. Hence $Q < Q'$ if the local ring $\mathfrak{o}_Q$ of Q on F is contained in the local ring $\mathfrak{o}_{Q'}$ of Q' on F', and $Q \equiv Q'$ if $\mathfrak{o}_Q = \mathfrak{o}_{Q'}$. We say that Q' *dominates* Q if $Q < Q'$ and that Q, Q' are *biregularly equivalent* points if $Q \equiv Q'$. *In this section we shall identify biregularly equivalent points.* With this identification, we denote by M the set of all points belonging to the various members F of B (note that any point of M is a simple point of any surface F which carries that point). The relation $<$ between points defines a partial ordering of M.

Let Q be a point of M, let F be a surface which carries that point and let T be a locally quadratic transformation of F with center Q. The set $T\{Q\}$ is a subset of M (Proposition I. 8. 2) and depends only on Q, being independent of the choice of F and T, and we have $Q < Q'$, $Q \neq Q'$, for any point Q' in $T\{Q\}$. Any point of $T\{Q\}$ will be called a *quadratic transform* *of Q*. A sequence (finite or infinite) of *successive quadratic transforms of a point* Q is a sequence $Q<Q_1<Q_2<\cdots$ such that each point Q_i of the sequence $(i\geqq 1)$ is a quadratic transform of its immediate predecessor Q_{i-1} $(Q_0=Q)$.

If $\mathfrak{p}$ is an algebraic place of Σ/k we denote by $M_\mathfrak{p}$ the set of all centers of $\mathfrak{p}$ on the various surfaces of the birational class B. If Q is any point of M we denote by M_Q the set of all points of M which dominate Q. Finally, if Q is any point of $M_\mathfrak{p}$ we denote by $M_{\mathfrak{p},Q}$ the intersection of $M_\mathfrak{p}$ with M_Q.

We observe that if $Q\in M_\mathfrak{p}$ then $\mathfrak{p}$ and Q determine a unique infinite sequence of successive quadratic transforms of Q such that all the points of the sequence belong to $M_{\mathfrak{p},Q}$. In fact, if T is a quadratic transformation with center Q then $\mathfrak{p}$ has a well defined center Q_1 in the set $T\{Q\}$. Similarly, if T_1 is a quadratic transformation with center Q_1 then $\mathfrak{p}$ has a well defined center Q_2 in the set $T_1\{Q\}$. Proceeding in this fashion we obtain the required infinite sequence $Q<Q_1<Q_2<\cdots$ of successive quadratic transforms of Q.

The following result will be essential for applications in the sequel:

THEOREM II. 1. 1. *If $\mathfrak{p}$ is any algebraic place of Σ/k and Q is any point of $M_\mathfrak{p}$ then $M_{\mathfrak{p},Q}$ coincides with the infinite, totally ordered simple sequence*

$$Q<Q_1<Q_2<\cdots<Q_n<\cdots \tag{1}$$

of successive quadratic transforms of Q which is determined by $\mathfrak{p}$ and Q, and the union of the local rings $\mathfrak{o}_{Q_i}$ is the valuation ring of the place $\mathfrak{p}$.

PROOF. We first prove that the union of the local rings $\mathfrak{o}_{Q_i}$ of the points of the sequence (1) is the valuation ring of the place $\mathfrak{p}$. (See our paper [25], Theorem 10, p. 681, where a corresponding stronger result is proved for sequences of normal points obtained by successive applications of quadratic transformations followed by normalization. For a suitable extension to varieties of higher dimension see our paper [33], p. 75). Let S be the union of the local rings $\mathfrak{o}_{Q_i}$. There exists a natural homomorphism π of S onto k. If S is not a valuation ring then we can find a quantity z in Σ such that neither z nor $1/z$ belongs to S. Since S is integrally closed we can extend the homomorphism π to a homomorphism π' of $S[z]$ onto $k[t]$, where t is a transcendental over k. We can then extend

π' to a place $\mathfrak{p}'$ of Σ/k, and the valuation v defined by this place will be necessarily of dimension 1.

If we express z as a quotient ξ/η, where ξ and η are elements of $\mathfrak{o}_Q$, then ξ and η must both belong to the maximal ideal $\mathfrak{m}$ of $\mathfrak{o}_Q$, and we will have $v(\xi)>0$, $v(\eta)>0$. Let x, y be local parameters at Q and let $\xi=ax+by$, $\eta=a'x+b'y$, where $a, b, a', b' \in \mathfrak{o}_Q \subset \mathfrak{o}_{Q_1}$. Since Q_1 is a quadratic transform of Q, either y/x or x/y belongs to $\mathfrak{o}_{Q_1}$. If, say, $y/x \in \mathfrak{o}_{Q_1}$, then the elements $\xi_1=\xi/x=a+by/x$ and $\eta_1=\eta/x=a'+b'y/x$ belong to $\mathfrak{o}_{Q_1}$, and since $v(x)>0$ we have $v(\xi)>v(\xi_1)$, $v(\eta)>v(\eta_1)$, and $z=\xi_1/\eta_1$. Proceeding in thisfash ion we obtain two infinite sequences $\{\xi, \xi_1, \xi_2, \ldots\}$, $\{\eta, \eta_1, \eta_2, \ldots\}$ of elements of Σ such that $v(\xi)>v(\xi_1)>v(\xi_2)>\cdots$ and $v(\eta)>v(\eta_1)>v(\eta_2)>\cdots$, and this is impossible since the $v(\xi_i)$ and $v(\eta_i)$ are non-negative integers.

In view of this result it follows that in order to prove that the sequence (1) coincides with $M_{\mathfrak{p},Q}$ it is only necessary to prove the following: *if Q' is any point of M such that $Q<Q'$ and $Q \neq Q'$, then there exists a quadratic transform Q^* of Q such that $Q^* \leq Q'$* (see our paper [29], "Lemma" on p. 538). We fix a place $\mathfrak{p}$ with center Q' and we denote by Q^* the center of $\mathfrak{p}$ in the set of $T\{Q\}$ of quadratic transforms of Q. Here $T: F \rightarrow F^*$ is a quadratic transformation of F, with center Q. We shall show that $Q^* \leq Q'$.

Assume the contrary: $Q^* \not\leq Q'$. Then, if F' is a surface in B which carries the point Q', the point Q' will be a fundamental point of the birational transformation $F' \rightarrow F^*$. Hence to Q' there must correspond on F^* at least one irreducible curve through Q^*. Any such curve must also correspond to the point Q under the quadratic transformation T since $\mathfrak{o}_Q \subset \mathfrak{o}_{Q'}$. However, the total transform $T\{Q\}$ is an irreducible curve Γ (I. 8, Example 2). Hence Γ is the only curve through Q^* which corresponds to the point Q' in the birational transformation $F' \rightarrow F^*$.

Let v be the one-dimensional valuation of Σ/k which is defined by the curve Γ, and let x, y be local parameters of the point Q. Without loss of generality we may assume that $y/x \in \mathfrak{o}_{Q^*}$. Then x can be taken as one of a pair of local parameters at the point Q^*, and the curve Γ is defined by the principal ideal (x) in the local ring $\mathfrak{o}_{Q^*}$ (I. 8, "Remark" preceding Example 1). Let $\mathfrak{m}$ and $\mathfrak{m}'$ be the maximal ideals of $\mathfrak{o}_Q$ and $\mathfrak{o}_{Q'}$ respectively. Since $v(x)=1$ and since every element of $\mathfrak{m}'$ has positive value in v (Q' being the center of v on F'), it follows that x does not belong to $\mathfrak{m}'^2$. Now, Q is a fundamental point of the birational transformation $F \rightarrow F'$ (since $Q<Q'$ and $Q \neq Q'$), and therefore there must exist on F' an irreducible curve Δ' through Q' which corresponds to Q. Since both x and y vanish on Δ', this curve must be defined by an irreducible element of $\mathfrak{o}_{Q'}$ which divides in $\mathfrak{o}_{Q'}$ both x and y. But x itself is an irreducible element of $\mathfrak{o}_{Q'}$, since $x \notin \mathfrak{m}'^2$. Hence x divides y in $\mathfrak{o}_{Q'}$, $y/x \in \mathfrak{o}_{Q'}$, showing that $\mathfrak{o}_{Q^*} \subset \mathfrak{o}_{Q'}$ and that our original assumption that Q' does not dominate Q^* was absurd. This completes the proof of Theorem II. 1. 1.

From the preceding theorem it follows that *given any two distinct points Q and Q' in M such that $Q<Q'$, there exists a finite sequence*

$$Q<Q_1<Q_2<\cdots<Q_n=Q' \tag{2}$$

of successive quadratic transforms of Q joining Q to Q', and this sequence is uniquely determined by Q and Q'. The existence of the sequence is obvious, and the uniqueness follows from the fact there cannot exist two distinct quadratic transforms of Q which are both dominated by Q'.

From the existence and unicity of the sequence (2) it follows that *every point Q in M has immediate successors in the partially ordered set M and that the set of immediate successors of Q is precisely the set of all quadratic transforms of Q.*

We now prove the following fundamental result which in the sequel will be the basis of many inductive proofs:

THEOREM II. 1. 2. (*The factorization theorem for antiregular birational transformations*). *Any antiregular (but not biregular) birational transformation $T: F \rightarrow F'$ ($F, F' \in B$) is a finite product of locally quadratic transformations.*

PROOF. If Q' is any point of F' and $Q=T^{-1}\{Q'\}$ is the corresponding point of F, we denote by $l(Q', F)$ the length of the sequence of successive quadratic transforms of Q joining Q to Q'. For any integer $n \geqq 1$ we denote by $C_n(F, F')$ the set of points Q' of F' such that $l(Q', F) \geqq n$. We assert that

(a) $C_n(F, F')$ *is an algebraic curve or an empty set.*

Let $P_1, P_2, \ldots, P_s$ be the fundamental points of T ($s \geqq 1$ since T is not biregular). It is clear that $C_1(F, F')$ is the union of the s curves $T\{P_i\}$, and thus (a) is true for $n=1$. We can therefore use induction on n. We first define inductively s surfaces F_i and, for each i $(1\leqq i\leqq s)$, a birational transformation $T_i: F_{i-1}\to F_i$, as follows: $F_0=F$; T_i is a locally quadratic transformation of F_{i-1} with center P_i. It is clear that the points $P_i, P_{i+1}, \ldots, P_s$ belong to F_{i-1}, and hence the definition of the T_i and the F_i makes sense. Since each point of $T\{P_i\}$ dominates some quadratic transform of P_i, the transformation

$$T'_i: F_i\to F'$$

is antiregular. The points $P_{i+1}, P_{i+2}, \ldots, P_s$ are among the fundamental points of T'_i (the other fundamental points of T'_i, if any, belong necessarily to the curves $(T_1 T_2 \cdots T_i)\{P_j\}$, $j=1, 2, \ldots, i$. It is now clear that if $n\geqq 2$ then

(b) $$C_n(F, F')=C_{n-1}(F_s, F'),$$

and this proves (a), since by our induction hypothesis the second term in (b) is an algebraic curve or an empty set.

We now assert that

(c) *if $C_n(F, F')$ is not empty then $C_{n+1}(F, F')$ is a proper subset of $C_n(F, F')$.*

For $n=1$ assertion (c) follows from the fact that $C_2(F, F')$ is the total T'_s-transform of the set of fundamental points of T'_s (or is the empty set, if T'_s is biregular) while $C_1(F, F')$ is the total T'_s-transform of the curve

$$\bigcup_{i=1}^{s} (T_1 T_2 \cdots T_s)\{P_i\},$$

this curve containing all the fundamental points of T'_s. For $n>1$ assertion (c) follows from (b) by induction on n.

From (a) and (c) it follows that $C_n(F, F')$ is empty if n is large. We denote by $N(F, F')$ the smallest of the integers n such that $C_n(F, F')$ is empty. We can now complete the proof of the theorem by induction on $N(F, F')$. We observe that $N(F, F')=N(F_s, F')+1$ and that if $N(F, F')=1$ then F_s and F' are biregularly equivalent surfaces. By our induction hypothesis, the birational transformation T'_s is a product of locally quadratic transformations or is a biregular transformation. Since $T=T_1 T_2 \cdots T_s T'_s$, the theorem is proved.[8)]

---

8) It can be easily established by induction that the number of quadratic factors of T depends only on T, being independent of the particular factorization of T.

COROLLARY II. 1. 3. *If a birational transformation $T:F\rightarrow F'$ is anti-regular at a point Q of F then T is biregularly equivalent at Q to a product of locally quadratic transformations.*

Obvious.

II. 2. Exceptional curves of the first and the second kind.

DEFINITION II. 2. 1. *An algebraic curve E (irreducible or not) on a normal (not necessarily non-singular) surface F is said to be an exceptional curve of F if there exists a birational transformation $T:F\rightarrow F'$ of F onto a surface F' (F'–not necessarily non-singular) such that E is the total T^{-1}-transform of a simple point of F'.*

DEFINITION II. 2. 2. *An exceptional curve E of F is said to be of the first kind if there exists a birational transformation T satisfying the condition stated in Definition* II. 2. 1 *and satisfying the further condition that it be regular at each point of E. In the contrary case E is said to be an exceptional curve of the second kind.*

THEOREM II. 2. 3. *Let E be an exceptional curve of a normal surface F, of the first kind, and let $f:F\rightarrow G$ be a birational transformation of F such that E is the total f^{-1} transform of a normal point A of G. If f is regular at each point of E then A is a simple point of G, and conversely. Furthermore, if A is a simple point of G then*

$$\mathfrak{o}_A=\bigcap_{P\in E}\mathfrak{o}_P, \tag{1}$$

and consequently, the point A—if it is simple—depends only on the curve E, being independent of the choice of the birational transformation f.

PROOF. By assumption, there exists a birational transformation $T:F\rightarrow F'$ and there exists a simple point P' of F' such that $E=T^{-1}\{P'\}$ and such that T is regular at each point of E (equivalently: T^{-1} is anti-regular at P'). Let T' denote the birational transformation $f^{-1}\circ T: G\rightarrow F'$. We have $T'=f^{-1}T$ at A (Lemma I. 7. 2) and hence $T'\{A\}=P'$. This implies that T' is regular at A, since A is a normal point. We have therefore

$$A>P'. \tag{2}$$

Now assume that f is regular at each point of E. We have then

$$\mathfrak{o}_A\subset\bigcap_{P\in E}\mathfrak{o}_P. \tag{3}$$

Since A is a normal point, we have, by known properties of the integral closure of a local domain (see ZARISKI [28], p. 513, proof of Theorem 9), that

$$\mathfrak{o}_A = \bigcap_{\mathfrak{p}} R_{\mathfrak{p}}, \tag{4}$$

where the intersection symbol is extended to all places $\mathfrak{p}$ of Σ/k which have center A on G and where $R_{\mathfrak{p}}$ denotes the valuation ring of $\mathfrak{p}$. Now if $\mathfrak{p}$ has center A on G its center Q on F belongs to E; hence $R_{\mathfrak{p}} \supset \mathfrak{o}_Q$ and, *a fortiori*,

$$R_{\mathfrak{p}} \supset \bigcap_{P \in E} \mathfrak{o}_P.$$

This implies, by (4) and (3), the equality (1).
By the same argument (since also T is regular at each point of E) we have

$$\mathfrak{o}_{P'} = \bigcap_{P \in E} \mathfrak{o}_P \tag{5}$$

and hence $\mathfrak{o}_A = \mathfrak{o}_{P'}$, showing that A is a simple point.

Conversely, assume that A is a simple point. If P is any point of E then there exists a place $\mathfrak{p}$ of Σ/k with center A and P on G and F respectively, and the center of $\mathfrak{p}$ on F' must be P' (since T is regular at P and sends P into P'). Therefore, in the notations of II. 1, we have that the points A, P, P' belong to $M_{\mathfrak{p}}$. Since $P \succ P'$ and since, by (2), also $A \succ P'$, we conclude that A and P belong to $M_{\mathfrak{p},P'}$. Consequently, by Theorem II. 1. 1, either $P \prec A$ or $A \prec P$. The first relation is impossible since f^{-1} is not regular at A (on the contrary, A is a fundamental point of f^{-1}). Hence $A \prec P$, i.e., f is regular at each point P of E. This completes the proof.

For an exceptional curve E of the 1st kind it is therefore true that there exists one and only one (simple) point in the set M of which E is the total transform. If P' is that point then E is, of course, a total *antiregular* transform of P'. The preceding theorem shows even more, namely that P' is the only normal point of which E is a total antiregular transform (note that in proof of this last conclusion no use has been made of Lemma II. 1. 1). However, it can be shown by examples that there can exist infinitely many (biregularly distinct) normal (non-simple) points of which E is a total transform if we do not insist on the condition of antiregularity.

For *any* exceptional curve E, of the 1st or of the 2nd kind, we call a simple point P' a *contraction of* E if E is a total transform of P'. If E is of the first kind then the contraction of E is uniquely determined by E.

One could widen the definition of an exceptional curve and call exceptional any curve on F which is a total transform of a normal point, such a point to be called a contraction of the curve. The exceptional curves in the sense of Definition II. 2. 1 would then be those exceptional curves which admit as contraction a simple point. The first part of the proof of Theorem II. 2. 3 makes no use of the fact that E is an exceptional curve of the 1st kind; it only uses the fact that E is a total antiregular transform of both point A and P' and that both these points are normal. Hence it remains true also for exceptional curves in the wider sense that any such curve has at most one contraction of which it is a total *antiregular* transform.

We state and prove the following theorem for exceptional curves in the wide sense, but we shall only have occasion of using it in the case of exceptional curves in the restricted sense of Definition II. 2. 1.

THEOREM II. 2. 4. *Let F be a non-singular surface, let E be an exceptional curve of F and assume that E admits a contraction P' such that E is a total antiregular transform of P'. Then there exists a surface G, birationally equivalent to F, which carries the point P', and there exists a birational transformation $f: F \to G$ such that*: (a) *f is regular at each point of E and is biregular on $F-E$*; (b) $E=f^{-1}\{P'\}$.

PROOF. By assumption, P' is a normal point of some surface F', and there exists a birational transformation $T: F \to F'$ such that T is regular at each point of E and such that $E=T^{-1}\{P'\}$. Without loss of generality we may assume that F' is a normal surface (replace F' by a derived normal model of F').

If T^{-1} has fundamental points other than P' we may eliminate these fundamental points by successive quadratic transformation (see ZARISKI [27], p. 592), and these transformations, which will replace F' by a new surface F'', will not affect the point P' nor the total transform of P' on F, while the resulting birational transformation $F \to F''$ will still be regular at each point of E. Hence we may assume that T^{-1} is regular at each point of F' other than P'.

Let L denote the defining linear system (on F) of the birational transformation T (I. 3). The system L may have base points (they are the fundamental points of T), but since T is regular at each point of E it follows that *no point of E is a base point of L.*

Since F is a non-singular surface it follows by a result which we have proved in [34, p. 167] that for large h the complete linear system $|hL|$ has no base points at all. We fix such a large integer h and we denote by M the system $|hL|$. Let $T': F \to F''$ be the rational transformation obtained by cutting M with hyperplanes (see I. 3). Since T is a birational transformation and since the linear system L is partially contained in the linear system M it follows immediately that also T' is a birational transformation. Since M has no base points and F is non-singular, T' is regular on F. Let G be a normalization of F'' and let $f = T' \circ g^{-1}$, where g is the natural projection of G onto F''. Then $f: F \to G$ is a birational transformation. We shall show that G and f satisfy the conditions (a) and (b) of the theorem.

We first show that if D is an irreducible curve on F which is not a component of E then D does not belong to the exceptional locus of T' (I. 2). In view of the " Remark " made at the end of I. 5 this amounts to showing that if M_1 denotes the trace of M on D then $\dim M_1 > 0$. Now, since M partially contains L, M_1 partially contains the trace L_1 of L on D, and therefore $\dim M_1 \geqq \dim L_1$. Since E is the exceptional locus of T, D is not an exceptional curve of T and hence $\dim L_1 > 0$, whence also $\dim M_1 > 0$, as was asserted.

We next show that *T' transforms the curve E into a point P'' of F''.* Let E_1 be any irreducible component of E and let L_1 be the trace of L on E_1. Since E_1 is an exceptional curve of T we have $\dim L_1 = 0$, i.e., the linear system L_1 on the curve E_1 consists of only one divisor. Were this divisor strictly positive then any point which is a component of this divisor would be a base point of L, and this is impossible since L has no base points on E. Hence L_1 consists only of the divisor zero. This signifies that the cycles D of L which do not contain E as component do not intersect E_1 at all. Hence the intersection number $(D \cdot E_1)$ is zero. Since any member D' of M is linearly equivalent to hD it follows that also $(D' \cdot E_1) = 0$ for any cycle D' in M. Therefore the cycles D' of M which do not contain E_1 as component do not intersect E_1, and thus the trace of

M on E_1 consists only of the divisor zero. This implies that E_1 is an exceptional curve of the birational transformation T', i.e., $T'\{E_1\}$ is a point of F''. Applying this argument to each of the irreducible components $E_1, E_2, \ldots, E_s$ of E we obtain s points P_i'' of F'', where $P_i''=T'\{E_i\}$. We now show that *the s points P_i'' coincide*. Consider the birational transformation $h=T^{-1}\circ T'$ $(=T^{-1}T')$ of F' onto F''. It is clear that $h\{P'\}$ is the set of s points P_i''. Since P' is a normal point (in view of our definition of contraction of an exceptional curve (in the wide sense)), it follows that $s=1$, as asserted.

Let us then denote by P'' the T'-transform of the curve E. It is clear that E is the total T'^{-1}-transform of P'', since T'^{-1} is not regular at P'' while T' is biregular at each point of F which does not belong to E. The foregoing considerations show that we have now a birational transformation of F'' onto the abstract normal surface $(F-E)+P'$ (in the sense of WEIL) which sends P'' into P' and which transforms in (1, 1) fashion $F''-P''$ onto $F-E$. Hence this abstract surface is a normalization G of F'' (and is therefore, in fact, a projective surface). It is now obvious that G and f satisfy the conditions (a) and (b) of the theorem.

REMARK. Actually it is not difficult to show that already F'' is itself a normal surface and that consequently already F'' and T' may replace G and f (Cf. p. 85, (3)). Since T' transforms biregularly $F-E$ onto $F''-P''$, we have only to show that P'' is a normal point. Since $h\{P'\}=P''$ and since P' is a normal point, we have $P'>P''$. We shall show now that $P''>P'$, and this will prove the normality of F'' (and also the fact that F'' carries the point P'). Let r and s be the dimensions of L and M respectively. We know that L and M contain subsystems L' and M', of dimension $r-1$ and $s-1$ respectively, such that every cycle belonging to these subsystems contains as component every irreducible component of E. We fix in L a cycle D, defined over k, which does not belong to L'. Then no irreducible component of E is a component of D, for in the contrary case that component would be a fixed component of L, and L has no fixed components. The rational functions f on F such that $(f)+D\in L$ form a k-module of dimension $r+1$. Let $\{1, f_1, f_2, \ldots, f_r\}$ be a k-basis of this module, where the f_i are defined over k and where we assume that $(f_i)+D\in L'$. Then the f_i are non-homogeneous coordinates of the general point of

F'/k. Since the f_i have trace 0 on each irreducible component of E it follows that, in the chosen system of non-homogeneous coordinates in the ambient space of F', the point P' is the origin. We now note that $hD \in M$, we consider the k-module (of dimension $s+1$) of the rational functions g on F such that $(g)+hD \in M$ and we choose a basis $\{1, g_1, g_2, \ldots, g_s\}$ of this module such that $(g_j)+hD \in M'$. We may also assume that for $i=1,2,\ldots,r$ we have $(g_i)+hD=(h-1)D+(f_i)+D$, i.e., $g_i=f_i$. Again, the g_j are non-homogeneous coordinates of the general point of F''/k, and with respect to this chosen system of non-homogeneous coordinates in the ambient space of F'' the point P'' is the origin. In these two systems of non-homogeneous coordinates the birational transformation $h^{-1}: F'' \to F'$ is therefore given by equations of the form $x'_i=f_i, i=1, 2, \ldots, r$, where on the right-hand side there appear functions which are regular at P''. Hence h^{-1} is regular at P'', and thus $P'' > P'$.

The following corollary is a generalization of the preceding theorem:

COROLLARY II. 2. 5. *Let F be a non-singular surface, let $E_1, E_2, \ldots, E_q$ be exceptional curves of F of the 1st kind, and let P'_i be the contraction of E_i. If no two of the curves E_i have common points then the set $F-(E_1+E_2+\cdots+E_q)+P'_1+P'_2+\cdots+P'_q$ (which is easily seen to be a complete abstract surface) is a projective surface.*

Since in the case $q=1$ we have a special case of the theorem, we may assume that $q>1$ and use induction with respect to q. The surface $F_1=F-E_1+P'_1$ is projective, by the case $q=1$, and non-singular (since P'_1 is a simple point). Since $E_1 \cap E_i$ is empty for $i \neq 1$, the birational transformation $F \to F_1$ is biregular at each point of $E_i, i=2, 3, \ldots, q$. By our induction hypothesis it follows then that $F_1-(E_2+E_3+\cdots+E_q)+P'_2+P'_3+\cdots+P'_q$ is a projective surface, and this establishes the corollary.

As an incidental consequence of Theorem II. 2. 4 and the above corollary we easily obtain the following result:

COROLLARY II. 2. 6. *Every abstract non-singular surface is projective* (i.e., *any such abstract surface is an open subset of a projective surface*).

Let $F=\{F_i-B_i, T_{ij}\}$ be a representation of the given abstract surface by means of a finite set of projective surfaces F_i, boundaries B_i and birational transformations $T_{ij}: F_i \to F_j$. We fix a *non-singular* projective surface F' which dominates each of the surfaces F_i and we denote by f_i the birational transformation of F_i onto F'. Let $B'_i=f_i\{B_i\}$ and B' be the intersection of the curves B'_i. Let f' be the birational transformation of F into F' induced by the f_i. It is clear that $f'\{F\}=F'-B'$ and that f'^{-1} is regular on $F'-B'$. Let $P_1, P_2, \ldots, P_q$ be the fundamental points of f' and let $E'_j=f'\{P_j\}$. Then each E'_j is an exceptional curve of F', of the first kind, and P_j is the contraction of E'_j. Furthermore, no two of the curves E'_j intersect. Hence the surface $G=F'-(E'_1+E'_2+\cdots+E'_q)+P_1+P_2+\cdots+P_q$ is projective. Since none of the curves E'_j intersects B' it follows that the algebraic locus B' can be regarded as embedded in G. Since F is biregularly equivalent to $G-B$, the corollary is proved.[9]

II. 3. Properties of exceptional curves of the first kind.

From now on the term " exceptional curve " will be used only in the restricted sense and will therefore be used only for exceptional curves of the first or second kind. We shall derive in this section some simple properties of exceptional curves of the first kind situated on non-singular surfaces. Since we wish to adopt the point of view of biregular equivalence

9) In view of Theorem II. 2. 4 it is clear that Corollary II. 2. 5 remains valid (and the proof remains the same) if one (and only one) of the curves E_i, say E_q, is allowed to be an exceptional curve in the wider sense, provided that it is an *antiregular* total transform of a normal point P_q (while the remaining E_i are exceptional curves in the strict sense, of the first kind). It follows that Corollary II. 2. 6 remains valid if it is assumed that the abstract surface, instead of being non-singular has *at most* one singular point. Hence if there exist abstract normal surfaces which cannot be embedded in a projective surface, one must look for them among surfaces which have *at least* two singular points. An example of such a surface has in fact been given recently by NAGATA and will be published in the Mem. Coll. Sci. Univ. Kyoto (In NAGATA's example, the surface is birationally equivalent to an elliptic cone, but we have a more general example of a non-projective abstract complete normal surface which is birationally equivalent to a preassigned ruled surface of arbitrary genus. By a remark which has been communicated to us by NAGATA it follows therefore that any surface which carries an irrational pencil is birationally equivalent to an abstract, complete, normal, non-projective surface). We point out also that the same argument which we have used in proof of Corollary II. 2. 6 establishes also the following more general result: *A normal abstract surface can be embedded in a projective surface if (and only if) there exists an affine surface which carries all the singular points of the abstract surface.*

we shall identify two exceptional curves of two surfaces F, F' if these curves correspond to each other under a biregular transformation of F onto F'. Hence, to some extent, we consider our exceptional curves as subsets of the big set M of all simple points of all projective models of our field Σ/k (see II. 1). However, some of the properties of an exceptional curve which are derived below make sense only if the exceptional curve is regarded as a subvariety of a surface F which carries that curve, because the set M by itself has no simple algebro-geometric structure.

The simplest exceptional curve of the first kind which has a *preassigned contraction* P' is the total transform of P' under a locally quadratic transformation with center P'. More precisely: we fix in our birational class B of non-singular projective models of Σ/k a surface F' which carries the point P' and we apply to F' a locally quadratic transformation $T_0 : F' \to F_0$ with center P'; then the total transform of P' on F_0 is an exceptional curve of the first kind, and this curve is irreducible, non-singular and rational (Proposition I. 8. 2). Up to biregular equivalence, this exceptional curve depends only on the point P', being independent of the choice of F' and T_0, and we shall denote it by $E_0(P')$; it is the set of all quadratic transforms of P', or also, the set of all points of the partially ordered set M which are immediate successors of P'.

As an irreducible curve of the non-singular (hence normal) surface F_0 the curve $E_0(P')$ determines a prime divisor of the field Σ/k whose center on F' is the point P' (whence, in the classical terminology, this prime divisor is of the "second kind" with respect of the surface F'). Again, this prime divisor depends only on the point P'; we shall call it the *principal P'-adic divisor* and we shall denote it by $\mathfrak{P}_0(P')$. The following determination of the valuation ring of $\mathfrak{P}_0(P')$ is well known and is easily derived from the definition of a quadratic transformation (Zariski [27], p. 588): if x and y are local parameters of the simple point P' then the valuation ring of $\mathfrak{P}_0(P')$ is the set of all quotients of the form $f(x, y)/g(x, y)$, where f and g are forms of like degree, with coefficients in the local ring of P', and where the coefficients of g do not all belong to the maximal ideal of that local ring. The prime divisor $\mathfrak{P}_0(P')$ maps y/x into a quantity t which is transcendental over k, and the residue field of the divisor is the purely transcendental extension $k(t)$.

It should be noted explicitly that two distinct points of M may very well have the same principal prime divisor. The following example will illustrate this possibility. Consider the CREMONA transformation $x'=x$ and $y'=x^2/y$ between two planes and let P and P' be the origin $x=y=0$ and $x'=y'=0$ respectively. The principal P-adic divisor $\mathfrak{P}_0(P)$ is defined by the principal ideal $(\bar{x})$ in the polynomial ring $k[\bar{x}, \bar{y}]$, where $\bar{x}=x$ and $\bar{y}=y/x$. Similarly, the prime divisor $\mathfrak{P}_0(P')$ is defined by the principal ideal $(\bar{x}')$ in the polynomial ring $k[\bar{x}', \bar{y}']$, where $\bar{x}'=x'$ and $\bar{y}'=y'/x'$. Since $\bar{x}=\bar{x}'$ and $\bar{y}=1/\bar{y}'$ it is obvious that $\mathfrak{P}_0(P)=\mathfrak{P}_0(P')$. On the other hand, it is clear that $P\neq P'$ (the quotient $x^2/y(=y')$ belongs to the local ring of P' but does not belong to the local ring of P).

LEMMA II. 3. 1. *If E is any exceptional curve of the first kind and if P' is the contraction of E, then E is a total antiregular transform of $E_0(P')$.*

PROOF. If F' is our surface which carries the point P' then we have $E_0(P')=T_0\{P'\}$ and $E=T\{P'\}$, where T_0 is a locally quadratic transformation with center P' and T is a birational transformation which is antiregular at P'. If we set $f=T_0^{-1}\circ T$ then $f=T_0^{-1}T$ (Lemma I. 7. 2) and hence $f\{E_0(P')\}=E$. Since every point of E dominates some quadratic transform of P' (II. 2), f is antiregular at each point of $E_0(P')$. Q.E.D.

COROLLARY II. 3. 2. *The curve $E_0(P')$ is the only irreducible exceptional curve of the first kind whose contraction is P'.*

For, in the notations of the proof of the preceding lemma, the curve E consists of the proper transform $f[E_0(P')]$ of $E_0(P')$ and of the exceptional curves of the transformation f^{-1} which correspond to the fundamental points of f on $E_0(P')$. Hence if E is irreducible then f has no fundamental points of $E_0(P')$ and is therefore biregular at each point of $E_0(P')$.

The notations and assumptions being the same as in the above lemma, we shall call the irreducible component $f[E_0(P')]$ the *principal component of E*. When E is regarded as embedded in some surface F each irreducible component of E determines a prime divisor of the field Σ/k (which depends only on that irreducible component and not on the choice of the surface). The principal component of E is that irreducible component of E for which the corresponding prime divisor is the principal P'-adic divisor $\mathfrak{P}_0(P')$, where P' is the contraction of E.

PROPOSITION II. 3. 3. *Let E be an exceptional curve of the first kind and let E_0 be the principal component of E. If $E \neq E_0$* (i.e., *if E is reducible; see Corollary* II. 3. 2) *then E has a decomposition of the form*

$$(1) \qquad E = E_0 \cup E_1 \cup E_2 \cup \ldots \cup E_s,$$

such that (a) *each E_i $(i=1, 2, \ldots, s)$ is an exceptional curve of the first kind;* (b) *$E_i \cap E_j$ is empty if $i \neq j$ $(i, j=1, 2, \ldots, s)$;* (c) *E_0 is not a component of any of the E_i, $i \neq 0$.*

PROOF. If, in the notations of the proof of Lemma II. 3. 1, we denote by $P'_1, P'_2, \ldots, P'_s$ the fundamental points of f on $E_0(P')$ then we obtain the decomposition (1) by setting $E_i = f\{P'_i\}$.

COROLLARY II. 3. 4. *Any exceptional curve E of the first kind is connected.*

Using induction with respect to the number of irreducible components of E we have that the curves $E_1, E_2, \ldots, E_s$ in (1) are connected. Since $E_i = f\{P'_i\}$, $E_0 = f[E_0(P')]$ and $P'_i \in E_0(P')$ it follows that each E_i intersects E_0. Since E_0 is irreducible, the corollary is proved.

The above corollary is, of course, also a consequence of the general "connectedness theorem" for birational transformation which we have proved in [31], but the latter is too deep a theorem to use in the present elementary connection.

COROLLARY II. 3. 5. *Let $T: F \rightarrow F'$ be any birational transformation of a surface F onto a surface F'* (*the surfaces need not be non-singular*) *and let W be any algebraic subvariety of F. If each point of W is simple for F and if W is connected then also $T\{W\}$ is connected.*

Assume first that F' is non-singular and that T is antiregular at each point of W. Using an indirect argument, assume that $T\{W\}$ is not connected and that it is therefore the union of two non-empty varieties W'_1 and W'_2 having no points in common. Then the union of the two varieties $W_i = T^{-1}\{W'_i\}$ $(i=1, 2)$ is W, and therefore W_1 and W_2 have a point P in common. The transformation T cannot be regular at P for otherwise $T\{P\}$ would consist of only one point, and this would have to belong to both W'_1 and W'_2. Hence P is a fundamental point of T, $T\{P\}$ is then an exceptional curve of the first kind (since T is antiregular at P), and this curve is the union of the two non-empty varieties $T\{P\} \cap W'_1$ and $T\{P\} \cap W'_2$ having no points in common, in contradiction with

Corollary II. 3. 4. In the general case we pass to a non-singular surface G which dominates both F and F' and we consider the birational transformations $f: F \rightarrow G$ and $f': F' \rightarrow G$. Then $T = ff'$ (Lemma I. 7. 2) and hence if $D = f\{W\}$ then $T\{W\} = f'\{D\}$. By the preceding part of the proof, D is connected, and since f' is single-valued it follows that also $f'\{D\}$ is connected.

COROLLARY II. 3. 6. *Exceptional curves of the second kind are also connected.*

Obvious.

LEMMA II. 3. 7. *If E_1 and E_2 are exceptional curves of the first kind belonging to a surface F and if P_1' and P_2' are the contractions of E_1 and E_2 respectively, then the relations $E_1 \supset E_2$ and $P_1' < P_2'$ are equivalent.*

PROOF. That the relation $E_1 \supset E_2$ implies the relation $P_1' < P_2'$ follows from Theorem II. 2. 3, relation (1). Conversely, if $P_1' < P_2'$ then any place with center P_2' has necessarily center P_1', and therefore E_2, which is the total transform of P_2' on F, is contained in the total transform E_1 of P_1'.

COROLLARY II. 3. 8. *The exceptional curves $E_i = f\{P_i'\}$ introduced in the course of the proof of Proposition* II. 3. 3 ($i = 1, 2, \ldots, s$) *are the maximal exceptional curves of the first kind which are properly contained in E.*

Since each point P_i' is an immediate successor of P' it follows from Lemma II. 3. 7 that the E_i are maximal elements in the set of all exceptional curves of the first kind which are properly contained in E. Let now $\bar{E}$ be any exceptional curve of the first kind which is properly contained in E and let $\bar{Q}$ be the contraction of $\bar{E}$. Then by the above lemma, the point $\bar{Q}$ dominates P' and is different from P'. Hence $\bar{Q}$ must dominate some point Q' of $E_0(P')$, and the total transform $f\{Q'\}$ must then contain the curve $\bar{E}$. It follows that Q' is one of the fundamental points P_i' of f on $E_0(P')$, $f\{Q'\}$ is one of the curves $E_1, E_2, \ldots, E_s$, showing that one of these curves must contain the curve $\bar{E}$.

COROLLARY II. 3. 9. *An exceptional curve E of the first kind does not admit a non-trivial decomposition into exceptional curves of the first kind.*

For any exceptional curve of the first kind which is properly contained in E is contained in one of the maximal exceptional curves E_i of the preceding corollary, and therefore does not contain the principal component E_0 of E.

COROLLARY II. 3. 10. *The decomposition* (1) *of Proposition* II. 3. 3 *is uniquely determined, and that is so even if condition* (c) *of that proposition is omitted and condition* (b) *is replaced by the following weaker condition:* (b′) $E_i \not\supset E_j$ *if* $i \neq j$ $(i, j = 1, 2, \dots, s)$.

For suppose that we have another decomposition

$$E = E_0 \cup \bar{E}_1 \cup \bar{E}_2 \cup \cdots \cup \bar{E}_h$$

satisfying condition (a) of Proposition II. 3. 3 and condition (b′) above (with $E_1, E_2, \cdots, E_s$ replaced with by $\bar{E}_1, \bar{E}_2, \cdots, \bar{E}_h$). Then condition (c) is automatically satisfied, for otherwise E would admit a non-trivial decomposition $\bar{E}_1 \cup \bar{E}_2 \cup \dots \cup \bar{E}_h$ into exceptional curves of the first kind, contrary to Corollary II. 3. 9. Hence the union of the $\bar{E}_j$ coincides with the union of the E_i. By Corollary II. 3. 8 each $\bar{E}_j$ is contained in some E_i. Since no two of the E_i have irreducible components in common it follows that each of the E_i is the union of a certain number of the $\bar{E}_j$, and hence, by Corollary II. 3. 9 and in view of (b′), each E_i coincides with some $\bar{E}_j$. Since the E_i are all the maximal exceptional curves of the first kind which are properly contained in E it follows by (b′) that $h = s$.

PROPOSITION II. 3. 11. *Each irreducible component D of an exceptional curve E of the first kind is the principal component of one and only one exceptional curve of the first kind contained in E. If $c(D)$ denotes the contraction of that exceptional curve and P' is the contraction of E itself, then the mapping $D \to c(D)$ is a* (1, 1) *mapping of the set of all irreducible components of E onto the set $N(E)$ of all points Q' of $M_{P'}$ which do not dominate any point of E.* (We recall that $M_{P'}$ denotes the set of all points of M which dominate P'; see II. 1).

PROOF. The first part of the proposition is easily established by induction with respect to the number of irreducible components of E. If D is the principal component E_0 of E then D does not belong to any exceptional curve of the first kind which is properly contained in E, for E_0 is not contained in any of the maximal exceptional curves $E_1, E_2, \dots, E_s$ which are properly contained in E (see Proposition II. 3. 3 and Corollary II. 3. 8). If $D \neq E_0$ then $D \subset E_i$ for some $i = 1, 2, \dots, s$, and the existence and unicity of the required exceptional curve now follow from our induction hypothesis and from the fact that the E_i are the only maximal exceptional curves properly contained in E.

It is clear that $c(D) \in N(E)$. If $\mathfrak{P}$ is the prime divisor of Σ/k which is defined by the irreducible curve D and if we set $Q' = c(D)$, then $\mathfrak{P}$ is the principal Q'-adic divisor. It follows that if D and D' are distinct irreducible components of E then $c(D) \neq c(D')$.

It remains to show that the mapping $D \rightarrow c(D)$ maps the set of irreducible components of E *onto* the set $N(E)$. If Q' is any point of $N(E)$ then the total transform W of Q' on the surface F which carries the curve E (i.e., the set of centers on F of all the places having center Q') is contained in E (since Q' dominates the contraction P' of E). The variety W does not consist of just one point, for in the contrary case Q' would dominate that point, contrary to our assumption that Q' belongs to the set $N(E)$. Furthermore, if Q is any point of W and if we fix a place $\mathfrak{p}$ having center Q and Q', then also P' is the center of $\mathfrak{p}$ (since Q dominates P'), and thus Q and Q' belong to the set $M_{\mathfrak{p},P}$ (see II. 1). Since this set is totally ordered (Lemma II. 1. 1) and since Q' does not dominate Q, it follows that Q dominates Q'. We have therefore shown that W is a total antiregular transform of Q' and is not a point. Therefore W is an exceptional curve of the first kind. If D is its principal component then D is an irreducible component of E and we have $Q' = c(D)$. This completes the proof of the proposition.

COROLLARY II. 3. 12. *Any exceptional curve E' contained in an exceptional curve E of the first kind is itself of the first kind.*

For if Q' is a contraction of E' then Q' dominates the contraction P' of E, and on the other hand the point Q' does not dominate any point of E. Hence Q' belongs to the set $N(E)$, and E' is therefore of the first kind.

PROPOSITION II. 3. 13. *The irreducible components of an exceptional curve E of the first kind are non-singular rational curves. The curve E has only ordinary double points, and the number of these is one less than the number of irreducible components of E. Any two irreducible components of E have at most one intersection.*

PROOF. Let P' be the contraction of E. If E is irreducible then $E = E_0(P')$ (Corollary II. 3. 2), hence is a rational non-singular curve. We can therefore use induction with respect to the number of irreducible components of E. The first part of the proposition follows then at once from the decomposition (1) of E (Proposition II. 3. 3). It follows also from our induction hypothesis, and from the fact no two of the curves E_i $(i=1,2,\ldots,s)$ intersect, that the curve $E_1 \cup E_2 \cup \ldots \cup E_s$ has only ordinary double points and that the number of its double points is equal to the number of irreducible components of E, diminished by $s+1$. Furthermore, any two irreducible components of E, different from E_0, have at most one intersection. Hence the proposition will be established if we prove that for each $i=1, 2, \ldots, s$ the curve E_i has only one intersection with E_0 and that that intersection is an ordinary double point of the curve $E_i \cup E_0$. In the notations of the proof of Lemma II. 3. 1 we have $E_i=f\{P'_i\}$ and $E_0=f[E_0(P')]$, where f is antiregular at P'. Now, $E_0(P')$ is a non-singular curve, and we can complete the proof of the proposition by establishing the following general result:

Lemma II. 3. 14. *Let P be a simple point of a surface F, let D be a curve on F which has at P a simple point and let $f: F \to F'$ be a birational transformation of F which is antiregular but not biregular at the point P. Then the proper transform $D'=f[D]$ has only one intersection with the exceptional curve $E'=f\{P\}$, and that intersection is an ordinary double point of the curve $D' \cup E'$.*

Proof. The lemma is obvious in the case in which f is biregularly equivalent at P to a locally quadratic transformation with center P, i.e., in the case in which E' is irreducible. We can therefore use induction with respect of the number of irreducible components of E'. Let $T: F \to F_1$ be a locally quadratic transformation of F, with center P, and let $f_1=T^{-1} \circ f$. Let $E_1=T\{P\}$ and $D_1=T[D]$. From the factorization theorem for antiregular quadratic transformations (Theorem II. 1. 2) it follows that f_1 is antiregular at each point of E_1. We also know that E_1 intersects D_1 in a single point, say P_1, and that this point P_1 is an ordinary double point of the curve $E_1 \cup D_1$. It is clear that $D'=f_1[D_1]$ and $E'=f_1\{E_1\}$. If P_1 is not a fundamental point of f_1 then the point $f_1\{P_1\}$ is the only intersection of D' and E', and this intersection is an ordinary double point of the curve $D' \cup E'$ (since f_1 is then biregular at P). If P_1 is a fundamental point of f_1, then by our induction hypothesis D' has only one intersection with the exceptional curve E'_1, where $E'_1=f_1\{P_1\}$, and this intersection is an ordinary double point of the curve $D' \cup E'_1$. On the other hand, E' is the union of E'_1 and the proper transform $E'_0=f_1[E_1]$ of

E_1, and D' does not intersect E_0' since D_1 and E_1 have at P_1 distinct tangents. Therefore the proof of the lemma (and also of Proposition II. 3. 13) is now complete.

II. 4. Exceptional curves of the second kind.

Exceptional curves of the second kind exist only on surfaces which are birationally equivalent to ruled surfaces. The proof of this fundamental result will be outlined in Part III of this work. The structure of an exceptional curve of the second kind may be much more complicated than of an exceptional curve of the first kind. For instance, an irreducible exceptional curve of the second kind may have very complicated singularities, in contrast with the fact that an irreducible exceptional curve of the first kind is always non-singular (see Corollary II. 3. 2 or also Proposition II. 3. 13). Our object in this section is not to study, for their own sake, many properties of exceptional curves of the second kind but to study only a few selected properties of these curves which we shall have to use in Part III.

PROPOSITION II. 4. 1. *Each irreducible component of an exceptional curve of the second kind is a rational curve.*

PROOF. Let E be an exceptional curve of the second kind on a surface F, let Q' be a contraction of E and let F' be a surface which carries the point Q'. Let G be a non-singular surface which dominates both F and F' and let f and f' denote the birational transformations $F \to G$ and $F' \to G$ respectively. It is clear that $f'\{Q'\}$ is a subset of $f\{E\}$. However, if D is any irreducible component of E then the prime divisor of the field Σ/k which is determined by the curve D on F, has center Q' on F' and hence its center $f[D]$ on G is contained in $f'\{Q'\}$. Since $f'\{Q'\}$ is an exceptional curve of the first kind, the irreducible component $f[D]$ of that curve is a rational curve (Proposition II. 3. 13). Hence also D is a rational curve.

PROPOSITION II. 4. 2. *Let E_1 and E_2 be exceptional curves of a surface F such that E_1 and E_2 have a non-empty intersection and $E_i \not\supset E_j$ if $i \neq j$. Let P_1' be a contraction of E_1, let P_2' be a contraction of E_2, let F_1' and F_2' be surfaces which carry the points P_1' and P_2' respectively, and assume that one of the common points of E_1 and E_2 dominates one of the points P_1', P_2'. If D_i' denotes the total transform of P_j' on F_i' $(i, j=1, 2;\ i \neq j)$, then D_1'*

and D'_2 *are exceptional curves of the second kind, and we have* $P'_i \in D'_i$. *Furthermore, if one of the exceptional curves* E_i, *say* E_1, *is of the first kind, then* D'_2 *is the total transform of* E_1.

PROOF. Let Q be a common point of E_1 and E_2 which dominates one of the points P'_i, say let Q dominate P'_1. There exists a place $\mathfrak{p}$ having centers Q and P'_2 (since Q belongs to the total transform E_2 of P'_2), and then necessarily also P'_1 will be a center of $\mathfrak{p}$ (since Q dominates P'_1). This shows that P'_1 and P'_2 are corresponding points of the birational correspondence between F'_1 and F'_2, and that $P'_i \in D'_i$. Neither of the points P'_1 and P'_2 dominates the other, since neither one of the curves E_1 and E_2 contains the other. Hence each of the curves D'_1 and D'_2 appears as a total, *but not antiregular*, transform of a *simple* point (namely of P'_2 and P'_1 respectively). Hence, by Theorem II. 2. 3, both curves D'_1 and D'_2 are exceptional curves of the *second* kind. It is clear that D'_2 is a subset of the total transform of E_1 and that if each point of E_1 dominates P'_1 (equivalently: if E_1 is of the first kind) then D'_2 coincides with the total transform of E_1. Q.E.D.

COROLLARY II. 4. 3. *If a surface F carries two intersecting exceptional curves of the first kind such that none of the two curves contains the other, then some surface which is birationally equivalent to F carries an exceptional curve of the second kind.*

For if in the above proposition one of the two exceptional curves E_1 and E_2 is of the first kind, say E_1, then each common point of E_1 and E_2 dominates the point P'_1.

LEMMA II. 4. 4. *If E is an exceptional curve of the second kind, the set of (simple) points (in M) which are contractions of E does not contain any infinite, strictly ascending sequence.*

PROOF. Assume that there exists an infinite, strictly ascending sequence

$$P' < P'' < \cdots < P^{(i)} < \cdots$$

such that each point $P^{(i)}$ is a contraction of E. Let $\mathfrak{o}_i$ be the local ring of $P^{(i)}$ and let $\mathfrak{o}$ be the union of the rings $\mathfrak{o}_i$. Let $\mathfrak{m}_i$ be the maximal ideal of $\mathfrak{o}_i$ and let $\mathfrak{m}$ be the union of the $\mathfrak{m}_i$. Then $\mathfrak{m}$ is the maximal ideal of $\mathfrak{o}$, and it is known that $\mathfrak{o}$ is the valuation ring of a zero dimensional valuation v' of the field Σ/k. On the other hand, let v be the one-dimensional valuation of Σ/k defined by some irreducible component of E and let R_v and M_v denote respectively the valuation ring and the maximal ideal of v. From the assumption that each point $P^{(i)}$ is a contraction of E follows that $R_v \supset \mathfrak{o}_i$ and $M_v \supset \mathfrak{m}_i$. Hence $R_v \supset \mathfrak{o}$ and $M_v \supset \mathfrak{m}$, and of course $R_v \neq \mathfrak{o}$. This is in contradiction with the well known fact that since v is composite with the valuation v' the maximal ideal M_v of v is a proper subset of the maximal ideal $\mathfrak{m}$ of v'.

PROPOSITION II. 4. 5. *If a surface F (in our birational class B of non-singular projective models of Σ/k) carries an exceptional curve of the second kind, then there exists a surface in the class B which carries an irreducible exceptional curve of the second kind.*

PROOF. Let E be an exceptional curve of F, of the second kind, and let q be the number of irreducible components of E. We assume that $q>1$ and we proceed to show that there exists a surface F' in B which carries an exceptional curve of second kind having less than q irreducible components.

In the set of contractions of E we fix a maximal element P^* (Lemma II. 4. 4) and we denote by E' the total transform $E_0(P^*)$ of P^* under a quadratic transformation with center P^* (II. 3). It is clear that E is the total transform of E'. Since E' is irreducible and E is reducible, the birational transformation of the carrier surface of E' onto the surface F must have fundamental point on E'. Let P' be a fundamental point on E' and let E_1 be the total transform of P' on F. Then E_1 is an exceptional curve of F, contained in E. It cannot be E itself, for in the contrary case the point P', which is a quadratic transform of P^*, would also be a contraction of E, in contradiction with the maximality of P^*. Hence E_1 is a proper subset of E and has therefore less than q irreducible components. If E_1 is of the second kind, the proof is complete. Suppose that E_1 is of the first kind. We pass then to the projective surface $F''=(F-E_1)+P'$ (Theorem II. 2. 4) and to the total transform E'' of E on F''.[10] Since the birational transformation $F \rightarrow F''$ is regular, E'' is the total transform of P^* on F'' (Lemma I. 7. 2) and is therefore an exceptional curve, with contraction P^*. Again, since the birational transformation $F \rightarrow F''$ is regular

10) The transition to the abstract surface F'' has been suggested to us by NAGATA; it simplifies our original proof.

and since E is not an antiregular transform of P^* (E being of the second kind), it follows *a fortiori* that also E'' is not an antiregular transform of its contraction P^*. Hence E'' is an exceptional curve of the second kind (Theorem II. 2. 3). Since the number of irreducible components of E'' is equal to the difference between the integer q and the number of irreducible components of E_1, the proof is complete.

In Part III we shall deal extensively with *relatively minimal models of the field Σ/k*, this being the term with which we designate those surfaces in the birational class B which are minimal elements of the partially ordered set B. A surface F in B is therefore a relatively minimal model of Σ/k if there exists no surface in B which is dominated by F (and is not biregularly equivalent to F). We note now that in the course of the above proof we have established the following result:

Corollary II. 4. 6. *If a relatively minimal model F of Σ/k carries an exceptional curve of the second kind then F also carries an irreducible exceptional curve of the second kind.*

For the exceptional curve E_1 of F must still be of the second kind, because in the contrary case we would have a surface F'' in B which is dominated by F and is not biregularly equivalent to F. Thus if F carries an exceptional curve E having q irreducible components it also carries an exceptional curve E_1 of the second kind having less than q irreducible components ($q>1$).

Proposition II. 4. 7. *Let E be an irreducible exceptional curve of the second kind on a surface F, let P' be a maximal contraction of E (Lemma II. 4. 4) and let P be a point of E which does not dominate the point P' (equivalently: P is a point of E which is a fundamental for the birational transformation of F onto a surface F' which carries the point P'; such a point P must exist since E is of the second kind). Let $T: F\to\bar{F}$ be a locally quadratic transformation of F, with center P. Then the proper T-transform $T[E]$ of E is an exceptional curve (not necessarily of the second kind) and P' is a maximal contraction of that curve.*

Proof. Let $T': F'\to\bar{F}'$ be a locally quadratic transformation of F', with center P', and let $\bar{E}'=T'\{P'\}$. Since $\bar{E}'$ and P' are total transforms of each other, it follows at once that E is a total transform of $\bar{E}'$. The birational transformation $f: \bar{F}'\to F$ can have no fundamental points on $\bar{E}'$, for a fundamental point $\bar{P}'$ on $\bar{E}'$ would then necessarily be a contraction of E, in contradiction with the maximality of P'. We assert that also the birational transformation $\bar{f}: \bar{F}' \rightarrow \bar{F}$ is regular at each point of $\bar{E}'$. For let $\bar{P}'$ be any point of $\bar{E}'$. If the point Q of E which corresponds to $\bar{P}'$ is different from P then $\bar{f}$ is regular at $\bar{P}'$ because f is regular at $\bar{P}'$ and the transformation $T: F \rightarrow \bar{F}$ is biregular at Q. If $Q=P$, then we observe that while we have $\bar{P}' > P$ (in view of the regularity of f at $\bar{P}'$) we cannot have $\bar{P}'=P$, for in the contrary case we would have $P > P'$, in contradiction with the fact that P does not dominate P' (it is here that we use the assumption that the exceptional curve E is of the second kind). Hence $\bar{P}'$ dominates some quadratic transform of P, and this proves that $\bar{f}$ is regular at $\bar{P}'$. From the regularity of $\bar{f}$ follows that $\bar{f}\{\bar{E}'\}=\bar{f}[\bar{E}']$. Now the prime divisor determined by the curve $\bar{E}'$ on $\bar{F}'$ coincides with the prime divisor determined by the curve E on F (since E is the total transform of $\bar{E}'$ and since the birational transformation f is regular on $\bar{E}'$). Hence $\bar{f}[\bar{E}']=T[E]$. Thus $T[E]$ is the total transform of the point P', showing that $T[E]$ is an exceptional curve having P' as contraction. That P' is a maximal contraction of $T[E]$ follows at once from the fact that P' is a maximal contraction of E. Q.E.D.

If $\bar{E}=T[E]$ is still of second kind we can repeat the process by applying to $\bar{F}$ a locally quadratic transformation whose center is a point of $\bar{E}$ which does not dominate P'. If this process were to continue indefinitely then it is easily seen that for a suitable choice of the point P on E (subject to the condition that P does not dominate P') we would obtain an infinite, strictly ascending sequence $P > \bar{P} > \bar{P}^{(1)} > \cdots > \bar{P}^{(n)} > \cdots$ of points belonging to surfaces $\bar{F}^{(i)}$, such that for each n the point $\bar{P}^{(n)}$ is a fundamental point of the birational transformation of $\bar{F}^{(n)}$ onto the fixed surface F'. Since this is impossible, we can state the following result:

PROPOSITION II. 4. 8. *If E is an irreducible exceptional curve of the second kind on a surface F, there exists an antiregular birational transformation $g: F \rightarrow G$ such that the proper transform $g[E]$ of E is an exceptional curve of the first kind. More precisely: if P' is a maximal contraction of E, then there exists a finite sequence of locally quadratic transformations $T_i: F_i \rightarrow F_{i+1}$ ($i=0, 1, \ldots, n-1$; $F_0=F$, $T_0=T$; $n \geqq 1$) such that if we define the curves E_i by induction, as follows: $E_0=E$, $E_{i+1}=T_i[E_i]$, then the $n+1$ curves E_i*

($i=0, 1, \ldots, n$) *are exceptional curves having* P' *as a maximal contraction, the first* n *of these exceptional curves are of the second kind, while* E_n *is of the first kind, and the center of each* T_i *is a point of* E_i *which does not dominate* P'.

COROLLARY II. 4. 9. *If* E *is an irreducible exceptional curve of the second kind, with maximal contraction* P', *and if* Q *is a singular point of* E, *then* Q *does not dominate* P'.

For in the contrary case, the product g of the n quadratic transformations T_i of the preceding proposition would be biregular at Q, and Q would then be a singular point of the irreducible exceptional curve E_n of the first kind, in contradiction with Corollary II. 3. 2.

The preceding corollary implies that if T is any antiregular transformation of F such that $T[E]$ is an exceptional curve of the first kind, then each singular point of E is a fundamental point of T. This, however, is obvious *a priori*, for an irreducible exceptional curve of the first kind has no singularities and therefore one of the effects of T is that of resolving the singularities of E.

Let us assume that E has a singular point P. From the preceding corollary and from Proposition II. 4. 7 it follows that we can take for T_0 a locally quadratic transformation with center P. If E_1 $(=T_0[E])$ still has singular points then E_1 is still of the second kind, and if P_1 is one of the singular points of E_1 then we can take for T_1 a locally quadratic transformation with center P_1. Proceeding in this fashion we can arrange the choice of the T_i so that for some integer $m \leqq n$ the following conditions are satisfied: (1) E_i has a singular point if $i<m$, while E_m is non-singular; (2) the center of T_i ($i=0, 1, \ldots, m-1$) is a singular point of E_i. Then E_m is an irreducible non-singular exceptional curve. It is clear that the integer m depends only on the curve E.

If $T'_0, T'_1, \ldots, T'_{m-1}$ is another sequence of locally quadratic transformations $T'_i: F'_i \rightarrow F'_{i+1}$ $(F'_0=F)$ satisfying conditions similar to (1) and (2) (with E_i being replaced by $E'_i=T'_{i-1}[E'_{i-1}]$, $i=1, 2, \ldots, m$; $E'_0=E$) then it is not difficult to see that the two antiregular transformations $f=T_0T_1\cdots T_{m-1}$ and $f'=T'_0T'_1\cdots T'_{m-1}$ of F are biregularly equivalent. This is proved by induction on m and by making use of the fact (whose proof is straightforward) that any two locally quadratic transformations having distinct centers commute (to within biregular equivalence). From the biregular equivalence of f and f' it follows that the model F_m and the non-singular exceptional curve E_m depend only on F and E (up to biregular equivalence). We shall refer to E_m as the *minimal resolution of* E.

The case $m=n$ is of particular interest; this is the case in which the minimal resolution of our irreducible exceptional curve E of the second kind is an exceptional curve *of the first kind.*

PROPOSITION II. 4. 10. *If an irreducible exceptional curve E of the second kind is such that its minimal resolution E' is of the first kind, then E has only one maximal contraction P', and every simple point of E dominates P'.*

PROOF. Any maximal contraction P' of E is also a contraction of E' (Proposition II. 4. 8), and since an exceptional curve of the first kind has only one contraction (Theorem II. 2. 3) the uniqueness of P' is established. Since every point of E' dominates P' and since the antiregular transformation f of F which transforms E into E' is biregular at each simple point of E (by the construction of f), the proposition is established.

COROLLARY II. 4. 11. *Let E be an irreducible exceptional curve of the second kind whose minimal resolution is of the first kind. If T is any birational transformation of the carrier surface F, such that E is the total T^{-1}-transform of a* (simple) *point, then T is regular at each simple point of E, while each singular point of E is fundamental for T.*

The first assertion follows from the preceding proposition: the second follows from Corollary II. 4. 9.

We shall see in II. 5. (Corollary II. 5. 15) that if the minimal resolution of E is of the second kind (in particular, if E itself is a *non-singular* irreducible exceptional curve of the second kind) then the conclusion of Proposition II. 4. 10 is false. We shall see that in that case for every simple point P of E there exists at least one contraction of E which is not dominated by P. Hence any simple point of E can and does occur as fundamental point of some birational transformation of F which contracts E to a (simple) point. It follows that if the minimal resolution of E is of the second kind then E has infinitely many maximal contractions.

II. 5. Antiregular transformations of divisorial cycles.

Let $T: F \to F'$ be an antiregular birational transformation of a non-singular surface F onto a non-singular surface F'. We shall study in this section the effect of T on divisorial cycles on F, i.e., on cycles X of the form

$$X = \sum m_i \Gamma_i, \tag{1}$$

where the Γ_i are distinct irreducible curves on F, finite in number and defined over k, and the m_i are integers. We shall continue to use the notation $\langle X \rangle$ (I. 3) for the carrier of the cycle X, i.e., for the union of the curves Γ_i such that $m_i \neq 0$.

If P is any point of F we set $m_i(P) = m_i$ if $P \in \Gamma_i$ and $m_i(P) = 0$ in the contrary case. We call the cycle $\sum m_i(P)\Gamma_i$ *the local P-component of X* and we denote it by X_P.

Since F is non-singular, every cycle is everywhere locally complete intersection, i.e., for every point P of F there exists an element g_P of the function field Σ $(=k(F))$ of F such that X_P coincides with the local P-component of the cycle (g_P) of the function g_P (see ZARISKI [30], Theorem 5, p. 22). The function g_P is uniquely determined by X and P to within a factor which is a unit in the local ring of P. We shall find it convenient to say that $g_P = 0$ is *a local equation of X at P.*

Let P' be any point of F' let P be the corresponding point of F, let X be a divisorial cycle on F and let $g_P = 0$ be a local equation of X at P. The function g_P defines also a divisorial cycle on F', and the local P'-component of this cycle depends only on X and P', since any unit in the local ring of P is also a unit in the local ring of P' (in view of the antiregularity of T). We denote the local P'-component of the cycle (g_P) by $X(P')$.

Let $\Delta_1', \Delta_2', \ldots, \Delta_s'$ be the irreducible exceptional curves of T^{-1} (I. 2) and let Q_j be the point of F which corresponds to Δ_j' (the points $Q_1, Q_2, \ldots, Q_s$ need not be distinct). Let $\Gamma_i' = T[\Gamma_i]$.

PROPOSITION II. 5. 1. *The local cycles $X(P')$ $(P' \in F')$ are the local components of a global cycle X' of F', of the form*

$$X' = \sum m_i \Gamma_i' + \sum \mu_j \Delta_j', \tag{2}$$

where the μ_j are integers. We have $\mu_j = 0$ if $Q_j \notin \langle X \rangle$. If X is a positive cycle and $Q_j \in \langle X \rangle$ then $\mu_j > 0$.

PROOF. Let D' be any prime divisorial cycle on F', let P' be any point of D' and let λ be the coefficient of D' in the cycle $X(P')$. We show that

λ depends only on D' and X, that $\lambda = m_i$ if $D' = \Gamma'_i$ and that $\lambda = 0$ if D' is different from any of the curves Γ'_i and Δ'_j. We denote by P the point of F which corresponds to P' under T^{-1}. Let $g_P = 0$ be a local equation of X at P. Assume first that D' is different from any of the Δ'_j. If D is the (irreducible) curve on F which corresponds to D' then D and D' define one and the same one-dimensional valuation v of the field Σ, and since $g_P = 0$ is also a local equation of $X(P')$ at P' it follows that $\lambda = v(g_P)$, i.e., the coefficient of D' in $X(P')$ is the same as the coefficient of D in X. Now assume that $D' = \Delta'_j$ for some $j = 1, 2, \ldots, s$. Then $P = Q_j$, whence the point P depends only on D' (being independent of the choice of the point P' on D'), and since (by the definition of $X(P')$) we have $\lambda = v_j(g_P)$, where v_j is the one-dimensional valuation defined by the curve Δ'_j, we conclude that also in the present case λ depends only on D' and X.

If $Q_j \notin \langle X \rangle$ then g_{Q_j} is a unit in the local ring of Q_j, whence $\mu_j = v_j(g_{Q_j}) = 0$. If X is a positive cycle and $Q_j \in \langle X \rangle$, then g_{Q_j} belongs to the local ring of Q_j and is a non-unit in that ring. Since Q_j is the center of v_j on F it follows that $\mu_j > 0$. This completes the proof.

DEFINITION II. 5. 2. *The divisorial cycle X' on F which is determined by its local components $X_{P'} = X(P')$ will be called the T-transform of the cycle X and will be denoted by $T(X)$. The cycle $\sum m_i \Gamma'_i$ will be called the proper T-transform of X and will be denoted by $T[X]$.*

Without fear of confusion we shall denote by the letter T not only the given antiregular birational transformation $F \to F'$ but also the cycle transformation $X \to X' = T(X)$.

We point out explicitly that in the course of the proof of Proposition II. 5. 1 we have proved that

$$\mu_j = v_j(g_{Q_j}), \tag{3}$$

where v_j is the one-dimensional valuation of Σ defined by the irreducible exceptional curve Δ'_j of T^{-1} and where $g_{Q_j} = 0$ is a local equation of X at the fundamental point Q_j of T.

PROPOSITION II. 5. 3. *The cycle transformation T is a monomorphism of the additive group of divisorial cycles on F into the additive group of divisorial cycles on F'. This monomorphism preserves linear equivalence and intersection numbers: if X and Y are divisorial cycles on F and $X' = X(T)$, $Y' = T(Y)$, then $(X' \cdot Y') = (X \cdot Y)$, and if $X \equiv Y$ then $X' \equiv Y'$. Furthermore, if $X' \equiv Y'$*

then $X \equiv Y$.

PROOF. That T is a monomorphism follows directly from the definition of the cycle transformation T, from Proposition II. 5. 1 and from (3). If X is linearly equivalent to zero then X is the cycle of a function g, and X' is obviously the cycle of the same function g, on F'. Hence T preserves linear equivalence. The same argument shows that if $X' \equiv 0$ then also $X \equiv 0$.

If none of the curves $\langle X \rangle$, $\langle Y \rangle$ contains any of the fundamental points of T and if furthermore these two curves have no common components, then the transformed cycles X' and Y' coincide with the proper transforms $T[X]$ and $T[Y]$. In this case, the equality $(X \cdot Y) = (X' \cdot Y')$ is obvious, since T is biregular at each common point of $\langle X \rangle$ and $\langle Y \rangle$. The general case is reduced to this special case by making use of the well known fact that every cycle X is linearly equivalent to a cycle $\bar{X}$ whose carrier does not contain any preassigned finite set of curves and points of F.[11)]

PROPOSITION II. 5. 4. *If X is any cycle on F and Δ'_j is any of the irreducible exceptional curves of T^{-1} on F', then $(T(X) \cdot \Delta'_j) = 0$.*

PROOF. If the fundamental point Q_j does not belong to $\langle X \rangle$ then $T(X)$ does not intersect Δ'_j at all. If Q_j does belong to $\langle X \rangle$ we replace X by a linearly equivalent cycle whose carrier does not pass through Q_j.

PROPOSITION II. 5. 5. *If $T: F \to F'$ is an antiregular birational transformation of a non-singular surface F onto a non-singular surface F' and $T': F' \to F''$ is an antiregular birational transformation of F' onto a non-singular surface F'', then we have for any divisorial cycle X on F: $(TT')(X) = T'(T(X))$.*

PROOF. Obvious.

Since any antiregular birational transformation of a non-singular surface onto a non-singular surface can be factored into locally quadratic transformations, Proposition II. 5. 5 furnishes the means of reducing the proofs of certain propositions concerning cycle transformations to the special case of a locally quadratic transformation (see, for instance, Proposition II. 5. 12 below).

---

11) Write X as a difference $X_1 - X_2$ of two positive cycles and observe that if C is a hyperplane section of F then the complete linear systems $|X_1 + hC|$, $|X_2 + hC|$ have no fixed components and no base points if h is large.

PROPOSITION II. 5. 6. *Let $T: F \to F'$ be a locally quadratic transformation of F, with center P, and let $E' = T\{P\}$. If K is a canonical divisor on F then $T(K)+E'$ is a canonical divisor on F'. Conversely, if K' is a canonical divisor on F' then $K'-E'$ is the T-transform of a canonical divisor on F.*

PROOF. Let K be a canonical divisor on F, i.e., K is the divisorial cycle $(\omega)^F$, on F, of a double differential ω of the field Σ. To prove the first part of the proposition it will be sufficient to consider the case in which $P \notin \langle K \rangle$. In that case the cycle $T(K)$ coincides with the proper T-transform $T[K]$ of K and does not contain E' as component (Proposition II. 5. 1). It is clear that the cycle $(\omega)^{F'}$ of ω on F' can differ from $T[K]$ only in the component E'. We shall show that E' appears in $(\omega)^{F'}$ with the coefficient $+1$. Let x, y be local parameters at P and let $\omega = A\,dx\,dy$, $A \in \Sigma$. Since $P \notin \langle K \rangle$, A is a unit in the local ring of P. Let $y' = y/x$. Then x and y' are local parameters, on F', of some point P' of E'. We have $\omega = A\,x\,dx\,dy'$. Since A is also a unit in the local ring of P and since $x=0$ is a local equation of E' at P' (see "Remark" just before Example 1, section I. 8) it follows that E' occurs in the cycle $(\omega)^{F'}$ with the coefficient $+1$, as asserted, showing that $(\omega)^{F'} = T(K)+E'$.

To prove the second part of the proposition, let K_1' be a canonical cycle on F', let $K_1' = (\omega_1)^{F'}$ and $K_1 = (\omega_1)^F$, where ω_1 is a double differential of Σ, and let ω and $K = (\omega)^F$ be as in the first part of the proof. Since $\omega_1/\omega \in \Sigma$ and since $K_1 - K = (\omega_1/\omega)^F$, it follows that $T(K_1) - T(K) = (\omega_1/\omega)^{F'}$. Since we have just proved that $(\omega)^{F'} = T(K)+E'$, it follows that $K_1' = (\omega_1)^{F'} = T(K_1)+E'$, and this completes the proof.

COROLLARY II. 5. 7. *If the canonical system $|K|$ on F exists, then the canonical system $|K'|$ on F' has E' as a fixed component, and the linear system $|K'-E'|$ is the transform of $|K|$* (i.e., *the members of that system are the transforms of the members of the system $|K|$*).

Obvious.

Let $\mathfrak{o}_P$ be the local ring of a point P of F and let $\mathfrak{m}$ be the maximal ideal of $\mathfrak{o}_P$. If X is a divisorial cycle on F and $g=0$ is a local equation of X at P, we write g in the form $g = g_1/g_2$, where $g_i \in \mathfrak{o}_P$. If r_i is the highest power of $\mathfrak{m}$ which contains g_i then we say that *P is an (r_1-r_2)-fold point of the cycle X*. The integer $r_1 - r_2$ depends only on X and P; it may be negative if X is not a positive cycle.

PROPOSITION II. 5. 8. *Let $T: F \to F'$ be a locally quadratic transformation of F, with center P, and let $E' = T\{P\}$. If a cycle X on F has at P an r-fold point then*

$$T(X) = T[X] + rE'. \tag{4}$$

PROOF. The proposition follows directly from Proposition II. 5. 1, from relation (3) and from the known determination (see II. 3) of the valuation ring of the principal P-adic divisor (the divisor defined by the curve E').

PROPOSITION II. 5. 9. *If E' is an irreducible exceptional curve of the first kind on a non-singular surface F', then $(E'^2) = -1$.*

PROOF. Let P be the (simple) contraction of E'. The abstract surface $F = F' - E' + P$ is projective (Theorem II. 2. 4) and the birational transformation $T: F \to F'$ is locally quadratic, with center P (Corollary II. 3. 2). We fix a prime divisorial cycle X on F which passes through P and has there a simple point. By Proposition II. 5. 8 we have $T(X) = T[X] + E'$, and by Proposition II. 5. 4 we have $(T[X] \cdot E') + (E'^2) = 0$. Since $(T[X] \cdot E') = +1$ (Lemma II. 3. 14), it follows that $(E'^2) = -1$.

The following converse of Proposition II. 5. 9 is well known and will be given here without proof:

THEOREM II. 5. 10. *If an irreducible curve E on a non-singular surface F is such that $p(E) = 0$ and $(E^2) = -1$, then E is an exceptional curve of the first kind.*

The proof, due to CASTELNUOVO and based on the theorem of RIEMANN-ROCH for algebraic surfaces, may be found on pp. 71–72 of our Ergebnisse monograph " Algebraic Surfaces " [23].

Propositions II. 5. 6–II. 5. 9 concern (directly or indirectly) locally quadratic transformations. We can now pass to the next object of this section: the proof of invariance of the arithmetic genus $p(X)$ of a cycle X under antiregular birational transformations (for the definition of $p(X)$ see our paper [32], pp. 580–583).

If X is a divisorial cycle on a non-singular surface F we set

$$\sigma(X) = -(X \cdot K), \tag{5}$$

where K is a canonical cycle. The integer $\sigma(X)$ is an additive character of the cycle X. We shall now prove the following relation:

$$\sigma(X) = (X^2) - 2p(X) + 2. \tag{6}$$

Let us denote the expression

$$\frac{(X)^2-\sigma(X)}{2}+1$$

by $\tilde{p}(X)$. We have to show that $\tilde{p}(X)=p(X)$. A direct calculation shows that if X and Y are any two divisorial cycles on F then

(7) $$\tilde{p}(X+Y)=\tilde{p}(X)+\tilde{p}(Y)+(X\cdot Y)-1.$$

We also have $p(X+Y)=p(X)+p(Y)+(X\cdot Y)-1$ (see ZARISKI [32], p. 581). If X is a non-singular hyperplane section of F, if G is a characteristic divisor on X and K is a canonical cycle which does not have X as component, then $K\cdot X+G$ is a canonical divisor on X (ZARISKI [32], pp. 588–589). Hence $(X^2)-\sigma(X)=2p(X)-2$ and this establishes (6) in the case under consideration.

Let now X be an arbitrary divisorial cycle on F. We denote by C a hyperplane section of F and we consider the linear system $L_h=|X+hC|$, h large. By (7) and by the preceding part of the proof, to establish the equality $\tilde{p}(X)=p(X)$ it will be sufficient to prove that $\tilde{p}(Y)=p(Y)$ for a general cycle of L_h. Now, if h is large then L_h is of positive dimension, has no base points and no fixed components. Hence L_h defines a rational *regular* transformation T_h of F. But then T_{h+1}, which is the graph of T_h, is a *biregular* birational transformation $F\to F'$. Since T_{h+1} transforms the general cycle Y of L_{h+1} biregularly onto a non-singular hyperplane section of F', the preceding part of the proof is applicable to Y, and (6) is proved.

We note that if E is an irreducible exceptional curve of the first kind on a non-singular surface F, then

(8) $$p(E)=0, \ \sigma(E)=1.$$

The first of these two relations is obvious (E is a rational non-singular curve); the second follows from Proposition II. 5. 9 and from (6).

Another proof of the equality $\sigma(E)=1$ is as follows:

We use the notations of the proof of Proposition II. 5. 9 and we consider a canonical cycle K' on F'. By the second part of Proposition II. 5. 6 we have $K'=T(K)+E'$, where K is a suitable canonical cycle on F. Since $(T(K)\cdot E')=0$ (Proposition II. 5. 4) we have $\sigma(E')=-(E'\cdot K')=-(E'^2)=1$ (Proposition II. 5. 9).

We can now prove the announced invariance of the arithmetic genus $p(X)$ under antiregular birational transformations:

THEOREM II. 5. 11. *Let $T: F \to F'$ be an antiregular birational transformation of a non-singular surface F onto a non-singular surface F' and let X be an arbitrary divisorial cycle on F. Then*

$$(9) \qquad \sigma(T(X)) = \sigma(X),$$

$$(9') \qquad p(T(X)) = p(X).$$

PROOF. By Proposition II. 5. 5 it is sufficient to consider the case of a locally quadratic transformation T. Let P be the center of T and let $E' = T\{P\}$. Let K be a canonical cycle on F and let $K' = T(K) + E'$, so that K' is a canonical cycle on F. We have $T((X) \cdot K') = (T(X) \cdot T(K)) + (T(X) \cdot E') = (X \cdot K)$ (Propositions II. 5. 3 and II. 5. 4). Hence $\sigma(T(X')) = \sigma(X)$. Relation (9′) follows from (9), (6) and Proposition II. 5. 3.

If X is a *prime* divisorial cycle on F we shall denote from now on by $\pi(X)$ the geometric (i.e., the effective) genus of X.

PROPOSITION II. 5. 12. *If X is a prime divisorial cycle on a non-singular surface F, then $p(X) \geqq \pi(X)$, with equality if and only if X has no singular points.*

PROOF. If X has no singular points then $p(X) = \pi(X)$. If X has a singular point P we apply to F a locally quadratic transformation $T: F \to F'$ with center P. If r is the multiplicity of the point P of X $(r > 1)$ then $T(X) = X' + r\,E'$, where X' is the irreducible curve $T[X]$, and $E' = T\{P\}$ (Proposition II. 5. 8). Since $\sigma(T(X)) = \sigma(X)$ and $\sigma(E') = 1$, we have

$$(10) \qquad \sigma(X) = \sigma(X') + r.$$

Since $(T(X) \cdot E') = 0$ (Proposition II. 5. 4) and $(E'^2) = -1$, we have

$$(11) \qquad (X' \cdot E') = r.$$

Finally, since $(T(X)^2) = (X^2)$ we find that $(X^2) = (X'^2) + 2r(X' \cdot E') + r^2(E'^2)$, i.e.,

$$(12) \qquad (X^2) = (X'^2) + r^2.$$

From (10) and (12) we conclude, in view of (6), that

$$(13) \qquad p(X) = p(X') + r(r-1)/2.$$

We can now use induction with respect to the number of quadratic transformations which are necessary in order to resolve the singularities of X. By our induction hypothesis we can assert that $p(X') \geqq \pi(X')$. Since $\pi(X) = \pi(X')$ and since $r > 1$ it follows from (13) that $p(X) > \pi(X)$.

We shall terminate this section by an application of formulae (10)–

(13) to irreducible exceptional curves of the second kind. This application will be essential for Part III.

PROPOSITION II. 5. 13. *Let E be an irreducible curve on a non-singular surface F and let $s_1, s_2, \ldots, s_m$ be the multiplicities of the singular points of E (including the infinitely near singular points of E). Then E is an exceptional curve if and only if* (a) *E is a rational curve and* (b) $(E^2) \geqq -1+\sum s_i^2$, *with equality if and only if the minimal resolution of E is of the first kind.*

PROOF. Assume that E is an exceptional curve. We know then that condition (a) is satisfied (Propositions II. 3. 13 and II. 4. 1). If E is of the first kind then it is non-singular (Proposition II. 3. 2), all the s_i are to be set equal to zero (or more precisely: the set of the s_i is empty), and condition (b) is satisfied, with strict equality (Proposition II. 5. 9).

Assume now that E is of the second kind. If E is non-singular and if it takes n successive quadratic transformations T_i to transform E into an exceptional curve E' of the first kind (Proposition II. 4. 8), then successive applications of (12) (where r is to be set equal to 1) yields $(E^2)=(E'^2)+n=n-1>-1$. If E has singular points and if E' denotes the minimal resolution of E (II. 4), then again successive applications of (12) yield the relation $(E^2)=(E'^2)+\sum s_i^2$, and hence, by the preceding case, we find that inequality (b) holds, with equality if and only if E' is of the first kind.

Assume now that conditions (a) and (b) are satisfied. We observe that if T is a locally quadratic transformation of F and if it is known that the curve $E'=T[E]$ is exceptional, then also E is exceptional (every contraction of E' is automatically a contraction of E). Now, if we set $n=(E^2)+1-\sum s_i^2$ ($n \geqq 0$) then (12) shows that after $m+n$ quadratic transformations the curve E is transformed into a curve E' such that $(E'^2)=-1$. It is understood that the first m of these transformations are so chosen as to lead to a minimal resolution of E. Hence E' is an exceptional curve of the first kind (Theorem II. 5. 12), and this completes the proof of the proposition.

COROLLARY II. 5. 14. *If E is an irreducible exceptional curve of the second kind on a non-singular surface F, then $(E^2) \geqq 0$, and if E has singularities then $(E^2)>0$. Furthermore, in the notations of Proposition* II. 5. 13 *we have*

$$(14)\qquad (E^2)=-1+\sum_{i=1}^{m} s_i^2+n, \quad n+m>0\,;\ n,\ m\geqq 0,$$

$$(15)\qquad p(E)=\sum s_i(s_i-1)/2,$$

$$(16)\qquad \sigma(E)=1+n+\sum s_i.$$

Obvious.

We note that after m quadratic transformations (with centers at the singular points of E and its successive proper transforms $E_1, E_2, \ldots, E_{m-1}$) we obtain a proper transform E_m which is non-singular and such that $(E_m^2)=n$. The curve E_m is a minimal resolution of E, and if $n\geqq 0$, i.e., if E_m is of the second kind, then the centers of the last n quadratic transformations can be taken arbitrarily on $E_m, E_{m+1}, \ldots, E_{m+n-1}$ respectively (since one only wishes to make sure that the self-intersection number of E_{m+n} is -1). It is clear that if P' is the contraction of E_{m+n} and if P_m denotes the point of E_m which has been used as center of a quadratic transformation, then P_m does not dominate P' (since the total transform of P_m on the carrier surface of E_{m+n} is not contained in E_{m+n}). Hence, referring to the considerations developed at the end of II. 4, we can state the following:

Corollary II. 5. 15. *If the minimal resolution of an irreducible exceptional curve E of the second kind is also of the second kind, then for any point P of E there exists a maximal contraction of E which is not dominated by P. Consequently, E has infinitely many maximal contractions.*

Finally we point out the following consequence of (16):

Corollary II. 5. 16. *If a non-singular surface F carries an exceptional curve E of the second kind, then all the plurigenera P_i of F are zero* $(i\geqq 1\,;\ P_1=p_g)$.

In view of the birational invariance of the plurigenera we may assume that E is irreducible (Proposition II. 4. 5). If K is a canonical cycle on F then we have by (16): $(E\cdot K)<0$. Now, $(E^2)\geqq 0$ (Corollary II. 5. 14) and hence $(E\cdot X)\geqq 0$ for any cycle $X\geqq 0$. Consequently the system $|iK|$ does not exist, for any $i\geqq 1$.

PART III

THE PROBLEM OF MINIMAL MODELS FOR ALGEBRAIC SURFACES

In this last part of the monograph we shall give a condensed account of our solution of the problem of minimal models for algebraic surfaces (over an algebraically closed field k of arbitrary characteristic p). The various propositions will be stated explicitly, but their proofs will be given only in outline or not at all. The complete proofs will be given in a paper which will appear in the American Journal of Mathematics ([35]).

III. 1. Statement of the main result.

Let B denote, as before, a birational class of non-singular surfaces over an algebraically closed ground field k, partially ordered in the usual way: $F \prec F'$ if the birational transformation of the projective model F onto the projective model F' is antiregular. We denote by Σ the common function field of the surfaces in B. By the theorem of reduction of singularities of algebraic surfaces (ZARISKI [25], [27]; ABHYANKAR [1]) the assumption that the surfaces are non-singular imposes no restriction on the field Σ.

DEFINITION III. 1. 1. *A surface F in B is called a relatively minimal model of the field Σ/k if F is a minimal element of the ordered set B*, i.e., *if there exists no surface F' in B such that F' is dominated by F ($F' \prec F$) and is different from F.*

(We shall make no distinction between biregularly equivalent surfaces).

DEFINITION III. 1. 2. *A surface F in B is called a minimal model of Σ/k if F is a lower bound of the ordered set B*, i.e., *if $F \prec F'$ for every F' in B.*

It is clear that if there exists a minimal model F in B then there exists only one (up to biregular equivalence), and F is then the only relatively minimal model of Σ/k.

The main result (which was known to the Italian geometers in the case k=field of complex numbers; see CASTELNUOVO and ENRIQUES [5];

also ENRIQUES [7], p. 374) is the following:

FUNDAMENTAL THEOREM. *If the field Σ/k does not possess a minimal model then Σ is the field of rational functions of a ruled surface (or equivalently: Σ is a simple transcendental extension $\Omega(t)$ of a field Ω/k of algebraic functions of one variable).*

Also the converse of this theorem is true, but that has to do with an elementary and well known property of ruled surfaces.

As to the existence of relatively minimal models the situation is quite different: *our class B always contains relatively minimal models.* As a matter of fact, the following stronger result holds for non-singular varieties of any dimension:

PROPOSITION III. 1. 3. *Each birational class of non-singular varieties satisfies the descending chain condition.*

In order to prove this proposition one makes use of the following well known result (see VAN DER WAERDEN [20], p. 154; ZARISKI [28], p. 532, Corollary to Theorem 17): if $T: V \to V'$ is an antiregular birational transformation of a *non-singular* variety V of dimension r onto a variety V' then the exceptional locus of T^{-1} on V' (see 1. 2) is pure, of dimension $r-1$. Using this result and applying the theorem of NERON-SEVERI (SEVERI [17, 18]; NERON [12]) one finds then easily that if we have $V < V'$ and V is non-singular then $\rho(V) < \rho(V')$, where ρ is the base number for algebraic divisors on a variety. This establishes the descending chain condition in any birational class of non-singular varieties. In the case of algebraic surfaces it is possible to give two other proofs of the descending chain condition. In one of these proofs, due to SERRE (See SERRE [15]) one observes that if $F < F' \neq F$ $(F, F' \in B)$ then $h^{1,1}(F) < h^{1,1}(F')$, where the notations are those of the theory of sheaves. The inequality is first proved directly in the case in which the birational transformation $F \to F'$ is quadratic, and then one uses the factorization theorem II. 1. 2.

A second proof is based on the consideration of the anticanonical system $|-K|$ of an algebraic surface. It can be shown that if we have an infinite strictly descending chain $F_1 > F_2 > F_3 > \cdots$ in B then $\dim |-K(F_i)| \to +\infty$ and this leads to a contradiction. We shall present this proof in a forthcoming note in the Mem. Col. Sci. of Kyoto University.

III. 2. A stronger formulation of the fundamental theorem.

From now on we shall assume that our birational class does *not* contain a minimal model. We fix a relatively minimal model F in B (Proposition III. 1. 3). It follows from Proposition II. 2. 4 that F does not contain exceptional curves of the first kind. Since F is not a minimal model there exists a surface F_1 in B such that $F \not< F_1$. By the descending chain condition in B there exists in B a relatively minimal model F' such that $F' < F_1$. Then F and F' are two distinct relatively minimal models, and the birational transformation between these two surfaces F and F' has exceptional curves on each of them. Thus F *carries some exceptional curve, which is then necessarily of the second kind.* By Corollary II. 4. 6, F must carry then also an *irreducible* exceptional curve of the second kind. We fix on F an irreducible exceptional curve E of the second kind.

We shall now give a stronger formulation of the fundamental theorem, in terms of exceptional curves of the second kind:

THEOREM III. 2. 1. *If a non-singular surface carries an irreducible exceptional curve* E *of the second kind, then* F *is birationally equivalent to a ruled surface, and if, furthermore,* $(E^2)>0$ *then* F *is a rational surface.*

Note that by (17), II. 5, we have $(E^2)\geqq 0$ for any irreducible exceptional curve E of the second kind, and that (E^2) is certainly *positive* if the curve E has singularities.

The case $(E^2)=0$ is relatively simple. In this case, using the RIEMANN-ROCH theorem for algebraic surfaces and the fact that $p(E)=0$ (E being non-singular and rational) one finds that $\dim|nE|=n+const.$, for n large. It follows that for n large the system $|nE|$ exists and has no fixed components. Any two cycles of $|nE|$ which pass through a given point of F must have a common component since the intersection number $(nE\cdot nE)$ is zero. Therefore, for large n, the system $|nE|$ is reducible. On the other hand, since $\dim|(n+1)E|=1+\dim|nE|$ (n-large) and $|nE|$ has no fixed components, while E is irreducible, it follows that the general cycle of $|(n+1)E|$ is unramified. Consequently, by BERTINI's theorem (Theorem I. 6. 3) the system $|nE|$ (n-large) is composite with a pencil H, which may be assumed to be irreducible. It is immediately seen that E itself is a member of this pencil. If Y is a general cycle of H/k we must have $p(Y)=0$, since we have $p(E)=0$ for a particular cycle E of H. Thus H is a pencil of curves which have arithmetic genus zero. By the theorem of NOETHER-ENRIQUES (or as a consequence of the purely algebraic theorem of TSEN [19]) it follows that F is birationally equivalent to a ruled surface.

We observe that the preceding proof is applicable also to the case $(E^2)>0$ provided E is *non-singular*, and in that case the conclusion is that *F is rational.* In fact, if $(E^2)=s>0$, then after s quadratic transformations with centers on E and on the successive transforms of E we obtain a surface F' such that E is transformed into an irreducible exceptional curve E' of F', of the second kind, and such that $(E'^2)=0$ (see the "if" part of Proposition II. 5. 13). To the pencil H' on F' of which E' is a member and whose existence has been established above, there corresponds on F a pencil H which has s base points. The pencil H, and therefore also the pencil H', is then necessarily rational (see SCHILLING-ZARISKI [14]). Hence F is a rational surface.

III. 3. The case $(E^2)>0$: vanishing of the arithmetic genus $p_a(F)$.

In this case we have to prove that F is rational. The first step of the proof consists in showing that *the arithmetic genus p_a of F is zero.*

DEFINITION III. 3. 1. *A variety V is said to be strongly minimal if every rational transformation of any variety V' into V (not necessarily onto V) is regular at each simple point of V'.*

We point out explicitly that a strongly minimal variety is not assumed to be necessarily non-singular.

Every abelian variety is strongly minimal (WEIL [22], Theorem 6, p. 27), and so it is every subvariety of a strongly minimal variety.

The following two propositions have been kindly communicated to me by MATSUSAKA:

PROPOSITION III. 3. 2. *If V is a strongly minimal variety and Σ' is a finite algebraic extension of the function field $k(V)$ of V/k, then the normalization V' of V in Σ' is also strongly minimal.*

The proof is straightforward and makes use of the well known fact that if a rational transformation of a normal variety V' is finitely valued at a given point P' of V' then it is regular at P' (see ZARISKI [28], pp. 512–

514). (The concept of normalization of V in the bigger field Σ' has been first defined in our paper [31], pp. 69–70).

PROPOSITION III. 3. 3. *If a surface F (over an algebraically closed field k) is not birationally equivalent to a strongly minimal surface, then F carries a pencil of genus equal to the irregularity q of F (irregularity=dimension of the* ALBANESE *variety of F).*

The proposition says nothing of significance if $q=0$. Assume $q>0$ and consider the canonical mapping f of F into its ALBANESE variety A. The image $f(F)$ cannot be a surface, for otherwise F would be birationally equivalent to the normalization of $f(F)$ in $k(F)$, and since this normalization is strongly minimal (by the preceding proposition) we would have a contradiction. Since $f(F)$ generates A and $q>0$, it follows that $f(F)$ is a curve, and for the same reason it follows that the genus of that curve is q. The curve $f(F)$ being a rational transform of F, the proposition is proved.

We now come to a proposition which is crucial for the proof of the vanishing of the arithmetic genus p_a of F.

PROPOSITION III. 3. 4. *Let P and P' be two corresponding simple points of two birationally equivalent surfaces F and F' (over k). If neither of these two points dominates the other* (i.e., if $\mathfrak{o}_P \not\subseteq \mathfrak{o}_{P'}$ and $\mathfrak{o}_{P'} \not\subseteq \mathfrak{o}_P$) *then the integral domain $\mathfrak{o}_P \cap \mathfrak{o}_{P'}$ has transcendence degree $\leqq 1$ over k.*

We have two proofs of this proposition: one is elementary and is based on the factorization theorem for antiregular birational transformations (Theorem II. 1. 2), the other is based on the consideration of the canonical mapping of a surface into its ALBANESE variety.

COROLLARY III. 3. 5. *Under the assumptions of the preceding proposition there exists no surface F^* (singular or non-singular) which is birationally equivalent to F and is such that $F^* < F$ and $F^* < F'$.*

Obvious.

COROLLARY III. 3. 6. *If the field Σ/k admits no minimal model then it admits no strongly minimal model.*

For, under our assumption, we can find in the birational class B two distinct relatively minimal models F and F' (see III. 2). Then there will exist points P and P' on F and F' respectively which satisfy the conditions of Proposition III. 3. 4. The corollary now follows from Corollary III.

3. 5 and from the definition of strongly minimal surfaces.

Combining Proposition III. 3. 3 and Corollary III. 3. 6 we conclude that our relatively minimal model F must carry a pencil H of genus equal to the irregularity q of F. Now suppose that $q>0$, so that the pencil H is irrational. Then we must have $(X^2)=0$ for any cycle X in H. On the other hand, the trace of H on our irreducible exceptional curve E must have dimension zero, for a rational curve carries no irrational involutions (theorem of LÜROTH). Hence E must be a component of some cycle X_0 in H. Let $X_0=hE+Y_0$, where Y_0 does not contain E as component. Since X_0 does not intersect a general cycle X of H, it follows that $(E\cdot X_0)=0$. Since $(E\cdot Y_0)\geqq 0$, we find that $(E^2)\leqq 0$, a contradiction. *This establishes the equality* $q=0$.

In the classical case, SEVERI has proved, by essentially algebraic methods, the completeness of the characteristic series of an ample complete continuous system of curves on any surface F whose geometric genus p_g is zero (see SEVERI [16], Anhang F; see also our monograph [23], pp. 84–86, where the proof and references can be found). In other words, SEVERI has established, in the classical case, the equality $q=p_g-p_a$ by algebraic methods, *if* $p_g=0$. This result has now been established also in the abstract case, and independent proofs have been given by AKIZUKI-MATSUMURA, by NAKAI and by MATSUMURA-NAGATA (see [2], [8], [10] and [11]). (NAKAI has even in [11] proved the following stronger result: $p_g-p_a\geqq q\geqq -p_a$). Since we know already that $p_g=0$ (Corollary II. 5. 16) and since we have just proved that $q=0$, the vanishing of p_a is established.

III. 4. The rationality of the surface F in the case $(E^2)>0$.

We have now established that our surface F has the following properties (under the assumption that F carries an irreducible exceptional curve E of the second kind such that $(E^2)>0$):

1) All the plurigenera of F are zero (Corollary II. 5. 16).
2) The arithmetic genus p_a of F is zero.
3) F carries an irreducible curve E such that $(K\cdot E)\leqq -2$, where K is a canonical divisor on F (see (16), II. 5).

In the proof of rationality of F we have to consider two cases:

Case A. $(K^2) \leqq 0$.

Case B. $(K^2) > 0$.

Our proof in the case B actually makes use only of the assumptions $P_2 = p_a = 0$ and therefore establishes, in that case, the criterion of rationality of a surface, due to CASTELNUOVO. Our proof is quite similar to a proof of CASTELNUOVO's criterion which has been given recently by KODAIRA (See SERRE [15]), except that at a certain stage of the proof KODAIRA has to make use of topological considerations, while our proof establishes CASTELNUOVO's criterion by purely algebraic methods (the case of characteristic $p=2$ is especially troublesome). In case A our proof is superceded by KODAIRA's proof which is much simpler than ours (also KODAIRA considers separately the two cases A and B). CASTELNUOVO's criterion of rationality (" If the bigenus P_2 and the arithmetic genus p_a of a surface F are zero then F is rational "; see CASTELNUOVO [4]; see also ENRIQUES [7], pp. 230–235.) is therefore now established in the abstract case.

By using CASTELNUOVO's criterion of rationality we can easily generalize to the abstract case CASTELNUOVO's theorem on the rationality of plane involutions (CASTELNUOVO [3]) in the following form:

Let $k(x, y)$ be a field of rational functions of two independent variables over an algebraically closed field k and let Σ be a field between k and $k(x, y)$, of transcendence degree 2 over k. If $k(x, y)$ is a separable extension of Σ, then also Σ is a purely transcendental extension of k.

It can be shown by examples that the condition of separability in the above theorem is necessary.

CORRECTIONS ADDED IN PROOF

(1) p. 4, lines 9–10.

The definition of $T[W]$ given in the text is not the correct one. One proves easily that there exists a subvariety G' of V' with the following properties: (a) every irreducible component of G'/k corresponds to W; (b) if an irreducible subvariety W' of V'/k corresponds to W then W' is contained in G'. It is this variety G' that must be defined as being the proper transform $T[W]$ of W. It remains true that $T[W]$ is contained in $T\{W\}$ and that each irreducible component of $T\{W\}/k$ which corresponds to W is an irreducible component of $T[W]/k$, but $T[W]/k$ may very well have irreducible components which are *properly* contained in irreducible components of $T\{W\}/k$.

(2) p. 18, lines 5–9 from bottom of page.

The proof of the absolute irreducibility of I is incomplete, since the field $F(P\times Z)$ is an not even an extension of the field $F(P\times\Gamma_1)$. The correct proof is as follows:

Since the cycles Γ_i form a complete set of conjugates over $F(Z)$ and since the variety $\langle\Gamma_i\rangle$ is defined over the field $F(\Gamma_i)$, it follows that all the varieties $\langle\Gamma_i\rangle$ are defined over the algebraic closure K of the field $F(Z)$. Since P is a general point of $\langle\Gamma_1\rangle/F(\Gamma_1)$, hence also of $\langle\Gamma_1\rangle/K$, it follows that P does not belong to $\langle\Gamma_i\rangle$ if $i\neq 1$. This shows that Γ_1 is the only specialization of Γ_1 over $F(P)$, i.e., that Γ_1 is purely inseparable over $F(P)$. Since F is quasi-maximally algebraic in $F(P)$, it follows that F is also quasi-maximally algebraic in $F(P\times\Gamma_1)$, showing that J is absolutely irreducible.

(3) p. 52, third line of "Remark".

The assertion that T transforms biregularly $F-E$ onto $F''-P''$ is not justified, and in fact may be false for low values of h. The correct proof is as follows:

For any integer i we denote by F_i'' the surface obtained from F by cutting out the system $|iL|$ with hyperplanes. We fix an integer h such that the system $|hL|$ has no base points. Then it is immediately seen that for any integer $m \geqq 1$ the surface F''_{mh} is obtained from F''_h by cutting with hyperplanes the complete system on F''_h containing the sections of F''_h with hypersurfaces of order m. Hence for large m, the surface F''_{mh} is a derived normal model of F''_h, and is therefore normal. This argument renders superfluous the special separate proof of the normality of the point P''.

REFERENCES

[1] S. Abhyankar. *Local uniformization of algebraic surfaces over ground fields of characteristic* $p \neq 0$. Ann. of Math. *63*, 1956, pp. 491–526.

[2] Y. Akizuki and H. Matsumura. *On the dimension of algebraic system of curves with nodes on a non-singular surface.* Mem. Coll. Sci. Univ. Kyoto, series A, *30*, n. 2, 1957, pp. 143–150.

[3] G. Castelnuovo. *Sulla razionalità delle involuzioni piane.* Math. Ann. *44*, 1894, pp. 125–155.

[4] G. Castelnuovo. *Sulle superficie di genere zero.* Mem. Soc. Ital. detta dei XL, III. s, *10*, 1896.

[5] G. Castelnuovo and F. Enriques. *Sopra alcune questioni fondamentali nella teoria delle superficie algebriche.* Ann. di Mat. pura appl., III. s, *6*, 1901.

[6] I. S. Cohen and O. Zariski. *A fundamental inequality in the theory of extensions of valuations*, Ill. Jour. of Math. *1*, 1957, pp. 1–8.

[7] F. Enriques. *Superficie algebriche.* Bologna, Zanichelli, 1949.

[8] H. Matsumura and M. Nagata. *On the algebraic theory of sheets of an algebraic variety.* Mem. Coll. Sci. Univ. Kyoto, series A, *30*, n. 2, 1957, pp. 157–164.

[9] T. Matsusaka. *The theorem of* Bertini *on linear systems.* Mem. Coll. Sci. Univ. Kyoto, series A, *26*, 1951, pp. 51–62.

[10] Y. Nakai. *A property of an ample linear system on a non-singular variety.* Mem. Coll. Sci. Univ. Kyoto, series A, n. 2, *30*, 1957, pp. 151–156.

[11] Y. Nakai. *On the characteristic linear systems of algebraic families.* Ill. Jour. of Math. *1*, 1957, pp. 552–561.

[12] A. Neron. *Problèmes arithmétiques et géométriques rattachés à la notion de rang d'une courbe algébrique dans un corps* (Thèse Fas. Sci. Univ. Paris), Bullet. Soc. Math. France *80*, 1952, pp. 101–166.

[13] P. Samuel. *Méthodes d'algèbre abstraite en géométrie algébrique.* Ergebnisse der Mathematik und ihrer Grenzgebiete, 1955, neue Folge, Heft 4.

[14] O. F. G. Schilling and O. Zariski. *On the linearity of pencils of curves on algebraic surfaces.* Amer. Jour. of Math. *60*, 1938, pp. 320–324.

[15] J. P. SERRE. *Critère de rationalité pour les surfaces algébriques (d'après un cours de* K. KODAIRA, *Princeton, Novembre*, 1956), Séminaire Bourbaki, Février, 1957.

[16] F. SEVERI. *Vorlesungen über algebraische Geometrie.* Teubner, Berlin, 1921.

[17] F. SEVERI. *Sulla totalità delle curve algebriche tracciate sopra una superficie algebrica.* Math. Ann. *62*, 1906, pp. 194–225.

[18] F. SEVERI. *Sui fondamenti della geometria numerativa e sulla teoria delle caratteristiche.* Atti R. Inst. Veneto, *75*, 1916.

[19] C. C. TSEN. *Divisionalgebren über Funktionenkörpern.* Nach. Ges. Wiss. Göttingen, 1933.

[20] B. L. VAN DER WAERDEN. *Algebraische Korrespondenzen und rationale Abbildungen.* Math. Ann. *110*, 1934, pp. 134–160.

[21] A. WEIL. *Foundations of Algebraic Geometry.* Amer. Math. Soc. Colloquium Publications, *29*, 1946.

[22] A. WEIL. *Variétés abéliennes et courbes algébriques.* Act. Sc. et Ind., n. 1064, Paris, 1948.

[23] O. ZARISKI. *Algebraic Surfaces.* Ergebnisse der Mathematik und ihrer Grenzgebiete *3*, n. 5, 1935.

[24] O. ZARISKI. *Some results in the arithmetic theory of algebraic varieties.* Amer. Jour. of Math. *61*, 1939, pp. 249–294.

[25] O. ZARISKI. *Reduction of the singularities of an algebraic surface.* Ann. of Math. *40*, 1939, pp. 639–689.

[26] O. ZARISKI. *Proof of the theorem of* BERTINI. Trans. Amer. Math. Soc. *50*, 1941, pp. 48–70.

[27] O. ZARISKI. *Simplified proof for the resolution of the singularities of an algebraic surface.* Ann. of Math. *43*, 1942, pp. 583–593.

[28] O. ZARISKI. *Foundations of a general theory of birational correspondences.* Trans. Amer. Math. Soc. *53*, 1943, pp. 490–542.

[29] O. ZARISKI. *Reduction of the singularities of algebraic three-dimensional varieties.* Ann. of Math. *45*, 1944, pp. 472–542.

[30] O. ZARISKI. *The concept of a simple point of an abstract algebraic variety.* Trans. Amer. Math. Soc. *62*, 1947, pp. 1–52.

[31] O. ZARISKI. *Theory and applications of holomorphic functions on algebraic varieties over arbitrary ground fields.* Mem. Amer. Math. Soc., n. 5, 1951.

[32] O. ZARISKI. *Complete linear systems on normal varieties and a generalization of a lemma of* ENRIQUES-SEVERI. Ann. of Math. *55*, 1952, pp. 552–592.

[33] O. Zariski. *Applicazioni geometriche della teoria delle valutazioni.* Rend. Mat. e delle sue applicazioni, serie V, *13*, Roma, 1955, pp. 51–88.

[34] O. Zariski. *Interprétations algébro-géométriques du quatorzième problème de* Hilbert. Bull. Sciences Mathématiques *78*, 1954, pp. 155–168.

[35] O. Zariski. *The problem of minimal models in the theory of algebraic surfaces.* Amer. J. Math. *80*, 1958, pp. 146–184.

THE PROBLEM OF MINIMAL MODELS IN THE THEORY OF ALGEBRAIC SURFACES.*

By Oscar Zariski.†

Introduction. The non-singular varieties, in a given birational class of varieties (over an algebraically closed field k), form—in a well-known fashion—a partially ordered set $\mathfrak{B}$. Any minimal element of $\mathfrak{B}$ is called a *relatively minimal model,* and the lower bound of $\mathfrak{B}$ (if it exists) is called a *minimal model* (we identify biregularly equivalent varieties). The object of the present paper is to prove, for ground fields of arbitrary characteristic, the following classical result (known to the Italian geometers in ths classical case):[1] *if a birational class $\mathfrak{B}$ of non-singular surfaces contains no minimal model then $\mathfrak{B}$ is the birational class of a ruled surface* (§ 3, Fundamental Theorem A). We prove actually a stronger result. It can be proved that if our birational class of non-singular surfaces contains no minimal model and if F is a relatively minimal model in the class (such models always exist even in the case of higher varieties, as long as our birational class contains non-singular models at all; see Corollary 1.8), then F carries an irreducible exceptional curve E of the second kind (Proposition 3.3, (a) and Proposition 2.6). It can also be shown that the self-intersection number (E^2) of an irreducible exceptional curve E of the second kind (on a non-singular surface F) is $\geqq 0$ (and that if E has singularities then $(E^2) > 0$); see Proposition 2.7. Then

* Received October 24, 1957.

† This research was supported by the United States Air Force through the Air Office of Scientific Research of the Air Research and Development Command under Contract No. AF18(600)-1503.

[1] See, for instance, F. Enriques [2], pp. 372-374. The proof in the classical case is due to Castelnuovo and Enriques, but that proof cannot be regarded as entirely rigorous. Furthermore, some steps of the proof, even those which could be accepted without objections in the classical case, would break down in the case of characteristic $p \neq 0$. Our proof differs from the classical proof also in some of its central ideas. Thus: (a) we do not have to presuppose that the Castelnuovo criterion of rationality $p_a = P_2 = 0$ has already been established (and in fact, the validity of that criterion does not result from our proof, except in some special cases; see, however, the concluding remarks of our introduction); (b) the crucial phases of the proof are based on specific properties of exceptional curves of the second kind.

what we actually prove is the following stronger result: *if a non-singular surface F carries an irreducible exceptional curve E of the second kind, then F is birationally equivalent to a ruled surface, and that if, furthermore, (E^2) is positive then F is a rational surface* (Fundamental Theorem B, § 3).

A preliminary study of the problem of minimal models is contained in our monograph "Introduction to the problem of minimal models in the theory of algebraic surfaces," now in course of publication in the "Publications of the Mathematical Society of Japan." In that monograph (based on a series of lectures given in Tokyo and Kyoto in the fall of 1956) we give an exposition of some background material concerning rational transformations and linear systems, then develop the theory of exceptional curves on a non-singular surface, and finally give an outline of the contents of the present paper. This monograph will be referred to as IP. Frequent and explicit references to various propositions proved in IP will be made in the present paper, especially in regard to the theory of exceptional curves.

In Section 1 we deal with varieties of any dimension and we prove that the class $\mathfrak{B}$ satisfies the descending chain condition (Proposition 1.5). Thus the existence of relatively minimal models as assured (as long as $\mathfrak{B}$ is non-empty; a question which is still open for varieties of dimension > 2 in the case of characteristic $\neq 0$, and for varieties of dimension > 3 in the case of characteristic zero). This result is a rather straightforward consequence of the theorem of Neron-Severi.[2] In Section 2 we review some of the results concerning exceptional curves which have been obtained in IP, and we derive a few other properties of exceptional curves. In Section 3 we give various criteria of existence of minimal models. In partilular, we show that if there does *not* exist a minimal model then we are dealing with surfaces all plurigenera of which are zero (Proposition 2.9). This result (above all, the vanishing of the geometric genus p_g and of the bigenus P_2) is repeatedly used in the rest of the paper. In Section 4 we prove the fundamental

[2] In the case of surfaces it is possible (as was pointed out by Serre) to prove the descending chain condition by using sheaf theory. Namely, one observes that if $F < F'$, where F and F' are non-singular surfaces, then $h^{1,1}(F) < h^{1,1}(F')$, where the notations are those of the theory of sheaves. This inequality is first proved directly in the case in which the birational transformation $F \to F'$ is locally quadratic, and then one uses our factorization theorem for antiregular birational transformations of non-singular surfaces (see IP, Theorem II. 1. 2; or also our paper [12], Lemma on p. 538).

A second proof (which we shall publish in the Mem. Col. Sci. of Kyoto University) is based on the consideration of the anticanonical system $|-K|$ of an algebraic surface. It can be shown that if we have an infinite strictly descending chain $F_1 > F_2 > \cdots$ of non-singular algebraic surfaces, then $\dim |-K(F_j)| \to +\infty$, and this can be shown to lead to a contradiction.

theorem B (stated in Section 3) under the assumption that $(E^2)=0$ and also in the case in which E is non-singular. The conditions $p(E)=0$ and $(E^2)\geqq 0$ characterize the irreducible *non-singular* exceptional curves of the second kind (Proposition 2.8), and what we are actually proving in Section 4 is that if a non-singular surface F carries an irreducible curve E such that $p(E)=0$ and $(E^2)\geqq 0$, then F is birationally equivalent to a ruled surface.

The remainder of the paper deals with the proof of the fundamental theorem B in the case $(E^2)>0$. In this case it is the rationality of F that has to be proved. A first step in that direction is the proof of the vanishing of the arithmetic genus p_a of F (Section 5). It will be noted that our present proof of the vanishing of p_a is much simpler than the proof which was outlined in IP (III, § 3). We now make no use of the concept of strongly minimal models (IP, Definition III. 3. 1), and—what is more important—we do not use a local result which was stated without proof in IP (Proposition III. 3. 4). The proof of this local result will be published elsewhere. What is common to both proofs is the recourse to Abelian varieties and the use of the equality $q=-p_a$, which holds on any surface of geometric genus zero.

The proof of the fundamental theorem in the case $(E^2)>0$ (and under the permissible assumption that F is a relatively minimal model; see remark at the end of Section 5) has to be divided into various cases, according to the value of (K^2), where K is a canonical divisor on F. A very short proof, given in Section 6, settles the case $(K^2)\leqq 0$. In Section 7 we develop some general properties of surfaces which satisfy the Castelnuovo criterion $p_a=P_2=0$ *and* the inequality $(K^2)>0$. In Section 8 we prove the Castelnuovo criterion of rationality in the case $(K^2)\geqq 3$. This leaves us only with two cases for completing the proof of the fundamental theorem: $(K_2)=1$ and $(K^2)=2$. The proof in the case $(K^2)=1$ (Section 9) is very short and simple. In the last section (Section 10) we deal with the case $(K^2)=2$, and here again, as in Section 8, we are proving not merely the fundamental theorem B but actually the Castelnuovo criterion of rationality $p_a=P_2=0$ in the case $(K^2)=2$ (the latter implies the former once the vanishing of p_a has been established in Section 5). We first give a very simple proof, but unfortunately this proof does not work for characteristic 2. We then give a second (and much longer) proof which is valid for any characteristic. This second proof is constructive in the sense that we construct the entire family of algebraic surfaces with $p_a=P_2=0$ and $(K^2)=2$. We show that the general member of this family is a rational surface, and we then prove the rationality of the given surface F by a specialization argument.

From the foregoing outline it is seen that the present investigation contains a proof of Castelnuovo's criterion of rationality of a surface ("a surface is rational if $p_a = P_2 = 0$") in the case $(K^2) \geqq 2$. A proof of Castelnuovo's criterion of rationality, in the classical case, was given by Kodaira in a lecture course at Princeton in 1956 (the existence of this proof, and the proof itself, because known to me in Paris through an exposé by Serre in a Bourbaki seminar in 1957). It appears clearly from that exposé that Kodaira's proof, *in the case* $(K^2) \leqq 0$, is algebraic in nature and is valid for any characteristic. [On the contrary, in the case $0 < (K^2) \leqq 9$, Kodaira's proof uses topological considerations; our proof in the case $(K^2) \geqq 3$ was presented in a Kyoto seminar in 1956 and is similar to, but somewhat simpler than, the proof of Kodaira in the case $(K^2) > 9$]. Thus the proof of Castelnuovo's criterion of rationality in the abstract case is complete, except for the case $(K^2) = 1$. We have a proof also in this case, but in order not to lengthen the present paper we do not include it here; it will be published elsewhere.

We point out that the proof of the fundamental theorem in the case $(K^2) \leqq 0$ which we gave in our Kyoto seminar was very long, while in the present paper the proof takes a few lines (Section 6). The reason for this is that in the Kyoto seminar we were essentially proving also the Castelnuovo criterion of rationality in the case $(K^2) \leqq 0$ (and not merely the fundamental theorem B). However, our Kyoto proof of Castelnuovo's criterion of rationality in the case $(K^2) \leqq 0$ was not as simple as Kodaira's proof and is therefore superseded by the latter.

Table of Contents.

1. Minimal and relatively minimal models. Let k be an algebraically closed field and let Σ be a field of algebraic functions of r independent variables over k, i.e., a finitely generated extension of k, of transcendence degree r. By a *projective model* of Σ/k we mean a pair (V, P), where V is a (projective) variety which is defined and irreducible over k, P is a general point of V/k, and $\Sigma = k(P)$. In the sequel we shall often refer to V itself as a projective model of Σ/k and shall write V instead of (V, P) without fear of confusion.

We shall denote by $\mathfrak{B}^*$ the totality of all normal projective models of Σ/k, and by $\mathfrak{B}$ the totality of all non-singular projective models of Σ/k.

Given two projective models $V = (V, P)$ and $V' = (V', P')$ of Σ/k, there is a unique birational transformation T of V onto V' such that $T(P) = P'$. We shall say that V *dominates* V', or in symbols: $V \succ V'$, if T is regular on V. The relation $\succ$ defines a partial ordering of the birational class $\mathfrak{B}^*$. As a rule we shall identify biregularly equivalent models, i.e., models V and V' such that $V \succ V'$ *and* $V' \succ V$.

Let V, V' be members of $\mathfrak{B}^*$ such that $V \succ V'$ and let $T: V \rightarrow V'$ be the regular transformation of V onto V'. To any irreducible subvariety W/k of V/k there corresponds a unique irreducible subvariety W'/k of V'/k, and we have $\dim W' \leqq \dim W$. We say that W is an exceptional variety of T if $\dim W' < \dim W$. It is known (see IP, Corollary I. 2. 4) that there exists a subvariety E of V, called the *exceptional locus* of T, having the following properties: 1) E is defined over k; 2) every irreducible component of E/k is an exceptional variety of T; 3) every exceptional variety of T is contained in E. The exceptional locus E of T is identional with the set of all points of V at which T is not biregular.

Let $\mathfrak{G}$ denote the group of divisorial cycles on V which is generated by the prime divisorial cycles which are rational over k. Let $\mathfrak{G}'$ have the similar meaning for the variety V'. We shall now associate with the regular transformation $T: V \rightarrow V'$ a mapping, denoted by the same letter T, of $\mathfrak{G}$ into $\mathfrak{G}'$.

If Z is a prime divisorial cycle on V, rational over k, and if the variety Z is not exceptional for T, then the corresponding variety on V' is of the same dimension as Z and defines a prime divisorial cycle Z', rational over k. We then set $T(Z) = Z'$. If, however, Z is an exceptional variety we set $T(Z) = 0$. We then extend the definition of the cycle transformation $Z \rightarrow T(Z)$ to the whole group $\mathfrak{G}$, by linearity. If we denote by $\mathfrak{G}(T)$ the subgroup of $\mathfrak{G}$ generated by prime divisorial cycles which are exceptional for T, then the cycle transformation T is a homomorphism of $\mathfrak{G}$ into $\mathfrak{G}'$, with kernel $\mathfrak{G}(T)$. It is clear that T is a mapping of $\mathfrak{G}$ onto $\mathfrak{G}'$, for every $(r-1)$-dimensional irreducible subvariety W'/k of V'/k corresponds, under T, to at least one irreducible subvariety W/k of V/k.

Let $\mathfrak{G}_a$ (resp. $\mathfrak{G}'_a$) be the subgroup of $\mathfrak{G}$ (resp. of $\mathfrak{G}'$) which is generated by the cycles which are algebraically equivalent to zero.

PROPOSITION 1.1. *Denoting by* $\mathfrak{E}(T)+\mathfrak{G}_a$ *the subgroup of* $\mathfrak{G}$ *which is generated by the subgroups* $\mathfrak{E}(T)$ *and* $\mathfrak{G}_a$, *we have*

$$T(\mathfrak{E}(T)+\mathfrak{G}_a)=\mathfrak{G}'_a.$$

Proof. We shall use notations and results from Weil's paper [10]. Any cycle in $\mathfrak{G}_a$ is of the form $Z(N)-Z(M)$, where Z is a divisorial cycle on the product $V\times C$ of V by a non-singular irreducible curve C/k and where M and N are points of C which are rational over k (Weil [10], Lemma 10). Here the divisorial cycle Z may be assumed to be a "reduced" one, i.e., free from prime components of the form $V\times W'$ or $W\times C$ ($W'\in V'$, $W\subset V$). As to the symbol $Z(Q)$, where Q is a point of C, it is defined by the relation $Z(Q)\times Q=Z\cdot V\times Q$. Since Z is a reduced divisorial cycle and both V and C are normal, it is true for any point Q of C that all the prime components of the cycle $Z\cdot V\times Q$ are proper intersections of the variety of the cycle Z with the variety $V\times Q$, i.e., they are all of dimension $r-1$ and are simple for $V\times C$ (and also for $V\times Q$). They are therefore all of the form $\Gamma\times Q$, where Γ is a prime divisorial cycle on V. Thus $Z(Q)$ is a divisorial cycle on V.

The birational transformation $T^*\colon V\times C\to V'\times C$ defined $T^*(A\times B)=T(A)\times B$ ($A\in V, B\in C$) is regular. By the preceding definition, we have an induced transformation, denoted by the same letter T^*, of the group of divisorial cycles of $V\times C$ onto the group of divisorial cycles of $V'\times C$. We set $Z'=T^*(Z)$. We now observe that if Q is a rational point of C, then the restriction to $V\times Q$ of the birational transformation $T^*\colon V\times C\to V'\times C$ is a birational transformation $T_Q\colon V\times Q\to V'\times Q$, defined over k, and it is immediate that for any divisorial cycle Γ on V we have

(1) $$T_Q(\Gamma\times Q)=T(\Gamma)\times Q.$$

We have the following relation:

(2) $$T^*(Z)\cdot V'\times Q=T_Q(Z(Q)\times Q).$$

In fact, let Γ be any prime divisorial cycle on V, rational over k. If Γ is not an exceptional variety of T, then T is biregular at a general point of Γ/k, T^* is biregular at a general point of $\Gamma\times Q/k$. In this case, the coefficient of $\Gamma\times Q$ in $Z\cdot V\times Q$ is the same as the coefficient of $T(\Gamma)\times Q$ in

$T^*(Z)\cdot V'\times Q$. On the other hand, the coefficient of $\Gamma\times Q$ in $Z\cdot V\times Q$ is the same as the coefficient of $T(\Gamma)\times Q$ in $T_Q(Z\cdot V\times Q)$, by the definition of the cycle transformation T_Q. Hence in this case, the cycle $T(\Gamma)\times Q$ occurs with the same coefficient in both sides of (2). Now, assume that Γ is an execptional variety of T. In that case, $\Gamma\times Q$ is also an exceptional variety of T_Q [by (1)], and thus $T_Q(\Gamma\times Q)=0$. Since any prime divisorial cycle of $V'\times Q$ is the T_Q-transform of a cycle of the form $\Gamma\times Q$, where Γ is a non-exceptional prime divisor cycle of V, relation (2) is established.

We now go back to our cycle $Z(N)-Z(M)$ in $\mathfrak{G}_a$. Let $Z(M)=\sum a_i\Gamma_i$, $Z(N)=\sum b_j\Delta_j$, where Γ_i and Δ_j are prime divisorial cycles on V, rational over k. We have

$$Z\cdot V\times M=\sum a_i(\Gamma_i\times M),\qquad Z\cdot V\times N=\sum b_j(\Delta_j\times N).$$

Hence, by (2),

$$Z'\cdot V'\times M=\sum a_i(T(\Gamma_i)\times M),\qquad Z'\cdot V'\times N=\sum b_j(T(\Delta_j)\times N),$$

where $Z'=T^*(Z)$. Consequently

$$Z'(M)-Z'(N)=T(Z(M)-Z(N)). \tag{3}$$

This establishes the inclusion $T(\mathfrak{G}_a)\subset\mathfrak{G}'_a$ and therefore also the inclusion $T(\mathfrak{E}(T)+\mathfrak{G}_a)\subset\mathfrak{G}'_a$, since $\mathfrak{E}(T)$ is the kernel of T. The opposite inclusion $\mathfrak{G}'_a\subset T(\mathfrak{G}_a)$ follows from the fact that given any reduced divisorial cycle Z' on $V'\times C$ there exists a reduced divisorial cycle Z on $V\times C$ such that $Z'=T^*(Z)$. This completes the proof of the proposition.

COROLLARY 1.2. *The group of Neron-Severi of the variety V' (i.e., the group $\mathfrak{G}'/\mathfrak{G}'_a$) is isomorphic to $\mathfrak{G}/(\mathfrak{E}(T)+\mathfrak{G}_a)$ and is therefore a homomorphic image of the group of Neron-Severi of the variety V.*

COROLLARY 1.3. *If the exceptional locus E of T has dimensional $r-1$ then the number of Picard $\rho(V')$ of $V'(\rho(V')=$ rank of $\mathfrak{G}'/\mathfrak{G}'_a)$ is less than the number of Picard $\rho(V)$ of V.*

For under our assumption, there exists a prime divisorial cycle Γ on V such that $T(\Gamma)=0$, and the corollary follows from the fact that no integral multiple $n\Gamma$ of Γ $(n\neq 0)$ can be algebraically equivalent to zero.

COROLLARY 1.4. *If V is either a non-singular variety or is a normal surface, and if the regular transformation $T\colon V\rightarrow V'$ is not biregular, then $\rho(V)>\rho(V')$.*

For in either case the exceptional locus E of T has dimension $r-1$

(van der Waerden, *Algebraische Korrespondenzen und rationale Abbildungen*, Math. Ann. *110*, 1934, p. 154; Zariski, *Foundations of a general theory of birational correspondences*, Trans. Amer. Math. Soc., *53*, 1943, p. 532, Corollary to Theorem 17).

As a consequence of Corollary 1.4 and the finiteness of the number ρ of Picard we have

PROPOSITION 1.5. *The birational class $\mathfrak{B}$ of non-singular projective models of Σ/k satisfies the descending chain condition. If $r = 2$ then also the birational class $\mathfrak{B}^*$ (of normal projective models of Σ/k) satisfies the descending chain condition.*

Note that in this proposition it is tacitly assumed that biregularly equivalent varieties have been identified.

DEFINITION 1.6. *By a relatively minimal model of Σ/k we mean a minimal element o fthe ordered set $\mathfrak{B}$, i.e., a non-singular projective model V of Σ/k which does not dominate (in the strict sense) any other non-singular projective model of Σ/k.*

DEFINITION 1.7. *By a minimal model of Σ/k we mean a lower bound of the ordered set $\mathfrak{B}$, i.e., a non-singular projective model V of Σ/k such that $V \prec V'$ for all non-singular projective models V' of Σ/k.*

COROLLARY 1.8. *If a field Σ/k of algebraic functions of r independent variables possesses at all non-singular projective models (in particular, if $r = 2$, or if $r = 3$ and the characteristic of k is 0) then Σ/k also possesses a relatively minimal model, and, in fact, given any non-singular projective model V of Σ/k there exists a relatively minimal model V' of Σ/k such that $V' \prec V$.*

This follows from the descending chain condition in $\mathfrak{B}$ (Proposition 1.5).

If there exists a minimal model of Σ/k, then there is only one such model. On the other hand, it follows from Corollary 1.8 that if there does not exist a minimal model of Σ/k and if the class $\mathfrak{B}$ is not empty, then there exist at least two distinct relatively minimal models of Σ/k (distinct = not biregularly equivalent).

2. Exceptional curves of the 1st and 2nd kind. The problem of minimal models for algebraic surfaces requires a preliminary study of exceptional curves on an algebraic surface. This preliminary work has been carried out in Part II of our paper IP and will be briefly reviewed here (without proofs).

Definition 2.1. *An algebraic curve E (reducible or not) on a normal surface F is said to be an exceptional curve of F if there exists a birational transformation $T: F \rightarrow F'$ of F onto a surface F' such that E is the* TOTAL *T^{-1}-transform of a* SIMPLE *point of F'*

Definition 2.2. *An exceptional curve E of a normal surface F is said to be* OF THE FIRST KIND *if there exists a birational transformation T satisfying the condition stated in the preceding definition and satisfying the further condition that it be regular at each point of E. In the contrary case E is said to be an exceptional curve* OF THE SECOND KIND.

Given a curve E (on F) and a birational transformation $T: F \rightarrow F'$ such that E is the total T^{-1}-transform of a *normal* point P' of F', we shall say, for the sake of brevity, that T *contracts* E into the point P'. This expression, when used in this paper, will therefore always mean a little more than what it actually says; it will mean that not only does every point of E correspond to the point P' in the birational transformation T, but also that no point of F which lies outside the curve E corresponds to P' (and it is also tacitly assumed that P' is a normal point of F').

If a birational transformation $T: F \rightarrow F'$ contracts a curve E into a (normal) point P' of F' then we shall say that *P' is a contraction of E.* If, furthermore, T is regular at each point of E then we say that *P' is a regular contraction of E.*

An exceptional curve on a normal surface F is therefore a curve which admits a *simple contraction* (i.e., a contraction P' which is a simple point), and an exceptional curve of the first kind is one which admits a *simple regular contraction.* On the other hand, if E is an exceptional curve of F, *of the second kind,* then *any* birational transformation T of F *w*hich contracts E to a *simple* point has necessarily at least one fundamental point *on the curve* E (since F is normal), and thus T "blows up" some points of E into a curve.

Proposition 2.3. *If P' is a simple regular contraction of an exceptional curve E, of the first kind, of a normal surface F, then P' is the only regular contraction of E (in the biregular sense; i.e., if P'' is any regular contraction of E then the local rings $\mathfrak{o}_{P'}$ and $\mathfrak{o}_{P''}$ coincide). Similarly, P' is the only simple contraction of E. The point P' is thus uniquely determined by the curve E, being independent of the choice of the birational transformation which contracts E into a simple point. The local ring $\mathfrak{o}_{P'}$ is the intersection of the local rings $\mathfrak{o}_P$, $P \in E$* (see IP, Theorem II.2.3).

It can be shown by examples that an exceptional curve of the first kind has infinitely many non-regular (and therefore necessarily non-simple, by Proposition 2.3) contractions. As to exceptional curves of the second kind, it can be shown that any such curve has infinitely many simple contractions (IP, Corollary II.5.15).

Let E be an exceptional curve of the first kind on a *non-singular* surface F, and let $T: F \rightarrow F'$ be a birational transformation of F which contracts E to a simple point P'. By Proposition 2.3, the birational transformation T is necessarily regular at each point of E, whence P' is the (uniquely determined) simple regular contraction of E. Nothing in our definition of exceptional curves of the first kind specifies the behavior of T at points of F which are not on E. Thus T may have fundamental points outside of E; T may also contract into points other curves of F (such curves will necessarily have no point in common with E, since E is the *total* T^{-1}-transform of P' and since T is regular, hence single-valued, at each point of E; and the contractions of these curves will necessarily all be different from P'). However, it can be shown that there exists another birational transformation of F which contracts E to P' *and which is biregular on* $F - E$ (see IP, Theorem II.2.4, where a more general result is proved). We can formulate this result in the following fashion:

We note first that $F - E + P'$ is easily seen to be an abstract surface in the sense of Weil. Then the above result can be stated as follows:

PROPOSITION 2.4. *If E is an exceptional curve, of the first kind, on a non-singular surface, and if P' is the (uniquely determined) simple (whence also regular) contraction of E, then the abstract surface $F - E + P'$ is a projective surface.*

By repeated applications of Proposition 2.4 we find the following generalization:

PROPOSITION 2.5. *If $E_1, E_2, \cdots, E_s$ are exceptional curves, of the first kind, on a non-singular surface F, such that no two of the curves E_i have points in common, and if P'_i is the simple contraction of E_i, then the abstract surface $F - \sum E_i + \sum P'_i$ is projective.*

The simplest exceptional curve of the first kind which has a preassigned contraction P', where P' is a simple point of a surface F', is the total transform of P' under a locally quadratic transformation of F' with center P'. The exceptional curve E which is thus obtained is irreducible, rational, non-singular, and its numerical characters $p(E)$ (arithmetic genus) and (E^2)

(virtual degree, or self-intersection number) are given by (see IP, Proposition II. 5. 9)

$$(1) \qquad p(E)=0, \qquad (E^2)=-1.$$

Conversely, every irreducible exceptional curve E of the first kind, on a non-singular surface, is the total quadratic transform of its simple construction P' (IP, Corollary II. 3. 2). A deeper result is the following: if an irreducible curve E on a non-singular surface is such that $p(E)=0$ and $(E^2)=-1$, then E is an exceptional curve of the first kind (see our Ergebnisse monograph "Algebraic surfaces" [11], pp. 71-72).

In IP we have derived a number of properties of reducible exceptional curves, of the first kind, on a non-singular surface (IP, Section II. 3), but we shall make no use of reducible exceptional curves of the first kind in the present paper.

Given a reducible exceptional curve E, of the first kind, on a non-singular surface F, it is possible to define a divisorial cycle $\mathcal{E}$ on F whose prime components are the irreducible components of E, counted to suitable (positive) multiplicities, in such way that relations similar to (1) be satisfied, i. e., so as to have $p(\mathcal{E})=0$ and $(\mathcal{E}^2)=-1$. For the definition and properties of these *exceptional cycles of the first kind* (which were not treated in IP) see our cited monograph [11], pp. 36-40; compare also our forthcoming note "Exceptional cycles of the first kind on a non-singular surface and the descending chain condition in birational classes of non-singular surfaces" in the Mem. Coll. Sci. Univ. Kyoto.

We terminate this section with listing a few properties of exceptional curves of the second kind, on a non-singular surface.

PROPOSITION 2. 6. *If a relatively minimal model carries an exceptional curve of the second kind, then it also carries an irreducible exceptional curve of the second kind.* (IP, Corollary II. 4. 6)

PROPOSITION 2. 7. *If E is an irreducible exception curve of the second kind, on a non-singular surface F, and if $s_1, s_2, \cdots, s_m$ are the multiplicities of its singular points (including its infinitely near singular points; if E is non-singular we set m and all the s_i equal to zero), then there exists a non-negative integer n such that*

$$(2) \qquad p(E)=\sum s_i(s_i-1)/2,$$

$$(3) \qquad (E^2)=-1+\sum s_i^2+n, \quad n+m>0;\ n, m\geqq 0.$$

It follows, in particular, that $(E^2)\geqq 0$ and that if E has singularities then $(E^2)>0$. (IP, Corollary II. 5. 14)

We now prove the converse of Proposition 2.7:

PROPOSITION 2.8. *Let E be an irreducible curve on a non-singular surface F, and let $s_1, s_2, \cdots, s_m$ be the multiplicities of its singular points (including its infinitely near singular points). If there exists an integer $n \geqq 0$ such that $n + m > 0$ and such that the numerical characters $p(E)$ and (E^2) of E are given by* (2) *and* (3), *then E is an exceptional curve of the second kind.*

Proof. We have only to show that there exists an *antiregular* birational transformation $T: F \rightarrow F'$ of F onto a non-singular surface F' (i.e., a *transformation T such that T^{-1} is regular*) such that the *proper* transform $E' = T[E]$ of E is an exceptional curve of the first kind. For assume that this has already been shown. Then, denoting by P'' the simple contraction of E', the antiregularity of T shows immediately that the total transform of P'' *on* F contains the curve E and no other *curves.* Since the proper transform of P'' must be pure one-dimensional, it follows that E is the proper transform of P'', whence E is an exceptional curve. Since $(E^2) \geqq 0$, by (3), E must be of the second kind.

Now, to obtain T we apply to F successive quadratic transformations whose centers are at the singular points of E and of the successive transforms of E, until we obtain a *non-singular* proper transform $\bar{E}$ of E. From (2) and (3) we obtain easily that $p(\bar{E}) = 0$, $(\bar{E}^2) = -1 + n$. We then choose n distinct points of $\bar{E}$ and we apply n quadratic transformations with these points as centers. The proper transform of $\bar{E}$ is then an irreducible curve E' such that $p(E') = 0$ and $(E')^2 = -1$, hence an exceptional curve of the first kind. Q. E. D.

If K is a canonical divisor on a non-singular surface F and Z is any divisorial cycle on F we set

$$\sigma(Z) = -(Z \cdot K). \tag{4}$$

It is known (see IP, II. 5, formula (6)) that

$$\sigma(Z) = (Z^2) - 2p(Z) + 2. \tag{5}$$

It follows from (2) and (3) that if E is an irreducible exceptional curve, of the second kind, on F, then

$$\sigma(E) = 1 + n + \sum s_i \geqq 2. \tag{6}$$

Thus $(E \cdot K) < 0$. On the other hand, we have $(E^2) \geqq 0$ and hence $(E \cdot Z) \geqq 0$ for any divisorial cycle $Z \stackrel{\bullet}{\geqq} 0$. Consequently, if i is any integer

$\geqq 1$ there cannot exist a divisorial cycle $\geqq 0$ which is linearly equivalent to iK. In other words, the multicanonical systems $|iK|$, $i = 1, 2, 3, \cdots$, do not exist. We have therefore

PROPOSITION 2.9. *If a non-singular surface F carries an irreducible exceptional curve E of the second kind, then all the plurigenera P_i of F $(i \geqq 1;\ P_1 = p_g)$ are zero.*

Remark. What matters in the proof of Proposition 2.9 is not that E is an exceptional curve of the second kind but the fact that E is an irreducible curve such that $(E^2) \geqq 0$ and $\sigma(E) > 0$. Consequently, already the presence of such an irreducible curve E implies the vanishing of all the plurigenera.

3. Preliminary criteria of existence of minimal models and statement of the main results. Let F be a member of our birational class $\mathfrak{B}$, i.e., F is a non-singular projective model of the function geld Σ/k. If F carries an exceptional curve E, of the first kind, and if P' is the simple contraction of E, then, by Proposition 2.4, the class $\mathfrak{B}$ contains the surface $F - E + P'$ which is strictly dominated by F, and hence F is not a relatively minimal model of Σ/k. Conversely, if F is not a relatively minimal model of Σ/k, then exists a surface F' in $\mathfrak{B}$ such that $F' < F \neq F'$, the birational transformation $T: F' \rightarrow F$ has then necessarily a fundamental point P' on F', and the total transform $T\{P'\}$ is an exception curve on F, of the first kind (since T^{-1} is regular on F, and P' is a simple point). We have therefore

PROPOSITION 3.1. *A necessary and sufficient condition that a surface F in the birational class $\mathfrak{B}$ be a relatively minimal model of Σ/k is that F be free from exceptional curves of the first kind.*

We now prove

PROPOSITION 3.2. *A necessary and sufficient condition that a non-singular projective model F of Σ/k be a minimal model of Σ/k is that F be free from exceptional curves* (of the first *or* of the second kind).

Proof. The condition is obviously sufficient. Suppose now that F carries an exception curve E. Let $T: F \rightarrow F'$ be a birational transformation of F which contracts E into a simple point P'. By the theorem of reduction of singularities there exists a non-singular projective model F'' of Σ/k such that $F' \prec F''$. Since P' is a simple point of F' we can choose the resolving birational transformation $F' \rightarrow F''$ of F' in such a way that it be biregular at P'. Then also the birational transformation $F \rightarrow F''$ contracts E to the point P'. Thus $F \not\prec F''$, and since $F'' \in \mathfrak{B}$, F is not a minimal model of Σ/k. This completes the proof.

Let E_1, E_2 be two exception curves, of the first kind, on a non-singular surface F, and let P' and P'' be the simple contractions of E_1 and E_2 respectively. If $E_1 \subset E_2$, then it follows from the last part of Proposition 2.3 that $\mathfrak{o}_{P'} \supset \mathfrak{o}_{P''}$. (Note that also the converse is true, for if $\mathfrak{o}_{P'} \supset \mathfrak{o}_{P''}$ then every zero-dimensional valuation with center P' has necessarily center P'', and thus the total transform E_1 of P' on F is contained in the total transform E_2 of P''). It follows that the projective surface $F - E_1 + P'$ strictly dominates the projective surface $F - E_2 + P''$ (we assume, of course, that $E_1 \neq E_2$). By the descending chain condition in $\mathfrak{B}$ (Proposition 1.5) it follows therefore that F *cannot carry a strictly ascending chain* $E_1 \subset E_2 \subset E_3 \cdots$ *of exceptional curves of the first kind.* If, then, we call an exceptional curve E of the first kind *maximal* when E is not a proper subset of any other exceptional curve of the first kind, then we conclude that *any exceptional curve, of the first kind, on a non-singular surface F is contained in at least one maximal exceptional curve of the first kind.*

PROPOSITION 3.3. *A necessary condition that the field Σ/k have a minimal model is that every non-singular projective model F of Σ/k have the following three properties:*

(a) *F carries no exceptional curves of the second kind.*

(b) *F has only a finite number of exceptional curves of the first kind.*

(c) *Any two distinct maximal exceptional curves of the first kind have no points in common.*

A sufficient condition that Σ/k have a minimal model is that there exist a non-singular projective model of Σ/k having properties (a), (b) *and* (c).

Proof. Assume that Σ/k has a minimal model, say F^*. Let F be any non-singular projective model of Σ/k. We first show that F can carry only exceptional curves *of the first kind.* Let E be an exceptional curve on F and let $T: F \rightarrow F'$ be a birational transformation of F which contracts E to a simple point P' ($P' \in F'$). Without loss of generality we may assume that F' is a non-singular surface (see proof of Proposition 3.2). Let P be any point of E. Since P and P' are corresponding points under T, there exists a zero-dimensional valuation v of Σ/k having centers P and P' on F and F' respectively. Let P^* be the center of v on F^*. Since F^* is a minimal model we have $F^* \prec F$ and $F^* \prec F'$, whence $\mathfrak{o}_{P^*} \subset \mathfrak{o}_P$ and $\mathfrak{o}_{P^*} \subset \mathfrak{o}_{P'}$. From these inclusions and from the fact that P, P' are simple centers of one and the same valuation it follows by a known result (see Zariski [12], Lemma on p. 538, or also IP, proof of Theorem II. 1. 1) that either $\mathfrak{o}_P \subset \mathfrak{o}_{P'}$ or $\mathfrak{o}_{P'} \subset \mathfrak{o}_P$. The first inclusion is impossible since T^{-1} is not regular at P' (the point P' is fundamental for T^{-1}). Hence $\mathfrak{o}_{P'} \subset \mathfrak{o}_P$, and thus T is regular at P. Since P was any point of E, the curve E is exceptional *of the first kind*, by Definition 2. 1.

We have thus shown that F has property (a). Now let $P_1^*, P_2^*, \cdots, P_s^*$ be the fundamental points of the birational transformation $f: F^* \to F$ and let $E_i = f\{P_i\}$. The s curves E_i on F are exceptional of the first kind (we assume, of course, that $F \neq F^*$, and that therefore $s \geqq 1$, for otherwise conditions (b) and (c) are trivially satisfied), and *no two of these curves have points in common*; both these assertions follow from the regularity of f. Let E be any exceptional curve on F. We know already that E is of the first kind. Let P' be the simple contraction of E, let $F' = F - E + P'$ and let P^* be *the* point of F^* which corresponds to P' in the birational transformation $F' \to F^*$. Since $\mathfrak{o}_{P^*} \subset \mathfrak{o}_{P'}$ and P' is fundamental for the birational transformation $F' \to F$, it follows that *a fortiori* also P^* is fundamental for $f: F^* \to F$. Hence P^* is one of the points P_i^*, say $P^* = P_1^*$. Since $\mathfrak{o}_{P^*} \subset \mathfrak{o}_{P'}$, it follows, by a remark made earlier, that the total transform E_1 of P^* on F contains the total transform E of P' on F. We have therefore shown that every exceptional curve E, of the first kind, on F is contained in one of the s exceptional curves E_i. Hence F carries exactly s maximal exceptional curves of the first kind, and this proves (b). Since no two of E_i intersect, the necessity of the condition is established.[3]

Assume now that there exists a non-singular projective model F of Σ/k having properties (a), (b) and (c). Let $E_1, E_2, \cdots, E_s$ be the maximal exceptional curves, of the first kind, on F and let P_i^* be the simple contraction of E_i. Since no two of E_i intersect, the abstract surface $F^* = F - \sum E_i + \sum P_i^*$ is projective (Proposition 2. 5). We shall show that F^* is a minimal model of Σ/k.

Let F' be any non-singular projective model of Σ/k and let $f: F' \to F$ and $f^*: F' \to F^*$ be the birational transformations of F' onto F and F^* respectively. Let P' be any point of F'. If f is regular at P' then also f^* is regular at P' since $F^* \prec F$. If f is not regular at P', let $E = f\{P'\}$ be the exceptional curve on F which is the total f-transform of the fundamental

[3] We observe that even if there does not exist a minimal model the following is true for any non-singular projective model F of Σ/k: there does not exist on F an infinite set of exceptional curves of the first kind such that no two curves in the set intersect. This is a straightforward consequence of the descending chain condition in the class $\mathfrak{B}$.

point P' of f. The exceptional curve E is of the first kind, since F has property (a), and E is contained in one of the s maximal exceptional curves E_i, say $E \subset E_1$. Since P' and P_1^* are the simple contractions of E and E_1 respectively, it follows from $E \subset E_1$ that $\mathfrak{o}_{P'} \supset \mathfrak{o}_{P_1^*}$. Thus also in this case the birational transformation $f^*: F' \rightarrow F^*$ is regular at the point P'. Since P' is any point of F', it follows that $F' \succ F^*$. Since F' is any non-singular projective model of Σ/k the assertion that F^* is a minimal model of Σ/k is established. This completes the proof of the proposition.

Our principal object in this paper is the characterization of those function fields Σ/k of transcendence degree 2 which have no minimal model. This characterization is given by the following theorem (due to the Italian geometers Castelnuovo and Enriques in the classical case):

FUNDAMENTAL THEOREM A. *If the field Σ/k does not possess a minimal model then Σ is the field of rational functions of a ruled surface (or equivalently: Σ is a simple transcendental extension $\Omega(t)$ of a field Ω/k of algebraic functions of one variable).*

The converse of the fundamental theorem A is also true, but is trivial. For, if $\Sigma = \Omega(t)$ and if C denotes a non-singular curve such that $k(C) = \Omega$, then the direct product $C \times L$, where L is a projective line, is a non-singular projective model of Σ/k (the Segre variety of the pairs of points of C and L). The curves E on $C \times L$, given by $Q \times L$ ($Q \in C$), are irreducible, non-singular, and we have $p(E) = 0$, $(E^2) = 0$. Hence each E is an exceptional curve of the second kind (Proposition 2.8), and therefore $\Omega(t)$ has no minimal model (Proposition 3.3).

We shall prove the fundamental theorem A in the following stronger form:

FUNDAMENTAL THEOREM B. *If a non-singular surface F carries an irreducible exceptional curve E of the second kind, then F is birationally equivalent to a ruled surface, and if, furthermore, $(E^2) > 0$ (in particular, if E has singular points; see Proposition 2.7) then F is a rational surface.*

The remainder of this paper is devoted to the proof of this theorem. In regard to this theorem we point out at this stage that if the theorem is true under the additional assumption that F is a relatively minimal model then it is true in general. For assume that the theorem has already been proved under the assumption that F is a relatively minimal model and let F any non-singular surface satisfying the conditions of the fundamental theorem B. By the descending chain condition in the birational class $\mathfrak{B}$ of F, there

11

exists a relatively minimal model F^* in B such that $F^* \prec F$. Let $T: F \rightarrow F^*$ be the (regular) birational transformation of F onto F^*. We assert that E is not an exceptional curve of T, i.e., that $T\{E\}$ is not a point. For assume the contrary, and let $T\{E\}$ be a point P^* of F^*. Let P' be a simple contraction of E and let F' be a non-singular surface which carries the point P'. From the regularity of T and from the fact that E is the total transform of P' it follows that P^* is the only point of F^* which corresponds to the point P' in the birational transformation $F' \rightarrow F^*$. Hence $\mathfrak{o}_{P^*} \subset \mathfrak{o}_{P'}$. Since we have also $\mathfrak{o}_{P^*} \subset \mathfrak{o}_P$ for any point P of E, the argument developed in the first part of the proof of Proposition 3.3 shows that $\mathfrak{o}_{P'} \subset \mathfrak{o}_P$. This being true for every point P of E, P' is a simple *regular* contraction of E, in contradiction with the fact that E is an exceptional curve of the *second* kind. [What we have proved here is that no irreducible component of an exceptional curve of the first kind (in the present case—of $T^{-1}\{P^*\}$) can be an exceptional curve of the second kind.]

Let then $T\{E\} = E^*$, where E^* is an irreducible curve on F^*. From the regularity of T and from the irreducibility of E it follows that also E^* is an exceptional curve (every simple contraction of E is also a contraction of E^*). The regularity of T implies that $(E^{*2}) \geqq (E^2)$ (IP, II 5, formula (12)). Since $(E^2) \geqq 0$ it follows that also $(E^{*2}) \geqq 0$, and consequently E^* is of the second kind. If, furthermore, $(E^2) > 0$ then also $(E^{*2}) > 0$, F^* is rational, and so is F.

The proof of the fundamental theorem will be divided in several cases, according as (E^2) is zero or positive, and in the latter case according to the value of (K^2), where K is a canonical divisor. In view of what was just said above, we may assume, in any given case, that F is a relatively minimal model. In some cases this assumption will play a crucial role in the proof, while in other cases that assumption is not required and will in fact not be made.

4. Case $(E^2) = 0$. *Beginning with this section, and throughout the rest of the paper, we assume that the field Σ/k has no minimal model.* We fix a relatively minimal model F of Σ/k. The surface F must then carry an *irreducible* exceptional curve E, of the second kind (Propositions 3.1, 3.2 and 2.6). In this section we assume that F carries an irreducible exceptional curve E, of the second kind, such that $(E^2) = 0$. Then E is non-singular and $p(E) = 0$ (Proposition 2.7), whence E is a non-singular rational curve.

We proceed to prove that *any non-singular surface F which carries an irreducible non-singular rational curve E such that* $(E^2) \geqq 0$ (such a curve is in fact exceptional of the second kind; see Proposition 2.8) *is birationally equivalent to a ruled surface.*

We first consider the case $(E^2) = 0$. By formula (5) of Section 2 we have $\sigma(E) = 2$ and consequently $\sigma(nE) = 2n$ for any integer n.

If Z is any divisorial cycle on F then by the Riemann-Roch theorem

$$(1) \qquad \dim |Z| \geqq \tfrac{1}{2}[\sigma(Z) + (Z^2)] + p_a - i(Z),$$

where $i(Z) = 1 + \dim |K - Z|$ and p_a is the arithmetic genus of F. If $Z \geqq 0$ then $i(Z) = 0$ since the canonical system dos not exist on our surface F (see *Remark* following Proposition 2.9). Applying (1) to the cycle $Z = nE$ we find

$$(2) \qquad \dim |nE| \geqq n + p_a, \qquad n \geqq 0.$$

For any integer $n \geqq 1$ we have only the following two possibilities: (a) either E is a fixed component of $|nE|$, and in that case $\dim |nE| = \dim |n-1)E$, or (b) E is not a fixed component of $|nE|$, and in that case $\dim |nE| = 1 + \dim |(n-1)E|$, since $(E^2) = 0$ and since therefore the trace of $|nE|$ on E consists only of the zero-cycle on E. Hence in both cases we have $0 \leqq \dim |nE| - \dim |(n-1)E| \leqq 1$, and from this it follows, in view of (2) that *for large* n we have

$$(3) \qquad \dim |nE| = n + c,$$

where c is a constant, and

$$(4) \qquad \dim |nE| = 1 + \dim |(n-1)E|.$$

Since E is irreducible, it follows from (4) that for large n the system $|nE|$ has no fixed components. On the other hand, if $n > 1$ the system $|nE|$ must be reducible since any two cycles in $|nE|$ which pass through a given point of F must have a common component (the intersection number $(nE \cdot nE)$ being zero)). Therefore, for large n, the system $|nE|$ is composite with a pencil, or else all the members of $|nE|$ are the p-folds of other cycles, where p is the characteristic of k (Theorem of Bertini; see IP, Theorem I.6.3). The second alternative is to be excluded, for if n is large, then also $|(n-1)E|$ has no fixed components and thus $|nE|$ contains a cycle of the form $E + D$, where $D \in |(n-1)E|$ and E is not a component of D, so that E is a simple component of that cycle. Therefore $|nE|$ is composite with a pencil (not neecssarily linear) and thus it is also composite with a (uniquely determined) irreducible pencil, say L_n. Since $nE \in |nE|$, some multiple of E, say s_nE, belongs to L_n. Then every cycle in $|nE|$ which contains E as component must contain E to multiplicity $\geqq s_n$. Since we have just seen that if n is large, then there exist cycles in $|nE|$ which contain E as a simple component, we conclude that $s_n = 1$, i.e., E itself is a member of the pencil L_n. Since E is irreducible and $(E^2) = 0$, E cannot belong to two distinct pencils. Hence L_n does not depend on n, and the systems $|nE|$, n large, are composite with one and the same irreducible pencil $L = L_n$.

We have thus found a pencil L on F one member of which, namely E, is a prime cycle whose arithmetic genus is zero. But then also the general cycle of L, over k, has arithmetic genus zero. By the theorem of Noether-Enriques (or as a consequence of the purely algebraic theorem of Tsen [8]),[4] it follows that F is birationally equivalent to a ruled surface.

The preceding proof is applicable also to the case $(E^2) > 0$ provided E is *non-singular*, and in that case the conclusion is that *the surface F is rational.* In fact, if $(E^2) = s > 0$, then after s quadratic transformations with center on E and the successive transforms of E we obtain a non-singular surface F' on which the proper transform of E is an irreducible, non-singular, rational surve E' such that $(E'^2) = 0$. By the preceding proof, E' is a member of a pencil L'. The corresponding pencil L on F has s base points $(s > 0)$ and is therefore a linear pencil (Schilling-Zariski [6]). Therefore also L' is a linear pencil, and F is a rational surface (the field Ω of footnote 4 is now a simple transcendental extension of k).

5. Case $(E) > 0$. The vanishing of $p_a(F)$. We now assume that our relatively minimal model F carries an irreducible exceptional curve E, of the second kind, such that the self-intersection number (E^2) of E *is positive.* Our object in the rest of the paper is to prove that F is a rational surfaces. In this section we shall carry out a first step of the proof, namely, we shall show that the arithmetic genus p_a *of F is zero.*

We shall need the following lemma:

LEMMA 5.1. *Let $T: G \to G'$ be a rational regular transformation of a non-singular surface G onto a (not necessarily non-singular) variety G' of*

[4] The pencil L determines a subfield Ω of Σ/k, of transcendence degree 1. The fact that the general member of L/k has arithmetic genus zero signifies that if Ω is taken as ground field then Σ—a field of algebraic functions of *one* variable over Ω—has genus zero. Thus Σ/Ω has divisors of degree 2 (the anticanonical divisors of Σ/Ω), and the Riemann-Roch theorem yields a projective model C of Σ/Ω which is a conic, defined over Ω. Since k is algebraically closed, the theorem of Tsen guarantees the existence of a rational point of C, over Ω, and consequently C is birationally equivalent, over Ω, to a projective line. Hence Σ is a purely transcendental extension of Ω.

positive dimension (whence G' is either a surface or a curve). If E is an irreducible curve on F such that $T\{E\}$ is a point then $(E^2) \leqq 0$.

Proof. Let L be the linear system on G, free from fixed components, which corresponds to the system of hyperplane sections of G' (see IP, I. 3 and I. 5). Since T is regular, L has no base points. Since E is exceptional for T, the linear system $Tr_E L$ has dimension zero, i.e., this linear system consists only of one divisor (on E). This divisor must be the null divisor, for L has no base points. Hence $(D \cdot L) = 0$ for any cycle D in L. Furthermore, if $\dim L = m$ (where necessarily $m \geqq 1$), then L contains a linear subsystem L_1, of dimension $m-1$, having E as fixed component. Fix a cycle $sE + \Delta$ in L_1, where $\Delta \geqq 0$, $s \geqq 1$ and where we assume that E is not a component of Δ. Then $(\Delta \cdot E) \geqq 0$, and since $(sE + \Delta \cdot E) = 0$, it follows that $(E^2) \leqq 0$, as asserted.[5]

We now go back to our relatively minimal model F.

PROPOSITION 5.2. *The irregularity q of F is zero ($q =$ dimension of the Albanese variety of F).*

Proof. Let A be the Albanese variety of F, let f be the canonical mapping of F into A and let $F' = f(F)$. Since F is non-singular it follows, by a well-known property of Abelian varieties (A. Weil, [9], Th. 6, p. 27) that the rational transformation f of F onto F' is regular. Since E is a rational curve and since A carries no rational curves, *the transform $f(E)$ of E must be a point* (for otherwise $f(E)$ would be a *curve*, which—by the theorem of Lüroth—would then be rational). Since $(E^2) > 0$, it follows from the preceding lemma that F' cannot have positive dimension. Hence F' is a point, and since F' generates A, also A is a point. Q. E. D.

In the classical case, Severi has proved, by an essentially algebraic method, the completeness of the characteristic series of any ample complete continuous system of curves on a surface F whose geometric genus p_g is zero (see Severi [7], Anhang F; see also our monograph [11], pp. 84-86, where the proof and references may be found). In other words, Severi has proved, in the classical case, the *equality* $q = p_g - p_a$ by algebraic methods, under the assumption $p_g = 0$. This equality, under the same assumption, has now been established also in the abstract case, and independent proofs have been given by Akizuki-Matsumura [1], Nakai [4], [5] and Matsumura-Nagata

[5] Note that if F' is a *surface* then one can conclude that $(E^2) < 0$. For in that case we have $m \geqq 2$, $\dim L_1 > 0$, and thus we can find a cycle $sE + \Delta$ in L_1 such that $\Delta > 0$. By the connectedness theorem (Zariski [13], Theorem 14, p. 77) it then follows that $(\Delta \cdot E) > 0$, whence $(E^2) < 0$.

[3]. Since we know that for our surface F we have $p_g = 0$ (Proposition 2.9), it follows from Proposition 5.2 that $p_a = 0$. This completes the preliminary step of our proof of rationality of F.

In view of the vanishing of p_a, the Riemann-Roch inequality on the surface F is now as follows:

$$\dim |Z| \geqq \tfrac{1}{2}[\sigma(Z) + (Z^2)] - i(Z), \tag{1}$$

where $i(Z) = 1 + \dim |K - Z|$ and $\sigma(Z) = -(Z \cdot K)$ (Z, a divisorial cycle; K, a canonical cycle). Since $p_g = 0$ we have $i(Z) = 0$ if $Z \geqq 0$.

We denote by Z' any cycle which is linearly equivalent to $Z + K$, whence $|Z'|$ is the so-called *adjoint system of* $|Z|$. We set

$$(K^2) = \nu. \tag{2}$$

Then

$$\sigma(Z') = \sigma(Z) - \nu, \quad (Z'^2) = (Z^2) + \nu - 2\sigma(Z), \quad i(Z') = 1 + \dim |-Z|.$$

Hence, applying (1) to Z' instead of to Z, and using the equality $\sigma(Z) = (Z^2) - 2p(Z) + 2$, we find

$$\dim |Z'| \geqq p(Z) - \dim |-Z| - 2, \tag{3}$$

and if $|Z|$ *contains positive cycles then*

$$\dim |Z'| \geqq p(Z) - 1, \qquad Z > 0. \tag{4}$$

6. The case $(E^2) > 0$, $(K^2) \leqq 0$. In this section we prove the fundamental theorem B under the assumptions that $(E^2) > 0$, $(K^2) \leqq 0$ and that F is a relatively minimal model.

We assert that *the linear system* $|E + nK|$ (*the n-th adjoint system of* $|E|$) *does not exist if n is large.* In fact, we have $\sigma(E) \geqq 2$ [see (6), § 2], i.e., $(E \cdot K) \leqq -2$, whence $(E + nK \cdot E) < 0$ if n is large. Hence, if the system $|E + nK|$ exists for large n, then that system would have the curve E (which is *irreducible*) as fixed component, in contradiction with the non-existence of $|nK|$ (Proposition 2.9).

Let m be the smallest integer ($m \geqq 0$) such that $|E + mK|$ exists and $|E + (m+1)K|$ does not exist. Since $(mK \cdot mK) = m^2(K^2) \leqq 0$ while $(E^2) > 0$, we have $E + mK \not\equiv 0$. Hence the system $|E + mK|$ consists of positive cycles. Let Z be a positive cycle in that system, and let D be any prime component of Z. Since $|Z'|$ does not exist, *a fortiori* $|D'|$ does not exist. Hence by (4), § 5, we must have $p(D) = 0$. On the other hand, since $\sigma(Z) = \sigma(E) - m(K^2) > 0$ we must have $\sigma(D) > 0$ for at least one prime component D of Z. For such a D we have therefore $(D^2)+2>0$ (since $p(D)=0$). The case $(D^2)=-1$ is excluded since F carries no exceptional curves of the first kind. Hence $(D^2)\geqq 0$, and thus F is birationally equivalent to a ruled surface (§ 4). Since $p_a(F)=0$, F is rational.

7. Some general properties of surfaces with $p_a=P_2=0$ and $(K^2)>0$. In this section we consider a-non-singular surface F for which the following conditions are satisfied: $p_a=P_2=0$ and $(K^2)>0$. We shall derive some general properties of such a surface and we shall use these properties in the rest of the paper.

We set $(K^2)=\nu$ and we denote by K_a any *anticanonical* cycle, i. e., any cycle of the form $-K$. We note first of all that

$$(1) \qquad p(K_a)=1.$$

This equality holds on *any* non-singular surface, for we have $\sigma(K_a)=-(K_a\cdot K)=(K_a{}^2)$, and (1) follows from (5), 2. From $\sigma(K_a)=\nu$ we find by (1), § 5:

$$(2) \qquad \dim|K_a|\geqq\nu\geqq 1,$$

since $i(K_a)=\dim|2K|+1=0$. Hence the anticanonical system $|K_a|$ exists and has positive dimension. *A fortiori*, all the multi-anticanonical systems $|nK_a|$ $(n>0)$ exist and have positive dimension. Therefore none of the multicanonical systems $|nK|$ $(n>0)$ exists, and thus all the plurigenera of F vanish.

Let Z be any divisorial cycle on F and let n be an integer >0. Assume that the system $|Z+nK|$ exists. Then if we fix a *positive* cycle K_a in $|K_a|$, the linear system $|Z|$ will contain a cycle which is $\geqq nK_a$. Since, for a given Z, this is not possible if n is large, it follows that *the system* $|Z+nK|$ *does not eixst if n is large.* Thus the process of successive adjunctions, applied to a cycle Z, always terminates after a finite number of steps.

LEMMA 7.1. *If D is an irreducible curve on F and $|D'|$ does not exist, then $p(D)=0$ (and hence D is a rational non-singular curve).*

Proof. By (4), § 5, the non-existence of $|D'|$ implies $p(D)\leqq 0$, and hence $p(D)=0$ since D is an irreducible curve.

LEMMA 7.2. *Let K_a be a positive anticanonical cycle. If K_a is not a prime cycle, then every prime component of K_a has arithmetic genus zero.*

Proof. Let *D be a prime component of K_a and let $K_a=D+\Delta$, $\Delta>0$.*

We have $|D'| = |K_a - \Delta + K| = |-\Delta|$. Hence $|D'|$ does not exist, and we can apply the preceding lemma.

PROPOSITION 7.3. *If $|K_a|$ contains a reducible linear subsystem L, of positive dimension, then F is a rational surface.*

Proof. The system L may have fixed components. Let B be the maximal fixed component of L and let L_1 be the system obtained from L by deleting the fixed cycle B (so that L_1 is a linear system of positive dimension and free from fixed components). Let the cycle A be defined as follows: 1) If L_1 is irreducible then A is a general cycle of L_1/k. 2) if L_1 is reducible, then L_1 is either composite with an irreducible pencil or the cycles of L_1 are all of the form p^eC, where C varies in an irreducible linear system M of positive dimension; in the first case, A shall be a general cycle of that pencil, in the second case it shall be a general cycle of the linear system M/k. In all cases, A is a prime cycle and is a component of a positive anticanonical cycle which is not prime. Therefore, by Lemma 7.2, we have $p(A) = 0$. We have thus on F either an irreducible linear system of positive dimension, whose general cycle A has arithmetic genus zero, in which case F is a rational surface, by the theorem of Noether-Enriques; or a pencil whose general member A has arithmetic genus zero. In the latter case, F must be birationally equivalent to a ruled surface, again by the theorem of NoetherEnriques. But since $p_a = 0$, this ruled surface is of genus zero, and so again F is rational.

8. Proof of Castelnuovo's criterion of rationality in the case $(K^2) \geqq 3$. We again assume that $p_a = P_2 = 0$ and we consider first the case $(K^2) = \nu \geqq 4$. We fix a rational point P of F, over k, and we denote by L the linear cystem of all cycles in $|K_a|$ which have at P a singular point (hence at least a double point). We have $\dim L \geqq \dim |K_a| - 3$, whence, by (2), §7: $\dim L \geqq \nu - 3 \geqq 1$. If L is reducible, then F is rational (Proposition 7.3). Assume that L is irreducible. Since $p(K_a) = 1$ it follows that the general member of L/k is a rational curve (which, however, as a cycle has arithmetic genus 1). Let $T: F \to \bar{F}$ be a quadratic transformation of F, with center P. For every cycle D in L we have $T(D) = \bar{D} + 2\bar{E}$, where $\bar{E} = T\{P\}$ is the irreducible exceptional curve of the first kind created by T and where $\bar{D}$ is a positive cycle. As D varies in L, $\bar{D}$ varies in a linear system $\bar{L}$, of positive dimension, and since $p(D) = 1$ it follows at once that $p(\bar{D}) = 0$. Thus $\bar{F}$ carries a linear system (of positive dimension) of cycles having arithmetic genus zero, the general cycle of the system being prime. Hence $\bar{F}$, and therefore also F, is rational.

We consider now the case $\nu = 3$. By Proposition 7.3 we may assume that $|K_a|$ is irreducible (since $\dim |K_a| \geqq 3$). If r denotes the dimension of $|K_a|$, the system $|K_a|$ defines a rational transformation $T: F \to F'$ of F onto an algebraic variety F' of positive dimension $\leqq 2$, immersed in a projective space of dimension r, whose system of hyperplane sections corresponds to $|K_a|$. Since $|K_a|$ is not composite with a pencil, F' is a surface (not a curve). Since the degree of $|K_a|$ is the prime integer 3 and the dimension is $\geqq 3$, the system $|K_a|$ cannot be composite with any unramified involution. Hence, the function field $\Sigma = k(F)$ either coincides with the function field $k(F')$ or is a purely inseparable extension of $k(F')$, of degree 3. In the latter case, F' would be a surface of *order* 1, i.e., a plane, which is impossible, since F' does not lie in a projective space of dimension $< r$, and $r \geqq 3$. Hence $k(F) = k(F')$, and F' is *a surface of* order $\geqq 3$, *birationally equivalent to* F (it could be a surface of order < 3, for *a priori* $|K_a|$ may have base points; see, however, Lemma 9.1 in the next section). If F' is a surface of order < 3, or if F' is a cubic surface and $r \geqq 4$, then F' is rational [if $r \geqq 4$ then a central projection of F' transforms F' birationally onto a quadratic surface; actually, it can be shown that $r = 3$ (see Lemma 9.1 in the next section)]. If F' is a cubic surface in S_3 $(r = 3)$ then F' cannot be an elliptic cubic cone since $p_a(F) = 0$ and since F and F' are birationally equivalent. Hence F' is necessarily a rational surface. This completes the proof of Castelnuovo's criterion in the case $\nu \geqq 3$.

9. Proof of the fundamental theorem B in the case $(K^2) = 1$. In this case we shall use the following result:

PROPOSITION 9.1. *Let F be a relatively minimal model of a field Σ/k of algebraic functions of two variables and assume that $p_a = P_2 = 0$ and that $(K^2) = \nu > 0$. If there exists on F a divisorial cycle which is not linearly equivalent to an integral multiple of K, F is rational.*

Proof. We first show that F *carries at least one irreducible curve having arithmetic genus zero.* Assume the contrary. Since there exist on F divisorial cycles which are not linearly equivalent to an integral multiple of K, there also exist positive divisorial cycles having this property. Fix one such positive cycle Z. Let D be a prime component of Z. Since, by assumption, $p(D) \geqq 1$, the adjoint system $|D'|$ exists (Lemma 7.1). *A fortiori*, also the adjoint system $|Z'|$ $(= |D' + Z - D|)$ exists. The system $|Z'|$ must contain positive cycles since $Z \not\equiv -K$. Obviously no cycle in $|Z'|$ can be linearly equivalent to an integral multiple of K. Therefore, replacing Z by a cycle belonging to $|Z'|$ and applying the same argument we deduce that also the second adjoint system $|Z''|$ $(=|Z'+K|)$ exists. Thus all the successive adjoint system $|Z+nK|$, $n=1,2,\cdots$, exist, and this is in contradiction with what has been shown in the beginning of Section 7.

The proposition will be established if we show that either F is rational or F carries an irreducible curve E such that $p(E)=0$ *and* $(E^2)\geqq 0$. For in the latter case F is birationally equivalent to a ruled surface (Section 4) and is therefore again rational, since $p_a=0$.

Assume that for *any* irreducible curve E on F such that $p(E)=0$ we have necessarily $(E^2)<0$, and therefore $(E^2)\leqq -2$ (since F—*a* relatively minimal model—carries no exceptional curves of the first kind (Proposition 3.1)). We fix an irreducible curve E on F such that $p(E)=0$. Since $(E^2)\leqq -2$, we have $\sigma(E)=(E^2)-2p(E)+2\leqq 0$, i.e., $(E\cdot K_a)\leqq 0$. If $(E\cdot K_a)<0$, then E is a fixed component of $|K_a|$, thus $|K_a|$ is reducible and F is rational (Proposition 7.3). We are trus led to the only remaining possibility: if E is any irreducible curve on F such that $p(E)=0$ then $(E^2)=-2$, whence $\sigma(E)=(K_a\cdot E)=0$. If we fix such a curve E, it is a fundamental curve of $|K_a|$ (in view of $(K_a\cdot E)=0$). Thus there exist a cycle $\bar{K}_a$ in $|K_a|$ which has E as component. For *every* irreducible component $\bar{E}$ of $\bar{K}_a$ we have $p(\bar{E})=0$ (Lemma 7.2), whence $\sigma(\bar{E})=0$. This is in contradiction with $\sigma(\bar{K}_a)=\nu>0$ and with the additive property of the numerical character σ. This completes the proof of the proposition.

Using the above proposition we shall now prove the fundamental theorem B in the case $(K^2)=1$ assuming—as we may—that F is a relatively minimal model. It will be sufficient to show that our (irreducible) exceptional curve E of the second kind cannot be linearly equivalent to an integral multiple of K. Assume the contrary: $E\equiv qK_a$, where q is necessarily a positive integer since $|K_a|$ contains positive cycles. Then $(E^2)=q^2$ and $\sigma(E)=q$. Using formulae (2) and (3) of Section 2 (Prop. 2.7) we find

$$q^2=-1+\sum_{i=1}^{h} s_i^2, \qquad q=1+\sum_{i=1}^{h} s_i,$$

where $h>0$, $s_i>0$ $(i=1,2,\cdots,h)$ and where, in the notations of Proposition 2.7, we have set $h=n+m$ and $s_{m+1}=s_{m+2}=\cdots=s_h=1$. Hence $(1+\sum_{i=1}^{h} s_i)^2=\sum_{i=1}^{h} s_i^2-1$, or $1+\sum_{i<j} s_is_j+\sum s_i=0$, which is impossible.

10. Proof of Castelnuovo's criterion of rationality in the case $(K^2) = 2$. We shall need the following lemma which deals with the general case of a non-singular surface F for which $p_a = P_2 = 0$ and ν $[= (K^2)]$ is an arbitrary positive integer.

LEMMA 10.1. *Let F be a non-singular surface such that $p_a = P_2 = 0$ and $\nu = (K^2) > 0$. Then either F is rational or the following is true: the linear system $|K_a|$ is irreducible, dim $|K_a| = \nu$ and, furthermore, if $\nu \geqq 2$, then $|K_a|$ has no base points and no fundamental curves.*

Proof. If $|K_a|$ is reducible, then F is rational (Proposition 7.3). Assume that $|K_a|$ is irreducible. We show that the assumption dim $|K_a| > \nu$ leads to a contradiction, and that consequently dim $|K_a| = \nu$, by (2), § 7. If dim $|K_a| > \nu$, then the characteristic series of $|K_a|$ on a general member $K_a{}^0$ of $|K_a|/k$ is a linear series of degree ν and of dimension $\geqq \nu$ (hence necessarily of dimension ν, so that dim $|K_a| = \nu + 1$). This implies that $K_a{}^0$ must have a singular point Q. But then the cycles of $|K_a|$ which pass through Q would cut out on $K_a{}^0$ a linear series of degree $\leqq \nu - 2$ and of dimension $\geqq \nu - 1$, which is impossible.

Assume now that $\nu \geqq 2$ (and that $|K_a|$ is irreducible). If $|K_a|$ has base points, then the characteristic series of $|K_a|$ on $K_a{}^0$, after the fixed cycle at the base points has been deleted, would be a linear series L of degree $< \nu$ and of dimension $\nu - 1$. Thus again $K_a{}^0$ would be of effective genus zero and would have to have a singular point Q. The point Q cannot be a base point, for in the contrary case the above series L would be actually of degree $\leqq \nu - 2$ (and of dimension $\nu - 1$), which is impossible. This being so, the cycles of $|K_a|$ which pass through Q would cut out on $K_a{}^0$, outside the fixed cycle at the base points *and* at Q, a series L_1 of degree $\leqq \nu - 3$ and of dimension $\nu - 2$, which again is impossible (note that L_1 is defined, since $\nu \geqq 2$ and since therefore dim $|K_a| \geqq 2$, so that there exist cycles in $|K_a|$, *different from* $K_a{}^0$, which pass through Q).

Assume now that $|K_a|$ has fundamental curves (always under the assumption that $\nu \geqq 2$) and that $|K_a|$ is irreducible, and let Γ be an irreducible fundamental curve of $|K_a|$. Then there exist cycles D in $|K_a|$ which contain Γ as component, and these cycles form a linear system of dimension $\nu - 1 > 0$. Therefore F is rational (Proposition 7.3). This completes the proof of the lemma.

The following is a generalization of the preceding lemma:

LEMMA 10.2. *Under the same assumptions as those of Lemma 8.1, either F is rational, or for any integer $n > 1$ the following is true: the system*

$|nK_a|$ is irreducible, has dimension $\frac{1}{2}[\nu n(n+1)]$, is free from base points, and furthermore it is also free from fundamental curves if $\nu \geqq 2$ or if F is a relatively minimal model.

Proof. We may assume that $|K_a|$ is irreducible (otherwise F is rational). Applying the Riemann-Roch inequality (1), Section 5, to the cycle $Z = nK_a$, we find

$$(1) \qquad \dim |nK_a| \geqq \tfrac{1}{2}[\nu n(n+1)].$$

From the irreducibility of $|K_a|$ and from the fact that $\dim |K_a| > 0$, it follows that $|nK_a|$ can mave no fixed component. If $\nu > 1$ that $\dim |K_a| \geqq 2$ and hence $|nK_a|$ cannot be composite with a pencil. If $\nu = 1$ then $|K_a|$ is a pencil, and this could be the only irreducible pencil with which $|nK_a|$ might be composite. But were $|nK_a|$ composite with the pencil $|K_a|$, then we would have $\dim |nK_a| \leqq n$, in contradiction with (1) which yields $\dim |nK_a| \geqq \frac{1}{2}[n(n+1)] > n$ if $n > 1$. Hence $|nK_a|$ is not composite with a pencil. It is also clear that not all the cycles in $|nK_a|$ are p-folds of other cycles, for $|nK_a|$ contains sums of n distinct primes cycles belonging to $|K_a|$. We therefore conclude that $|nK_a|$ is irreducible.

We shall now prove, by induction on n, that $\dim |nK_a| = \nu n(n+1)/2$. Since this equality is valid for $n = 1$ (by the above Lemma 10.1), we assume $n \geqq 2$. We denote by r the dimension of the linear series $g = Tr_C |nK_a|$, where C is a general member of $|K_a|/k$. Then g is a series of degree $n\nu$, and we have $\dim |nK_a| = 1 + r + \dim |(n-1)K_a|$, whence, by induction hypothesis, $\dim |nK_a| = 1 + r + \frac{1}{2}[\nu n(n-1)]$. By (1) we must have $r \geqq \nu n - 1$. Were $\dim |nK_a| > \frac{1}{2}[\nu n(n+1)]$, we would have $r \geqq \nu n$, and the linear series g would have degree νn *and* dimension νn. That would imply that C has effective genus zero. Since $p(C) = 1$ [see (1), § 7], C would have to have a double point Q. The linear series cut out on C by the cycles of $|nK_a|$ passing through Q would then have degree $\leqq \nu n - 2$ and dimension $\geqq \nu n - 1$, which is impossible.[6] This proves the equality $\dim |nK_a| = \nu n(n+1)/2$.

If $\nu \geqq 2$, then $|K_a|$ has no base points or fundamental curves, and this implies that also $|nK_a|$ has this same property. In the case $\nu = 1$ we use an indirect argument. Assume that $|nK_a|$ has a base point A $(n \geqq 2)$.

[6] Note that the cycles in $|nK_a|$ which pass through the point Q form a linear system of dimension $\geqq \nu n(n+1)/2 - 1$, while the cycles in $|nK_a|$ which contain C as component form a linear system of dimension $\nu n(n-1)/2$. If $n \geqq 2$ the former dimension is greater than the latter, and thus there exist cycles in $|nK_a|$ which pass through Q and do not contain C as component. Therefore the linear series in question is defined.

Then A is also a base point of $|K_a|$, $A \in C$, and the series $g = Tr_C |nK_a|$, outside of a fixed cycle at A, has degree $\leqq n-1$. Since the dimension r of g is $n-1$, C is of effective genus zero and must have a double point Q. The point Q is different from A, for otherwise the above series g, outside a fixed cycle at A, would have degree $\leqq n-2$ and dimension $n-1$, which is impossible. The cycles of $|nK_a|$ passing through Q would cut out on C a series of degree $\leqq n-3$ and of dimension $n-2$, which is impossible.

For any irreducible fundamental curve E of $|nK_a|$, $n > 1$, we must have $(nK_a \cdot E) = 0$, since $|nK_a|$ has no base points. This implies that E is not linearly equivalent to any integral multiple of K_a. Hence if F is relatively minimal model, F is rational (Proposition 9.1).

This completes the proof of the lemma. Note that Proposition 9.1 has been used only at the end of the proof, and hence only in the case $\nu = 1$, since in the case $\nu \geqq 2$ the non-existence of fundamental curves of $|nK_a|$ has already been established in the earlier part of the proof.

After these preliminaries we begin to treat the case $\nu = 2$. We assume that we have a surface F which is a relatively minimal model of Σ/k and that we have $p_a = P_2 = 0$, $(K^2) = 2$. We also assume that $|K_a|$ is irreducible (Proposition 7.3). Then $|K_a|$ has no base points and has dimension 2 (see proof of Lemma 10.1). We may also assume that $|K_a|$ has no fundamental curves, for otherwise F is rational (Lemma 10.1) and there is nothing to prove.

Under these assumptions a very short proof of the rationality of F can be given, *provided the characteristic p is different from* 2. The proof is as follows:

The irreducible net $|K_a|$ defines a rational transformation $T: F \rightarrow S_2$ of F onto a projective plane, the T-transform of $|K_a|$ being the net of lines in S_2. Since $|K_a|$ has no base points, T is regular on F. Since $|K_a|$ has degree 2, the functions field Σ $(= k(F))$ is an algebraic extension, of degree 2, of the subfield $\Omega = k(S_2)$. Since $|K_a|$ has no fundamental curves, T^{-1} has no fundamental points in S_2. Since F is non-singular, it follows that F *is the normalization of S_2 in the field Σ* (F is a "double plane"). Since $p \neq 2$, the quadratic extension $\Sigma/k(S_2)$ is separable, and if x, y are affine coördinates in S_2, then the portion of F which lies over the affine (x, y)-plane is biregularly equivalent to a surface given by an equation of the form $z^2 = f(x, y)$, where $f(x, y)$ is a polynomial *free from multiple factors.* The curve $f(x, y) = 0$ is the affine part of the branch curve Δ of the double plane F. It follows from the equation $z^2 = f(x, y)$ that if l is a line in S_2 and C is the corresponding cycle in $|K_a|$ then a singular point of C necessarily corresponds to a multiple intersection of l with Δ. It follows that the general cycle in $|K_a|$ is a *non-singular* curve and is therefore of effective genus 1 (since $p(K_a)=1$). Since in an elliptic function field any function of order 2 has 4 branch points (provided the characteristic p is $\neq 2$), it follows that the branch curve Δ is of order 4. There exists at least one line l_0 in S_2 *which has no simple intersections with* Δ (this is an elementary fact from the theory of plane quartic curves). Then the corresponding cycle C_0 of $|K_a|$ will either have two double points or a tacnode (according as l_0 has two distinct intersections with C_0, both of multiplicity 2, or a single intersection of multiplicity 4); in both cases C_0 has the "equivalent" of *two* double points and therefore degenerates, i.e., C_0 is not a prime cycle. A prime component of C_0 cannot be linearly equivalent to an integral multiple of C_0, and therefore F is rational, by Proposition 9.1.

The above proof is not applicable to the case $p=2$. We shall now give another proof which is valid for any characteristic. This second proof is "constructive," for in it the surface F is explicitly constructed as a member of an algebraic system of surfaces having $p_a=P_2=0$ and $(K^2)=2$.

By Lemma 10.2 we have

$$\text{(2)} \qquad \dim|2K_a|=6, \qquad \dim|4K_a|=20.$$

We fix a prime cycle C_0 in $|K_a|$, defined over k. The functions ξ in Σ such that $(\xi)+C_0>0$ form a vector space (over k) of dimension 3 (since $\dim|K_a|=2$), and 1 is one such function. We fix two elements x, y in Σ such that $\{1, x, y\}$ is a basis of that vector space. From what was said above in the course of the proof of rationality of F in the case $p\neq 2$, Σ is a quadratic extension of $k(x,y)$, and x, y are affine coördinates of the projective plane S_2 of which F is the normalization (in Σ). Clearly, Σ is a simple extension of $k(x,y)$. We shall now introduce a particular primitive element of $\Sigma/k(x,y)$.

The functions ξ in Σ such that $(\xi)+2C_0>0$ form a vector space (over k) of dimension 7 (since $\dim|2K_a|=6$, by (2)). The 6 functions $1, x, y, x^2, xy, y^2$ belong to that space and are linearly independent over k. Let z be a function in that vector space which is linearly independent of the above 6 functions. *We assert that z is a primitive element of $\Sigma/k(x,y)$.*

We first show that *z is an integral function of x and y.* For this purpose we have only to show that if $\mathfrak{p}$ is any algebraic place of Σ/k such that $\mathfrak{p}(x)$ and $\mathfrak{p}(y)$ are both finite then also $\mathfrak{p}(z)$ is finite. Let P be the center of $\mathfrak{p}$ on F. It will be sufficient to show that $P\notin C_0$, since C_0 is the only polar prime divisor of the function z. Assume the contrary: $P\in C_0$. Let

$(x) = C_1 - C_0$, $(y) = C_2 - C_0$, where $C_1, C_2 \in |K_a|$. Since $\mathfrak{p}(x) \neq \infty$, $\mathfrak{p}(y) \neq \infty$ and since $P \in C_0$, P necessarily belongs to C_1 and C_2. Since C_0, C_1, C_2 are three linearly independent cycles of the net $|K_a|$, P necessarily belongs to every cycle in $|K_a|$, i.e., P is a base point of $|K_a|$, in contradiction with the fact that $|K_a|$ has no base points.

To show that z is a primitive element of $\Sigma/k(x, y)$ we have only to show that $z \notin k(x, y)$. Assume the contrary. Then z is a polynomial $f(x, y)$ in x, y, since z is an integral function of x, y. Let n be the degree of the polynomial f. *We shall show that* $n \leqq 2$, and this will produce the desired contradiction, since the 7 functions $1, x, y, x^2, xy, y^2, z$ are linearly independent over k (by our choice of z).

Assume $n > 2$ and let $f_n(x, y)$ be the sum of terms of $f(x, y)$ which are of degree n. Since $f(x, y) - f_n(x, y)$ has order $\geqq -(n-1)$ on C_0 and $f(x, y)$ has order -2 on C_0, the quotient $f_n(x, y)/x^n$ must have positive order at the prime divisor C_0 (since $n > 2$). Hence, if we denote by c the C_0-residue of y/x then $f_n(1, c) = 0$, i.e., c is algebraic over k, and so $c \in k$ (note that $c \neq 0, \infty$ since both x and y have order -1 on C_0). The function $y - cx$ does not, therefore, become infinite at the prime divisor C_0 and thus has no polar prime divisors at all. Hence $y - cx \in k$, in contradiction with the linear independence of $1, x, y$ over k.

We have thus established that z is a primitive element of $\Sigma/k(x, y)$ (and, moreover, that it is an integral function of x, y).

To find the irreducible equation for z over $k(x, y)$ we consider the 22 monomials $x^m y^n$ $(0 \leqq m + n \leqq 4)$, $x^m y^n z$ $(0 \leqq m + n \leqq 2)$ and z^2, where m and n are non-negative integers. If ω is any of these monomials then $(\omega) + 4C_0 > 0$. Since $\dim |4K_a| = 20$ and since therefore the vector space of the functions ξ such that $(\xi) + 4C_0 > 0$ has dimension 21, it follows that *the above 22 monomials are linearly dependent over k.* The first 21 of those monomials (i.e., those which are different from z^2) are linearly independent over k (since x and y are algebraically independent over k and since $z \notin k(x, y)$). Hence the relation of linear dependence between the above 22 monomials has the following form:

$$z^2 + g_2(x, y) z + g_4(x, y) = 0, \tag{3}$$

where g_2 and g_4 are polynomials in x, y, of degree not greater than 2 and 4 respectively. The equation (3) is irreducible (since $z \notin k(x, y)$). We shall denote by g the polynomial on the left-hand side of (3) as well as the (projective) surface in S_3 defined by the equation (3). The surface g is birationally equivalent to F since $\Sigma = k(x, y, z)$. *We have therefore to prove the rationality of the surface g.*

We shall now make three remarks.

Remark 1. We denote by g_a the affine part of the surface, i.e., the set of points of g which are at finite distance in the (x, y, z)-space. We assert that *the affine surface g_a is normal.* To see this we have to show that $k[x, y, z]$ is the integral closure of $k[x, y]$ in Σ $(=k(x, y, z))$; or equivalently, that $\{1, z\}$ form an integral basis, over $k[x, y]$, of the integral closure of $k[x, y]$ in $k(x, y, z)$. Let ω be any function in $k(x, y, z)$ which is integral over $k[x, y]$. Since C_0 is the only polar curve of x and y on F, C_0 is also the only polar curve of ω on F. Hence there exists an integer $n \geqq 1$ such that $(\omega) + nC_0 \geqq 0$ (on F). Now, we have $\dim |nK_a| = n(n+1)$ [see (1)], and hence the vector space V of all functions ω in Σ such that $(\omega) +)nC_0 \geqq 0$ (on F) has dimension $n(n+1)+1$. On the other hand, the functions ζ of the form $f_n(x, y) + zf(x, y)$, where f_a and f_{n-2} are polynomials in x, y, of degree not greater than n and $n-2$ respectively, belong to that vector space. Since x, y are algebraically independent and since $z \notin k(x, y)$, the dimension of the space of functions ζ is $(n+1)(n+2)/2 + (n-1)n/2 = n(n+1)+1$. Hence this vector space coincides with V. Thus ω has the above form $f_n(x, y) + zf_{n-2}(x, y)$, and this proves our assertion.

Thus g_a is the normalization of the affine (x, y)-plane in the field Σ. Since F is the normalization of the *projective* (x, y)-plane in Σ, it follows that—up to biregular equivalence—the affine surface g_a can be identified with the part of the surface F which consists of those points P at which x and y are both finite (i.e., the point P of F such that $x, y \in \mathfrak{o}_P$). Since $(x) = C_1 - C_0$, $(y) = C_2 - C_0$, it follows that the part of F in question is $F - C_0$. *Thus g_a can be identified with $F - C_0$.*

If we now set $x' = 1/x$, $y' = y/x$, $z' = z/x^2$, then we find $(x') = C_0 - C_1$, $(y') = C_2 - C_1$, $(z') + 2C_1 > 0$. The three functions $x', y', z'$are connected by the equation

$$(3') \qquad z'^2 + g'_2(x', y')z' + g'_4(x', y') = 0,$$

where $g'_2(x', y') = x'^2 g_2(1/x', y'/x')$, $g'_4(x', y') = x'^4 g_4(1/x', y'/x')$. If we denote by g' the (projective) surface defined by (3′) and by g'_a the affine part of that surface (in the affine (x, y, z)-space), then we have, by a similar argument, that *the affine surface g'_a can be identified with $F - C_1$.*

Finally, if we set $x'' = x/y$, $y'' = 1/y$, $z'' = z/y^2$, then x'', y'', z'' are connected by the equation

$$(3'') \qquad z''^2 + g''_2(x'', y'')z'' + g''_4(x'', y'') = 0,$$

where $g''_2(x'', y'') = y''^2 g_2(x''/y'', 1/y'')$, $g''_4(x'', y'') = y''^4 g_4(x''/y'', 1/y'')$, and the *affine* surface g''_a defined by (3″) can be identified with $F - C_2$. Since C_0, C_1, C_2 have no common points, it follows that *the three affine surfaces* g_a, g'_a, g''_a *cover* F.

Remark 2. The surface g is of order 4, unless g_4 is of degree $\leqq 3$ in which case g is of order $\leqq 3$. In the latter case the surface g is certainly rational since it cannot be an elliptic cubic cone (the birationally equivalent surface F having arithmetic genus zero). If g is a quartic surface, the point O at infinity on the z-axis is a double point of g. The plane sections of g passing through O have a tacnode at O (i.e., they have a double point infinitely near the double point O). This is easily seen by a direct analysis of the singular point O. [In the case of characteristic zero the general plane section of g through O is an elliptic quartic curve, since it corresponds birationally to a general anticanonical curve K_a on F. Since g has no singularities at finite distance (in view of the inclusion $g_a \subset F$) the general plane section of F has no singular points outside of O, and this shows again that O must be a tacnode of the section.] In other words, O is a tacnodal point of the surface g. The plane at infinity is the tangent plane of g at O, and the section of g with that plane decomposes into 4 lines through O (these are obtained by equating to zero the highest degree homogeneous component g_{44} of the polynomial $g_4(x, y)$). Conversely, it is easily seen that if a quartic surface in S_3 has a tacnodal point O then the equation of that surface takes the form (3) provided we take O as the point at infinity on the z-axis and take the tangent plane of the surface at O as the plane at infinity. The proof developed in this section will therefore establish the fact that *any quartic surface in* S_3 *which has a tacnodal point is either rational or is birationally equivalent to the ruled surface of genus* 1 (*say, an elliptic cubic cone*).

Remark 3. The linear system L on F which corresponds to the system of plane sections of g is defined by the 4 functions $1, x, y, z$. Since we have on F: $(x) = C_1 - C_0$, $(y) = C_2 - C_0$ and $(z) = D - 2C_2$, where $D \in |2K_a|$, L is generated by the 4 (linearly independent) cycles $2C_0, C_1 + C_0, C_2 + C_0, D$. (Note that L has no fixed components, since the *irreducible* curve C_0 cannot be a component of D; otherwise D would be of the form $C_0 + C_3$, where $C_3 \in |K_a|$, and D would belong to the system generated by $2C_0, C_1 + C_0$, $C_2 + C_0$, and this is not the case.) The base points of L are the 4 (generally distinct) intersections of C_0 and D. These intersections are the fundamental points of the birational transformation $T: F \to g$. To each of these points corresponds on g one of 4 lines in which the plane at infinity cuts g. The

12

irreducible curve C_0 is a fundamental curve of L (since C_0 is a fixed component of the two-dimensional subsystem of L which is generated by the cycles $2C_0, C_1 + C_0, C_2 + C_0$). This curve C_0 is the only exceptional curve of T and to it corresponds on g the point O, which is thus a fundamental point of T^{-1}.. It is easily checked that O is the only fundamental point of T^{-1} and that the intersections of C_0 with D are the only fundamental points of T.

Note that the above considerations show that with our choice of C_0 (*C_0-irreducible*) the surface g must be of order exactly 4 (and not less), for L—a subsystem of the system $|2K_a|$, which has degree 8—has exactly 4 base points (or the equivalent of 4 base points if C_0 and D have some multiple intersections). The same conclusion holds true if C_0 is reducible, provided C_0 and D have no common components. Suppose, however, that C_0 is reducible and that one of its irreducible components, say Δ, is also a component of D. Then Δ is a fixed component of L, and the defining linear system of the birational transformation $T: F \rightarrow g$ is not L but the system L_1 obtained from L by deleting the fixed component Δ from all the cycles in L (we assume here, for simplicity, that L_1 has no fixed component). In this case it is easily found that L_1 has degree < 4, whence g is a surface of order $\leqq 3$. In this case, then, the rationality of F follows at once. We observe, however, that since we have assumed F to be a relatively minimal model, the rationality of F follows already, in view of Lemma 9.1, from the existence of a cycle $C_0 \in |K_a|$ which is not prime. (End of the remarks.)

As was stated at the end of Remark 2, we propose to prove in this section the following proposition:

PROPOSITION 10.3. *Every irreducible quartic surface g given by an equation of the form (2) is birationally equivalent to a cubic surface in S_3 (or to a surface of order < 3).*

This will establish Castelnuovo's criterion of rationality in the case $(K^2) = 2$, since every cubic surface, other than an elliptic cubic cone, is rational and since F—which has arithmetic genus zero—cannot be birationally equivalent to an elliptic cubic cone.

The intersection of the surface g with a plane is a cycle of degree 4. The proof of Proposition 10.3 will be based on the following lemma:

LEMMA 10.4. *If there exists a plane π through the tacnodal point O of g different from the plane at infinity, such that the intersection $g \cdot \pi$ splits into two cycles of degree 2, both passing through O, then the surface g is birationally equivalent to a surface of order $\leqq 3$.*

Proof. Let p_1, p_2, p_3, p_4 be the 4 lines in which the surface g is cut by the plane at infinity (the lines p_i need not be distinct). Subject to an affine change of the nonhomogeneous coördinates x and y, we may assume that π is the plane $x=0$. Setting $g_2(0,y)=h_2(y)$, $g_4(0,y)=h_4(y)$, we see that the intersection cycle $g\cdot\pi$ consists of the cycle Γ defined by the equation

$$z^2+h_2(y)z+h_4(y)=0 \tag{4}$$

and perhaps of one (and only one) of the line p_i (counted to a suitable multiplicity). If $h_4(y)$ is of degree >2 then equation (4) must be reducible, for otherwise Γ would be a prime cycle of degree $\geqq 3$, and the intersection cycle $g\cdot\pi$ could not split into two cycles of degree 2. Assume now that $h_4(y)$ is of degree $\leqq 2$. Then Γ is a cycle of degree 2, and Γ *does not pass through the point* O (the point at infinity on the z-axis). The intersection cycle $g\cdot\pi$ is then of the form $\Gamma+2p_1$ (where p_1 is one of the lines p_i). Since $O\notin\Gamma$, $g\cdot\pi$ could not be a sum of two cycles of degree 2, both passing through O, unless Γ splits into two lines q_1, q_2 (in which case $g\cdot\pi$ has the desired decomposition $(q_1+p_1)+(q_2+p_1)$). Hence also in this case equation (4) must be reducible. Let then

$$z^2+h_2(y)z+h_4(y)=(z+\psi(y))(z+\psi'(y)),$$

where $\psi(y)$ and $\psi'(y)$ are polynomials of degree $\leqq 2$. We now replace z by the element $z+\psi(y)$ (this amounts to a birational transformation $x'=x$, $y'=y$, $z'=z+\psi(y)$).[7] Then the equation (3) takes the form

$$z^2+\bar g_2(x,y)z+x\bar g_3(x,y)=0,$$

where $\bar g_2(x,y)$ and $\bar g_3(x,y)$ are polynomials of degree at most 2 and 3 respectively. Thus we may assume that *in the original equation* (3) *the polynomial* $g_4(x,y)$ *is divisible by* x. But then the birationally equivalent surface g' defined by equation (3′) (Remark 1) is a surface of degree $\leqq 3$ (since $g'_4(x',y')$ has now degree $\leqq 3$). This establishes Lemma 10.4.

The surface g belongs to an irreducible algebraic system N/k of quartic surfaces in S_3 whose general member G is defined by an equation of the form similar to (3):

$$G\colon z^2+G_2(x,y)z+G_4(x,y)=0,$$

[7] We observe—going back to the original surface F—that the conditions imposed on the element z (i. e., the divisor of the function z to be a mutiple of $-2C_0$ and z to be linearly independent of the functions $1, x, y, x^2, xy, y^2$) do not determine z uniquely. Any function of the form $cz+\psi(x,y)$, where $0\neq c\in k$ and $\psi(x,y)$ is any polynomial of degree $\leqq 2$, will satisfy the same conditions.

where $G_2(x,y)$ and $G_4(x,y)$ are polynomials of degree 2 and 4 respectively, with coefficients which are algebraically independent over k. Note that $\dim N = 21$.

We shall prove that *there exists a plane* σ *such that* σ *and the intersection cycle* $G\cdot\sigma$ *satisfy the conditions of Lemma* 10.4 (*with* g *and* π *replaced by* G *and* σ). We assume for a moment that this assertion has already been proved and we show how Projosition 10.3 can be derived from this assertion.

The surface g is a specialization of G, over k. Let $(G,\sigma) \xrightarrow{k} (g,\pi)$ be an extension of the specialization $G \xrightarrow{k} g$. Then π is a plane through O, and since $G\cdot\sigma$ is a sum of two cycles of degree 2 passing through O, the same is true of the intersection cycle $g\cdot\pi$ (note that O is rational point over k). If π is not the plane at infinity, then Proposition 10.3 is established, in view of Lemma 10.4. If, however, π is the plane at infinity, then we replace both G and g by their birational transforms G' and g' under the transformation $x' = 1/x$, $y' = y/x$, $z' = z/x^2$. Then also G' is a general member of N/k. Furthermore, if the equation of σ is $ax + by + c = 0$ we denote by σ' the plane $a + by' + cx' = 0$ (the transform of σ under the transformation $x' = 1/x$, $y' = y/x$) and by π' the plane $x' = 0$. It is clear that σ' and G' are in the same relationship to each other as are σ and G (i.e., the conditions of Lemma 10.4 are satisfied when g and π are replaced by G' and σ') and that (g',π') is a specialization of (G',σ') over k. Since π' is not the plane at infinity, g'—and therefore also g—is birationally equivalent to a surface of order $\leqq 3$. This completes the proof of Proposition 10.3.

We now proceed to the proof of the assertion that there exists a plane σ such that σ and the cycle $G\cdot\sigma$ satisfy the conditions of Lemma 10.4 (with g and π replaced by G and σ respectively).

The algebraic system N, of which G is a general member over k, contains an irreducible subsystem N' whose general member over k is the surface H defined by

$$(5) \qquad H: [z + \phi_2(x,y)]^2 + A_2(x,y)[z + \phi_2(x,y)] + A_1(x,y)A_3(x,y) = 0,$$

where A_1, A_2, A_3 and ϕ_2 are polynomials in x, y, of degree 1, 2, 3 and 2 respectively, with coefficients which are algebraically independent quantities over k. We shall denote also by H the left-hand side of the equation (5). Then

$$(6) \qquad H = z^2 + H_2(x,y)z + H_4(x,y),$$

where

$$H_2(x,y) = A_2(x,y) + 2\phi_2(x,y), \tag{7}$$

$$H_4(x,y) = A_1(x,y)A_3(x,y) + A_2(x,y)\phi_2(x,y) + [\phi_2(x,y)]^2, \tag{8}$$

which shows that H is indeed a specialization of G over k and that consequently $N' \subset N$. Now, the surface H satisfies the conditions of Lemma 10.4 (with g replaced by H and π replaced by the plane $A_1(x,y) = 0$), for the intersection of H with the plane $A_1(x,y) = 0$ splits into two cycles of degree 2, defined respectively by the equations $z + \phi_2(x,y) = 0$ and $z + \phi_2(x,y) + A_2(x,y) = 0$, and both these cycles pass through the tacnodal point O. So everything will be proved if we show that *also the surface H is (as well as G) a general member of N/k* (and that therefore N' actually coincides with N).

From now on we shall denote by G^* (resp., H^*) the point, in the affine space of dimension 21, whose (non-homogeneous) coördinates are the coefficients of $G_2(x,y)$ and $G_4(x,y)$ (resp., of $H_2(x,y)$ and $H_4(x,y)$). We have $\dim G^*/k = 21$, *and what we have to prove is that also*

$$\dim H^*/k = 21. \tag{9}$$

We shall denote by A^* (resp., ϕ^*) the point in the affine space of dimension 25 (resp., of dimension 6) whose (non-homogeneous) coördinates are the coefficients of the polynomials $A_1(x,y)$, $A_2(x,y)$, $A_3(x,y)$, $\phi_2(x,y)$ (resp., of the polynomial $\phi_2(x,y)$). We have $\dim A^*/k = 25$, and we also know, by (7) and (8), that the point H^* is rational over the field $k(A^*)$. *Hence, in order to prove* (9), *we have to show* that

$$\dim A^*/k(H^*) = 4. \tag{10}$$

Since we have at any rate $\dim H^*/k \leqq 21$, it follows that $\dim A^*/k(H^*) \geqq 4$. So *all we have to prove is that*

$$\dim A^*/k(H^*) \leqq 4. \tag{11}$$

Let $A_1{}^*$ denote the point, in the affine 3-space, whose (non-homogeneous) coördinates are the coefficients of the linear polynomial $A_1(x,y)$ (note that this point is also rational over $k(A^*)$). We show first that

$$\dim A^*/k(H^*, A_1{}^*) \leqq 3. \tag{12}$$

We fix a (finite) specialization $\bar{A}^*$ of A^* over $k(H^*, A_1{}^*)$ *which is algebraic over $k(H^*, A_1{}^*)$*. The coördinates of the point $\bar{A}^*$ are the coefficients of certain polynomials $\bar{A}_1(x,y)$, $\bar{A}_2(x,y)$, $\bar{A}_3(x,y)$ and $\bar{\phi}_2(x,y)$ (of indicated degrees). We have $\bar{A}_1(x,y) = A_1(x,y)$ since $\bar{A}^*$ is a specialization of A^* over $k(A_1^*)$. By (7) and (8) we have

$$H_2(x,y) = \bar{A}_2(x,y) + 2\bar{\phi}_2(x,y),$$

$$(13) \qquad H_4(x,y) = A_1(x,y)\bar{A}_3(x,y) + \bar{A}_2(x,y)\bar{\phi}_2(x,y) + [\bar{\phi}_2(x,y)]^2,$$

whence upon setting

$$(14) \qquad q_2(x,y) = \bar{\phi}_2(x,y) - \phi_2(x,y),$$

we find, again by (7) and (8):

$$(15) \qquad A_2(x,y) = \bar{A}_2(x,y) + 2q_2(x,y),$$

$$(16) \qquad A_1(x,y)A_3(x,y) = A_1(x,y)\bar{A}_3(x,y) + \bar{A}_2(x,y)q_2(x,y) + [q_2(x,y)]^2.$$

From (16) it follows that $q_2(x,y)$ must satisfy the congruence

$$\bar{A}_2(x,y)q_2(x,y) + [q_2(x,y)]^2 \equiv 0 \quad (\operatorname{mod} A_1(x,y)).$$

Since q_2 must be a polynomial of degree $\leqq 2$ this congruence has only two general solutions q_2^*; they are given by $q_2^*(x,y) = A_1(x,y)p_1(x,y)$ and $q_2^*(x,y) = -\bar{A}_2(x,y) + A_1(x,y)p_1(x,y)$ respectively, where $p_1(x,y)$ is the general linear polynomial in x, y. Each of these solutions depends on three arbitrary parameters; more precisely, we have $\dim q_2^*/k(H^*, A_1^*) = 3$ (note that the coefficients of $\bar{A}_2(x,y)$ are algebraic over $k(H^*, A_1^*)$, by our choice of the specialiaztion $\bar{A}^*$). From (14), (15) and (16) we see that the coefficients of ϕ_2, A_2 and A_3 are rational over $k(\bar{A}^*, q_2^*)$. Since $\bar{A}^*$ is algebraic over $k(H^*, A_1^*)$, it follows that A^* is algebraic over $k(H^*, A_1^*, q_2^*)$. The inequality (12) now follows from the equality $\dim q_2^*/k(H^*, A_1^*) = 3$.

In view of (12), the inequality (11) will be established if we show that

$$(17) \qquad \dim A_1^*/k(H^*) \leqq 1.$$

It will be sufficient to show that the number of *planes* $\bar{A}_1(x,y) = 0$ (in the ambient 3-space of the surface H) such that the $\bar{A}_1(x,y)$ is a (finite) specialization (over $k(H^*)$, of the linear polynomial $A_1(x,y)$), is finite.

Now, any such plane shares with the plane $A_1(x,y) = 0$ the property that its intersection with the surface H splits into two cycles of degree 2, both passing through the tacnodal point 0 of H. Now, let $\bar{A}_1(x,y)$ be such a plane. The quadratic polynomial H in z factors, $\operatorname{mod}\bar{A}_1$, into a product $(z+\bar{\psi})(z+\bar{\psi}')$, where $\bar{\psi}$ and $\bar{\psi}'$ are polynomials in x, y, of degree 2 (see proof of Lemma 10.4). We have seen in the proof of Lemma 10.4 that the birational transformation

$$(18) \qquad \bar{x} = x/\bar{A}_1, \qquad \bar{y} = y/\bar{A}_1, \qquad \bar{z} = z + \psi/\bar{A}_1^2$$

transforms H into the cubic surface

$$C: \bar{z}^2 + C_2(\bar{x}, \bar{y})z + C_3(\bar{x}, \bar{y}) = 0,$$

where C_2 and C_3 are polynomials of degree 2 and 3 respectively. From the fact that A_1, A_2, A_3 and ϕ_2 are polynomials whose coefficients are algebraically independent over k it follows at once that C is the *general cubic surface* in S_3 (referred to a special system of coödinates). Now, it is clear that in the birational transformation (18) the system of plane section of H through O corresponds to the system of plane sections of C passing through the point at infinity on the $\bar{z}$-axis. If a plane π passing through O, different from the plane at infinity and the plane $\bar{A}_1 = 0$, intersects H in a sum of two cycles, of degree 2, both passing through O, then the section of C with the corresponding plane $\alpha\bar{x} + \beta\bar{y} + \gamma = 0$ is reducible and therefore contains a line of the cubic surface. Since the general cubic surface contains only a finite number of lines, the number of planes π with the above indicated property is finite. This completes the proof.

HARVARD UNIVERSITY.

REFERENCES.

[1] Y. Akizuki and H. Matsumura, "On the dimension of algebraic systems of curves with nodes on a non-singular surface," *Mem. Coll. Sci. Univ. Kyoto*, series A, vol. 30, n. 2 (1957), pp. 143-150.

[2] F. Enriques, *Le superficie algebriche*, Bologna, Zanichelli, 1949.

[3] H. Matsumura and M. Nagata, "On the algebraic theory of sheets of an algebraic variety," *Mem. Coll. Sci. Univ. Kyoto*, series A, vol. 30, n. 2 (1957), pp. 157-164.

[4] Y. Nakai, "A property of an ample linear system on a non-singular variety," *ibid.*, series A, vol. 30, n. 2 (1957), pp. 151-156.

[5] ———, "On the characteristic linear systems of algebraic families," forthcoming in the *Illinois Journal of Mathematics*.

[6] O. F. G. Schilling and O. Zariski, "On the linearity of pencils of curves on algebraic surfaces," *American Journal of Mathematics*, vol. 60 (1938), pp. 320-324.

[7] F. Severi, *Vorlesungen über algebraische Geometrie*, Teubner, Berlin, 1921.

[8] C. C. Tsen, "Divisionalgebren über Funktionenkörpern," *Nach. Ges. Wiss.*, Göttingen,1933.

[9] A. Weil, "Variétés abéliennes et courbes algébriques," *Actual. Scient. et Indust.*, n. 1064, Paris, 1948.

[10] ———, "Sur les critères d'équivalence en géométrie algébrique," *Math. Ann.*, vol. 128 (1954), pp. 95-127.

[11] O. Zariski, "Algebraic surfaces," *Ergebnisse der Mathematik und ihrer Grenzgebiete* 3, n. 5 (1935).

[12] ———, "Reduction of the singularities of algebraic three-dimensional varieties," *Annals of Mathematics*, vol. 45 (1944), pp. 472-542.

[13] ———, "Theory and applications of holomorphic functions on algebraic varieties over arbitrary ground fields," Mem. Amer. Math. Soc., n. 5 (1951).

[14] ———, "Introduction to the problem of minimal models in the theory of algebraic surfaces," forthcoming in the *Publications of the Mathematical Society of Japan* (referred to as IP).

Reprinted from Illinois Journal of Mathematics
Vol. 2, No. 3, September 1958
Printed in U.S.A.

ON CASTELNUOVO'S CRITERION OF RATIONALITY $p_a = P_2 = 0$ OF AN ALGEBRAIC SURFACE

To Emil Artin on his sixtieth birthday

BY
Oscar Zariski[1]

1. Introduction

Let F be a nonsingular (irreducible) algebraic surface over an algebraically closed ground field k. A theorem of Castelnuovo asserts that if the arithmetic genus p_a and the bigenus P_2 of F are both zero then F is a rational surface. This theorem has now been proved for fields k of arbitrary characteristic p, except in the case $(K^2) = 1$, where K is a canonical divisor on F.[2] In our cited paper MM (see footnote 2) we have stated that we have also a proof for the case $(K^2) = 1$, and in the present paper we shall give this proof.

An immediate consequence of Castelnuovo's criterion of rationality is the well-known theorem of Castelnuovo on the rationality of plane involution. This theorem, in the case of arbitrary characteristic, is to be stated as follows:

Let $k(x, y)$ be a purely transcendental extension of an algebraically closed field k, of transcendence degree 2, and let Σ be a field between k and $k(x, y)$, also of transcendence degree 2 over k.[3] *If $k(x, y)$ is a separable extension oj Σ, then Σ is a pure transcendental extension of k.*

We shall show by an example that the condition of separability of $k(x, y)/\Sigma$ is essential.

2

We shall make use of results established in MM for the case of surfaces F for which $P_a = P_2 = 0$ and $(K^2) > 0$. If $(K^2) = 1$, then the Riemann-Roch inequality shows that the dimension of the anticanonical system $| K_a | (= | -K |)$ is $\geqq 1$. If $| K_a |$ is reducible, then F is rational, by Proposition 7.3 of MM. *We shall therefore assume that* $| K_a |$ *is irreducible.* In that case we have $\dim | K_a | = 1$ (MM, Lemma 10.1), i.e., $| K_a |$ is a pencil; it has a single base point O, every member K_a of $| K_a |$ has a simple point at O, and

Received February 5, 1958.

[1] This research was supported in part by the United States Air Force through the Office of Scientific Research of the Air Research and Development Command.

[2] See our recent paper *The problem of minimal models in the theory of algebraic surfaces*, Amer. J. Math., vol. 80 (1958), pp. 146–184. This paper will be referred to in the sequel as MM.

[3] The theorem is also true if Σ/k has transcendence degree 1 (without any assumption on separability), but in that case the theorem is an easy consequence of the theorem of of Lüroth.

two distinct K_a's in $|K_a|$ have distinct tangents at O (all this follows from $(K^2) = 1$).

From the irreducibility of $|K_a|$ follows (see MM, Lemma 10.2) that for each $n > 1$ the system $|nK_a|$ is irreducible, is free from base points, and has dimensions $n(n+1)/2$. We shall use the systems $|2K_a|$, $|3K_a|$ and $|6K_a|$. We have

$$\dim |2K_a| = 3, \tag{1}$$

$$\dim |3K_a| = 6, \tag{2}$$

$$\dim |6K_a| = 21. \tag{3}$$

An irreducible curve D on F will be a fundamental curve of $|nK_a|$, $n > 1$, if and only if $(K_a \cdot D) = 0$ (since $|nK_a|$, $n > 1$, has no base points). There will then exist a member K_a^0 of the pencil $|K_a|$ such that D is a component of K_a^0. We cannot have $K_a^0 = D$ since $(K_a^0 \cdot K_a) = 1$ while $(D \cdot K_a) = 0$. Hence K_a^0 is not a prime cycle. Now we prove the following:

PROPOSITION 1. *If K_a^0 is a member of $|K_a|$ which is not a prime cycle, then some prime component of K_a^0 is an exceptional curve of the first kind.*

Proof. For every prime component E of K_a^0 we must have $p(E) = 0$ (MM, Lemma 7.2). Since $|K_a|$ is irreducible (whence E is not a fixed component of $|K_a|$), we have $(K_a \cdot E) \geqq 0$. Since $(K_a^2) = 1$, it follows that there exists one and only one prime component E of K_a^0 such that $(K_a \cdot E) > 0$, and for that component E we must have $(K_a \cdot E) = 1$. Since $p(E) = 0$ and since $(X^2) - 2p(X) + 2 = (K_a \cdot X)$ for every cycle X on F, it follows that $(E^2) = -1$. Thus E is an exceptional curve of the first kind. QED.

The presence of an irreducible exceptional curve E of the first kind implies that F can be transformed birationally into a surface F' which is strictly dominated by F and on which the self-intersection number of a canonical divisor is 2. Hence F' (and therefore also F) is rational, by the case $(K^2) = 2$.

We may therefore assume that the systems $|nK_a|$, $n > 1$, are free from fundamental curves. By Proposition 1, this is equivalent to assuming that *each member of $|K_a|$ is a prime cycle.*

We summarize our assumptions concerning the nonsingular surface F:

(A) $p_a = P_2 = 0$; $(K^2) = 1$.

(B) *Every member of $|K_a|$ is a prime cycle.*

Our proof of the rationality of F will consist in showing that *under the assumptions* (A) *and* (B) *the surface F carries an exceptional curve of the first kind* (whence F is not a relatively minimal model). The rationality of F follows then by the case $(K^2) \geqq 2$. As in the case $(K^2) = 2$ (see MM, §10), so also in the present case, our method of proof will consist in constructing the entire algebraic family of surfaces satisfying assumptions (A) and (B).

3

Let t be a parameter of the pencil $| K_a |$:

$$(t) = C - C_0, \tag{4}$$

where $C, C_0 \in | K_a |$. For any $n \geqq 1$ we denote by $\mathfrak{L}(nC_0)$ the space of functions ξ in $k(F)$ such that $(\xi) + nC_0 \geqq 0$. We have $1, t, t^2 \in \mathfrak{L}(2C_0)$, and by (1), we have $\dim \mathfrak{L}(2C_0) = 4$. We choose an element x in $\mathfrak{L}(2C_0)$ such that $\{1, t, t^2, x\}$ is a basis of $\mathfrak{L}(2C_0)$ over k. Since $| 2K_a |$ is irreducible (therefore not composite with a pencil), the field $k(x, t)$, generated over k by the homogeneous coordinates of the point $(1, t, t^2, x)$ in S_3, cannot be of transcendence degree 1. *Hence x and t are algebraically independent over k*, and $k(F)$ is an algebraic extension of $k(x, t)$.

The locus, over k, of the point $(z_0, z_1, z_2, z_3) = (1, t, t^2, x)$ is the cone W: $z_0 z_2 - z_1^2 = 0$, and this cone is a rational transform of F, the plane sections of W corresponding to the cycles of $| 2K_a |$. Since W has order 2 and $| 2K_a |$ has degree 4, it follows that[4]

$$[k(F):k(x, t)] = 2. \tag{5}$$

The 6 elements $1, t, t^2, t^3, x, xt$ belong to $\mathfrak{L}(3C_0)$, and by (2) we have $\dim \mathfrak{L}(3C_0) = 7$. We can therefore find an element y in $\mathfrak{L}(3C_0)$ such that $\{1, t, t^2, t^3, x, xt, y\}$ is a basis of $\mathfrak{L}(3C_0)$ over k.

---

[4] It is at this stage that a short proof of the rationality of F can be given, provided that the characteristic p of k is different from 2. We shall outline here this proof.

Since $| 2K_a |$ has no base points and no fundamental curves and since F is nonsingular (hence normal), F is a normalization of the cone W in the field $k(F)$ (F is a double covering of the cone W). The pencil $| K_a |$ on F corresponds to the pencil of generators of the cone, each K_a being a double covering (*a priori*, not necessarily a normalization) of the corresponding generator. Let D be the branch curve, on W, of the double covering $W \rightarrow F$. It can be shown that D does not pass through the vertex of W. The general generator g of the cone W cannot have a contact P with D such that the intersection multiplicity of D and g at P is > 2, for in the contrary case it is easily seen that the general cycle K_a would not be prime (one must remember that $p(K_a) = 1$). It cannot have a contact with intersection multiplicity 2 since $p \neq 2$. From this it follows that the general K_a is *nonsingular and hence is elliptic*. The general generator g of the cone W must carry 4 branch points of the double covering $g \rightarrow K_a$. It is not difficult to see that one of the branch points is at the vertex O of the cone W (it is an isolated branch point of the covering $W \rightarrow F$, since $O \notin D$). Hence $(g \cdot D) = 3$, and therefore *D is a curve of order 6*. It can be shown that D is in fact complete intersection of W with a cubic surface. By using this fact it is possible to derive the existence of a tritangent plane π of D. The cycle in $| 2K_a |$ which corresponds to the plane section $W \cdot \pi$ splits then into two prime cycles D, E, neither one of which is a member of $| K_a |$. This shows that there exist cycles on F which are not linearly equivalent to an integral multiple of K_a, and thus the rationality of F follows from Proposition 9.1 of MM. As a matter of fact, the curves D, E are necessarily exceptional curves of the first kind; this is proved in the beginning of §7. Thus F is not a relatively minimal surface.

If $\mathfrak{p}$ is an (algebraic) place of $k(F)/k$ such that $t\mathfrak{p}$ and $x\mathfrak{p}$ are finite, and if P is the center of $\mathfrak{p}$ on F, then P cannot belong to C_0, for in the contrary case P would have to lie on each of the cycles $(t) + 2C_0$, $(t^2) + 2C_0$, $(x) + 2C_0$, and thus P would be a base point of $|2K_a|$. Since $P \notin C_0$ it follows that also $y\mathfrak{p} \neq \infty$. *Hence y is integral function of t and x.*

I assert that $y \notin k(x, t)$ (whence y is a primitive element of $k(F)/k(x, t)$; see (5)). For in the contrary case y would be a polynomial in x, t, say $y = f(x, t)$. Let $c_0 t^m + c_1 t^{m-2}x + c_2 t^{m-4}x^2 + \cdots$ be the sum of terms ct^ix^j in $f(x, t)$ for which $m = i + 2j$ is maximum. Were $m > 3$, then there would have to be at least two such terms (since t^ix^j is infinite on C_0 to the order $i + 2j$, while y is infinite to the order 3 on C_0), and the C_0-residue α of x/t^2 would have to satisfy the equation $c_0 + c_1\alpha + c_2\alpha^2 + \cdots = 0$. Thus $\alpha \in k$, and $x - \alpha t^2$ would belong to $\mathfrak{L}(C_0)$, i.e., $x - \alpha t^2$ would be linearly dependent on $1, t$, in contradiction with the linear independence of $1, t, t^2, x$ over k. Hence $m = 3$ and $y = f_3(t) + xf_1(t)$, where f_1 and f_3 are polynomials of degree 1 and 3 respectively. This contradicts the linear independence of $1, t, t^2, t^3, x, xt, y$ over k.

To find the irreducible equation (of degree 2) for y over $k(x, t)$, we observe that the 23 functions

$$(6) \qquad \omega_\nu = t^qx^ry^s, \qquad 0 \leqq q + 2r + 3s \leqq 6 \quad (0 \leqq q, r, s),$$

belong to the vector space $\mathfrak{L}(6C_0)$, and that, by (3), this space has dimension 22. Hence the above 23 functions are linearly dependent. A relation of linear dependence between these functions yields a relation of algebraic dependence between t, x, and y, of degree $\leqq 2$ in y. Since $y \notin k(t, x)$, y^2 must be present in the relation, and thus we find that the equation of algebraic (and integral) dependence for y over $k[x, t]$ has the following form:

$$(7) \qquad g = y^2 + [g_3(t) + xg_1(t)]y + [g_6(t) + g_4(t)x + g_2(t)x^2 + g_0x^3] = 0,$$

where $g_i(t)$ is a polynomial of degree $\leqq i$ (with coefficients in k). In particular, the coefficient g_0 of x^3 is a constant. It is important to observe that

$$(8) \qquad g_0 \neq 0.$$

To see this we note that if ω is any of the monomials ω_ν in (6) *other than* y^2 and x^3, then either C_0 or C is a component of the positive cycle $(\omega) + 6C_0$, and hence the base point O of the pencil $|K_a|$ belongs to this cycle. Were x^3 missing in (7) it would then follow that O belongs also to the cycle $(y) + 3C_0$. Since O also belongs to the cycles $(t^i) + 3C_0$ $(i = 0, 1, 2, 3)$, $(x) + 3C_0$, $(xt) + 3C_0$, it would then follow that O belongs to each cycle in $|3K_a|$, in contradiction with the fact that $|3K_a|$ has no base points. In the sequel we shall set $g_0 = 1$ (this amounts to replacing x by a constant multiple of x).

Since equation (7) is irreducible, it is the only relation of linear dependence between the monomials ω_ν in (6). It follows that these monomials span the entire space $\mathfrak{L}(6C_0)$. We denote by G the locus, over k, of the point in the projective space S_{22} whose homogeneous coördinates are the ω_ν (actually G lies in S_{21}, since the ω_ν are linearly independent). Thus G is a rational transform of our surface F, and is obtained by having the linear system $|6K_a|$ cut out by hyperplanes. The surface G is a *birational* transform of F, since $k(t, x, y) = k(F)$. Since $|6K_a|$ has no base points and no fundamental curves and since F is nonsingular (hence normal), it follows that *F is a normalization of G.* We shall show, however, that *G is itself normal*, and from this it will follow that G and F are biregularly equivalent surfaces and that, consequently, G is *nonsingular*. To show the normality of G we shall show that for any $n \geqq 1$ it is true that the linear system L_n cut out on G by the hypersurfaces of order n, in the ambient space S_{21} of G, coincides with the complete system $|6nK_a|$ (and hence G is arithmetically normal; note that $L_1 = |6K_a|$, by the definition of G, and that consequently $L_n \subset |6nK_a|$). If ρ_n is the number of linearly independent monomials ω' of the form $t^\alpha x^\beta y^\gamma$ such that $\alpha + 2\beta + 3\gamma \leqq 6n$ (α, β, γ nonnegative integers), then $\dim L_n = \rho_n - 1$. Since y^2 is linearly dependent on the monomials $t^q x^r y^s$ such that $q + 2r + 3s \leqq 6$ and $s = 0, 1$, it follows that the monomials ω' are linearly dependent on these particular monomials ω' for which γ is either zero or 1. An easy computation shows that the number of these particular monomials ω' is $3n(6n + 1) + 1$, and these monomials are linearly independent since t and x are algebraically independent over k and since $y \notin k(t, x)$. Hence $\dim L_n = 3n(6n + 1)$, i.e., $\dim L_n = \dim |6nK_a|$, and this proves our assertion.

4

In this section we shall retrace in reverse the procedure of the preceding section. We shall start with an affine surface g in S_3 defined by an equation of the form (7). We denote by G the projective surface in S_{22} which is the locus, over k, of the point whose homogeneous coördinates are the monomial ω_ν given in (6) (actually G lies in an S_{21}). We make the following two assumptions:

(A′) The coefficient g_0 of x^3 in (7) is different from zero (and we shall assume that $g_0 = 1$).

(B′) The surface G is nonsingular.

Proposition 2. *Under assumptions* (A′) *and* (B′), *the surface G satisfies conditions* (A) *and* (B) *of Section* 2.

Proof. If L_n denotes the system cut out on G by the hypersurfaces of order n, then—as was shown in the preceding section—we have $\dim L_n = 18n^2 + 3n$. Since the constant term of this quadratic polynomial is zero, it follows that *the arithmetic genus of G is zero.*

For the rest of the proof (and also for other applications which will be made in subsequent sections) we shall exhibit a suitable covering of G by three affine representatives.

We number the monomials ω_ν in (6) so as to have $\omega_0 = 1$, $\omega_1 = t^6$, $\omega_2 = y^2$. Let G_j $(j = 0, 1, 2)$ be the affine representative of G on which all the quotients ω_ν/ω_j are regular. It is clear that G_0 can be identified with the given affine surface g in S_3. We set

$$(9) \quad t' = 1/t \ (= t^5/\omega_1), \qquad x' = x/t^2 \ (= xt^4/\omega_1), \qquad y' = y/t^3 \ (= yt^3/\omega_1).$$

Then t', x', y' are among the quotients ω_ν/ω_1, and we see at once that all the quotients ω_ν/ω_1 are monomials in t', x', and y'. Hence G_1 can be identified with the affine surface g' in S_3 which is the locus, over k, of the point (t', x', y'). The equation of g' has the same form as that of g:

$$g' = y'^2 + [g_3'(t') + x'g_1'(t')]y' + [g_6'(t') + g_4'(t')x' + g_2'(t')x'^2 + x'^3] = 0,$$

where

$$g_i'(t') = t'^i g_i(1/t'), \qquad 1 \leqq i \leqq 6.$$

As to the affine representative G_2 of G, we introduce the functions

$$(10) \quad \tau = xt/y \ (= xty/\omega_2), \qquad \xi = x/y \ (= xy/\omega_2), \qquad \eta = x^3/y^2 \ (= x^3/\omega_2),$$

and we denote by g'' the locus of (τ, ξ, η) over k. Since τ, ξ, η are regular on G_2, the affine surface g'' is a regular (and obviously birational) transform of G_2. On the other hand, we have, for all nonnegative integers q, r and s such that $q + 2r + 3s \leqq 6$,

$$t^q x^r y^s/y^2 = \tau^q \xi^{6-q-2r-3s} \eta^{r-2+s}.$$

Hence the functions which are regular on G_2 are also regular at all points of g'' where $\eta \neq 0$. These points of g'' form an open subset which we shall denote by g_0''. Thus g_0'' can be identified with a part of G_2.

We note that g'' has the following equation:

$$(11) \quad \begin{aligned} g'' = \eta^3 + [1 + g_1''(\tau, \xi) + g_2''(\tau, \xi)]\eta^2 \\ + [g_3''(\tau, \xi) + g_4''(\tau, \xi)]\eta + g_6''(\tau, \xi) = 0, \end{aligned}$$

where

$$g_i''(\tau, \xi) = \xi^i g_i(\tau/\xi), \qquad 1 \leqq i \leqq 6.$$

We shall show that

$$(12) \qquad G = g \cup g' \cup g_0''.$$

In fact, *we shall show that the point* $\xi = \tau = 0$, $\eta = -1$ *is the only point of* g_0'' *which is not covered by* $g \cup g'$.

Let v be any valuation of $k(G)/k$ whose center on G belongs neither to G_0 nor to G_1, i.e., let

$$\min \{v(t), v(x), v(y)\} < 0,$$

$$\min \{v(t'), v(x'), v(y')\} < 0.$$

If $v(x) \geqq 0$, then necessarily $v(t) < 0$, for otherwise we would have $v(y) \geqq 0$ since y is an integral function of x and t. But then, by (9), $v(t') > 0, v(x') > 0$, and hence also $v(y') \geqq 0$, in contradiction with our assumption. Thus we have necessarily $v(x) < 0$, and similarly $v(x') < 0$.

Since x is an integral function of t and y, and $v(x) < 0$, we have either $v(t) < 0$ or $v(y) < 0$. Similarly, from $v(x') < 0$ follows that either $v(t') < 0$ or $v(y') < 0$. Were $v(y) \geqq 0$, we would have $v(t) < 0$, and consequently, by (9), $v(t') > 0$, $v(y') > 0$, which is impossible. Hence we have necessarily $v(y) < 0$, and similarly $v(y') < 0$.

Thus our valuation v is necessarily such that

$$v(x), \quad v(y), \quad v(x'), \quad v(y')$$

are all negative.

We note that the expressions of τ, ξ, and η in terms of t', x', and y' are similar to their expressions (10) in terms of t, x, and y, with τ and ξ interchanged, namely

$$\tau = x'/y', \qquad \xi = x't'/y', \qquad \eta = x'^3/y'^2.$$

Hence we may assume that $v(t) \geqq 0$ (if not, then $v(t') \geqq 0$).

It is clear that in equation (7) of the surface g the terms y^2 and x^3 are the only possible terms which have minimum v-value. Hence we must have $v(y^2) = v(x^3)$, $v(y^2/x^3 + 1) > 0$, and $0 < v(y) < v(x)$.

This implies, by (10), that $v(\tau) > 0$, $v(\xi) > 0$, and $v(\eta + 1) > 0$, showing that the center of v on g'' is the point $\tau = \xi = 0$, $\eta = -1$, as asserted. We shall denote this point by A.

Thus the covering (12) of G is established.

We now show that the *pencil* $\{C\}:t =$ const. *consists of anticanonical curves on G and that A is an ordinary simple base point of that pencil.* In fact, consider the differential $\omega = dtdx/(\partial g/\partial y)$. We have $\omega = -dt'dx'/(t'\partial g'/\partial y')$. If we take into account the fact that both affine surfaces g and g' are nonsingular and that the part of G which is not covered by $g \cup g'$ reduces to the point A, we conclude that the divisor (ω) of ω on G is equal to $-C_0$, where C_0 is the member of the pencil $\{C\}$ which corresponds to the value ∞ of the parameter t. This shows that the C's are anticanonical cycles. This, of course, implies already that *all the plurigenera of G are zero.* Furthermore, the pencil $\{C\}$ has only one base point on G, namely the point A. Equation (11) shows that τ and ξ are uniformizing parameters at A, and since $t = \tau/\xi$ [by (10)], it follows that any two C's have a simple intersection at A. Consequently $(K_a^2) = 1$ (where K_a denotes any anticanonical cycle on F).

The presence of the terms y^2 and x^3 in the equation (7) of the surface and the fact that the coefficient of y is of degree $\leqq 1$ in x imply that every cycle $t =$ const. of the pencil $\{C\}$ is prime. This completes the proof of Proposition 2.

5

From now on we always assume that our surface G is nonsingular.

PROPOSITION 3. *If the equation $g(X, Y, t) = 0$, regarded as an equation in X, Y, over $k(t)$, has a solution of the form*

$$X = \psi_2(t), \qquad Y = \psi_3(t), \tag{12'}$$

where ψ_i is a polynomial in $k[t]$, of degree $\leqq i$, then the surface G is rational. Furthermore, the irreducible curve E defined by (12′) is an exceptional curve on G, of the first kind.

Proof. By assumption, the polynomial $g(t, \psi_2(t), Y)$, of degree 2 in Y, has a root $Y = \psi_3(t)$. Therefore, it has a second root $Y = \varphi_3(t)$, where $\varphi_3(t)$ is also a polynomial of degree $\leqq 3$. We denote by D the irreducible curve defined by

$$X = \psi_2(t), \qquad Y = \varphi_3(t).$$

The replacing of X and Y by $X - \psi_2(t)$ and $Y - \varphi_3(t)$ respectively amounts to a linear transformation of the homogeneous coördinates in the ambient space of G, and, furthermore, this transformation sends g into a polynomial of the same type [see (7)]. Hence we may assume that $\psi_2(t) = \varphi_3(t) = 0$. Then in (7), $g_6(t)$ is zero, and the equations of the surfaces g and g' have the following form:

$$g\colon y(y - \psi_3(t)) + x[g_1(t)y + g_4(t) + g_2(t)x + x^2] = 0,$$

$$g'\colon y'(y' - \psi_3'(t')) + x'[g_1'(t')y' + g_4'(t') + g_2'(t')x' + x'^2] = 0,$$

where $\psi_3'(t') = t'^3\psi_3(1/t')$, $g_i'(t') = t'^i g_i(1/t')$.

Let v and v_1 be the prime divisors of $k(G)$ defined by the curves E and D respectively. We observe that $\psi_3(t)$ is not the constant zero, for if it were and if we denote by t_0 a root of $g_4(t)$, then the point $t = t_0$, $x = y = 0$ would be a singular point of g. (If $g_4(t)$ is a constant, different from zero, then $g_4'(t')$ is a constant multiple of t'^4, and the point $t' = x' = y' = 0$ would be a singular point of g'.) Since $v(x) > 0$ and $v(y - \psi_3(t)) > 0$, it follows from $\psi_3(t) \neq 0$ that $v(y) = 0$. Hence $v(\eta) = v(x^2/y^3) > 0$, showing that the point $A\colon \tau = \xi = 0, \eta = -1$ of G does not belong to E. The preceding argument about the nonvanishing of $\psi_3(t)$ shows also that $g_4(t)$ is not zero; in fact, it also shows that $\psi_3(t)$ and $g_4(t)$ have no common factor. The presence of the terms $g_4(t)x - \psi_3(t)y$ in g implies that $v_1(x) = v_1(y)$, since $v_1(x) > 0$ and $v_1(y) > 0$. Hence $v_1(\eta) > 0$, showing that also the curve D does not pass through A. We have therefore proved that

$$E + D \subset g \cup g'.$$

Since $g_6(t)$ is zero, equation (11) of the surface g'' (after division by η) has the form

$$(13)\qquad g'': \eta^2 + [1 + g_1''(\tau, \xi) + g_2''(\tau, \xi)]\eta + g_3''(\tau, \xi) + g_4''(\tau, \xi) = 0.$$

Equations (10) define a birational transformation T of G onto g''. Since $v(x) > 0$ and $v(y - \psi_3(t)) > 0$, while $\psi_3(t) \neq 0$, it follows that $v(y) = v(t) = 0$. Hence $v(t) > 0$, $v(\xi) > 0$, and $v(\eta) > 0$, showing that the center, on g'', of the prime divisor v is the origin O. Thus E is an exceptional curve of T and corresponds to the point O.

We assert that E is the total T^{-1}-transform of O. For let w be any zero-dimensional valuation of $k(G)$ having center O on g''. Since O, $A \in g''$ and $A \neq O$, it follows that the center P of w on G belongs to $g \cup g'$. Because of the symmetric roles played by g and g', we may assume that $P \in g$, i.e., that $w(x) \geqq 0$, $w(y) \geqq 0$, and $w(t) \geqq 0$. Since $\xi = x/y$ and $w(\xi) > 0$, it follows that $w(x) > 0$. Hence $P \in E + D$. Suppose $P \in D$, whence $w(y) > 0$. Since $w(x) > w(y)$, division of the equation $g = 0$ by y shows that $w(\psi_3(t)) = 0$, i.e., P is the point $t = t_0$, $x = y = 0$, where t_0 is a root of $\psi_3(t)$. But then P belongs also to E, and this proves our assertion.

We also assert that T is regular at each point of E. We have only to show that τ, ξ, and η are regular at any point P of G such that $P \in E$. We may assume that $P \in g$ (since $E \subset g \cup g'$). Let $t = t_0$, $x = 0$, $y = y_0 = \psi_3(t_0)$ at P. If $y_0 \neq 0$, then (10) shows that τ, ξ, and η are regular at P. Assume $y_0 = 0$, whence t_0 is a root of $\psi_3(t)$. It was pointed out above that $\psi_3(t)$ and $g_4(t)$ can have no common root t_0 (for otherwise the point $(t_0, 0, 0)$ would be a singular point of g). Hence $g_4(t_0) \neq 0$, and $g_4(t) + g_2(t)x + x^2$ is a unit ε in the local ring $\mathfrak{o}_P$ of P. Therefore $x/y = -(y - \psi_3 + xg_1)/\varepsilon \in \mathfrak{o}_P$, showing that

$$\xi = x/y \in \mathfrak{o}_P, \qquad \tau = \xi t \in \mathfrak{o}_P, \quad \text{and} \quad \eta = \xi^2 x \in \mathfrak{o}_P,$$

as asserted.

Since it is obvious from (13) that O is a simple point of g'', we have therefore established the second part of our proposition: *E is an irreducible exceptional curve, of the first kind.* It follows that G is not a relatively minimal model, and thus the rationality of G follows from the case $(K^2) \geqq 2$ and from Proposition 2.

Note. Clearly, also, D is an exceptional curve of the first kind. Furthermore, assuming, as we may, that $\psi_3(t)$ is exactly of degree 3 in t (if not, pass to g' and $\psi_3'(t')$), we see at once that the common points of E and D are the points $P(t_0, 0, 0)$, where t_0 is any root of $\psi_3(t)$, and that the intersection multiplicity of E and D at P is the multiplicity of t_0 as a root of $\psi_3(t)$. Hence $(E \cdot D) = 3$.

6

If, in (7), we allow the coefficients of the polynomials $g_i(t)$ $(0 \leqq i \leqq 6, i \neq 5)$ to vary arbitrarily in the universal domain, we obtain an irreducible algebraic system M of surfaces in S_3, of dimension 22. It is easily verified that any *general* member g of M/k is an absolutely irreducible, nonsingular (in the absolute sense) surface, and that the projective surface G, defined in terms of g as in Section 4, satisfies conditions (A) and (B) of Section 2.

PROPOSITION 4. *The general surface g of M/k carries a curve $X = \psi_2(t)$, $Y = \psi_3(t)$, where ψ_2 and ψ_3 are polynomials of degree* 2 *and* 3 *respectively (with coefficients in the universal domain).*

Proof. We consider the most general surface h in M whose equation $h(X, Y, t) = 0$ has a rational solution $X = \psi_2(t)$, $Y = \psi_3(t)$ of the above indicated form. Then

$$h(X, Y, t) = [Y - \psi_3(t)][Y - \varphi_3(t)] + \lambda[X - \psi_2(t)][h_1(t)y + x^2 + h_2(t)x + h_4(t)], \tag{14}$$

where λ and the 21 coefficients of the polynomials ψ_3, φ_3, ψ_2, and h_i $(i = 1, 2, 4)$ are indeterminates (the subscript indicates in each case the degree of the polynomial). We have to show that the surface h is a general member of M/k. Let N be the irreducible subsystem of M which is the locus of h over k. We have to show that dim $N = 22$. We denote by h^* the point in the affine space A_{22}, of dimension 22, whose coördinates are λ and the coefficients of ψ_3, φ_3, ψ_2, h_i $(i = 1, 2, 4)$. The locus, over k, of the pair (h^*, h) is a rational transformation of A_{22} onto N. To prove that dim $N = 22$ we have to show that dim $T^{-1}\{h\} = 0$, or—equivalently—that dim $h^*/K = 0$, where K denotes here the field generated over k by the coefficients of the polynomial $h(X, Y, t)$. Let E and D denote the curves $X = \psi_2(t)$, $Y = \psi_3(t)$ and $X = \psi_2(t)$, $Y = \varphi_3(t)$ respectively. Since both E and D are exceptional curves of the first kind on the surface h, neither one can vary on h in an algebraic system of positive dimension. Hence the coefficients of ψ_2, ψ_3, and φ_3 are algebraic over K. From the identity (14) it follows at once that λ and the coefficients of $h_1(t)$, $h_2(t)$, and $h_4(t)$ belong to the field generated over K by the coefficients of $\psi_2(t)$, $\psi_3(t)$, and $\varphi_3(t)$. This shows the point h^* is algebraic over K, and the proposition is proved.

7

We shall now proceed to the proof of the results which has been announced at the end of Section 2, namely, that the surface F, under the assumptions (A) and (B), carries an exceptional curve of the first kind. We may replace F by the biregularly equivalent surface G which lies in a projective space S_{21} (see Section 3). This surface has order 36, since it is the image of the complete system $|6K_a|$ (which has degree 36). By Proposition 3, it will be sufficient to show that the equation (7) has two rational solutions $X = \psi_2(t)$, $Y = \psi_3(t)$ and $X = \psi_2(t)$, $Y = \varphi_3(t)$ of the type stated in Proposition 3. Let us interpret the existence of such a pair of solutions. They would correspond to two irreducible exceptional curves E and D on G, of the first kind, and these two curves would be the zeros (and the only zeros) on G of the function $x - \psi_2(t)$. Hence $E + D$ is a cycle in $|2K_a|$.[5] It is a composite cycle, consisting of two (not necessarily distinct) prime cycles E and D. *Moreover, neither one of these two prime cycles is a member of the pencil* $|K_a|$ *defined by* $t = const.$ Conversely, assume that the system $|2K_a|$ contains a cycle Γ which is not prime but is not a sum of two cycles of $|K_a|$.[6] Since $(\Gamma \cdot K_a) = 2$ and since by assumption (A) of Section 2 we have $(\Delta \cdot K_a) > 0$ for every prime cycle Δ on G, it follows that $\Gamma = E + D$, where E and D are prime cycles, and that $(E \cdot K_a) = (D \cdot K_a) = 1$. Neither E nor D can pass through the base point of the pencil $|K_a|$, for in the contrary case the cycles in $|K_a|$ would have no variable intersections with E (or D), and thus E (or D) would be a component of some cycle of $|K_a|$, and this, in view of assumption (A), is impossible, since $E \notin |K_a|$ and $D \notin |K_a|$. It follows that the trace of $|K_a|$ on E is a linear series of degree 1 and of dimension 1. Hence E is a rational curve, and similarly for D. The curve E has no singular points, since $(E \cdot K_a) = 1$ and since $E \notin |K_a|$. Hence $p(E) = 0$. Similarly $p(D) = 0$. Since $1 = (E \cdot K_a) = (E^2) - 2p(E) + 2$, it follows that $(E^2) = -1$, and thus E is an exceptional curve of the first kind. Similarly D is an exceptional curve of the first kind.

We therefore have only to show that *the system* $|2K_a|$ *on* G *contains a cycle which is not prime and which is not a sum of two cycles of* $|K_a|$.

Now, let us consider the affine surface h in A_3 which is the general member of the system M/k [see (14)]. This surface h defines a projective surface H in S_{21}, in the same way as G is defined by the affine surface g. From Proposition 2 it follows that also H is a surface of order 36. Since g is a specialization of h over k, it follows that G is a specialization of H over k. By Proposition 3, the surface H carries two prime cycles E and D which are exceptional curves of the first kind and such that $(E \cdot D) = 3$ [see Note at the end of Section 5]. Moreover, $E + D$ is the null cycle, on H, of the function $x - \psi_2(t)$. By the specialization $H \xrightarrow{h} G$ we find on G a composite cycle $\bar{E} + \bar{D}$ which is the null cycle of a function of the form $\lambda x - \psi_2(t)$, whence at any rate $\bar{E} + \bar{D} \in |2K_a|$. Now, since $(E \cdot D) = 3$, we have also $(\bar{E} \cdot \bar{D}) = 3$, and consequently $\bar{E} \notin |K_a|$, $\bar{D} \notin |K_a|$, since $(K_a^2) = 1$. This completes the proof of our original assertion made in Section 2.

[5] Just in the way of (redundant) checking, we observe that since $(E^2) = (D^2) = -1$ and $(E \cdot D) = 3$ [see Note at the end of section 5], the self-intersection number of $E + D$ is equal to 4, while $p(E + D) = p(E) + p(D) + (E \cdot D) - 1 = 2$. This checks with the characters of $|2K_a|$.

[6] Under this assumption it would already follow from Proposition 9.1 of our paper MM that G is either rational or is not a relatively minimal model (hence again rational, by the case $(K^2) \geqq 2$), since G would then carry a cycle which is not linearly equivalent to an integral multiple of K_a. However, we have assigned ourselves the task of proving not only the rationality of G but also the assertion that G is not a relatively minimal model.

8

We now turn to the Castelnuovo theorem of rationality of plane involutions, as formulated in Section 1. We fix a nonsingular projective model F of Σ/k, and we choose a nonsingular model F' of $k(x, y)/k$ such that F' dominates the normalization of F in $k(x, y)$. The rational transformation $f: F' \to F$ is regular, and the inverse transformation $f^{-1}: F \to F'$ has only a finite number of fundamental points on F. The multicanonical systems $| nK |$ on F can be defined by regular differential forms $\omega = A(dxdy)^n$, of weight n, on F. The separability of $k(x, y)/\Sigma$ implies that if $\omega \neq 0$ then the inverse image ω' of ω by f is also different from zero, and is of course also regular on F'. Since F' is rational, it follows that we cannot have $\omega \neq 0$. *Thus all the plurigenera of F are zero.*

The rational mapping f of F' onto F defines a rational mapping of the Albanese variety of F' *onto* the Albanese variety of F. Since the former is a point, it follows that also the Albanese variety of F reduces to a point, i.e., the irregularity q of F is zero. Since $p_g = 0$, it follows that $q = -p_a$, whence $p_a = 0$ (see, for instance, Y. NAKAI, *On the characteristic linear systems of algebraic families*, Illinois J. Math., vol. 1 (1957), pp. 552–561). Thus we have shown that $P_2 = p_a = 0$ for F, and consequently F is a rational surface.

Let us call a surface F *unirational* if it is a rational transform of a rational surface F'. The theorem of Castelnuovo asserts that under the separability assumption concerning $k(F')/k(F)$ a unirational surface F is in fact rational. Now consider, for characteristic $p \neq 0$, any surface F in S_3 given by an equation of the form

$$z^p = f(x, y), \tag{15}$$

where $f(x, y)$ is a polynomial. Any such surface F is unirational, for $k(x, y, z) \subset k(x^{1/p}, y^{1/p})$. Now we shall find a surface F of this type such that the geometric genus p_g of F is > 0, and this will show that the condition of separability in Castelnuovo's theorem is essential.

The following is well known, and is true for arbitrary characteristic: if a surface F in S_3 is such that its only singularities are isolated double points, and if each double point of F is no worse than a biplanar point (i.e., the tangent quadric cone at the point is either irreducible or splits into two *distinct* planes), then every surface in S_3 is an adjoint surface of F.[7] Thus, if such a surface F has order $\geqq 4$, the geometric genus of F will be positive. This being so, let the characteristic p be different from 2, and let us consider the surface F defined by the equation

$$F(x, y, z) = z^p + x^{p+1} + y^{p+1} - (x^2 + y^2)/2 = 0, \tag{16}$$

[7] In other words, if $f(x, y, z) = 0$ is the irreducible equation of the surface F, and if n is the degree of f, then (under the assumption that f is a separable polynomial in z) every double differential of the form $(\phi_{n-4}(x, y, z)/f_z') \, dx \, dy$, where ϕ is an arbitrary polynomial of degree $\leqq n - 4$, is regular (not only on F but also) on every nonsingular model of the field $k(x, y, z)$.

which is of type (15). One finds at once that the only singular points of the surface are the points $x = m$, $y = n$, $z = \{(n^2 + m^2)/2\}^{1/p}$, where m and n are arbitrary elements of the prime field of characteristic p (there are no singular points at infinity). It is also immediately seen that these p^2 singular points are biplanar double points. Since the order of the surface F is $p + 1 \geqq 4$, its geometric genus is positive.[8]

HARVARD UNIVERSITY
CAMBRIDGE, MASSACHUSETTS

[8] An example, of similar nature, could also be given for $p = 2$. It would be similar to (16), but the degree in z would have to be a power 2^n of 2, $n > 1$.

421 LINEAR SYSTEMS

MEMOIRS OF THE COLLEGE OF SCIENCE, UNIVERSITY OF KYOTO, SERIES A
Vol. XXXII, Mathematics No. 1, 1959.

Proof that any birational class of non-singular surfaces satisfies the descending chain condition.

By

Oscar ZARISKI*

Received January 28, 1959

(Communicated by Prof. Akizuki)

In our monograph "Introduction to the problem of minimal models in the theory of algebraic surfaces" (Publications of the Mathematical Society of Japan, no. 4; this monograph will be referred to as IMM) we have stated the proposition that *each birational class of non-singular varieties satisfies the descending chain condition* (see IMM, Proposition III. 1.3, p. 79), it being understood that the underlying partial ordering of the class is the one in which $V<V'$ if V' dominates V. In the quoted monograph we gave a proof based on the theorem of Neron-Severi. We have also mentioned the existence, in the case of surfaces, of a sheaf-theoretic proof due to Serre (a similar sheaf-theoretic proof has been given recently by Matsumura in an unpublished paper). Finally we have alluded in IMM to a forthcoming note in Mem. Col. Sci. of Kyoto University in which we proposed to prove the above descending chain condition for algebraic surfaces by elementary algebro-geometric considerations, using properties of *exceptional cycles* and the *anticanonical system* $|-K|$. This is the note in which we propose to give this proof.

§ 1. Exceptional cycles of the first kind.

Let F be a non-singular surface (over an algebraically closed

---

* This research was supported by the United States Air Force through the Air Office of Scientific Research of the Air Research and Development Command under Contract No. AF18 (600)-1503.

ground field k) and let E be an exceptional curve of the 1st kind on F (E may be reducible). We shall associate with E a well-defined positive divisorial cycle $\mathcal{E}$ whose components are the irreducible components of E, counted to suitable (positive) multiplicities.

Let P be the (simple) contraction of E and let $\mathfrak{m}_P$ be the maximal ideal of the local ring $\mathfrak{o}_P$ of the point P. If v is any valuation of the function field $k(F)$ of F and if v is non-negative on $\mathfrak{o}_P$ (i.e., if $v(z) \geqq 0$ for all z in $\mathfrak{o}_P$) then from the fact that $\mathfrak{m}_P$ has a finite basis it follows that min $\{v(z),\ z \in \mathfrak{m}_P\}$ exists. We denote this minimum by $v(\mathfrak{m}_P)$. It is clear that $v(\mathfrak{m}_P) \geqq 0$ and that $v(\mathfrak{m}_P) > 0$ if and only if P is the center of v (on the surface which carries the point P).

Let now $\Gamma_1, \Gamma_2, \cdots, \Gamma_h$ be the irreducible components of E and let v_{Γ_i} be the divisorial (discrete) valuation of $k(F)$ defined by the irreducible curve Γ_i. Since P is the contraction of E, P is the center of v_{Γ_i}, and thus $v_{\Gamma_i}(\mathfrak{m}_P)$ is defined (and is a *positive* integer). We set

$$\mathcal{E} = \textstyle\sum_{i=1}^{h} v_{\Gamma_i}(\mathfrak{m}_P)\Gamma_i\,, \tag{1}$$

and we refer to $\mathcal{E}$ as the *cycle associated with the exceptional curve* E. We say that a divisorial cycle on F is an *exceptional cycle* (of the first kind) if it is the cycle associated with an exceptional curve E of the first kind. If P is the (simple) contraction of E we shall also refer to P as the contraction of the exceptional cycle $\mathcal{E}$.

Proposition 1. *If E is an irreducible exceptional curve then the exceptional cycle $\mathcal{E}$ associated with E is E itself.*

Proof. If E is irreducible then v_E is the principal P-adic divisor (IMM, p. 55 and Corollary II. 3. 2, p. 56), and hence $v_E(\mathfrak{m}_P) = 1$.

If X is any divisorial cycle on F we denote by $\langle X \rangle$ the *support* of X, i.e., the curve whose irreducible components are the prime components of X.

Proposition 2. *Let $\mathcal{E}$ be an exceptional cycle, let P be the contraction of $\mathcal{E}$ and let Q be a point of the support $\langle \mathcal{E} \rangle$ of $\mathcal{E}$. Then the ideal $\mathfrak{o}_Q\mathfrak{m}_P$ is principal, and if g is a generator of this ideal then $g=0$ is a local equation of $\mathcal{E}$ at Q.*

Proof. Let x and y be regular parameters of the local ring $\mathfrak{o}_P$. Since $Q>P$ and $Q\neq P$, it follows that either y/x or x/y belongs $\mathfrak{o}_Q$ (IMM, Theorem II. 1. 2, p. 46). If, say $y/x\in\mathfrak{o}_Q$ then $\mathfrak{o}_Q\cdot\mathfrak{m}_P=\mathfrak{o}_Q\cdot x$. An irreducible curve Γ on F is a component of E if and only if $v_\Gamma(\mathfrak{m}_P)>0$, hence (and assuming furthermore that $Q\in\Gamma$) if and only if $v_\Gamma(x)>0$; and for any such curve Γ we have $v_\Gamma(\mathfrak{m}_P)=v_\Gamma(x)$. Hence $x=0$ is a local equation of $\mathscr{E}$ at Q. QED.

Let $T:G\rightarrow F$ be an antiregular birational transformation of a non-singular surface G onto a non-singular surface F. Let P be a fundamental point of T (on G). If $E=T\{P\}$ is the total T-transform of P (whence E is an exceptional curve on F, with contraction P), we denote by $T(P)$ the exceptional cycle $\mathscr{E}$ associated with E.

Proposition 3. *Let $T_1:H\rightarrow G$ and $T_2:G\rightarrow F$ be antiregular birational transformations, the surfaces H, G, F being non-singular, and let $T=T_1T_2$: $H\rightarrow F$. If P is a fundamental point of T_1 then $T(P)=T_2(T_1(P))$ [here $T_2(T_1(P))$ denotes the T_2-transform of the divisorial cycle $T_1(P)$].*

Proof. Since $T_1(P)$ is a positive cycle, the support of the T_2-transform of $T_1(P)$ coincides with the total T_2-transform of the support of $T_1(P)$ (IMM, Proposition II. 5. 1, p. 69). Hence $T(P)$ and $T_2(T_1(P))$ have the same support. Let now R be any point of $\langle T(P)\rangle$, let Q be the point of $\langle T_1(P)\rangle$ which corresponds to R, let x, y be uniformizing parameters at P and let, say, $y/x\in\mathfrak{o}_Q$. By Proposition 2, $x=0$ is a local equation of $T(P)$ at R. By the same proposition, $x=0$ is also a local equation of $T_1(P)$ at Q, and hence, by the definition of the T_2-transform of a cycle on G (IMM, Definition II. 5. 2, p. 70), $x=0$ is also the local equation of $T_2(T_1(P))$ at R. Thus $T(P)$ and $T_2(T_1(P))$ have the same local equation at each point R of their common support. QED.

If P and Q are points of birationally equivalent surfaces, we write $P<Q$ if $\mathfrak{o}_P$ is a *proper* subring of $\mathfrak{o}_Q$; and if X and Y are two divisorial cycles on a surface F we write $X<Y$ if Y-X is a *strictly positive* cycle. In the latter case we say that X is a *proper sub-cycle* of Y.

Proposition 4. *Let $\mathscr{E}_1$ and $\mathscr{E}_2$ be exceptional cycles of the 1st kind on a non-singular surface F and let P_1, P_2 be their contractions. Then the following relations are equivalent.*

(a) $\mathcal{E}_1 < \mathcal{E}_2$;
(b) $\langle\mathcal{E}_1\rangle < \langle\mathcal{E}_2\rangle$;
(c) $P_1 > P_2$.

Proof. If (a) is satisfied, then clearly $\langle\mathcal{E}_1\rangle \subset \langle\mathcal{E}_2\rangle$, and we cannot have $\langle\mathcal{E}_1\rangle = \langle\mathcal{E}_2\rangle$ since any exceptional curve determines uniquely the exceptional cycle associated with it. Thus (a) implies (b).

That (b) and (c) are equivalent has been proved in IMM (Lemma II. 3. 7, p. 58).

Now assume (c). If Γ is any prime component of $\mathcal{E}_1$ then P_1 is a center of v_Γ, and hence also P_2 is a center of v_Γ. Furthermore, we have $v_\Gamma(\mathfrak{m}_{P_1}) \leqq v_\Gamma(\mathfrak{m}_{P_2})$ since $\mathfrak{m}_{P_2} \subset \mathfrak{m}_{P_1}$. This shows that $\mathcal{E}_1 \leqq \mathcal{E}_2$, and since equality is clearly impossible, the proof is complete.

With the notations of Proposition 4 we say that $\mathcal{E}_1$ is a *maximal exceptional sub-cycle of* $\mathcal{E}_2$ if $\mathcal{E}_1 < \mathcal{E}_2$ and if there exist no exceptional cycles $\mathcal{E}$ such that $\mathcal{E}_1 < \mathcal{E} < \mathcal{E}_2$.

Corollary 4. 1. *$\mathcal{E}_1$ is a maximal exceptional sub-cycle of $\mathcal{E}_2$ if and only if P_1 is a quadratic transform of P_2.*

This follows from Proposition 4 and Theorem II. 1. 1 of IMM, p. 44.

Corollary 4. 2. *Let $T: H \to F$ be an antiregular birational transformation of an non-singular surface H onto a non-singular surface F, let P be a fundamental point of T and let $\mathcal{E} = T(P)$. Let $T_1: H \to G$ be a locally quadratic transformation of H, with center P, let E_0 be the (irreducible) curve $T_1\{P\}$ on G, and let $T_2: G \to F$ be the antiregular birational transformation of G such that $T_1 T_2 = T$. If $\mathcal{E}$ is not a prime cycle (or equivalently: if T_2 has fundamental points on E_0) and if $P_1', P_2', \cdots, P_g'$ denote the fundamental points of T_2 on E_0, then the g exceptional cycles $T_2(P_i')$ are the only maximal exceptional sub-cycles of $\mathcal{E}$.*

Obvious.

Proposition 5. *If $\mathcal{E}$ is an exceptional cycle on a non-singular surface F then $p(\mathcal{E}) = 0$ and $(\mathcal{E}^2) = -1$. If Γ is any prime component of $\mathcal{E}$, different from the principal component of $\langle\mathcal{E}\rangle$, then $(\mathcal{E}\cdot\Gamma) = 0$.*

Proof. Let P be the contraction of $\mathcal{E}$. There exists a non-singular surface H which carries the point P and such that $\mathcal{E} = T(P)$, where $T: H \to F$ is an anti-regular birational transformation of H

onto F (for instance, take $H=F-\langle\mathscr{E}\rangle+P$). Let $T_1: H\rightarrow G$ and $T_2: G\rightarrow F$ have the same meaning as in Corollary 4.2. Using the notations of that corollary, we have, by Proposition 3: $\mathscr{E}=T_2(E_0)$. Since $p(E_0)=0$ and $(E_0^2)=-1$ (E_0 being an *irreducible* exceptional curve of the first kind) and since anti-regular transformations preserve the arithmetic genus and the self-intersection number of any divisorial cycle, it follows that also $p(\mathscr{E})=0$ and $(\mathscr{E}^2)=-1$.

To prove the second part of the proposition we fix some *proper* exceptional sub-cycle $\mathscr{E}_1$ of $\mathscr{E}$ such that Γ is a component of $\mathscr{E}_1$ (the existence of $\mathscr{E}_1$ follows from IMM, Proposition II. 3. 3, p. 57). We replace in the preceding part of the proof $\mathscr{E}$ by $\mathscr{E}_1$. Let P', H', T' have the same meaning in relation to $\mathscr{E}_1$ as P, H and T had in relation to $\mathscr{E}$. Since $\mathscr{E}_1$ is a proper exceptional sub-cycle of $\mathscr{E}$, we have $H<H'$ (assuming, as we may, that $H=F-\langle\mathscr{E}\rangle+P$, $H'=F-\langle\mathscr{E}_1\rangle+P'$). Let $\mathscr{E}'$ be the exceptional cycle on H' which is the transform of the point P. By Proposition 3 (as applied to the surfaces H, H', F) we have $\mathscr{E}=T'(\mathscr{E}')$. From Proposition II. 5. 4 of IMM, p. 71, it now follows directly that $(\mathscr{E}.\Gamma)=0$.

Corollary 5. 1. *If* $\mathscr{E}_1$ *and* $\mathscr{E}_2$ *are distinct exceptional sub-cycles of* $\mathscr{E}$ *then* $(\mathscr{E}_1\cdot\mathscr{E}_2)=0$.

By Proposition 5 it is sufficient to consider the case in which both $\mathscr{E}_1$ and $\mathscr{E}_2$ are proper sub-cycles of $\mathscr{E}$, for if, say, $\mathscr{E}_1=\mathscr{E}$ then $\mathscr{E}_2$ is a proper exceptional sub-cycle of $\mathscr{E}$ and therefore no prime component of $\mathscr{E}_2$ is the principal component of $\mathscr{E}$ (IMM, Corollary II. 3. 8, p. 58). Since the corollary is vacuous if $\langle\mathscr{E}\rangle$ is irreducible, we use induction with respect to the number of prime components of $\mathscr{E}$. We use the notations of the proof of Proposition 5 and we denote by $P_1', P_2' \cdots, P_g'$ the fundamental points of T_2 on E_0. By Corollary 4. 1, each of the exceptional cycles $\mathscr{E}_1, \mathscr{E}_2$ is a sub-cycle of one of the exceptional cycles $T_2(P_i')$. Let, say $\mathscr{E}_1$ be a sub-cycle of $T_2(P_\alpha')$ and $\mathscr{E}_2$ a sub-cycle of $T_2(P_\beta')$. If $\alpha\neq\beta$, then $\langle T_2(P_\alpha')\rangle$ and $\langle T_2(P_\beta')\rangle$ have no common points, and the relation $(\mathscr{E}_1\cdot\mathscr{E}_2)=0$ is proved. If $\alpha=\beta$, then we observe that the number of prime components of $T_2(P_\alpha')$ is less than that of $\mathscr{E}$, and hence $(\mathscr{E}_1\cdot\mathscr{E}_2)=0$, by our induction hypothesis.

If $T: H\rightarrow F$ is an antiregular birational transformation of a non-singular surface H onto a non-singular surface F and if $X=\sum_{i=1}^{q} m_i\Gamma_i$ is any divisorial cycle on H whose distinct prime components are $\Gamma_1, \Gamma_2, \cdots, \Gamma_q$, then we denote by $T[X]$ the divisorial cycle $\sum_{i=1}^{q} m_i T[\Gamma_i]$, where $T[\Gamma_i]$ denotes the proper T-transform of Γ_i. This cycle $T[X]$ does not, in general, coincide with the T-transform $T(X)$ of X as defined in IMM, p. 70.

Proposition 6. *Let $T: H \to F$ be an anti-regular birational transformation of a non-singular surface H onto a non-singular surface F, let $P_1, P_2, \cdots, P_h$ be the fundamental points of T (on H) and let $\mathcal{E}_i = T(P_i)$, $i=1, 2, \cdots, h$. If $\{\mathcal{E}_{i,1}, \mathcal{E}_{i,2}, \cdots, \mathcal{E}_{i,s_i}\}$ is the set of all proper exceptional sub-cycles of $\mathcal{E}_i$, then for any divisorial cycle X on H we have*

$$T(X) = T[X] + \sum_{i=1}^{h} \lambda_i \mathcal{E}_i + \sum_{i=1}^{h} \sum_{j_i=1}^{s_i} \lambda_{i,j_i} \mathcal{E}_{i,j_i},$$

where the coefficients λ_i, λ_{i,j_i} are integers and where λ_i is the multiplicity of X at P_i. If $X > 0$ then the λ_i, λ_{i,j_i} are non-negative.

Proof. It is obviously sufficient to prove the proposition under the assumption that $h=1$. The transformation T has in that case only one fnndamental point P. We set $\mathcal{E} = T(P)$ and we let $\{\mathcal{E}_1, \mathcal{E}_2, \cdots, \mathcal{E}_3\}$ be the set of all proper exceptional sub-cycles of $\mathcal{E}$. By IMM, Proposition II. 5. 8 (p. 73) the proposition is true if T is a locally quadratic transformation. We shall therefore use induction with respect to the number of prime components of $\mathcal{E}$. We use the notations of Corollary 4. 2. We have, by IMM, Proposition II. 5. 8 (p. 73),

$$T_1(X) = T_1[X] + \lambda E_0,$$

where λ is the multiplicity of X at P, and hence, by Proposition 3,

$$T(X) = T_2(T_1(X)) = T_2(T_1[X]) + \lambda \mathcal{E}.$$

The fundamental points of T_2 are the points $P_1', P_2', \cdots, P_g'$, and their T_2-transforms represent all the maximal exceptional sub-cycles of $\mathcal{E}$ (Corollary 4. 2). Hence, by our induction hypothesis, the proposition is applicable to T_2 and to any divisorial cycle on G (and, in particular, to the cycle $T_1[X]$). Since $T_2[T_1[X]] = T[X]$, the proposition is proved.

Corollary 6. 1. *Let $\mathcal{E}$ be an exceptional cycle of the first kind on a non-singular surface and let E_1 be the principal component of $\mathcal{E}$. If $\mathcal{E}_1, \mathcal{E}_2, \cdots, \mathcal{E}_s$ are the proper exceptional sub-cycles of $\mathcal{E}$, then*

$$\mathcal{E} = E_1 + \sum_{i=1}^{s} \lambda_i \mathcal{E}_i,$$

where the λ_i are non-negative integers and λ_i is positive if $\mathcal{E}_i$ is a maximal exceptional sub-cycle of $\mathcal{E}$.

In the notation of the proof of Proposition 6 we have $\mathcal{E}=T(P) = T_2(E_0)$, and the corollary follows by applying Proposition 6 to the transformation T_2 and the cycle E_0.

§ 2. The anticanonical system and exceptional cycles

Proposition 1. *If K is a canonical divisor on a non-singular surface F and $\mathcal{E}$ is an exceptional cycle on F, of the first kind, then $(K\cdot\mathcal{E})=-1$.*

Proof. The proposition follows directly from Proposition 5, § 1, in view of the equality $(K\cdot X)=2p(X)-2-(X^2)$ which holds for any divisorial cycle X on F.

Proposition 2. *If an anti-regular birational transformation $T: F'\rightarrow F$ of a non-singular surface F' onto a non-singular surface F is a product of n quadratic transformations then the dimension of the anticanonical system $|-K'|$ on F' satisfies the inequality*

$$\dim|-K'|\geq n+(K^2)+P_a\,;$$

where K is a canonical divisor on F and where P_a is the arithmetic genus of F' (and of F).

Proof. Assume first that $n=1$. Let P' be the center of the locally quadratic transformation T and let $E=T\{P'\}$. Then it is known (see IMM, Proposition II. 5. 6, p. 72) that $T(K')+E$ is a canonical divisor K on F. Hence $(K^2)=(K'^2)-1$, since $(T(X')\cdot E) =0$ for any divisorial cycle X' of F' (Proposition II. 5. 5, IMM, p. 71) and since $(E^2)=-1$. By induction with respect to n we find that if T is a product of n quadratic transformations then $(K'^2)=(K^2)+n$. Since $p(-K')=1$ our proposition follows from the Riemann-Roch theorem on F'.

Theorem 1. *On a non-singular surface F there cannot exist an infinite strictly ascending chain $\mathcal{E}_1<\mathcal{E}_2<\cdots<\mathcal{E}_n<\cdots$ of exceptional cycles of the first kind.*

Proof. We shall assume that such a chain exists and we shall show that this assumption leads to a contradiction. Let $F_i=(F-\mathcal{E}_i)+P_i$, where P_i is the contraction of $\mathcal{E}_i$. Then F_i is a non-singular surface and we have $F>F_1>F_2>\cdots>F_n>\cdots$. We also have $P_1 > P_2 > \cdots > P_n > \cdots$, and each F_n carries an infinite strictly ascending chain of exceptional cycles of the first kind: namely, the cycles on F_n which are the transforms of the points $P_{n+1}, P_{n+2}, \cdots$ form such a chain. We therefore may replace in our proof the surface F by any of the surfaces F_n. Since the anti-regular birational transformation of F_n onto F is the product of at least n locally quadratic transformations, the dimension of the anticanonical system $|-K_n|$ on F_n satisfies the inequality: $\dim|-K_n| \geqq n+(K^2)+P_a$, where K is a canonical divisor on F and P_a is the arithmetic genus of F. Thus $\dim|-K_n|\geqq 1$ if n is sufficiently large, and *we may therefore assume* that $\dim|-K|\geqq 1$.

Let E_i be the principal component of $\mathcal{E}_i$. Then E_i is not a component of $\mathcal{E}_j$, $j<i$ (IMM, Corollary II. 3. 8, p. 58), and hence the irreducible curves $E_1, E_2, \cdots, E_n, \cdots$ are distinct.

By Corollary 6. 1, § 1, we have that $\mathcal{E}_i$ is the sum of E_i and a certain number ν_i of exceptional cycles of the first kind. Here $\nu_i \geqq 1$ except if $\mathcal{E}_i$ is a prime cycle (which can happen only for $i=1$). Hence we may assume that $\nu_i \geqq 1$ for all i. By Proposition 1 it follows that

$$(1) \qquad (-K\cdot E_i) = 1-\nu_i \leqq 0\,.$$

Let N be an integer such that *no* E_i, $i \geqq N$, *is a fixed component of the linear system* $|-K|$. Then $(-K\cdot E_i)\geqq 0$ if $i\geqq N$, and hence by (1) we conclude that

$$(2) \qquad (-K\cdot E_i) = 0\,, \qquad i \geqq N\,.$$

This shows that each E_i, $i\geqq N$, is a fundamental curve of $|-K|$, i.e., that the cycles in $|-K|$ which have E_i as component form a (linear) subsystem L_i of $|-K|$ the dimension of which is one less than the dimension of $|-K|$. Thus $|-K|$ has infinitely many fundamental curves. This implies that the rational transformation of F which is defined by the linear system $|-K|$ (IMM, p. 10) is necessarily a curve. In other words, if we denote by B the fixed cycle of $|-K|$, then the linear system obtained by deleting B from the members of $|-K|$ is composite with some irreducible pencil H. Since H contains at most a finite number of cycles which are not prime and since each E_i, $i\geqq N$, is a component of some member of H, it follows that some E_i is a member of H (actually, all but a finite number of the E_i must be members of

H). However, *we now show that* (E_i^2) *is negative* and therefore no E_i can be a member of a pencil. This contradiction will complete the proof.

Let, then, quite generally, $\mathcal{E}$ be an exceptional cycle of the first kind and let E_1 be the principal component of $\mathcal{E}$. We have then, by Corollary 6.1, §1:

$$(3) \qquad \mathcal{E} = E_1 + \sum_{i=1}^{s} \lambda_i \mathcal{E}_i ,$$

where the λ_i are non-negative integers and the $\mathcal{E}_i$ are proper exceptional sub-cycles of $\mathcal{E}$. Since $(\mathcal{E}\cdot\mathcal{E}_i)=0$ (Corollary 5.1, §1) and $(\mathcal{E}^2)=-1$, it follows from (3) that

$$(4) \qquad (\mathcal{E}\cdot E_1) = -1 .$$

For a fixed j, we intersect both sides of (3) with $\mathcal{E}_j$, and we note that $(\mathcal{E}_i\cdot\mathcal{E}_j)=0$ if $i \neq j$ (Corollary 5.1, §1). We thus obtain

$$(5) \qquad (\mathcal{E}_j\cdot E_1) = \lambda_j .$$

Intersecting both sides of (3) with E_1 we find in view of (4) and (5): $(E_1^2)=-1-\sum_{i=1}^{s}\lambda_i<0$. This completes the proof of the theorem.

Remark. After the relation (2) has been obtained, the rest of the proof admits another variation. From (1) and (2) it follows that $\nu_i=1$ if $i\geqq N$. It follows therefore from Corollary 6.1, §1, that each $\mathcal{E}_i$, $i\geqq N$, has only one maximal exceptional sub-cycle, say $\mathcal{E}_i'$, and that $\mathcal{E}_i=E_i+\mathcal{E}_i'$. By a refinement of the original sequence $\mathcal{E}_1<\mathcal{E}_2<\cdots$ we may arrange matters so that each $\mathcal{E}_i$ is a maximal exceptional sub-cycle of its successor $\mathcal{E}_{i+1}$. Then $\mathcal{E}'_{i+1}=\mathcal{E}_i$. We have then $0=(\mathcal{E}_i\cdot\mathcal{E}_{i+1})=(E_i\cdot E_{i+1})+(E_i\cdot\mathcal{E}_i)+(\mathcal{E}_{i-1}\cdot\mathcal{E}_{i+1})=(E_i\cdot E_{i+1})+(E_i\cdot\mathcal{E}_i)$. By relation (4), applied to $\mathcal{E}=\mathcal{E}_i$, we have $(E_i\cdot\mathcal{E}_i)=-1$. Hence $(E_i\cdot E_{i+1})=1$. Now let $i\geqq N$ and let L_i be the above sub-system of $|-K|$ whose members contain E_i as component. If D is any member of L_i, $D=E_i+D_i$, then $0=(D\cdot E_{i+1})=1+(D_i\cdot E_{i+1})$, i.e., $(D_i\cdot E_{i+1})=-1$. Hence E_{i+1} is a component of D_i, and if we set $D_i=E_{i+1}+D_{i+1}$, then from $(E_i\cdot E_{i+2})\geqq 0$, $(E_{i+1}\cdot E_{i+2})=1$ and $(D\cdot E_{i+2})=0$ follows at once that $(D_{i+1}\cdot E_{i+2})<0$ and that consequently E_{i+2} is a component of D_{i+1}. Proceeding in this fashion we see that all the curves $E_i, E_{i+1}, \cdots$ are components of D, and this is absurd.

§ 3. The descending chain condition in the birational class of F

We now come to our main object, i.e. to the proof of the following theorem:

Every strictly descending chain $F > F_1 > F_2 > \cdots$ of birationally equivalent non-singular surfaces is necessarily finite.

Proof. We shall assume that there exists an infinite strictly descending chain

$$F > F_1 > F_2 > \cdots > F_n > \cdots$$

of non-singular surfaces (each F_i dominating its successor F_{i+1}) and we shall show that this assumption leads to a contradiction.

We fix on each F_n $(n \geqq 1)$ a fundamental point P_n of the antiregular birational transformation of F_n onto F_{n-1} and we denote by $\mathcal{E}_n$ the exceptional cycle of the first kind on F which corresponds to the point P_n in the antiregular birational transformation of F_n onto F. By Theorem 1, § 2, the infinite set $\{\mathcal{E}_1, \mathcal{E}_2, \cdots, \mathcal{E}_n, \cdots\}$ contains an infinite subset $\{\mathcal{E}_{i_1}, \mathcal{E}_{i_2}, \cdots, \mathcal{E}_{i_n}, \cdots\}$ consisting of maximal elements of the set. I assert that $\langle \mathcal{E}_{i_\alpha} \rangle \cap \langle \mathcal{E}_{i_\beta} \rangle = \emptyset$ *if* $\alpha \neq \beta$. For assume the contrary and let Q be a common point of $\langle \mathcal{E}_{i_\alpha} \rangle$ and $\langle \mathcal{E}_{i_\beta} \rangle$. Then Q corresponds to both points P_{i_α}, P_{i_β}, and since $Q > P_{i_\alpha}$ and $Q > P_{i_\beta}$ it follows that P_{i_α} and P_{i_β} are corresponding points in the birational transformation between F_{i_α} and F_{i_β}. If, say, $\alpha < \beta$, then it follows that $P_{i_\alpha} > P_{i_\beta}$, whence $\mathcal{E}_{i_\alpha} < \mathcal{E}_{i_\beta}$, which is impossible. This proves our above assertion.

Any *minimal* exceptional sub-cycle of an exceptional cycle of the first kind is a prime cycle. We fix a minimal exceptional sub-cycle E_α of $\mathcal{E}_{i_\alpha}$, for each α. Then the E_α are *irreducible* exceptional curves of the first kind, and

$$(1) \qquad E_\alpha \cap E_\beta = \emptyset .$$

We may assume that the anticanonical system $|-K|$ on F has dimension $\geqq 2$ (see § 2). We fix a linear subsystem L of $|-K|$ which has dimension 2. If D is any member of L we have

$$(2) \qquad (D \cdot E_\alpha) = 1 , \quad \text{all } \alpha .$$

Let B be the fixed cycle of L (if such a cycle exists) and let $L_1 = L - B$. If B meets a given E_α then $(D_1 \cdot E_\alpha) = 0$ for any D_1 in L_1 in view of (2), and thus E_α is a fundamental curve of L_1.

The assumption that L has infinitely many fundamental curves E_α would lead to the same contradiction as was reached in the proof of Theorem 1 (in view of $(E_\alpha^2)=-1$). Hence B meets at most a finite number of E_α. Omitting if necessary a finite number of the E_α we may therefore assume that B meets no E_α and that consequently $(D_1 \cdot E_\alpha)=1$ for all α. We replace L by L_1, and we may therefore assume that L has no fixed components without violating (2), and also that no E_α is fundamental for L. Since $\dim L=2$, it follows from (2) that for each α there exists one and only one cycle D_α in L such that E_α is a component of D_α. This cycle D_α cannot be E_α itself since (E_α^2) is negative. Hence D_α is not a prime cycle. Let M/k be the smallest algebraic sub-system of L/k which contains all the cycles D_α and let N be an irreducible component of M/k such that N contains infinitely many of the cycles D_α. Then it is clear that if D^* is a general member of N/k (the coördinates of the Chow point of D^* belonging to a universal domain), infinitely many of the curves E_α (regarded as cycles) will be specializations, over k, of one and the same prime component of D^*. Since this is a contradiction with the fact that the E_α have a negative self-intersection number, the proof of the theorem is complete.

Harvard University

ON THE SUPERABUNDANCE OF THE COMPLETE LINEAR SYSTEMS $|nD|$ (n — *large*) FOR AN ARBITRARY DIVISOR D ON AN ALGEBRAIC SURFACE (*)

By OSCAR ZARISKI, Cambridge. Mass.

§ 1. THE PROBLEM.

Let F be a non-singular irreducible algebraic surface, defined over an algebraically closed ground field k of arbitrary characteristic, and let

$$D = \lambda_1 D_1 + \lambda_2 D_2 + \ldots + \lambda_q D_q \tag{1}$$

be a divisorial cycle on F. Here $D_1, D_2, \ldots, D_q$ are distinct prime cycles, and the λ_i are integers. *Our ploblem is to study the dimension of the complete linear system* $|nD|$ *as a function of* n, *for large values of* n. If K is a canonical cycle on F and if we set, for any divisorial cycle Z:

$$\varrho(Z) = \frac{Z(Z-K)}{2}, \tag{2}$$

$$i(Z) = 1 + \dim |K - Z|,$$

then (1) the theorem of RIEMANN-ROCH on F gives us only the

(*) The results presented in these lectures were obtained in the course of work supported by the AF Office of Scientific Research of the Air Research and Development Command, under Contract n° AF 49 (638) - 494.

The full proofs will appear in a separate paper.

(1) Note that $1 + \varrho(Z)$ is equal to the arithmetic genus $p_a(-Z)$ of the cycle $-Z$.

inequality

$$\dim |nD| \geqq \varrho(nD) + p_a - i(nD) \ ,$$

where p_a is the arithmetic genus of F. The difference

$$s(nD) = \dim |nD| - [\varrho(nD) + p_a - i(nD)] \ ,$$

called the *superabundance* of $|nD|$, is thus a non-negative integer. For our purposes we may regard $i(nD)$ as a known function of n. In fact, if $|nD|$ is empty for all n, then there is no problem, since in that case we have $\dim |nD| = -1$, for all n. Suppose, however, that for some $n = n_0$ the system $|n_0 D|$ is non-empty, and let Δ be a member of that system (whence Δ is an effective cycle). If $\Delta \neq$ zero cycle and if C is a hyperplane section of F, then the intersection number $(\Delta \cdot C)$ is positive. If for any integer $n > 0$ we set then $n = mn_0 + j$, where $0 \leqq j < n_0$, then $(nD \cdot C) = = m(\Delta \cdot C) + j(D \cdot C) \to +\infty$ as $n \to +\infty$, $(K - nD \cdot C) < 0$ for all large n and therefore $i(nD) = 0$ if n is large. On the other hand, if $\Delta = 0$ then $i(nD) = i(jD)$, i.e., $i(nD)$ can take only n_0 distinct values (at most).

Thus our problem is equivalent to finding $s(nD)$ as a function of n, for large values of n.

We shall discuss here this problem *only under the assumption that D is an effective cycle.* However, we shall indicate now how the general case can be reduced to the case of an effective D. In the above notations, we may assume that $|n_0 D|$ contains an *effective* cycle Δ. Then $\dim |nD| = \dim |m\Delta + jD|$, where m is large and j assumes the values $0, 1, ..., n_0 - 1$. Thus we can achieve a reduction to the case of an effective cycle D, provided we generalize the terms of our problem and study the dimension of the complete linear system $|nD + X|$ as a function of n, for large n, *where now D is an effective cycle and X is a fixed cycle.* We have precise information also on the solution of this slightly more general problem, but we shall not discuss it here.

From the sheaf-theoretic point of view our problem can be formulated as that of finding $h^0(nD)$ and $h^1(nD)$ as functions of n, for large n. (We use the customary notations of sheaf theory). In fact, we have $h^0(nD) = 1 + \dim |nD|$ and $h^1(nD) = s(nD)$. Note that $h^2(nD) = h^0(K - nD) = i(nD)$, and that the EULER-POINCARÉ characteristic $h^0(nD) - h^1(nD) + h^2(nD)$ of the cycle nD is equal to the polynomial (of degree $\leqq 2$) $\varrho(nD) + p_a + 1 = p_a(-nD) + p_a$. In the case of divisorial cycles D on a non-singular variety V of any dimension, the problem of finding the functions $h^i(nD)$ $(0 \leqq i \leqq r)$ is completely open.

§ 2. A RING-THEORETIC ASPECT OF THE PROBLEM.

In this section we shall deal with a non-singular variety V of arbitrary dimension r and with a divisorial cycle D on V. We denote by Σ the function field $k(V)$ and by L_n the finite-dimensional vector space (over k) consisting of the elements f of Σ such that the cycle $(f) + nD$ is effective (in symbols: $(f) + nD > 0$). Here (f) denotes the cycle on V determined by f (we include in L_n the element zero of Σ). Let t be a transcendental over Σ. In the polynomial ring $\Sigma[t]$ we consider the polynomials of the form

$$\Sigma a_i t^i , \qquad a_i \in L_i .$$

Since $L_i L_j \subset L_{i+j}$, these polynomials form a subring of $\Sigma[t]$. We denote this subring by $R^*[D]$. We have a natural grading of the ring $R^*[D]$, the homogeneous component $R_i^*[D]$, of degree i, being given by $L_i t^i$. We also have: $R_0^*[D] = k$ and

$$\dim_k R_i^*[D] = \dim_k L_i = 1 + \dim |iD| .$$

In the case in which V is a surface, the results which we obtain (and which will be stated later on) contain, in particular, the following conclusion: *there exists a finite number of polynomials $g_1(x), g_2(x), \ldots, g_h(x)$ such that for all large n the dimension of $|nD|$ is equal to the value which one of these polynomials assumes for $x = n$* (2). This conclusion would follow automatically from a result due to P. SAMUEL (see [3]) *if the graded ring $R^*[D]$ were finitely generated over k.* Unfortunately, this is not always the case (already in the case of surfaces), as we shall now show by exhibiting a very general class of counter-examples.

(2) Our results also include the additional information to the effect that the polynomials $g_i(x)$ differ only in their constant terms and that they are of degree $\leqq 2$.

Let us denote by B_n the fixed component of the linear system $|nD|$. We make the following preliminary observation: *the ring $R^*[D]$ is certainly not finitely generated over k if the following two conditions are satisfied:*

a) $B_n > 0$ *for all* n;

b) B_n *is bounded* (from above).

In fact, assume that $R^*[D]$ is finitely generated over k. Then there exists an integer N such that for all $n \geqq N$ the vector space L_n is spanned by the monomials $\Pi_{i=1}^{N} u_i^{\nu_i}$ such that $u_i \in L_i$ and $\Sigma i\nu_i = n$.

From this it follows that

$$B_n > \min \{ \Sigma_{i=1}^{N} \nu_i B_i \mid \Sigma i\nu_i = n \} .$$

If *a)* holds then the cycles $B_1, B_2, \ldots, B_N$ are all strictly positive, and hence condition *b)* is not satisfied, since

$$\max \{\nu_1, \nu_2, \ldots, \nu_N\} \to +\infty \text{ as } n \to +\infty .$$

With this observation in mind, let us consider an arbitrary non-singular surface F'. We denote by C' a hyperplane section of F' and we fix on F' an irreducible non-singular curve E', of genus $g \geqq 1$. We fix an integer h such that the following conditions are satisfied:

1) The system $|nC'|$ is regular (i.e., $i(nC') = s(nC') = 0$), for all $n \geqq h$.

2) the intersection number $m = (hC' + E' \cdot E')$ is greater than $2g$.

3) $(hC' \cdot E') > 2g - 2$.

We fix on E' m distinct points $P_1', P_2', \ldots, P_m'$ satisfying the following condition: *if g_m denotes the linear series cut out on E' by $|hC' + E'|$, then for all integers $n \geqq 1$*

$$(3) \qquad n(P_1' + P_2' + \ldots + P_m') \notin |ng_m| .$$

Such a set of m points exists, by 2). We now apply to F' successive locally quadratic transformations, with centers at $P_1', P_2', \ldots, P_m'$, and we denote by F the new (non-singular) surface thus obtained.

We now fix an irreducible non-singular curve Γ' in $|hC'|$ which does not pass through any of the m points P_i' and we denote by Γ and E the irreducible curves on F which correspond to Γ' and E' respectively (i.e., the proper transforms of Γ' and E'). We set $D=\Gamma+E$, *and we assert that the fixed component* B_n *of* $|nD|$ *is precisely* E. Thus B_n satisfies conditions a) and b) above, and $R^*|D|$ is not finitely generated over k.

The equality $B_n=E$ is proved as follows. The non-singular curves E and E' are birationally equivalent, and the divisor class determined by the cycle D on the curve E corresponds to the divisor class $|G_m-P_1'-P_2'-\ldots-P_m'|$ on E', where G_m is any divisor in g_m. The degree of this divisor class is zero, and by (3) it follows that the divisor *class* determined on E by nD contains no effective divisors. Therefore E *must be part of the fixed component* B_n, *for any* n. The systems $|n\Gamma|$ are regular, for all n (condition 1). Using condition 3), one finds that $\varrho(n\Gamma+iE)>\varrho(n\Gamma+(i-1)E)$ if $i<n$, where $\varrho(Z)$ is defined by (2). Thus, *assuming* that $|n\Gamma+(i-1)E|$ is regular (*for a given* $i<n$), it follows that $\dim|n\Gamma+iE|>\dim|n\Gamma+(i-1)E|$, and that consequently E *is not a fixed component of* $|n\Gamma+iE|$. Furthermore, again by condition 3), the intersection number $(n\Gamma+iE\cdot E)$ is greater than $2g-2$, and thus the linear series trace of $|n\Gamma+iE|$ on E is non-special. Hence $s(n\Gamma+iE)=0$, since, by assumption, $s(n\Gamma+(i-1)E)=0$. Therefore, $|n\Gamma+iE|$ is also a regular system. This inductive reasoning shows, in particular, that E is not a fixed component of $|n\Gamma+(n-1)E|$. Since $|nD|=|n\Gamma+nE|$, the assertion $B_n=E$ is proved.

§ 3. Connection with the generalized 14th problem of Hilbert.

In this section we still deal with an arbitrary non-singular variety V, of dimension r, and with a divisorial cycle D on V. The notations being the same as § 2, we note that $L_n\subset L_{n-1}$ and that the set-theoretic sum $\cup_{n=1}^{\infty}L_n$ is a subring of the function fiield Σ of V. We denote this ring by $R[D]$. We have proved in [5] that if $r=2$ then $R[D]$ is finitely generated over k (even if V is only a *normal* surface, not necessarily non-singular), and we have proposed the question whether this is still true for $r>2$.

This question was answered in the negative by REES in [2]. In the counter-example of REES, the variety V is a suitable birational transform of the direct product of an elliptic curve and a projective plane. We shall show (in outline) that under suitable conditions (specified below) the ring $R^*[D]$, defined in § 2, can be interpreted as a ring $R[\tilde{D}]$, where $\tilde{D}$ is a suitable divisorial cycle on some birational transform $\tilde{V}$ of the direct product $V \times \mathbb{P}_1$, where $\mathbb{P}_1$ is the projective line. Thus, the counter example (in dimension 2), which we have exhibited in § 2, gives also a general class of counter examples, in dimension 3, to our original generalization of the 14th problem of HILBERT.

We shall assume that D is effective and has no multiple components:

$$D = D_1 + D_2 + \ldots + D_q \ .$$

We consider the direct product

$$\bar{V} = V \times \mathbb{P}_1 \ .$$

We fix two distinct points P_∞, P_0, on $\mathbb{P}_1$, and on the variety $\bar{V}$ we consider the divisorial cycle

$$\bar{D} = D \times \mathbb{P}_1 + V \times P_\infty \ .$$

If t is a function on $\mathbb{P}_1$ such that $(t) = P_0 - P_\infty$, then one sees immediately that

$$R[\bar{D}] = R[D][t] \ ,$$

(and $R\,[\bar{D}]$ is therefore finitely generated over k if $R[D]$ is finitely generated). On the other hand, we have $R^*[D] \subset R[\bar{D}]$.

We set

$$\bar{\Gamma}_i = D_i \times P_0 \ , \qquad i = 1, 2, \ldots, q.$$

Then $\bar{\Gamma}_i$ is an irreducible subvariety of $\bar{V}$, of codimension 2. Since $\bar{V}$ is non-singular, the local ring $\mathfrak{o}_i$ of $\bar{\Gamma}_i$ on $\bar{V}$ is regular, of dimension 2. If $\mathfrak{m}_i$ denotes the maximal ideal of $\mathfrak{o}_i$, we denote by v_i the $\mathfrak{m}_i$ — adic divisorial valuation ([3]) of $k(\bar{V})$ associated with $\bar{\Gamma}_i$.

The following can be easily proved: *an element* ξ *of the ring* $R[\bar{D}]$ *belongs to the subring* $R^*[D]$ *if* and only if

$$(4) \qquad v_i(\xi) \geq 0 \; , \qquad i=1, 2, \ldots, q.$$

Now, let us furthermore assume that each of the prime cycles D_i is a non-singular variety and that any singular point P of the variety $D_1+D_2+\ldots+D_q$ is a normal crossing of tho (and only two) of the varieties D_i. Under these assumptions, a sequence of consecutive monoidal transformations of $\bar{V}$, with centers at $\bar{\Gamma}_1, \bar{\Gamma}_2, \ldots, \bar{\Gamma}_q$ (in some order), will lead to a non-singular birational transform $\tilde{V}$ of $\bar{V}$ which satisfies the following three conditions:

1) $\tilde{V}$ dominates $\bar{V}$.

2) Each of the divisorial valuations $v_i (i=1, 2, \ldots, q)$ is of the first kind with respenct to $\tilde{V}$ (i.e, the center of v_i on $\tilde{V}$ has codimension 1).

3) Every prime divisorial cycle on $\tilde{V}$ is either the proper transform of a prime divisorial cycle on $\bar{V}$ or is the center of one of the q divisorial valuations v_i.

It follows at once from these properties of $\tilde{V}$ and from the above characterization of the subring $R^*[D]$ of $R[\bar{D}]$ (see 4)) that if $\tilde{D}$ denotes the proper transform of $\bar{D}$ on the variety $\tilde{V}$, then

$$R^*[D] = R[\tilde{D}] \; .$$

We note that the various conditions that we have imposed on the cycle D and on its prime components D_i are easily satisfied in the counter example of § 2. In fact, in that counter example, the cycle D was the sum $\Gamma+E$ of two distinct non-singular

([3]) We recall that v_i is characterized by the following stipulation: if $z \in \mathfrak{m}_i^n$, $z \notin \mathfrak{m}^{n+1}$, then $v_i(z) = n$.

curves Γ, E, and it is permissible to assume that Γ and E have only simple intersections (by taking for Γ' a generic member of $|hC'+E'|$).

§ 4. Some preliminary results on linear systems free from base points.

The results which we proceed to state in this section will be needed later on only in the case in which the underlying variety V is either a non-singular curve or a normal surface, but for the sake of completeness we shall deal with the general case of a normal variety V, of any dimension r.

If H and G are two linear systems on V (not necessarily complete), we denote by $H+G$ the linear subsystem of $|H+G|$ which is spanned by thr cycles $X+Y$, where $X \in H$ and $Y \in G$ ($H+G=$ *minimal sum* of H and G).

Let $\{H_{i_1 i_2 \ldots i_m}\}$ be an infinite m-fold sequence of linear systems on $V(i_\alpha=0, 1, \ldots; \alpha=1, 2, \ldots, m)$ *not necessarily complete.* We denote by $H^1, H^2, \ldots, H^m$ the linear systems $H_{10\ldots0}, H_{010\ldots0}, H_{00\ldots1}$ respectively, while by $H_{0\ldots0}$ we means the system consisting of the zero-cycle only.

We make the following assumptions:

$$A) \qquad H_{i_1 i_2 \ldots i_m} + H_{j_1 j_2 \ldots j_m} \subset H_{i_1+j_1,\, i_2+j_2, \ldots, i_m+j_m}.$$

B) The linear systems $H^1, H^2, \ldots, H^m$ are free from base points.

We fix a cycle D^α in H^α. By *A*), the cycle $i_1 D^1 + i_2 D^2 + \ldots + i_m D^m$ belongs to $H_{i_1 i_2 \ldots i_m}$. We set:

$$L_{i_1 i_2 \ldots i_m} = \{x \in k(V) \mid (x) + i_1 D^1 + i_2 D^2 + \ldots + i_m D^m \in H_{i_1 i_2 \ldots i_m}\}.$$

Then $L_{(i)}$, where (i) stands fort he m-tuple $(i_1, i_2, \ldots, i_m)$, contains the field k and is a finite dimensional vector space over k, and by *A*) we have $L_{(i)} \cdot L_{(j)} \subset L_{(i+j)}$. We can therefore introduce the m-fold graded ring

$$R^* = \Sigma_{(i) \geqq 0} L_{i_1 i_2 \ldots i_m} t_1^{i_1} t_2^{i_2} \ldots t_m^{i_m},$$

where $t_1, t_2, \ldots, t_m$ are algebraically independent transcendentals over $k(V)$ and where the summation symbol stands for the weak direct sum (every element of R^* is a finite sum). We denote by $L^1, L^2, \ldots, L^m$ the vector spaces $L_{10\ldots0}, L_{010\ldots}, \ldots, L_{00\ldots1}$, respectively. Note that $L_{00\ldots0}=k$.

Let $\eta_1^a, \eta_2^a, \ldots, \eta^a_{s_a}$ be a minimal set of elements L^a such that the cycles $D^a, D^a+(\eta_1^a), \ldots, D^a+(\eta_s^a)$ have no common points. We note that from $B)$ it follows easily that $\eta_1^a, \eta_2^a, \ldots, \eta^a_{s_a}$ are algebraically independent elements of $k(V)$, over k, and that s_a is in fact the transcendence degree of the field $k(L^a)$, over k. We set

$$y_0^a=t_a, \; y_1^a=t_a\eta_1^a, \; \ldots, \; y_{s_a}^a=t_a\eta_{s_a}^a \; .$$

and we consider the ring

$$k[(y^1), (y^2), \ldots, (y^m)] \; (=k[y_0^1, \ldots, y_{s_1}^1 ; y_0^2, \ldots, y_{s_2}^2 ; \ldots; y_0^m \ldots, y_{s_m}^m]).$$

This ring has a natural m-fold grading, and its elements belong to the polynomial ring (over $k(V)$)

$$k(V)[t_1, t_2, \ldots, t_m] \; .$$

Note that we have $k[(y^1), (y^2), \ldots, (y^m)] \subset R^* \subset k(V)[t_1, t_2, \ldots, t_m]$.

THEOREM 1. *Under assumptions A) and B), the elements of R^* are integrally dependent on the ring $k[(y^1), (y^2), \ldots, (y^m)]$ (and therefore R^* is a finite module over $k[(y^1), (y^2), \ldots, (y^m)]$ and is finitely generated over k). Furthermore, there exists an integer N such that*

$$L_{i_1, i_2, \ldots, i_a+1, \ldots, i_m} = L_{i_1, i_2, \ldots, i_m} + L^a \; , \tag{5}$$

for all $i_a \geqq N$ and for all $i_1, i_2, \ldots, i_{a-1}, i_{a+1}, \ldots, i_m \geqq 0$ $(a=1, 2, \ldots, m)$.

The ring R^* coincides with the integral closure of $k[(y^1), (y^2), \ldots, (y^m)]$ in $k(V)(t_1, t_2, \ldots, t_m)$ if and only if all the linear systems $H_{(i)}$ are complete.

From the fact that R^* is a finitely generated multigraded module over the multigraded ring $k[(y^1), (y^2), \ldots, (y^m)]$ follows (as in the case of graded modules over polynomials rings) the existence of a characteristic polynomial of R^*. Thus, taking into account that $k[(y^1), (y^2), \ldots, (y^m)] \subset R^*$, we find the following.

COROLLARY 1.1. *Under assumptions A) and B), there exists a polynomial* $\varphi(X_1, X_2, \ldots, X_m)$, *of degree* s_α *in* X_α, *and there exists an integer* N, *such that*

$$\dim L_{i_1 i_2 \ldots i_m} (= 1 + \dim H_{i_1 i_2 \ldots i_m}) = \varphi(i_1, i_2, \ldots i_m)$$

if $i_1 \geqq N$, $i_2 \geqq N$, ..., $i_m \geqq N$.

Applying this corollary to the special case $m=1$, $H_i = |iD|$, where D is an effective divisorial cycle on V, we find

COROLLARY 1.2. *If* $|D|$ *is a complete linear system on* V, *free from base points* (4), *and if* s *is the dimension of* $k(L_1)$, *then we have, for all sufficiently large values of* n:

$$(6) \qquad 1 + \dim |nD| = a_0 \binom{n}{s} + a_1 \binom{n}{s-1} + \ldots + a_s. \qquad (a \neq 0)$$

where the coefficients a_i *are constants (in fact, integers).* (We recall the meaning of L_1: it is the set of all $x \in k(V)$ such that $(x) + D > 0$).

We can go a step further. In the case $m=1$, $H_i = |iD|$, the ring R^* is the ring $R^*[D]$ introduced in §2 (after replacing t_1 by t). By Theorem 1, this ring is the integral closure, in $k(V)(t)$, of the ring $k[y_0, y_1, \ldots, y_s]$ (here $y_0 = t$ and y_j^1 has been replaced by y_j). It follows, by well-known rerults on arithmetically normal varieties, that for *sufficiently large* h the ring $R^*[hD]$ is the homogeneous coördinate ring of an arithmetically normal variety V_h^* of dimension s. The varieties V_h^* obtained in this way are all biregularly equivalent: each of them is a normalization of the projective space (with general point $(y_0, y_1, \ldots, y_s)$), relative to the algebraic closure of $k(y_1/y_0, \ldots, y_s/y_0)$ in $k(V)$. We note also that V_h^* is a regular transform of V, since the system $|hD|$ has no base points, and that to the linear system $|hnD|$ there corresponds

(4) The absence of base points implies, in particular, that $|D|$ has no fixed components (and has positive dimension). It follows also that $s \geqq 1$.

on V_h^* the complete system cut out on V_h^* by the hypersurfaces of order n.

Denoting by V^* any of the biregularly equivalent varieties V_h^* and by D^* the cycle which corresponds on V^* to the cycle D, we find (see [4]), for large n:

$$(7) \qquad 1+\dim |nD| = (-1)^s \{p_a(V^*)+p_a(-nD^*)\} .$$

Now, let us suppose that V is a surface. The fact that $|D|$ has no fixed components implies, at any rate, that $(D^2)\geqq 0$. If $(D^2)>0$ then necessarily $s=2$ (since, by the definition of s, $s+1$ is the smallest number of cycles in $|D|$ which have no common points). If $(D^2)=0$, then necessarily $s=1$.

Assume $s=2$. In this case, the regular transformation $T:V\rightarrow V^*$ is birational, and is in fact biregular at all points of a generic member of $|D|$. Therefore $p_a(-n\,D^*)=p_a(-n\,D)$, and it follows from (7) that

$$(8) \qquad s(nD)=p_a(V^*)-p_a(V)$$

for large n.

Assume now $s=1$. In this case V^* is curve, and all the systems $|n\,D|$ are composite with an irreducible pencil, free from base points. The genus π of that pencil is the genus $p_a(V^*)$ of the curve V^*. If g denotes the genus of the generic member of the pencil and if D is the sum of h members of the pencil, then one obtains easily from (7) that for large n we have:

$$(9) \qquad \dim |nD| = hn-\pi$$

$$(10) \qquad s(nD)=hgn+\pi-p_a(V)=[p_a(D)+h-1]n+\pi-p_a(V) .$$

The assumption that D has no base points plays an essential role in Corollary 1.2. However, in the case of *non-singular* varieties, we can weaken this assumption, in view of the following

THEOREM 2. - *If V is a non-singular variety and $|D|$ is complete linear system on V which has only a finite number of base points, then $|nD|$ has no base points, for large n.*

Proof. Let C be a generic hyperplane section of V and let

$$H_{ij} = Tr_C |iC + jD| .$$

The family of linear systems H_{ij} on the non-singular variety C satisfies conditions $A)$ and $B)$, since C does not pass through any of the base points of $|D|$. Therefore, by Theorem 1 (relation (5)) we have

$$H_{ij} = H_{i,j-1} + H_{0,1}, \qquad j \geqq N (i\text{-arbitrary}).$$

This implies that

$$Tr_C |iC + jD| = Tr_C \{ |iC + (j-1) D| + |D| \} (j \geqq N;\ i\text{-arbitrary}) .$$

It follows that $|iC + Dj|$ is spanned by its two subsystems

$$(11) \qquad |iC + (j-1) D| + |D| , \quad |(i-1) C + jD| + C ,$$

where the second system has C as fixed component. *For fixed j, the system $|iC + jD|$ has no base points if i is sufficienty large. Let then j be a fixed but arbitrary integer $\geqq N$* and let i be such that $|iC + jD|$ has no base points. If P is any base point of $|D|$, then P is also a base point of the first of the two systems (11). It follows that P is *not* a base point of $|(i-1) C + jD|$. Applying the same argument to this last system (which is permissible, since $j \geqq N$) we conclude that P is not a base point of $|(i-2) C + jD|$. Ultimately we reach the conclusion that P is not a base point of $|jD|$, if $j \geqq N$. Q.E.D.

COROLLARY 2.1. *If V is a non-singular variety, then the conclusion of Corollary 1.2 continues to hold if $|D|$ has only a finite number of base points, provided that in the definition of the integer s we replace the space L_1 by the space L_j, where j is such that $|jD|$ has no base points. Similarly,* (7) *continues to hold (without any changes whatsoever; V^* is still one of the varieties V_h^*, h large).*

The proof of this corollary is straightforward.

COROLLARY 2.2. *If a complete linear system* $|D|$ *on a non-singular surface* F *has no fixed components, then* $|D|$ *has no base points if* n *is large* (5).

Obvious.

From these last two corollaries, and in view of (8) and (10), one reaches the following final conclusion we were aiming at in this section, in the case of non-singular surfaces:

THEOREM 3. *If a complete linear system* $|D|$ *on a non-singular surface* F *is such that some multiple* $|jD|$ *of this system has no fixed components, then for large* n *we have* $s(nD) = \text{const.}$, if $(D^2) > 0$, *and* $s(nD) = an + b$, *with* $a > 0$, if $(D^2) = 0$.

We note that the condition that $|jD|$ have no fixed components for some j is always satisfied in the following case: *D is an irreducible curve and* $(D^2) > 0$. For if $(D^2) > 0$ then the RIEMANN-ROCH inequality shows that $\dim |nD| \to +\infty$ as $n \to +\infty$, and thus there exist (infinitely many) integers j such that $\dim |jD| > \dim |(j-1)D|$. If D is an irreducible curve, then it is clear that for any such integer j the curve D is not a fixed component of $|jD|$.

More generally, the conclusion of Theorem 3 is still applicable if *all* the prime components of D have *positive* self-intersection number (in which case it is still true that $s(nD) = \text{const.}$ for large n, and it is easily seen in this case that $|nD|$ is irreducible, for n large). The difficulties begin when some of the prime components of D have self-intersection number $\leqq 0$; and the most serious difficulties are caused by prime components which have *negative* self-intersection number. Thus, Theorem 3 does not take us very far in as much as our general problem is concerned. Nevertheless, this theorem plays an essential role in the proofs of the various results which give the complete solution of our problem and which we now proceed to state (without proofs).

§ 5. STATEMENT OF THE RESULTS.

Let F be a non-singular surface and let D be a divisorial cycle on F.

---

(5) This is the correct version of a theorem which we have stated in [5].

NOTATION: If $D_1, D_2, \dots, D_q$ are the prime components of D, we denote by φ_D the quadratic form $\Sigma_{ij}(D_i \cdot D_j)\, x_i x_j$.

We say that φ_D is of type (s, t) if in the canonical form $\Sigma a_\alpha u_\alpha^2$ of φ_D there are s positive and t negative coefficients a_α. *It is well known that we always have* $s \leqq 1$ (this is a property of non-singular surfaces, proved by HODGE; see also GROTHENDIECK [1]). We write $\varphi_D < 0$ if φ_D is of type $(0, q)$; $\varphi_D \leqq 0$ if φ_D is of type $(0, t,)$ $t < q$. Thus, the only other possible case is the one in which φ_D is of type $(1, t)$, $t \leqq q - 1$.

DEFINITION *The cycle D is said to be* ARITHMETICALLY EFFECTIVE *if* $(D.Z) \geqq 0$ *for every effective cycle Z.*

THEOREM 4. *If $D > 0$, then there exists one and only one effective cycle ε, with* RATIONAL *coefficients, such that the following conditions are satisfied:*

1) *If $\varepsilon \neq 0$, then $\varphi_\varepsilon < 0$.*
2) *$D - \varepsilon$ is arithmetically effective.*
3) *If $\varepsilon \neq 0$, then $(D - \varepsilon \cdot E) = 0$ for every prime component E of ε.*

Furthermore, this unique cycle ε is necessarily such that $D - \varepsilon$ is effective.

COROLLARY 4.1. $\varepsilon = 0$ *if and only if D is arithmetically effective.*

If $\varepsilon = r_1 E_1 + r_2 E_2 + \dots + r_s E_s$ is an effective cycle with rational coefficients, then we denote by $[\varepsilon]$ the integral part $[r_1] E_1 + [r_2] E_2 + \dots + [r_s] E_s$ of ε.

From now on D denotes an effective cycle and ε is the corresponding (unique) cycle introduced in Theorem 4.

THEOREM 5. *If B_n is the fixed component of $|nD|$, then we have for each n: $[n\varepsilon] < B_n$. Furthermore, the effective cycle $B_n - [n\varepsilon]$ is bounded from above.*

COROLLARY 5.1. *B_n is bounded from above if and only if D is arithmetically effective.*

Let m be the smallest positive integer such that $m\varepsilon$ has integral coefficients, and let $\Delta = m(D - \varepsilon)$ (whence Δ is an effective and arithmetically effective cycle (Theorem 4), with integral coefficients). It follows from Theorem 5, that for any integer n' the cycle $n'm\varepsilon$ is parts of the fixed component of $n'mD$ and that $|n'mD| = n'm\varepsilon + |n'\Delta|$, $\dim |n'mD| = \dim |n'\Delta|$. From this follows, by a simple computation.

COROLLARY 5.2. *If* $n = mn' + r$, $0 \leqq r < m$, *then*

$$s(nD) = -\varrho(n\varepsilon) + s(n'\Delta + rD) \quad (12)$$

where $\varrho(\ )$ *is defined in* (1), §1.

Formula (12) reduces our problem essentially to the case of an arithmetically effective cycle D. In this case it is clear that the quadratic form φ_D is either of type $(1, t,)$ or we must have $\varphi_D \leqq 0$. The case in which φ_D is of type $(1, t,)$ is the more important of the two. In this case we have:

THEOREM 6. *If* $D > 0$ *is an arithmetically effective cycle and if* φ_D *is of type* $(1, t,)$, *then* $s(nD)$ *is bounded from above.*

Note. It is not difficult to show that if $D > 0$ is arithmetically effective then φ_D is of type $(1, t,)$ if and only if $(D^2) > 0$.

If, however, $\varphi_D \leqq 0$ we have the following.

THEOREM 7. *Let* $D > 0$ *be an arithmetically effective cycle such that* $\varphi_D \leqq 0$. *Then there are two possibilities:*

1) $\dim |nD| = 0$ *for all* n *(this can occur only if* $p_a(D) \geqq 1$). *In this case we have, for large* n:

$$s(nD) = [p_a(D) - 1]n - p_a(F) \quad (13)$$

2) $|nD| - B_n$ *is composite with an irreducible pencil of degree zero. If* π *is the genus of the pencil, then, for* n *large:*

$$\dim |nD| = \frac{n - r}{h} - \pi \quad (14)$$

$$s(nD) = \left[p_a(D) - 1 + \frac{1}{h}\right]n - \frac{r}{h} - \pi - p_a(F) \quad (15)$$

where h *is some fixed integer* $\geqq 1$ *and where* $n \equiv r \pmod{h}$, $0 \leqq r < h$.

From these theorems one concludes with the following general result:

THEOREM 8. *If D is any effective cycle on F, then there exists a finite number of polynomials* $\varphi_1(X), \varphi_2(X), \ldots, \varphi_N(X)$, *of* degree $\leqq 2$, *which differ only in their constant terms, such that for large n the superabundance* $s(nD)$ *is equal to* $\varphi_i(n)$ *for some i. With the only exceptions which are specified below, the polynomials* $\varphi_i(X)$ *are of degree zero (or — equivalently —* $s(nD)$ *is bounded from above) if and only if D is arithmetically effective (or — also — if and only if* B_n *is bounded from above).*

The exceptions are as follows:

a) *D is as in Theorem 7, part* 1), *and* $p_a(D) > 1$.

b) *D is as in Theorem 7, part* 2), *and* $p_a(D) > 1$.

or $p_a(D) = 0$ *and* $h > 1$.

As a special case of arithmetically effective cycles one has the *arithmetically positive* cycles, i.e. cycles D such that $(D \cdot Z) > 0$ if $Z > 0$. It is clear that if $D > 0$ is arithmetically positive then $D \neq 0$ and hence $(D^2) > 0$ (whence φ_D is of type $(1, t)$). In this case one finds that $|nD|$ *is ample, if n is large* (whence $s(nD) = 0$ if n is large). This result has also been proved by NAKAI [6]. Most of the difficulties one encounters in the general case of arithmetically effective cycles are absent in the case of arithmetically positive cycles.

REFERENCES

[1] A. Grothendieck, *Sur une note de Mattuck-Tate*, « J. für r. und angew. Math. », v. 200 (1958), pp. 208-215.

[2] P. Rees, *On a problem of Zariski*, « Ill. J. of Math. », v. 2 (1958), pp. 145-149.

[3] P. Samuel, *Sur les anneaux gradués*, « C. R. Acad. Cien. Brasil » (1958).

[4] O. Zariski, *Complete linear systems on normal varieties and a generalization of a lemma of Enriques-Severi*, « Ann. of Math. », v. 55 (1952), pp. 552-592.

[5] — —, *Intérpretations algébro-géométriques du quatorzième problème de Hilbert*, « Bull. Sc. Math. », v. 78 (1954), pp. 155-168.

[6] Yoshikazu Nakai, *Non-degenerate divisors on an algebraic surface*. « Journal of the Science of the Hiroshima University », v. 24, no. 1 (1960).

ANNALS OF MATHEMATICS
Vol. 76, No. 3, November, 1962
Printed in Japan

THE THEOREM OF RIEMANN-ROCH FOR HIGH MULTIPLES OF AN EFFECTIVE DIVISOR ON AN ALGEBRAIC SURFACE.

BY OSCAR ZARISKI*

(WITH AN APPENDIX BY DAVID MUMFORD)
(Received Feburary 13, 1962)

CONTENTS

PART I. THE PROBLEM AND ITS BACKGROUND

1. Formulation of the problem

Let F be a projective non-singular irreducible algebraic surface F, defined over an algebraically closed ground field k (of arbitrary characteristic). Let D be an arbitrary effective divisorial cycle on F. Our problem is *to find the dimension of the complete linear system* $|nD|$ *as a function of* n, *for large values of* n. If K is a canonical divisor on F and if we set for any divisorial cycle Z on F:

$$(1) \qquad \rho(Z) = \frac{Z(Z-K)}{2}, \qquad i(Z) = 1 + \dim |K - Z| ,$$

then[1] the Riemann-Roch theorem on F gives us only the inequality

* This research was supported by the Air Force Office of Scientific Research.

[1] Note that $\rho(Z) = p(-Z) - 1 = (Z^2) - p(Z) + 1$, where $p(-Z)$ is the arithmetic genus of the cycle $-Z$ (see Zariski [9, p. 592]).

$$\dim |nD| \geqq \rho(nD) + p_a - i(nD) , \tag{2}$$

where p_a is the arithmetic genus of F. The difference

$$s(nD) = \dim |nD| - [\rho(nD) + p_a - i(nD)] , \tag{3}$$

called *the superabundance of* $|nD|$, is thus a non-negative integer. Clearly, $i(nD) = 0$ if $D > 0$ and n is large. *Thus our problem is equivalent to finding* $s(nD)$ *as a function of* n, *for large* n.

From the sheaf-theoretic point of view our problem can be formulated as that of finding $h^0(nD)$ and $h^1(nD)$ as functions of n for large n (we use the customary notations of sheaf theory). In fact, we have $h^0(nD) = 1 + \dim |nD|$ and $h^1(nD) = s(nD)$. Note that $h^2(nD) = h^0(K - nD) = i(nD)$ and that the Euler-Poincaré characteristic $h^0(nD) - h^1(nD) + h^2(nD)$ is equal to the polynomial (of degree $\leqq 2$) $\rho(nD) + p_a + 1$.

2. A ring-theoretic aspect of the problem

In this section we shall deal with a non-singular variety V of any dimension r and with an effective divisorial cycle D on V. We denote by K the function field $k(V)$ and by L_n the finite-dimensional vector space (over k) consisting of the elements f of K such that the cycle $(f) + nD$ is effective (in symbols: $(f) + nD \succ 0$). Here (f) denotes the cycle of the function f (we include in L_n also the element zero of K). Let t be a transcendental over K and let $R^*[D]$ be the set of all elements of the polynomial ring $K[t]$ which are of the form $\sum a_i t^i$, $a_i \in L_i$. Since $L_i L_j \subset L_{i+j}$, $R^*[D]$ is a ring. This ring has a natural grading, the homogeneous component $R_i^*[D]$ of degree i being $L_i t^i$. We have $R_0^*[D] = k$, and

$$\dim_k R_i^*[D] = \dim_k L_i = 1 + \dim |iD| .$$

In the case of a *surface* V one of the results which we obtain in this paper will show that *there exists a finite number of polynomials* $g_1(X)$, $g_2(X), \cdots, g_h(X)$ *of one variable* X *such that for all large* n *the dimension of* $|nD|$ *is equal to the value which one of these polynomials assumes for* $X = n$. This particular statement[2] would have followed automatically by a well-known result on graded rings (see, for instance, the note [5] by P. Samuel) *if it were true that ring* $R^*[D]$ *is finitely generated over* k. However, this is not always true, already in the case in which V is a surface, as we shall now show by constructing a very general class of counter-examples.

[2] Our results are much more explicit than the above statement. For instance, it turns out that the polynomials $g_i(X)$ differ from each other only in their constant term and are of degree $\leqq 2$.

Let us denote by B_n the fixed component of $|nD|$. We make the following observation: *the ring $R^*[D]$ is certainly not finitely generated over k if the following two conditions are satisfied*:

(a) $B_n \neq 0$ *for all* n;

(b) B_n *is bounded (from above)*.

In fact, assume that $R^*[D]$ is finitely generated over k. Then there exists an integer N such that for $n \geqq N$ the vector space L_n is spanned by the monomials

$$\prod_{i=1}^{N} u_i^{\nu_i} \mid u_i \in L_i \, , \qquad \sum i\nu_i = n \, .$$

From this it follows at once that B_n dominates the cycle

$$\text{g.l.b. } \{\textstyle\sum_{i=1}^{N} \nu_i B_i \mid \sum i\nu_i = n\} \, .$$

It is obvious that we have for any $m, n : mB_n > B_{mn}$. Therefore each of the N cycles $(N!/i)B_i (i = 1, 2, \cdots, N)$ dominates the cycle $B_{N!}$, which, by (a), is strictly positive. It follows that $B_1, B_2, \cdots, B_N$ have at least one prime component in common, and hence condition (b) is not satisfied, since $\sum \nu_i$ approaches $+\infty$ as $n \to +\infty$.

With this observation in mind, let us consider an arbitrary non-singular surface F'. We fix on F' an arbitrary irreducible non-singular curve E', *of positive genus* g. We denote by C' a hyperplane section of F' and we fix an integer $h \geqq 1$ such that the following conditions are satisfied:

(1) The systems $|nC'|$ are regular (i.e., $i(nC') = s(nC') = 0$) for all $n \geqq h$.

(2) The intersection number $m = (hC' + E' \cdot E')$ is greater than $2g$.

(3) $(hC' \cdot E') > 2g - 2$.

We denote by $\mathfrak{A}$ the divisor class (of degree m) determined on E' by the cycle $hC' + E'$ and we fix on E' a set of m distinct points $P_1', P_2', \cdots, P_m'$ such that the following condition is satisfied *for each integer* $n \geqq 1$:

$$(4) \qquad n(P_1' + P_2' + \cdots + P_m') \notin n\mathfrak{A} \, .$$

Such a set of m points exists in view of (2) and in view of our assumption that $g > 0$.[3]

[3] *Proof. We must assume that k is not absolutely algebraic, of characteristic* $\neq 0$. We can find then on the Jacobian variety of E' a point of infinite order. We denote by $\mathfrak{B}_0$ the corresponding divisor class (of degree zero) on E'. The divisor class $\mathfrak{A} - \mathfrak{B}_0$ is of degree m, and the set of effective divisors belonging to that class is a complete linear series L of degree m and dimension $m - g$ (since $m > 2g - 2$). Since $m \geqq 2g + 1$, it is well-known that this series is not composite with an involution. In other words, there exists a curve E, of order m, in a projective space of dimension $m - g (\geqq g + 1 \geqq 2)$, which is birationally equivalent to E' and on which the hyperplanes cut out the series L. It follows that L contains sets consisting of m distinct points. Let $P_1', P_2', \cdots, P_m'$ be such a set. Then, for any $n \geqq 1$, we have $n(P_1' + \cdots + P_m') \in n\mathfrak{A} - n\mathfrak{B}_0$, and since $n\mathfrak{B}_0 \neq 0$, (4) follows.

We now apply to F' successive, locally quadratic transformations, with centers at $P'_1, P'_2, \cdots, P'_m$, and we denote by F the new (non-singular) surface, thus obtained. We fix an irreducible curve Γ' in $|hC'|$ which does not pass through any of the m points P'_i and we denote by Γ and E the irreducible curves on F which are the proper transforms of Γ' and E'. We set $D = \Gamma + E$ *and we assert that the fixed component* B_n *of* $|nD|$ *is precisely* E (n-arbitrary). Thus B_n satisfies conditions (a) and (b) above, and $R^*[D]$ is therefore not finitely generated over k.

The curve E is irreducible, non-singular and is birationally equivalent to E'. To prove our assertion that $B_n = E$ we observe that the divisor class determined on E by the cycle D corresponds (in the biregular correspondence between E and E') to the divisor class $\mathfrak{A} - \{P'_1 + P'_2 + \cdots + P'_m\}$ (where $\{P'_1 + P'_2 + \cdots + P'_m\}$ denotes the divisor class containing the divisor $P'_1 + P'_2 + \cdots + P'_m$). In view of (4) it follows that the divisor class (of degree zero) determined on E by nD does not contain the divisor zero. *Therefore* E *must be part of the fixed component* B_n *of* $|nD|$, for any n. On the other hand, we shall now show that the linear system $|n\Gamma + (n-1)E|$ has no fixed component. This will complete the proof of the equality $B_n = E$.

For this purpose it will be necessary to prove something more: namely, we prove that if $0 \leqq i < n$ *then the linear system* $|n\Gamma + iE|$ *is regular and has no fixed components.* We note that $|n\Gamma|$ corresponds to the linear system $|nhC'|$ on F', which is regular by condition (1). Since both F and F' are non-singular, it follows that also $|n\Gamma|$ is regular. Thus the above statement is true for $i = 0$. We shall therefore proceed by induction with respect to i, assuming that the statement is true for $i \leqq q$, where q is a given integer, $0 \leqq q < n-1$. By assumption, $|n\Gamma + qE|$ is regular, whence

$$(5) \qquad \dim |n\Gamma + qE| = \rho(n\Gamma + qE) + p_a ,$$

where $\rho(Z)$ is defined by (1). For any two cycles X, Y on F one verifies immediately the following equality:

$$(6) \qquad \rho(X + Y) = \rho(X) + \rho(Y) + (X \cdot Y) .$$

Hence

$$(7) \qquad \rho\bigl(n\Gamma + (q+1)E\bigr) = \rho(n\Gamma + qE) + \rho(E) + (n\Gamma + qE \cdot E) .$$

Since $(\Gamma + E \cdot E) = (D \cdot E) = 0$, we have $(E^2) = -(\Gamma \cdot E) = -(\Gamma' \cdot E')$, and since $\rho(E) = (E^2) - g + 1$ (see footnote[1]), we find from (7) that

$$(8) \quad \rho\bigl(n\Gamma + (q+1)E\bigr) = \rho(n\Gamma + qE) + (n - q - 1)(\Gamma' \cdot E') - g + 1 .$$

Now, $(\Gamma' \cdot E') = (hC' \cdot E') > 2g - 2$, by condition (3), and $q + 1 < n$. Therefore $(n - q - 1)(\Gamma' \cdot E') - g + 1 > g - 1 \geqq 0$. Thus, by (8):

$$(9) \qquad \rho(n\Gamma + (q+1)E) > \rho(n\Gamma + qE)\,.$$

Since $i(n\Gamma + qE) = 0$, we have *a fortiori* $i(n\Gamma + (q+1)E) = 0$, whence $\dim |n\Gamma + (q+1)E| \geqq \rho(n\Gamma + (q+1)E) + p_a$. From this and in view of (9), we deduce that $\dim |n\Gamma + (q+1)E| > \dim |n\Gamma + qE|$; in other words, *E is not a fixed* component of the system $|n\Gamma + (q+1)E|$, and thus—obviously—this system has no fixed components at all.

To prove that regularity of the system $|n\Gamma + (q+1)E|$ we have only to show that the superabundance $s(n\Gamma + (q+1)E)$ is zero, since the equality $i(n\Gamma + (q+1)E) = 0$ follows trivially from the equality $i(n\Gamma) = 0$. For this purpose we shall make use of a well-known result from the theory of linear systems on a non-singular algebraic surface F (this result will also be used repeatedly in the rest of the paper).

For any divisorial cycle X on F define *the irregularity of* X as follows:

$$(10) \qquad \operatorname{irreg}(X) = s(X) - i(X)\,.$$

Let Y be an irreducible curve on F. We denote by $\{X + Y\} \cdot Y$ the divisor class defined by $X + Y$ on the curve Y, and by $\mathrm{Tr}_Y |X + Y|$ the linear series cut out on Y by $|X + Y|$. The series $\mathrm{Tr}_Y |X + Y|$ is empty if and only if $|X + Y|$ does not exist or if Y is a fixed component of $|X + Y|$; if it is not empty then it is contained in the complete series $|\{X + Y\} \cdot Y|$. The curve Y may have singularities, and if it has, we consider it, *with its singularities*, as a curve of (virtual) genus $p(Y)$, where $p(Y)$ is the arithmetic genus of the cycle $Y (p(Y) \geqq$ effective genus of Y, with equality if and only if Y is a non-singular curve). Linear equivalence of divisors on Y is defined accordingly (see Rosenlicht [4′]), and in this sense we speak of complete linear series and of the Riemann-Roch theorem on Y. The series $\mathrm{Tr}_Y |X + Y|$ may not be complete (in the indicated sense), and we denote by $\delta_Y(X + Y)$ its deficiency:

$$(11) \qquad \delta_Y(X + Y) = \dim |\{X + Y\} \cdot Y| - \dim \mathrm{Tr}_Y |X + Y|\,.$$

It should be understood that in (11) we must set $\dim \mathrm{Tr}_Y |X + Y| = -1$ if the series $\mathrm{Tr}_Y |X + Y|$ does not exist; similarly we must set $\dim |\{X + Y\} \cdot Y| = -1$ if the divisor class determined on Y by $X + Y$ contains no effective divisors. We now denote by $i_Y(X + Y)$ the *index of specialty* of the divisor class $\{X + Y\} \cdot Y$ (the index of specialty is in the sense of the generalized definition of the canonical class on a curve Y with singularities; see Rosenlicht [4′]). Then the fact that we shall make repeated use of is the following equality[4]:

(12) $$\text{irreg}\,(X+Y) = \text{irreg}\,(X) + i_Y(X+Y) - \delta_Y(X+Y)\,.$$

Since $i(X+Y) \leqq i(X)$ and $\delta_Y(X+Y) \geqq 0$, (12) implies that

(13) $$s(X+Y) \leqq s(X) + i_Y(X+Y)\,.$$

By the Riemann-Roch theorem on the curve Y (of virtual genus $p(Y)$), we have $i_Y(X+Y) = 0$ if $(X+Y) > 2p(Y) - 2$. Therefore

(14) $$s(X+Y) \leqq s(X)\,, \qquad [\text{if } (X+Y\cdot Y) > 2p(Y) - 2]\,.$$

The assumptions under which (14) is valid are satisfied if $Y = E$ and $X = n\Gamma + qE$, where $q < n - 1$, since $(n\Gamma + (q+1)E \cdot E) = (n-q-1)(\Gamma' \cdot E) = (\Gamma' \cdot E') > 2g - 2$. Hence $s(n\Gamma + (q+1)E) \leqq s(n\Gamma + qE)$, and consequently $s(n\Gamma + (q+1)E) = 0$ since $s(n\Gamma + qE) = 0$. This completes the proof.

Note. Relation (12) is an immediate consequence of the following relations: $\dim|X+Y| = \rho(X+Y) + p_a + \text{irreg}\,(X+Y)$, $\dim|X| = \rho(X) + p_a + \text{irreg}\,(X)$ [see (3), § 1], $\dim|X+Y| = \dim|X| + \dim \text{Tr}_Y|X+Y| + 1$, $\rho(X+Y) = \rho(X) + \rho(Y) + (X\cdot Y)$, $\rho(Y) = (Y^2) - p(Y) + 1$ *and of the Riemann-Roch theorem* on Y: $\dim|\{X+Y\}\cdot Y| = (X+Y\cdot Y) - p(Y) + i_Y(X+Y)$. In sheaf-theory, relation (12) follows from the consideration of the exact sequence $0 \to \mathcal{L}(F, X'-Y) \to \mathcal{L}(F, X') \to \mathcal{L}(Y, X'\cdot Y) \to 0$, where $\mathcal{L}(F, Z)$ denotes the sheaf of F defined by a divisor Z, X' is a divisor linearly equivalent to $X+Y$, not having Y as component and not passing through the singular points of Y, and $\mathcal{L}(Y, X'\cdot Y)$ is the sheaf on Y defined by the divisor $X'\cdot Y$.

In the present context, the curve Y is our *non-singular* curve E. However, later on we shall have to apply relation (12) to the case in which the curve Y has singular points. In that case, the above derivation of (12) requires the knowledge of the Riemann-Roch theorem on curves Y *with singularities* (the genus of Y being then computed by regarding these singularities as virtually non-existent). However, we shall actually only need in our applications the inequality (14), and for this inequality one only needs the proof of the following fact: *if* $|D|$ *is a complete linear system which does not have* Y *as fixed component and if* $(D\cdot Y) > 2p(Y) - 2$, *then* $\dim \text{Tr}_Y|D| = (D\cdot Y) - p(Y) - \sigma$, *with* $\sigma \geqq 0$. For the convenience of the reader we give here a direct proof of this assertion.

We shall use induction with respect to the difference $p(Y) - \pi$, where π is the effective genus of Y, since our assertion is true if $p(Y) = \pi$ (in

[4] See O. Zariski [7, p. 70]. See also Note at the end of this section.

which case the curve Y has no singularities). Let P be a singular point of Y and let $T:F \to F'$ be the locally quadratic transformation of F, with center P. Let $E' = T(P)$ be the exceptional curve on F' introduced by T, let $D' = T(D)$ the T-transform of a cycle D in $|D|$ and let $Y' = T[Y]$ be the *proper* T-transform of Y. If $p = p(Y)$, $p' = p(Y')$ and P is an r-fold point of $Y(r > 1)$, then $p = p' + \{r(r-1)/2\}$. Since $\dim \mathrm{Tr}_{Y'} |D'| = \dim \mathrm{Tr}_Y |D|$, we can write

$$\text{(a)}\quad \dim \mathrm{Tr}_{Y'} |D'| = (D \cdot Y) - p' - \frac{r(r-1)}{2} - \sigma\,.$$

From $(D' \cdot E') = 0$ and $(E'^2) = -1$ follows that $(D' - hE' \cdot E') = h$, and therefore we obtain at once the following inequality

$$\text{(b)}\quad \dim |D'| \leqq \dim |D' - (r-1)E'| + \frac{r(r-1)}{2}\,.$$

We have $(Y' \cdot E') = r$ and hence $(D' - Y' - hE' \cdot E') = h - r < 0$ if $h < r$. From this it follows that hE' is a fixed component of $|D' - Y'|$ if $h \leqq r$. Therefore, for $h = r - 1$, we find

$$\text{(c)}\quad \dim |D' - Y'| = \dim |D' - (r-1)E' - Y'|\,.$$

Combining (b) and (c) we find that

$$\text{(d)}\quad \dim \mathrm{Tr}_{Y'} |D'| \leqq \dim |D' - (r-1)E'| - \dim |D' - (r-1)E' - Y'| + \frac{r(r-1)}{2} - 1$$
$$= \dim \mathrm{Tr}_{Y'} |D' - (r-1)E'| + \frac{r(r-1)}{2}\,.$$

Now, $(D' - (r-1)E' \cdot Y') = (D \cdot Y) - (r-1)r > 2p - 2 - (r-1)r = 2p' - 2$, and since $p(Y') - \pi(Y') = p' - \pi < p - \pi$, our induction hypothesis implies that $\dim \mathrm{Tr}_{Y'} |D' - (r-1)E'| \leqq (D \cdot Y) - (r-1)r - p'$. Consequently, by (d), $\dim \mathrm{Tr}_{Y'} |D'| \leqq (D \cdot Y) - p' - \{(r-1)r/2\}$. This, in view of (a), yields the desired inequality $\sigma \geqq 0$.

3. Connection with the generalized 14th problem of Hilbert

In this section we shall still deal with an arbitrary non-singular variety V, of dimension r, and with an effective divisorial cycle D on V. The notations being the same as in § 2, we note that $L_n \subset L_{n+1}$ and that the set-theoretic sum $\bigcup_{n=1}^{\infty} L_n$ is a subring of the function field K of V/k. We denote this ring by $R[D]$. We have proved in [10] that if $r=2$ then $R[D]$ is finitely generated over k (even if V is only a *normal* surface, not necessarily non-singular), and we have proposed there the question (which is a generalization of Hilbert's 14th problem) whether this is still true if $r > 2$. This question was answered in the negative by Rees in [4]. In the counter-example of Rees the variety V is a suitable birational transform of the direct product of an elliptic curve and of the projective plane. We shall now show that under suitable conditions *the ring* $R^*[D]$ *defined in* § 2 *can be interpreted as a ring* $R[\tilde{D}]$, *where* $\tilde{D}$ *is a suitable divisorial cycle on some birational transform of the direct product* $V \times \mathbf{P}_1$ *of* V *and the projective line* $\mathbf{P}_1$. Thus, the counter-example in dimension 2 which we have constructed in § 2 gives also a general class of counter-examples, in dimension 3, to the question which was solved negatively by Rees.

We shall assume that D has no multiple prime components:

$$D = D_1 + D_2 + \cdots + D_h ,$$

where the D_i are then distinct prime divisorial cycles on V. We consider the direct product

$$\bar{V} = V \times \mathbf{P}_1 .$$

We fix two distinct points P_∞, P_0 on $\mathbf{P}_1$, and we consider on the variety $\bar{V}$ the following divisorial cycle:

$$\bar{D} = D \times \mathbf{P}_1 + V \times P_\infty .$$

If t is a function on $\mathbf{P}_1$ such that $(t) = P_0 - P_\infty$, then $k(\bar{V}) = K(t)$, and it is seen immediately that the subring $R[\bar{D}]$ of $K(t)$ is given by

$$R[\bar{D}] = R[D][t] . \tag{15}$$

[Thus $R[\bar{D}]$ is finitely generated over k if $R[D]$ is finitely generated over k, and this is centainly the case if $r = 2$]. It is also clear from (15) that

$$R^*[D] \subset R[\bar{D}] .$$

We set

$$\bar{\Gamma}_i = D_i \times P_0 , \qquad i = 1, 2, \cdots, h . \tag{16}$$

Each $\bar{\Gamma}_i$ is an irreducible subvariety of $\bar{V}$, of codimension 2. Since $\bar{V}$ is non-singular, the local ring $\bar{\mathfrak{o}}_i$ of $\bar{\Gamma}_i$ on $\bar{V}$ is regular, of dimension 2. If $\bar{\mathfrak{m}}_i$ denotes the maximal ideal of $\bar{\mathfrak{o}}_i$, we denote by $\bar{v}_i$ the $\bar{\mathfrak{m}}_i$-adic divisorial valuation of $k(\bar{V})$ (we recall that $\bar{v}_i$ is characterized by the following stipulation: if $z \in \bar{\mathfrak{m}}_i^n$, $z \notin \bar{\mathfrak{m}}_i^{n+1}$ then $\bar{v}_i(z) = n$).

We now prove the following:

An element ξ *of of the ring* $R[\bar{D}]$ *belongs to the subring* $R^*[D]$ *if and only if*

$$\bar{v}_i(\xi) \geqq 0 , \qquad i = 1, 2, \cdots, h . \tag{17}$$

PROOF. Let v_i be the valuation of $k(V)$ defined by the prime divisorial cycle D_i and let $\mathfrak{m}_i$ be the maximal ideal of the valuation ring $\mathfrak{o}_i$ of v_i ($\mathfrak{o}_i =$ local ring of D_i on V). Let $\mathfrak{m}_i = \mathfrak{o}_i\tau_i$. From the definition (16) of $\bar{\Gamma}_i$ it is clear that $\bar{\mathfrak{o}}_i$ is the ring of quotients of $\mathfrak{o}_i[t]$ with respect to the prime ideal generated in $\mathfrak{o}_i[t]$ by τ_i and t, and that therefore τ_i and t are regular parameters of $\bar{\mathfrak{o}}_i$. If $\xi = \sum_n a_n t^n$ is an element of $R[\bar{D}]$ ($a_n \in R[D]$ and if $v_i(a_n) = s_n$, then from the definition of the $\bar{\mathfrak{m}}_i$-adic valuation $\bar{v}_i$ it follows that $\bar{v}_i(\xi) = \min_n \{s_n + n\}$. Hence $\bar{v}_i(\xi) \geqq 0$ if and only if $v_i(a_n) \geqq -n$, all n. Since $a_n \in R[D]$, it follows that $\bar{v}_i(\xi) \geqq 0$ for $i = 1, 2, \cdots, h$ if and only if $(a_n) + nD \succ 0$, i.e., if and only if $\xi \in R^*[D]$. This proves our assertion.

It follows that the elements ξ of $R^*[D]$ are those and only those elements of $k(\bar{V})$ which satisfy conditions (17) *and* the following condition:

$$v_{\bar{\Gamma}}(\xi) \geqq 0\ , \tag{18}$$

for any prime divisorial cycle $\bar{\Gamma}$ on $\bar{V}$ which is not a component of $\bar{D}$ (here $v_{\bar{\Gamma}}$ denotes the valuation of $k(\bar{V})$ defined by $\bar{\Gamma}$).

Now, let us furthermore assume that each of the prime cycles D_i is a variety *without singular points* and that each singular point of the variety $D_1 + D_2 + \cdots + D_h$ is a normal crossing of two (and only two) of the D_i. Under these assumptions it is well-known (and is easily seen) that the sequence of h consecutive monoidal transformations of $\bar{V}$, with centers at $\bar{\Gamma}_1, \bar{\Gamma}_2, \cdots, \bar{\Gamma}_h$ (in some arbitrary order) leads to a non-singular variety $\tilde{V}$ which satisfies the following conditions:

(a) $\tilde{V}$ dominates $\bar{V}$;

(b) each of the valuations $\bar{v}_i$ is of the first kind with respect to $\tilde{V}$ and has therefore on $\tilde{V}$ a center of codimension 1.

(c) every prime divisorial cycle on $\tilde{V}$ is either the proper transform of a prime divisorial cycle on $\bar{V}$ or is the center of one of the $\bar{v}_i$.

It follows from these properties of $\tilde{V}$ that if we denote by $\tilde{D}$ the proper transform of $\bar{D}$ on $\tilde{V}$, then the conditions (17) and (18) are equivalent to the following condition: $v_{\tilde{\Gamma}}(\xi) \geqq 0$ *for any prime divisorial cycle* $\tilde{\Gamma}$ *on* $\tilde{V}$ *which is not a prime component of* $\tilde{D}$. This shows that $R^*[D] = R[\tilde{D}]$, which is the result we were aiming at.

We note that the various conditions which we have imposed on the cycle D and its prime components D_i are easily satisfied in the counter-example of § 2. In fact, in that counter-example we had $D = \Gamma + E$. Since E' was a non-singular curve on F', it is sufficient to take for Γ' a generic member of $|hC'|$ and take h sufficiently high.

Part II. Preliminary Results on Linear Systems Free from Base Points

In this part of the paper we shall deal with the general case of a normal variety V of any dimension r (over an algebraically closed field k). In the last section of Part II the results obtained will be applied to the case in which V is a non-singular surface.

4. On certain graded subrings in function fields

Let H be a linear system on V (we exclude the trivial case in which H consists only of the cycle 0). We fix a cycle D in H and we set

$$L = \{x \in k(V) \mid (x) + D \in H\} .$$

Here $k(V)$ denotes the field of rational functions on V and (x) denotes the divisor of x on V (we agree to include in L the element zero of $k(V)$).

Lemma 4.1. *If H has no base points and if s denotes the transcendence degree of the field $k(L)/k$, then $s > 0$ and any s cycles in H have a common point, but there exist s cycles $D_1, D_2, \cdots, D_s$ in H such that the $s+1$ cycles $D, D_1, D_2, \cdots, D_s$ have no common point. Furthermore, if we fix, for such a set of cycles $D_1, D_2, \cdots, D_s$, a set of s elements $\eta_1, \eta_2, \cdots, \eta_s$ in L such that $(\eta_h) + D = D_h$, then the s elements η_h constitute a transcendence* basis of $k(L)/k$.

Proof. The assumption that H has no base points implies that $\dim H > 0$. Hence $L \neq k$, and thus $s > 0$. The vector space L/k determines (up to a non-singular projective transformation) a projective model V' of the field $k(L)/k$, the homogeneous coordinates of a general point of V'/k being given by the basis elements of L/k. There is a natural rational transformation T of V onto V' such that the T-image of the linear system H on V is the system of hyperplane sections of V' (see Zariski [11, p. 10]), and we have $\dim V' = s > 0$. Since, by assumption, H has no base points, *T is a regular map*, and therefore, given a set of cycles in H, these cycles will have a common point on V if and only if the corresponding hyperplane sections of V' have a common point on V'. Since $\dim V' = s$, this shows that any s cycles in H have a common point on V. On the other hand, since for any hyperplane section Z of V' one can find other s hyperplane sections which have no common point on Z, the assertion concerning the existence of the s cycles $D_1, D_2, \cdots, D_s$ is proved.

The cycles $D, D_1, \cdots, D_s$ span a linear subsystem H_0 of H, and also H_0 is free from base points (by our choice of the cycles D_h). There is a corresponding subspace L_0 of L, namely $L_0 = k + k\eta_1 + k\eta_2 + \cdots + k\eta_s$, and a corresponding projective model V_0 of $k(L_0)/k$, and $1, \eta_1, \eta_2, \cdots, \eta_s$ are homogeneous coordinates of a general point of V_0/k. Also V_0 is a regular image of V, the transform of the linear system H_0 being the system of hyperplane sections of V_0. By the regularity of the map $V \rightarrow V_0$ and in view of the fact that any s cycles in H_0 have a common point on V, it follows that any s hyperplane sections of V_0 have a common point on V_0. This implies that $\dim V_0 \geqq s$, and since $k(V_0) = k(\eta_1, \eta_2, \cdots, \eta_s)$, the proof of the lemma is now complete.

If H and G are two (not necessarily complete) linear systems on V we shall denote by $H + G$ the linear subsystem of $|H + G|$ which contains all the cycles of the form $X + Y$, $X \in H$, $Y \in G$ ($H + G$ = *minimal sum of* H and G).

Let $\{H_{i_1 i_2 \cdots i_m}\}$ be an infinite m-fold sequence of linear systems on V ($i_\alpha = 0, 1, 2, \cdots; \alpha = 1, 2, \cdots, m$). We shall denote by $H^1, H^2, \cdots, H^m$ the linear systems $H_{10\cdots 0}, H_{01\cdots 0}, \cdots, H_{00\cdots 1}$ respectively. *We make the following assumptions*:

(A) $H_{i_1 i_2 \cdots i_m} + H_{j_1 j_2 \cdots j_m} \subset H_{i_1 + j_1, i_2 + j_2, \cdots, i_m + j_m}$.

(B) *The linear systems* $H^1, H^2, \cdots, H^m$ *are free from base points.*

Note that (A) implies that $H_{00\cdots 0}$ consists only of the divisorial cycle 0. On the other hand, (B) implies that $\dim H^\alpha > 0 (\alpha = 1, 2, \cdots, m)$.

We fix a cycle D_α in H^α. By (A), the cycle $i_1 D_1 + i_2 D_2 + \cdots + i_m D_m$ belongs to $H_{i_1 i_2 \cdots i_m}$. We set

$$L_{\{i\}} = L_{i_1 i_2 \cdots i_m} \tag{19}$$
$$= \{x \in k(V) \mid (x) + i_1 D_1 + i_2 D_2 + \cdots + i_m D_m \in H_{i_1 i_2 \cdots i_m}\} .$$

The set $L_{\{i\}}$ contains the ground field k and is a finite dimensional vector space over k. By (A) we have $L_{\{i\}} \cdot L_{\{j\}} \subset L_{\{i+j\}}$. Therefore, if we adjoin to $k(V)$ a set of m indeterminates $t_1, t_2, \cdots, t_m$ and consider the following subset R^* of the polynomial ring $k(V)[t_1, t_2, \cdots, t_m]$:

$$R^* = \{\textstyle\sum_{\{i\}} x_{\{i\}} t_1^{i_1} t_2^{i_2} \cdots t_m^{i_m} \mid x_{\{i\}} \in L_{\{i\}}\} , \tag{20}$$

then R^* is a subring of $k(V)[t_1, t_2, \cdots, t_m]$ and has a natural m-fold grading, the homogeneous component $R^*_{\{i\}}$, of degree $\{i_1, i_2 \cdots i_m\}$, being given by

$$R^*_{\{i\}} = L_{i_1 i_2 \cdots i_m} t_1^{i_1} t_2^{i_2} \cdots t_m^{i_m} . \tag{21}$$

We have $R^*_{\{0\}} = k$. We set $L^1 = L_{10\cdots 0}, L^2 = L_{01\cdots 0}, \cdots, L^m = L_{00\cdots 1}$, and we denote by $R^{*1}, R^{*2}, \cdots, R^{*m}$ the homogeneous components $R^*_{10\cdots 0}, R^*_{01\cdots 0}, \cdots, R^*_{00\cdots 1}$ of R^*, respectively, so that we have

$$R^{*\alpha} = L^\alpha t_\alpha\,, \qquad \alpha = 1, 2, \cdots, m\,.$$

We note that $L^\alpha \neq k$ since $\dim H^\alpha > 0$.

Let s_α be the transcendence degree of the field $k(L^\alpha)/k$. In view of assumption (B), the preceding Lemma 4.1 is applicable to each of the m linear systems H^α. Let then $D_{\alpha 1}, D_{\alpha 2}, \cdots, D_{\alpha s_\alpha}$ be s_α cycles in H^α which have no common point on D_α and let $\eta_{\alpha h}$ be an element of L^α such that $(\eta_{\alpha h}) + D_\alpha = D_{\alpha h}(h = 1, 2, \cdots, s_\alpha)$. By Lemma 4.1, the s_α elements $\eta_{\alpha h}$ are algebraically independent over k and constitute a transcendence basis of $k(L^\alpha)/k$. We set

$$y_0^{(\alpha)} = t_\alpha\,, \qquad y_h^{(\alpha)} = t_\alpha \eta_{\alpha h}$$

and we consider the ring

$$R = k[\{y^{(1)}\}, \{y^{(2)}\}, \cdots, \{y^{(m)}\}]\,, \tag{22}$$

generated over k by the $(s_1 + 1) + (s_2 + 1) + \cdots + (s_m + 1)$ elements $y_{h_\alpha}^{(\alpha)}(h_\alpha = 0, 1, \cdots, s_\alpha; \alpha = 1, 2, \cdots, m)$. This ring is a graded subring of R^* [see (20)], and the homogeneous component $R_{\{i\}}$ [$= R \cap R^*_{\{i\}}$; see (21)] of R, of degree $\{i_1, i_2, \cdots, i_m\}$, is the set of all elements of the form $\varphi(\{y^{(1)}\}, \{y^{(2)}\}, \cdots, \{y^{(m)}\})$, where φ is a multi-homogeneous polynomial (with coefficients in k) in the m sets of elements $\{y^{(1)}\}, \{y^{(2)}\}, \cdots, \{y^{(m)}\}$, of degree i_α in the $s_\alpha + 1$ elements of the set $\{y^{(\alpha)}\}$. We note that the $s_\alpha + 1$ elements of the set $\{y^{(\alpha)}\}$ are algebraically independent over k; this follows from the fact that $\eta_{\alpha 1}, \eta_{\alpha 2}, \cdots, \eta_{\alpha s_\alpha}$ are algebraically independent over k and that t_α is a transcendental over $k(V)$.

Let $\bar{R}$ be the integral closure of R in the field $k(V)(t_1, t_2, \cdots, t_m)$. Since $R \subset k(V)[t_1, t_2, \cdots, t_m]$, we have also $\bar{R} \subset k(V)[t_1, t_2, \cdots, t_m]$. It is known that $\bar{R}$ has a natural m-fold grading which is an extension of the grading of R (see Zariski-Samuel [6, v. 2, Ch. VII, Th. 11, p. 157], for the case of rings with a simple grading; the proof for multigraded rings is quite similar). Furthermore, $\bar{R}$ is a *finite* (graded) R-module.

THEOREM 4.2. *Let $\bar{H}_{\{i\}}$ $(=\bar{H}_{i_1 i_2 \cdots i_m})$ be the complete linear system containing $H_{\{i\}}$ and let*

$$\bar{L}_{\{i\}} = \{x \in k(V) \mid (x) + i_1 D_1 + i_2 D_2 + \cdots + i_m D_m \in \bar{H}_{\{i\}}\}$$

[*whence $\bar{L}_{\{i\}} \supset L_{\{i\}}$; see* (19)]. *Then:*

(1) *Then union of the sets $\bar{L}_{\{i\}}$ is the integral closure of $k[\eta_{11}, \eta_{12}, \cdots, \eta_{1s_1}; \eta_{21}, \eta_{22}, \cdots, \eta_{2s_2}; \cdots; \eta_{m1}, \eta_{m2}, \cdots, \eta_{ms_m}]$ in the field $k(V)$.*

(2) $\bar{R} = \{\sum_{\{i\}} x_{\{i\}} t_1^{i_1} t_2^{i_2} \cdots t_m^{i_m} \mid x_{\{i\}} \in \bar{L}_{\{i\}}\}$, *whence $R^* \subset \bar{R}$ and also R^* is a finite R-module.*

(3) *There exists an integer N such that for each* $\alpha = 1, 2, \cdots, m$ *we have*

$$L_{i_1 i_2 \cdots, i_\alpha + 1, \cdots, i_m} = L_{i_1 i_2 \cdots i_m} L^\alpha , \tag{23}$$

or — equivalently:

$$H_{i_1 i_2 \cdots, i_\alpha + 1, \cdots, i_m} = H_{i_1 i_2 \cdots i_m} + H^\alpha , \tag{24}$$

whenever $i_\alpha \geqq N$ *(while the remaining* $m - 1$ *indices* i_β *are arbitrary).*

Proof. To avoid typographically cumbersome notations we shall prove the theorems only in the case $m = 2$; for arbitrary m the proof is the same (actually, we shall only need this theorem in the cases $m = 1, 2$).

We shall need a lemma which we now proceed to state.

Let $R = \sum R_{\{i\}}$ be a ring with an m-fold grading; here $\{i\}$ stands for a set of m non-negative indices $i_1, i_2, \cdots, i_m$. Let $\Omega = \sum \Omega_{\{i\}}$ be a graded R-module (whence $R_{\{i\}}\Omega_{\{j\}} \subset \Omega_{\{i+j\}}$). We write $\{i\} \geqq N$ if $i_\alpha \geqq N$ for $\alpha = 1, 2, \cdots, m$. Furthermore, we set $R^1 = R_{10\cdots 0}$, $R^2 = R_{01\cdots 0}, \cdots, R^m = R_{00\cdots 1}$.

We shall say that Ω *has the property* **(a)** if there exists an integer N such that

(a) $\Omega_{\{i+j\}} = R_{\{j\}}\Omega_{\{i\}}$ *for all* $\{i\} \geqq N$ *and all* $\{j\}$.

We shall say that Ω *has the property* **($\mathbf{b}_\alpha$)** $(1 \leqq \alpha \leqq m)$ if there exists an integer N such that

($\mathbf{b}_\alpha$) $\Omega_{i_1 i_2 \cdots, i_\alpha + 1, \cdots i_m} = R^\alpha \Omega_{i_1 i_2 \cdots i_m}$, *for all* $i_\alpha \geqq N$ *(the remaining* $m - 1$ *indices* i_β *being arbitrary)*[5].

Lemma 4.3. *Assume that there exists an integer M such that* Ω *is generated over R by the homogeneous components* $\Omega_{\{\nu\}}$ *such that* $\{\nu\} \leqq M$. *Then*:

(1) *If R has property* **(a)**, *so does* Ω.

(2) *If R has property* **($\mathbf{b}_\alpha$)**, *so does* Ω.

Proof of the Lemma (in the case $m = 2$).

(1) By assumption we have

$$\Omega_{i+s, j+t} = \sum\nolimits_{\nu, \mu \leqq M} R_{i+s-\nu, j+t-\mu}\Omega_{\nu\mu} , \tag{25}$$

where we set $R_{qr} = 0$ if at least one of the indices q, r is negative. If R has property **(a)** then there exists an integer N such that

$$R_{i+s-\nu, j+t-\mu} = R_{st}R_{i-\nu, j-\mu}$$

if $i, j \geqq M + N$ (and $\nu, \mu \leqq M$). Substituting into (25) we find:

$$\Omega_{i+s, j+t} = R_{st}\Omega_{ij} , \qquad \text{if } j \geqq M + N. \tag{26}$$

This proves the first part of the lemma.

[5] For $m = 1$, properties **(a)** and **(b)** are equivalent.

(2) We prove the second part of the lemma for $\alpha = 1$. We set

$$\begin{aligned} R^1 &= \textstyle\sum_i R_{i0}\,, & R_i^1 &= R_{i0}\,. \\ R^{1,j} &= \textstyle\sum_i R_{ij}\,, & R_i^{1,j} &= R_{ij}\,. \\ \Omega^{1,j} &= \textstyle\sum_i \Omega_{ij}\,, & \Omega_i^{1,j} &= \Omega_{ij}\,. \end{aligned}$$

Then R^1 is a (simply) graded ring, and, for any j, $R^{1,j}$, $\Omega^{1,j}$ are graded R^1-modules. If R has property $(\mathbf{b}_1)$ there exists an integer N such that

$$R_{i+1}^{1,j} = R_1^1 R_i^{1,j}\,, \qquad \text{if } i \geqq N\,. \tag{27}$$

In (27) we may also allow j to be negative, provided we set $R_i^{1,j} = 0$ if $j < 0$. By repeated application of (27) we deduce that the R^1-module $R^{1,j}$ is generated by its homogeneous components $R_\mu^{1,j}$, $0 \leqq \mu \leqq N$, and we have therefore

$$R_i^{1,j} = \textstyle\sum_{\mu=0}^N R_{i-\mu}^1 R_\mu^{1,j} = \sum_{\mu=0}^N R_{i-\mu}^1 R_{\mu j}\,, \tag{28}$$

for *all* i and j (always with the stipulation that $R_q^1 = 0$ if $q < 0$). We have by (25)

$$\Omega_i^{1,j} = \textstyle\sum_{\mu,\nu=0}^M R_{i-\mu}^{1,j-\nu}\Omega_{\mu\nu}\,,$$

and hence, by (28)

$$\begin{aligned} \Omega_i^{1,j} &= \textstyle\sum_{\mu,\nu=0}^M \sum_{q=0}^N R_{i-\mu-q}^1 R_{q,j-\nu}\Omega_{\mu\nu} \\ &= \textstyle\sum_{\mu=0}^M \sum_{q=0}^N R_{i-\mu-q}^1 \Omega_{q+\mu,j} \\ &= \textstyle\sum_{h=0}^{M+N} R_{i-h}^1 \Omega_h^{1,j}\,. \end{aligned}$$

This shows that the R^1-module $\Omega^{1,j}$ is generated by its homogeneous components of degree $\leqq M + N$. On the other hand, relation (27) shows (in the case $j = 0$) that R^1 has property (**a**). Therefore, by the first part of the lemma (as applied to simply graded modules), and using the specific estimate given in (26), we find that

$$\Omega_{i+1}^{1,j} = R_1^1 \Omega_i^{1,j}\,, \qquad \text{if } i \geqq M + 2N\,.$$

This completes the proof of the lemma.

We now proceed with the proof of the theorem (in the case $m = 2$).

(1) Let x be any element of $\bar{L}_{ij}$. We consider any valuation v of $k(V)/k$ which is non-negative on the ring $k[\eta_{11}, \eta_{12}, \cdots, \eta_{1s_1}, \eta_{21}, \cdots, \eta_{2s_2}]$. We assert that the center P of v on V does not belong to any of the two cycles D_1, D_2. For, assuming the contrary, let say, $P \in D_1$. Since the (effective) cycles $D_1, D_1 + (\eta_{11}), \cdots, D_1 + (\eta_{1s_1})$ have no common points, we may assume that, say, $P \notin D_1 + (\eta_{11})$. Then P belongs to the polar cycle of the element η_{11} but does not belong to its cycle of zeros. This implies that $v(\eta_{11}) < 0$,

a contradiction.

From the fact that $P \notin D_\alpha(\alpha = 1, 2)$ and that the polar cycle of x is dominated by $iD_1 + jD_2$ (since $x \in \bar{L}_{ij}$) it follows that $v(x) \geqq 0$. Since this inequality holds for any valuation v which is non-negative on the ring $k[\eta_{11}, \eta_{12}, \cdots, \eta_{2s_2}]$, it follows that x is integrally dependent on that ring.

Conversely, if an element x of $k(V)$ is integrally dependent on the ring $k[\eta_{11}, \eta_{12}, \cdots, \eta_{2s_2}]$, then it is clear that any prime component of the polar cycle of x must be a component of one of the cycles D_1, D_2. Therefore x belongs to one of the sets $\bar{L}_{ij}$. This completes the proof of the first part of the theorem.

(2) The homogeneous elements of $\bar{R}$ are of the form $\omega = xt_1^i t_2^j$, $x \in k(V)$. To prove the second part of the theorem we have to show an element ω of the form $\omega = xt_1^i t_2^j$ belongs to $\bar{R}$ if and only if $x \in \bar{L}_{ij}$, i.e., if and only if $(x) + iD_1 + jD_2 \succ 0$.

Assume first that $\omega \in \bar{R}$. We have then an equation of integral dependence for ω over R, whose coefficients may be assumed to be homogeneous elements of R:

$$\omega^n + \varphi_1(\{y^{(1)}\}, \{y^{(2)}\})\omega^{n-1} + \cdots + \varphi_n(\{y^{(1)}\}, \{y^{(2)}\}) = 0 ,$$

where φ_q is a doubly homogeneous polynomial in the two sets of elements $\{y^{(1)}\}, \{y^{(2)}\}$, of degrees qi, qj respectively. Dividing through this equation by $t_1^{in} t_2^{jn} (= y_0^{(1)in} y_0^{(2)jn})$, we find

$$x^n + \psi_1(\eta_{11}, \eta_{12}, \cdots, \eta_{2s_2})x^{n-1} + \cdots + \psi_n(\eta_{11}, \eta_{12}, \cdots, \eta_{2s_2}) = 0 ,$$

where ψ_q is a polynomial, of degree $\leqq qi$ in the elements $\eta_{11}, \eta_{12}, \cdots, \eta_{1s_1}$, and of degree $\leqq qj$ in the elements $\eta_{21}, \eta_{22}, \cdots, \eta_{2s_2}$. Since this is an equation of integral dependence on the ring $k[\eta_{11}, \eta_{12}, \cdots, \eta_{2s_2}]$, the first part of the theorem tells us that x belongs to some $\bar{L}_{\mu\nu}$. If a prime component Γ of D_α occurs in D_α with multiplicity $h_\alpha(\alpha = 1, 2)$, then $\psi_q(\eta_{11}, \eta_{12}, \cdots, \eta_{2s_2})$ becomes infinite on Γ at most to the order $qih_1 + qjh_2$. Therefore x becomes infinite on Γ at most to the order $ih_1 + jh_2$. This shows that $x \in \bar{L}_{ij}$, as asserted.

Conversely assume that $x \in \bar{L}_{ij}$ (where $\omega = xt_1^i t_2^j$). By the first part of the theorem, x is integrally dependent on the ring $k[\eta_{11}, \eta_{12}, \cdots, \eta_{2s_2}]$. From this it follows at once (since $\eta_{1h} = y_h^{(1)}/y_0^{(1)}$ and $\eta_{2h} = y_h^{(2)}/y_0^{(2)}$) that for all sufficiently high integers q_1, q_2 the product

$$x(y_0^{(1)})^{q_1}(y_0^{(2)})^{q_2}$$

is integrally dependent on the ring R. Since $y_0^{(1)} = t_1, y_0^{(2)} = t_2$ and $\omega = xt_1^i t_2^j$, we can assert that for all sufficiently high integers q_1, q_2 the product

$$\omega(y_0^{(1)})^{q_1}(y_0^{(2)})^{q_2}$$

is integrally dependent on R. A similar argument shows that if we fix an element $y_\mu^{(1)}$ in the set $\{y_0^{(1)}, y_1^{(1)}, \cdots, y_{s_1}^{(1)}\}$ and an element $y_\nu^{(2)}$ in the set $\{y_0^{(2)}, y_1^{(2)}, \cdots, y_{s_2}^{(2)}\}$, then for all sufficiently high integers q_1, q_2 the product

$$\omega(y_\mu^{(1)})^{q_1}(y_\nu^{(2)})^{q_2}$$

is integrally dependent on R. It follows that for all sufficiently high integers q we have

$$\omega R_{qq} \subset \bar{R}_{q+i,q+j} \,. \tag{29}$$

We now apply Lemma 4.3 to the doubly graded ring R and to the graded R-module $\bar{R}$. Since R is generated over k by the homogeneous elements $y_h^{(1)}$ of degree (1, 0) ($h = 1, 2, \cdots, s_1$) and the homogeneous elements $y_h^{(2)}$ of degree (0, 1) ($h = 1, 2, \cdots, s_2$), it follows that $R_{i+\mu,j+\nu} = R_{ij}R_{\mu\nu}$. Thus R (regarded as a module over itself) certainly has property (**a**) and also property ($\mathbf{b}_\alpha$) ($\alpha = 1, 2$). Since $\bar{R}$ is finitely generated over R, it follows from the first part of Lemma 4.3 that also $\bar{R}$ has property (**a**). Therefore, there exists an integer N such that

$$\bar{R}_{q+i,q+j} = R_{ij}\bar{R}_{qq} \,, \tag{30}$$

for all $i, j \geqq 0$ and all $q \geqq N$. By (29) and (30), we find, for q sufficiently large:

$$\omega\bar{R}_{qq} = \omega R_{q-N,q-N}\bar{R}_{NN} \subset \bar{R}_{q-N+i,q-N+j}\bar{R}_{NN} \subset \bar{R}_{q+i,q+j} = R_{ij}\bar{R}_{qq} \,,$$

i.e.,

$$\omega\bar{R}_{qq} \subset R_{ij}\bar{R}_{qq} \,.$$

Since $\bar{R}_{qq}$ has a finite basis over k, this last relation shows that ω is integral over R. This completes the proof of the second part of the theorem.

(3) Recalling that R has property ($\mathbf{b}_\alpha$) ($\alpha = 1, 2$) we apply the second part of Lemma 4.3 to the graded (*finite*) R-module R^*. There exists therefore an integer N such that

$$R^*_{i+1,j} = R_{10}R^*_{ij} = R^*_{10}R^*_{ij} \,, \qquad \text{if } i \geqq N \,;$$
$$R^*_{i,j+1} = R_{01}R^*_{ij} = R^*_{01}R^*_{ij} \,, \qquad \text{if } j \geqq N \,.$$

In view of the definition (21) of R^*_{ij}, these relations yield precisely relations (23).

This completes the proof of the theorem.

5. On certain graded R^*-modules defined by linear systems

We now assume that in addition to the infinite m-fold sequence $\{H_{(i)}\}$

of linear systems considered in the preceding section (and satisfying conditions (A) and (B)), we have also another m-fold sequence $\{G_{\{i\}}\}$ of linear systems on V satisfying the following condition:

$$G_{\{i\}} + H_{\{i\}} \subset G_{\{i+j\}} \, . \tag{31}$$

We fix a cycle Δ in $G_{00\cdots 0}$. By (31) we have $\Delta + i_1D_1 + \cdots + i_mD_m \in G_{\{i\}}$. We set

$$M_{\{i\}} = \{x \in k(V) \mid (x) + \Delta + i_1D_1 + i_2D_2 + \cdots + i_mD_m \in G_{\{i\}}\} \, .$$

Then $M_{\{i\}}$ is a finite-dimensional vector space over k. We set

$$S^* = \{\textstyle\sum_{\{i\}} x_{\{i\}} t_1^{i_1} t_2^{i_2} \cdots t_m^{i_m} \mid x_{\{i\}} \in M_{\{i\}}\} \, . \tag{32}$$

By (31), S^* is a graded R^*-module (where R^* is defined in (20), § 4).

THEOREM 5.1. *S^* is a finite R^*-module.*

PROOF. Again, we shall only consider the case $m = 2$. We first make three preliminary observations.

Observation 1. Let $\bar{H}_{ij}$ and $\bar{G}_{ij}$ be the complete linear systems $|H_{ij}|$ and $|G_{ij}|$ respectively; let $\bar{L}_{ij} = \{x \in k(V) \mid (x) + iD_1 + jD_2 \in \bar{H}_{ij}\}$, $\bar{M}_{ij} = \{x \in k(V) \mid (x) + \Delta + iD_1 + jD_2 \in \bar{G}_{ij}\}$, and let

$$\bar{R}^* = \{\textstyle\sum_{ij} x_{ij} t_1^i t_2^j \mid x_{ij} \in \bar{L}_{ij}\} \, ,$$
$$\bar{S}^* = \{\textstyle\sum_{ij} x_{ij} t_1^i t_2^j \mid x_{ij} \in \bar{M}_{ij}\} \, .$$

Then $\bar{S}^*$ is an $\bar{R}^*$-module, and therefore also an R^*-module. Furthermore, S^* is a submodule of the R^*-module $\bar{S}^*$. Assume that the theorem is true for $\bar{S}^*$ and $\bar{R}^*$. It will then follow that $\bar{S}^*$ is also a finite R^*-module, since $\bar{R}^*$ is finite over R^* (Theorem 4.2). Since R^* is noetherian, the submodule S^* of $\bar{S}^*$ is then also finite over R^*. Thus, *it is sufficient to prove the theorem under the assumption that the linear systems H_{ij} and G_{ij} are complete.*

Observation 2. Assume that all the systems G_{ij} are complete, and let Δ' be a divisorial cycle on V such that $\Delta \prec \Delta'$. Upon replacing in the definition (32) of S^* the linear systems $G_{ij} = |\Delta + iD_1 + jD_2|$ by the linear systems $G'_{ij} = |\Delta' + iD_1 + jD_2|$ we obtain an R^*-module S'^*, and it is clear that S^* is a submodule of S'^*. Thus, *it is sufficient to prove the theorem in the case in which $G_{ij} = |qC + iD_1 + jD_2|$, where C is a hyperplane section of V and q is a sufficiently high integer.*

Observation 3. Our theorem is equivalent with the following assertion: there exists an integer N such that

$$\begin{cases} G_{i+1,j} = G_{ij} + H_{10} \, , & \text{if } i \geqq N, j \geqq 0 \\ G_{i,j+1} = G_{ij} + H_{01} \, , & \text{if } i \geqq 0, \ j \geqq N \, . \end{cases} \tag{32'}$$

For, (32′) is equivalent to

$$(32'') \qquad \begin{cases} S^*_{i+1,j} = S^*_{ij}R^*_{10}\,, & \text{if } i \geqq N, j \geqq 0 \\ S^*_{i,j+1} = S^*_{ij}R^*_{01}\,, & \text{if } i \geqq 0, j \geqq N\,. \end{cases}$$

Clearly, if (32″) holds true, then S^* is generated over R^* by its homogeneous components S^*_{ij} such that $i \leqq N$ *and* $j \leqq N$. Since each homogeneous component of S^* has a finite basis over k, it follows that S^* is a finite R^*-module. Conversely, if S^* is finite over R^*, then, by Lemma 4.3, S^* has property $(\mathbf{b}_\alpha)$ ($\alpha = 1, 2$) (since R^* — as a finite R-module — has the property $(\mathbf{b}_\alpha)$), and thus (32″) is valid.

With these observations in mind we now prove our theorem by induction with respect to the dimension r of V. Consider the case $r = 1$. By *observation* 1 we may assume that the systems G_{ij} and H_{ij} are complete. If $D_1 = 0$ then $G_{ij} = G_{0j}$ and H_{q0} consists of the cycle 0 only, and the first of the two relations (32′) holds, with $N = 0$. Assume therefore that $D_1 \neq 0$, whence $\deg D_1 > 0$. Let q be an integer such that $qD_1 - \Delta$ is linearly equivalent to an effective divisor Y and let $\Delta' = \Delta + Y$. Then $|\Delta' + iD_1 + jD_2| = |(q+i)D_1 + jD_2|$, and therefore — in the notations of *observation* 2 — we have $S'^* \subset t_1^{-q}R^*$. This shows that S'^* is finite over R^* (since R^* is noetherian), and hence — by *observation* 2 — also S^* is finite over R^*.

Let now r be arbitrary and assume that the theorem is true for varieties of dimension $r - 1$. By *observation* 1 we may assume that $H_{ij} = |iD_1 + jD_2|$ and that $G_{ij} = |C + iD_1 + jD_2|$, where C is a hyperplane section of V and where we may also assume that C is irreducible and normal. We denote by $\tilde{G}_{ij}$ and $\tilde{H}_{ij}$ the traces, on C, of G_{ij} and H_{ij} respectively. It is clear that the linear systems $\tilde{H}_{ij}$ satisfy conditions similar to (A) and (B) of §4, and that $\tilde{G}_{ij} + \tilde{H}_{st} \subset \tilde{G}_{i+s,j+t}$. Therefore, by our induction hypothesis, and by *observation* 3, there exists an integer N such that

$$(33) \qquad \begin{aligned} \tilde{G}_{i+1,j} &= \tilde{G}_{ij} + \tilde{H}_{10}\,, && \text{if } i \geqq N, j \geqq 0\,. \\ \tilde{G}_{i,j+1} &= \tilde{G}_{ij} + \tilde{H}_{01}\,, && \text{if } i \geqq 0, j \geqq N\,. \end{aligned}$$

The first of these relations implies that $G_{i+1,j}$ is spanned by its following two subsystems:

$$G_{ij} + H_{10}, \quad C + |(i+1)D_1 + jD_2|\,,$$

where the second system has C as fixed component. Now, it is clear that we can replace N by any larger integer, without affecting the validity of (33). By Theorem 4.2, we may assume then that $|(i+1)D_1 + jD_2| = |iD_1 + jD_2| + |D_1|$, if $i \geqq N$ and $j \geqq 0$. Then we will have that $C + |(i+1)D_1 + jD_2| \subset |C + iD_1 + jD_2| + |D_1| = G_{ij} + H_{10}$, which yields the first of the two relations (32′). In a similar way, the second of these relations is established. This completes the proof of the theorem.

Our next theorem concerns the existence of a *characteristic polynomial* for the multigraded R^*-module S^* (and hence — in particular — for R^* itself) and the degree of this polynomial. We recall that s_α denotes the transcendence degree of $k(L^\alpha)/k$, where $L^1 = L_{10\cdots 0}$, $L^2 = L_{010\cdots 0}$, $\cdots$, $L^m = L_{00\cdots 1}$.

THEOREM 5.2. *There exists a polynomial $f(X_1, X_2, \cdots, X_m)$ in m variables, of degree s_α in X_α, such that*

$$\dim_k S^*_{\{i\}} (= \dim_k M_{\{i\}} = 1 + \dim G_{\{i\}}) = f(i_1, i_2, \cdots, i_m)$$

for all sufficiently high values of $i_1, i_2, \cdots, i_m$.

PROOF. As S^* is a finite R^*-module and R^* is a finite R-module [where R is defined in (22), § 4], S^* is a finite R-module. Now, R is the residue class ring, modulo a multi-homogeneous ideal $\mathfrak{A}$, of the polynomial ring $A = k[\{Y^{(1)}\}, \{Y^{(2)}\}, \cdots, \{Y^{(m)}\}]$ in m sets of indeterminates $\{Y^{(\alpha)}\}$ $(=\{Y_0^{(\alpha)}, Y_1^{(\alpha)}, \cdots, Y_{s_\alpha}^{(\alpha)}\})$. Thus S^* is also a finite graded A-module, and the existence of a characteristic polynomial $f(X_1, X_2, \cdots, X_m)$ of S^* is therefore well known (for the case $m = 1$ see Zariski-Samuel [6, v. 2, Ch. VII, Th. 41, p. 232]; the proof in the case $m > 1$ is the same). It remains to show that f is of degree s_α in X_α.

The integers $s_1, s_2, \cdots, s_m$ are the projective dimensions of the multihomogeneous ideal $\mathfrak{A}$(see Lemma 4.1). As a finite graded A-module, also R has a characteristic polynomial $g(X_1, X_2, \cdots, X_m)$, and it is well known that g is of degree s_α in X_α (for the case $m = 1$ see Zariski-Samuel [6, v. 2, Ch. VII, Th. 42, p. 234]). Now, we have $L_{\{i\}} \subset M_{\{i\}}$ (since $k \subset M_{00\cdots 0}$) and thus $L_{\{i\}} = kL_{\{i\}} \subset M_{00\cdots 0}L_{\{i\}} \subset M_{\{i\}}$. Therefore $R^*_{\{i\}} \subset S^*_{\{i\}}$. On the other hand, $R_{\{i\}} \subset R^*_{\{i\}}$. Therefore $\dim_k R_{\{i\}} \leqq \dim_k S^*_{\{i\}}$ for all $\{i\}$. It follows that

$$(34) \qquad g(i_1, i_2, \cdots, i_m) \leqq f(i_1, i_2, \cdots, i_m)$$

for all sufficiently high values of $i_1, i_2, \cdots, i_m$.

On the other hand, since S^* is a finite graded R-module it follows from the first part of Lemma 4.3 that there exists an integer N such that

$$S^*_{i_1 i_2 \cdots i_m} = R_{i_1 - q, i_2 - q, \cdots, i_m - q} S^*_{qq\cdots q}$$

for all $q \geqq N$ and for all $i_\alpha \geqq q (\alpha = 1, 2, \cdots, m)$. Consequently

$$f(i_1, i_2, \cdots, i_m) \leqq cg(i_1 - N, i_2 - N, \cdots, i_m - N)$$

for all $i_1, i_2, \cdots, i_m \geqq N$; here c is constant: $c = \dim_k S^*_{NN\cdots N}$. This inequality, combined with (34), shows that f and g have the same degree in each variable X_α. q.e.d.

The proof of the next theorem is similar to that of Theorem 4.2 but requires some additional material from the theory of complete modules (see Zariski-Samuel [6, v. 2, Appendix 4]). We shall omit the proof, also because we shall not make use of the theorem.

THEOREM 5.3. *The notations and assumptions being the same as those introduced in the beginning of this section, let* $\bar{G}_{\{j\}}$ *be the complete system containing* $G_{\{j\}}$, *and let* $\bar{M}_{\{i\}}$ *and* $\bar{S}^*$ *have the similar meaning as* $M_{\{i\}}$ *and* S^* *respectively, with* $G_{\{j\}}$ *being replaced by* $\bar{G}_{\{j\}}$. *Then the module* $\bar{S}^*$ *is the integral closure of* S^* *in the field* $k(V)(t_1, t_2, \cdots, t_m)$.

6. Applications to non-singular algebraic surfaces

Beginning with this section we leave the general case of a normal variety V and confine ourselves to the case in which V is a non-singular surface F. The results of the preceding two sections are applicable, without further ado, to the case of complete linear systems of the form $G_n = |\Delta + nD|$ and $H_n = |nD|$ on F, *provided* $|D|$ *is free from base points.* What we shall do in this section is to

(1) re-interpret and complete the previous results in this case, for the purpose of obtaining more explicit information about the characteristic polynomials $\dim G_i$, $\dim H_i$ (i — large): and

(2) extend the application to the case in which *some* multiple $|qD|$ of $|D|$ (rather than $|D|$ itself) is free from base points. In this last application we shall make use of the following theorem which we have proved elsewhere (see Zariski [10]):

THEOREM 6.1. *If a complete linear system* $|D|$ *on a non-singular surface* F *has no fixed components then all sufficiently high multiples* $|nD|$ *of* $|D|$ *are free from base points.*

For convenience of the reader we shall prove this theorem here, and in a somewhat more general form. We observe that in the case of surfaces a linear system $|D|$ has no fixed components if and only if $|D|$ has only a finite number of base points. Therefore, Theorem 6.1 is a special case of the following theorem:

THEOREM 6.2. *If* V *is a non-singular variety and* $|D|$ *is a complete linear system on* V *which has only a finite number of base points, then* $|nD|$ *has no base points if* n *is sufficiently large.*

PROOF. Let C be a generic (hence irreducible and non-singular) hyperplane section of V and let

$$H_{ij} = \mathrm{Tr}_C \,|\, iC + jD \,| \,.$$

The family of linear systems H_{ij} on the non-singular variety C satisfies conditions (A) and (B) of § 4, since we may assume that C does not pass through any of the base points of $|D|$. Therefore, by Theorem 4.2 [relation (24)] we have, for a suitable integer N:

$$H_{ij} = H_{i,j-1} + H_{01} \qquad (j \geqq N; i \geqq 0).$$

This implies that

$$\mathrm{Tr}_C \,|\, iC + jD \,| = \mathrm{Tr}_C \{|\, iC + (j-1)D \,| + |\, D \,|\} \,, \qquad (j \geqq N; i \geqq 0).$$

It follows that $|\, iC + jD \,|$ is spanned by its following two subsystems:

$$|\, iC + (j-1)D \,| + |\, D \,| \,, \qquad |\, (i-1)C + jD \,| + C \,, \tag{35}$$

where the second system has C as fixed component. *For fixed* j, the system $|\, iC + jD \,|$ has no base points if i is sufficiently high. Let then j be *a fixed integer* $\geqq N$ and let i be such that $|\, iC + jD \,|$ has no base points. If P is any base point of $|\, D \,|$ then P is also a base point of the first of the two systems (35) (since we have dealing here with the *minimal* sum of $|\, iC + (j-1)D \,|$ and $|\, D \,|$). Therefore P is *not* a base point of $|\, (i-1)C + jD \,|$. Applying the same argument to this last system (if $i - 1 > 0$) we find that P is not a base point of $|\, (i-2)C + jD \,|$. Ultimately we reach the conclusion that P is not a base point of $|\, jD \,|$ (if $j \geqq N$). This completes the proof.

We shall now introduce a convenient notation. If D is a divisorial cycle on a normal variety V such that the complete system $|\, D \,|$ is of positive dimension, we shall denote by $H(D)$ the projective variety which is a rational transform of V and which is defined — to within projective equivalence — by the condition that the transform of $|\, D \,|$ is the system of hyperplane sections of $H(D)$. This transition from V to $H(D)$ has been described explicitly in the course of the proof of Lemma 4.1 (for any — not only complete — linear systems on V). We note that $H(D)$ depends only on $|\, D \,|$, and not on the choice of the divisor D in its divisorial class. The trivial case in which $|\, D \,|$ consists only of one divisor and therefore $\dim |\, D \,| = 0$ (in which case $H(D)$ is a point), will be generally excluded.

We set $H_n = |\, nD \,|$ and we use the notations L_n, R^* of § 4 (relative to the case $m = 1$):

$$L_n = \{x \in k(V) \,|\, (x) + nD \succ 0\} \,, \qquad n \geqq 0 \,, \tag{36}$$

where D is a preassigned member of $|\, D \,|$, and

$$R^* = R^*[D] = \{\textstyle\sum x_n t^n \,|\, x_n \in L_n\} \,.$$

We denote by T_n the natural rational map of V onto $H(nD)$. We note that if $\omega_0, \omega_1, \cdots$, is a k-basis of L_n, then $\omega_0 t^n, \omega_1 t^n, \cdots$, is a k-basis of the homogeneous component R_n^*, of degree n, of R^*, and the elements of the basis of R_n^* are strictly homogeneous coordinates of the general point of $H(nD)$. Thus R_n^* can be identified with the vector space of homogeneous elements of degree 1 of the homogeneous coordinate ring of $H(nD)$, and this latter ring can be identified with $k[R_n^*](=k[\omega_0 t^n, \omega_1 t^n, \cdots])$.

Let us now assume that some multiple $|qD|$ *of* $|D|$ *has no base points.* Then for each n, the rational map $T_{nq}: V \to H(nqD)$ is regular (see proof of Lemma 4.1). We now apply Theorem 4.2 to the case in which $m = 1$ and the systems H_i are the complete linear system $|iqD|$ (whence the L_i in Theorem 4.2 must now be replaced by L_{iq}, where L_n is defined by (36)). We now have $L_{iq} = \bar{L}_{iq}$, and thus the ring $\bar{R}$ of Theorem 4.2 is now the ring $R^{(q)*} = \sum_{i=0}^{\infty} R_{iq}^*$ (provided we replace in Theorem 4.2 the indeterminate t by t^q). The ring R of Theorem 4.2 [see (22)] is a subring of the homogeneous coordinate ring $k[R_q^*]$ of $H(qD)$: it is generated, over k, by a suitable set of $s + 1$ algebraically independent elements of R_q^*, where $s = \dim H(qD)$ (see Lemma 4.1); thus R is, in fact, a polynomial ring in $s + 1$ variables. Since $R \subset k[R_q^*] \subset \bar{R}$, and $\bar{R}$ is the integral closure of R in $k(V)(t^q)$ (Theorem 4.2), $\bar{R}$ is also the integral closure of $k[R_q^*]$ in $k(V)(t^q)$. It follows, by known results on arithmetically normal varieties, that *for all sufficiently large* n *the variety* $H(nqD)$ *is arithmetically normal and is a derived normal model of* $H(qD)$ *with respect to the extension field* $k(V)$ *of* $k(H(qD))$[6]. Note therefore that the varieties $H(nqD)$ are biregularly equivalent (if n is large), namely $T_{nq}T_{n'q}^{-1}$ is a biregular map if both n and n' are sufficiently large.

Let now q' be another integer such that $|q'D|$ has no base points, whence also the varieties $H(nq'D)$ are arithmetically normal and biregularly equivalent to each other, if n is sufficiently large. Since the two classes of varieties $\{H(nqD)\}$ and $H(nq'D)$ have in common the varieties $H(nqq'D)$, it follows that also the varieties $H(nqD)$ and $H(nq'D)$ are biregularly equivalent (for large n). We have therefore associated with the given complete linear system $|D|$ (*under the assumption that some multiples* $|qD|$ *of* $|D|$ *has no base points*) a class of biregularly equivalent arithmetically normal varieties which we shall denote by $\mathrm{Cl}(D)$. For each

[6] That $H(nqD)$ is arithmetically normal for large n follows also directly from Theorem 4.2, relation (23). This relation tells us, in the present case, that $L_{(n+1)q} = L_{nq}L_q$; i.e., $R^*_{(n+1)q} = R^*_{nq}R^*_q$, for large n and all i. This shows that the homogeneous coordinate ring $k[R^*_{nq}]$ of $H(nqD)$ is the ring $R^{(nq)*} = \sum_i R^*_{inq}$, and this latter ring is integrally closed, since the ring $R^{(q)*}$ is integrally closed.

variety V^* in that class we have a canonical regular map $T_{V^*}: V \to V^*$, such that if V^*, V'^* are any two varieties in the class, then $T^{-1}T$ is a biregular map of V^* onto V'^*. Any member V^* of the class $\mathrm{Cl}(D)$ will be denoted by V_D^*, unless we wish to specify a definite projective choice $H(nqD)$ for V_D^*.

If V is a surface then the dimension s of V_D^* is either 2 or 1. If $s = 2$, then T_{V^*} is a *birational transformation*, since $H(nqD)$ is a derived normal model of a rational transform $H(qD)$ of V, with respect to the field $k(V)$, and the function field of such a model is always maximally algebraic in $k(V)$. If $s = 1$, i.e., if V_D^* is a curve (necessarily non-singular, since it is normal), then the inverse images of the points of V_D^* (under $T_{V^*}^{-1}$) form a pencil $\{\Gamma\}$, of genus equal to the genus of V_D^*, and since $k(V_D^*)$ is maximally algebraic in $k(V)$ the generic number Γ of this pencil is an absolutely irreducible curve (see S. Lang, *Abelian varieties*, p. 23, Lemma 3). Each system $|nqD|$ is composite with the pencil $\{\Gamma\}$. Since $|nqD|$ has no base points, the same must be true for the pencil $\{\Gamma\}$, and hence two distinct members of $\{\Gamma\}$ do not meet. We have therefore $(\Gamma^2) = 0$, whence also $(D^2) = 0$. On the other hand, if $s = 2$, we have necessarily $(D^2) > 0$ (see Lemma 4.1).

THEOREM 6.3. *Let F be a non-singular surface and $|D|$ a complete linear system on F, free from base points. Then:*

(a) *If $(D^2) > 0$ the superabundance $s(nD)$ is constant, for large n, and namely*

$$s(nD) = p_a(V_D^*) - p_a(F) . \tag{37}$$

(b) *If $(D^2) = 0$, the systems $|nD|$ are composite with an irreducible pencil $\{\Gamma\}$, of genus $\pi = p_a(V_D^*)$ (where V_D^* is now a non-singular curve), and we have for large n:*

$$s(nD) = gmn - \pi - p_a(F) , \tag{38}$$

$$\dim |nD| = mn - \pi , \tag{38'}$$

where $g = p_a(\Gamma)$ and m denotes the number of cycles Γ whose sum is equal to D.

PROOF. (a) We consider the case $(D^2) > 0$. Since $|D|$ has no base points (and therefore also no fixed components) and since $(D^2) > 0$, it follows, by the theorem of Bertini (see Zariski [8]) that $|D|$ is irreducible. If in the inequality (14) of §1 we identify Y with a generic (hence irreducible) member D of $|D|$ and X with nD, and observing that the assumption $(X + Y \cdot Y) > 2p(Y) - 2$ is now satisfied if n is large (in view of $(D^2) > 0$), we find that $s((n+1)D) \leq s(nD)$, for large n. This shows that $s(nD)$

remains constant for large n.

We now consider the surface V_D^* (which is a birationally equivalent to and is a regular map of V). We fix an integer h such that the surface $V^* = H(hD)$ belongs to the class $\mathrm{Cl}(D)$ (whence V^* is arithmetically normal). We denote by C^* a hyperplane section of V^*. We know that the regular map $T_h\colon V \to V^*$ transforms $|nhD|$ onto $|nC^*|$. Hence $\dim|nhD| = \dim|nC^*|$, and therefore, by known results (see Zariski [9])

$$\dim|nhD| = p_a(V^*) + p_a(-nC^*) - 1 . \tag{39}$$

Let X^* be a generic member of $|nC^*|$. Then X^* does not pass through any of the fundamental points of T_h^{-1}, and consequently T_h^{-1} is biregular at each point of X^*. Therefore, if X is the member of $|nhD|$ which corresponds to X^*, we will have $(X^2) = (X^{*2})$ and $p_a(X) = p_a(X^*)$ (note that (X^{*2}) is defined, even if V^* has singular points, since any singular point of V^* is necessarily fundamental for T_h^{-1}). Since $p_a(-X) = (X^2) - p_a(X) + 2$ (see footnote (1), § 1) and similarly $p_a(-X^*) = (X^{*2}) - p_a(X^*) - 2$, it follows that $p_a(-X) = p_a(-X^*)$; i.e., $p_a(-nhD) = p_a(-nC^*)$. Therefore, by (39), we have for large n:

$$\dim|nhD| = p_a(V^*) + p_a(-nhD) - 1 . \tag{40}$$

Hence, $s(nhD) = p_a(V_D^*) - p_a(F)$ for large n, and this, in conjunction with the fact that $s(nD) = \text{const}$, for large n, establishes (37).

(b) The case $(D^2) = 0$ is elementary. By the Riemann-Roch theorem on the curve V_D^*, it is clear that $\dim|nD| = mn - \pi$, for large n. Since $(\Gamma^2) = 0$, we have $(\Gamma \cdot K) = 2p(\Gamma) - 2 = 2g - 2$, and hence $(nD \cdot K) = 2mn(g - 1)$. Therefore $\rho(nD) = -mn(g - 1)$ [see (1), § 1], and $s(nD) = \dim|nD| - p_a(F) - \rho(nD) = gmn - \pi - p_a(F)$, for large n. This completes the proof of the theorem.

THEOREM 6.4. *Let F be a non-singular surface and let $|D|$ be a complete linear system on F, free from base points, and such that $(D^2) > 0$. Let Δ be an effective divisorial cycle on F. Then*

(a) *$s(\Delta + nD) = const$, for large n;*

(b) *if B_n is the fixed component of $|\Delta + nD|$ then $B_n = B =$ a constant cycle, for large n; every prime component Γ of B is an exceptional curve of the regular map $T\colon F \to V_D^*$ (in other words, we have $(\Gamma \cdot D) = 0$), and we have*

$$s(\Delta + nD) \leqq p_a(V_D^*) - p_a(F) - \rho(B) . \tag{41}$$

PROOF. The proof of (a) is the same as the proof of the corresponding part of the preceding theorem.

It is clear that $B_n \succ B_{n+1}$, for all n (this follows from the fact that $|D|$ has no fixed components). Hence the assertion $B_n = B =$ a constant cycle is obvious. We also point out the self-evident fact that we must have $\Delta \succ B$.

Let Γ be a prime component of B. By part (a) of the theorem, let $s(\Delta - B + nD) = s_0, n -$ large. We set $\Delta' = \Delta - B$. We have, then, for large n,

$$\dim |\Delta' + nD| = (1/2)(\Delta' + nD \cdot \Delta' + nD - K) + p_a + s_0 ,$$

i.e.,

$$\dim |\Delta' + nD| = (1/2)(D^2)n^2 + (D \cdot \Delta')n - \left(\frac{D \cdot K}{2}\right)n + \text{const} . \tag{42}$$

On the other hand, the inequality (n — large)

$$\dim |\Gamma + \Delta' + nD| \geqq \rho(\Gamma + \Delta' + nD) + p_a$$

is equivalent with

$$\dim|\Gamma + \Delta' + nD| \geqq (1/2)(D^2)n^2 + \left[D\cdot(\Delta' + \Gamma) - \left(\frac{D \cdot K}{2}\right)\right]n + \text{const} . \tag{43}$$

Were $(\Gamma \cdot D) > 0$, comparison of (42) and (43) would lead to the conclusion that $\dim |\Gamma + \Delta' + nD| > \dim |\Delta' + nD|$, for large n, in contradiction with the fact that B is a fixed component of $|B + \Delta' + nD|$.

It remains to prove (41). Let us first assume that $B = 0$; i.e., that $|\Delta + nD|$ has no fixed component, for large n. In this case we have to show that $s(\Delta + nD) \leqq p_a(V_D^*) = p_a(F)$, $n -$ large. We fix a value n_0 of n such that $|\Delta + n_0D|$ has no fixed components. Now, it is easy to see that $|\Delta + nD - B_n|$ is not composite with a pencil, and hence is irreducible (since it has no fixed components). In fact, set $L_n = \{x \mid (x) + nD \succ 0\}$ and $M_n = \{x \mid (x) + nD + \Delta \succ 0\} = \{x \mid (x) + nD + \Delta - B_n \succ 0\}$. Then tr.d. $k(L_n)/k = 2$, since $(D^2) > 0$ and $|D|$ has no base points (and since therefore $|D|$ — and *a fortiori* $|nD|$ — is not composite with a pencil). Since $L_n \subset M_n$, also tr.d. $k(M_n)/k = 2$, and this proves our assertion. In particular, then, $|\Delta + n_0D|$ is irreducible. We fix in $|\Delta + n_0D|$ a generic (hence prime) cycle Δ_0. It is clear that $(\Delta_0 \cdot D) > 0$. Therefore, by (14), § 1, we have for large n: $s(\Delta_0 + nD) \leqq s(nD)$; i.e., by Theorem 6.3, part (a), $s(\Delta_0 + nD) \leqq p_a(V_D^*) - p_a(F)$. Since $|\Delta + nD| = |\Delta_0 + (n - n_0)D|$, our assertion follows.

In the general case, we set $\Delta_1 = \Delta - B(\Delta_1 \succ 0)$. By the previous proof we have

$$s(\Delta_1 + nD) \leqq p_a(V_D^*) - p_a(F) , \qquad n - \text{large} . \tag{44}$$

Now, $\dim|\Delta + nD| = \dim|\Delta_1 + nD|$, if n is large. Hence, for large n, $s(\Delta+nD) = \dim|\Delta+nD| - \rho(\Delta+nD) - p_a = \dim|\Delta_1+nD| - \rho(\Delta_1+nD) - p_a - \rho(B) - (B \cdot \Delta_1 + nD) = s(\Delta_1 + nD) - \rho(B) - (B \cdot \Delta_1 + nD)$. Since $|\Delta_1 + nD|$ has no fixed component, we have $(B \cdot \Delta_1 + nD) \geqq 0$, and this establishes (41), since we have already shown that $s(\Delta_1 + nD) \leqq p_a(V_D^*) - p_a(F)$[7].

THEOREM 6.5. *Let $|D|$ be a complete linear system on a non-singular surface F, such that some multiple of $|D|$ has no fixed components. Then the ring $R^*[D]$ (see § 2) is finitely generated over k. If we assume furthermore that $(D^2) > 0$, then*

(a) *the superabundance $s(nD)$ is bounded from above and is, in fact, a periodic function of n (for large n);*

(b) *if all sufficiently high multiples of $|D|$ are free from fixed components then $s(nD) = p_a(V_D^*) - p_a(F)$ for large n.*

PROOF. By Theorem 6.2, some multiple $|qD|$ has no base points. We know then that the ring $R^{(q)*} = \sum_{j=0}^{\infty} R_{jq}^*$ is finitely generated over k (Theorem 4.2), and we know also that for each $i = 0, 1, \cdots, q-1$, $\sum_{j=0}^{\infty} R_{i+jq}$ is a finite $R^{(q)*}$-module (Theorem 5.1). Hence the ring $R^*[D] = \sum_{n=0}^{\infty} R_n^*$ is finitely generated over k.

Now assume that $(D^2) > 0$. By the preceding theorem we have for each $i = 0, 1, \cdots, q-1$: $s(iD + nqD) =$ const, for large n. This proves part (a) of the theorem.

By Theorem 6.2, we can find an integer q' such that $(q, q') = 1$ and such that also $|q'D|$ is free from base points. We know then that $s(nq'D) = p_a(V_D^*) - p_a(F)$ for large n. For a given i, $0 \leqq i \leqq q$, the congruence $i + nq \equiv 0(q')$ has infinitely many solutions n. Since $s(iD + nqD)$ is constant for n large, it follows that $s(iD + nqD) = p_a(V_D^*) - p_a(F)$, for large n. This completes the proof.

Note. Under the assumptions stated in the above theorem the superabundance $s(nD)$ is a periodic function of n, for large n, with period q, where q is any integer such that $|qD|$ has no base points.

For our next — and final — theorem of this section we shall need a well-known theorem of Hodge (see [2]; see also Grothendieck [1, Th. 1.1], where a purely algebraic and elementary proof of Hodge's theorem can be found) which we shall frequently use in the sequel.

LEMMA 6.6. *If D is a divisorial (not necessarily effective) cycle on a non-singular surface F such that $(D^2) > 0$, and if E is any cycle on F,*

[7] It can be easily shown that $\rho(B)$ is always negative (if $B \neq 0$). It can also be shown by examples that in (41) we may have strict inequality, even if $B = 0$.

then $(D \cdot E) = 0$ *implies that either* $(E^2) < 0$ *or that some multiple of* E *is algebraically equivalent to zero*;

PROOF. Since $(D^2) > 0$, no rational multiple $\lambda D (\lambda \neq 0)$ of D is algebraically equivalent to zero. Therefore, there exists a set of divisorial cycles $D_1, D_2, \cdots, D_q$ such that $(D, D_1, D_2, \cdots, D_q)$ is an independent basis (over the rationals) of the group of divisorial cycles on F modulo algebraic equivalence, and such that $(D_i \cdot D_j) = 0$ if $i \neq j$ (here $D_0 = D$). It is then known that $(D_i^2) < 0$ for $i = 1, 2, \cdots, q$ (this is the theorem of Hodge). Now, let E be algebraically equivalent to $\lambda D + \lambda_1 D_1 + \lambda_2 D_2 + \cdots + \lambda_q D_q$ (the λ's being rational numbers). If $(D \cdot E) = 0$ then $\lambda = 0$, and hence $(E^2) = \lambda_1^2(D_1^2) + \cdots + \lambda_q^2(D_q^2) \leqq 0$. Hence either $\lambda_1 = \lambda_2 = \cdots = \lambda_q = 0$ or $(E^2) < 0$. q.e.d.

THEOREM 6.7. *Let D be an effective divisorial cycle on a non-singular surface F and assume that D satisfies the following conditions*:

(1) $(D^2) > 0$;

(2) *every prime component of D has a non-negative self-intersection number. Then all sufficiently high multiples $|nD|$ of $|D|$ are free from fixed components, and hence $s(nD) = p_a(V_D^*) - p_a(F)$ for all large n (see Theorem* 6.5, *part* (b)).

PROOF. Let us first consider the case in which D is a prime cycle. Then, the argument used in the proof of part (a) of Theorem 6.3 shows that $s(nD) =$ const, for large n, say $s(nD) = s_0$. Thus, $\dim |nD| = ((D^2)/2)n^2 - ((D \cdot K)/2)n + p_a + s_0$, for large n. Since $(D^2) > 0$ it follows that $\dim |(n+1)D| > \dim |nD|$ for large n, and this shows that D is not a fixed component of $|nD|$, if n is large. Hence $|nD|$ is free from fixed components if n is sufficiently large.

In the general case, let $D = \sum_{i=1}^{h} \lambda_i \Gamma_i$, where $\Gamma_1, \Gamma_2, \cdots, \Gamma_h$ are the distinct prime components of D, and $\lambda_i > 0$. Since $(D^2) > 0$, the inequality $\dim |nD| \geqq ((D^2)/2)n^2 - ((D \cdot K)/2)n + p_a$ (n — large) shows that, for large n, $|nD|$ has positive dimension, and that the complete linear system H obtained from $|nD|$ by deleting the fixed component of $|nD|$ is not composite with a pencil. Thus, if Δ is the generic member of H, Δ is irreducible, and we have $(\Delta^2) > 0$. On the other hand, the prime components of the fixed component of $|nD|$ must be in the set $\{\Gamma_1, \Gamma_2, \cdots, \Gamma_h\}$. We therefore can fix a sufficiently large integer q such that qD is linearly eqivalent to a cycle of the form $\Delta + \sum \mu_i \Gamma_i$, with $(\Delta^2) > 0$, Δ — a prime cycle, and $\mu_i \geqq 0$ $(i = 1, 2, \cdots, h)$.

By assumption, we have $(\Gamma_i^2) \geqq 0$, $i = 1, 2, \cdots, h$. Hence, by the preceding lemma,

$$(45) \qquad (\Delta \cdot \Gamma_i) > 0 , \qquad i = 1, 2, \cdots, h .$$

By the first part of the proof we have

(46) $$s(n\Delta) = s_0 = \text{const}\,, \qquad \text{if } n \text{ is large}\,.$$

We now fix an integer N satisfying the following conditions:

(47) $$N\cdot(\Delta\cdot\Gamma_i) > \max\{2p(\Gamma_i) - 2,\, s_0 - \rho(\Gamma_i)\}\,,$$

(48) $$i(n\Delta) = 0,\, s(n\Delta) = s_0\,, \qquad \text{if } n \geqq N\,.$$

Such an integer exists, in view of (45) and (46). *We now assert that if* $m_1, m_2, \cdots, m_h$ *are arbitrary non-negative integers and* $n \geqq N$ *then* $|n\Delta + m_1\Gamma_1 + \cdots + m_h\Gamma_h|$ *has no fixed components.* To see this, we first observe that we have

(49) $$s(n\Delta + m_1\Gamma_1 + \cdots + m_h\Gamma_h) \leqq s_0\,, \qquad \text{if } n \geqq N\,.$$

This follows at once, by induction with respect to $m_1 + \cdots + m_h$, taking into account (48), the inequalities $N\cdot(\Delta\cdot\Gamma_i) > 2p(\Gamma_i) - 2$ [see (47)], $(\Gamma_i\cdot\Gamma_j) \geqq 0$, and using inequality (14) of § 2. It is clear that the only possible prime fixed components of the system $|n\Delta + m_1\Gamma_1 + \cdots + m_h\Gamma_h|$ are $\Gamma_1, \Gamma_2, \cdots, \Gamma_h$. If $m_1 = 0$, Γ_1 is not a fixed component of that system. Suppose that $m_1 > 1$. Then we have, setting $X = n\Delta + (m_1 - 1)\Gamma_1 + m_2\Gamma_2 + \cdots + m_h\Gamma_h$, for $n \geqq N$:

$$\begin{aligned}\dim|n\Delta + m_1\Gamma_1 + \cdots + m_h\Gamma_h| &= \dim|X + \Gamma_1| \geqq \rho(X + \Gamma_1) + p_a\\ &= \rho(X) + \rho(\Gamma_1) + (X\cdot\Gamma_1) + p_a \geqq \rho(X) + p_a + n(\Delta\cdot\Gamma_1) + \rho(\Gamma_1)\,.\end{aligned}$$

By (49), we have $\rho(X) + p_a \geqq \dim|X| - s_0$, if $n \geqq N$ (always using the fact that, by the first of the two relations (48), $n\Delta$ — and hence also X — is non-special). Therefore

$$\dim|X + \Gamma_1| \geqq \dim|X| + \big(n(\Delta\cdot\Gamma_1) - s_0 + \rho(\Gamma_1)\big)\,,$$

and hence, by (47): $\dim|X + \Gamma_1| > \dim|X|$, showing that Γ_1 is not a fixed component of $X + \Gamma_1$. The same argument can be repeated for $\Gamma_2, \cdots, \Gamma_h$ and establishes the above assertion.

Now, we complete the proof of our theorem by showing that if $n \geqq Nq$ then $|nD|$ has no fixed components. We write $n = mq + r$, $0 \leqq r < q$. Then necessarily $m \geqq N$ if $n \geqq Nq$. Now

$$|nD| = |m\Delta + (m\mu_1 + r\lambda_1)\Gamma_1 + \cdots + (m\mu_h + r\lambda_h)\Gamma_h|\,,$$

and since $m \geqq N$, this establishes our assertion that $|nD|$ has no fixed components if $n \geqq Nq$.

Remark. It is not difficult to see that under the assumptions made in the theorem it is also true that $|nD|$ *has no base points if n is sufficiently* *large.* In fact, let S be the set of all positive integers n such that $|nD|$ has no base points. Since $|nD|$ has no fixed components if n is large, it follows from Theorem 6.2 that S is non-empty. On the other hand, S is closed under addition. Hence if d denotes the highest common divisor of all the integers $n \in S$, then S contains all sufficiently high multiples of d. So we have only to show that $d = 1$. We fix two integers q, q' which are relatively prime and are such that $|qD|$ and $|qD'|$ have no fixed components. There exists an integer N such that $|nqD|$ and $|nq'D|$ have no base points, for all $n \geqq N$. Then all integers of the form nq and $n'q'$, with $n \geqq N$ and $n' \geqq N$, belong to S. This shows that $d = 1$.

Part III. Solution of the General Problem

7. The arithmetically negative component of an effective cycle

Given a finite set of divisorial cycles $\{Z_1, Z_2, \cdots, Z_m\} = \{Z\}$, we denote by $\varphi\{Z\}$ the quadratic form

$$\sum (Z_i \cdot Z_j) x_i x_j \ .$$

We have, then, that for any $x_1, x_2, \cdots, x_m$ (*the* x_i *will be assumed, as a rule, to be rational numbers*), $\varphi\{Z\}(x_1, x_2, \cdots, x_m)$ is the self-intersection number of the cycle $x_1 Z_1 + x_2 Z_2 + \cdots + x_m Z_m$. It is known — and we have already used this fact in the proof of Lemma 6.6 — that in the canonical form $a_1 U_1^2 + \cdots + a_h U_h^2$ $(h \leqq m, a_i \neq 0)$ of $\varphi\{Z\}$ at most one of the coefficients a_i can be positive. We shall say that $\varphi\{Z\}$ is of type (s, t) if s of the coefficients a_i are positive and t of the coefficients are negative; here $s + t = h \leqq m$. We therefore have that s can only have the values 0 or 1.

If D is a divisorial cycle and if $\Gamma_1, \Gamma_2, \cdots, \Gamma_m$ are the prime components of D we shall also denote the quadratic form $\varphi\{\Gamma\}$ by φ_D.

In the sequel we shall deal with divisorial cycles $\sum x_i \Gamma_i$, where, unless otherwise specified, the x_i are rational numbers. We shall say that the cycle is *integral* if the x_i are integers.

Lemma 7.1. *Let* $E_1, E_2, \cdots, E_q$ *be distinct prime divisorial cycles such that the quadratic form* $\varphi\{E\}$ *is of type* $(0, q)$ (*hence is definite negative*) *and let* $X = x_1 E_1 + x_2 E_2 + \cdots + x_q E_q$. *If* $(X \cdot E_i) \leqq 0$ *for* $i = 1, 2, \cdots, q$, *then the* x_i *are non-negative* (*whence* X *is an effective cycle*).

Proof. Write X in the form $X = A - B$, where A and B are effective cycles, without common components. We have $(X \cdot B) \leqq 0$, by assumption, hence $(A \cdot B) - (B^2) \leqq 0$. Since $(A \cdot B) \geqq 0$ and $(B^2) \leqq 0$, it follows that $(B^2) = 0$, and hence $B = 0$ since $\varphi\{E\}$ is definite negative. This completes the proof.

Note. The assumption that the E_i are prime cycles is not essential. This assumption can be replaced by the assumption $(E_i \cdot E_j) \geqq 0$ if $i \neq j$, and the cycles E_i need not be assumed to be effective. The same proof can be repeated (except that, in writing X in the form $A - B$, $A = \sum a_i E_i$, $B = \sum b_i E_i$, we should assume that $a_i b_i = 0$ for all i) and leads to the same conclusion, namely that all the x_i are non-negative.

COROLLARY 7.2. *Let* $\mathcal{E} = x_1 E_1 + x_2 E_2 + \cdots + x_q E_q$, *where the* $E_1, E_2, \cdots, E_q$ *are the prime cycles of Lemma* 7.1, *and let* D *be an effective cycle. If* $(D - \mathcal{E} \cdot E_i) \leqq 0$ *for* $i = 1, 2, \cdots, q$, *then* $D - \mathcal{E} \succ 0$.

We write D in the form $D = X + D_1$, where X, D_1 are effective cycles, the prime components of X belong to the set $\{E_1, E_2, \cdots, E_q\}$, while no E_i is a component of D_1. Then $(D_1 \cdot E_i) \geqq 0$, whence $(X \cdot E_i) \leqq (D \cdot E_i) \leqq (\mathcal{E} \cdot E_i)$. Therefore, by Lemma 7.1, $X - \mathcal{E}$ is effective, and hence, *a fortiori*, $D - \mathcal{E}$ is effective.

Given a finite set of *distinct* prime divisorial cycles $E_1, E_2, \cdots, E_q$, we denote by $V(E_1, E_2, \cdots, E_q)$ the vector space, over the rationals, formed by all the cycles $\sum x_i E_i$, where the x_i are rational numbers. The *singular space* of the quadratic form $\varphi\{E\}$ (or — what is the same — of the associated symmetric bilinear function) is then the subspace of $V(E_1, E_2, \cdots, E_q)$ consisting of those cycles $Z = \sum x_i E_i$ for which $(Z \cdot E_i) = 0$, $i = 1, 2, \cdots, q$.

LEMMA 7.3. *If* $E_1, E_2, \cdots, E_q$ *are distinct prime divisorial cycles such that* $\varphi\{E\}$ *is of type* $(0, r)$, $r < q$, *then the singular space* W *of* $\varphi\{E\}$ *(of dimension* $q - r$*) has a basis of strictly positive cycles.*

PROOF. We may assume that $W, E_1, E_2, \cdots, E_r$ span $V(E_1, E_2, \cdots, E_q)$. Let $W' = V(E_1, E_2, \cdots, E_r)$. It is clear that the quadratic form $\varphi\{E_1, E_2, \cdots, E_r\}$ is definite negative. For each $j = 1, 2, \cdots, q - r$, let $E_{r+j} = -\mathcal{E}'_j + \mathcal{E}_{r+j}$, where $\mathcal{E}'_j \in W'$ and $\mathcal{E}_{r+j} \in W$. Then it is clear that $E_1, E_2, \cdots, E_r, \mathcal{E}_{r+1}, \cdots, \mathcal{E}_q$ form a basis of $V(E_1, \cdots, E_q)$, and hence $\mathcal{E}_{r+1}, \cdots, \mathcal{E}_q$ form a basis of the singular space W. On the other hand, we have, for $j = 1, 2, \cdots, q - r$ and $\nu = 1, 2, \cdots, r$: $(\mathcal{E}'_j \cdot E_\nu) = -(E_{r+j} \cdot E_\nu) \leqq 0$. Therefore, by Lemma 7.1, $\mathcal{E}'_j \succ 0$ $(j = 1, 2, \cdots, r)$, and hence $\mathcal{E}_{r+j} = E_{r+j} + \mathcal{E}'_j$ is also effective (and necessarily positive, since the $\mathcal{E}_{r+j}$ form a basis of W). This completes the proof.

LEMMA 7.4. *Let* $E_1, E_2, \cdots, E_q$ *be distinct prime divisorial cycles and let* D *be an effective cycle. If* $(D \cdot E_i) < 0$ *for* $i = 1, 2, \cdots, q$, *then the quadratic form* $\varphi\{E\}$ *is definite negative.*

PROOF. We first show that $\varphi\{E\}$ must be of type $(0, r)$, $r \leqq q$. Assume the contrary. Then there exists a cycle $\mathcal{E}$ in $V(E_1, E_2, \cdots, E_q)$ such that $(\mathcal{E}^2) > 0$, and we may assume that $\mathcal{E}$ is effective (if $\mathcal{E} = \mathcal{E}_1 - \mathcal{E}_2$, where

$\mathcal{E}_i \in V(E_1, E_2, \cdots, E_q)$, $\mathcal{E}_i > 0$, and where $\mathcal{E}_1$, $\mathcal{E}_2$ have no prime components in common, then $(\mathcal{E}^2) > 0$ implies that either $(\mathcal{E}_1^2) > 0$ or $(\mathcal{E}_2^2) > 0$). Consider the complete linear system $|n\mathcal{E}|$. If B_n is the fixed component of this system then for n large we will have $|n\mathcal{E}| = H + B_n$, where H is an irreducible linear system, of positive dimension. Set $\Delta_0 = n\mathcal{E} - B_n$. Then Δ_0 is a *positive* cycle and $\Delta_0 \in V(E_1, E_2, \cdots, E_q)$. Since $(D \cdot E_i) < 0$ for $i = 1, 2, \cdots, q$, it follows that $(D \cdot \Delta_0) < 0$ and thus $(D \cdot \Delta) < 0$ for every Δ in H. This is impossible since H has no fixed components and D is effective.

So we now know that $\varphi\{E\}$ is of type $(0, r)$. Suppose $r < q$. Then, by Lemma 7.3, there exists a strictly positive cycle Z in the singular space of $\varphi\{E\}$. Since $(D \cdot E_i) < 0$ for $i = 1, 2, \cdots, q$ and Z is strictly positive, it follows that $(D \cdot Z) < 0$, in contradiction with the fact that Z belongs to the singular space of $\varphi\{E\}$ and that consequently $(X \cdot Z) \geqq 0$ for every $X \succ 0$. This completes the proof of the lemma.

LEMMA 7.5. *Let D be a strictly positive cycle, let $E_1, E_2, \cdots, E_q$ be distinct prime cycles ($q \geqq 1$), and assume that $(D \cdot E_i) \leqq 0$ for $i = 1, 2, \cdots, q$. If the quadratic form φ_D is not definite negative then the quadratic form $\varphi\{E\}$ is of type $(0, r)$ ($r \leqq q$). If the quadratic form φ_D is of type $(1, t)$ then $\varphi\{E\}$ is definite negative* (i.e., $r = q$).

PROOF. To prove the first assertion we repeat the indirect argument of the first part of the proof of the preceding lemma, with the following modification: We now have $(D \cdot \Delta) \leqq 0$, and hence necessarily $(D \cdot \Delta) = 0$, for every Δ in H. Since H has no fixed components, it follows that if $D_1, D_2, \cdots, D_h$ are the prime components of D then $(D_i \cdot \Delta) = 0, i = 1, 2, \cdots, h$. Therefore if $x_1, x_2, \cdots, x_h$ are arbitrary rational numbers and if we set $X = x_1 D_1 + x_2 D_2 + \cdots + x_h D_h$, then $(X \cdot \Delta) = 0$. Since $(\Delta^2) > 0$, it follows from Lemma 6.6 that either $(X^2) < 0$ or some multiple of X is algebraically equivalent to zero. This implies already that $(X^2) < 0$ if X is a strictly positive cycle, and from this — by writing X as the difference of two effective cycles, without common components — it follows that $(X^2) < 0$ if $X \neq 0$. Thus φ_D is definite negative, a contradiction.

To prove the second assertion of the lemma, we repeat the argument given in the second part of the proof of the preceding lemma, with the following modifications: We now have $(D \cdot Z) \leqq 0$ and hence necessarily $(D \cdot Z) = 0$, as was pointed out already in the proof of the preceding lemma. Therefore $(D_i \cdot Z) = 0$ for $i = 1, 2, \cdots, h$ (always by the same remark that $(Y \cdot Z) \geqq 0$ for every effective cycle Y), and thus $(X \cdot Z) = 0$ for every cycle X of the form $x_1 D_1 + x_2 D_2 + \cdots + x_h D_h$. By hypothesis, we have $(X^2) > 0$ for some X, and we have $(Z^2) = 0$. Therefore, by Lemma 6.6,

Z must be algebraically equivalent to zero, which is impossible since Z is a strictly positive cycle. This completes the proof.

DEFINITION 7.6. *A divisorial cycle D is said to be arithmetically effective* (a.e.) *if $(D \cdot X) \geqq 0$ for every effective cycle X.*[8]

The reduction of our problem to the case of arithmetically effective cycles D is based on the following theorem:

THEOREM 7.7. *If D is an effective divisorial cycle then there exists one and only one effective cycle $\mathcal{E}$ having the following properties:*

(1) *Either $\mathcal{E} = 0$ or $\varphi_{\mathcal{E}}$ is definite negative.*

(2) *$D - \mathcal{E}$ is arithmetically effective.*

(3) *$(D - \mathcal{E} \cdot E) = 0$ for every prime component of $\mathcal{E}$.*

Furthermore, we have necessarily $D - \mathcal{E} \succ 0$. (The cycle $\mathcal{E}$ will be referred to as the arithmetically negative component of D).

PROOF. We first prove the unicity of $\mathcal{E}$ (in this part of the proof the assumption that $D \succ 0$ is irrelevant). Suppose $\mathcal{E}$ and $\mathcal{E}'$ are two effective cycles having properties (1), (2) and (3). Let $\mathcal{E} = \sum m(E)E$, $\mathcal{E}' = \sum m'(E)E$, where the summation symbols range over the totality of all prime divisorial cycles E on our surface F.

If S denotes the set of prime cycles E which are components of both $\mathcal{E}$ and $\mathcal{E}'$ (whence $m(E)$ and $m'(E)$ are both positive if and only if $E \in S$) we set $\mathcal{E}_1 = \sum_{E \in S} m(E)E$, $\mathcal{E}_1' = \sum_{E \in S} m'(E)E$, $\mathcal{E} = \mathcal{E}_1 + \mathcal{E}_2$, $\mathcal{E}' = \mathcal{E}_1' + \mathcal{E}_2'$. Let E be any prime component of $\mathcal{E}$. Then E is not a component of $\mathcal{E}_2'$, thus $(\mathcal{E}_2' \cdot E) \geqq 0$, and hence

$$(\mathcal{E}_1' \cdot E) \leqq (\mathcal{E}' \cdot E) \leqq (D \cdot E) = (\mathcal{E} \cdot E) .$$

(The second inequality follows from the fact that $D - \mathcal{E}'$ is a.e., and the last equality follows from condition (3)). We have therefore that the cycle $\mathcal{E}_1' - \mathcal{E}$, which is a linear combination of the prime components of $\mathcal{E}$, has the property $(\mathcal{E}_1' - \mathcal{E} \cdot E) \leqq 0$, for every prime component E of $\mathcal{E}$. Since $\varphi\{E\}$ is definite negative, it follows from Lemma 7.1 that $\mathcal{E}_1' \succ \mathcal{E}$, and *a fortiori* $\mathcal{E}' \succ \mathcal{E}$. In a similar way it follows that $\mathcal{E} \succ \mathcal{E}'$, whence $\mathcal{E} = \mathcal{E}'$.

[8] An important example of arithmetically effective cycles is given by the canonical cycles K on a non-singular surface F *free from exceptional curves of the first kind* ($F =$ a relatively minimal model) *and not belonging to the class of ruled surfaces.* In fact, it is known that if there exists in F a prime cycle Γ such that $(K \cdot \Gamma) < 0$, i.e., $2p(\Gamma) - 2 < (\Gamma^2)$, (or equivalently if the process of adjuntion on F terminates after a finite number of steps; see Zariski [7, p. 73]), then F is birationally equivalent to a ruled surface. (In the case of characteristic $p \neq 0$ the classical proof of this result encounters a new difficulty, owing to the possibility that the map of F into its Albanese variety may not be separable. However, a proof which is valid in arbitrary characteristic was recently found by D. Mumford (unpublished)).

Assuming that $\mathcal{E}$ exists, the assertion $D-\mathcal{E}\succ 0$ is a direct consequence of properties (1), (3) of $\mathcal{E}$ and of Corollary 7.2.

We now turn to the proof of the existence of $\mathcal{E}$. We first observe that if D is a.e. then we can take for $\mathcal{E}$ the cycle 0. We also observe that if φ_D is definite negative then we can set $\mathcal{E}=D$. We shall therefore assume that φ_D is not definite negative, and that D is not a.e.

Let $E_1, E_2, \cdots, E_q$ be the prime components of D such that $(D\cdot E_i)\leqq 0, i=1,2,\cdots,q$. Here q is at least 1, since D is not a.e. We shall now prove that *there exists an effective cycle* $\mathcal{E}_0$ *in* $V(E_1, E_2, \cdots, E_q)$ *having the following properties:* $\varphi_{\mathcal{E}_0}$ *is definite negative,* $D-\mathcal{E}_0$ *is effective and* $(D\cdot E_i)=(\mathcal{E}_0\cdot E_i)$ *for* $i=1,2,\cdots,q$. Once this assertion is proved, the proof of the existence of $\mathcal{E}$ can be rapidly completed as follows:

We set $D'=D-\mathcal{E}_0$, whence D' is an effective cycle and

$$(D'\cdot E_i)=0\ , \qquad i=1,2,\cdots,q\ . \tag{50}$$

Let n be the number of distinct prime components of D (whence $n\geqq q$). If $n=q$, then $(D'\cdot E)=0$ for every prime component E of D' (since every such E is also a prime component of D and is therefore one of the cycles $E_1, E_2, \cdots, E_q$), and hence D' is a.e. Therefore in this case we can take for $\mathcal{E}$ the cycle $\mathcal{E}_0$. We now use induction with respect to the difference $n-q$ (i.e., with respect to the number of prime components E of D such that $(D\cdot E)>0$). We shall denote this numerical character $n-q$ of D by $\mu(D)$. In view of (50), we have $\mu(D')\leqq\mu(D)$, and if the equality holds then D' is a.e., in which case we can take, as above, $\mathcal{E}=\mathcal{E}_0$. We may therefore assume that $\mu(D')<\mu(D)$. By our induction hypothesis, there exists then an effective cycle $\mathcal{E}'$ such that conditions (1), (2) and (3) of the theorem are satisfied, with D and $\mathcal{E}$ being replaced by D' and $\mathcal{E}'$. We now set $\mathcal{E}=\mathcal{E}_0+\mathcal{E}'$. Then condition (2) is satisfied for D and $\mathcal{E}$ since $D-\mathcal{E}=D'-\mathcal{E}'$. It is also clear that condition (3) is satisfied if E is a prime component of $\mathcal{E}'$. Let now E be a prime component of $\mathcal{E}$ which is not a component of $\mathcal{E}'$ and is therefore a component of $\mathcal{E}_0$. Then E is one of the cycles $E_1, E_2, \cdots, E_q$ and therefore, by (50), $(D'\cdot E)=0$. On the other hand, we have $(D'\cdot E)\geqq(\mathcal{E}'\cdot E,)$ since $D'-\mathcal{E}'$ is a.e. Hence $(\mathcal{E}'\cdot E)\leqq 0$, and since E is not a component of the *effective* cycle $\mathcal{E}'$, it follows that $(\mathcal{E}'\cdot E)=0$. We have thus shown that both $(D'\cdot E)$ and $(\mathcal{E}'\cdot E)$ are zero; consequently $(D-\mathcal{E}\cdot E)=0$ (since $D-\mathcal{E}=D'-\mathcal{E}'$). This shows that $\mathcal{E}$ has property (3).

By the induction hypothesis, the quadratic form $\varphi_{\mathcal{E}'}$ is definite negative. The set S of prime components of $\mathcal{E}$ which are *not* components of $\mathcal{E}'$ are components of $\mathcal{E}_0$, and since $\varphi_{\mathcal{E}_0}$ is definite negative it follows that the quadratic form associated with the set S is also definite negative. We have just shown that $(\mathcal{E}' \cdot E) = 0$ if $E \in S$, and this implies that $(E' \cdot E) = 0$ if E' is a prime component of $\mathcal{E}'$ and $E \in S$. This shows that also $\varphi_{\mathcal{E}}$ is definite negative.

Thus, to complete the proof of the theorem we have only to prove the existence of the cycle $\mathcal{E}_0$ having the above stated properties. In the proof we shall consider separately two cases, according as φ_D is of type $(1, t)$ or $(0, t)$.

First case. φ_D *is of type* $(1, t)$.

By Lemma 7.5, the quadratic form $\varphi\{E_1, E_2, \cdots, E_q\}$ is definite negative. As this form is therefore, at any rate, non-singular, we can find a cycle $\mathcal{E}_0$ in $V(E_1, E_2, \cdots, E_q)$ such that $(D \cdot E_i) = (\mathcal{E}_0 \cdot E_i)$, $i = 1, 2, \cdots, q$. Since $(D \cdot E_i) \leqq 0$, it follows from Lemma 7.1 that $\mathcal{E}_0$ is an effective cycle. It is strictly positive because the $(D \cdot E_i)$ are not all zero (since D is not a.e.). Hence $\varphi_{\mathcal{E}_0}$ is definite negative. Since $(D - \mathcal{E}_0 \cdot E_i) = 0$ for $i = 1, \cdots, q$, it follows by Corollary 7.2 that $D - \mathcal{E}_0 \succ 0$.

Second case. φ_D *is of type* $(0, t)$.

In this case we can only assert, by Lemma 7.5, that the quadratic form $\varphi\{E_1, E_2, \cdots, E_q\}$ is type $(0, r)$, $r \leqq q$, and the proof of the existence of $\mathcal{E}_0$ requires some additional considerations. We use Lemma 7.3 and the specific information acquired in the course of the proof of that lemma. In the notations introduced in that proof, we have

$$\left.\begin{array}{lll}(51) & E_{r+j} = -\mathcal{E}'_j + \mathcal{E}_{r+j}\,, \\ (52) & 0 \prec \mathcal{E}'_j \in V(E_1, E_2, \cdots, E_r)\,, \\ (53) & \mathcal{E}_{r+j} \in V(E_1, E_2, \cdots, E_q) \\ (54) & (\mathcal{E}_{r+j} \cdot E_i) = 0,\ i = 1, 2, \cdots, q\,,\end{array}\right\} \quad j = 1, 2, \cdots, q - r.$$

Since $(D \cdot E_i) \leqq 0$ and $\mathcal{E}'_j \succ 0$, it follows that $(D \cdot \mathcal{E}'_j) \leqq 0$. From (53) and (54) it follows that $\mathcal{E}_{r+j}$ is a.e., whence also $(D \cdot \mathcal{E}_{r+j}) \geqq 0$. Since, by (51), we have $(D \cdot E_{r+j}) = -(D \cdot \mathcal{E}'_j) + (D \cdot \mathcal{E}_{r+j})$, and since $(D \cdot E_{r+j}) \leqq 0$, it follows that

$$\left.\begin{array}{ll}(55) & (D \cdot \mathcal{E}'_j) = 0 \\ (56) & (D \cdot \mathcal{E}_{r+j}) = 0\end{array}\right\} \quad j = 1, 2, \cdots, q - r.$$

From (56) it follows at once (since $\mathcal{E}_{r+j}$ is a.e.) that

$$(57) \qquad (\Gamma \cdot \mathcal{E}_{:+j}) = 0\,, \qquad \text{if } \Gamma \text{ is any prime component of } D.$$

The quadratic form $\varphi\{E_1, E_2, \cdots, E_r\}$ is definite negative, and therefore, at any rate, non-singular. We take for $\mathcal{E}_0$ the unique cycle in $V(E_1, E_2, \cdots, E_r)$ which is determined by the conditions

$$(\mathcal{E}_0 \cdot E_\nu) = (D \cdot E_\nu) , \qquad \nu = 1, 2, \cdots, r . \tag{58}$$

Since $(D \cdot E_\nu) \leqq 0$ and $\varphi\{E_1, E_2, \cdots, E_r\}$ is definite negative, it follows from Lemma 7.1 that $\mathcal{E}_0 \succ 0$, and from Corollary 7.2 that $D - \mathcal{E}_0 \succ 0$. Furthermore, $\varphi_{\mathcal{E}_0}$ is definite negative (since $\mathcal{E}_0 \in V(E_1, E_2, \cdots, E_r)$) or $\mathcal{E}_0 = 0$. It remains to show that

$$(\mathcal{E}_0 \cdot E_{r+j}) = (D \cdot E_{r+j}) , \qquad j = 1, 2, \cdots, q - r , \tag{59}$$

and if that is shown, it will also follow that $\mathcal{E}_0 \neq 0$ (since D is not a.e.). We set $D' = D - \mathcal{E}_0$. Since $0 \prec D' \prec D$, it follows from (57) that $(D' \cdot \mathcal{E}_{r+j}) = 0$, for $j = 1, 2, \cdots, q - r$. Since, by (58), we have $(D' \cdot E_\nu) = 0$ ($\nu = 1, 2, \cdots, r$), it follows that $(D' \cdot \mathcal{E}'_j) = 0$, for $j = 1, 2, \cdots, q - r$. This, in conjunction with $(D' \cdot \mathcal{E}_{r+j}) = 0$ yields, by (51), the relations $(D' \cdot E_{r+j}) = 0$. This establishes (59) and completes the proof of the theorem.

The cycle $\mathcal{E}$ (the arithmetically negative component of D), which according to our theorem is uniquely determined by D, shall be denoted by $N(D)$. This cycle $N(D)$ is defined if D is effective. We point out explicitly that $N(D)$ is not necessarily an integral cycle (even if D is integral).

COROLLARY 7.8. *$N(D) = 0$ if and only if D is* a.e.

8. Reduction to the case of arithmetically effective cycles

Let D be an arbitrary strictly positive *integral* cycle and let $\mathcal{E} = N(D)$ be the a.n. (arithmetically negative) component of D. If $X = x_1\Gamma_1 + x_2\Gamma_2 + \cdots + x_h\Gamma_h$ is any effective cycle, where $\Gamma_1, \Gamma_2, \cdots, \Gamma_h$ are the distinct prime components of X, we denote by $[X]$ the integral cycle $[x_1]\Gamma_1 + [x_2]\Gamma_2 + \cdots + [x_h]\Gamma_h$, where $[x]$ denotes the integral part of x.

THEOREM 8.1. *If B_n denotes the fixed component of $|nD|$ then $B_n \succ [n\mathcal{E}]$.*

PROOF. Let $E_1, E_2, \cdots, E_q$ be the distinct prime components of $\mathcal{E}$. We have $(D - \mathcal{E} \cdot E_i) = 0$, and hence $(D_n - n\mathcal{E} \cdot E_i) = 0$, for any cycle D_n in $|nD|$ and for $i = 1, 2, \cdots, q$. Hence, by Corollary 7.2, $D_n \succ n\mathcal{E}$, and since $n\mathcal{E} \succ [n\mathcal{E}]$, the theorem is proved.

If φ_D is definite negative then $D = \mathcal{E}$ and $\dim |nD| = 0$ for all n. Hence in this case, we have

$$s(nD) = -\rho(nD) - p_a = -\rho(n\mathcal{E}) - p_a . \tag{60}$$

In the next theorem we assume that φ_D is not definite negative.

THEOREM 8.2. *Assume that φ_D is not definite negative and let m be the smallest positive integer such that $m\mathcal{E}$ is an integral cycle. Let $\Delta = m(D - \mathcal{E})$ (whence Δ is an effective and* a.e. *integral cycle). If n is a* *positive integer and* $n = n'm + r$, *where* $0 \leqq r < m$, *then for large* n

$$s(nD) = s(n'\Delta + rD) - \rho(n\mathcal{E}) + \rho(r\mathcal{E}) . \tag{61}$$

Here ρ *is the expression* (1) *of* § 1, i.e.,

$$\rho(n\mathcal{E}) = \frac{(\mathcal{E}^2)}{2}n^2 - \frac{(\mathcal{E} \cdot K)}{2}n . \tag{62}$$

PROOF. From Theorem 8.1 it follows that the integral cycle $n'm\mathcal{E}$ is part of the fixed component of the complete linear system $|nD|$. Since

$$nD = n'm\mathcal{E} + n'\Delta + rD , \tag{63}$$

it follows that $\dim |nD| = \dim |n'\Delta + rD|$. On the other hand, since φ_D is not definite negative, Δ is a strictly positive cycle, and therefore $i(nD) = i(n\Delta) = 0$ for large n. We find therefore, for large n:

$$\begin{aligned} s(nD) &= \dim |n'\Delta + rD| - \rho(nD) - p_a \\ &= \rho(n'\Delta + rD) - \rho(nD) + s(n'\Delta + rD) . \end{aligned} \tag{64}$$

Now, by (63), we have $\rho(nD) = \rho(n'\Delta + rD) + \rho(n'm\mathcal{E}) + (nD - n'm\mathcal{E} \cdot n'm\mathcal{E})$, and $(nD - n'm\mathcal{E} \cdot n'm\mathcal{E}) = (n(D - \mathcal{E}) + r\mathcal{E} \cdot n'm\mathcal{E})$. Since $(D - \mathcal{E} \cdot \mathcal{E}) = 0$, we have $\rho(nD) = \rho(n'\Delta + rD) + \rho(n'm\mathcal{E}) + rn'm(\mathcal{E}^2)$. On the other hand, we have $\rho(n\mathcal{E}) = \rho(n'm\mathcal{E} + r\mathcal{E}) = \rho(n'm\mathcal{E}) + \rho(r\mathcal{E}) + rn'm(\mathcal{E}^2)$. Hence

$$\rho(nD) = \rho(n'\Delta + rD) + \rho(n\mathcal{E}) - \rho(r\mathcal{E}) , \tag{65}$$

and substituting into (64) we find (61).

We note that (61) is false if φ_D is definite negative, for in that case we have $D = \mathcal{E}$, $\Delta = 0$, $r = 0$, and the right hand of (61) is equal to $-\rho(n\mathcal{E}) + s(0) = -\rho(n\mathcal{E}) + p_g - p_a$, while, by (60), $s(nD) = -\rho(n\mathcal{E}) - p_a$.

Formula (61) reduces the problem of finding $s(nD)$ to that of finding $s(n'\Delta + rD)$. Here Δ is arithmetically effective and r is a fixed integer, $0 \leqq r < m$. The difference $s(nD) - s(n'\Delta + rD)$, for given r, is a quadratic polynomial in n, with *positive* leading coefficient $-(\mathcal{E}^2)/2$.

We shall therefore proceed to the study of the systems $|nD|$, where D is an effective and arithmetically effective cycle, and we shall later on apply the results to the systems $|X + nD|$, where X is a fixed effective cycle and D is effective and a.e. The interesting case is the one in which φ_D is of type $(1, t)$, and it is with this case that we shall be concerned in most of the remainder of this part of the paper (§§ 9 and 10).

9. A theorem on arithmetically effective cycles

Let D be an effective cycle. We denote by B_n the fixed component of

$|nD|$ and by $\mathscr{B}(D)$ the set of prime cycles E which have the following property: there exist infinitely many integers n such that E is a prime component of B_n. It is clear that each cycle E in $\mathscr{B}(D)$ is a prime component of D and that $\mathscr{B}(D)$ is therefore a finite set. Let $\mathscr{B}(D) = \{E_1, E_2, \cdots, E_q\}$. *For all sufficiently large n*, the set of prime components of B_n is a subset of the set $\mathscr{B}(D)$, and for each $i = 1, 2, \cdots, q$ it happens for infinitely many values of n that E_i is a prime component of B_n.

For any E_i, let M_i denote the set of all positive integers n such that E_i is *not* a component of B_n. The set M_i, if not empty, is closed under addition. Assume that for a given i, M_i is non-empty. Let h_i be the highest common divisor of the integers n belonging to M_i; thus every integer belonging to M_i is divisible by h_i. *We assert that all sufficiently high multiples of h_i belong to M_i.* For, let $n_1, n_2, \cdots, n_s$ be elements of M_i which have h_i as their highest common divisor. It is a straightforward matter to show that all sufficiently high multiples of h_i are linear combinations of $n_1, n_2, \cdots, n_s$, with coefficients which are non-negative integers, and from this our assertion follows. The integer h_i is defined only if M_i is non-empty, and in that case we must have $h_i > 1$ (since $E_i \in \mathscr{B}(D)$). We note the following consequence: *there exists an integer N such that for any $n \geqq N$, the set of prime components of B_n coincides with $\mathscr{B}(D)$ if and only if n is not divisible by any of the h_i.*

THEOREM 9.1. *If D is arithmetically effective then $(D \cdot E_i) = 0$ $(i = 1, 2, \cdots, q)$.*

PROOF. The theorem is trivial if φ_D is of type $(0, t)$, and in this case we have in fact $(D \cdot E) = 0$ for *each prime component of D*. To see this, we observe that since D is a.e. we must have $(D^2) \geqq 0$, and therefore $(D^2) = 0$ (since φ_D is of type $(0, t)$). Since $(D \cdot E) \geqq 0$ (always because D is a.e.) and since D is an effective cycle, the equality $(D \cdot E) = 0$ follows.

We therefore assume from now on that φ_D is of type $(1, t)$. We shall also assume, of course, that $\mathscr{B}(D)$ is a non-empty set. We shall assume that there is an element E in the set $\mathscr{B}(D)$ such that $(D \cdot E) > 0$ and we shall show that this assumption leads to an absurd conclusion. Let, say, $E = E_1$.

Let S denote the (finite) set of all pairs (Δ, Γ) of cycles Δ, Γ such that $0 \prec \Delta \prec D$ and Γ is a prime component of $D - \Delta$. We know that there exists an integer N such that either E_1 is a component of B_n for all $n \geqq N$ (this is the case if the set M_1 introduced earlier is empty) or E_1 is a component of B_n for all integers n which are $\geqq N$ and are *not* divisible by a certain integer $h_1 > 1$ (this is the case in which M_1 is not empty). By increasing N, if necessary, we may assume that also the following condition is satisfied:

(66) $$(ND - \Delta \cdot \Gamma) > 2p(\Gamma) - 2, \qquad \text{for all } (\Delta, \Gamma) \in S$$

such that $(D \cdot \Gamma) > 0$. We set

(67) $$\sigma = \max\{p(\Gamma), p(\Gamma) - 1 + (\Delta \cdot \Gamma)\}, \qquad \text{for all } (\Delta, \Gamma) \in S.$$

We also fix a sequence $(\Delta_1, \Gamma_1), (\Delta_2, \Gamma_2), \cdots, (\Delta_m, \Gamma_m)$ in S such that:

$$\Delta_1 = 0, \Gamma_1 = E_1;$$
$$\Delta_{i+1} = \Delta_i + \Gamma_i, \qquad 1 \leqq i \leqq m - 2,$$
$$D = \Delta_m + \Gamma_m.$$

Consider any element (Δ, Γ) in S. If $(D \cdot \Gamma) > 0$ then it follows from (66) and (14) (§2) that $s(nD - \Delta) \leqq s(nD - \Delta - \Gamma)$ for all $n \geqq N$. On the other hand, if $(D \cdot \Gamma) = 0$ then $(nD - \Delta \cdot \Gamma) = -(\Delta \cdot \Gamma)$, and the index of specialty $i_\Gamma(nD - \Delta)$ is not greater than σ, by (67). Hence in this case we have, by (13), § 2: $s(nD - \Delta) \leqq s(nD - \Delta - \Gamma) + \sigma$. Since $\sigma \geqq 0$, it follows that

(68) $$s(nD - \Delta) \leqq s(nD - \Delta - \Gamma) + \sigma, \qquad \text{for all } (\Delta, \Gamma) \in S \text{ and all } n \geqq N.$$

Applying this inequality successively to the pairs $(\Delta_1, \Gamma_1), (\Delta_2, \Gamma_2), \cdots, (\Delta_m, \Gamma_m)$ we find

(69) $$s(nD) \leqq s((n-1)D) + m\sigma, \qquad \text{for all } n \geqq N.$$

Here m and σ are fixed integers, independent of n. The inequality (69) can, however, be improved if E_1 is a fixed component of $|nD|$, for in that case the linear series $\mathrm{Tr}_{E_1}|nD|$ does not exist and hence the deficiency $\delta_{E_1}(nD)$ is equal to $\dim|\{nD\} \cdot E_1| + 1$. Since, by (66), as applied to $\Delta = \Delta_1 = 0$, $\Gamma = \Gamma_1 = E_1$, we have $(nD \cdot E_1) > 2p(E_1) - 2$ if $n \geqq N$, it follows that $\delta_{E_1}(nD) = n(D \cdot E_1) - p(E_1) + 1$. Therefore, by (12) (§ 2), we find

(70) $$s(nD) \leqq s(nD - E_1) - n(D \cdot E_1) + p(E_1) - 1.$$

If we now apply (68) successively to the pairs $(\Delta_2, \Gamma_2), \cdots, (\Delta_m, \Gamma_m)$ and combine with (70) we find

(71) $$s(nD) \leqq s((n-1)D) - n(D \cdot E_1) + p(E_1) - 1 + (m-1)\sigma,$$
$$\text{(if } E_1 \text{ is a component of } B_n \text{ and } n \geqq N.)$$

If M_1 is empty, then E_1 is a component of B_n for all n, and since $(D \cdot E_1) > 0$ it follows from (71) that if n is sufficiently large the *non-negative* integers $s(nD), s((n+1)D), \cdots$ form an infinite, *strictly decreasing* sequence, which is absurd. If M_1 is not empty, then for all $n \geqq N$ (71) holds true if n is not divisible by h_1, while (69) holds true for all $n \geqq N$. It follows at once also in this case that if n is sufficiently large then the non-negative integers $s(nD)$, $s((n+1)D)$, $s((n+2)D)$, $\cdots$ form an infinite, strictly decreasing sequence. This completes the proof of the theorem.

COROLLARY 9.2. *Assume that* D *is* a.e., *that the quadratic form* φ_D *is of type* $(1, t)$ *and that* $\mathscr{B}(D)$ $(=\{E_1, E_2, \cdots, E_q\})$ *is non-empty. Then the quadratic form* $\varphi\{E_1, E_2, \cdots, E_q\}$ *is definite negative.*

The corollary is a direct consequence of the preceding theorem and of Lemma 7.5.

Note. The assumption that D is arithmetically effective is essential for the validity of the above theorem; if this assumption is not satisfied, then we may very well have $(D \cdot E) > 0$ for some i. The following is an example.

Let Γ, E_1, E_2 be irreducible curves such that $(\Gamma^2) > 0$, $(\Gamma \cdot E_1) = (\Gamma \cdot E_2) = 0$ and such that $\varphi\{E_1, E_2\}$ is definite negative. Let $D = \Gamma + E_1 + E_2$. Then $\mathscr{E} = N(D) = E_1 + E_2$, and hence $n\mathscr{E}$ is part of the fixed component of $|nD|$, and — in fact — for n large the fixed component B_n of precisely nD (since $(\Gamma^2) > 0$ and Γ is irreducible). It follows that $\mathscr{B}(D) = \{E_1, E_2\}$. Let $(E_1^2) = -\alpha$, $(E_2^2) = -\beta$, $(E_1, E_2) = \gamma$, and *assume* that $\alpha > \gamma > \beta$. Then $(D \cdot E_2) = \gamma - \beta > 0$. The conditions imposed on Γ, E_1, E_2 can be realized, for instance, as follows. Let F' be the surface $xyz + x^3 + y^3 + z^4 = 0$. This has at the origin O', a triple point. After one quadratic transformation $T: F' \to F''$, with center at O', the singularity O' is resolved and is replaced by an irreducible curve E'' having an ordinary double point O'' and such that $(E'')^2 = -3$. Another quadratic transformation with center O'' yields a surface F on which the proper transform E_1 of E'' is such that $(E_1^2) = -\alpha = -7$. Furthermore, if E_2 is the exceptional curve which corresponds to O'' then $(E_2^2) = -\beta = -1$ and $(E_1 \cdot E_2) = \gamma = 2$. Clearly $\varphi\{E_1, E_2\}$ is definite negative. We take for Γ the proper transform of a generic plane section of the initial surface F'.

A special case of arithmetically effective cycles is given by the *arithmetically positive* cycles; i.e., cycles D such that $(D \cdot Z) > 0$ for every strictly positive cycles Z. In this case we have

COROLLARY 9.3. *If* D *is an effective, arithmetically positive cycle, then* $s(nD) = 0$ *for large* n, *and* $|nD|$ *has no base points if* n *is large. Furthermore, the system* $|nD|$ *is ample for large* n, i.e., *the rational transformation* $F \to V_D^*$ *defined by* $|nD|$ (see § 6) *is biregular.*

In the first place it follows from Theorem 9.1 that $|nD|$ has no fixed components for large n. Hence $|nD|$ has no base points for large n. By Theorem 6.3 it follows then that $s(nD) = p_a(V_D^*) - p_a(F)$. If the regular map $F \to V_D^*$ were not biregular, we would have curves E on F which are contracted to points of V_D^*, and for any such curve E we would have

$(D \cdot E) = 0$, in contradiction with the assumption that D is arithmetically positive. Hence $F \to V_D^*$ is a biregular map and $p_a(V_D^*) = p_a(F)$.

The assumption that D is arithmetically positive is a very strong one and makes it possible to give a proof of the above corollary which avoids all the difficulties of the proof of the general Theorem 9.1 (see Nakai [3]).

10. Arithmetically effective cycles D such that φ_D is of type $(1, t)$: boundedness of the superabundance and of the fixed component of $|nD|$.

We use the notations of the preceding section; in particular, B_n denotes the fixed component of $|nD|$, where D is a given effective cycle.

THEOREM 10.1. *A necessary condition that $s(nD)$ or B_n be bounded (from above) is that D be arithmetically effective. If φ_D is of type $(1, t)$ then this condition is also sufficient.*

PROOF. The first part of the theorem is a direct consequence of Theorems 8.1 and 8.2, since in (65) the leading coefficient of the second degree polynomial $\rho(n\mathcal{E})$ is negative. Now suppose that D is arithmetically effective and that φ_D is of type $(1, t)$. We observe that these assumptions imply that $(D^2) > 0$. For, we certainly have $(D^2) \geqq 0$ (since D is a.e.). If $(D^2) = 0$ and if $D_1, D_2, \cdots, D_m$ are the prime components of D, then it would follow that $(D \cdot D_i) = 0$, for $i = 1, 2, \cdots, m$ (always because D is a.e), and hence, by Lemma 7.5, the quadratic form φ_D would be definite negative, a contradiction. It follows that if we write $|mD| = B_m + H_m$ then, for all m sufficiently large, the complete linear system H_m is irreducible, has positive dimension, and if Γ is a generic member of H_m then $(\Gamma^2) > 0$ and Γ is a prime cycle (see proof of Theorem 6.7). We fix such an integer m and we impose on m the further condition: $m \not\equiv 0 \pmod{h_i}$, for all i such that M_i is non-empty (see §9). Then $E_1, E_2, \cdots, E_q$ are precisely the prime components of B_m. Given any integer n, we write $n = mn' + r$, $0 = \leqq r < m$, and we note that we have obviously $B_n \prec B_{n'm} + rD$. Therefore, in order to prove that the B_n are bounded it is sufficient to prove that the $B_{n'm}$ are bounded. We therefore may replace D by mD; i.e., we may assume that if $|D| = B_1 + H_1$ then $\dim H_1 > 0$, that the generic member Γ of H_1 is a prime cycle Γ, that $(\Gamma^2) > 0$ and that $\mathcal{B}(D)$, i.e., the set $\{E_1, E_2, \cdots, E_q\}$, is precisely the set of prime components of B_1.

Let s_0 be the constant superabundance of $|n\Gamma|$, n-large (see Theorem 6.7), say $s(n\Gamma) = s_0$ for $n \geqq N$. For any $n \geqq N$ there exist then cycles Z_n having the following properties:

(a) $n\Gamma \prec Z_n \prec n(B_1 + \Gamma)$;

(b) $|Z_n|$ is irreducible;

(c) $s(Z_n) \leqq s_0$ (for instance, $Z_n = n\Gamma$ is such a cycle).

For each given $n \geqq N$ we fix a maximal element C_n in the (finite) set of such cycles Z_n. We note that by (a) we have $\dim|C_n| > 0$ (since $\dim|n\Gamma| > 0$). Hence, by (b), $|C_n|$ has no fixed components. We set

$$B'_n = n(B_1 + \Gamma) - C_n .$$

Then it is clear that $B_n \prec B'_n \prec nB_1$ (since $C_n + B'_n \in |nD|$ and since $n\Gamma \prec C_n$). It will therefore be sufficient to prove that the B'_n are bounded (from above), and in proving this we will take into account the relation $B'_n \prec nB_1$ and the fact that $E_1, E_2, \cdots, E_q$ are the prime components of B_1; these imply that the prime components of B'_n are in the set $\{E_1, E_2, \cdots, E_q\}$.

We assert that for any $i = 1, 2, \cdots, q$ and for all n sufficiently large we have

$$(72) \qquad (C_n \cdot E_i) \leqq \max\{s_0 - \rho(E_i), (E_i \cdot K), 0\} .$$

This inequality is obvious if E_i is not a component of B'_n, for in that case we have $(B'_n \cdot E_i) \geqq 0$ and hence $(C_n \cdot E_i) \leqq 0$, since $C_n + B'_n \equiv nD$ and $(D \cdot E_i) = 0$ (Theorem 9.1). Assume now that E_i is a component of B'_n. We have $\dim|C_n| \leqq \rho(C_n) + p_a + s_0$, since $s(C_n) \leqq s_0$ and $i(C_n) \leqq i(n\Gamma) = 0$ (if n is large). On the other hand, $\dim|C_n + E_i| \geqq \rho(C_n + E_i) + p_a = \rho(C_n) + \rho(E_i) + (C_n \cdot E_i) + p_a$. Therefore, if $(C_n \cdot E_i) > s_0 - \rho(E_i)$, then $\dim|C_n + E_i| > \dim|C_n|$, showing that E_i is not a fixed component of $|C_n + E_i|$. Hence $|C_n + E_i|$ is irreducible. If, furthermore, $(C_n \cdot E_i) > (E_i \cdot K)$, then $(C_n + E_i \cdot E_i) > 2p(E_i) - 2$, and therefore [see (14), § 2] $s(C_n + E_i) \leqq s(C_n)$; i.e., $s(C_n + E_i) \leqq s_0$. Finally, since E_i is a component of B'_n it follows from the definition of B'_n that $C_n + E_i \prec n(\Gamma + B_1)$. Thus, if (72) is false, then $C_n + E_i$ has the properties (a), (b) and (c) above, in contradiction with the maximality of C_n.

Since $C_n + B'_n \equiv nD$, and since $(D \cdot E_i) = 0$ (Theorem 9.1) we have from (72):

$$(73) \qquad (B'_n \cdot E_i) \geqq -\sigma_i ,$$

where we have set

$$\sigma_i = \max\{s_0 - \rho(E_i), (E_i \cdot K), 0\} .$$

Since $\varphi\{E_1, E_2, \cdots, E_q\}$ is definite negative, hence — at any rate — non-singular, we can find a cycle $\mathcal{E}$ in $V(E_1, E_2, \cdots, E_q)$ such that $(\mathcal{E} \cdot E_i) = -\sigma_i$. Since $B'_n \in V(E_1, E_2, \cdots, E_q)$, it follows from (73) and Lemma 7.1 that $B'_n \prec \mathcal{E}$, for large n. This shows that the B'_n, and hence also the B_n, are bounded (from above).

To show that also the superabundances $s(nD)$ are bounded (from above) we need the following lemma:

LEMMA 10.2. *Let* $\{|D_n|\}$ *be an infinite sequence of complete linear systems and let* X *be an effective divisorial cycle. If the sequence of irregularities* $\{\text{irreg}(X_i)\}$ [*see* (10), § 2] *is bounded from above and if for each prime component* Γ *of* X *the sequence of intersection numbers* $(D_n \cdot \Gamma)$ *is bounded from below, then also the sequence* $\{\text{irreg}(X + D_n)\}$ *is bounded from above.*

PROOF OF THE LEMMA. If $X = \sum r_i \Gamma_i$, we shall use induction with respect to the integer $\sum r_i$ (if $\sum r_i = 0$; i.e., if $X = 0$, there is nothing to prove). Let Γ be a prime component of X and $X' = X - \Gamma$. By our induction hypothesis, the sequence $\{\text{irreg}(X' + D_n)\}$ is bounded from above, say $\text{irreg}(X' + D_n) \leqq \sigma$, for all n. By (12) (§ 2), we have $\text{irreg}(X + D_n) \leqq \text{irreg}(X' + D_n) + i_\Gamma(X + D_n) \leqq \sigma + i_\Gamma(X + D_n)$. By assumption, the intersection numbers $(X + D_n \cdot \Gamma)$ are bounded from below, say $(X + D_n \cdot \Gamma) \geqq \tau$, for all n. Then $i_\Gamma(X + D_n) \leqq \rho = \max\{0, 2p(\Gamma) - 2 - \tau\}$. Hence $\text{irreg}(X + D_n) \leqq \sigma + \rho$, all n. q.e.d.

This lemma can be applied to the systems $|C_n|$, since $s(C_n) \leqq s_0$ and $i(C_n) = 0$ for large n. Furthermore, we have $(C_n \cdot E_i) \geqq 0$ for n large and $i = 1, 2, \cdots, q$ (since $|C_n|$ has no fixed components for large n). We have just shown that $B'_n \prec \mathcal{E}$ for large n. Now, the preceding lemma tells us that if X ranges over the (*finite*) set of all effective cycles such that $0 \prec X \prec E$ then the superabundances $s(X + C_n)$ are bounded. Thus, in particular, the superabundances $s(B'_n + C_n)$, i.e., $s(nD)$, are bounded. This completes the proof of the theorem.

COROLLARY 10.3. *If an effective divisorial cycle* D *is arithmetically effective and if the quadratic form* φ_D *is of type* $(1, t)$, *then a necessary and sufficient condition that the ring* $R^*[D]$ (see §2) *be finitely generated over* k *is that for some* n *the complete linear system* $|nD|$ *have no fixed components.*

The necessity of the condition follows from the boundedness of the fixed components B_n of $|nD|$ and from an observation made in § 2 ("if $B_n \neq 0$ for all n and B_n is bounded, then $R^*[D]$ is not finitely generated over k"). The sufficiency follows directly from Theorem 6.5 (without *any* conditions on the effective cycle D).

THEOREM 10.4. *Let* D *be an effective, arithmetically effective divisorial cycle, such that* φ_D *is of type* $(1, t)$, *and let* X *be an arbitrary effective divisorial cycle. Then the superabundance* $s(X + nD)$ *is bounded* (*from above*).

PROOF. The truth of the theorem follows from the boundedness of $s(nD)$ (Theorem 10.1) and from Lemma 10.2.

THEOREM 10.5. *If D is an arbitrary effective cycle such that φ_D is of type $(1, t)$ and if $\mathcal{E} = N(D)$ is the arithmetically negative component of D (see § 7), then there exists a set of integers $\{b_1, b_2, \cdots, b_\mu\}$ such that, for n large, we have*

$$s(nD) = -\rho(n\mathcal{E}) + b_{\lambda(n)} ,$$

where $\lambda(n)$ depends on n $(1 \leqq \lambda(n) \leqq \mu)$.

PROOF. The theorem is a direct consequence of Theorem 8.2 and of Theorem 10.4.

THEOREM 10.6. *If D is an effective divisorial cycle such that φ_D is of type $(1, t)$ and $\mathcal{E} = N(D)$ is the arithmetically negative component of D, then a necessary and sufficient condition that the ring $R^*[D]$ be finitely generated over k is that for some n the complete linear system $|n(D-\mathcal{E})|$ has no fixed components.*

PROOF. To prove the sufficiency of the condition, we set $L_n = \{x \in k(F) \mid (x) + nD \succ 0\}$ and we set $\Delta_0 = m_0(D-\mathcal{E})$, where m_0 is a positive integer satisfying the following two conditions:

(a) Δ_0 is an integral cycle;

(b) $|\Delta_0|$ is free from base points.

Such an integer m_0 exists in view of the assumption made in the statement of the theorem and also in view of Theorem 6.1. The ring $R^*[\Delta_0]$ $(=R^*[m_0D - m_0\mathcal{E}])$ can be identified with the subring $\sum_{n \geqq 0} L_{nm_0} t^{nm_0}$ of $R^*[D]$, since $nm_0\mathcal{E}$ is part of the fixed component of $|nm_0D|$ and since therefore $L_{nm_0} = \{x \in k(F) \mid (x) + nm_0D - nm_0\mathcal{E} \succ 0\}$. For any integer r, $0 \leqq r < m_0$, we set $S^{(r)} = \sum_{n \geqq 0} L_{nm_0+r} t^{nm_0}$. Then $S^{(0)} = R^*[\Delta_0]$ and $R^*[D] = S^{(0)} + S^{(1)}t + \cdots + S^{(m_0-1)}t^{m_0-1}$. By Theorem 5.1, $S^{(r)}$ is a finite $S^{(0)}$-module. Since the ring $S^{(0)}$ is finitely generated over k (Theorem 6.5, or also — directly — Theorem 4.2), it follows that $R^*[D]$ is finitely generated over k.

Conversely, assume that $R^*[D]$ is finitely generated over k. Then it is clear that for any integer q also $R^*[qD]$ is finitely generated over q. In particular, if m is an integer such that $m(D-\mathcal{E})$ is an integral cycle, then we know that $R^*[mD] = R^*[\Delta]$, where $\Delta = m(D-\mathcal{E})$, and so $R^*[\Delta]$ is finitely generated over k. Since Δ is a.e. and since it is clear that also φ_Δ is of type $(1, t')$, it follows from Corollary 10.3 that for some n the linear system $|n\Delta|$ has no fixed components. This completes the proof of the theorem.

11. The case $\varphi_D \leqq 0$

Again, as in the preceding section, the case in which D is not arithmetically effective will be easily settled once we solve the question for arithmetically effective cycles D. We assume therefore that our effective cycle D is arithmetically effective and is such that the quadratic form φ_D is $\leqq 0$; i.e., that φ_D is of type $(0, t)$, where $t < q =$ number of distinct prime components $E_1, E_2, \cdots, E_q$ of D. We shall postpone the discussion of the case in which $\dim |nD| = 0$ for all n. We therefore assume now that $\dim |nD| > 0$ for large n (and that consequently $\dim |nD| \to +\infty$ as $n \to +\infty$). Since D is arithmetically effective, we must have $(D^2) \geqq 0$, and hence $(D^2) = 0$, since $\varphi_D \leqq 0$. Furthermore, if Δ is any effective cycle such that $nD \succ \Delta$ for some n, we must have $(\Delta^2) \leqq 0$ (since Δ is then a linear combination of $E_1, E_2, \cdots, E_q$). It follows that if we denote by B_n the fixed component of the complete linear system $|nD|$ and set $|nD| = B_n + |\Delta_n|$, where $\Delta_n = nD - B_n$, then for all n sufficiently large (and, in fact, for all n such that $\dim |nD| > 0$) we will have $(\Delta_n^2) = 0$ and that consequently the complete linear system $|\Delta_n|$ is composite with some irreducible pencil $\{\Gamma\}$, free from base points (and therefore such that $(\Gamma^2) = 0$). *This pencil* $\{\Gamma\}$ *is independent of* n. In fact, let L_n be the usual vector space determined by $|nD|$: $L_n = \{x \in k(F) \mid (x) + nD \succ 0\}$ and let Σ be the subfield of $k(F)$ (of transcendence degree 1 over k) determined by the pencil $\{\Gamma\}$. The fact that $\{\Gamma\}$ is irreducible implies that Σ is maximally algebraic on $k(F)$, and the fact that $|nD|$ is composite with $\{\Gamma\}$ implies that Σ is an algebraic extension of $k(L_n)$. Thus tr.d. $k(L_n)/k = 1$, and Σ is the algebraic closure of $k(L_n)$ in $k(F)$. Since $k(L_n) \subset k(L_{n+1})$ and both these fields have the same transcendence degree 1 over k, they have the same (relative) algebraic closure in $k(F)$, and this proves that $\{\Gamma\}$ is independent of n.

Since D is arithmetically effective we have $(D \cdot E_i) \geqq 0$, $i = 1, 2, \cdots, q$. On the other hand, if $D = \lambda_1 E_1 + \lambda_2 E_2 + \cdots + \lambda_q E_q$, we have $0 = (D^2) = \sum_{i=1}^{q} \lambda_i (D \cdot E_i)$. Since the λ_i are positive, it follows that

$$(D \cdot E_i) = 0\,, \qquad i = 1, 2, \cdots, q\,. \tag{74}$$

For any n, Δ_n is a sum of cycles Γ_i which are members of the pencil $\{\Gamma\}$ (from now on we tacitly assume that we deal only with such (sufficiently large) values of n for which $\dim |nD| > 0$). Since $nD \succ \Delta_n$, *it follows that* $[D]$ *contains some* $[\Gamma_0]$, where $[Z]$ denotes the support of a cycle $[Z]$ and where Γ_0 is a member of $\{\Gamma\}$. It follows therefore from (74) that $(D \cdot \Gamma_0) = 0$, whence $(D \cdot \Gamma) = 0$ for every Γ in $\{\Gamma\}$. We thus have $\sum \lambda_i (E_i \cdot \Gamma) = 0$, and since $(E_i \cdot \Gamma) \geqq 0$ ($\{\Gamma\}$ being irreducible) we conclude that

(75) $$(E_i \cdot \Gamma) = 0\,, \qquad i = 1, 2, \cdots, q\,.$$

There is a member Γ_i of the pencil $\{\Gamma\}$ which passes through a given point of E_i. In view of (75) this particular cycle Γ_i must have E_i as a prime component. Thus, *each E_i is a prime component of some member Γ_i of the pencil $\{\Gamma\}$.*

We now assert the following: *if a given member Γ_0 of $\{\Gamma\}$ is such that $[D]$ and $[\Gamma_0]$ have a non-empty intersection then $[D] \supset [\Gamma_0]$.* In fact, since each E_i is a prime component of some member Γ_i of $\{\Gamma\}$, and since two distinct Γ's do not intersect, the assumption that $[D] \cap [\Gamma_0]$ is non-empty implies that $[D] \cap [\Gamma_0] = \bigcup_{i=1}^{m} X_i$, where $m \geqq 1$ and $X_1, X_2, \cdots, X_m$ are among the prime components of $[\Gamma_0]$ and belong also to the set $\{E_1, E_2, \cdots, E_q\}$. Write Γ_0 in the form

$$\Gamma_0 = X + X'\,,$$

where $X = \sum_{i=1}^{m} r_i X_i$ $(r_i > 0)$ and where $X' \succ 0$ and none of the h prime cycles $X_1, X_2, \cdots, X_m$ is a component of X'. Assume $X' \neq 0$. Then $[X] \cap [X'] \neq \varnothing$, since Γ_0 is connected. Since $[X] \subset [D]$ and since D and X' have no prime components in common, it follows that $(D \cdot X') > 0$. Since $(D \cdot \Gamma_0) = 0$ (as was pointed out above), we find that $(D \cdot X) < 0$, in contradiction with the fact that D is arithmetically effective. We thus conclude that $X' = 0$; i.e., that $[D] \supset [\Gamma_0]$, as asserted.

It follows that $[D]$ is the union of the supports of cycles Γ_i in $\{\Gamma\}$, say

(76) $$[D] = [\Gamma_1] \cup [\Gamma_2] \cup \cdots \cup [\Gamma_h]\,,$$

where $\Gamma_1, \Gamma_2, \cdots, \Gamma_h$ are distinct members of the pencil $\{\Gamma\}$. To this decomposition of $[D]$ into maximal connected components there corresponds a decomposition of the cycle D:

$$D = D_1 + D_2 + \cdots + D_h\,,$$

where $[D_i] = [\Gamma_i]$, $i = 1, 2, \cdots, h$. *We assert that D_i is a rational multiple of Γ_i.* To see this, we observe that each D_i is arithmetically effective. In fact, for any prime component E_j of D we have either $[D_i] \cap [E_j] = \varnothing$ (if E_j is not a component of D_i), or $(D_i \cdot E_j) = (D \cdot E_j)$ (if E_j is a component of D_i). Thus in either case, we have $(D_i \cdot E_j) = 0$, showing that D_i is arithmetically effective. We furthermore observe that we have obviously $\varphi_{D_i} \leqq 0$. To complete the proof we need the following lemma:

LEMMA 11.1. *Let $D \succ 0$ be a connected, arithmetically effective cycle and let $\varphi_D \leqq 0$. If D has q distinct prime components then φ_D is of type $(0, q-1)$.*

PROOF. Let φ_D be of type $(0, r)$, where, by assumption, $r < q$. We shall assume that $q - r > 1$ and we shall show that this assumption contradicts the connectedness of D. We use the notations of the proof of Lemma 7.3. The cycle $\mathcal{E}_{r+1}$ depends only on $E_1, E_2, \cdots, E_r, E_{r+1}$ and is effective, strictly positive. Since $(\mathcal{E}_{r+1} \cdot E_j) = 0$ for $j = 1, 2, \cdots, q$, it follows that $(E_i \cdot E_j) = 0$ if E_i belongs to the set A of prime components of $\mathcal{E}_{r+1}$ and E_j is in the complement of A in $\{E_1, E_2, \cdots, E_q\}$. Since A is non-empty and since also the complement of A is non-empty if $q > r + 1$, we have our contradiction.

The notations and assumptions still being the same in the lemma, it follows that any cycle $X = a_1E_1 + \cdots + a_qE_q$ such that $(X \cdot E_i) = 0$ for $i = 1, 2, \cdots, q$ must be a rational multiple of $\mathcal{E}_{r+1}$. Therefore any two such cycles X and X' must be rational multiples of each other (assuming, of course, that $X \neq 0$ *and* $X' \neq 0$).

We now apply the lemma to the cycles D_i and Γ_i. Since $[D_i] = [\Gamma_i]$, they have the same prime component. We know that $(D_i \cdot E_j) = 0$ for any prime components E_j of D (in particular, for any prime component of D_i). Since $\Gamma_i \in \{\Gamma\}$ we have also $(\Gamma_i \cdot E_j) = 0$ for all j. Therefore D_i and Γ_i are rational multiples of each other, as asserted.

We can write therefore

(77) $D = \tau_1\Gamma_1 + \tau_2\Gamma_2 + \cdots + \tau_h\Gamma_h$, τ_i — positive rational members.

We set

$$\nu_{n,i} = [\tau_i n], \; r_{n,i} = \tau_i n - \nu_{n,i}$$
$$\nu_n = \nu_{n,1} + \nu_{n,2} + \cdots + \nu_{n,h}$$
$$r_n = r_{n,1} + r_{n,2} + \cdots + r_{n,h} .$$

The number r_n, as function of n, is rationally valued and periodic, with period m, where m is the least common multiple of the denominators of the rational numbers $\tau_1, \tau_2, \cdots, \tau_h$, and we have $r_n < h$. We set

(78) $$\tau = \tau_1 + \tau_2 + \cdots + \tau_h .$$

Then τ is strictly positive and we have

(79) $$\nu_n = \tau n - r_n .$$

It is clear that

(80) $$\Delta_n = \nu_{n,1}\Gamma_1 + \nu_{n,2}\Gamma_2 + \cdots + \nu_{n,h}\Gamma_h .$$

If we set, for each $j = 0, 1, \cdots, m - 1$:

(81) $$B'_j = r_{j,1}\Gamma_1 + r_{j,2}\Gamma_2 + \cdots + r_{j,h}\Gamma_h ,$$

then the fixed component B_n of $|nD|$ is given by

(82) $$B_n = B'_j\,, \qquad \text{if } n \equiv j (\text{mod } m)\,.$$

From the expression (80) of Δ_n and from the Riemann-Roch theorem in the pencil $\{\Gamma\}$ it follows that

$$\dim |nD| = \nu_n - \pi\,, \qquad n - \text{large},$$

where π is the genus of the pencil $\{\Gamma\}$. Therefore, the superabundance $s(nD)$ is given by

(83) $$s(nD) = \nu_n - \pi - \rho(nD) - p_a(F)\,, \qquad n - \text{large}.$$

Since $(\Gamma^2)=0$, it follows from (77) that $\rho(nD)=n\tau\rho(\Gamma)$, where τ is defined in (78). Hence, by (81) and (79), we find

(84) $$s(nD) = an + b_n\,,$$

where

(85) $$a = \tau(1 - \rho(\Gamma))$$

(86) $$b_n = -(r_n + \pi + p_a(F))\,.$$

Here, again, b_n is a rationally valued function of n, with period m. As to the coefficient a of n in (84), we observe that $\rho(\Gamma) = 1 - p(\Gamma)$, whence $\rho(\Gamma) \leqq 1$. Hence, by (85), a is strictly positive, *except when* $p(\Gamma) = 0$. It is well known (see Zariski [11]) that in this exceptional case the surface F belongs to the class of ruled surface (of genus π) and that every member of the pencil $\{\Gamma\}$ is irreducible (and non-singular, of genus zero). In this case then we have $a = 0$, and furthermore the coefficients τ_i in (77) are integers, the numbers $r_{n,i}$, r_n are zero, and therefore we find in this exceptional case, from (84), (86), (81) and (82) that

$$\left.\begin{aligned} s(nD) &= -\pi - p_a(F) = \text{const.} \\ B_n &= 0 \end{aligned}\right\} \qquad n - \text{large}.$$

This completes the discussion of the case in which $\dim |nD| > 0$ for large n.

We now assume that $\dim |nD| = 0$ for all n, but maintain our previous assumption that D is arithmetically effective (and strictly positive). In this case we have, for n large,

$$s(nD) = -\rho(D) \cdot n - p_a(F)\,,$$

i.e., $s(nD)$ is a polynomial in n, of degree 1, *unless* $\rho(D) = 0$, in which case $s(nD) = \text{const.}$ $(= -p_a(F)$, for n large). *We shall now investigate the case* $\rho(D) = 0$.

Since $(D^2) = 0$, we have $\rho(D) = (-1/2)(D \cdot K) = (-1/2)\sum_{j=1}^{q} \lambda_j(E_j \cdot K)$; i.e.,

$$\rho(D) = (-1/2)\sum_{j=1}^{q} \lambda_j(2p(E_j) - 2 - (E_j^2))\,.$$

Since $(E_j^2) \leqq 0$ and $p(E_j) \geqq 0$, and also since $\rho(E_j) \leqq 0$ (since $\dim |nE_j| = 0$, all n) we have $2p(E_j) - 2 - (E_j^2) > 0$, except in the following cases:

(a) $p(E_j) = 1$, $(E_j^2) = 0 \ (\Rightarrow (E_j \cdot K) = 0)$;

(b) $p(E_j) = 0$, $(E_j^2) = -1 \ (\Rightarrow (E_j \cdot K) = -1)$;

(c) $p(E_j) = 0$, $(E_j^2) = -2 \ (\Rightarrow (E_j \cdot K) = 0)$.

Therefore at least one of the prime components of D must be of one of these three types. The conclusions which we will reach will concern the nature of each maximal connected component D_i of D. It is clear that each D_i satisfies the same conditions that we have imposed on D; i.e., we have that D_i is arithmetically effective, (since $(D_i \cdot E) = (D \cdot E) = 0$ for each prime component of D_i), that $\dim |nD_i| = 0$ for all n, and that $\varphi_{D_i} \leqq 0$ (since $(D_i^2) = 0$ and $\varphi_D \leqq 0$). We shall therefore assume temporarily (and in order to simplify the notation) that D is connected.

Assume that D has a prime component of type (a), and let E_1 be such a component. There exists a cycle $\mathcal{E}(\neq 0)$ of the form $a_1E_1 + \cdots + a_qE_q$ such that $(\mathcal{E} \cdot E_j) = 0, j = 1, 2, \cdots, q$. Since D is connected, it is clear that none of the a_i is zero, and hence each a_i is positive. Since $(E_1^2) = 0$, it follows from $(\mathcal{E} \cdot E_1) = 0$ and from the fact that the a_i are positive that $(E_j \cdot E_1) = 0, j > 1$. Since D is connected, this implies that $q = 1$. *We have thus shown that if D (is connected and) contains a prime component E_1 of type* (a) *then D is simply an integral multiple of E_1.*

Assume now that D contains no prime components of type (a). We note that the prime cycles of type (b) are exceptional curves of the first kind. Thus, if the surface F carries no exceptional curve of the first kind, then *all the prime components of D must be of type* (c),

Assume now that some prime component of D, say E_1, is exceptional, of the first kind: we can then contract E_1 to a simple point P' of a non-singular surface F' such that F dominates F' and such that $F' - P'$ and $F - E_1$ are bi-regularly equivalent. Let D' be the transform of D by the regular map $f: F \rightarrow F'$. Since $(D \cdot E_1) = 0$ it follows that D is the f^{-1}-transform of the cycle D' (see Zariski [11, p. 71]). We have $(D'^2) = (D^2)$ (Zariski [11, p. 70]) and cleary $\dim |nD'| = \dim |nD|$ for all n. Thus $(D'^2) = 0$, $\dim |nD'| = 0$, for all n. It is also clear that D' is connected. Furthermore, if X' is any effective cycle on F' and if we set $X = f^{-1}(X')$, then X is an effective cycle. Since $(D' \cdot X') = (D \cdot X)$, we have $(D' \cdot X') \geqq 0$, which shows that D' is arithmetically effective. Finally, if $E_2', \cdots, E_q'$ are the f-transforms of $E_2, \cdots, E_q$ respectively and if Z' is a cycle of the form $Z' = b_2E_2' + \cdots + b_qE_q'$, then $f^{-1}(Z')$ is a cycle Z of the form $Z = b_2E_2 + \cdots + b_qE_q + cE_1$, where c is an integer, and since $(Z'^2) = (Z^2)$ it follows that $(Z'^2) \leqq 0$. Thus $\varphi_{D'} \leqq 0$. We have therefore shown that D' satisfies the same condition that D does (including the condition of being connected). If D' still has a prime component which is exceptional of the first kind, we contract that component to a simple point of a surface F'', and thus we continue, until we get ultimately a regular map $\tilde{f}: F \to \tilde{F}$ of f onto a non-singular surface such that the $\tilde{f}$-transform $\tilde{D}$ of D contains no prime components which are exceptional of the first kind and such that $D = \tilde{f}^{-1}(\tilde{D})$. To this new cycle $\tilde{D}$ we can apply the preceding considerations and we may conclude that *either* $\tilde{D}$ is an integral multiple of a prime cycle $\tilde{E}$ such that $(\tilde{E}^2) = 0$ and $p(\tilde{E}) = 1$, or all prime components of $\tilde{D}$ are curves $\tilde{E}$ such that $(\tilde{E}^2) = -2$ and $p(\tilde{E}) = 0$.

We have therefore proved the following results:

PROPOSITION 11.2. *Let D be an effective and arithmetically effective cycle such that $\varphi_D \leqq 0$ and such that* $\dim|nD| > 0$ *for large n. Let B_n be the fixed component of $|nD|$. Then there exists a positive integer m, and there exist rational numbers $a, b'_1, \cdots, b'_m$ and effective cycle $B'_1, \cdots, B'_m$, such that, for large n, we have*

$$\left.\begin{aligned} s(nD) &= an + b'_i \\ B_n &= B'_i \end{aligned}\right\} \qquad \textit{if } n \equiv i (\operatorname{mod} m) .$$

Here a is an integer $\geqq 0$, with equality only in the following case: the surface F carries a pencil of genus π, of curves Γ such that $(\Gamma^2) = p(\Gamma) = 0$ (whence F belongs to the class of ruled surfaces) and D is a sum of curves Γ of the pencil. In this special case, we have, for large n: $s(nD) = -\pi - p_a(F)$, $B_n = 0$. In all other cases, $s(nD) \to +\infty$ as $n \to +\infty$.

PROPOSITION 11.3. *Let D be an effective and arithmetically effective cycle such that $\varphi_D \leqq 0$ and such that* $\dim|nD| = 0$ *for all n. Then we have, for large n:*

$$s(nD) = -\rho(D)n - p_a(F) .$$

Here $\rho(D) \leqq 0$, with equality only in the following cases: there exists a birational regular map $f: F \to \tilde{F}$ of F onto a non-singular surface $\tilde{F}$ and there exists an effective, arithmetically effective cycle $\tilde{D}$ on $\tilde{F}$ such that $D = f^{-1}\{\tilde{D}\}$ and such that the cycle $\tilde{D}$ has the following properties:

(a) $\varphi_{\tilde{D}} \leqq 0$;

(b) $\dim|n\tilde{D}| = 0$, *for all n;*

(c) *each maximal connected component of $\tilde{D}$ is either an integral multiple of a prime cycle $\tilde{\Gamma}$ such that $p(\tilde{\Gamma}) = 1$, $(\tilde{\Gamma}^2) = 0$, or all its prime components $\tilde{E}_j$ satisfy the conditions $p(\tilde{E}_j) = 0$ and $(\tilde{E}_j^2) = -2$.*

To complete the discussion of the case $\varphi_D \leqq 0$ we have only to consider now the case in which D is not arithmetically effective. Let $\mathcal{E}$ be the arithmetically negative part of D and let μ be the smallest integer such that $\mu\mathcal{E}$ is an integral cycle. Let $\Delta = \mu D - \mu\mathcal{E}$. Then Δ is arithmetically effective and we have $\varphi_\Delta \leqq 0$, Furthermore, we know (see Theorem 8.1) that if we set $n = \mu n' + r, 0 \leqq r < \mu$, then $n'\mu\mathcal{E}$ is part of the fixed component of $|nD|$, and thus

$$\dim |nD| = \dim |rD + n'\Delta| .$$

We have two cases to consider:

(a) $\dim |n'\Delta| > 0$ for some n';

(b) $\dim |n'\Delta| = 0$ for all n'.

The case (b) is straightforward. In this case we have for all n, such that $n \equiv 0 (\mathrm{mod}\, \mu)$, $\dim |nD| = \dim |n'\Delta|$, where $n = n'\mu$, and hence $\dim |nD| = 0$. Therefore, $\dim |nD| = 0$ for *all* n, and

$$s(nD) = -\rho(nD) - p_a(F) .$$

Thus $s(nD)$ is a polynomial in n. *This polynomial is definitely of degree* 2. This can be seen in two different ways. We either observe that, by Theorem 8.2 (formula (64)), $s(nD) \geqq -\rho(n\mathcal{E}) + \rho(r\mathcal{E})$, and $\rho(n\mathcal{E})$ is of degree 2 in n since $(\mathcal{E}^2) < 0$. Or we observe that since $D - \mathcal{E}$ is an arithmetically effective (rational) cycle and since $\varphi_{D-\mathcal{E}} = \varphi_\Delta \leqq 0$, we have $((D - \mathcal{E})^2) = 0$. Since $(D - \mathcal{E} \cdot \mathcal{E}) = 0$ it follows that $(D^2) = (\mathcal{E}^2) < 0$.

We consider now the case (a). We know that $|n\Delta|$ is composite with an irreducible pencil $\{\Gamma\}$ and that $s(n\Delta)$ is given by (84), where a and b are given in (85) and (86). *We assert that every prime component E of D is a component of some member of the pencil* $\{\Gamma\}$, To see this, we observe that either E is a component of $\mathcal{E}$ or is a component of Δ. In the first case, we have $(\Delta \cdot E) = 0$ (see Theorem 7.7), and our assertion follows from the fact that Δ is a linear combination (with rational coefficients) of members $\Gamma_1, \Gamma_2, \cdots, \Gamma_h$, of $\{\Gamma\}$ (see (77)). In the second case our assertion is trivial, and follows already from the fact that the support of Δ is the union of the supports of $\Gamma_1, \Gamma_2, \cdots, \Gamma_h$ [see (76)].

From this we can draw the following consequences. In the first place it follows that, apart from a fixed component, $|nD|$ is composite with the pencil $\{\Gamma\}$. Let again $n = n'\mu + r$ $(0 \leqq r < \mu)$. Using the notations of Proposition 11.2, where we now replace D by Δ and n by n', let $n' \equiv i (\mathrm{mod}\, m)$ $(0 \leqq i < m)$. We have that B_i' is the fixed component of $|n'\Delta|$ (n' — large). Let g_n be the greatest integer such that $rD + B_i'$ dominates the sum of g_n members of the pencil $\{\Gamma\}$; here g_n has, therefore, only a finite number of possible values. Then nD dominates exactly a sum of $\nu_n + g_n$ members of the pencil $\{\Gamma\}$. Hence, for n large, $\dim |rD + n'\Delta| - \dim |n'\Delta|$ is bounded from above (since $\dim|rD + n'\Delta| = \nu_n + g_n - \pi$, $\dim|n'\Delta| = \nu_n - \pi$). On the other hand, we have $(\Delta^2) = 0$ and $(D - \mathcal{E}\cdot\mathcal{E}) = 0$, and since $\Delta = \mu(D - \mathcal{E})$, it follows that $(D\cdot\Delta) = 0$. This shows that $\rho(rD + n'\Delta) - \rho(n'\Delta)$ depends only on $r\ (r = 0, 1, \cdots, \mu - 1)$. *Consequently, $s(rD + n'\Delta) - s(n'\Delta)$ assumes only a finite number of values (depending on r and the smallest non-negative residue i of n' modulo m).* We know that $s(n'\Delta) = an + b'_i$, for n large, where $n \equiv i (\mathrm{mod}\ m)$, $0 \leqq i < m$ (Proposition 11.2). We know also that $s(nD) = -\rho(n\mathcal{E}) + \rho(r\mathcal{E}) + s(r\Delta + n'D)$ (Theorem 8.2). *Therefore, we can conclude that*

$$s(nD) = -\rho(n\mathcal{E}) + an + b_{r,i}\,, \qquad (n - \text{large}),$$

where $b_{r,i}$ depends only on r and i. This yields for $s(nD)$ a finite number of polynomials, *of degree exactly* 2, which differ only in their constant terms. We note that $b_{r,i}$, as a function of n, is periodic, with period $m\mu$. We note also that the fixed component B_n of $|nD|$ is given by $n'\mu\mathcal{E} + B''_n$, where the cycle B''_n depends only on the cycle $rD + B'_i$ and is therefore a periodic function of n, with period $m\mu$. We have therefore proved the following result:

THEOREM 11.4. *Let D be an arbitrary effective cycle such that $\varphi_D \leqq 0$. Let $\mathcal{E}$ be the arithmetically negative component of D and let B_n be the fixed component of $|nD|$. Then there exist an integer M and $M + 1$ integers $a, b_1, b_2, \cdots, b_M$ such that*

$$s(nD) = -\rho(n\mathcal{E}) + an + b_j\,, \qquad \text{if } n \equiv j (\mathrm{mod}\ M).$$

Furthermore, there exist M cycles $\bar{B}_1, \bar{B}_2, \cdots, \bar{B}_M$ (having rational coefficients) such that

$$B_n = n\mathcal{E} + \bar{B}_j\,.$$

If $\dim|nD| > 0$ *for some n, then $nD - B_n$ is composite with a pencil $\{\Gamma\}$.*

As our final proposition we have the following:

PROPOSITION 11.5. *If the cycle D is as in Proposition* 11.4, *then the ring $R^*[D]$ is finitely generated over k.*

PROOF. The proposition is trivial if $\dim|nD| = 0$ for all n, for in that case $R^*[D] = k[t]$. We shall assume then that $\dim|nD| > 0$ for some n.

Assume first that D is arithmetically effective. From the expression (77) of D it follows that some multiple eD of D is a sum of a certain number, say d, of members Γ of the pencil. If π is the genus of the pencil, then it follows that if $n'e \geqq 2\pi$ then $|n'eD|$ will have no fixed components. The proposition follows then from the first part of Theorem 6.5.

If D is not arithmetically effective we repeat the argument given in the proof of Theorem 10.6.

12. A summary of principal results

We shall now give a brief summary of the principal results which were obtained in this part of the paper (§§ 7–11, and also § 6), or are directly obtainable by combining theorems explicitly stated in those sections. In what follows, *D denotes an effective cycle and n a sufficiently high integer.*

(1) There exists a finite number of polynomials $f_1(x), f_2(x), \cdots, f_m(x)$, in one variable x, such that $s(nD) = f_{\lambda(n)}(n)$, where $\lambda(n)$ is one of the integers $1, 2, \cdots, m$ and depends on n. These polynomials $f_\lambda(x)$ are of degree $\leqq 2$ and differ only in their constant terms. (Theorem 10.5; Theorem 11.4).

(2) If some multiple $|hD|$ of $|D|$ has no fixed components or if the quadratic form φ_D (introduced in the beginning of § 7) is of type $(0, t)$, then $\lambda(n)$ is a periodic function of n (Theorem 6.5, part (a); Theorem 11.4). (It is an open question whether $\lambda(n)$ is always a periodic function of n.)

(3) The superabundance $s(nD)$ is bounded from above (or, equivalently, the polynomials $f_\lambda(x)$ are of degree zero) if and only if one of the following conditions is satisfied:

(3a) D is arithmetically effective (see Definition 7.6) and the quadratic form φ_D is of type $(1, t)$ (Theorem 10.1);

(3b) D is arithmetically effective and D is a sum of members of a pencil of curves Γ such that $(\Gamma^2) = p(\Gamma) = 0$ (Proposition 11.2; in this case, the surface F is biregularly equivalent to a ruled surface);

(3c) D is arithmetically effective, $\dim |nD| = 0$ for all n, and each maximal connected component of D is either an integral multiple of a prime cycle Γ such that $(\Gamma^2) = 0$, $p(\Gamma) = 1$, or is a sum of prime cycles Γ such that $(\Gamma^2) = -2$ and $p(\Gamma) = 0$ (Proposition 11.3);

(3d) D is the inverse image $f^{-1}(\tilde{D})$ of a cycle $\tilde{D}$ satisfying condition (3c), f being a regular birational map of F onto a non-singular surface $\tilde{F}$ (Proposition 11.3).

(4) In the cases (3b), (3c) and (3d) above, $s(nD)$ is constant. Another case in which $s(nD)$ is constant is the following: $(D^2) > 0$ and *all* sufficiently high multiples $|nD|$ of $|D|$ have no fixed components (Theorem 6.5, part (b)).

(5) The polynomials $f_\lambda(x)$ referred to in (1) are of degree 1, if and only if D is arithmetically effective, φ_D is of type $(0, t)$ and D is not of the type described in (3). (Theorem 8.2; Theorem 8.1; Theorem 11.4)

(6) If B_n denotes the fixed components of $|nD|$ and $\mathcal{E}$ is the arithmetic-

ally negative component of D (see Theorem 7.7), then, under the assumption that $\dim |hD| > 0$ for some h, there exists an integer m and there exist m effective cycles $\bar{B}_1, \bar{B}_2, \cdots, \bar{B}_m$ (with rational coefficients) such that

$$B_n = n\mathcal{E} + \bar{B}_{\nu(n)} ,$$

where $\nu(n)$ is one of the integers $1, 2, \cdots, m$ (Theorem 8.1; Theorem 10.1; Theorem 11.4). Thus, under the assumption that $\dim |hD| > 0$ for some h, B_n is bounded (from above) if and only if D is arithmetically effective.[9]

(7) If the quadratic form φ_D is of type $(0, t)$, the function $\nu(n)$ which occurs in (6) is periodic (Theorem 11.4).

(8) The ring $R^*[D]$ (introduced in § 2) is finitely generated over k if and only if one of the following conditions is satisfied:

(8a) φ_D is of type $(1, t)$ and some multiple $|h(D-\mathcal{E})|$ ($\mathcal{E}$ = arithmetically negative component of D) has no fixed components;

(8b) φ_D is of type $(0, t)$ (Theorem 10.6; Proposition 11.5).

In the case (8b), $R^*[D]$ has transcendence degree ≤ 1 over k.

APPENDIX

THE CANONICAL RING OF AN ALGEBRAIC SURFACE

BY DAVID MUMFORD

In this appendix we wish to examine how the general theory developed by O. Zariski applies to the canonical divisor class. To be precise, suppose F is a non-singular algebraic surface over an algebraically closed field k, which

(a) is not birationally equivalent to a ruled surface, and

(b) is minimal [11].

Moreover let K be the canonical divisor class. We set

$$R = \bigoplus_{n=0}^{\infty} H^0(\mathfrak{o}_F(nK)) ,$$

and we call R the canonical ring of the surface ($R = R^*[K]$ in Zariski's notation). There are three essentially different cases to consider according as (K^2) is negative, zero, or positive. We assert:

---

[9] The assumption that $\dim |hD| > 0$ for some h is necessary. Thus, it is possible to have a prime cycle E such that $(E^2) = 0$ (and which is therefore arithmetically effective, whence $\mathcal{E} = 0$) and such that $\dim |nE| = 0$ or all n (whence $B_n = nE$, and B_n is not bounded). To obtain such a cycle E, we use the construction of §2, with the following modifications: we take for E' a generic plane section of F', we take for $\mathfrak{A}$ the divisor class determined on E' by $|E'|$ (i.e., we take for h the integer 0) and we determine $P'_1, P'_2, \cdots, P'_m$ (where $m = (E'^2)$) by the condition (4). Then it is immediate that the proper transform E of E' satisfies the desired conditions.

THEOREM. (i) *Under the above hypotheses,* $(K^2) < 0$ *is impossible.*

(ii) *If* $(K^2) = 0$, *then for some* n *either* $nK \equiv 0$, *or* $|nK|$ *is a linear system without base points, composite with a pencil. Therefore* R *is a finitely generated ring of dimension* 1 *or* 2.

(iii) *If* $(K^2) > 0$, *then for sufficiently large* n, $|nK|$ *is a linear system without base points. Therefore* R *is a finitely generated ring of dimension* 3. *Moreover* $s(nK) = \dim H^1(\mathfrak{o}_F(nK)) = 0$ *for sufficiently large* n.

Only the proof of (iii) will be given in this Appendix, since the proof of (i) and (ii) is rather long and will be published elsewhere. If the characteristic is 0, the latter proof depends chiefly upon Enriques' theorem: *if* F *is a relatively minimal non-singular algebraic surface, and* $|nK|$ *is empty for all* n, *then* F *is ruled.* The first complete proof of this in characteristic 0 (and of its refinement: if $|12K|$ is empty, then F is ruled) was obtained several years ago by K. Kodaira (unpublished). In characteristic p, new difficulties arise, but Enriques' result and parts (i) and (ii) of the theorem can still be proved.

We shall now establish (iii). Notice first that by the Riemann-Roch theorem, either $|nK|$ or $|-nK|$ is non-empty for large n. The latter case is impossible. For suppose q is the irregularity of F (= dimension of the Picard variety), and ρ is the base number (= rank of the Neron-Severi group). Then by Noether's formula for $p_a(F)$ and by Igusa's inequality,[10] we see that

$$12(p_a(F) + 1) = (K^2) - \deg(c_2) > 2 - 4q + \rho .$$

But since $p_g(F) = 0$, it follows that $p_a(F) = -q$.[11] Therefore:

$$8(1 - q) \geqq \rho - 1 .$$

But if $q = 0$, then F is rational, by Castelnuovo's criterion[12] which we have excluded; and if $q = 1$, then the Albanese map is a regular map onto a curve, and $\rho \geqq 2$. Therefore this last inequality cannot be fulfilled.

Therefore $|nK|$ is at least non-empty for sufficiently large n. Let D be any irreducible curve on F. Suppose $(D \cdot K) \leqq 0$. Then by Hodge's index theorem, since $(K^2) > 0$, it follows that $(D^2) < 0$. But also $-2 \leqq 2p_a(D) - 2 = (D^2) + (D \cdot K)$. Therefore $p_a(D) = 0$; i.e., D is a non-singular rational curve, and (D^2) equals -1 or -2. In the first case, D would be

[10] See J. I. Igusa, *Betti and Picard numbers of abstract algebraic surfaces*, Proc. Nat. Acad. Sci. U.S.A., 46 (1960), p. 724

[11] See Y. Nakai, *On the characteristic linear systems of algebraic families*, Ill. J. Math., 1 (1957), p. 552.

[12] See O. Zariski, *On Castelnuovo's criterion of rationality* $p_a = p_2 = 0$ *of an algebraic* surface, Ill. J. of Math., 2 (1958), p. 303.

exceptional of the first kind and F would not be minimal [11]. Therefore, we conclude:

(*) *If D is an irreducible curve, and $(D \cdot K) \leqq 0$, then D is a non-singular rational curve, $(D^2) = -2$, and $(D \cdot K) = 0$.*

Notice that there can be at most a finite number of such irreducible curves D. In fact, by the Riemann-Roch theorem, there is an m such that $\dim |mK| \geqq 2$. Then every curve D such that $(D \cdot K) = 0$ is either a fixed component of the linear system $|mK|$ or else is disjoint from every divisor of $|mK|$. In either case, there is only a finite set of such irreducible curves.

Let $E_1, E_2, \cdots, E_n$ be the set of all irreducible curves D such that $(D \cdot K) = 0$. Then by a very beautiful theorem of M. Artin,[13] which is the central point of this proof, there is a normal surface F^*, and a regular birational map $f: F \to F^*$, with the following five properties:

(i) f is biregular on $F - \bigcup E_i$,
(ii) f maps each E_i to one point,
(iii) the canonical divisor K^* on F^* is a Cartier divisor,
(iv) $f^{-1}(K^*) = K$,
(v) $p_a(F) = p_a(F^*)$.

By (iv), the linear systems $|nK|$ (on F), and $|nK^*|$ (on F^*) are canonically isomorphic. The proof that for sufficiently large n, $|nK^*|$ has no base points proceeds in three steps:

Step I. For all sufficiently large n, $|nK|$ (and hence $|nK^*|$) is non-empty. This is a corollary of the Riemann-Roch theorem.

Step II. For all sufficiently large n, $|nK^*|$ has no fixed components. For let k and l be relatively prime integers such that $|kK|$ and $|lK|$ are non-empty. Then by Theorem 9.1 above, for all sufficiently large n, the only fixed components of $|nkK|$ and $|nlK|$ are the irreducible curves E_i. But since all sufficiently large integers are of the form $nk + n'l$, for "sufficiently" large n and n', it follows that the only fixed components of $|nK|$ for sufficiently large n are the curves E_i. Hence by (ii) and (iv), the corresponding linear system $|nK^*|$ has no fixed components.

Step III. For sufficiently large n, $|nK|$ has no base points at all. The proof of this depends on a slight extension of Theorem 6.2 above. Namely, notice that this theorem, together with the proof of that theorem (§ 6), are equally valid whenever (in the notation of that theorem) V is a normal surface, and D is a Cartier divisor. Now let k and l be relatively prime integers such that $|kK^*|$ and $|lK^*|$ have no fixed components. By (iii),

[13] See M. Artin, *Some numerical criteria for contractability of curves on an algebraic surface*, Amer. J. Math., forthcoming, Th. (2.7).

and this extension of Theorem 6.2, for all sufficiently large n, the linear systems $|nkK^*|$ and $|nlK^*|$ have no base points. Hence just as before, for all sufficiently large n, $|nK^*|$ (and hence $|nK|$) has no base points.

The result on the superabundance follows from (v) and Theorem 6.5 above, once one observes that for sufficiently large n, the linear system $|nK|$ must define a regular map of F into projective space, with image F^*.

Finally, one sees that R is finitely generated as follows: Let k and l be relatively prime integers such that $|kK|$ and $|lK|$ have no base points. Then by Theorem 6.5, the rings $R^*[kK]$, and $R^*[lK]$ are finitely generated. But these two rings together generate a subring of R that contains all but a finite number of its homogeneous components. As each component of R is a finite dimensional vector space, R itself is therefore finitely generated. q.e.d.

HARVARD UNIVERSITY

REFERENCES

1. A. GROTHENDIECK, *Sur une note de Mattuck-Tate*, J. für reine und angew. Math., 20 (1958), 208-215.
2. W. V. D. HODGE, *Note on the theory of the base for curves on an algebraic surface*, J. London Math. Soc., 12 (1937), 58-63.
3. Y. NAKAI, *Non-degenerate divisors on an algebraic surface*, J. Science Hiroshima Univ., 24 (1960).
4. D. REES, *On a problem of Zariski*, I. J. of Math., 2 (1958), 145-149.
4′. M. ROSENLICHT, *Equivalence relations on algebraic curves*, Ann. of Math., 56 (1952), 169-191.
5. P. SAMUEL, *Sur les anneaux gradués*, C. R. Acad. Cien. Brasil (1958).
6. O. ZARISKI and P. SAMUEL, Commutative Algebra, volumes 1 and 2 (1958 and 1960), D. van Nostrand Company, Princeton.
7. O. ZARISKI, Algebraic Surfaces, Ergebnisse der Mathematik und ihrer Grenzgebiete, 3 (1935), n. 5.
8. ———, *Proof of a theorem of Bertini*, Trans. Amer. Math. Soc., 50 (1941), 48-70.
9. ———, *Complete linear systems on normal varieties and a generalization of a lemma of Enriques-Severi*, Ann. of Math., 55 (1952), 552-592.
10. ———, *Interprétations algébro-géometriques du quatorzième problème de Hilbert*, Bull. Soc. Math., 78 (1954), 155-168.
11. ———, Introduction to the problem of minimal models in the theory of algebraic surfaces, Publications of the Mathematical Society of Japan 4 (1958).